McGraw-Hill

Dictionary of Computing & Communications

McGraw-Hill

New York Chicago San Francisco Lisbon London Madrid
Mexico City Milan New Delhi San Juan Seoul Singapore
Sydney Toronto

The **McGraw-Hill** Companies

1 2 3 4 5 6 7 8 9 0 DOC/DOC 0 9 8 7 6 5 4 3

ISBN 0-07-142178-5

This book is printed on recycled, acid-free paper containing a minimum of 50% recycled, de-inked fiber.

This book was set in Helvetica Bold and Novarese Book by TechBooks, Fairfax, Virginia. It was printed and bound by RR Donnelley, The Lakeside Press.

McGraw-Hill books are available at special quantity discounts to use as premiums and sales promotions, or for use in corporate training programs. For more information, please write to the Director of Special Sales, Professional Publishing, McGraw-Hill, Two Penn Plaza, New York, NY 10121-2298. Or contact your local bookstore.

Library of Congress Cataloging-in-Publication Data

McGraw-Hill dictionary of computing and communications/McGraw-Hill.
p. cm.
ISBN 0-07-142178-5
1. Computer science—Dictionaries. 2. Telecommunication–Dictionaries.
3. Engineering—Dictionaries. I. The McGraw-Hill Companies, Inc.

QA76.15D52634 2003
004'.03—dc21 2003051209

Contents

Preface

The *McGraw-Hill Dictionary of Computing & Communications* provides a compendium of more than 11,000 terms that are not only central to these fields but are relevant in virtually every area of science, engineering, and commerce. The coverage includes terminology in computer science, telecommunications, aerospace and other fields of engineering, control systems, electromagnetism, electronics, navigation, optics, and systems engineering.

The definitions are drawn from the *McGraw-Hill Dictionary of Scientific and Technical Terms*, 6th edition (2003). Each one is classified according to the field with which it is primarily associated. The pronunciation of each term is provided along with synonyms, acronyms, and abbreviations where appropriate. A guide to the use of the *Dictionary* is included, explaining the alphabetical organization of terms, the format of the book, cross referencing, and how synonyms, variant spellings, abbreviations, and similar information are handled. A pronunciation key is also provided to assist the reader. An appendix provides conversion tables for commonly used scientific and technical units as well as charts, a "family tree" of programming languages, and listings of useful mathematical, technical, and scientific data.

Many of the terms used in computing and communications are found in specialized dictionaries and glossaries; this dictionary, however, aims to provide the user with the convenience of a single, comprehensive reference. It is the editors' hope that it will serve the needs of scientists, engineers, specialists in information technology, students, teachers, librarians, and writers for high-quality information, and that it will contribute to scientific literacy and communication.

Mark D. Licker
Publisher

Staff

Mark D. Licker, Publisher—Science

Elizabeth Geller, Managing Editor
Jonathan Weil, Senior Staff Editor
David Blumel, Staff Editor
Alyssa Rappaport, Staff Editor
Charles Wagner, Digital Content Manager
Renee Taylor, Editorial Assistant

Roger Kasunic, Vice President—Editing, Design, and Production

Joe Faulk, Editing Manager
Frank Kotowski, Jr., Senior Editing Supervisor

Ron Lane, Art Director

Thomas G. Kowalczyk, Production Manager
Pamela A. Pelton, Senior Production Supervisor

Henry F. Beechhold, Pronunciation Editor
Professor Emeritus of English
Former Chairman, Linguistics Program
The College of New Jersey
Trenton, New Jersey

How to Use the Dictionary

ALPHABETIZATION. The terms in the *McGraw-Hill Dictionary of Computing & Communications* are alphabetized on a letter-by-letter basis; word spacing, hyphen, comma, and solidus in a term are ignored in the sequencing. For example, an ordering of terms would be:

absolute-value computer
absolute vector
accuracy control system
ac/dc receiver
airborne radar
air navigation

FORMAT. The basic format for a defining entry provides the term in boldface, the field is small capitals, and the single definition in lightface:

term [FIELD] Definition.

A field may be followed by multiple definitions, each introduced by a boldface number:

term [FIELD] **1.** Definition. **2.** Definition. **3.** Definition.

A term may have definitions in two or more fields:

term [COMMUN] Definition. [COMPUT SCI] Definition.

A simple cross-reference entry appears as:

term *See* another term.

A cross reference may also appear in combination with definitions:

term [COMMUN] Definition. [COMPUT SCI] *See* another term.

CROSS REFERENCING. A cross-reference entry directs the user to the defining entry. For example, the user looking up "chroma band-pass amplifier" finds:

chroma band-pass amplifier *See* burst amplifier.

The user then turns to the "B" terms for the definition. Cross references are also made from variant spellings, acronyms, abbreviations, and symbols.

ACK *See* acknowledge character.
A-O-I gate *See* AND-OR-INVERT gate.
bps *See* bit per second.
chip *See* microchip.

ALSO KNOWN AS ..., etc. A definition may conclude with a mention of a synonym of the term, a variant spelling, an abbreviation for the term, or other such information, introduced by "Also known as ...," "Also spelled ...," "Abbreviated ...," "Symbolized ...," "Derived from ...." When a term has more than one definition, the positioning of any of these phrases conveys the extent of applicability. For example:

term [COMPUT SCI] **1.** Definition. Also known as synonym. **2.** Definition. Symbolized T.

In the above arrangement, "Also known as ..." applies only to the first definition; "Symbolized ..." applies only to the second definition.

term [COMMUN] **1.** Definition. **2.** Definition. [COMPUT SCI] Definition. Also known as synonym.

In the above arrangement, "Also known as ..." applies only to the second field.

term [COMPUT SCI] Also known as synonym. **1.** Definition. **2.** Definition. [COMMUN] Definition.

In the above arrangement, "Also known as ..." applies only to both definitions in the first field.

term Also known as synonym. [COMMUN] **1.** Definition. **2.** Definition. [COMPUT SCI] Definition.

In the above arrangement, "Also known as ..." applies to all definitions in both fields.

Fields and Their Scope

[AERO ENG] **aerospace engineering**—The branch of engineering pertaining to the design and construction of aircraft and space vehicles and of power units, and dealing with the special problems of flight in both the earth's atmosphere and space, such as in the flight of air vehicles and the launching, guidance, and control of missiles, earth satellites, and space vehicles and probes.

[COMMUN] **communications**—The science and technology by which information is collected from an originating source; converted into a form suitable for transmission; transmitted over a pathway such as a satellite channel, underwater acoustic channel, telephone cable, or fiber-optic link; and reconverted into a form suitable for interpretation by a receiver.

[COMPUT SCI] **computer science**—The study of computing, including computer hardware, software, programming, networking, database systems, information technology, interactive systems, and security.

[CONT SYS] **control systems**—The study of those systems in which one or more outputs are forced to change in a desired manner as time progresses.

[ELEC] **electricity**—The science of physical phenomena involving electric charges and their effects when at rest and when in motion.

[ELECTROMAG] **electromagnetism**—The branch of physics dealing with the observations and laws relating electricity to magnetism, and with magnetism produced by an electric current.

[ELECTR] **electronics**—The technological area involving the manipulation of voltages and electric currents through the use of various devices for the purpose of performing some useful action with the currents and voltages; this field is generally divided into analog electronics, in which the signals to be manipulated take the form of continuous currents or voltages, and digital electronics, in which signals are represented by a finite set of states.

[ENG] **engineering**—The science by which the properties of matter and the sources of power in nature are made useful to humans in structures, machines, and products.

[ENG ACOUS] **engineering acoustics**—The field of acoustics that deals with the production, detection, and control of sound by electrical devices, including the study, design, and construction of such things as microphones, loudspeakers, sound recorders and reproducers, and public address sytems.

[FL MECH] **fluid mechanics**—The science concerned with fluids, either at rest or in motion, and dealing with pressures, velocities, and accelerations in the fluid, including fluid deformation and compression or expansion.

[GEOPHYS] **geophysics**—The branch of geology in which the principles and practices of physics are used to study the earth and its environment, that is, earth, air, and (by extension) space.

[GRAPHICS] **graphic arts**—The fine and applied arts of representation, decoration, and writing or printing on flat surfaces together with the techniques and crafts associated with each: includes painting, drawing, engraving, etching, lithography, photography, and printing arts.

[IND ENG] **industrial engineering**—A branch of engineering dealing with the design, development, and implementation of integrated systems of humans, machines, and information resources to provide products and services.

[MATH] **mathematics**—The deductive study of shape, quantity, and dependence; the two main areas are applied mathematics and pure mathematics, the former arising from the study of physical phenomena, the latter involving the intrinsic study of mathematical structures.

[NAV] **navigation**—The science or art of directing the movement of a craft, such as a ship, small marine craft, underwater vehicle, land vehicle, aircraft, missile, or spacecraft, from one place to another with the assistance of onboard equipment, objects, or devices, or of systems external to the craft.

[OPTICS] **optics**—The study of phenomena associated with the generation, transmission, and detection of electromagnetic radiation in the spectral range extending from the long-wave edge of the x-ray region to the short-wave edge of the radio region; and the science of light.

[PHYS] **physics**—The science concerned with those aspects of nature which can be understood in terms of elementary principles and laws.

[SOLID STATE] **solid-state physics**—The branch of physics centering on the physical properties of solid materials; it is usually concerned with the properties of crystalline materials only, but it is sometimes extended to include the properties of glasses or polymers.

[STAT] **statistics**—The science dealing with the collection, analysis, interpretation, and presentation of masses of numerical data.

[SYS ENG] **systems engineering**—The branch of engineering dealing with the design of a complex interconnection of many elements (a system) to maximize an agreed-upon measure of system performance.

Pronunciation Key

Vowels

a as in b**a**t, th**a**t
ā as in b**ai**t, cr**a**te
ä as in b**o**ther, f**a**ther
e as in b**e**t, n**e**t
ē as in b**ee**t, tr**ea**t
i as in b**i**t, sk**i**t
ī as in b**i**te, l**igh**t
ō as in b**oa**t, n**o**te
ȯ as in b**ough**t, t**au**t
u̇ as in b**oo**k, p**u**ll
ü as in b**oo**t, p**oo**l
ə as in b**u**t, sof**a**
au̇ as in cr**ow**d, p**ow**er
ȯi as in b**oi**l, sp**oi**l
yə as in form**u**la, spectac**u**lar
yü as in f**ue**l, m**u**le

Semivowels/Semiconsonants

w as in **w**ind, t**w**in
y as in **y**et, on**io**n

Stress (Accent)

ˈ precedes syllable with primary stress

ˌ precedes syllable with secondary stress

¦ precedes syllable with variable or indeterminate primary/secondary stress

Consonants

b as in **b**i**b**, dri**bb**le
ch as in **ch**arge, stre**tch**
d as in **d**og, ba**d**
f as in **f**ix, sa**f**e
g as in **g**ood, si**g**nal
h as in **h**and, be**h**ind
j as in **j**oint, di**g**it
k as in **c**ast, bri**ck**
k̲ as in Ba**ch** (used rarely)
l as in **l**oud, be**ll**
m as in **m**ild, su**mm**er
n as in **n**ew, de**n**t
n̲ indicates nasalization of preceding vowel
ŋ as in ri**ng**, si**ng**le
p as in **p**ier, sli**p**
r as in **r**ed, sca**r**
s as in **s**ign, po**s**t
sh as in **s**ugar, **sh**oe
t as in **t**imid, ca**t**
th as in **th**in, brea**th**
t̲h̲ as in **th**en, brea**th**e
v as in **v**eil, wea**v**e
z as in **z**oo, crui**s**e
zh as in bei**g**e, trea**s**ure

Syllabication

· Indicates syllable boundary when following syllable is unstressed

A

A AND NOT B gate *See* AND NOT gate. { 'ā an nöt 'bē ˌgāt }

abbreviated dialing [COMMUN] A feature which requires less than the usual number of dialing operations to connect two or more subscribers. { ə'brē·vē·ād·əd 'dī·liŋ }

ABC *See* automatic brightness control.

abend [COMPUT SCI] An unplanned program termination that occurs when a computer is directed to execute an instruction or to process information that it cannot recognize. Also known as blow up; bomb; crash. { 'ab·end }

able [COMPUT SCI] A name for the hexadecimal digit whose decimal equivalent is 10. { 'ā·bəl }

abnormal propagation [COMMUN] Phenomena of unstable or changing atmospheric or ionospheric conditions acting upon transmitted radio waves, preventing such waves from following their normal path, thereby causing difficulties and disruptions of communications. { ab'nȯr·məl ˌpräp·ə'gā·shən }

abnormal reflections [ELECTROMAG] Sharply defined reflections of substantial intensity at frequencies greater than the critical frequency of the ionized layer of the ionosphere. { ab'nȯr·məl re'flek·shənz }

abnormal statement [COMPUT SCI] An element of a FORTRAN V (UNIVAC) program which specifies that certain function subroutines must be called every time they are referred to. { ab'nȯr·məl 'stāt·mənt }

abort [COMPUT SCI] To terminate a procedure, such as the running of a computer program or the printing of a document, while it is still in progress. { ə'bȯrt }

abrupt junction [ELECTR] A *pn* junction in which the concentration of impurities changes suddenly from acceptors to donors. { ə'brəpt 'jəŋk·shən }

abs [COMPUT SCI] A special function occurring in ALGOL, which yields the absolute value, or modulus, of its argument.

absolute address [COMPUT SCI] The numerical identification of each storage location which is wired permanently into a computer by the manufacturer. { 'ab·səˌlüt ə'dres }

absolute addressing [COMPUT SCI] The identification of storage locations in a computer program by their physical addresses. { 'ab·sə ˌlüt ə'dres·iŋ }

absolute category rating mean opinion score [COMMUN] Methodology for subjectively testing audio quality where participants are presented with sound samples, one at a time, and are asked to grade them on a 5-point scale. For the NRSC FM IBOC tests, the MOS scale used was 5 = excellent, 4 = good, 3 = fair, 2 = poor, 1 = bad. Abbreviated ACR-MOS. { ¦ab·səˌlut kad·əˌgȯr·ē rād·iŋ mēn 'ə·'pin·yən ˌskȯr }

absolute cell reference [COMPUT SCI] A cell reference used in a formula in a spreadsheet program that does not change when the formula is copied or moved. { ¦ab·səˌlüt 'sel ˌref·rəns }

absolute code [COMPUT SCI] A code used when the addresses in a program are to be written in machine language exactly as they will appear when the instructions are executed by the control circuits. { 'ab·səˌlüt 'kōd }

absolute gain of an antenna [ELECTROMAG] Gain in a given direction when the reference antenna is an isotropic antenna isolated in space. Also known as isotropic gain of an antenna. { 'ab·səˌlüt ˌgān əv ən an'ten·ə }

absolute index of refraction *See* index of refraction. { 'ab·səˌlüt 'inˌdeks əv ri'frak·shən }

absolute instruction [COMPUT SCI] A computer instruction in its final form, in which it can be executed. { 'ab·səˌlüt in'strək·shən }

absolute programming [COMPUT SCI] Programming with the use of absolute code. { 'ab·səˌlüt 'prō·gram·iŋ }

absolute refractive constant *See* index of refraction. { 'ab·səˌlüt ri'frak·tiv 'kän·stənt }

absolute-value computer [COMPUT SCI] A computer that processes the values of the variables rather than their increments. { 'ab·səˌlüt 'val·yü kəm'pyüd·ər }

absolute vector [COMPUT SCI] In computer graphics, a vector whose end points are given in absolute coordinates. { 'ab·səˌlüt 'vek·tər }

absorption circuit [ELECTR] A series-resonant circuit used to absorb power at an unwanted signal frequency by providing a low impedance to ground at this frequency. { əb'sȯrp·shən 'sər·kət }

absorption control *See* absorption modulation. { əb'sȯrp·shən kən'trōl }

absorption fading [COMMUN] Slow type of fading, primarily caused by variations in the absorption rate along the radio path. { əb'sȯrp·shən 'fād·iŋ }

absorption loss [COMMUN] That part of the transmission loss due to the dissipation or conversion of either sound energy or electromagnetic energy into other forms of energy, either within the medium or attendant upon a reflection. { əb'sȯrp·shən ˌlȯs }

absorption modulation [ELECTR] A system of amplitude modulation in which a variable-impedance device is inserted in or coupled to the output circuit of the transmitter. Also known as absorption control; loss modulation. { əb'sȯrp·shən mäd·yü'lā·shən }

abstract automata theory [COMPUT SCI] The mathematical theory which characterizes automata by three sets: input signals, internal states, and output signals; and two functions: input functions and output functions. { 'abz·trakt ȯ'tam·ə·tə 'thē·ə·rē }

abstract data type [COMPUT SCI] A mathematical model which may be used to capture the essentials of a problem domain in order to translate it into a computer program; examples include queues, lists, stacks, trees, graphs, and sets. Abbreviated ADT. { 'abz·trakt 'dad·ə ˌtīp }

ACAS *See* airborne collision avoidance system.

accelerated graphics port [COMPUT SCI] A personal computer graphics bus that transfers data at a greater rate than a PCI bus. { akˌsel·əˌrād·əd 'graf·iks ˌpȯrt }

acceleration time [COMPUT SCI] The time required for a magnetic tape transport or any other mechanical device to attain its operating speed. { akˌsel·ə'rā·shən ˌtīm }

accentuation [ELECTR] The enhancement of signal amplitudes in selected frequency bands with respect to other signals. { akˌsen·chə'wā·shən }

accentuator [ELECTR] A circuit that provides for the first part of a process for increasing the strength of certain audio frequencies with respect to others, to help these frequencies override noise or to reduce distortion. Also known as accentuator circuit. { ak'sen·chəˌwād·ər }

accentuator circuit *See* accentuator. { ak'sen·chəˌwād·ər 'sər·kət }

accept [COMPUT SCI] A data transmission statement which is used in FORTRAN when the computer is in conversational mode, and which enables the programmer to input, through the teletypewriter, data the programmer wishes stored in memory. { ak'sept }

access [COMPUT SCI] The reading of data from storage or the writing of data into storage. { 'ak ˌses }

access arm [COMPUT SCI] The mechanical device which positions the read/write head on a magnetic storage unit. { 'akˌses ˌärm }

access code [COMMUN] **1.** Numeric identification for internetwork or facility switching. **2.** The preliminary digits that a user must dial to be connected through an automatic PBX to the serving switching center. [COMPUT SCI] A sequence of characters which a user must enter into a terminal in order to use a computer system. { 'akˌses ˌkōd }

access control [COMPUT SCI] A restriction on the operations that a user of a computer system may perform on files and other resources of the system. { 'akˌses kənˌtrōl }

access-control list [COMPUT SCI] A column of an access matrix, containing the access rights of various users of a computer system to a given file or other resource of the system. { 'akˌses kənˌtrōl ˌlist }

access-control mechanism *See* reference monitor. { ¦akˌses kən'trōl ¦me·kə·ni·zəm }

access-control register [COMPUT SCI] A storage device which controls the word-by-word transmission over a given channel. { 'akˌses kən'trōl ˌrej·ə·stər }

access-control words [COMPUT SCI] Permanently wired instructions channeling transmitted words into reserved locations. { 'akˌses kən 'trōl ˌwərdz }

access gap *See* memory gap. { 'akˌses ˌgap }

access line [COMMUN] Four-wire circuit between a subscriber or a local PBX to the serving switching center. { 'akˌses ˌlīn }

access management [COMPUT SCI] The use of techniques to allow various components of a computer's operating system to be used only by authorized personnel. { 'akˌses ˌman·ij·mənt }

access matrix [COMPUT SCI] A method of representing discretionary authorization information, with rows representing subjects or users of the system, columns corresponding to objects or resources of the system, and cells (intersections of rows and columns) composed of allowable operations that a subject may apply to an object. { 'akˌses ˌmā·triks }

access mechanism [COMPUT SCI] The mechanism of positioning reading or writing heads onto the required tracks of a magnetic disk. { 'akˌses 'mek·əˌniz·əm }

access method [COMMUN] The procedures required to obtain access to a communications network. [COMPUT SCI] A set of programming routines which links programs and the data that these programs transfer into and out of memory. { 'akˌses ˌmeth·əd }

access mode [COMPUT SCI] A programming clause in COBOL which is required when using a random-access device so that a specific record may be read out of or written into a mass storage bin. { 'akˌses ˌmōd }

access privileges [COMPUT SCI] The extent to which a user of a computer in a network is allowed to use and read, write to, and execute files in other computers in the network. { 'ak ˌses ˌpriv·ə·ləj·əs }

access protocol [COMMUN] A set of rules observed by all nodes in a local-area network so that one node can get the attention of another and its data packet can be transferred, and so that no two data packets can be simultaneously transmitted over the same medium. { 'akˌses ˌprōd·əˌkȯl }

access provider *See* service provider. { 'akˌses prəˌvīd·ər }

access time [COMPUT SCI] The time period required for reading out of or writing into the computer memory. { 'ak·ses ,tīm }

access type [COMPUT SCI] One of the allowable operations that a given user of a computer system governed by access controls may perform on a file or other resource of the system, such as own, read, write, or execute. { 'ak·ses ,tīp }

accounting package [COMPUT SCI] A set of special routines that allow collection of information about the usage level of various components of a computer system by each production program. { ə'kau̇nt·iŋ 'pak·ij }

accumulator [COMPUT SCI] A specific register, in the arithmetic unit of a computer, in which the result of an arithmetic or logical operation is formed; here numbers are added or subtracted, and certain operations such as sensing, shifting, and complementing are performed. Also known as accumulator register; counter. { ə'kyü·myə ,lād·ər }

accumulator jump instruction [COMPUT SCI] An instruction which programs a computer to ignore the previously established program sequence depending on the status of the accumulator. Also known as accumulator transfer instruction. { ə'kyü·myə,lād·ər ,jəmp in'strək·shən }

accumulator register *See* accumulator. { ə'kyü·myə,lād·ər 'rej·ə·stər }

accumulator shift instruction [COMPUT SCI] A computer instruction which causes the word in a register to be displaced a specified number of bit positions to the left or right. { ə'kyü·myə ,lād·ər 'shift in'strək·shən }

accumulator transfer instruction *See* accumulator jump instruction. { ə'kyü·myə,lād·ər 'trans·fər in'strək·shən }

accuracy control system [COMPUT SCI] Any method which attempts error detection and control, such as random sampling and squaring. { 'ak·yə·rə·sē kən'trōl ,sis·təm }

ACK *See* acknowledge character.

acknowledge character [COMPUT SCI] A signal that a receiving station transmits in order to indicate that a block of information has been received and that its validity has been checked. Also known as acknowledgement. Abbreviated ACK. { ak'nä·lij 'kar·ək·tər }

acknowledgement *See* acknowledge character. { ak'nä·lij·mənt }

acoustic convolver *See* convolver. { ə'küs·tik kən'välv·ər }

acoustic filter *See* filter. { ə'küs·tik 'fil·tər }

acoustooptic interaction [OPTICS] A way to influence the propagation characteristics of an optical wave by applying a low-frequency acoustical field to the medium through which the wave passes. { ə¦küs·tō¦äp·tik ,in·tə'rak·shən }

acoustooptic modulator [OPTICS] A device utilizing acoustooptic interaction ultrasonically to vary the amplitude or the phase of a light beam. Also known as Bragg cell. { ə¦küs·tō¦äp·tik 'mäd·yə,lād·ər }

acoustooptics [OPTICS] The science that deals with interactions between acoustic waves and light. { ə¦küs·tō¦äp·tiks }

acquisition [ELECTR] Also known as target acquisition. **1.** Of acquisition radars, the process of detecting and locating a target so as to permit reliable tracking and possible identification of it or other determinations about it. **2.** Of precision tracking radars, the detecting and tracking of a target designated to it by another radar or other initial data source to support continued intended action. [ENG] The process of pointing an antenna or a telescope so that it is properly oriented to allow gathering of tracking and telemetry data from a satellite or space probe. { ,ak·wə'zish·ən }

acquisition tone [COMPUT SCI] An audible tone that verifies entry into a computer. { ,ak·wə 'zish·ən ,tōn }

ACR-MOS *See* absolute category rating mean opinion score.

action entries [COMPUT SCI] The lower right-hand portion of a decision table, indicating which of the various possible actions result from each of the various possible conditions. { 'ak·shən ,en·trēz }

action portion [COMPUT SCI] The lower portion of a decision table, comprising the action stub and action entries. { 'ak·shən ,pȯr·shən }

action stub [COMPUT SCI] The lower left-hand portion of a decision table, consisting of a single column listing the various possible actions (transformations to be done on data and materials). { 'ak·shən ,stəb }

activation record [COMPUT SCI] A variable part of a program module, such as data and control information, that may vary with different instances of execution. { ,ak·tə'vā·shən 'rek·ərd }

active array [ELECTROMAG] A radar antenna composed of many radiating elements, each of which contains an amplifier, generally solid state in nature, for the final amplification of the signal transmitted; when the elements are also phased controlled for electronic beam steering, the term active phased array is used. { ¦ak·tiv ə'rā }

active balance [COMMUN] Summation of all return currents, in telephone repeater operation, at a terminal network balanced against the impedance of the local circuit or drop. { 'ak·tiv 'bal·əns }

active cell [COMPUT SCI] The cell that continues the value being used or modified in a spreadsheet program, and that is highlighted by the cell pointer. Also known as current cell. { ¦ak·tiv 'sel }

active communications satellite [AERO ENG] Satellite which receives, regenerates, and retransmits signals between stations. { 'ak·tiv kə ,myü·nə'kā·shənz 'sad·ə,līt }

active computer [COMPUT SCI] When two or more computers are installed, the one that is on-line and processing data. { 'ak·tiv kəm'pyüd·ər }

active device [ELECTR] A component, such as an electron tube or transistor, that is capable of amplifying the current or voltage in a circuit. { 'ak·tiv di'vīs }

active electronic countermeasures [ELECTR] The major subdivision of electronic countermeasures that concerns electronic jamming and electronic deceptions. { 'ak·tiv ə,lek'trän·ik 'kau̇nt·ər,mezh·ərz }

active file [COMPUT SCI] A collection of records that is currently being used or is available for use. { 'ak·tiv 'fīl }

active filter [ELECTR] A filter that uses an amplifier with conventional passive filter elements to provide a desired fixed or tunable pass or rejection characteristic. { 'ak·tiv 'fil·tər }

active jamming *See* jamming. { 'ak·tiv 'jam·iŋ }

active logic [ELECTR] Logic that incorporates active components which provide such functions as level restoration, pulse shaping, pulse inversion, and power gain. { 'ak·tiv 'läj·ik }

active master file [COMPUT SCI] A relatively active computer master file, as determined by usage data. { 'ak·tiv 'mas·tər 'fīl }

active master item [COMPUT SCI] A relatively active item in a computer master file, as determined by usage data. { 'ak·tiv 'mas·tər 'ī·təm }

active-matrix liquid-crystal display [ELEC] A liquid-crystal display that has an active element, such as a transistor or diode, on every picture element. Abbreviated AMLCD. { ¦ak·tiv ¦mā·triks ¦lik·wid 'kris·təl di,splā }

active-RC filter [ELEC] An active filter whose frequency-sensitive mechanism is the charging of a capacitor (C) through a resistor (R), giving a characteristic frequency at which the impedances of the resistor and the capacitor are equal. { ¦ak·tiv ¦är¦sē 'fil·tər }

active region [ELECTR] The region in which amplifying, rectifying, light emitting, or other dynamic action occurs in a semiconductor device. { 'ak·tiv 'rē·jən }

active-RLC filter [ELEC] An integrated-circuit filter that uses both inductors (L), made as spirals of metallization on the top layer, and amplifiers, connected to simulate negative resistors (R), that enhance the performance of the inductors as well as capacitors (C). { ¦ak·tiv ¦är¦el'sē fil·tər }

active satellite [AERO ENG] A satellite which transmits a signal. { 'ak·tiv 'sad·ə,līt }

active sonar [ENG] A system consisting of one or more transducers to send and receive sound, equipment for the generation and detection of the electrical impulses to and from the transducer, and a display or recorder system for the observation of the received signals. { 'ak·tiv 'sō,när }

active termination [COMPUT SCI] A means of ending a chain of peripheral devices connected to a small computer system interface (SCSI) port, suitable for longer chains, where it can reduce electrical interference. { ¦ak·tiv ,tər·mə'nā·shən }

active window [COMPUT SCI] In a windowing environment, the window in which the user is currently working and which receives keyboard input. { ¦ak·tiv 'win,dō }

activity [COMPUT SCI] The use or modification of information contained in a file. { ,ak'tiv·əd·ē }

activity level [COMPUT SCI] **1.** The value assumed by a structural variable during the solution of a programming problem. **2.** A measure of the number of times that use or modification is made of the information contained in a file. { ,ak'tiv·əd·ē 'lev·əl }

activity ratio [COMPUT SCI] The ratio between used or modified records and the total number of records in a file. { ,ak'tiv·əd·ē ,rā·shō }

activity sequence method [COMPUT SCI] A method of organizing records in a file so that the records most frequently used are located where they can be found most quickly. { ak'tiv·əd·ē 'sē·kwəns ,meth·əd }

actual argument [COMPUT SCI] The variable which replaces a dummy argument when a procedure or macroinstruction is called up. { 'ak·chə·wəl 'är·gyə·mənt }

actual decimal point [COMPUT SCI] The period appearing on a printed report as opposed to the virtual point defined only by the data structure within the computer. { 'ak·chə·wəl 'des·məl 'pȯint }

actual instruction *See* effective instruction. { 'ak·chə·wəl in'strək·shən }

actual key [COMPUT SCI] A data item in COBOL computer language which can be used as an address. { 'ak·chə·wəl 'kē }

acyclic feeding [COMPUT SCI] A method employed by alphanumeric readers in which the trailing edge or some other document characteristic is used to activate the feeding of the succeeding document. { ā'sik·lik 'fēd·iŋ }

Ada [COMPUT SCI] A computer language that was chosen by the United States Department of Defense to support the development of embedded systems, and uses the language Pascal as a base to meet the reliablity and efficiency requirements imposed by these systems. { 'ā·də }

adapter [COMPUT SCI] A device which converts bits of information received serially into parallel bit form for use in the inquiry buffer unit. { ə'dap·tər }

adaptive antenna [ELECTROMAG] An antenna that adjusts its pattern automatically to be the inverse to any nonuniform distribution in angle of offending interference sources, tending to "whiten" or make appear uniform the noise in angle and minimizing the effects of strong jamming. { ə'dap·tiv an'ten·ə }

adaptive communications [COMMUN] A communications system capable of automatic change to meet changing inputs or changing characteristics of the device or process being controlled. Also known as self-adjusting communications; self-optimizing communications. { ə'dap·tiv kə,myü·nə'kā·shənz }

adaptive control [CONT SYS] A control method in which one or more parameters are sensed and used to vary the feedback control signals in order to satisfy the performance criteria. { ə'dap·tiv kən'trōl }

adaptive differential pulse-code modulation [COMMUN] A method of compressing speech and music signals in which the transmitted signals represent differences between input signals and predicted signals, and these predicted signals are synthesized by predictors with response functions representative of the short- and long-term correlation inherent in the signal. Abbreviated ADPCM. { ə¦dap·tiv ˌdif·ə¦ren·chəl ˈpəls ˌcōd ˌmäj·əˌlā·shən }

adaptive equalization [COMMUN] A signal-processing technique designed to compensate for impairments in received signals over a communications channel resulting from imperfect transmission characteristics. { əˈdap·tiv ˌē·kwə·ləˌzā·shən }

adaptive filter [ELECTR] An electric filter whose frequency response varies with time, as a function of the input signal. { əˌdap·tiv ˈfil·tər }

adaptive robot [CONT SYS] A robot that can alter its responses according to changes in the environment. { əˈdap·tiv ˈrōˌbät }

adaptive signal processing [COMMUN] The design of adaptive systems for signal-processing applications. { ə¦dap·tiv ¦sig·nəl ˈprä·sə·siŋ }

adaptive system [SYS ENG] A system that can change itself in response to changes in its environment in such a way that its performance improves through a continuing interaction with its surroundings. { əˈdap·tiv ˈsis·təm }

adaptive system theory [COMPUT SCI] The branch of automata theory dealing with adaptive, or self-organizing, systems. { əˈdap·tiv ˈsis·təm ˌthe·ə·rē }

adaptor [COMPUT SCI] A printed circuit board that is plugged into an expansion slot in a computer to communicate with an external peripheral device. { əˈdap·tər }

Adcock antenna [ELECTROMAG] A pair of vertical antennas separated by a distance of one-half wavelength or less and connected in phase opposition to produce a radiation pattern having the shape of a figure eight. { ˈad·käk ˌanˈten·ə }

Adcock direction finder [NAV] A radio direction finder utilizing one or more pairs of Adcock antennas. { ˈad·käk dəˈrek·shən ˌfīn·dər }

ADCON *See* address constant. { ˈadˌkän }

add *See* add operation. { ad }

adder [COMPUT SCI] A computer device that can form the sum of two or more numbers or quantities. [ELECTR] A circuit in which two or more signals are combined to give an output-signal amplitude that is proportional to the sum of the input-signal amplitudes. Also known as adder circuit. { ˈad·ər }

adder circuit *See* adder. { ˈad·ər ˌsər·kət }

add-in [COMPUT SCI] An electronic component that can be placed on a printed circuit board already installed in a computer to enhance the computer's capability. { ˈad ˌin }

adding circuit [ELECTR] A circuit that performs the mathematical operation of addition. { ˈad·iŋ ˈsər·kət }

adding machine [COMPUT SCI] A device which performs the arithmetical operation of addition and subtraction. { ˈad·iŋ məˌshēn }

add-in program [COMPUT SCI] A computer program that enhances the capabilities of a particular application. { ˈadˌin ˌprō·grəm }

addition item [COMPUT SCI] An item which is to be filed in its proper place in a computer. { əˈdi·shən ˈīd·əm }

addition record [COMPUT SCI] A new record inserted into an updated master file. { əˈdi·shən ˌrek·ərd }

addition table [COMPUT SCI] The part of memory that holds the table of numbers used in addition in a computer employing table look-up techniques to carry out this operation. { əˈdi·shən ˌtā·bəl }

additive white Gaussian noise Noise that contains equal energy per frequency across the spectrum of the noise employed. Also known as white noise. Abbreviated AWGN. { ˈad·əd·iv wīt ¦gau̇·sē·ən ˈnȯiz }

add-on [COMPUT SCI] A peripheral device, such as a printer or disk drive, that is added to a basic computer. { ˈadˌȯn }

add-on memory [COMPUT SCI] Computer storage that is added to the original main storage to enhance the computer's processing capability. { ˈadˌȯn ˈmem·rē }

add operation [COMPUT SCI] An operation in computer processing in which the sum of two or more numbers is placed in a storage location previously occupied by one of the original numbers. Also known as add. { ˈad ˌäp·əˌrā·shən }

address [COMPUT SCI] The number or name that uniquely identifies a register, memory location, or storage device in a computer. { ˈad·res }

addressable [COMPUT SCI] Capable of being located by a computer through an addressing technique. { əˈdres·ə·bəl }

addressable cursor [COMPUT SCI] A cursor that can be moved by software or keyboard controls to any point on the screen. { əˈdres·ə·bəl ˈkər·sər }

address book [COMPUT SCI] A feature in an e-mail program for storing e-mail addresses. { ˈad·rəs ˌbu̇k }

address bus [COMPUT SCI] An internal computer communications channel that carries addresses from the central processing unit to components under the unit's control. { ˈad·res ˌbəs }

address computation [COMPUT SCI] The modification by a computer of an address within an instruction, or of an instruction based on results obtained so far. Also known as address modification. { ˈad·res ˌkäm·pyəˈtā·shən }

address constant [COMPUT SCI] A value, or its expression, used in the calculation of storage addresses from relative addresses for computers. Abbreviated ADCON. Also known as base address; presumptive address; reference address. { ˈad·res ˌkän·stənt }

address conversion [COMPUT SCI] The use of an assembly program to translate symbolic or relative computer addresses. { ˈad·res kən ˌvər·zhən }

address counter [COMPUT SCI] A counter which increments an initial memory address as a block of data is being transferred into the memory locations indicated by the counter. { 'ad·res ˌkaůnt·ər }

address field [COMPUT SCI] The portion of a computer program instruction which specifies where a particular piece of information is located in the computer memory. { 'ad·res ˌfēld }

address format [COMPUT SCI] A description of the number of addresses included in a computer instruction. { 'ad·res ˌfór·mat }

address-free program [COMPUT SCI] A computer program in which all addresses are represented as displacements from the expected contents of a base register. { 'ad·res ¦frē 'prō·grəm }

address generation [COMPUT SCI] An addressing technique which facilitates addressing large storages and implementing dynamic program relocation; the effective main storage address is obtained by adding together the contents of the base register of the index register and of the displacement field. { 'ad·res ˌjen·ə'rā·shən }

addressing [COMPUT SCI] **1.** The methods of locating and gaining access to information in a computer's storage. **2.** The methods of selecting a particular peripheral device from several that are available at a given time. { ə'dres·iŋ }

addressing mode [COMPUT SCI] The specific technique by means of which a memory reference instruction will be spelled out if the computer word is too small to contain the memory address. { ə'dres·iŋ ˌmōd }

addressing system [COMPUT SCI] A labeling technique used to identify storage locations within a computer system. { ə'dres·iŋ ˌsis·təm }

address interleaving [COMPUT SCI] The assignment of consecutive addresses to physically separate modules of a computer memory, making possible the very-high-speed access of a sequence of contiguously addressed words, since all modules operate nearly simultaneously. { 'ad·res ˌin·tər'lēv·iŋ }

addressless instruction format *See* zero-address instruction format. { ə'dres·ləs ˌin 'strək·shən 'fór·mat }

address modification *See* address computation. { 'ad·res ˌmäd·ə·fə'kā·shən }

address part [COMPUT SCI] That part of a computer instruction which contains the address of the operand, of the result, or of the next instruction. { 'ad·res ˌpärt }

address register [COMPUT SCI] A register wherein the address part of an instruction is stored by a computer. { 'ad·res ˌrej·ə·stər }

address resolution [COMPUT SCI] **1.** The process of obtaining the actual machine address needed to perform an operation. **2.** The process by which the address used to identify a workstation on a local-area network is translated to an address that can be handled on the Internet. { 'ad·res ˌrez·əˌlü·shən }

address sort routine [COMPUT SCI] A debugging routine which scans all instructions of the program being checked for a given address. { 'ad·res 'sórt ˌrü'tēn }

address space [COMPUT SCI] The number of storage locations available to a computer program. { 'ad·rəs ˌspās }

address track [COMPUT SCI] A path on a magnetic tape, drum, or disk on which are recorded addresses used in the retrieval of data stored on other tracks. { 'ad·res ˌtrak }

address translation [COMPUT SCI] The assignment of actual locations in a computer memory to virtual addresses in a computer program. { 'ad·res tranz'lā·shən }

add-subtract time [COMPUT SCI] The time required to perform an addition or subtraction, exclusive of the time required to obtain the quantities from storage and put the sum or difference back into storage. { 'ad səb'trakt ˌtīm }

add time [COMPUT SCI] The time required by a computer to perform an addition, not including the time needed to obtain the addends from storage and put the sum back into storage. { 'ad ˌtīme }

add-to-memory technique [COMPUT SCI] In direct-memory-access systems, a technique which adds a data word to a memory location; permits linear operations such as data averaging on process data. { ¦ad tə ¦mem·rē 'tek·nēk }

ad hoc inquiry [COMPUT SCI] A single request for a piece of information, such as a report. { 'ad ¦häk in'kwī·rē }

A-display [ELECTR] A radar display in cartesian coordinates; the targets appear as vertical deflection lines; their Y coordinates are proportional to signal intensity; their X coordinates are proportional to distance to targets. Also known as A-indicator; A-scan; A-scope. { 'ā diˌsplā }

adjacency [COMPUT SCI] A condition in character recognition in which two consecutive graphic characters are separated by less than a specified distance. { ə'jās·ən·sē }

adjacent-channel interference [COMMUN] Interference that is caused by a transmitter operating in an adjacent channel. Also known as A-scan; A-scope. { ə'jās·ənt 'chan·əl in·tər 'fir·əns }

adjacent-channel selectivity [ELECTR] The ability of a radio receiver to respond to the desired signal and to reject signals in adjacent frequency channels. { ə'jās·ənt 'chan·əl səˌlek'tiv·əd·ē }

adjusted decibel [ELECTR] A unit used to show the relationship between the interfering effect of a noise frequency, or band of noise frequencies, and a reference noise power level of −85 dBm. Abbreviated dBa. Also known as decibel adjusted. { ə'jəs·təd 'des·əˌbel }

ADP *See* automatic data processing.

ADPCM *See* adaptive differential pulse-code modulation.

ADSEL *See* Mode S.

ADSL *See* asymmetric digital subscriber line; asynchronous digital subscriber loop. { a·dē·es 'el *or* 'ad·səl }

ADT *See* abstract data type.

Advanced Research Projects Agency Network [COMPUT SCI] The computer network developed by the U.S. Department of Defense in 1969 from which the Internet originated. Abbreviated ARPANET. { əd,vanst ri'sərch ,prä,jeks ,ā·jən·sē ,net,wərk }

advanced signal-processing system [COMPUT SCI] A portable data-processing system for military use; its complete configuration may consist of the analyzer unit, a postprocessing unit (for data-processing and control tasks), and an advanced signal-processing display unit. Also known as Proteus. { əd'vanst 'sig·nəl 'präs·əs·iŋ ,sis·təm }

Advanced Television Technology Center A private, nonprofit corporation organized by members of the television broadcasting and consumer products industries to test and recommend technologies for the delivery and recaption of new U.S. digital services. Abbreviated ATTC. { əd'vanst 'tel·ə,vizh·ən tek'näl·ə·jē sen·tər }

aerial *See* antenna. { 'e·rē·əl }

aeronautical mobile satellite service [COMMUN] A mobile satellite service in which the mobile earth stations are located on board aircraft. Abbreviated AMSS. { ,er·ə¦nȯd·ə·kəl ,mō·bəl 'sad·əl,īt ,sər·vəs }

aeronautical mobile service [COMMUN] A mobile service between aircraft stations and land stations, or between aircraft stations, in which survival craft stations may also participate. { ,er·ə¦nȯd·ə·kəl ¦mō·bəl 'sər·vəs }

aerophare *See* radio beacon. { 'e·rə'fer }

AFC *See* automatic frequency control.

affinity [COMPUT SCI] A specific relationship between data processing elements that requires one to be used with the other, where a choice might otherwise exist. { ə'fin·əd·ē }

a format [COMPUT SCI] A nonexecutable statement in FORTRAN which permits alphanumeric characters to be transmitted in a manner similar to numeric data. { 'ā 'fȯr,mat }

AGC *See* automatic gain control.

agenda [COMPUT SCI] **1.** The sequence of control statements required to carry out the solution of a computer problem. **2.** A collection of programs used for manipulating a matrix in the solution of a problem in linear programming. { ə'jen·də }

aggregate data type *See* scalar data type. { 'ag·rə·gət 'da·də tīp }

aggregate function [COMPUT SCI] A command in a database management program that performs an arithmetic operation on the values in a specified column or field in all the records in the database, such as computing their sum or average or counting the number of records that satisfy particular criteria. { ¦ag·rə·gət 'fəŋk·shən }

aggressive device [COMPUT SCI] A unit of a computer that can initiate a request for communication with another device. { ə'gres·iv di'vīs }

AGP *See* accelerated graphics port.

aid to navigation [NAV] A device external to a craft, designed to assist in determination of position of the craft, or of a safe course, or to warn of dangers or obstructions. { ād tə ,nav·ə'gā·shən }

A-indicator *See* A-display. { 'ā ,in·də,kād·ər }

airborne collision avoidance system [NAV] A navigation system for preventing collisions between aircraft that relies primarily on equipment carried on the aircraft itself, but which may make use of equipment already employed in the ground-based air-traffic control system. Abbreviated ACAS. { 'er,bȯrn kə'lizh·ən ə'vȯid·əns ,sis·təm }

airborne collision warning system [ENG] A system such as a radar set or radio receiver carried by an aircraft to warn of the danger of possible collision. { 'er,bȯrn kə'lizh·ən 'wȯrn·iŋ ,sis·təm }

airborne profile recorder [ENG] An electronic instrument that emits a pulsed-type radar signal from an aircraft to measure vertical distances between the aircraft and the earth's surface. Abbreviated APR. Also known as terrain profile recorder (TPR). { 'er,bȯrn 'prō,fīl ri,kȯrd·ər }

airborne radar [ENG] Radar equipment carried by aircraft to assist in navigation by pilotage, to determine drift, and to locate weather disturbances; a very important use is locating other aircraft either for avoidance or attack. { 'er,bȯrn 'rā,där }

airborne self-protection jammer [ELECTR] An electronic system carried by an aircraft to prevent detection by enemy radar by emitting signals that deceive the radar, causing confusion and uncertainty. { 'er,bȯrn ¦self·prə'tek·shən ,jam·ər }

air-control center [COMMUN] An area set aside in a submarine for the control of aircraft; it is the equivalent of a combat information center on an aircraft or a ship. { 'er kən'trōl ,sent·ər }

aircraft antenna [ELECTR] An airborne device used to detect or radiate electromagnetic waves. { 'er,kraft an'ten·ə }

air-ground communication [COMMUN] Two-way communication between aircraft and stations on the ground. { ¦er ¦graůnd kə,myü·nə'kā·shən }

air navigation [NAV] The process of directing and monitoring the progress of an aircraft between selected geographic points or with respect to some predetermined plan. Also known as avigation. { ¦er ,nav·ə'gā·shən }

airport surface detection equipment [ENG] Radar and other equipment specifically designed to assist in the control of aircraft and the many other vehicles that must use taxiways and other surface routes in the airport area. Also known as surface movement radar. { 'er,pȯrt 'sər·fəs di'tek·shən i,kwip·mənt }

airport surveillance radar [ENG] Radar designed for air surveillance and to assist in air traffic management in the area of airports; designated as ASR in the United States nomenclature; usually composed of both primary and secondary radars. { 'er,pȯrt sər'vā·ləns ,rā,där }

air-route surveillance radar [ENG] Radar designed for air surveillance along established air routes to assist, through netted data operation, in air traffic management. Often in rather remote locations, such radars are designed for minimum on-site operator and maintenance attention. Abbreviated ARSR. { ¦er ˌrüt sər¦vā·ləns ˌrāˌdär }

air-spaced coax [ELECTROMAG] Coaxial cable in which air is basically the dielectric material; the conductor may be centered by means of a spirally wound synthetic filament, beads, or braided filaments. { ˈer ˌspāst ˈkōˌaks }

air-traffic control radar beacon system [NAV] A system adopted by the Federal Aviation Agency for use in controlling air traffic over the United States; the aircraft carry identification transponders designed to transmit an airplane identity code, altitude, and additional message when interrogated by an air-traffic controller's equipment. Abbreviated ATCRBS. { ˈer ¦traf·ik kən'trōl ¦rāˌdär ˈbē·kən ˌsis·təm }

airwave [ELECTR] A radio wave used in radio and television broadcasting. { ˈerˌwāv }

alarm signal [ELECTR] The international radiotelegraph alarm signal transmitted to actuate automatic devices that sound an alarm indicating that a distress message is about to be broadcast. { əˈlärm ˌsig·nəl }

ALC *See* automatic level control.

alert box [COMPUT SCI] A dialog box that warns of an existing condition or the consequences of a command that has been given, or explains why a command cannot be executed. { əˈlərt ˌbäks }

alerting signal [COMMUN] Specific signal that is applied to subscriber access lines to indicate an incoming call. { əˈlərt·iŋ ˈsig·nəl }

Alexanderson antenna [ELECTROMAG] An antenna, used at low or very low frequencies, consisting of several base-loaded vertical radiators connected together at the top and fed at the bottom of one radiator. { ˌal·igˈzan·dər·sən an ˌten·ə }

Alford loop [ELECTROMAG] An antenna utilizing multielements which usually are contained in the same horizontal plane and adjusted so that the antenna has approximately equal and in-phase currents uniformly distributed along each of its peripheral elements and produces a substantially circular radiation pattern in the plane of polarization; it is known for its purity of polarization. { ˈȯl·fərd ˌlüp }

algebraic computation system *See* symbolic system. { ˌal·jəˌbrā·ik ˌkäm·pyəˈtā·shən ˌsis·təm }

algebraic manipulation language [COMPUT SCI] A programming language used in the solution of analytic problems by symbolic computation. { ¦al·jə¦brā·ik mə·ni·pyəˈlā·shən ˌlaŋ·gwij }

Algol [COMPUT SCI] An algorithmic and procedure-oriented computer language used principally in the programming of scientific problems. { ˈalˌgȯl }

algorithmic error [COMPUT SCI] An error in computer processing resulting from imprecision in the method used to carry out mathematical computations, usually associated with either rounding or truncation of numbers. { ¦al·gə ¦rith·mik ˈer·ər }

algorithmic language [COMPUT SCI] A language in which a procedure or scheme of calculations can be expressed accurately. { ¦al·gə¦rith·mik ˈlaŋ·gwij }

algorithm translation [COMPUT SCI] A step-by-step computerized method of translating one programming language into another programming language. { ˈal·gəˌrith·əm tranzˈlā·shən }

alias [COMPUT SCI] **1.** An alternative entry point in a computer subroutine at which its execution may begin, if so instructed by another routine. **2.** An alternative name for a file or device. { ˈā·lē·əs }

aliasing [COMPUT SCI] In computer graphics, the jagged appearance of diagonal lines on printouts and on video monitors. { ˈāl·yəs·iŋ }

alignment [ELECTR] The process of adjusting components of a system for proper interrelationship, including the adjustment of tuned circuits for proper frequency response and the time synchronization of the components of a system. [ENG] Placing of surveying points along a straight line. { əˈlīn·mənt }

all-channel tuning [COMMUN] The ability of a television set to receive ultra-high-frequency as well as very-high-frequency channels. { ˈȯl ˌchan·əl ˈtün·iŋ }

all-diffused monolithic integrated circuit [ELECTR] Microcircuit consisting of a silicon substrate into which all of the circuit parts (both active and passive elements) are fabricated by diffusion and related processes. { ¦ȯl də¦fyüzd ˌmän·əˈlith·ik ˈin·təˌgrād·əd ˈsər·kət }

all-digital AM IBOC [COMMUN] The final mode of the AM IBOC system approved by the Federal Communications Commission for use in the United States that increases data capacity by increasing signal power and adjusting the bandwidth of the digital sidebands to minimize adjacent channel interference; uses four frequency partitions and no analog carrier. In this mode, the digital audio data rate can change from 40 to 60 kbits/s, and the corresponding ancillary data rate will remain at 0.4 kbits/s. { ¦ȯl dij·əd·əl ˈāˌem ˈīˌbäk }

all-digital FM IBOC [COMMUN] The third of three modes in the FM IBOC system approved by the Federal Communications Commission for use in the United States that increases data capacity by adding additional digital carriers; uses four frequency partitions and no analog carrier. In this mode, the digital audio data rate can range from 64 to 96 kbits/s, and the corresponding ancillary data rate can range from 213 kbits/s for 64-kbits/s audio to 181 kbits/s for 96-kbits/s audio. { ¦ȯl dij·əd·əl ˈefˌem ˈīˌbäk }

allocate [COMPUT SCI] To place a portion of a computer memory or a peripheral unit under control of a computer program, through the action of an operator, program instruction, or executive program. { ˈa·lōˌkāt }

allotter [COMMUN] A telephone term referring to a distributor, which allots an idle line-finder in preparation for an additional call. { əˈläd·ər }

all-pass network [ELECTR] A network designed to introduce a phase shift in a signal without introducing an appreciable reduction in energy of the signal at any frequency. { ¦ȯl ¦pas 'net ˌwərk }

all-wave receiver [ELECTR] A radio receiver capable of being tuned from about 535 kilohertz to at least 20 megahertz; some go above 100 megahertz and thus cover the FM band also. { ¦ȯl ¦wāv ri'sē·vər }

aloha [COMMUN] A radio-channel random-access technique that depends on positive acknowledgement of correct receipt for error control. { ə'lō·ə }

alphabetic character [COMPUT SCI] A letter or other symbol used to form data, other than a digit. { ¦al·fə¦bed·ik 'kar·ik·tər }

alphabetic coding [COMPUT SCI] **1.** Abbreviation of words for computer input. **2.** A system of coding with a number system of base 26, the letters of the alphabet being used instead of the cardinal numbers. { ¦al·fə¦bed·ik 'kōd·iŋ }

alphabetic string *See* character string. { ¦al·fə ¦bed·ik 'striŋ }

alphageometric technique *See* alphamosaic technique. { ¦al·fəˌjē·ə'me·trik ˌtekˌnēk }

alphameric characters *See* alphanumeric characters. { ¦al·fə¦mer·ik 'kar·ik·tərz }

alphameric typebar [COMPUT SCI] A metal bar containing the alphabet, the ten numerical characters, and the ampersand, in use in electromechanical accounting machines. { ¦al·fə¦mer·ik 'tīpˌbär }

alphamosaic technique [COMPUT SCI] In computer graphics, a technique for displaying very-low-resolution images by constructing them from a set of elementary graphics characters. Also known as alphageometric technique. { ¦al·fə·mō'zā·ik ˌtekˌnēk }

alphanumeric characters [COMPUT SCI] All characters used by a computer, including letters, numerals, punctuation marks, and such signs as $, @, and #. Also known as alphameric characters. { ¦al·fə·nü¦mer·ik 'kar·ik·tərz }

alphanumeric display device [ELECTR] A device which visibly represents alphanumeric output information from some signal source. { ¦al·fə·nü ¦mer·ik dis'plādiˌvīs }

alphanumeric instruction [COMPUT SCI] The name given to instructions which can be read equally well with alphabetic or numeric kinds of fields of data. { ¦al·fə·nü¦mer·ik in'strək·shən }

alphanumeric pager [COMMUN] A receiver in a radio paging system that contains a device which can display text or numeric messages. { ¦al·fə·nü¦mer·ik 'pā·jər }

alphanumeric reader [ELECTR] A device capable of reading alphabetic, numeric, and special characters and punctuation marks. { ¦al·fə·nü ¦mer·ik 'rēd·ər }

alpha test [COMPUT SCI] A test of software carried out at the user's location and using actual data. { 'al·fə ˌtest }

alpha test site [COMPUT SCI] A place where a complete computer system is tested with actual data and transactions. { 'al·fə ˌtest ˌsīt }

alternate-channel interference [COMMUN] Interference that is caused in one communications channel by a transmitter operating in the next channel beyond an adjacent channel. Also known as second-channel interference. { ¦ȯl·tər·nət ¦chan·əl in·tər'fir·əns }

alternate index *See* secondary index. { 'ȯl·tər·nət 'inˌdeks }

alternate key [COMPUT SCI] A key on a computer keyboard that does not itself generate a character but changes the nature of the character generated by another key when depressed simultaneously with it; similar to the control and shift keys. Abbreviated ALT key. { 'ȯl·tər·nət ˌkē }

alternate routing [COMMUN] The operation of a switching center when all circuits are found busy in a programmed route to the destination, and the call is offered to another programmed route. { 'ȯl·tər·nət 'rüt·iŋ }

alternate track [COMPUT SCI] The disk track used if, after a disk volume is initialized, a defective track is sensed by the system. { 'ȯl·tər·nət 'trak }

alternating-current dump [ELECTR] The removal of all alternating-current power from a computer intentionally, accidentally, or conditionally. { ¦ȯl·tərˌnād·iŋ ¦kər·ənt 'dəmp }

altitude delay [ELECTR] Synchronization delay introduced between the time of transmission of the radar pulse and the start of the trace on the indicator to eliminate the altitude/height hole on the plan position indicator-type display. { 'al·tə ˌtüd di'lā }

altitude hole [ELECTR] The blank area in the center of a plan position indicator-type radarscope display caused by the time interval between transmission of a pulse and the receipt of the first ground return. { 'al·təˌtüd ˌhōl }

ALT key *See* alternate key. { 'ȯlt 'kē }

ALU *See* arithmetical unit.

AM *See* amplitude modulation.

amateur bands [COMMUN] Bands of frequencies assigned to licensed radio amateurs. { 'a·mə·chər ˌbanz }

amateur radio [ELECTR] A radio used for two-way radio communications by private individuals as leisure-time activity. Also known as ham radio. { 'a·mə·chər 'rād·ēˌō }

ambiguity [ELECTR] The condition in which a synchro system or servosystem seeks more than one null position. [ELECTROMAG] In radar, the consequence of using a periodic waveform in estimating a target's range and, in coherent radar, its radial velocity by Doppler sensing; deliberate change of periodicity is used to help resolve these ambiguities. [NAV] The condition in which navigation coordinates derived from a navigational instrument define more than one point, direction, line of position, or surface of position. { ˌam·bə'gyü·əd·ē }

ambiguity error [COMPUT SCI] An error in reading a number represented in a digital display that can occur when this representation is changing; for example, the number 699 changing to 700 might be read as 799 because of imprecise synchronization in the changing of digits. { ˌam·bə'gyü·əd·ē ˌer·ər }

ambiguous name [COMPUT SCI] A name of a file or other item which is only partially specified; it is useful in conducting a search of all the items to which it might apply. { am'big·yə·wəs 'nām }

AMC *See* automatic modulation control.

Amdahl's law [COMPUT SCI] A law stating that the speed-up that can be achieved by distributing a computer program over p processors cannot exceed $1/\{f + [1 - f]/p\}$, where f is the fraction of the work of the program that must be done in serial mode. { 'am,dälz ,lȯ }

amendment record *See* change record. { ə'mend·mənt ,rek·ərd }

American Morse code *See* Morse code. { ə'mer·ə·kən 'mȯrs 'kōd }

American Standard Code for Information Interchange [COMMUN] Coded character set to be used for the general interchange of information among information-processing systems, communications systems, and associated equipment; the standard code, comprising characters 0 through 127, includes control codes, upper- and lower-case letters, numerals, punctuation marks, and commonly used symbols; an additional set is known as extended ASCII. Abbreviated ASCII. { ə'mer·ə·kən 'stan·dərd 'kōd fər in·fər'mā·shən 'in·tər,chānj }

AML *See* automatic modulation limiting.

AMLCD *See* active-matrix liquid-crystal display.

amorphous memory array [COMPUT SCI] An array of memory switches made of amorphous material. { ə'mȯr·fəs 'mem·rē ə,rā }

amplifier [ENG] A device capable of increasing the magnitude or power level of a physical quantity, such as an electric current or a hydraulic mechanical force, that is varying with time, without distorting the wave shape of the quantity. { 'am·plə,fī·ər }

amplifying delay line [ELECTR] Delay line used in pulse-compression systems to amplify delayed signals in the super-high-frequency region. { 'am·plə,fī·iŋ di'lā,līn }

amplitude distortion *See* frequency distortion. { 'am·plə,tüd di'stȯr·shən }

amplitude fading [COMMUN] Fading in which the amplitudes of frequency components of a modulated carrier wave are uniformly attenuated. { 'am·plə,tüd ,fād·iŋ }

amplitude-frequency distortion *See* frequency distortion. { 'am·plə,tüd 'frē·kwən·sē di'stȯr·shən }

amplitude-frequency response *See* frequency response. { 'am·plə,tüd 'frē·kwən·sē ri'späns }

amplitude-modulated indicator [ENG] A general class of radar indicators, in which the sweep of the electron beam is deflected vertically or horizontally from a base line to indicate the existence of an echo from a target. Also known as deflection-modulated indicator; intensity-modulated indicator. { 'am·plə,tüd ¦mäj·ə,lād·əd ¦in·də,kād·ər }

amplitude modulation [ELECTR] Abbreviated AM. **1.** Modulation in which the aplitude of a wave is the characteristic varied in accordance with the intelligence to be transmitted. **2.** In telemetry, those systems of modulation in which each component frequency f of the transmitted intelligence produces a pair of sideband frequencies at carrier frequency plus f and carrier minus f. { 'am·plə,tüd ,maj·ə'lā·shən }

amplitude-modulation noise [COMMUN] Noise produced by undesirable amplitude variations of a radio-frequency signal. { 'am·plə,tüd ,maj·ə'lā·shən ,nȯiz }

amplitude-modulation radio [COMMUN] Also known as AM radio. **1.** The system of radio communication employing amplitude modulation of a radio-frequency carrier to convey the intelligence. **2.** A receiver used in such a system. { 'am·plə,tüd ,maj·ə'lā·shən 'rād·ē,ō }

amplitude modulator [PHYS] Any device which imposes amplitude modulation upon a carrier wave in accordance with a desired program. { 'am·plə,tüd 'maj·ə,lād·ər }

amplitude noise [ELECTROMAG] Effect on radar accuracy of the fluctuations in the amplitude of the signal returned by the target; these fluctuations are caused by any change in aspect if the target is not a point source. { 'am·plə,tüd ,nȯiz }

amplitude resonance [PHYS] The frequency at which a given sinusoidal excitation produces the maximum amplitude of oscillation in a resonant system. { 'am·plə,tüd 'rez·ə·nəns }

amplitude response [ELECTR] The maximum output amplitude obtainable at various points over the frequency range of an instrument operating under rated conditions. { 'am·plə,tüd ri'späns }

amplitude selector *See* pulse-height selector. { 'am·plə,tüd si,lek·tər }

amplitude separator [ELECTR] A circuit used to isolate the portion of a waveform with amplitudes above or below a given value or between two given values. { 'am·plə,tüd 'sep·ə,rād·ər }

amplitude shift keying [COMMUN] A method of transmitting binary coded messages in which a sinusoidal carrier is pulsed so that one of the binary states is represented by the presence of the carrier while the other is represented by its absence. Abbreviated ASK. { 'am·plə,tüd ¦shift ,kē·iŋ }

amplitude suppression ratio [ELECTR] Ratio, in frequency modulation, of the undesired output to the desired output of a frequency-modulated receiver when the applied signal has simultaneous amplitude and frequency modulation. { 'am·plə,tüd sə'presh·ən ,rā·shō }

amplitude-versus-frequency distortion [ELECTR] The distortion caused by the nonuniform attenuation or gain of the system, with respect to frequency under specified terminal conditions. { 'am·plə,tüd ¦vər·səs ¦frē·kwən·sē di'stȯr·shən }

AM radio *See* amplitude-modulation radio. { ¦ā¦em 'rād·ē,ō }

AM signature [COMMUN] A graphic representation of the significant identifying characteristics of an amplitude-modulated signal. { ¦ā¦em 'sig·nə·chər }

AMSS *See* aeronautical mobile satellite service.

analog [ELECTR] **1.** A physical variable which remains similar to another variable insofar as the proportional relationships are the same over some specified range; for example, a temperature may be represented by a voltage which is its analog. **2.** Pertaining to devices, data, circuits, or systems that operate with variables which are represented by continuously measured voltages or other quantities. { 'an·əl,äg }

analog adder [ELECTR] A device with one output voltage which is a weighted sum of two input voltages. { 'an·əl,äg 'ad·ər }

analog channel [ELECTR] A channel on which the information transmitted can have any value between the channel limits, such as a voice channel. { 'an·əl,äg 'chan·əl }

analog communications [COMMUN] System of telecommunications employing a nominally continuous electric signal that varies in frequency, amplitude, or other characteristic, in some direct correlation to nonelectrical information (sound, light, and so on) impressed on a transducer. { 'an·əl,äg kə,myü·nə'kā·shənz }

analog comparator [ELECTR] **1.** A comparator that checks digital values to determine whether they are within predetermined upper and lower limits. **2.** A comparator that produces high and low digital output signals when the sum of two analog voltages is positive and negative, respectively. { 'an·əl,äg kəm'par·əd·ər }

analog computer [COMPUT SCI] A computer is which quantities are represented by physical variables; problem parameters are translated into equivalent mechanical or electrical circuits as an analog for the physical phenomenon being investigated. { 'an·əl,äg kəm'pyüd·ər }

analog data [COMPUT SCI] Data represented in a continuous form, as contrasted with digital data having discrete values. { 'an·əl,äg 'dad·ə }

analog-digital computer *See* hybrid computer. { 'an·əl,äg 'dij·ə·təl kəm,pyüd·ər }

analog monitor [ELECTR] A display unit that accepts only analog signals, which must be converted from digital signals by the computer's video display board. { 'an·əl·äg ,män·əd·ər }

analog multiplexer [ELECTR] A multiplexer that provides switching of analog input signals to allow use of a common analog-to-digital converter. { 'an·əl,äg 'məl·tə,plek·sər }

analog multiplier [ELECTR] A device that accepts two or more inputs in analog form and then produces an output proportional to the product of the input quantities. { 'an·əl,äg 'məl·tə ,plī·ər }

analog simulation [COMPUT SCI] The representation of physical systems and phenomena by variables such as translation, rotation, resistance, and voltage. { 'an·əl,äg ,sim·yə'lā·shən }

analog-to-digital converter [ELECTR] A device which translates continuous analog signals into proportional discrete digital signals. { ¦an·əl,äg tə ¦dij·ət·əl kən'vərd·ər }

analysis by synthesis [COMMUN] A method of determining the parameters of a speech coder in which the consequence of choosing a particular value of a coder parameter is evaluated by locally decoding the signal and comparing it to the original input signal. { ə¦nal·ə·sis ,bī 'sin·thə·səs }

analytical engine [COMPUT SCI] An early-19th-century form of mechanically operated digital computer. { ,an·əl'id·ə·kəl 'en·jən }

analytical function generator [ELECTR] An analog computer device in which the dependence of an output variable on one or more input variables is given by a function that also appears in a physical law. Also known as natural function generator; natural law function generator. { ,an·əl'id·ə·kəl 'fəŋk·shən ,jen·ə,rād·ər }

analytic hierarchy [MATH] A systematic procedure for representing the elements of any problem which breaks down the problem into its smaller constituents and then calls for only simple pairwise comparison judgments to develop priorities at each level. { ,an·əl'id·ik 'hī·ər,är·kē }

analyzer [COMPUT SCI] **1.** A routine for the checking of a program. **2.** One of several types of computers used to solve differential equations. { 'an·ə,līz·ər }

anchor [COMPUT SCI] A tag that indicates either the source or destination of a hyperlink; for example, HTML anchors are used to create links within a document or to another document. { 'aŋ·kər }

anchored graphic [COMPUT SCI] A picture or graph that remains at a fixed position on a page of a document rather than being attached to the text. { ¦aŋ·kərd 'graf·ik }

anchor frame [COMMUN] In MPEG-2, a video frame that is used for prediction. I-frames and P-frames are generally used as anchor frames, but B-frames are never anchor frames. { 'aŋ·kər ,frām }

AND circuit *See* AND gate. { 'and ,sər·kət }

AND gate [ELECTR] A circuit which has two or more input-signal ports and which delivers an output only if and when every input signal port is simultaneously energized. Also known as AND circuit; passive AND gate. { 'and ,gāt }

AND/NOR gate [ELECTR] A single logic element whose operation is equivalent to that of two AND gates with outputs feeding into a NOR gate. { ¦and ¦nȯr ,gāt }

AND NOT gate [ELECTR] A coincidence circuit that performs the logic operation AND NOT, under which a result is true only if statement A is true and statement B is not. Also known as A AND NOT B gate. { ¦and ¦nät ,gāt }

AND-OR circuit [ELECTR] Gating circuit that produces a prescribed output condition when several possible combined input signals are applied; exhibits the characteristics of the AND gate and the OR gate. { ¦and ¦ȯr ,sər·kət }

AND-OR-INVERT gate [ELECTR] A logic circuit with four inputs, a_1, a_2, b_1, and b_2, whose output is 0 only if either a_1 and a_2 or b_1 and b_2 are 1. Abbreviated A-O-I gate. { ¦and ¦ȯr in'vərt ˌgāt }

anechoic chamber [ENG] **1.** A test room in which all surfaces are lined with a sound-absorbing material to reduce reflections of sound to a minimum. Also known as dead room; free-field room. **2.** A room well shielded from interference and lined with signal-absorbing material within which measurements of radar and radio performance can be reliably made. { ¦an·ə¦kō·ik 'chām·bər }

angel echo [ENG] A radar echo from a region where there are no visible targets; may be caused by insects, birds, or refractive index variations in the atmosphere. { 'ān·jəl ˌek·ō }

angle diversity [COMMUN] Diversity reception in which beyond-the-horizon tropospheric scatter signals are received at slightly different angles, equivalent to paths through different scatter volumes in the troposphere. { 'aŋ·gəl də'vər·səd·ē }

angle jamming [ELECTR] Electronic countermeasures used to introduce large errors in angle-measuring radars; methods involve producing a false echo with pulse-to-pulse modulation that is inverse to that otherwise produced by a radar using conical scanning, or the generation of multiple interfering signals that may confuse monopulse radars. { 'aŋ·gəl ˌjam·iŋ }

angle marker *See* azimuth marker. { 'aŋ·gəl ˌmärk·ər }

angle modulation [ELECTR] The variation in the angle of a sine-wave carrier; particular forms are phase modulation and frequency modulation. Also known as sinusoidal angular modulation. { 'aŋ·gəl mäj·ə'lā·shən }

angle tracking noise [ELECTR] Deviation of the tracking axis or other angle estimate from the true angle of a radar target; it results from target reflective behavior and propagation path characteristics (such as fluctuation, glint, and scintillation) and also from the radar's own receiver, mechanical or computational noise. { 'aŋ·gəl ¦trak·iŋ ˌnȯiz }

ANL *See* automatic noise limiter.

annotation [COMPUT SCI] Any comment or note included in a program or flow chart in order to clarify some point at issue. { ˌan·ə'tā·shən }

anode [ELECTR] **1.** The collector of electrons in an electron tube. Also known as plate; positive electrode. **2.** In a semiconductor diode, the terminal toward which forward current flows from the external circuit. { 'aˌnōd }

anomaly detection [COMPUT SCI] The technology that seeks to identify an attack on a computer system by looking for behavior that is out of the norm. { ə'näm·ə·lē diˌtek·shən }

anonymous FTP [COMPUT SCI] A public FTP (file transfer protocol) site at which users can log in and download documents by entering "anonymous" as their user ID, and their e-mail address as password. { əˌnän·ə·məs ¦ef¦tē'pē }

A-N radio range [NAV] A type of radio beacon station whose signals provide definite track guidance for aircraft by establishing four radial lines of position which can be identified by a continuous-tone signal made up of keyed pulses of equal amplitude representing the Morse code letters A and N. { ¦ā¦en 'rād·ēˌō ˌrānj }

answer back [COMPUT SCI] The ability of a device such as a computer or terminal to automatically identify itself when it is contacted by another communicating device. { 'an·sər ¦bak }

answer-only modem [COMMUN] A modem that can answer but not initiate a call. { ¦an·sər ˌōn·lē 'mōˌdem }

antenna [ELECTROMAG] A device used for radiating or receiving radio waves. Also known as aerial; radio antenna. { an'ten·ə }

antenna amplifier [ELECTROMAG] One or more stages of wide-band electronic amplification placed within or physically close to a receiving antenna to improve signal-to-noise ratio and mutually isolate various devices receiving their feed from the antenna. { an'ten·ə 'am·pləˌfī·ər }

antenna coil [ELECTROMAG] Coil through which antenna current flows. { an'ten·ə ˌkȯil }

antenna coupler [ELECTROMAG] A radio-frequency transformer, tuned line, or other device used to transfer energy efficiently from a transmitter to a transmission line or from a transmission line to a receiver. { an'ten·ə ˌkəp·lər }

antenna crosstalk [ELECTROMAG] The ratio or the logarithm of the ratio of the undesired power received by one antenna from another to the power transmitted by the other. { an' ten·ə'krȯs ˌtȯk }

antenna directivity diagram [ELECTROMAG] Curve representing, in polar or cartesian coordinates, a quantity proportional to the gain of an antenna in the various directions in a particular plane or cone. { an'ten·ə di·rek'tiv·əd·ē 'dī·əˌgram }

antenna effect [ELECTROMAG] A distortion of the directional properties of a loop antenna caused by an input to the direction-finding receiver which is generated between the loop and ground, in contrast to that which is generated between the two terminals of the loop. Also known as electrostatic error; vertical component effect. { an'ten·ə i'fekt }

antenna efficiency [ELECTROMAG] The ratio of the amount of power radiated into space by an antenna to the total energy received by the antenna. { an'ten·ə iˌfish·ən·sē }

antenna field [ELECTROMAG] A group of antennas placed in a geometric configuration. { an'ten·ə ˌfēld }

antenna gain [ELECTROMAG] A measure of the effectiveness of a directional antenna as compared to a standard nondirectional antenna. Also known as gain. { an'ten·ə ˌgān }

antenna matching [ELECTROMAG] Process of adjusting impedances so that the impedance of an antenna equals the characteristic impedance of its transmission line. { an'ten·ə ˌmach·iŋ }

antenna pair [ELECTROMAG] Two antennas located on a base line of accurately surveyed length, sometimes arranged so that the array may be rotated around an axis at the center of the base line; used to produce directional patterns and in direction finding. { an'ten·ə ˌper }

antenna pattern See radiation pattern. { an'ten·ə ˌpad·ərn }

antenna polarization [ELECTROMAG] The orientation of the electric field lines in the electromagnetic field radiated or received by the antenna. { an'ten·ə ˌpō·lə·rə'zā·shən }

antenna power [ELECTROMAG] Radio-frequency power delivered to an antenna. { an'ten·ə ˌpaů·ər }

antenna power gain [ELECTROMAG] The ratio of the spatial power density on transmit, or sensitivity on receive, experienced at a distant point for using an actual directive antenna, as in radar, to that density or sensitivity experienced had an imaginary isotropic antenna been used. Power gain will, then, be slightly less than directive gain, differing by the insertion loss of the actual antenna, and is the gain actually measured in constructed antennas and used in most calculations about radar performance. { an'ten·ə 'paů·ər ˌgān }

antenna resistance [ELECTROMAG] The power supplied to an entire antenna divided by the square of the effective antenna current measured at the point where power is supplied to the antenna. { an'ten·ə riˌzis·təns }

antenna scanner [ELECTROMAG] A microwave feed horn which moves in such a way as to illuminate sequentially different reflecting elements of an antenna array and thus produce the desired field pattern. { an'ten·ə ˌskan·ər }

antenna tilt error [ENG] Angular difference between the tilt angle of a radar antenna shown on a mechanical indicator, and the electrical center of the radar beam. { an'ten·ə 'tilt ˌer·ər }

antennna directive gain [ELECTROMAG] The ratio of the spatial power density on transmit, or sensitivity on receive, experienced at a distant point for using an idealized (lossless) directive antenna, as in radar, to that density of sensitivity experienced had an imaginary isotropic antenna been used. { ¦an'ten·ə 'də'rek·tiv ˌgān }

antialiasing technique [COMPUT SCI] In computer graphics, a technique for smoothing the jagged appearance of diagonal lines on printouts and on video monitors. { ˌan·tē'āl·ē·əs·iŋ ˌtek ˌnēk }

anticipatory staging [COMPUT SCI] Moving blocks of data from one storage device to another prior to the actual request for them by the program. { 'an'tis·ə·pəˌtȯr·ē 'stāj·iŋ }

anticoincidence circuit [ELECTR] Circuit that produces a specified output pulse when one (frequently predesignated) of two inputs receives a pulse and the other receives no pulse within an assigned time interval. { ˌan·tēˌkō'in·sə·dəns ˌsər·kət }

anticollision radar [ENG] A radar set designed to give warning of possible collisions during movements of ships or aircraft. { ˌan·tē·kə'li·zhən ˌrā ˌdär }

antiglare shield [COMPUT SCI] A sheet of nonreflective material placed over the screen of an electronic display to reduce the amount of light reflected from the screen. { 'an·tēˌgler 'shēld }

anti-intrusion technology [COMPUT SCI] One of the different ways in which an attack on a computer system can be detected and countered, including prevention, deterrence, detection, deflection, and diminution. { ˌan·tēˌin'trü·zhən ˌtek¦näl·ə·jē }

antijamming [ELECTR] Any system or technique used to counteract the jamming of communications or of radar operation; part of electronic protection. { ˌan·tē'jam·iŋ }

antiradar coating [ENG] A surface treatment used to reduce the reflection of electromagnetic waves so as to avoid radar detection. { ˌan·tē'rā ˌdär ¦kōd·iŋ }

antistatic mat [COMPUT SCI] A floor mat placed in front of a device such as a tape drive that is sensitive to discharges of static electricity to safeguard against loss of data from such discharges during human handling of the device. { ¦an·tē¦stad·ik 'mat }

antivirus software [COMPUT SCI] Software that is designed to protect against computer viruses. { ¦an·tēˌvī·rəs 'sȯfˌwer }

A-O-I gate See AND-OR-INVERT gate. { ˌāˌō'ī ˌgāt }

APC See automatic phase control.

aperiodic antenna [ELECTROMAG] Antenna designed to have constant impedance over a wide range of frequencies because of the suppression of reflections within the antenna system; includes terminated wave and rhombic antennas. { ¦āˌpir·ē¦äd·ik an'ten·ə }

aperture antenna [ELECTROMAG] Antenna in which the beam width is determined by the dimensions of a horn, lens, or reflector. { 'ap·ə ˌchər an'ten·ə }

aperture grill picture tube [ELECTR] An in-line gun-type picture tube in which the shadow mask is perforated by long, vertical stripes and the screen is coated with vertical phosphor stripes. { 'ap·əˌchər ˌgril 'pik·chər ˌtüb }

aperture mask See shadow mask. { 'ap·əˌchər ˌmask }

aperture synthesis [ELECTROMAG] The use of one or more pairs of instruments of relatively small aperture, acting as interferometers, to obtain the information-gathering capability of a telescope of much larger aperture. { 'ap·əˌchər ˌsin·thə·səs }

API See application program interface.

APL [COMPUT SCI] An interactive computer language whose operators accept and produce arrays with homogeneous elements of type number or character.

apodization [ELECTR] A technique for modifying the response of a surface acoustic wave filter by varying the overlap between adjacent electrodes of the interdigital transducer. { ˌa·pə·də'zā·shən }

applet [COMPUT SCI] A small program, typically written in Java. { 'ap·lət }

Appleton layer *See* F_2 layer. { 'ap·əl·tən ˌlā·ər }

application [COMPUT SCI] A computer program that performs a specific task, for example, a word processor, a Web browser, or a spread sheet. { ˌap·lə'kā·shən }

application development language [COMPUT SCI] A very-high-level programming language that generates coding in a conventional programming language or provides the user of a data-base management system with a programming language that is easier to implement than conventional programming languages. { ˌap·lə'kā·shən di'vel·əp·mənt ˌlaŋ·gwij }

application development system [COMPUT SCI] An integrated group of software products used to assist in the efficient development of computer programs and systems. { ˌap·lə'kā·shən di'vel·əp·mənt ˌsis·təm }

application generator [COMPUT SCI] A commercially prepared software package used to create applications programs or parts of such programs. { ˌap·lə'kā·shən ˌjen·əˌrād·ər }

application package [COMPUT SCI] A combination of required hardware, including remote inputs and outputs, plus programming of the computer memory to produce the specified results. { ˌap·lə'kā·shən ˌpak·ij }

application processor [COMPUT SCI] A computer that processes data. { ˌap·lə¦kā·shən 'prä ˌses·ər }

application program [COMPUT SCI] A program written to solve a specific problem, produce a specific report, or update a specific file. { ˌap·lə'kā·shən ˌprō·grəm }

application program interface [COMPUT SCI] A language that enables communication between computer programs, in particular between application programs and control programs. Abbreviated API. { ˌap·lə¦kā·shən ¦prō·grəm 'in·tərˌfās }

application server [COMPUT SCI] A computer that executes commands requested by a Web server to fetch data from databases. Also known as app server. { ˌap·lə'kā·shən ˌser·vər }

application-specific integrated circuit [ELECTR] An integrated circuit that is designed for a particular application by integrating standard cells from a library, making possible short design times and rapid production cycles. Abbreviated ASIC. { ˌap·ləˌkā·shən spi¦sif·ik ˌint·iˌgrād·əd 'sər·kət }

application study [COMPUT SCI] The detailed process of determining a system or set of procedures for using a computer for definite functions of operations, and establishing specifications to be used as a base for the selection of equipment suitable to the specific needs. { ap·lə'kā·shən ˌstəd·ē }

application system [COMPUT SCI] A group of related applications programs designed to perform a specific function. { ˌap·lə'kā·shən ˌsis·təm }

application window [COMPUT SCI] In a graphical user interface, the chief window of an application program, with a title bar, a menu bar, and a work area. { ˌap·lə'kā·shən ˌwinˌdō }

applicative language [COMPUT SCI] A programming language in which functions are repeatedly applied to the results of other functions and, in its pure form, there are no statements, only expressions without side effects. { 'ap·ləˌkād·iv 'laŋ·gwij }

applied epistemology [COMPUT SCI] The use of machines or other models to simulate processes such as perception, recognition, learning, and selective recall, or the application of principles assumed to hold for human categorization, perception, storage, search, and so on, to the design of machines, machine programs, scanning, storage, and retrieval systems. { ə'plīd i¦pis·tə ¦mäl·ə·jē }

approach vector [CONT SYS] A vector that describes the orientation of a robot gripper and points in the direction from which the gripper approaches a workpiece. { ə'prōch ˌvek·tər }

app server *See* application server. { 'ap ˌsər·vər }

APT *See* Automatic Programming Tool.

arbiter [COMPUT SCI] A computer unit that determines the priority sequence in which two or more processor inputs are connected to a single functional unit such as a multiplier or memory. { 'är·bəd·ər }

arbitrary function generator *See* general-purpose function generator. { 'är·bəˌtrer·ē 'fəŋk·shən ˌjen·əˌrād·ər }

arbitration [COMPUT SCI] The set of rules in a computer's operating system for allocating the resources of the computer, such as its peripheral devices or memory, to more than one program or user. { ˌar·bə'trā·shən }

Archie [COMPUT SCI] A system of file servers that searches for specific files that are publicly available in File Transfer Protocol archives on the Internet. { 'är·chē }

archival storage [COMPUT SCI] Storage of infrequently used or backup information that cannot be readily or immediately accessed by a computer system. { 'ärˌkīv·əl 'stȯr·ij }

archiving [COMPUT SCI] The storage of files in auxiliary storage media for very long periods, in the event it is necessary to regenerate the file due to subsequent errors introduced. { 'ärˌkīv·iŋ }

area [COMPUT SCI] A section of a computer memory assigned by a computer program or by the hardware to hold data of a particular type. { 'er·ē·ə }

area code [COMMUN] A three-digit prefix used in dialing long-distance telephone calls in the United States and Canada. { 'er·ē·ə ˌkōd }

areal density [COMPUT SCI] The amount of data that can be stored on a unit area of the surface of a hard disk, floppy disk, or other storage device. { ˌer·ē·əl 'den·səd·ē }

area navigation [NAV] An aircraft navigation system in which radial and distance information from a Vortac station or distance information from two or more Vortacs is used to fix aircraft position, so that navigation is not restricted to airways or direct routes between stations. Abbreviated RNAV. { ēr·ē·ə ˌnav·ə'gā·shən }

area search [COMPUT SCI] A computer search that examines only those records which satisfy some broad criteria. { 'er·ē·ə ˌsərch }

A register See arithmetic register. { 'ā ˌrej·ə·stər }

ARGOS system [COMMUN] An operational international satellite system to collect, locate, and disseminate environmental data. { 'är·gos ˌsis·təm }

argument [COMPUT SCI] A value applied to a procedure, subroutine, or macroinstruction which is required in order to evaluate any of these. { 'är·gyə·mənt }

argument separator [COMPUT SCI] A comma or other punctuation mark that separates successive arguments in a command or statement in a computer program. { 'är·gyü·mənt ˌsep·əˌrād·ər }

arithmetic address [COMPUT SCI] An address in a computer program that results from performing an arithmetic operation on another address. { ¦a·rith¦med·ik ə'dres }

arithmetical element See arithmetical unit. { ¦a·rith¦med·ə·kəl 'el·ə·mənt }

arithmetical instruction [COMPUT SCI] An instruction in a computer program that directs the computer to perform an arithmetical operation (addition, subtraction, multiplication, or division) upon specified items of data. { ¦a·rith ¦med·ə·kəl ˌin'strək·shən }

arithmetical operation [COMPUT SCI] A digital computer operation in which numerical quantities are added, subtracted, multiplied, divided, or compared. { ¦a·rith¦med·ə·kəl ˌäp·ə'rā·shən }

arithmetical unit [COMPUT SCI] The section of the computer which carries out all arithmetic and logic operations. Also known as arithmetical element; arithmetic-logic unit (ALU); arithmetic section; logic-arithmetic unit; logic section. { ¦a·rith¦med·ə·kəl 'yü·nət }

arithmetic check [COMPUT SCI] The verification of an arithmetical operation or series of operations by another such process; for example, the multiplication of 73 by 21 to check the result of multiplying 21 by 73. { ə'rith·məˌtik ˌchek }

arithmetic circuitry [COMPUT SCI] The section of the computer circuitry which carries out the arithmetic operations. { ¦a·rith¦med·ik 'sər·kə·trē }

arithmetic coding [COMMUN] A method of data compression in which a long character string is represented by a single number whose value is obtained by repeatedly partitioning the range of possible values in proportion to the probabilities of the characters. { ¦a·rith¦med·ik 'cōd·iŋ }

arithmetic-logic unit See arithmetical unit. { ə¦rith·məˌtik 'läj·ik ˌyü·nət }

arithmetic processor See numeric processor extension. { ə'rith·məˌtik ˌpräsˌes·ər }

arithmetic register [COMPUT SCI] A specific memory location reserved for intermediate results of arithmetic operations. Also known as A register. { ¦a·rith¦med·ik 'rej·ə·stər }

arithmetic scan [COMPUT SCI] The procedure for examining arithmetic expressions and determining the order of execution of operators, in the process of compilation into machine-executable code of a program written in a higher-level language. { ¦a·rith¦med·ik ˌskan }

arithmetic section See arithmetical unit. { ¦a·rith¦med·ik ˌsek·shən }

arithmetic shift [COMPUT SCI] A shift of the digits of a number, expressed in a positional notation system, in the register without changing the sign of the number. { ¦a·rith¦med·ik 'shift }

ARPA See automated radar plotting aid. { 'är·pə }

ARQ See automatic repeat request.

array [COMPUT SCI] A collection of data items with each identified by a subscript or key and arranged in such a way that a computer can examine the collection and retrieve data from these items associated with a particular subscript or key. [ELECTR] A group of components such as antennas, reflectors, or directors arranged to provide a desired variation of radiation transmission or reception with direction. { ə'rā }

array element [COMPUT SCI] A single data item in an array. { ə'rāˌ el·ə·mənt }

array processor [COMPUT SCI] A multiprocessor composed of a set of identical central processing units acting synchronously under the control of a common unit. { ə'rā 'präsˌes·ər }

array radar [ENG] A radar incorporating a multiplicity of phased antenna elements. { ə'rā 'rā ˌdär }

array sonar [ENG] A sonar system incorporating a phased array of radiating and receiving transducers. { ə'rā 'sōˌnär }

ARSR See air-route surveillance radar.

articulation [COMMUN] The percentage of speech units understood correctly by a listener in a communications system; it generally applies to unrelated words, as in code messages, in distinction to intelligibility. [CONT SYS] The manner and actions of joining components of a robot with connecting parts or links that allow motion. { ärˌtik·yə'lā·shən }

articulation equivalent [COMMUN] Of a complete telephone connection, a measure of the articulation of speech reproduced over it, expressed numerically in terms of the trunk loss of a working reference system when the latter is adjusted to give equal articulation. { ärˌtik·yə'lā·shən i'kwiv·ə·lənt }

artifact [COMMUN] Any component of a signal that is extraneous to the variable represented by the signal. { 'ärd·əˌfakt }

artificial antenna See dummy antenna. { ¦ärd·ə ¦fish·əl an'ten·ə }

artificial delay line See delay line. { ¦ärd·ə ¦fish·əl di'lāˌlīn }

artificial intelligence [COMPUT SCI] The property of a machine capable of reason by which it can learn functions normally associated with human intelligence. { ¦ärd·ə¦fish·əl in'tel·ə·jəns }

artificial ionization [COMMUN] Introduction of an artificial reflecting or scattering layer into the atmosphere to permit beyond-the-horizon communications. { ¦ärd·ə¦fish·əl ˌī·ə·nə'zā·shən }

artificial language [COMPUT SCI] A computer language that is specifically designed to facilitate communication in a particular field, but is not yet natural to that field; opposite of a natural language, which evolves through long usage. { ¦ärd·ə¦fish·əl 'laŋ·gwij }

artificial radio aurora [COMMUN] Modification of the ionosphere by high-power high-frequency radio transmitters to improve scatter and auroral long-distance communication. Also known as radio aurora. { ¦ärd·ə¦fish·əl 'rād·ē‚ō ə'rȯr·ə }

artificial reality *See* virtual reality. { ˌärd·ə'fish·əl rē'al·əd·ē }

A-scan *See* A-display. { 'ā ˌskan }

ascending sort [COMPUT SCI] The arrangement of records or other data into a sequence running from the lowest to the highest in a specified field. { ə'send·iŋ 'sȯrt }

ASCII *See* American Standard Code for Information Interchange. { 'as‚kē }

ASCII file [COMPUT SCI] A data or text file that contains only codes that constitute the 128-character ASCII set. { ¦as‚kē ¦fīl }

ASCII protocol [COMMUN] A protocol for the simplest mode of transmitting ASCII data, with little or no error checking. { ¦as‚kē ¦prōd·ə‚kȯl }

ASCII sort order [COMPUT SCI] A sort order determined by the numbering of characters in the American Standard Code for Information Interchange. { ¦as‚kē 'sȯrt ˌȯrd·ər }

A-scope *See* A-display. { 'ā skōp }

ASIC *See* application-specific integrated circuit. { 'ā‚sik *or* ¦ā¦es¦ī'sē }

ASK *See* amplitude shift keying.

aspect ratio [COMPUT SCI] In computer graphics, the ratio between the width and height of an image. [DES ENG] The ratio of frame width to frame height in television; in the United States and Britain it is 4:3 for standard television and 16:9 for high-definition television. { 'a‚spekt ˌrā·shō }

assembler [COMPUT SCI] A program designed to convert symbolic instruction into a form suitable for execution on a computer. Also known as assembly program; assembly routine. { ə'sem·blər }

assembler directive [COMPUT SCI] A statement in an assembly-language program that gives instructions to the assembler and does not generate machine language. { ə'sem·blər di‚rek·tiv }

assembler language *See* assembly language. { ə'sem·blər ˌlaŋ·gwij }

assembler program [COMPUT SCI] A program that is written in assembly language. { ə'sem·blər ˌprō·grəm }

assembly [COMPUT SCI] The automatic translation into machine language of a computer program written in symbolic language. { ə'sem·blē }

assembly language [COMPUT SCI] A symbolic, nonbinary format for instructions (human-readable version of machine language) that allows mnemonic names to be used for instructions and data; for example, the instruction to add the number 39321 to the contents of register D1 in the central processing unit might be written as ADD#39321, D1 in assembly language, as opposed to a string of 0's and 1's in machine language. { ə'sem·blē ˌlaŋ·gwij }

assembly list [COMPUT SCI] A printed list which is the by-product of an assembly procedure; it lists in logical instruction sequence all details of a routine, showing the coded and symbolic notation next to the actual notations established by the assembly procedure; this listing is highly useful in the debugging of a routine. { ə'sem·blē ˌlist }

assembly program *See* assembler. { ə'sem·blē 'prō·grəm }

assembly robot [COMPUT SCI] A robot that positions, mates, fits, and assembles components or parts and adjusts the finished product to function as intended. { ə'sem·blē ˌrō‚bät }

assembly routine *See* assembler. { ə'sem·blē rü'tēn }

assembly system [COMPUT SCI] An automatic programming software system with a programming language and machine-language programs that aid the programmer by performing different functions such as checkout and updating. { ə'sem·blē ˌsis·təm }

assembly unit [COMPUT SCI] **1.** A device which performs the function of associating and joining several parts or piecing together a program. **2.** A portion of a program which is capable of being assembled into a larger program. { ə'sem·blē ˌyü·nət }

assign [COMPUT SCI] A control statement in FORTRAN which assigns a computed value *i* to a variable *k*, the latter representing the number of the statement to which control is then transferred. { ə'sīn }

assignment problem [COMPUT SCI] A special case of the transportation problem in a linear program, in which the number of sources (assignees) equals the number of designations (assignments) and each supply and each demand equals 1. { ə'sīn·mənt 'präb·ləm }

assignment statement [COMPUT SCI] A statement in a computer program that assigns a value to a variable. { ə'sīn·mənt ˌstāt·mənt }

assisted panel [COMPUT SCI] In an interactive system, a screen that explains a question the computer has asked, the available options, the expected format, and so forth. { ə'sis·təd 'pan·əl }

associated document [COMPUT SCI] A file that is linked to the application program in which it was created, so that the application can be started by choosing such a file. { ə,sō·sē,ād·əd 'däk·yə·mənt }

association trail [COMPUT SCI] A linkage between two or more documents or items of information, discerned during the process of their examination and recorded with the aid of an information retrieval system. { ə,sō·sē'ā·shən ,trāl }

associative dimensioning system [COMPUT SCI] A system for making automatic changes in the dimensions of workpieces manufactured by machine tools. { ə'sō·sē,ād·iv di'men·shən·iŋ 'sis·təm }

associative key [COMPUT SCI] In a computer system with an associative memory, a field used to reference items through comparing the value of the field with corresponding fields in each memory cell and retrieving the contents of matching cells. { ə'sō·sē,ād·iv 'kē }

associative memory [COMPUT SCI] A data-storage device in which a location is identified by its informational content rather than by names, addresses, or relative positions, and from which the data may be retrieved. Also known as associative storage. { ə'sō·sē,ād·iv 'mem·rē }

associative processor [COMPUT SCI] A digital computer that consists of a content-addressable memory and means for searching rapidly changing random digital data stored within, at speeds up to 1000 times faster than conventional digital computers. { ə'sō·sē,ād·iv 'präs,es·ər }

associative storage *See* associative memory. { ə'sō·sē,ād·iv 'stȯr·ij }

associator [COMPUT SCI] A device for bringing like entities into conjunction or juxtaposition. { ə'sō·sē,ād·ər }

assumed decimal point [COMPUT SCI] For a decimal number stored in a computer or appearing on a printout, a position in the number at which place values change from positive to negative powers of 10, but to which no location is assigned or at which no printed character appears, as opposed to an actual decimal point. Also known as virtual decimal point. { ə'sümd 'des·məl ,pȯint }

astable multivibrator [ELECTR] A multivibrator in which each active device alternately conducts and is cut off for intervals of time determined by circuit constants, without use of external triggers. Also known as free-running multivibrator. { ā'stā·bəl ,məlt·i'vī,brād·ər }

A station [NAV] In loran, the designation applied to one transmitting station of a pair, the signal of which always occurs less than half a repetition period after the preceding signal and more than half a repetition period before the succeeding signal of the other station, designated a B station. { 'ā ¦stā·shən }

astigmatism [ELECTR] In an electron-beam tube, a focus defect in which electrons in different axial planes come to focus at different points. { ə'stig·mə,tiz·əm }

asymmetrical modem [COMMUN] A modem that simultaneously transmits and receives data, but at different speeds. { ,ā·si¦me·trə·kəl 'mō ,dem }

asymmetrical-sideband transmission *See* vestigial-sideband transmission. { ¦ā·sə¦me·tri·kəl 'sīd,band ,tranz'mish·ən }

asymmetric digital subscriber line [COMMUN] A broadband communication technology designed for use on conventional telephone lines, which reserves more bandwidth for receiving data than for sending data. Abbreviated ADSL. { ¦ā·sə'me·trik ¦dij·ə·dəl ,səb'skrī·bər ,līn }

asynchronous [COMPUT SCI] Operating at a speed determined by the circuit functions rather than by timing signals. { ā'siŋ·krə·nəs }

asynchronous communications [COMMUN] The transmission and recognition of a single character at a time. { ā'siŋ·krə·nəs kə,myü·nə'kā·shənz }

asynchronous communications adaptor [COMPUT SCI] A device connected to a computer to allow it to carry out asynchronous communications over a telephone line. { ā'siŋ·krə·nəs kə ,myü·nə'kā·shənz ə,dap·tər }

asynchronous computer [COMPUT SCI] A computer in which the performance of any operation starts as a result of a signal that the previous operation has been completed, rather than on a signal from a master clock. { ā'siŋ·krə·nəs kəm'pyüd·ər }

asynchronous control [CONT SYS] A method of control in which the time allotted for performing an operation depends on the time actually required for the operation, rather than on a predetermined fraction of a fixed machine cycle. { ā'siŋ·krə·nəs kən'trōl }

asynchronous data [COMPUT SCI] Information which is sampled at irregular intervals with respect to another operation. { ā'siŋ·krə·nəs 'dad·ə }

asynchronous digital subscriber loop *See* asymmetric digital subscriber line. { ā'siŋ·krə·nəs 'dij·əd·əl səb'skrīb·ər ,lüp }

asynchronous input/output [COMPUT SCI] The ability to receive input data while simultaneously outputting data. { ā'siŋ·krə·nəs 'in,pu̇t 'au̇t ,pu̇t }

asynchronous logic [ELECTR] A logic network in which the speed of operation depends only on the signal propagation through the network. { ā'siŋ·krə·nəs 'läj·ik }

asynchronous operation [ELECTR] An operation that is started by a completion signal from a previous operation, proceeds at the maximum speed of the circuits until finished, and then generates its own completion signal. { ā'siŋ·krə·nəs ,äp·ə'rā·shən }

asynchronous time-division multiplexing [COMMUN] A data-transmission technique in which several users utilize a single channel by means of a system which assigns time slots only to active channels. { ā'siŋ·krə·nəs 'tīm də'vi·zhən 'məlt·i ,pleks·iŋ }

asynchronous transfer mode [COMMUN] A high-speed packet-switching technology based on cell-oriented switching and multiplexing that uses 53-byte packets to transfer different types of information, such as voice, video, and data, over the same communications network at different speeds. Abbreviated ATM. { ˌā¦siŋ·krə·nəs 'tranz·fər ˌmōd }

asynchronous transmission [COMMUN] Data transmission in which each character contains its own start and stop pulses and there is no control over the time between characters. { ā'siŋ·krə·nəs ˌtranz'mish·ən }

asynchronous working [COMPUT SCI] The mode of operation of a computer in which an operation is performed only at the end of the preceding operation. { ā'siŋ·krə·nəs 'wərk·iŋ }

asyndetic [COMPUT SCI] **1.** Omitting conjunctions or connectives. **2.** Pertaining to a catalog without cross references. { ¦as·ən¦ded·ik }

ATCRBS *See* air-traffic control radar beacon system.

ATDM *See* asynchronous time-division multiplexing.

ATM *See* asynchronous transfer mode; automatic teller machine.

atmospheric attenuation [GEOPHYS] The loss of radar or radio signals sent through earth's (or other) atmosphere due to the thermal agitation of various gas molecules as the electromagnetic wave passes through; oxygen and water vapor are the two most sensitive gases in the microwave region, with severity generally, but very linearly, increasing with frequency. { ¦at·mə¦sfir·ik əˌten·yə'wā·shən }

atmospheric noise [ELECTR] Noise heard during radio reception due to atmospheric interference. { ¦at·mə¦sfir·ik 'nȯiz }

atmospheric radio wave [ELECTROMAG] Radio wave that is propagated by reflection in the atmosphere; may include either the ionospheric wave or the tropospheric wave, or both. { ¦at·mə ¦sfir·ik 'rād·ē·ō ˌwāv }

atom [COMPUT SCI] A primitive data element in a data structure. { 'ad·əm }

atomic operation [COMPUT SCI] An operation that cannot be broken up into smaller parts that could be performed by different processors. { ə'täm·ik ˌäp·ə'rā·shən }

attached processing [COMPUT SCI] A method of data processing in which several relatively inexpensive computers dedicated to specific tasks are connected together to provide a greater processing capability. { ə'tacht 'präsˌes·iŋ }

attached processor [COMPUT SCI] A computer that is electronically connected to and operates under the control of another computer. { ə'tacht 'präsˌes·ər }

attachment [COMPUT SCI] An additional file sent with an e-mail message. { ə'tach·mənt }

attachment unit interface [COMMUN] A 15-pin connector on an Ethernet card for connecting a network cable. Abbreviated AUI. { ə¦tach·mənt ˌyü·nət 'in·tərˌfās }

attack director [COMPUT SCI] An electromechanical analog computer which is designed for surface antisubmarine use and which computes continuous solution of several lines of submarine attack; it is part of several antisubmarine fire control systems. { ə'tak di'rek·tər }

ATTC *See* Advanced Television Technology Center.

attendant's switchboard [COMMUN] Switchboard of one or more positions in a central-office location which permits the central-office operator to receive, transmit, or cut in on a call to or from one of the lines which the office services. { ə'ten·dəns 'swichˌbȯrd }

attended time [COMPUT SCI] The time in which a computer is either switched on and capable of normal operation (including time during which it is temporarily idle but still watched over by computer personnel) or out of service for maintenance work. { ə'tend·əd ¦tīm }

attenuation constant [PHYS] A rating for a line or medium through which a plane wave is being transmitted, equal to the relative rate of decrease of an amplitude of a field component, voltage, or current in the direction of propagation, in nepers per unit length. { əˌten·yə'wā·shən ˌkän·stənt }

attenuation distortion [COMMUN] **1.** In a circuit or system, departure from uniform amplification or attenuation over the frequency range required for transmission. **2.** The effect of such departure on a transmitted signal. { əˌten·yə'wā·shən dis ˌtȯr·shən }

attenuator [ELECTR] An adjustable or fixed transducer for reducing the amplitude of a wave without introducing appreciable distortion. { ə'ten·yəˌwād·ər }

attribute [COMPUT SCI] **1.** A data item containing information about a variable. **2.** A characteristic of computer-generated characters, such as underline, boldface, or reverse image. { 'a·trə ˌbyüt }

audible feedback [COMPUT SCI] A feature of a computer keyboard that generates sound each time a key is depressed sufficiently to generate a character on the screen. { ˌȯd·ə·bəl 'fēdˌbak }

audio adapter *See* sound board. { ˌȯd·ē·ō ə'dap·tər }

audio amplifier *See* audio-frequency amplifier. { 'ȯd·ē·ō 'am·pləˌfī·ər }

audio-frequency amplifier [ELECTR] An electronic circuit for amplification of signals within, and in some cases above, the audible range of frequencies in equipment used to record and reproduce sound. Also known as audio amplifier. { 'ȯd·ē·ō ¦frē·kwən·sē ¦am·pləˌfī·ər }

audio-frequency shift modulation [COMMUN] System of facsimile transmission over radio, in which the frequency shift required is applied through a change in audio signal, rather than shifting the radio transmitter frequency; the radio signal is modulated by the shifting audio signal, usually at 1500 to 2300 hertz. { 'ȯd·ē·ō ¦frē·kwən·sē ˌshift mäj·ə'lā·shən }

audio response [COMMUN] A form of computer output in which prerecorded spoken syllables, words, or messages are selected and put together by a computer as the appropriate verbal response to a keyboarded inquiry on a time-shared on-line information system. { 'òd·ē·ō ri'späns }

audio response unit [COMMUN] A system that provides voice response to an inquiry; the inquiry is typically made using the dual-tone multifrequency (DTMF) dial on a telephone set. { 'òd·ē·ō ri'späns ˌyü·nət }

audiovisual [COMMUN] Pertaining to methods of education and training that make use of both hearing and sight. { ¦òd·ē·ō¦vizh·ə·wəl }

audit [COMPUT SCI] The operations developed to corroborate the evidence as regards authenticity and validity of the data that are introduced into the data-processing problem or system. { 'òd·ət }

audit total [COMPUT SCI] A count or sum of a known quantity, calculated in order to verify data. { 'òd·ət ˌtōd·əl }

audit trail [COMPUT SCI] A system that provides a means for tracing items of data from processing step to step, particularly from a machine-produced report or other machine output back to the original source data. { 'òd·ət ˌtrāl }

augmented operation code [COMPUT SCI] An operation code which is further defined by information from another portion of an instruction. { 'òg·men·təd äp·ə'rā·shən ˌkōd }

AUI *See* attachment unit interface.

aural transmitter [COMMUN] Radio equipment used for transmitting aural (sound) signals from a television broadcast station. { 'òr·əl ˌtranz'mid·ər }

auroral propagation [COMMUN] The propagation of radio waves that are reflected from the aurora in the presence of unusual solar activity. { ə'ròr·əl ˌpräp·ə'gā·shən }

authentication [COMMUN] Security measure designed to protect a communications system against fraudulent transmissions and establish the authenticity of a message. { əˌthent·ə'kā·shən }

authenticator [COMMUN] Letter, numeral, or groups of letters or numerals attesting to the authenticity of a message or transmission. { ə'thent·əˌkād·ər }

authoring language [COMPUT SCI] A programming language designed to be convenient for authors of computer-based learning materials. { 'ò·thər·iŋ 'laŋ·gwij }

authorization code [COMPUT SCI] A password or identifying number that is used to gain access to a computer system. { ˌòth·ə·rə'zā·shən ˌkōd }

authorized carrier frequency [COMMUN] A specific carrier frequency authorized for use, from which the actual carrier frequency is permitted to deviate, solely because of frequency instability, by an amount not to exceed the frequency tolerance. { 'ò·thəˌrīzd 'kar·ē·ər ˌfrē·kwən·sē }

authorized library [COMPUT SCI] A group of authorized programs. { 'ò·thəˌrīzed 'līˌbrer·ē }

authorized program [COMPUT SCI] A computer program that can alter the fundamental operation or status of a computer system. { 'ò·thə ˌrizd 'prō·grəm }

auto-abstract [COMPUT SCI] **1.** To select key words from a document, commonly by an automatic or machine method, for the purpose of forming an abstract of the document. **2.** The material abstracted from a document by machine methods. { ¦òd·ō 'abˌstrakt }

autoadaptivity [CONT SYS] The ability of an advanced robot to sense the environment, accept commands, and analyze and execute operations. { ¦òd·ō·ˌəˌdap'tiv·əd·ē }

autoalarm *See* automatic alarm receiver. { 'òd·ō·əˌlärm }

auto answer [COMMUN] The feature of a modem that receives the telephone ring for an incoming call and accepts the call to establish a connection. { ¦òd·ō 'an·sər }

auto bypass [COMPUT SCI] The ability of a computer network to bypass a terminal or other device if it fails, allowing other devices connected to the network to continue operation. { ¦òd·ō 'bīˌpas }

autocall [COMPUT SCI] The automatic placing of a telephone call by a computer or a computer-controlled modem. Also known as automatic call origination. { 'òd·ōˌkòl }

autocode [COMPUT SCI] The process of using a computer to convert automatically a symbolic code into a machine code. Also known as automatic code. { 'òd·ōˌkōd }

autocoder [COMPUT SCI] A person or machine producing or using autocode as a part or the whole of a task. { 'òd·ōˌkōd·ər }

autodecrement addressing [COMPUT SCI] An addressing mode of computers in which the register is first decremented and then used as a pointer. { ˌòd·ō'dek·rə·mənt ə'dres·iŋ }

auto dial [COMMUN] The feature of a modem that automatically opens a telephone line and dials the telephone of a receiving computer to establish a connection. { ¦òd·ō 'dīl }

autodyne reception [COMMUN] System of heterodyne reception through the use of a device which is both an oscillator and a detector. { 'òd·ōˌdīn ri'sep·shən }

autoincrement addressing [COMPUT SCI] An addressing mode of minicomputers in which the operand address is gotten from the specified register which is then incremented. { ¦òd·ō'iŋ·krə·mənt ə'dres·iŋ }

autoindexing *See* automatic indexing. { ¦òd·ō 'inˌdeks·iŋ }

automata theory [MATH] A theory concerned with models used to simulate objects and processes such as computers, digital circuits, nervous systems, cellular growth and reproduction. { ò'täm·əd·ə 'thē·ə·rē }

automated decision making [COMPUT SCI] The use of computers to carry out tasks requiring the generation or selection of options. { ¦òd·ə ˌmād·əd di'sizh·ən ˌmāk·iŋ }

automated guided vehicle system [CONT SYS] A computer-controlled system that uses pallets and other interface equipment to transport workpieces to numerically controlled machine tools and other equipment in a flexible manufacturing system, moving in a predetermined pattern to ensure automatic, accurate, and rapid work-machine contact. { 'ȯd·ə,mād·əd ¦gīd·əd 'vē·ə·kəl ,sis·təm }

automated identification system [COMPUT SCI] In a data processing system, the use of a technology such as bar coding, image recognition, or voice recognition instead of keyboarding for data entry. { ,ȯd·ə¦mād·əd ī,den·tə·fə'ka·shən ,sis·təm }

automated radar plotting aid [NAV] A marine computer-based anticollision system that automatically processes time coordinates of radar echo signals into space coordinates in digital form, determines consecutive coordinates and motion parameters of targets, calculates the predicted closest point of approach and time to closest point of approach and presents them in graphic or alphanumeric form on the radar display, and switches on alarms if there is a danger of collision. { 'ȯd·ə,mād·əd ¦rā,där 'pläd·iŋ ,ād }

automated tape library [COMPUT SCI] A computer storage system consisting of several thousand magnetic tapes and equipment under computer control which automatically brings the tapes from storage, mounts them on tape drives, dismounts the tapes when the job is completed, and returns them to storage. { 'ȯd·ə,mād·əd 'tāp ¦lī,brer·ē }

automatic abstracting [COMPUT SCI] Techniques whereby, on the basis of statistical properties, a subset of the sentences in a document is selected as representative of the general content of that document. { ¦ȯd·ə ¦mad·ik 'ab,strakt·iŋ }

automatic acceleration *See* dynamic resolution. { ¦ȯd·ə¦mad·ik ik,sel·ə'rā·shən }

automatic alarm receiver [ELECTR] A complete receiving, selecting, and warning device capable of being actuated automatically by intercepted radio-frequency waves forming the international automatic alarm signal. Also known as autoalarm. { ¦ȯd·ə¦mad·ik ə'lärm ri,sē·vər }

automatic-alarm-signal keying device [COMMUN] A device capable of automatically keying the radiotelegraph transmitter on board a vessel to transmit the international automatic-alarm signal, or to respond to receipt of an internationally agreed-upon distress signal and wake up the radio operator on ships not having a 24-hour radio watch. { ¦ȯd·ə¦mad·ik ə'lärm ,sig·nəl 'kē·iŋ di,vīs }

automatic back bias [ELECTR] Radar technique which consists of one or more automatic gain control loops to prevent overloading of a receiver by large signals, whether jamming or actual radar echoes. { ¦ȯd·ə¦mad·ik 'bak ,bī·əs }

automatic background control *See* automatic brightness control. { ¦ȯd·ə¦mad·ik 'bak,graủnd kən,trōl }

automatic bass compensation [ELECTR] A circuit related to the volume control in some radio receivers and audio amplifiers to make bass notes sound properly balanced, in the audio spectrum, at low volume-control settings. { ¦ȯd·ə¦mad·ik 'bās käm·pən'sā·shən }

automatic brightness control [ELECTR] A circuit used in an analog television receiver to keep the average brightness of the reproduced image essentially constant. Abbreviated ABC. Also known as automatic background control. { ¦ȯd·ə¦mad·ik 'brīt·nəs kən,trōl }

automatic calling unit [COMPUT SCI] A device that enables a business machine or computer to automatically dial calls over a communications network. { ¦ȯd·ə¦mad·ik 'kȯl·iŋ ,yü·nət }

automatic call origination *See* autocall. { ¦ȯd·ə¦mad·ik 'kȯl ə,rij·ə'nā·shən }

automatic carriage [COMPUT SCI] Any mechanism designed to feed continuous paper or plastic forms through a printing or writing device, often using sprockets to engage holes in the paper. { ¦ȯd·ə¦mad·ik 'kar·ij }

automatic character recognition [COMPUT SCI] The technology of using special machine systems to identify human-readable symbols, most often alphanumeric, and then to utilize this data. { ¦ȯd·ə¦mad·ik 'kar·ik·tər ,rek·ig'nish·ən }

automatic check [COMPUT SCI] An error-detecting procedure performed by a computer as an integral part of the normal operation of a device, with no human attention required unless an error is actually detected. { ¦ȯd·ə¦mad·ik 'chek }

automatic chroma control *See* automatic color control. { ¦ȯd·ə¦mad·ik 'krōm·ə kən,trōl }

automatic chrominance control *See* automatic color control. { ¦ȯd·ə¦mad·ik 'krōm·ə·nəns kən ,trōl }

automatic code *See* autocode. { ¦ȯd·ə¦mad·ik 'kōd }

automatic coding [COMPUT SCI] Any technique in which a computer is used to help bridge the gap between some intellectual and manual form of describing the steps to be followed in solving a given problem, and some final coding of the same problem for a given computer. { ¦ȯd·ə¦mad·ik 'kōd·iŋ }

automatic color control [ELECTR] A circuit used in an analog color television receiver to keep color intensity levels essentially constant despite variations in the strength of the received color signal; control is usually achieved by varying the gain of the chrominance band-pass amplifier. Also known as automatic chroma control; automatic chrominance control. { ¦ȯd·ə¦mad·ik 'kəl·ər kən,trōl }

automatic computer [COMPUT SCI] A computer which can carry out a special set of operations without human intervention. { ¦ȯd·ə¦mad·ik kəm'pyüd·ər }

automatic contrast control [ELECTR] A circuit that varies the gain of the radio-frequency and video intermediate-frequency amplifiers in such a way that the contrast of the television picture is maintained at a constant average level. { ¦ȯd·ə¦mad·ik 'kän,trast kən,trōl }

automatic control [CONT SYS] Control in which regulating and switching operations are performed automatically in response to predetermined conditions. Also known as automatic regulation. { ¦ȯd·ə¦mad·ik kən,trōl }

automatic controller [CONT SYS] An instrument that continuously measures the value of a variable quantity or condition and then automatically acts on the controlled equipment to correct any deviation from a desired preset value. Also known as automatic regulator; controller. { ¦ȯd·ə¦mad·ik kən¦trōl·ər }

automatic data processing [ENG] The machine performance, with little or no human assistance, of any of a variety of tasks involving informational data; examples include automatic and responsive reading, computation, writing, speaking, directing artillery, and the running of an entire factory. Abbreviated ADP. { ¦ȯd·ə¦mad·ik ¦dad·ə 'präs,əs·iŋ }

automatic degausser [ELECTR] An arrangement of degaussing coils mounted around a color television picture tube, combined with a circuit that energizes these coils only while the set is warming up; demagnetizes any parts of the receiver that have been affected by the magnetic field of the earth or of any nearby devices. { ¦ȯd·ə¦mad·ik dē'gau̇s·ər }

automatic detection [ELECTR] A computer-based process in radar wherin the receiver's output video is examined, compared to appropriate thresholds and contacts (detections) reported; augments or replaces the similar role played by the human operator viewing an analog display of the video in more elementary radar. { ¦ȯd·ə¦mad·ik di'tek·shən }

automatic dialer [ELECTR] A device in which a telephone number up to some maximum number of digits can be stored in a memory and then activated, directly into the line, by the caller's pressing a button. { ¦ȯd·ə¦mad·ik 'dīl·ər }

automatic dictionary [COMPUT SCI] Any table within a computer memory which establishes a one-to-one correspondence between two sets of characters. { ¦ȯd·ə¦mad·ik 'dik·shə,ner·ē }

automatic error correction [COMMUN] A technique, usually requiring the use of special codes or automatic retransmission, which detects and corrects errors occurring in transmission; the degree of correction depends upon coding and equipment configuration. { ¦ȯd·ə¦mad·ik 'er·ər kə'rek·shən }

automatic exchange [ELECTR] A telephone, teletypewriter, or data-transmission exchange in which communication between subscribers is effected, without the intervention of an operator, by devices set in operation by the originating subscriber's instrument (for example, the dial n a telephone). Also known as automatic switching system; machine switching system. { ¦ȯd·ə ¦mad·ik iks'chanj }

automatic fine-tuning control [ELECTR] A circuit used in a color television receiver to maintain the correct oscillator frequency in the tuner for best reception by compensating for drift and incorrect tuning. { ¦ȯd·ə¦mad·ik ,fīn 'tün·iŋ kən ,trōl }

automatic frequency control [ELECTR] Abbreviated AFC. **1.** A circuit used to maintain the frequency of an oscillator within specified limits, as in a transmitter. **2.** A circuit used to keep a superheterodyne receiver tuned accurately to a given frequency by controlling its local oscillator, as in an FM receiver. **3.** A circuit used in radar superheterodyne receivers to vary the local oscillator frequency so as to compensate for changes in the frequency of the received echo signal. **4.** A circuit used in television receivers to make the frequency of a sweep oscillator correspond to the frequency of the synchronizing pulses in the received signal. { ¦ȯd·ə¦mad·ik 'frē·kwən·sē kən,trōl }

automatic gain control [ELECTR] A control circuit that automatically changes the gain (amplification) of a receiver or other piece of equipment so that the desired output signal remains essentially constant despite variations in input signal strength. Abbreviated AGC. { ¦ȯd·ə¦mad·ik 'gān kən,trōl }

automatic head parking [COMPUT SCI] A feature that moves the read/write head of a hard disk over the landing zone whenever power is shut off to ensure against a head crash. { ¦ȯd·ə¦mad·ik 'hed ,pärk·iŋ }

automatic indexing [COMPUT SCI] Selection of key words from a document by computer for use as index entries. Also known as autoindexing. [CONT SYS] The procedure for determining the orientation and position of a workpiece with respect to an automatically controlled machine, such as a robot manipulator, that is to perform an operation on it. { ¦ȯd·ə¦mad·ik 'in,deks·iŋ }

automatic intercept [COMMUN] Telephone service that automatically records messages a caller may leave when the called party is away from his telephone. This may be an answering machine or a function provided by an automatic exchange. { ¦ȯd·ə¦mad·ik 'in·tər,sept }

automatic interrupt [COMPUT SCI] Interruption of a computer program brought about by a hardware device or executive program acting as a result of some event which has occurred independently of the interrupted program. { ¦ȯd·ə¦mad·ik 'in·tə,rəpt }

automatic level compensation [COMMUN] System which automatically compensates for amplitude variations in a circuit. { ¦ȯd·ə¦mad·ik 'lev·əl ,käm·pen'sā·shən }

automatic level control [ELECTR] A circuit that keeps the output of a radio transmitter, tape recorder, or other device essentially constant, even in the presence of large changes in the input amplitude. Abbreviated ALC. { ¦ȯd·ə¦mad·ik 'lev·əl kən,trōl }

automatic mathematical translator [COMPUT SCI] An automatic-programming computer capable of receiving a mathematical equation from a remote input and returning an immediate solution. { ¦ȯd·ə¦mad·ik ,math·ə'mad·ə·kəl 'tranz,lād·ər }

automatic message accounting [COMMUN] System whereby toll calls are automatically recorded and timed. { ¦ȯd·ə¦mad·ik 'mes·ij ə,kaunt·iŋ }

automatic message-switching center [COMMUN] A center in which messages are automatically routed according to information in them. { ¦ȯd·ə¦mad·ik 'mes·ij ,swich·iŋ ,sen·tər }

automatic modulation control [ELECTR] A transmitter circuit that reduces the gain for excessively strong audio input signals without affecting the strength of normal signals, thereby permitting higher average modulation without overmodulation. Abbreviated AMC. { ¦ȯd·ə¦mad·ik ,mäj·ə'lā·shən kən,trōl }

automatic modulation limiting [COMMUN] A circuit that prevents overmodulation in some citizen-band radio transmitters by reducing the gain of one or more audio amplifier stages when the voice signal becomes stronger. Abbreviated AML. { ¦ȯd·ə¦mad·ik mäj·ə'lā·shən ,lim·əd·iŋ }

automatic noise limiter [ELECTR] A circuit that clips impulse and static noise peaks, and sets the level of limiting or clipping according to the strength of the incoming signal, so that the desired signal is not affected. Abbreviated ANL. { ¦ȯd·ə¦mad·ik 'nȯiz ,lim·əd·ər }

automatic phase control [ELECTR] **1.** A circuit used in color television receivers to reinsert a 3.58-megahertz carrier signal with exactly the correct phase and frequency by synchronizing it with the transmitted color-burst signal. **2.** An automatic frequency-control circuit in which the difference between two frequency sources is fed to a phase detector that produces the required control signal. Abbreviated APC. { ¦ȯd·ə¦mad·ik 'fāz kən,trōl }

automatic programming [COMPUT SCI] The preparation of machine-language instructions by use of a computer. { ¦ȯd·ə¦mad·ik 'prō,gram·iŋ }

Automatic Programming Tool [COMPUT SCI] A computer language used to program numerically controlled machine tools. Abbreviated APT. { ¦ȯd·ə¦mad·ik 'prō,gram·iŋ ,tül }

automatic regulation *See* automatic control. { ¦ȯd·ə¦mad·ik ,reg·yə'lā·shən }

automatic regulator *See* automatic controller. { ¦ȯd·ə¦mad·ik 'reg·yə,lād·ər }

automatic relay [COMMUN] Means of selective switching which causes automatic equipment to record and retransmit communications. { ¦ȯd·ə¦mad·ik 'rē,lā }

automatic repeat request [COMPUT SCI] A request from a receiving device to retransmit the most recent block of data. Abbreviated ARQ. { ¦ȯd·ə¦mad·ik ri'pēt ri,kwest }

automatic routine [COMPUT SCI] A routine that is executed independently of manual operations, but only if certain conditions occur within a program or record, or during some other process. { ¦ȯd·ə¦mad·ik rü'tēn }

automatic scanning receiver [ELECTR] A receiver which can automatically and continuously sweep across a preselected frequency, either to stop when a signal is found or to plot signal occupancy within the frequency spectrum being swept. { ¦ȯd·ə¦mad·ik 'skan·iŋ ri,sē·vər }

automatic sensitivity control [ELECTR] Circuit used for automatically maintaining receiver sensitivity at a predetermined level; it is similar to automatic gain control, but it affects the receiver constantly rather than during the brief interval selected by the range gate. { ¦ȯd·ə¦mad·ik sen·sə'tiv·əd·ē kən,trōl }

automatic sequences [COMPUT SCI] The characteristic of a computer that can perform successive operations without human intervention. { ¦ȯd·ə¦mad·ik 'sē·kwən·səs }

automatic shutdown [COMPUT SCI] A procedure whereby a network or computer system stops work in an orderly fashion with as little data loss and other damage as possible when the system's software determines that it has encountered unacceptable conditions. { ¦ȯd·ə¦mad·ik 'shət ,daun }

automatic speed sensing [COMPUT SCI] The capability of a modem to automatically determine the maximum rate of data transfer over a connection. { ¦ȯd·ə¦mad·ik 'spēd ,sen·siŋ }

automatic stop [COMPUT SCI] An automatic halting of a computer processing operation as the result of an error detected by built-in checking devices. { ¦ȯd·ə¦mad·ik 'stäp }

automatic switchboard [COMMUN] Telephone switchboard in which the connections are made by using remotely controlled switches. { ¦ȯd·ə¦mad·ik 'swich,bȯrd }

automatic switching system *See* automatic exchange. { ¦ȯd·ə¦mad·ik 'swich·iŋ ,sis·təm }

automatic teller machine [COMPUT SCI] A banking terminal that is activated by inserting a magnetic card containing the user's account number, and that accepts deposits, dispenses cash, provides information about current balances, and may perform other services such as making payments and transfers and providing account statements. Abbreviated ATM. { ¦ȯd·ə¦mad·ik 'tel·ər mə,shēn }

automatic threshold variation [ELECTR] Constant false-alarm rate scheme that is an open-loop of automatic gain control in which the decision threshold is varied continuously in proportion to the incoming intermediate frequency and video noise level. { ¦ȯd·ə¦mad·ik 'thresh ,hōld ,ver·ē'ā·shən }

automatic tint control [ELECTR] A circuit used in color television receivers to maintain correct flesh tones by correcting phase errors before the chroma signal is demodulated. { ¦ȯd·ə¦mad·ik 'tint kən,trōl }

automatic tracking [ELECTR] A computer-based process in radar wherein successive contacts (detections) are associated and tracks of targets are estimated and updated with further observations. [NAV] **1.** Tracking in which a servomechanism autpmatically follows some characteristic of the signal; specifically, a process by which tracking or data-acquisition systems are enabled to keep their antennas continuously directed at a moving target without manual operation. **2.** An instrument which displays the actual course made good through the use of navigation derived from several sources. { ¦ȯd·ə¦mad·ik 'trak·iŋ }

automatic tuning system [CONT SYS] An electrical, mechanical, or electromechanical system that tunes a radio receiver or transmitter automatically to a predetermined frequency when a button or lever is pressed, a knob turned, or a telephone-type dial operated. { ¦ȯd·ə¦mad·ik 'tün·iŋ ˌsis·təm }

automatic video noise leveling [ELECTR] Constant false-alarm rate scheme in which the video noise level at the output of the receiver is sampled at the end of each range sweep and the receiver gain is readjusted accordingly to maintain a constant video noise level at the output. { ¦ȯd·ə¦mad·ik ¦vid·ē·ō 'nȯiz ˌlev·əl·iŋ }

automatic volume control [ELECTR] An automatic gain control that keeps the output volume of a radio receiver essentially constant despite variations in input-signal strength during fading or when tuning from station to station. Abbreviated AVC. { ¦ȯd·ə¦mad·ik 'väl·yəm kən,trōl }

automation [ENG] **1.** The use of technology to ease human labor or extend the mental or physical capabilities of humans. **2.** The mechanisms, machines, and systems that save or eliminate labor, or imitate actions typically associated with human beings. { ˌȯd·ə'mā·shən }

automaton [COMPUT SCI] A robot which functions without step-by-step guidance by a human operator. { ȯ'täm·ə,tän }

automechanism [CONT SYS] A machine or other device that operates automatically or under control of a servomechanism. { ¦ȯd·ō'mek·əˌniz·əm }

automonitor [COMPUT SCI] A computer program used in debugging which instructs a computer to make a record of its own operations. { ¦ȯd·ō¦män·əd·ər }

autonomous channel operation [COMPUT SCI] The rapid transfer of data between computer peripherals and the main store in which an entire block of data is transferred, word by word; the cycles of storage time for the word transfer are stolen from those available to the central processing unit. { ȯ'tän·ə·məs ¦chan·əl ˌäp·ə'rā·shən }

autonomous robot [ENG] A robot that not only can maintain its own stability as it moves, but also can plan its movements. { ȯ¦tän·ə·məs 'rō ˌbät }

autonomous vehicle [ENG] A vehicle that is able to plan its path and to execute its plan without human intervention. { ȯ¦tän·ə·məs 'vē·ə·kəl }

autopatch [ELECTR] A device for connecting radio transceivers to telephone lines by remote control, generally through the use of repeaters. { 'ȯd·ō,pach }

autoplotter [COMPUT SCI] A machine which automatically draws a graph from input data. { 'ȯd·ō,pläd·ər }

autostability [CONT SYS] The ability of a device (such as a servomechanism) to hold a steady position, either by virtue of its shape and proportions, or by control by a servomechanism. { ¦ȯd·ō·stə'bil·əd·ē }

autostart routine [COMPUT SCI] A set of instructions that is permanently stored in a computer memory and activated when the computer is turned on, to perform diagnostic tests and then load the operating system. { 'ȯd·ō,stärt rü,tēn }

autotest program [COMPUT SCI] A computer program within the operating system that aids in testing and debugging programs. { 'ȯd·ō ˌtest 'prō·grəm }

autotrace [COMPUT SCI] A routine that locates outlines of raster graphics images and transforms them into vector graphics, usually at higher resolution. { 'ȯd·ō,trās }

auxiliary channel [COMMUN] A secondary path for low-speed communication that uses the same circuit as a higher-speed stream of data. { ȯg'zil·yə·rē 'chan·əl }

auxiliary equipment *See* off-line equipment. { ȯg'zil·yə·rē ə'kwip·mənt }

auxiliary instruction buffer [COMPUT SCI] A section of storage in the instruction unit, 16 bytes in length, used to hold prefetched instructions. { ȯg'zil·yə·rē in'strək·shən ˌbəf·ər }

auxiliary memory [COMPUT SCI] **1.** A high-speed memory that is in a large main frame or supercomputer, is not directly addressable by the central processing unit, and is connected to the main memory by a high-speed data channel. **2.** *See* auxiliary storage. { ȯg¦zil·yə·rē 'mem·rē }

auxiliary operation [COMPUT SCI] An operation performed by equipment not under continuous control of the central processing unit of a computer. { ȯg'zil·yə·rē ˌäp·ə'rā·shən }

auxiliary processor [COMPUT SCI] Any equipment which performs an auxiliary operation in a computer. { ȯg'zil·yə·rē 'präs,es·ər }

auxiliary routine [COMPUT SCI] A routine designed to assist in the operation of the computer and in debugging other routines. { ȯg'zil·yə·rē rü'tēn }

auxiliary storage [COMPUT SCI] Storage device in addition to the main storage of a computer; for example, magnetic tape, magnetic or optical disk, or magnetic drum. Also known as auxiliary memory. { ȯg'zil·yə·rē 'stȯr·ij }

availability [COMPUT SCI] Of data, data channels, and input-output devices in computers, the condition of being ready for use and not immediately committed to other tasks. [NAV] The probability that a navigational system will function and provide required levels of accuracy, integrity, and continuity. [SYS ENG] The probability that a system is operating satisfactorily at any point in time, excluding times when the system is under repair. { ə,väl·ə'bil·ə·dē }

available line [ELECTR] Portion of the length of the scanning line which can be used specifically for picture signals in a facsimile system. { ə'väl·ə·bəl 'līn }

available space list [COMPUT SCI] A pool of inactive memory cells, available for use in a list-processing system, to which cells containing items deleted from data lists are added, and from which cells needed for newly inserted data items are removed. { ə'väl·ə·bəl 'spās ˌlist }

available time See up time. { ə'väl·ə·bəl 'tīm }

avalanche breakdown [ELECTR] Nondestructive breakdown in a semiconductor diode when the electric field across the barrier region is strong enough so that current carriers collide with valence electrons to produce ionization and cumulative multiplication of carriers. { 'av·ə ˌlanch 'brākˌdau̇n }

avalanche diode [ELECTR] A semiconductor breakdown diode, usually made of silicon, in which avalanche breakdown occurs across the entire *pn* junction and voltage drop is then essentially constant and independent of current; the two most important types are IMPATT and TRAPATT diodes. { 'av·əˌlanch 'dīˌōd }

avalanche-induced migration [ELECTR] A technique of forming interconnections in a field-programmable logic array by applying appropriate voltages for shorting selected base-emitter junctions. { 'av·əˌlanch in¦düsd ˌmī'grā·shən }

avalanche oscillator [ELECTR] An oscillator that uses an avalanche diode as a negative resistance to achieve one-step conversion from direct-current to microwave outputs in the gigahertz range. { 'av·əˌlanch ¦äs·əˌlād·ər }

avalanche photodiode [ELECTR] A photodiode operated in the avalanche breakdown region to achieve internal photocurrent multiplication, thereby providing rapid light-controlled switching operation. { 'av·əˌlanch ˌfōd·ō'dīˌōd }

avatar [COMPUT SCI] A virtual representation of a person or a person's interactions with others in a virtual environment, conveying a sense of someone's presence (known as telepresence) by providing the location (position and orientation) and identity; examples include the graphical human figure model, the talking head, and the real-time reproduction of a three-dimesional human image. { 'av·əˌtär }

AVC See automatic volume control.

average-calculating operation [COMPUT SCI] A common or typical calculating operation longer than an addition and shorter than a multiplication; often taken as the mean of nine additions and one multiplication. { 'av·rij ¦kal·kyəˌlād·iŋ ˌäp·əˌrā·shən }

average-edge line [COMPUT SCI] The imaginary line which traces or smooths the shape of any written or printed character to be recognized by a computer through optical, magnetic, or other means. { 'av·rij ¦ej ˌlīn }

average effectiveness level See effectiveness level. { 'av·rij i'fek·tiv·nəs ˌlev·əl }

average information content [COMMUN] The average of the information content per symbol emitted from a source. { 'av·rij ˌin·fər'mā·shən ˌkän·tent }

average power output [ELECTR] Radio-frequency power, in an audio-modulation transmitter, delivered to the transmitter output terminals, averaged over a modulation cycle. { 'av·rij 'pau̇·ər 'au̇tˌpu̇t }

avigation See air navigation. { ˌa·və'gā·shən }

AWGN See additive white Gaussian noise.

axial ratio [ELECTR] The ratio of the major axis to the minor axis of the polarization ellipse of a waveguide. Also known as ellipticity. { 'ak·sē·əl 'rā·shō }

azel display [ELECTR] Modified type of plan position indicator presentation showing two separate radar displays on one cathode-ray screen; one display presents bearing information and the other shows elevation. { 'az·el disˌplā }

azimuth blanking [ELECTR] Blanking (disabling) either the radar receiver or transmitter or both in selected azimuth regions, to reduce interference or lessen radiation hazards. { 'az·ə·məth ˌblaŋk·iŋ }

azimuth error [ENG] An error in the indicated azimuth of a target detected by radar. { 'az·ə·məth ˌer·ər }

azimuth gating [ELECTR] The practice of selectively brightening and enhancing the gain-desired sectors of a radar plan position indicator display, usually by applying a step waveform to the automatic gain control circuit, or similar data separation by sectors in more automated systems. { 'az·ə·məth ˌgād·iŋ }

azimuth indicator [ENG] An approach-radar scope which displays azimuth information. { 'az·ə·məth ˌin·dəˌkād·ər }

azimuth marker [ELECTR] On a radar plan position indicator, a bright rotatable radial line used for bearing determination. Also known as angle marker; bearing marker. { 'az·ə·məth ˌmär·kər }

azimuth resolution [ELECTROMAG] Angle or distance by which two targets must be separated in azimuth to be distinguished by a radar set, when the targets are at the same range. { 'az·ə·məth ˌrez·ə'lü·shən }

azimuth-stabilized plan position indicator [ENG] A north-upward plan position indicator (PPI), a radarscope, which is stabilized by a gyrocompass so that either true or magnetic north is always at the top of the scope regardless of vehicle orientation. { 'az·ə·məth ¦sta·bəˌlīzd 'plan pə'zish·ən 'in·dəˌkād·ər }

azimuth versus amplitude [ELECTR] Electronic protection technique using a plan position indicator to display strobes due to jamming sources, particularly useful in making passive fixes when two or more radar sites operate together. { 'az·ə·məth ˌvər·səs 'am·pləˌtüd }

B

babble [COMMUN] **1.** Aggregate crosstalk from a large number of channels. **2.** Unwanted disturbing sounds in a carrier or other multiple-channel system which result from the aggregate crosstalk or mutual interference from other channels. { 'bab·əl }

babs *See* blind approach beacon system. { babz }

backbone [COMPUT SCI] The portion of a communication network that handles the largest volume of traffic, usually employing a high-speed, high-capacity medium designed to transmit data over long distances. { 'bak,bōn }

back echo [ELECTROMAG] An echo signal produced on a radar screen by one of the minor back lobes of a search radar beam. { 'bak ,ek·ō }

back-echo reflection [ELECTR] A radar echo produced by radiation reflected to the target by a large, fixed obstruction; that is, the ray path is from the antenna to obstruction to target and back similarly, giving a false indication of target position; an indirect-path echo. { 'bak ,ek·ō ri'flek·shən }

back-end system [COMPUT SCI] A computer that operates on data which have been previously processed by another computer system. { 'bak ¦end ,sis·təm }

backfire antenna [ELECTROMAG] An antenna which exhibits significant gain in a direction 180° from its principal lobe. { 'bak,fīr an'ten·ə }

background [COMMUN] **1.** Picture white of the facsimile copy being scanned when the picture is black and white only. **2.** Undesired printing in the recorded facsimile copy of the picture being transmitted, resulting in shading of the background area. **3.** Noise heard during radio reception caused by atmospheric interference or the operation of the receiver at such high gain that inherent circuit noises become noticeable. { 'bak,graủnd }

background ink [COMPUT SCI] In optical character recognition, a highly reflective ink used to print the parts of a document that are to be ignored by the scanner. { 'bak,graủnd ,iŋk }

background noise [ENG] The undesired signals that are always present in an electronic or other system, independent of whether or not the desired signal is present. { 'bak,graủnd ,nȯiz }

background processing [COMPUT SCI] **1.** The execution of lower-priority programs when higher-priority programs are not being handled by a data-processing system. **2.** Computer processing that is not interactive or visible on the display screen. { 'bak,graủnd 'prä·ses·iŋ }

background program [COMPUT SCI] A computer program that has low priority in a multiprogramming system. { 'bak,graủnd 'prō·grəm }

background reflectance [COMPUT SCI] The reflectance, relative to a standard, of the surface on which a printed or handwritten character has been inscribed in optical character recognition. { 'bak,graủnd ri'flek·təns }

background returns [ENG] **1.** Signals on a radar screen from objects which are of no interest. **2.** *See* clutter. { 'bak,graủnd ri'tərnz }

backhaul [COMMUN] Point-to-point satellite transmission of video from a remote site to a network distribution center in real time. { 'bak ,hȯl }

backing storage [COMPUT SCI] A computer storage device whose capacity is larger, but whose access time is slower, than that of the computer's main storage or immediate access storage; usually slower than main storage. Also known as bulk storage. { 'bak·iŋ ,stȯr·ij }

back lobe [ELECTROMAG] The three-dimensional portion of the radiation pattern of a directional antenna that is directed away from the intended direction. { 'bak ,lōb }

backout [COMPUT SCI] To remove a change that was previously made in a computer program. { 'bak,aủt }

backplane [ELECTR] A wiring board, usually constructed as a printed circuit, used in computers to provide the required connections between logic, memory, input/output modules, and other printed circuit boards which plug into it at right angles. { 'bak,plān }

back porch [ELECTR] The period of time in a television circuit immediately following a synchronizing pulse during which the signal is held at the instantaneous amplitude corresponding to a black area in the received picture. { 'bak¦pȯrch }

back radiation *See* backscattering. { 'bak ,rād·ē'ā·shən }

backscattering [COMMUN] Propagation of extraneous signals by F- or E-region reflection in addition to the desired ionospheric scatter mode; the undesired signal enters the antenna through the back lobes. [ELECTROMAG] **1.** Radar echoes from a target. **2.** Undesired radiation of energy to the rear by a directional antenna. Also known as back radiation; backward scattering. { 'bak¦skad·ə·riŋ }

backspace [COMPUT SCI] To move a recording medium one unit in the reverse or background direction. { 'bak,spās }

backtalk [COMPUT SCI] Passage of information from a standby computer to the active computer. { 'bak,tȯk }

backtracking [COMPUT SCI] A method of solving problems automatically by a systematic search of the possible solutions; the invalid solutions are eliminated and are not retried. { 'bak,trak·iŋ }

backup [COMPUT SCI] **1.** Logical or physical facilities to aid the process of restarting a computer system and recovering the information in it following a failure. **2.** The provision of such facilities. { 'bak,əp }

backup arrangement *See* cascade. { 'bak,əp ,ə 'rānj·mənt }

Backus-Naur form [COMPUT SCI] A metalanguage that specifies which sequences of symbols constitute a syntactically valid program language. Abbreviated BNF. { ¦bäk·əs ¦naůr ,fȯrm }

backward-acting regulator [ELECTR] Transmission regulator in which the adjustment made by the regulator affects the quantity which caused the adjustment. { 'bak·wərd 'ak·tiŋ 'reg·yə,lād·ər }

backward chaining [COMPUT SCI] In artificial intelligence, a method of reasoning which starts with the problem to be solved and repeatedly breaks this goal into subgoals that are more readily solvable with the relevant data and the system's rules of inference. { ¦bak·wərd 'chān·iŋ }

backward compatibility *See* downward compatibility. { ¦bak·wərd kəm,pad·ə'bil·əd·ē }

backward error analysis [COMPUT SCI] A form of error analysis which seeks to replace all errors made in the course of solving a problem by an equivalent perturbation of the original problem. { 'bak·wərd 'er·ər ə,nal·ə·səs }

backward read [COMPUT SCI] The transfer of data from a magnetic tape to computer storage when the tape is running in reverse. { 'bak·wərd 'rēd }

backward scattering *See* backscattering. { 'bak·wərd ¦skad·ə·riŋ }

backward search [COMPUT SCI] A search of a document or database that starts at the cursor's location and moves backwards toward the beginning of the document or database. { ¦bak·wərd 'sərch }

backward wave [ELECTROMAG] An electromagnetic wave traveling opposite to the direction of motion of some other physical quantity in an electronic device such as a traveling-wave tube or mismatched transmission line. { 'bak·wərd ,wāv }

bad branch [COMPUT SCI] An error in which execution of a computer program jumps to an incorrect instruction, usually as a result of errors in the program. { ¦bad 'branch }

bad page break [COMPUT SCI] A soft page break at an inappropriate location in a document, such as one that splits a table or leaves a single line of text at the top or bottom of a page. { ¦bad 'pāj ,brāk }

bad sector [COMPUT SCI] An area of disk storage that does not record data reliably and therefore is not used. { ,bad 'sek·tər }

bad track [COMPUT SCI] A disk track that contains a bad sector. { ,bad 'trak }

bad track table [COMPUT SCI] A listing of the bad sectors on a disk, which is packaged with or attached to a disk. { ¦bad 'trak ,tā·bəl }

balanced detector [ELECTR] A detector used in frequency-modulation receivers; in one form the audio output is the rectified difference between voltages produced across two resonant circuits, one being tuned slightly above the carrier frequency and one slightly below. { 'bal·ənst di 'tek·tər }

balanced merge [COMPUT SCI] A merge or sort operation in which the data involved are divided equally between the available storage devices. { 'bal·ənst 'mərj }

balanced modulator [ELECTR] A modulator in which the carrier and modulating signal are introduced in such a way that the output contains the two sidebands without the carrier. { 'bal·ənst 'maj·ə,lād·ər }

balanced-tree [COMPUT SCI] A system of indexes that keeps track of stored data, and in which data keys are stored in a hierarchy that is continually modified in order to minimize access times. Abbreviated B-tree. { 'bal·ənst 'trē }

balance error [COMPUT SCI] An error voltage that arises at the output of analog adders in an analog computer and is directly proportional to the drift error. { 'bal·əns ,er·ər }

balancing [COMPUT SCI] The distribution of workload among computing resources to optimize performance. { 'bal·əns·iŋ }

ballistic tracking *See* dynamic resolution. { bə ,lis·tik 'trak·iŋ }

band [COMMUN] A range of electromagnetic-wave frequencies between definite limits, such as that assigned to a particular type of radio service. [COMPUT SCI] A set of circular or cyclic recording tracks on a storage device such as a magnetic drum, disk, or tape loop. { band }

band-elimination filter *See* band-stop filter. { ¦band i,lim·ə'nā·shən 'fil·tər }

band-pass [ELECTR] A range, in hertz or kilohertz, expressing the difference between the limiting frequencies at which a desired fraction (usually half power) of the maximum output is obtained. { 'band ,pas }

band-pass amplifier [ELECTR] An amplifier designed to pass a definite band of frequencies with essentially uniform response. { 'band ,pas ¦am·plə,fī·ər }

band-pass filter [ELECTR] An electric filter which transmits more or less uniformly in a certain band, outside of which the frequency components are attenuated. { 'band ,pas ,fil·tər }

band-pass response [ELECTR] Response characteristics in which a definite band of frequencies is transmitted uniformly. Also known as flat-top response. { 'band ,pas ri'späns }

band printer [COMPUT SCI] A line printer that uses a band of type characters as its printing mechanism. { 'band ¦print·ər }

band-rejection filter *See* band-stop filter. { 'band ri'jek·shən ,fil·tər }

band selector [ELECTR] A switch that selects any of the bands in which a receiver, signal generator, or transmitter is designed to operate and usually has two or more sections to make the required changes in all tuning circuits simultaneously. Also known as band switch. { 'band sə'lek·tər }

band spreading [COMMUN] Method of double-sideband transmission in which the frequency band of the modulating wave is shifted upward in frequency so that the sidebands produced by modulation are separated in frequency from the carrier by an amount at least equal to the bandwidth of the original modulating wave, and second-order distortion products may be filtered from the demodulator output. { 'band ,spred·iŋ }

band-spread tuning control [ELECTR] A tuning control provided on some shortwave receivers to spread the stations in a single band of frequencies over an entire tuning dial. { 'band ,spred 'tün·iŋ kən'trōl }

band-stop filter [ELECTR] An electric filter which transmits more or less uniformly at all frequencies of interest except for a band within which frequency components are largely attenuated. Also known as band-elimination filter; band-rejection filter. { 'band ,stäp ,fil·tər }

band switch *See* band selector. { 'band ,swich }

bandwidth [COMMUN] **1.** The difference between the frequency limits of a band containing the useful frequency components of a signal. **2.** A measure of the amount of data that can travel a communications path in a given time, usually expressed as thousands of bits per second (kbps) or millions of bits per second (Mbps). { 'band ,width }

bang-bang circuit [ELECTR] An operational amplifier with double feedback limiters that drive a high-speed relay (1–2 milliseconds) in an analog computer; involved in signal-controlled programming. { ¦baŋ ¦baŋ ,sər·kət }

bang-bang control [COMPUT SCI] Control of programming in an analog computer through a bang-bang circuit. [CONT SYS] A type of automatic control system in which the applied control signals assume either their maximum or minimum values. { ¦baŋ ¦baŋ kən'trōl }

bang-bang-off control *See* bang-zero-bang control. { ¦baŋ ¦baŋ 'ȯf kən,trōl }

bang-bang robot [CONT SYS] A simple robot that can make only two types of motions. { ¦baŋ ¦baŋ 'rō,bät }

bang-zero-bang control [CONT SYS] A type of control in which the control values are at their maximum, zero, or minimum. Also known as bang-bang-off control. { ¦baŋ ,zir·ō 'baŋ kən ,trōl }

bank select [COMPUT SCI] To activate and deactivate blocks of memory or other internal system components using electronic control signals. Also known as bank switch. { 'baŋk si,lekt }

bank selected memory [COMPUT SCI] Auxiliary blocks of memory in a microcomputer that can be switched in to replace some or all of the internal memory by software-controlled switches located outside the microprocessor. { 'baŋk si¦lek·təd 'mem·rē }

bank switch *See* bank select. { 'baŋk ,swich }

bar code [COMPUT SCI] The representation of alphanumeric characters by series of adjacent stripes of various widths, for example, the universal product code. { 'bär ,kōd }

bar-code reader *See* bar-code scanner. { 'bär ,kōd 'rēd·ər }

bar-code scanner [COMPUT SCI] An optical scanning device that reads texts which have been converted into a special bar code. Also known as bar-code reader. { 'bär ,kōd 'skan·ər }

bare disk [ELECTR] A floppy-disk drive without electronic control circuits. { ¦ber 'disk }

bar generator [ELECTR] Generator of pulses or repeating waveforms that are equally separated in time; these pulses are synchronized by the synchronizing pulses of a television system, so that they can produce a stationary bar pattern on a television screen. { 'bär ¦jen·ə,rād·er }

BARITT diode *See* barrier injection transit-time diode. { 'bar·ət ¦dī,ōd }

Barkhausen interference [COMMUN] Interference caused by Barkhausen oscillations. { 'bärk,hau̇z·ən in·tər'fir·əns }

Barkhausen oscillation [ELECTR] Undesired oscillation in the horizontal output tube of a television receiver, causing one or more ragged dark vertical lines on the left side of the picture. { 'bärk,hau̇z·ən ,äs·ə'lā·shən }

bar pattern [ELECTR] Pattern of repeating lines or bars on a television screen. { 'bär ,pad·ərn }

bar printer [COMPUT SCI] An impact printer in which the character heads are mounted on type bars. { 'bär ,print·ər }

barrage jamming [COMMUN] The simultaneous jamming of a number of radio frequencies or even multiple radar bands of frequencies. { bə'räzh ,jam·iŋ }

barrel printer [COMPUT SCI] A computer printer in which the entire set of characters is placed around a rapidly rotating cylinder at each print position; computer-controlled print hammers opposite each print position strike the paper and press it against an inked ribbon between the paper and the cylinder when the appropriate character reaches a position opposite the print hammer. { 'bar·əl ,prin·tər }

barrier injection transit-time diode [ELECTR] A microwave diode in which the carriers that traverse the drift region are generated by minority carrier injection from a forward-biased junction instead of being extracted from the plasma of an avalanche region. Abbreviated BARITT diode. { 'bar·ē·ər in'jek·shən 'trans·ət ,tīm 'dī,ōd }

barrier layer *See* depletion layer. { 'bar·ē·ər ˌlā·ər }

barrier-layer cell *See* photovoltaic cell. { 'bar·ē·ər ˌlā·ər ˌsel }

barrier-layer photocell *See* photovoltaic cell. { 'bar·ē·ər ˌlā·ər 'fōd·ōˌsel }

base [COMPUT SCI] *See* root. [ELECTR] **1.** The region that lies between an emitter and a collector of a transistor and into which minority carriers are injected. **2.** The part of an electron tube that has the pins, leads, or other terminals to which external connections are made either directly or through a socket. **3.** The plastic, ceramic, or other insulating board that supports a printed wiring pattern. { bās }

base address *See* address constant. { bās ə'dres }

baseband [COMMUN] The band of frequencies occupied by all transmitted signals used to modulate the radio wave. { 'bāsˌband }

baseband frequency response [COMMUN] Frequency response characteristics of the frequency band occupied by all of the signals used to modulate a transmitted carrier. { 'bāsˌband 'frē·kwən·sē ri'späns }

baseband system [COMMUN] A communications system in which information is transmitted over a single unmodulated band of frequencies. { 'bāsˌband ˌsis·təm }

base-displacement [COMPUT SCI] In machine-language programming, a technique in which addresses are specified relative to a base address where the beginning of the program is stored. { 'bās disˌplās·mənt }

base font [COMPUT SCI] The font used in a document if none other is specified. { 'bās ˌfänt }

base language [COMPUT SCI] The component of an extensible language which provides a complete but minimal set of primitive facilities, such as elementary data types, and simple operations and control constructs. { 'bās 'laŋ·gwij }

baseline [ENG] The geographic line between transmitter and receiver locations in bistatic radar, or between pairs of radars or radio receivers in a network, used in calculations relative to the data. Abbreviated BL. { 'bās ˌlīn }

base-line break [ELECTR] Technique in radar which uses the characteristic break in the base line on an A-scope display due to a pulse signal of significant strength in noise jamming. { 'bās ˌlīn ˌbrāk }

base-loaded antenna [ELECTROMAG] Vertical antenna having an impedance in series at the base for loading the antenna to secure a desired electrical length. { 'bās ˌlōd·əd an'ten·ə }

base modulation [ELECTR] Amplitude modulation produced by applying the modulating voltage to the base of a transistor amplifier. { 'bās ˌmäj·ə'lā·shən }

base rate area [COMMUN] Area within which service is given without mileage charges. { 'bās ˌrāt ¦er·ē·ə }

base register *See* index register. { 'bās ˌrej·ə·stər }

base station [COMMUN] **1.** A land station, in the land mobile service, carrying on a service with land mobile stations (a base station may secondarily communicate with other base stations incident to communications with land mobile stations). **2.** A station in a land mobile system which remains in a fixed location and communicates with the mobile stations. { 'bās ˌstā·shən }

base system [COMPUT SCI] A computer system containing only program modules that carry out basic functions. { 'bās ˌsis·təm }

BASIC [COMPUT SCI] A procedure-level computer language designed to be easily learned and used by nonprofessionals, and well suited for an interactive, conversational mode of operation. Derived from Beginners All-purpose Symbolic Instruction Code. { 'bā·sik }

basic batch [COMPUT SCI] The least complex level of computer processing, in which application systems are normally made up of small programs that are run through the computer one at a time and that can process transactions only from sequential files. { 'bā·sik 'bach }

basic disk operating system [COMPUT SCI] The part of a computer's operating system that handles the transfer of data between programs and disk units and the control of files. Abbreviated BDOS. { 'bā·sik ¦disk ˌäp·ə'rād·iŋ 'sis·təm }

basic input/output system [COMPUT SCI] The part of a computer's operating system that handles communications between a program and external devices such as printers and electronic displays. Abbreviated BIOS. { 'bā·sik 'inˌput 'au̇tˌpu̇t ˌsis·təm }

basic instruction [COMPUT SCI] An instruction in a computer program which is systematically changed by the program to obtain the instructions which are actually carried out. Also known as presumptive instruction; unmodified instruction. { 'bā·sik in'strək·shən }

basic linkage [COMPUT SCI] Computer coding that provides a standard means of connecting a given routine or program with other routines and that can be used repeatedly according to the same rules. { 'bā·sik 'liŋ·kij }

basic processing unit [COMMUN] Principal controller and data processor within the communications system. { 'bā·sik 'präsˌes·iŋ ˌyü·nət }

basic software [COMPUT SCI] Software requirements that are taken into account in the design of the data-processing hardware and usually are provided by the original equipment manufacturer. { 'bā·sik 'sȯftˌwer }

basic telecommunications access method [COMPUT SCI] A method of controlling data transmission between a computer's main storage and its terminals and of providing applications programs with the capability of communicating with printers, terminals, and other devices. Abbreviated BTAM. { 'bā·sik ˌtel·ə·kəˌmyü·nə¦kā·shənz 'akˌses ˌmeth·əd }

basic variables [COMPUT SCI] The *m* variables in a basic feasible solution for a linear programming model. { 'bā·sik 'ver·ē·ə·bəlz }

bass compensation [ELECTR] A circuit that emphasizes the low-frequency response of an audio amplifier at low volume levels to offset the lower sensitivity of the human ear to weak low frequencies. { 'bās ,käm·pən'sā·shən }

batch [COMPUT SCI] A set of items, records, or documents to be processed as a single unit. { bach }

batch-and-forward system [COMPUT SCI] A data-processing system in which data are collected for a time and then transmitted as a unit to a computer. { 'bach ən 'fȯr·wərd ,sis·təm }

batching [COMPUT SCI] Grouping records for the purpose of processing them in a computer. { 'bach·iŋ }

batch job [COMPUT SCI] One of a group of jobs that are executed together by batch-processing techniques. { 'bach ,jäb }

batch-oriented applications [COMMUN] Applications of data communications that involve the transfer of thousands or even millions of bytes of data and are usually point-to-point and computer-to-computer. { 'bach ,ȯr·ē ¦ent·əd ,ap·lə'kā·shənz }

batch processing [COMPUT SCI] A technique that uses a single program loading to process many individual jobs, tasks, or requests for service. { 'bach ,präs·es·iŋ }

batch stream [COMPUT SCI] A group of batch processing programs that are scheduled to run on a computer. { 'bach ,strēm }

batch system [COMPUT SCI] A computer system that uses batch processing. { 'bach ,sis·təm }

batch total [COMPUT SCI] The total for a specified constituent quantity in a batch; used to verify the accuracy of operations on the batch. { 'bach ¦tōd·əl }

battery, overvoltage, ringing, supervision, coding, hybrid and test access *See* BORSCHT. { 'bad·ə·rē ¦ō·vər¦vōl·tij 'riŋ·iŋ ,sü·pər'vizh·ən 'kōd·iŋ 'hī·brid ən 'test ,ak,ses }

baud [COMMUN] A unit of telegraph signaling speed equal to the number of code elements (pulses and spaces) per second or twice the number of pulses per second. { bȯd }

Baudot code [COMMUN] A teleprinter code that uses a combination of five or six marking and spacing intervals of equal duration for each character; no longer in extensive use since it has been replace by ASCII code. { bȯ'dō ,kōd }

bay [COMPUT SCI] *See* drive bay. [ELECTROMAG] One segment of an antenna array. { bā }

B box *See* index register. { 'bē ,bäks }

BBS *See* bulletin board system.

BCAS *See* beacon collision avoidance system.

BCD system *See* binary coded decimal system. { ¦bē¦sē'dē ,sis·təm }

B-display [ELECTR] The presentation of radar output data in rectangular coordinates in which range and azimuth are plotted on the coordinate axes. Also known as B-indicator; B-scan; B-scope; range-bearing display. { 'bē dis'plā }

BDOS *See* basic disk operating system. { 'bē ,dȯs }

beacon [ELECTR] A radio transmitter and antenna used to indicate its location or that of the vehicle carrying it; a beacon that responds to an interrogation, as in secondary radar, is more properly called a transponder. [NAV] **1.** A light, group of lights, electronic apparatus, or other device that emits identifying signals related to their position so that the information so produced can be used by the navigator or pilots of aircraft and ships for guidance orientation or warning. **2.** A structure where such a device is mounted or located. { 'bē·kən }

beacon collision avoidance system [NAV] An airborne collision avoidance system that makes use of the air-traffic control radio beacon system (ATCRBS) transponders. Abbreviated BCAS. { 'bē·kən kə'lizh·ən ə'vȯid·əns ,sis·təm }

beacon delay [ELECTR] The amount of transponding delay within a beacon, that is, the time between the arrival of a signal and the response of the beacon. { 'bē·kən di'lā }

beacon presentation [ELECTR] The radar display resulting from receipt of signals from a beacon. { 'bē·kən ,prē·zən'tā·shən }

beacon skipping [ELECTR] A condition where transponder return pulses from a beacon are missing at the interrogating radar. { 'bē·kən ,skip·iŋ }

beacon stealing [ELECTR] Loss of beacon tracking by one radar due to stronger signals from other beacons, transponders, or interfering radars. { 'bē·kən ,stēl·iŋ }

beacon tracking [ENG] The tracking of a moving object by means of signals emitted from a transmitter or transponder within or attached to the object. { 'bē·kən ,trak·iŋ }

beacon-tracking radar [NAV] Radar equipment used in air-traffic control facilities for beacon tracking. { 'bē·kən ,trak·iŋ ¦rā,där }

bead [COMPUT SCI] A small subroutine. [ELECTROMAG] A glass, ceramic, or plastic insulator through which passes the inner conductor of a coaxial transmission line and by means of which the inner conductor is supported in a position coaxial with the outer conductor. { bēd }

beaded transmission line [ELECTROMAG] Line using beads to support the inner conductor in coaxial transmission lines. { 'bēd·əd tranz'mish·ən ,līn }

beam angle *See* beam width. { 'bēm ,aŋ·gəl }

beam antenna [ELECTROMAG] An antenna that concentrates its radiation into a narrow beam in a definite direction. { 'bēm an'ten·ə }

beam approach beacon system *See* blind approach beacon system. { 'bēm ə'prōch 'bē·kən ,sis·təm }

beam blank *See* blank. { 'bēm blaŋk }

beam efficiency [ELECTROMAG] The fraction of the total radiated energy from an antenna contained in a single beam. { 'bēm i,fish·ən·sē }

beam-forming electrode [ELECTR] Electron-beam focusing elements in power tetrodes and cathode-ray tubes. { 'bēm ,fȯrm·iŋ i'lek,trōd }

beamguide [ELECTROMAG] A set of elements arranged and spaced so as to form and conduct a beam of electromagnetic radiation. { 'bēm,gīd }

beam-indexing tube [ELECTR] A single-beam color television picture tube in which the color phosphor strips are arranged in groups of red, green, and blue. { ¦bēm 'in,dek·siŋ ,tüb }

beam magnet *See* convergence magnet. { 'bēm ,mag·nət }

beam power tube [ELECTR] A vacuum tube, most often an amplifier, used in radar and other microwave transmitters in which the electrons travel from the cathode in a well-focused beam, to interact with the electromagnetic signal being amplified. { 'bēm ¦pau̇·ər ,tüb }

beam splitting [ELECTR] Process for increasing angle accuracy in locating targets by radar by noting the azimuths at which one radar scan first discloses a target and at which the echoes cease, revealing the azimuth center, or by similarly intended algorithms in more automated systems. { 'bēm ,splid·iŋ }

beam steering [ELECTR] Changing the direction of the major lobe of a radiation pattern, usually by switching antenna elements. { 'bēm ,stir·iŋ }

beam switching [ELECTR] Method of obtaining more accurately the bearing or elevation of an object by comparing the signals received when the beam is in directions differing slightly in bearing or elevation; when these signals are equal, the object lies midway between the beam axes. Also known as lobe switching. { 'bēm ,swich·iŋ }

beam width [ELECTROMAG] The angle, measured in a horizontal plane, between the directions at which the intensity of an electromagnetic beam, such as a radar or radio beam, is one-half its maximum value. Also known as beam angle. { 'bēm ,width }

bearing cursor [ENG] Of a radar set, the radial line inscribed on a transparent disk which can be rotated manually about an axis coincident with the center of the plan position indicator; used for bearing determination. Also known as mechanical bearing cursor. { 'ber·iŋ ,kər·sər }

bearing marker *See* azimuth marker. { 'ber·iŋ ,märk·ər }

bearing resolution [ELECTR] Minimum angular separation in a horizontal plane between two targets at the same range that will allow an operator to obtain data on either target. { 'ber·iŋ ,rez·ə ,lü·shən }

beat frequency [ELECTR] The frequency of a signal equal to the difference in frequencies of two signals which produce the signal when they are combined in a nonlinear circuit. { 'bēt ,frē·kwən·sē }

beat-frequency oscillator [ELECTR] An oscillator in which a desired signal frequency, such as an audio frequency, is obtained as the beat frequency produced by combining two different signal frequencies, such as two different radio frequencies. Abbreviated BFO. Also known as heterodyne oscillator. { 'bēt ,frē·kwən·sē 'äs·ə ,lād·ər }

beat reception *See* heterodyne reception. { 'bēt ri'sep·shən }

beat-time programming [COMPUT SCI] A type of programming which requires that data be made available to the computer during some ongoing process prior to a particular point in time. { ¦bēt 'tīm 'prō,gram·iŋ }

beavertail [ELECTROMAG] Fan-shaped radar beam, wide in the horizontal plane and narrow in the vertical plane, which is swept up and down for height finding. { 'be·vər,tāl }

bedspring array *See* billboard array. { 'bed ,spriŋ ə'rā }

BEGIN [COMPUT SCI] An enclosing statement of ALGOL used to indicate the beginning of a block; any variable in a block enclosed by BEGIN and END is normally local to this block. { bi'gin }

beginning-of-information marker [COMPUT SCI] A section of magnetic tape covered with reflective material that indicates the beginning of the area on which information is to be recorded. { bi'gin·iŋ əv ,in·fər'mā·shən ,mär·kər }

bell character [COMPUT SCI] A control character that activates a bell, alarm, or other audio device to get someone's attention. { 'bel ,kar·ik·tər }

bells and whistles [COMPUT SCI] Special hardware features that are likely to attract attention but may not be important or even practical. { 'belz ən 'wis·əlz }

belt printer [COMPUT SCI] A type of impact printer similar to a chain printer in which the characters are carried on a moving belt rather than a chain. { 'belt ,print·ər }

benchmark problem [COMPUT SCI] A problem to be run on computers to evaluate their performances relative to one another. { 'bench,märk ,präb·ləm }

benchmark test [COMPUT SCI] A test of computer software or hardware that is generally run on a number of products to compare their performance. { 'bench,märk ,test }

bent-pipe system [COMMUN] A transponder on board a communications satellite that performs no signal processing other than heterodyning (frequency-changing) the uplink frequency bands to those of the downlinks. { ,bent 'pīp ,sis·təm }

beta [ELECTR] The current gain of a transistor that is connected as a grounded-emitter amplifier, expressed as the ratio of change in collector current to resulting change in base current, the collector voltage being constant. { 'bād·ə }

beta rule *See* reduction rule. { 'bād·ə ,rül }

beta software [COMPUT SCI] An application or program that is in development and undergoing testing. Also known as beta version; betaware. { ,bād·ə 'sȯf,wer }

beta test [COMPUT SCI] The first test of a computer system outside the laboratory, in its actual working environment. { 'bād·ə ,test }

beta test site [COMPUT SCI] An organization or company that tests a software or hardware product under actual working conditions and reports the results to the vendor. { 'bād·ə ¦test ,sīt }

beta version *See* beta software. { 'bād·ə ˌvər·zhən }

betaware *See* beta software. { 'bād·əˌwer }

Beverage antenna *See* wave antenna. { 'bev·rij an'ten·ə }

beyond-the-horizon communication *See* scatter propagation. { bə'yänd thə hə'rīz·ən kə ˌmyü·nə'kā·shən }

Bézier curve [COMPUT SCI] A curve in a drawing program that is defined mathematically, and whose shape can be altered by dragging either of its two interior determining points with a mouse. { ¦bāz·yā'kərv }

BFL *See* buffered FET logic.

BFO *See* beat-frequency oscillator.

B-frames *See* bidirectional pictures. { 'bē ˌfrāmz }

bias [ELECTR] A direct-current voltage applied to a transistor control electrode to establish the desired operating point. { 'bī·əs }

biased automatic gain control *See* delayed automatic gain control. { 'bī·əst ȯd·ə'mad·ik 'gān kənˌtrōl }

bias meter [COMMUN] A meter used in teletypewriter work for measuring signal bias directly in percent; a positive reading indicates a marking signal bias; a negative reading, a spacing signal bias. { 'bī·əs ˌmēd·ər }

bias register [COMPUT SCI] A computer device that stores a number that is added to the memory address each time the computer memory is referenced by the program, thus offsetting the program addresses by a fixed amount. { 'bī·əs ˌrej·ə·stər }

BiCMOS technology [ELECTR] An integrated circuit technology that combines bipolar transistors and CMOS devices on the same chip. { ¦bī'sē ˌmōs tekˌnäl·ə·jē }

biconditional gate *See* equivalence gate. { ˌbī·kən'dish·ən·əl 'gāt }

biconical antenna [ELECTROMAG] An antenna consisting of two metal cones having a common axis with their vertices coinciding or adjacent and with coaxial-cable or waveguide feed to the vertices. { bī'kän·ə·kəl an'ten·ə }

bidirectional antenna [ELECTROMAG] An antenna that radiates or receives most of its energy in only two directions. { ˌbī·də'rek·shən·əl an'ten·ə }

bidirectional counter *See* forward-backward counter. { ˌbī·də'rek·shən·əl 'kau̇n·tər }

bidirectional data bus [COMPUT SCI] A channel over which data can be transmitted in either direction within a computer system. { ˌbī·də'rek·shən·əl 'dad·ə ˌbəs }

bidirectional parallel port [COMPUT SCI] A parallel port that can transfer data in both directions, and at speeds much greater than a standard parallel port. { ˌbī·dəˌrek·shən·əl ˌpar·əˌlel 'pȯrt }

bidirectional pictures [COMMUN] In MPEG-2, pictures that use both future and past pictures as a reference. This technique is termed bidirectional prediction; bidirectional pictures provide the most compression and do not propagate coding errors as they are never used as a reference. Also known as B-frames; B-pictures. { ˌbī·də'rek·shən·əl 'pik·chərz }

bidirectional printer [COMPUT SCI] A printer in which printing can be done in both a left-to-right and a right-to-left direction. { ˌbī·də'rek·shən·əl 'print·ər }

bidirectional pulse-amplitude modulation *See* double-polarity pulse-amplitude modulation. { ˌbī·də'rek·shən·əl ¦pəls ¦am·pləˌtüd ˌmäj·ə'lā·shən }

bidirectional transistor [ELECTR] A transistor that provides switching action in either direction of signal flow through a circuit; widely used in telephone switching circuits. { ˌbī·də'rek·shən·əl tran'zis·tər }

bigit *See* bit. { 'bij·ət }

big LEO system [COMMUN] A system of relatively large satellites in low earth orbit (LEO) to provide global mobile handheld telephony and other services. { ˌbig 'lē·ōˌsis·təm }

big M method [COMPUT SCI] A technique for solving linear programming problems in which artificial variables are assigned cost coefficients which are a very large number M, say, $M = 10^{35}$. { ˌbig 'em ˌmeth·əd }

bilateral amplifier [ELECTR] An amplifier capable of receiving as well as transmitting signals; used primarily in transceivers. { bī'lad·ə·rəl 'am·pləˌfī·ər }

bilateral antenna [ELECTROMAG] An antenna having maximum response in exactly opposite directions, 180° apart, such as a loop. { bī'lad·ə·rəl an'ten·ə }

billboard array [ELECTROMAG] A broadside antenna array consisting of stacked dipoles spaced one-fourth to three-fourths wavelength apart in front of a large sheet-metal reflector. Also known as bedspring array; mattress array. { 'bilˌbȯrd ə'rā }

bin [COMPUT SCI] A magnetic-tape memory in which a number of tapes are stored in a single housing. [ENG] An enclosed space, box, or frame for the storage of bulk substance. { bin }

binary [COMPUT SCI] Possessing a property for which there exists two choices or conditions, one choice excluding the other. { 'bīn·ə·rē }

binary arithmetic operation [COMPUT SCI] An arithmetical operation in which the operands are in the form of binary numbers. Also known as binary operation. { 'bīn·ə·rē ˌar·ith'med·ik äp·ə'rā·shən }

binary cell [COMPUT SCI] An elementary unit of computer storage that can have one or the other of two stable states and can thus store one bit of information. { 'bīn·ə·rē ¦sel }

binary chain [COMPUT SCI] A series of binary circuit elements so arranged that each can change the state of the one following it. { 'bīn·ə·rē ¦chān }

binary chop *See* binary search. { 'bīn·ə·rē 'chäp }

binary code [COMPUT SCI] A code in which each allowable position has one of two possible states, commonly 0 and 1; the binary number system is one of many binary codes. { 'bīn·ə·rē ¦kōd }

binary coded character [COMPUT SCI] One element of a notation system representing alphanumeric characters such as decimal digits, alphabetic letters, and punctuation marks by a predetermined configuration of consecutive binary digits. { 'bīn·ə·rē ˌkōd·əd 'kar·ik·tər }

binary coded decimal system [COMPUT SCI] A system of number representation in which each digit of a decimal number is represented by a binary number. Abbreviated BCD system. { 'bīn·ə·rē ˌkōd·əd 'des·məl ˌsis·təm }

binary coded decimal-to-decimal converter [COMPUT SCI] A computer circuit which selects one of ten outputs corresponding to a four-bit binary coded decimal input, placing it in the 0 state and the other nine outputs in the 1 state. { 'bīn·ə·rē ˌkōd·əd 'des·məl tə 'des·məl kən'vərd·ər }

binary coded octal system [COMPUT SCI] Octal numbering system in which each octal digit is represented by a three-place binary number. { 'bīn·ə·rē ˌkōd·əd 'äk·təl ˌsis·tem }

binary component [ELECTR] An electronic component that can be in either of two conditions at any given time. Also known as binary device. { 'bīn·ə·rē kəm'pō·nənt }

binary conversion [COMPUT SCI] Converting a number written in binary notation to a number system with another base, such as decimal, octal, or hexadecimal. { 'bīn·ə·rē kən'vər·zhən }

binary counter *See* binary scaler. { 'bīn·ə·rē 'kau̇nt·ər }

binary decision [COMPUT SCI] A decision between only two alternatives. { 'bīn·ə·rē di 'sizh·ən }

binary device *See* binary component. { 'bīn·ə·rē di'vīs }

binary digit *See* bit. { 'bīn·ə·rē 'dij·ət }

binary dump [COMPUT SCI] The operation of copying the contents of a computer memory in binary form onto an external storage device. { 'bīn·ə·rē ¦dəmp }

binary encoder [ELECTR] An encoder that changes angular, linear, or other forms of input data into binary coded output characters. { 'bīn·ə·rē en'kōd·ər }

binary field [COMPUT SCI] A field that contains data in the form of binary numbers. { 'bīn·ə·rē 'fēld }

binary file [COMPUT SCI] A computer program in machine language that can be directly executed by the computer. { 'bīn·ə·rē 'fīl }

binary incremental representation [COMPUT SCI] A type of incremental representation in which the value of change in a variable is represented by one binary digit which is set equal to 1 if there is an increase in the variable and to 0 if there is a decrease. { 'bīn·ə·rē ˌiŋ·krə'men·təl ˌrep·riˌzen'tā·shən }

binary large object [COMPUT SCI] In a database management system, a file-storage system used most often for multimedia files (large files). Abbreviated BLOB. { ¦bīn·ə·rē ¦lärj 'äbˌjekt }

binary loader [COMPUT SCI] A computer program which transfers to main memory an exact image of the binary pattern of a program held in a storage or input device. { 'bīn·ə·rē ¦lōd·ər }

binary logic [ELECTR] An assembly of digital logic elements which operate with two distinct states. { 'bīn·ə·rē 'läj·ik }

binary operation *See* binary arithmetic operation. { 'bīn·ə·rē äp·ə'rā·shən }

binary phase-shift keying [COMMUN] Keying of binary data or Morse code dots and dashes by ±90° phase deviation of the carrier. Abbreviated BPSK. { 'bīn·ə·rē 'fāz ˌshift 'kē·iŋ }

binary point [COMPUT SCI] The character, or the location of an implied symbol, that separates the integral part of a numerical expression from its fractional part in binary notation. { 'bīn·ə·rē 'pȯint }

binary scaler [ELECTR] A scaler that produces one output pulse for every two input pulses. Also known as binary counter; scale-of-two circuit. { 'bīn·ə·rē ¦skā·lər }

binary search [COMPUT SCI] A dichotomizing search in which the set of items to be searched is divided at each step into two equal, or nearly equal, parts. Also known as binary chop. { 'bīn·ə·rē 'sərch }

binary signal [ELECTR] A voltage or current which carries information by varying between two possible values, corresponding to 0 and 1 in the binary system. { 'bīn·ə·rē 'sig·nəl }

binary word [COMPUT SCI] A group of bits which occupies one storage address and is treated by the computer as a unit. { 'bīn·ə·rē ¦wərd }

B-indicator *See* B-display. { ¦bē ¦in·dəˌkād·ər }

binding time [COMPUT SCI] **1.** The instant when a symbolic expression in a computer program is reduced to a form which is directly interpretable by the hardware. **2.** The instant when a variable is assigned its data type, such as integer or string. { 'bīn·diŋ ˌtīm }

binomial array antenna [ELECTROMAG] Directional antenna array for reducing minor lobes and providing maximum response in two opposite directions. { bī'nō·mē·əl ə'rā an'ten·ə }

biochip [ELECTR] An experimental type of integrated circuit whose basic components are organic molecules. { 'bī·ōˌchip }

bioinformatics [COMPUT SCI] The use of computers to study biological systems. { ˌbī·ōˌin·fər'mad·iks }

biometric device [COMPUT SCI] A device that identifies persons seeking access to a computing system by determining their physical characteristics through fingerprints, voice recognition, retina patterns, pictures, weight, or other means. { ˌbī·ō¦me·trik di¦vīs }

BIOS *See* basic input/output system.

bipolar amplifier [ELECTR] An amplifier capable of supplying a pair of output signals corresponding to the positive or negative polarity of the input signal. { bī'pō·lər 'am·pləˌfī·ər }

bipolar circuit [ELECTR] A logic circuit in which zeros and ones are treated in a symmetric or bipolar manner, rather than by the presence or absence of a signal; for example, a balanced arrangement in a square-loop-ferrite magnetic circuit. { bīˈpō·lər ˈsər·kət }

bipolar format [COMPUT SCI] A method of representing binary data in which 0 bits have zero voltage and each 1 bit has a polarity opposite that of the preceding 1 bit. { bīˈpō·lər ˈfȯr,mat }

bipolar integrated circuit [ELECTR] An integrated circuit in which the principal element is the bipolar junction transistor. { bīˈpō·lər ˈin·tə,grā d·əd ˈsər·kət }

bipolar junction transistor [ELECTR] A bipolar transistor that is composed entirely of one type of semiconductor, silicon. Abbreviated BJT. Also known as silicon homojunction. { ¦bī,pōl·ər ,jəŋk·shən tranˈzis·tər }

bipolar memory [COMPUT SCI] A computer memory employing integrated-circuit bipolar junction transistors as bistable memory cells. { bīˈpō·lər ˈmem·rē }

bipolar signal [COMMUN] A signal in which different logical states are represented by electrical voltages of opposite polarity. { bīˈpō·lər ˈsig·nəl }

bipolar spin device *See* magnetic switch. { ¦bī ,pō·lər ˈspin di,vīs }

bipolar spin switch *See* magnetic switch. { ¦bī ,pō·lər ˈspin ,swich }

bipolar transistor [ELECTR] A transistor that uses both positive and negative charge carriers. { bīˈpō·lər tranzˈis·tər }

bipolar video *See* coherent video. { bīˈpō·lər ˈvid·ē·ō }

biquartic filter [ELECTR] An active filter that uses operational amplifiers in combination with resistors and capacitors to provide infinite values of Q and simple adjustments for band-pass and center frequency. { ¦bī¦kwȯrd·ik ˈfil·tər }

birefringence [OPTICS] **1.** Splitting of a light beam into two components, which travel at different velocities, by a material. **2.** For a light beam that has been split into two components by a material, the difference in the indices of refraction of the components within the material. Also known as double refraction. { ,bī·riˈfrin·jəns }

bistable circuit [ELECTR] A circuit with two stable states such that the transition between the states cannot be accomplished by self-triggering. { ¦bī¦stā·bəl ,sar·kət }

bistable multivibrator [ELECTR] A multivibrator in which either of the two active devices may remain conducting, with the other nonconducting, until the application of an external pulse. Also known as Eccles-Jordan circuit; Eccles-Jordan multivibrator; flip-flop circuit; trigger circuit. { ¦bī¦stā·bəl məl·tiˈvī,brād·ər }

bistable optical device [OPTICS] A device which can be in either of two stable states of optical transmission for a single value of the input light intensity. { ¦bī¦stā·bəl ˈäp·tə·kəl diˈvīs }

bistable unit [ENG] A physical element that can be made to assume either of two stable states; a binary cell is an example. { ¦bī¦stā·bəl ˈyü·nət }

bistatic radar [ENG] Radar in which the transmitter and receiver are not located in the same place; the line between their positions is called the baseline. { ˈbī,stad·ik ˈrā,där }

bisynchronous transmission [COMMUN] A set of procedures for handling synchronous transmission of data and, in particular, for handling a block of data, called a message format, that is transmitted in a single operation. { bīˈsiŋ·krə·nəs tranzˈmish·ən }

bit [COMPUT SCI] **1.** A unit of information content equal to one binary decision or the designation of one of two possible and equally likely values or states of anything used to store or convey information. **2.** A dimensionless unit of storage capacity specifying that the capacity of a storage device is expressed by the logarithm to the base 2 of the number of possible states of the device. { bit }

bitblt *See* bit block transfer.

bit block transfer [COMPUT SCI] In computer graphics, a hardware function that moves a rectangular block of bits from the main memory to the display memory at high speed. Abbreviated bitblt. { ¦bit ,bläk ˈtranz·fər }

bit buffer unit [COMMUN] A unit that terminates bit-serial communications lines coming from and going to technical control. { ¦bit ˈbəf·ər ,yü·nət }

bit count appendage [COMPUT SCI] One of the two-byte elements replacing the parity bit stripped off each byte transferred from main storage to disk volume (the other element is the cyclic check); these two elements are appended to the block during the write operation; on a subsequent read operation these elements are calculated and compared to the appended elements for accuracy. { ˈbit ,kau̇nt əˈpen·dij }

bit density [COMPUT SCI] Number of bits which can be placed, per unit length, area, or volume, on a storage medium; for example, bits per inch of magnetic tape. Also known as record density. { ˈbit ˈden·səd·ē }

bit depth [COMPUT SCI] In a digital file, the number of colors for an image; calculated as 2 to the power of the bit depth; for example, a bit depth of 8 supports up to 256 colors, and a bit depth of 24 supports up to 16 million colors. { ˈbit ,depth }

bit flipping *See* bit manipulation. { ˈbit ,flip·iŋ }

bit location [COMPUT SCI] Storage position on a record capable of storing one bit. { ˈbit lōˈkā·shən }

bit manipulation [COMPUT SCI] Changing bits from one state to the other, usually to influence the operation of a computer program. Also known as bit flipping. { ˈbit mə,nip·yəˈlā,shən }

bit-mapped font [COMPUT SCI] A font that is specified by a complete set of dot patterns for each character and symbol. { ¦bit ,mapt ˈfänt }

bit-mapped graphics *See* raster graphics. { ¦bit ,mapt ˈgraf·iks }

bit mapping [COMPUT SCI] The assignment of each location in a computer's storage to a physical location on an electronic display. { ˈbit ˈmap·iŋ }

bit-oriented protocol [COMMUN] A communications protocol in which individual bits within a

byte are used as control codes. { ¦bit ˌȯr·ēˌent·əd ˈprōd·əˌkȯl }
bit pattern [COMPUT SCI] A combination of binary digits arranged in a sequence. { ˈbit ˌpad·ərn }
bit per second [COMMUN] A unit specifying the instantaneous speed at which a device or channel transmits data. Abbreviated bps. { ˈbit pər ˈsek·ənd }
bit position [COMPUT SCI] The position of a binary digit in a word, generally numbered from the least significant bit. { ˈbit pəˈzish·ən }
bit rate [COMMUN] Quantity, per unit time, of binary digits (or pulses representing them) which will pass a given point on a communications line or channel in a continuous stream. { ˈbit ˌrāt }
bit serial [COMMUN] Sequential transmission of character-forming bits. { ¦bit ˈsir·ē·əl }
bit-sliced microprocessor [COMPUT SCI] A microprocessor in which the major logic of the central processor is partitioned into a set of large-scale-integration circuits, as opposed to being placed on a single chip. { ˈbit ˌslīst ˌmī·krō ˈpräs·əs·ər }
bit stream [COMPUT SCI] **1.** A consecutive line of bits transmitted over a circuit in a transmission method in which character separation is accomplished by the terminal equipment. **2.** A binary signal without regard to grouping by character. { ˈbit ˌstrēm }
bit-stream generator [COMMUN] An algorithmic procedure for producing an unending sequence of binary digits to implement a stream. { ˈbit ˌstrēm ˈjen·əˌrād·ər }
bit string [COMPUT SCI] A set of consecutive binary digits representing data in coded form, in which the significance of each bit is determined by its position in the sequence and its relation to the other bits. { ˈbit ˌstriŋ }
bit stuffing [COMMUN] The insertion of extra bits in a transmitted message in order to fill a frame to a fixed size or to break up a pattern of bits that could be mistaken for control codes. { ˈbit ˌstəf·iŋ }
bit synchronization [COMMUN] Element of a message header used to synchronize all of the bits and characters that follow. { ˈbit ˌsiŋ·krə·nəˈzā·shən }
bit test [COMPUT SCI] A check by a computer program to determine the status of a particular bit. { ˈbit ˌtest }
bit zone [COMPUT SCI] **1.** One of the two leftmost bits in a commonly used system in which six bits are used for each character; related to overpunch. **2.** Any bit in a group of bit positions that are used to indicate a specific class of items; for example, numbers, letters, special signs, and commands. { ˈbit ˌzōn }
BJT *See* bipolar junction transistor.
BL *See* base line.
black *See* black signal. { blak }
black-and-white television *See* monochrome television. { ¦blak ən ¦wīt ˈtel·əˌvizh·ən }
blacker-than-black level [COMMUN] In television, a level of greater instantaneous amplitude than the black level, used for synchronization and control signals. { ˈblak·ər t͟hən ˈblak ˌlev·əl }
black hole *See* stale link. { ¦blak ˈhōl }
black level [ELECTR] The level of the television picture signal corresponding to the maximum limit of black peaks. { ˈblak ˌlev·əl }
blackout *See* radio blackout. { ˈblakˌau̇t }
black peak [COMMUN] A peak excursion of the television picture signal in the black direction. { ˈblak ˌpēk }
black signal [COMMUN] Signal at any point in a facsimile system produced by the scanning of a maximum density area of the subject copy. Also known as black; picture black. { ˈblak ˌsig·nəl }
black transmission [COMMUN] The amplitude-modulated transmission of facsimile signals in which the maximum signal amplitude corresponds to the greatest copy density or darkest shade. { ˈblak tranzˈmish·ən }
blank [ELECTR] To cut off the electron beam of a television picture tube or camera tube during the process of retrace by applying a rectangular pulse voltage to the grid or cathode during each retrace interval. Also known as beam blank. { blaŋk }
blank cell [COMPUT SCI] A cell of a spreadsheet that contains no text or numeric values, and for which no formatting is specified other than the global formats of the spreadsheet. { ˈblaŋk ˌsel }
blank character [COMPUT SCI] A character, either printed or appearing as a blank, used to denote a blank space among printed characters. Also known as space character. { ˈblaŋk ˈkar·ik·tər }
blanketing [COMMUN] Interference due to a nearby transmitter whose signals are so strong that they override other signals over a wide band of frequencies. { ˈblaŋ·kəd·iŋ }
blank form *See* blank medium. { ˈblaŋk ˌfȯrm }
blanking [ELECTR] The act, useful in adapting a radar to its environment, of disabling selected apparatus at specified times or of deleting certain data from further treatment. [ENG] **1.** The closing off of flow through a liquid-containing process pipe by the insertion of solid disks at joints or unions; used during maintenance and repair work as a safety precaution. Also known as blinding. **2.** Cutting of metal or plastic sheets into shapes by striking with a punch. Also known as die cutting. { ˈblaŋk·iŋ }
blanking circuit [ELECTR] A circuit preventing the transmission of brightness variations during the horizontal and vertical retrace intervals in television scanning. { ˈblaŋk·iŋ ˌsər·kət }
blanking level [ELECTR] The level that separates picture information from synchronizing information in a composite television picture signal; coincides with the level of the base of the synchronizing pulses. Also known as pedestal; pedestal level. { ˈblaŋk·iŋ ˌlev·əl }
blanking pulse [ELECTR] A control pulse used to switch off a part of a television or radar set electronically for a predetermined length of time. { ˈblaŋk·iŋ ˌpəls }
blanking signal [ELECTR] The signal rendering the return trace invisible on the picture tube of a television receiver. { ˈblaŋk·iŋ ˌsig·nəl }

blanking time [ELECTR] The length of time that the electron beam of a cathode-ray tube is shut off. { 'blaŋk·iŋ ,tīm }

blank medium [COMPUT SCI] An empty position on the medium concerned, such as a column without holes on a punch tape, used to indicate a blank character. Also known as blank form. { ¦blaŋk 'mēd·ē·əm }

blank tape [COMPUT SCI] A portion of a paper tape having sprocket holes only, to indicate a blank character. { ¦blaŋk 'tāp }

blank tape halting problem [COMPUT SCI] The problem of finding an algorithm that, for any Turing machine, decides whether the machine eventually stops if it started on an empty tape; it has been proved that no such algorithm exists. { ¦blaŋk 'tāp 'hȯl·tiŋ ,präb·ləm }

blast [COMPUT SCI] To release internal or external memory areas from the control of a computer program in the course of dynamic storage allocation, making these areas available for reallocation to other programs. { blast }

bleed [COMPUT SCI] In optical character recognition, the flow of ink in printed characters beyond the limits specified for their recognition by a character reader. { blēd }

blend to analog [COMMUN] The point at which the block error rate of an AM/FM IBOC receiver falls below some predefined threshold and the digital audio is faded out while simultaneously the analog audio is faded in, preventing the received audio from simply muting when the digital signal is lost. The receiver audio will also blend to digital upon reacquisition of the digital signal. { ¦blend tə 'an·əl,äg }

blend to mono [COMMUN] The process of progressively attenuating the left-right component of a stereo decoded signal as the received radio frequency signal decreases, with the net result of lowering the audible noise. { ¦blend tə 'män·ō }

BLER *See* block error rate.

blind approach beacon system [NAV] A pulse-type, ground-based navigation beacon used for runway approach at airports, which sends out signals that produce range and runway position information on the L-scan cathode-ray indicator of an aircraft making an instrument approach. Also known as beam approach beacon system (British usage). Abbreviated babs. { ¦blīnd ə'prōch 'bē·kən ,sis·təm }

blind zone [COMMUN] Area from which echoes cannot be received; generally, an area shielded from the transmitter by some natural obstruction and therefore from which there can be no return. { 'blīnd ,zōn }

B line *See* index register. { 'bē ,līn }

blinking [COMMUN] Method of providing information in pulse systems by modifying the signal at its source so that signal presentation on the display scope alternately appears and disappears; in loran, this indicates that a station is malfunctioning. [ELECTR] Electronic-attack technique employed by two aircraft separated by a short distance and not resolved in azimuth so as to appear as one target to a tracking radar; the two aircraft alternately spot-jam, causing the radar system to oscillate from one place to another, greatly degrading the fire-control accuracy. [NAV] Regular shifting right and left or alternate appearance and disappearance of a loran signal to indicate that the signals of a pair of stations are out of synchronization. { 'bliŋ·kiŋ }

blip [ELECTR] **1.** The display of a received pulse on the screen of a cathode-ray tube. Also known as pip. **2.** An ideal infrared radiation detector that detects with unit quantum efficiency all of the radiation in the signal for which the detector was designed, and responds only to the background radiation noise that comes from the field of view of the detector. { blip }

blip-scan ratio [ELECTR] The ratio of the number of times a target is detected (a contact generated, or a display clearly evident) to the number of times of opportunities to do so provided by the radar routine; provides a rough estimate of the probability of detection occurring during the detection process. { 'blip ,skan 'rā·shō }

bloatware *See* fatware. { 'blōt,wer }

BLOB *See* binary large object. { bläb *or* ¦bē¦el¦ō 'bē }

block [COMMUN] An 8-by-8 array of pel values or discrete cosine transform coefficients representing luminance or chrominance information. [COMPUT SCI] A group of information units (such as records, words, characters, or digits) that are transported or considered as a single unit by virtue of their being stored in successive storage locations; for example, a group of logical records constituting a physical record. { bläk }

block body [COMPUT SCI] A list of statements that follows the block head in a computer program with block structure. { 'bläk ,bäd·ē }

block chaining *See* chained block encryption. { 'bläk ,chān·iŋ }

block check character [COMMUN] A character that is added to a block of data to check its accuracy, and consists of parity bits each of which is set by observing a specified set of bits in the block. { 'bläk ¦chek ,kar·ik·tər }

block cipher [COMMUN] A cipher that transforms a string of input bits of fixed length into a string of output bits of fixed length. { 'bläk ,sī·fər }

block code [COMMUN] An error-correcting code generated by an encoder that produces a fixed-length code word with each incoming fixed-length message block. { 'bläk ,kōd }

block data [COMPUT SCI] A statement in FORTRAN which declares that the program following is a data specification subprogram. { 'bläk ,dad·ə }

blocked F-format data set *See* FB data set. { 'bläkt ¦ef'fȯr,mat 'dad·ə ,set }

blocked process [COMPUT SCI] A program that is running on a computer but is temporarily prevented from making progress because it requires some resource (such as a printer or user input) that is not immediately available. { ¦bläkt 'prä,ses }

block encryption [COMMUN] The use of a block cipher, usually employing the data encryption standard (DES), in which each 64-bit block of data is enciphered or deciphered separately, and every bit in a given output block depends on every bit in its respective input block and on every bit in the key, but on no other bits. Also known as electronic codebook mode (ECB). { 'bläk en'krip·shən }

block error rate [COMMUN] A ratio of the number of data blocks received with atleast one uncorrectable bit to the total number of blocks received. Abbreviated BLER. { 'bläk 'er·ər ,rāt }

blockette [COMPUT SCI] A subdivision of a group of consecutive machine words transferred as a unit, particularly with reference to input and output. { blä'ket }

block head [COMPUT SCI] A list of declarations at the beginning of a computer program with block structure. { 'bläk ¦hed }

block identifier [COMPUT SCI] A means of identifying an area of storage in FORTRAN so that this area may be shared by a program and its subprograms. { 'bläk ī'den·tə,fī·ər }

block ignore character [COMPUT SCI] A character associated with a block which indicates the presence of errors in the block. { 'bläk ig'nór ,kar·ik·tər }

blocking [COMPUT SCI] Combining two or more computer records into one block. [ELECTR] Overloading a receiver by an unwanted signal so that the automatic gain control reduces the response to a desired signal. { 'bläk·iŋ }

blocking factor [COMPUT SCI] The largest possible number of records of a given size that can be contained within a single block. { 'bläk·iŋ ,fak·tər }

blocking layer *See* depletion layer. { 'bläk·iŋ ,lā·ər }

blocking oscillator [ELECTR] A relaxation oscillator that generates a short-time-duration pulse by using a single transistor or electron tube and associated circuitry. Also known as squegger; squegging oscillator. { 'bläk·iŋ 'äs·ə,lād·ər }

block input [COMPUT SCI] **1.** A block of computer words considered as a unit and intended or destined to be transferred from an internal storage medium to an external destination. **2.** *See* output area. { 'bläk 'in,pu̇t }

block length [COMPUT SCI] The total number of records, words, or characters contained in one block. { 'bläk ,leŋkth }

block loading [COMPUT SCI] A program loading technique in which the control sections of a program or program segment are loaded into contiguous positions in main memory. { 'bläk ,lōd·iŋ }

block mark [COMPUT SCI] A special character that indicates the end of a block. { 'bläk ,märk }

block move *See* cut and paste. { ¦bläk 'müv }

block multiplexor channel [COMPUT SCI] A transmission channel in a computer system that can simultaneously transmit blocks of data from several high-speed input/output devices by interleaving the data. { 'bläk ¦məlt·i,plek·sər ,chan·əl }

block operation [COMPUT SCI] An editing or formatting procedure that is carried out on a selected block of text in a word-processing document. { 'bläk ,äp·ə¦rā·shən }

block parity [COMMUN] An error-checking technique involving the comparison of a transmitted block check character with one calculated by the receiving device. { 'bläk 'par· əd·ē }

block protection [COMPUT SCI] An instruction in a word-processing or page-layout program that prevents a soft page break from being inserted in a specified block of text, ensuring against a bad page break. { 'bläk prə,tek·shən }

block standby [COMPUT SCI] Locations always set aside in storage for communication with buffers in order to make more efficient use of such buffers. { 'bläk ,stand,bī }

block structure [COMPUT SCI] In computer programming, a conceptual tool used to group sequences of statements into single compound statements and to allow the programmer explicit control over the scope of the program variables. { 'bläk ,strək·chər }

block transfer [COMPUT SCI] The movement of data in blocks instead of by individual records. { 'bläk ¦trans·fər }

blooming [ELECTR] **1.** Defocusing of television picture areas where excessive brightness results in enlargement of spot size and halation of the fluorescent screen. **2.** An increase in radar display spot size due to a particularly strong signal exciting the phosphorus material. **3.** The wide spatial dispersion of chaff after being dispensed in small bundles. { 'blüm·iŋ }

blow [COMPUT SCI] To write data or code into a programmable read-only memory chip by melting the fuse links corresponding to bits that are to be zero. { blō }

blow up *See* abend. { 'blō ,əp }

Bluetooth [COMMUN] A technical specification for the wireless connection over short distances of digital devices, such as cellular telephones, portable computers, and computer peripheral equipment, utilizing the unlicensed 2.4GHz radio frequency spectrum.

BNF *See* Backus-Naur form.

bobbing [ELECTR] Fluctuation of the strength of a radar echoand its display, due to alternate constructive and destructive interference of the received signal as in a multipath propagation situation. { 'bäb·iŋ }

bobtail curtain antenna [ELECTROMAG] A bidirectional, vertically polarized, phased-array antenna that has two horizontal sections, each 0.5 electrical wavelength long, that connect three vertical sections, each 0.25 electrical wavelength long. { 'bäb,tāl 'kərt·ən an,ten·ə }

boiler plate [COMPUT SCI] A commonly used expression or phrase that is stored in memory and can be copied into a word-processing document as needed. { 'bȯil·ər ,plāt }

bomb *See* abend. { bäm }

Book A *See* DVD-read-only. { ¦bu̇k 'ā }

Book B *See* DVD-video. { ¦bu̇k 'bē }

Book D *See* DVD-write once. { ¦bu̇k 'dē }

Book E *See* DVD-rewritable. { ¦bu̇k 'ē }

bookkeeping operation [COMPUT SCI] A computer operation which does not directly contribute to the result, that is, arithmetical, logical, and transfer operations used in modifying the address section of other instructions in counting cycles and in rearranging data. Also known as red-tape operation. { 'bu̇k,kēp·iŋ äp·ə'rā·shən }

bookmark [COMPUT SCI] **1.** Any method of halting the processing of a transaction and holding it, as far as it has been completed, until processing resumes. **2.** A code that is inserted at a particular place in a document or that is associated with a particular document so that the user can easily return to the specified insertion point or document. **3.** A Web page location (URL) which is saved by a user for quick reference. { 'bu̇k ,märk }

boolean [COMPUT SCI] A scalar declaration in ALGOL defining variables similar to FORTRAN's logical variables. { 'bü·lē·ən }

Boolean algebra [MATH] An algebraic system with two binary operations and one unary operation important in representing a two-valued logic. { 'bü·lē·ən 'al·jə·brə }

Boolean calculus [MATH] Boolean algebra modified to include the element of time. { 'bü·lē·ən 'kal·kyə·ləs }

Boolean data type *See* logical data type. { 'bü·lē·ən 'dad·ə ,tīp }

Boolean determinant [MATH] A function defined on Boolean matrices which depends on the elements of the matrix in a manner analogous to the manner in which an ordinary determinant depends on the elements of an ordinary matrix, with the operation of multiplication replaced by intersection and the operation of addition replaced by union. { ¦bül·ē·ən di'tər·mə·nənt }

Boolean function [MATH] A function $f(x,y,\ldots,z)$ assembled by the application of the operations AND, OR, NOT on the variables $x, y, \ldots, z$ and elements whose common domain is a Boolean algebra. { 'bü·lē·ən 'fəŋk·shən }

Boolean matrix [MATH] A rectangular array of elements each of which is a member of a Boolean algebra. { ¦bül·ē·ən 'mā,triks }

Boolean operation table [MATH] A table which indicates, for a particular operation on a Boolean algebra, the values that result for all possible combination of values of the operands; used particularly with Boolean algebras of two elements which may be interpreted as "true" and "false." { ¦bül·ē·ən ,äp·ə'rā·shən ,tā·bəl }

Boolean operator [MATH] A logic operator that is one of the operators AND, OR, or NOT, or can be expressed as a combination of these three operators. { ¦bül·ē·ən 'äp·ə,rād·ər }

Boolean ring [MATH] A commutative ring with the property that for every element a of the ring, $a \times a$ and $a + a = 0$; it can be shown to be equivalent to a Boolean algebra. { ¦bül·ē·ən 'riŋ }

Boolean search [COMPUT SCI] A search for selected information, that is, information satisfying conditions that can be expressed by AND, OR, and NOT functions. { 'bü·lē·ən 'sərch }

booster [ELECTR] **1.** A separate radio-frequency amplifier connected between an antenna and a television receiver to amplify weak signals. **2.** A radio-frequency amplifier that amplifies and rebroadcasts a received communications service radio carrier frequency for reception by the general public. { 'büs·tər }

booster voltage [ELECTR] The additional voltage supplied by the damper circuit to the horizontal output, horizontal oscillator, and vertical output circuits of a television receiver to give greater sawtooth sweep output. { 'büs·tər ,vōl·tij }

boot [COMPUT SCI] To load the operating system into a computer after it has been switched on; usually applied to small computers. { büt }

boot button *See* bootstrap button. { 'büt ,bət·ən }

boot record [COMPUT SCI] A special area of a floppy diskette or hard drive which is used by the computer during system startup. { 'büt ,rek·ərd }

bootstrap [COMPUT SCI] The procedures for making a computer or a program function through its own actions. { 'büt,strap }

bootstrap button [COMPUT SCI] The first button pressed when a computer is turned on, causing the operating system to be loaded into memory. Also known as boot button; initial program load button; IPL button. { 'büt,strap ,bət·ən }

bootstrap instructor technique [COMPUT SCI] A technique permitting a system to bring itself into an operational state by means of its own action. Also known as bootstrap technique. { 'büt ,strap in'strək·tər tek'nēk }

bootstrap loader [COMPUT SCI] A very short program loading routine, used for loading other loaders in a computer; often implemented in a read-only memory. { 'büt,strap 'lōd·ər }

bootstrap memory [COMPUT SCI] A device that provides for the automatic input of new programs without erasing the basic instructions in the computer. { 'büt,strap 'mem·rē }

bootstrap program *See* loading program. { 'büt,strap ,prō·grəm }

bootstrap technique *See* bootstrap instructor technique. { 'büt,strap tek'nēk }

boot virus [COMPUT SCI] A virus that infects the boot records on floppy diskettes and hard drives and is designed to self-replicate from one disk to another. { 'büt ,vī·rəs }

BORSCHT [COMMUN] An interface circuit between ordinary telephone lines carrying analog voice signals and digital time-division multiplex facilities, which digitizes voice signals, assigns them time slots, and then multiplexes them. Acronym for battery, overvoltage, ringing, supervision, coding, hybrid and test access. { bȯrsht }

bottleneck analysis [COMPUT SCI] A detailed study of the manner in which elements of a computer system are related to find out where bottlenecks arise, so that the system's performance can be improved. { 'bäd·əl,nek ə ,nal·ə·səs }

bottom [COMPUT SCI] The termination of a file. { 'bäd·əm }

bottom-up analysis [COMPUT SCI] A reductive method of syntactic analysis which attempts to reduce a string to a root symbol. { ¦bäd·əm·əp ¦ə'nal·ə·səs }

bounced message [COMPUT SCI] An electronic mail message that is returned to sender because attempts to deliver it have been unsuccessful. { ,baúnst 'mes·ij }

boundary-layer photocell *See* photovoltaic cell. { 'baún·drē ,lā·ər 'fō·dō,sel }

bounds register [COMPUT SCI] A device which stores the upper and lower bounds on addresses in the memory of a given computer program in a time-sharing system. { 'baúnz ,rej·ə·stər }

Bourne shell [COMPUT SCI] The original Unix shell. { 'búrn ,shel }

bowtie antenna [ELECTROMAG] An antenna that consists of two triangular pieces of stiff wire or two triangular flat metal plates, arranged in the configuration of a bowtie, with the feed point at the gap between the apexes of the triangles. { 'bō,tī an,ten·ə }

boxcar [COMMUN] One of a series of long signal-wave pulses which are separated by very short intervals of time. { 'bäks,kär }

boxcar circuit [ELECTR] A circuit used in radar for sampling voltage waveforms and storing the latest value sampled; the term is derived from the flat, steplike segments of the output voltage waveform. { 'bäks,kär ,sər·kət }

B-pictures *See* bidirectional pictures. { 'bē 'pik·chərz }

bps *See* bit per second.

BPSK *See* binary phase-shift keying.

Bragg cell *See* acoustooptic modulator. { 'brag ,sel }

branch [COMPUT SCI] **1.** Any one of a number of instruction sequences in a program to which computer control is passed, depending upon the status of one or more variables. **2.** *See* jump. { branch }

branch-circuit distribution center [COMMUN] Distribution center at which branch circuits are supplied. { ¦branch ¦sər·kət dis·trə'byü·shən ,sen·tər }

branch gain *See* branch transmittance. { 'branch ,gān }

branching [COMPUT SCI] The selection, under control of a computer program, of one of two or more branches. { 'branch·iŋ }

branch instruction [COMPUT SCI] An instruction that makes the computer choose between alternative subprograms, depending on the conditions determined by the computer during the execution of the program. { 'branch in'strək·shən }

branch point [COMPUT SCI] A point in a computer program at which there is a branch instruction. { 'branch ,póint }

branch prediction [COMPUT SCI] A method whereby a processor guesses the outcome of a branch instruction so that it can prepare in advance to carry out the instructions that follow the predicted outcome. { 'branch prə,dik·shən }

branch transmittance [CONT SYS] The amplification of current or voltage in a branch of an electrical network; used in the representation of such a network by a signal-flow graph. Also known as branch gain. { ¦branch trans'mit·əns }

Braun tube *See* cathode-ray tube. { 'braún ,tüb }

breadboarding [ELECTR] Assembling an electronic circuit in the most convenient manner, without regard for final locations of components, to prove the feasibility of the circuit and to facilitate changes when necessary. { 'bred,bórd·iŋ }

break [COMPUT SCI] **1.** To interrupt processing by a computer, usually by depressing a key. **2.** A place in a file of records where one or more of the values in the records change. { brāk }

break frequency [CONT SYS] The frequency at which a graph of the logarithm of the amplitude of the frequency response versus the logarithm of the frequency has an abrupt change in slope. Also known as corner frequency; knee frequency. { 'brāk ,frē·kwən·sē }

break-in operation [COMMUN] A method of radio communication in which it is possible for the receiving operator to interrupt or break into the transmission. { 'brā¦kin ,äp·ə,rā·shən }

break key [COMPUT SCI] A key on a computer keyboard whose depression causes processing to be interrupted. { 'brāk ,kē }

breakout box [ELECTR] A device connected to a multiconductor cable that provides terminal connections to test the signals in a transmission. { 'brāk,aút ,bäks }

breakoutput [COMPUT SCI] An ALGOL procedure which causes all bytes in a device buffer to be sent to the device rather than wait until the buffer is full. { ¦brā¦kaút,pút }

break period [COMMUN] Of a rotary dial telephone, the time interval during which the circuit contacts are open. { 'brāk ,pir·ē·əd }

breakpoint [COMPUT SCI] A point in a program where an instruction, instruction digit, or other condition enables a programmer to interrupt the run by external intervention or by a monitor routine. { 'brāk,póint }

breakpoint switch [COMPUT SCI] A manually operated switch which controls conditional operation at breakpoints, used primarily in debugging. { 'brāk,póint ,swich }

breakpoint symbol [COMPUT SCI] A symbol which may be optionally included in an instruction, as an indication, tag, or flag, to designate it as a breakpoint. { 'brāk,póint ,sim·bəl }

breakthrough [COMPUT SCI] An interruption in the intended character stroke in optical character recognition. { 'brāk,thrü }

B register *See* index register. { 'bē ,rej·ə·stər }

brevity code [COMMUN] Code which has as its sole purpose the shortening of messages rather than the concealment of their content. { 'brev·əd·ē ,kōd }

bridge [COMMUN] A device that joins two networks of the same type. { brij }

bridge hybrid *See* hybrid junction. { 'brij 'hī·brəd }

bridgeware [COMPUT SCI] Software or hardware that translates programs or converts data from one format to another. { 'brij,wer }

bridging connection [ELECTR] Parallel connection by means of which some of the signal energy in a circuit may be withdrawn frequently, with imperceptible effect on the normal operation of the circuit. { 'brij·iŋ kə,nek·shən }

bridging loss [ELECTR] Loss resulting from bridging an impedance across a transmission system; quantitatively, the ratio of the signal power delivered to that part of the system following the bridging point, and measured before the bridging, to the signal power delivered to the same part after the bridging. { 'brij·iŋ ,lós }

brightness control [ELECTR] A control that varies the luminance of the fluorescent screen of a cathode-ray tube, for a given input signal, by changing the grid bias of the tube and hence the beam current. Also known as brilliance control; intensity control. { 'brīt·nəs kən'trōl }

brilliance [ELECTR] **1.** The degree of brightness and clarity of the display of a cathode-ray tube. **2.** The degree to which the higher audio frequencies of an input sound are reproduced by a sound system. { 'bril·yəns }

brilliance control *See* brightness control. { 'bril·yəns kən'trōl }

broadband [COMMUN] A band with a wide range of frequencies. { 'bród,band }

broadband amplifier [ELECTR] An amplifier having essentially flat response over a wide range of frequencies. { 'bród,band ¦am·plə,fī·ər }

broadband antenna [ELECTROMAG] An antenna that functions satisfactorily over a wide range of frequencies, such as for all 12 very-high-frequency television channels. { 'bród,band an'ten·ə }

broadband channel [COMMUN] A data transmission channel that can handle frequencies higher than the normal voice-grade line limit of 3 to 4 kilohertz; it can carry many voice or data channels simultaneously or can be used for high-speed single-channel data transmission. { 'bród,band ¦chan·əl }

broadband path [COMMUN] A path having a bandwidth of 20 kilohertz or greater. { 'bród ,band ,path }

broadcast [COMMUN] A television, radio, or data transmission intended for public reception. { 'bród,kast }

broadcast band [COMMUN] The band of frequencies extending from 535 to 1605 kilohertz, corresponding to assigned radio carrier frequencies that increase in multiples of 10 kHz between 540 and 1600 kHz for the United States. Also known as standard broadcast band. { 'bród,kast ,band }

broadcast message [COMMUN] A message that is sent to all users of a computer network when they log on to the network. { ¦bród,kast ¦mes·ij }

broadcast station [COMMUN] A television or radio station used for transmitting programs to the general public. Also known as station. { 'bród ,kast ,stā·shən }

broadcast transmitter [ELECTR] A transmitter designed for use in a commercial amplitude-modulation, frequency-modulation, or television broadcast channel. { 'bród,kast tranz'mid·ər }

broadside array [ELECTROMAG] An antenna array whose direction of maximum radiation is perpendicular to the line or plane of the array. { 'bród,sīd ə'rā }

browse mode [COMPUT SCI] A mode of operation in which data in a document or database are conveniently displayed for rapid, on-screen review. { 'braúz ,mōd }

browser [COMPUT SCI] An interactive program (client) that requests, retrieves, and displays pages from the World Wide Web. { 'braúz·ər }

brute force attack [COMPUT SCI] An attempt to gain unauthorized access to a computing system by generating and trying all possible passwords. { ¦brüt ¦fórs ə'tak }

brute-force technique [COMPUT SCI] Any method that relies chiefly on the advanced processing capabilities of a large computer to accomplish a task. { ¦brüt ,fórs tek'nēk }

B-scan *See* B-display. { 'bē ,skan }

B-scope *See* B-display. { 'bē ,skōp }

b-spline [COMPUT SCI] A curve that is generated by a computer-graphics program, guided by a mathematical formula which ensures that it will be continuous with other such curves; it is mathematically more complex but easier to blend than a Bézier curve. { 'bē,splīn }

B station [NAV] In loran, the designation applied to one transmitting station of a pair, the signal of which always occurs more than half a repetition period after the succeeding signal and less than half a repetition period before the preceding signal from the other station of the pair, designated an A station. { 'bē ,stā·shən }

B store *See* index register. { 'bē ,stór }

B-tree *See* balanced-tree. { 'bē ,trē }

B+-tree [COMPUT SCI] A version of the balanced-tree that maintains a hierarchy of indexes while linking the data sequentially. { ¦bē ¦pləs ,trē }

bubble [COMPUT SCI] A circle that represents data in a data flow diagram. { 'bəb·əl }

bubble chart *See* data flow diagram. { 'bəb·əl ,chärt }

bubble memory [COMPUT SCI] A computer memory in which the presence or absence of a magnetic bubble in a localized region of a thin magnetic film designates a 1 or 0; storage capacity can be well over 1 megabit per cubic inch. Also known as magnetic bubble memory. { 'bəb·əl ¦mem·rē }

bubble sort [COMPUT SCI] A procedure for sorting a set of items that begins by sequencing the first and second items, then the second and third, and so on, until the end of the set is reached, and then repeats this process until all items are correctly sequenced. { 'bəb·əl ,sórt }

bucket [COMPUT SCI] A name usually reserved for a storage cell in which data may be accumulated. { 'bək·ət }

buffer [COMPUT SCI] *See* buffer storage. [ELECTR] **1.** An isolating circuit in an electronic computer used to prevent the action of a driven circuit from affecting the corresponding driving circuit. **2.** *See* buffer amplifier. { 'bəf·ər }

buffer amplifier [ELECTR] An amplifier used after an oscillator or other critical stage to isolate it from the effects of load impedance variations in subsequent stages. Also known as buffer; buffer stage. { ¦bəf·ər 'am·plə,fī·ər }

buffered computer [COMPUT SCI] A computer having a temporary storage device to compensate for differences in transmission speeds. { 'bəf·ərd kəm'pyüd·ər }

buffered device [COMPUT SCI] A piece of peripheral equipment, such as a printer, that is equipped with a buffer storage so that it can accept information more rapidly than it can process it. { 'bəf·ərd di'vīs }

buffered FET logic [ELECTR] A logic gate configuration used with gallium-arsenide field-effect transistors operating in the depletion mode, in which the level shifting required to make the input and output voltage levels compatible is achieved with Schottky barrier diodes. Abbreviated BFL. { 'bəf·ərd ¦ef¦ē¦tē 'läj·ik }

buffered I/O channel [COMPUT SCI] A storage device located between input/output (I/O) channels and main storage control to free the channels for use by other operations. { 'bəf·ərd ¦ī¦ō,chan·əl }

buffered terminal [COMPUT SCI] A computer terminal which contains storage equipment so that the rate at which it sends or receives data over its line does not need to agree exactly with the rate at which the data are entered or printed. { 'bəf·ərd 'tər·mən·əl }

buffer pooling [COMPUT SCI] A technique for receiving data in an input/output control system in which a number of buffers are available to the system; when a record is produced, a buffer is taken from the pool, used to hold the data, and returned to the pool after data transmission. { 'bəf·ər ,pül·iŋ }

buffer stage *See* buffer amplifier. { 'bəf·ər ,stāj }

buffer storage [COMPUT SCI] A synchronizing element used between two different forms of storage in a computer; computation continues while transfers take place between buffer storage and the secondary or internal storage. Also known as buffer. { 'bəf·ər ,stór·ij }

buffer zone [COMPUT SCI] An area of main memory set aside for temporary storage. { 'bəf·ər ,zōn }

bug [COMPUT SCI] A defect in a program code or in designing a routine or a computer. [ELECTR] **1.** A semiautomatic code-sending telegraph key in which movement of a lever to one side produces a series of correctly spaced dots and movement to the other side produces a single dash. **2.** An electronic listening device, generally concealed, used for commercial or military espionage. { bəg }

built-in antenna [ELECTROMAG] An antenna that is located inside the cabinet of a radio or television receiver. { 'bilt,in an'ten·ə }

built-in check [COMPUT SCI] A hardware device which controls the accuracy of data either moved or stored within the computer system. { 'bilt,in 'chek }

built-in function [COMPUT SCI] A function that is available through a simple reference and specification of arguments in a given higher-level programming language. Also known as built-in procedure; intrinsic procedure; standard function. { 'bilt,in 'fəŋk·shən }

built-in pointing device [COMPUT SCI] A trackball or pointing stick that is built into the case of a portable computer and used to move an on-screen pointer. { ¦bilt,in 'póint·iŋ di,vīs }

built-in procedure *See* built-in function. { 'bilt ,in prə'sēj·ər }

bulk diode [ELECTR] A semiconductor microwave diode that uses the bulk effect, such as Gunn diodes and diodes operating in limited space-charge-accumulation modes. { ¦bəlk 'dī,ōd }

bulk effect [ELECTR] An effect that occurs within the entire bulk of a semiconductor material rather than in a localized region or junction. { 'bəlk i'fekt }

bulk-effect device [ELECTR] A semiconductor device that depends on a bulk effect, as in Gunn and avalanche devices. { 'bəlk i'fekt di'vīs }

bulk memory [COMPUT SCI] A high-capacity memory used in connection with a computer for bulk storage of large quantities of data. { ¦bəlk 'mem·rē }

bulk storage *See* backing storage. { ¦bəlk 'stór·ij }

bulletin board [COMPUT SCI] A collection of information that is stored in a computer system and can be accessed either by a specified group of people or the general public, usually by dialing a number on the public telephone system. { 'búl·ət·ən ,bórd }

bulletin board system [COMPUT SCI] A computer system that enables its users, usually members of a particular interest group, to leave messages and to share information and software. Abbreviated BBS. { 'búl·ət·ən ,bórd ,sis·təm }

bundled program [COMPUT SCI] A computer program written, maintained, and updated by the computer manufacturer, and included in the price of the hardware. { ¦bənd·əld 'prō·grəm }

bundling [COMMUN] The provision of a combination of services, such as cable television and telephone service, over a single communications system. [COMPUT SCI] The provision of hardware and software as a single product or the combination of different software packages for sale as a single unit. { 'bən·dliŋ }

burnthrough [ELECTR] An electronic-protection effort by a radar to overcome the obscuration effect of jamming signals by using the highest energy transmission and longest possible dwell in the direction of the jamming or other direction of specific interest being affected. { 'bərn ˌthrü }

burst [COMMUN] **1.** A sudden increase in the strength of a signal being received from beyond line-of-sight range. **2.** A group of bits of characters that are transmitted together as a unit. **3.** A group of errors that occur together in a communication and alter its content. **4.** *See* color burst. [COMPUT SCI] **1.** To separate a continuous roll of paper into stacks of individual sheets by means of a burster. **2.** The transfer of a collection of records in a storage device, leaving an interval in which data for other requirements can be obtained from or entered into the device. **3.** A sequence of signals regarded as a unit in data transmission. { bərst }

burst amplifier [COMMUN] An amplifier stage in an analog color television receiver that is keyed into conduction and amplification by a horizontal pulse at the instant of each arrival of the color burst. Also known as chroma band-pass amplifier. { 'bərst ˌam·pləˌfī·ər }

burster [COMPUT SCI] An off-line device in a computer system used to separate the continuous roll of paper produced as output from a printer into individual sheets, generally along perforations in the roll. { 'bər·stər }

burst mode [COMPUT SCI] A method of transferring data between a peripheral unit and a control processing unit in a computer system in which the peripheral unit sends the central processor a signal to receive data until the peripheral unit signals that the transfer is completed. { 'bərst ˌmōd }

burst pedestal [COMMUN] Rectangular pulselike analog television signal which may be part of the color burst; the amplitude of the color burst pedestal is measured from the alternating-current axis of the sine-wave portion to the horizontal pedestal. { 'bərst ˌped·ə·stəl }

burst separator [ELECTR] The circuit in a color television receiver that separates the color burst from the composite video signal. { 'bərst sep·ə'rād·ər }

bus [COMPUT SCI] The circuitry and wiring connecting the various components of a computer through which data are transmitted; for example, in a personal computer the system bus interconnects the CPU, memory, and input/output devices. [ELECTR] One or more conductors in a computer along which information is transmitted from any of several sources to any of several destinations. { bəs }

bus architecture [COMPUT SCI] A structure for handling data transmission in a computer system or network, in which components are all linked to a common bus. { 'bəs 'är·kəˌtek·chər }

bus cable [ELECTR] An electrical conductor that can be attached to a bus to extend it outside the computer housing or join it to another bus within the same computer. { 'bəs ˌkā·bəl }

bus cycle [COMPUT SCI] A single transaction between the main memory and the CPU. { 'bəs ˌsī·kəl }

bus extender [ELECTR] A printed circuit board that can be joined to a bus to increase its capacity. { 'bəs ikˌsten·dər }

bus mouse [COMPUT SCI] A mouse that is plugged into a printed circuit board inserted into the computer's bus. { 'bəs ˌmau̇s }

bus network [COMMUN] A communications network whose components are joined together by a single cable. { 'bəs 'netˌwərk }

busy test [COMMUN] A test, in telephony, made to find out whether certain facilities which may be desired, such as a subscriber line or trunk, are available for use. { 'biz·ē ˌtest }

busy tone [COMMUN] Interrupted low tone returned to the subscriber as an indication that the party's line is busy. { 'biz·ē ˌtōn }

butterfly network [COMPUT SCI] A scheme that connects the units of a multiprocessing system and needs n stages to connect 2^n processors; at each stage a switch is thrown, depending on a particular bit in the addresses of the processors being connected. { ¦bəd·ərˌflī ¦netˌwərk }

Butterworth filter [ELECTR] An electric filter whose pass band (graph of transmission versus frequency) has a maximally flat shape. { 'bəd·ər ˌwərth 'fil·tər }

button [COMPUT SCI] A small circle or rectangle on a graphical user interface, such that moving the pointer to it and clicking the mouse initiates some action. { 'bət·ən }

buzz [CONT SYS] *See* dither. [ELECTR] The condition of a combinatorial circuit with feedback that has undergone a transition, caused by the inputs, from an unstable state to a new state that is also unstable. { bəz }

bypass [COMMUN] The use of alternative systems, such as satellite and microwave, to transmit data and voice signals, avoiding use of the communication lines of the local telephone company. { 'bīˌpas }

bypass filter [ELECTR] Filter which provides a low-attenuation path around some other equipment, such as a carrier frequency filter used to bypass a physical telephone repeater station. { 'bīˌpas ˌfil·tər }

byte [COMPUT SCI] A sequence of adjacent binary digits operated upon as a unit in a computer and usually shorter than a word. { bīt }

byte addressable computer [COMPUT SCI] A computer in which each byte of memory can be addressed independently of the others. { ¦bīt ə¦dres·ə·bəl kəm'pyüd·ər }

byte-aligned [COMMUN] A bit in a coded bit stream is byte-aligned if its position is a multiple of 8 bits from the first bit in the stream. { 'bīt ə'līnd }

bytecode [COMPUT SCI] Compiled Java programs that can be transferred across a network and executed by the Java virtual machine. { 'bīt ˌkōd }

byte mode [COMPUT SCI] A method of transferring data between a peripheral unit and a central processor in which one byte is transferred at a time. { bīt ˌmōd }

byte multiplexor channel [COMPUT SCI] A transmission channel in a computer system that can transmit data simultaneously from several devices and only one byte at a time. { ˈbīt ˈməlt·i ˌplek·sər ˌchan·əl }

byte-oriented protocol [COMPUT SCI] A communications protocol in which full bytes are used as control codes. Also known as character-oriented protocol. { ¦bīt ˌȯr·ē‚ent·əd ˈprōd·əˌkȯl }

C

C [COMPUT SCI] A programming language designed to implement the Unix operating system.

C++ [COMPUT SCI] An object-oriented language that was created as an extension to the C language. { 'sē,pləs,pləs }

C² *See* command and control. { 'sē'tü }

C³ *See* command, control, and communications. { 'sē'thrē }

cable code *See* Morse cable code. { 'kā·bəl ,kōd }

cable delay [COMPUT SCI] The time required for one bit of data to go through a cable, about 1.5 nanoseconds per foot of cable. { 'kā·bəl di'lā }

cable television [COMMUN] A television program distribution system in which signals from all local stations and usually a number of distant stations and program services are picked up by one or more high-gain antennas amplified on individual channels, then fed directly to individual receivers of subscribers by overhead or underground coaxial cable. Also known as community antenna television (CATV). { 'kā·bəl 'tel·ə,vizh·ən }

cabletext [COMMUN] Any videotex service that uses coaxial cable. { 'kā·bəl,tekst }

cache [COMPUT SCI] A small, fast storage buffer integrated in the central processing unit of some large computers. { kash }

CAD *See* computer-aided design. { kad }

CADD *See* computer-aided design and drafting. { kad }

caddy [COMPUT SCI] In certain types of disk drives, a plastic tray in which a CD-ROM disk is placed before loading. { 'kad·ē }

cage antenna [ELECTROMAG] Broad-band dipole antenna in which each pole consists of a cage of wires whose overall shape resembles that of a cylinder or a cone. { 'kāj an'ten·ə }

CAI *See* computer-assisted instruction.

CAL [COMPUT SCI] A higher-level language, developed especially for time-sharing purposes, in which a user at a remote console typewriter is directly connected to the computer and can work out problems on-line with considerable help from the computer. Derived from Conversational Algebraic Language. { kal }

calculated address *See* generated address. { 'kal·kyə,lād·əd 'ad,res }

calculating machine *See* calculator. { 'kal·kyə ,lād·iŋ mə'shēn }

calculator [COMPUT SCI] A device that performs logic and arithmetic digital operations based on numerical data which are entered by pressing numerical and control keys. Also known as calculating machine. { 'kal·kyə,lād·ər }

calculus of enlargement *See* calculus of finite differences. { 'kal·kyə·ləs əv in'lärj·mənt }

calculus of finite differences [MATH] A method of interpolation that makes use of formal relations between difference operators which are, in turn, defined in terms of the values of a function on a set of equally spaced points. Also known as calculus of enlargement. { 'kal·kyə·ləs əv 'fī,nīt 'dif·rən·səs }

call [COMPUT SCI] **1.** To transfer control to a specified closed subroutine. **2.** A statement in a computer program that references a closed subroutine or program. { kȯl }

call by location [COMPUT SCI] A method of transferring arguments from a calling program to a subprogram in which the referencing program provides to the subprogram the memory location at which the value of the argument can be found, rather than the value itself. Also known as call by reference. { 'kȯl bī ,lō'kā·shən }

call by name [COMPUT SCI] A method of transferring arguments from a calling program to a subprogram in which the actual expression is passed to the subprogram. { 'kȯl bī 'nām }

call by reference *See* call by location. { 'kȯl bī 'ref·rəns }

call by value [COMPUT SCI] A method of transferring arguments from a calling program to a subprogram in which the subprogram is provided with the values of the argument and on path leads back to the referencing program. { 'kȯl bī 'val·yü }

called routine [COMPUT SCI] A subroutine that is accessed by a call or branch instruction in a computer program. { 'kȯld rü,tēn }

call forwarding [COMMUN] A telephone service that automatically transfers incoming calls to a designated number. { 'kȯl 'fȯr·wərd·iŋ }

call in [COMPUT SCI] To transfer control of a digital computer, temporarily, from a main routine to a subroutine that is inserted in the sequence of calculating operations, to fulfill an ancillary purpose. { 'kȯl ,in }

calling device [ELECTR] Apparatus which generates signals, either dual-tone multifrequency (DTMF) or the pulse required for establishing connections in an automatic telephone switching system. { 'kȯl·iŋ di'vīs }

calling program [COMPUT SCI] A computer program that initiates a call to another program. { 'kȯl·iŋ ˌprō·grəm }

calling routine [COMPUT SCI] A subroutine that initiates a call to another subroutine. { 'kȯl·iŋ rüˌtēn }

calling sequence [COMPUT SCI] A specific set of instructions to set up and call a given subroutine, make available the data required by it, and tell the computer where to return after the subroutine is executed. { 'kȯl·iŋ ˌsē·kwəns }

call letters [COMMUN] Identifying letters, sometimes including numerals, assigned to radio and television stations by the Federal Communications Commission and other regulatory authorities throughout the world. Also known as call sign. { 'kȯl ˌled·ərz }

call number [COMPUT SCI] In computer operations, a set of characters identifying a subroutine, and containing information concerning parameters to be inserted in the subroutine, or information to be used in generating the subroutine, or information related to the operands. { 'kȯl ˌnəm·bər }

call setup time [COMMUN] The period of time between the lifting of a handset to make a telephone call and the start of voice or data transmission. { 'kȯl 'sedˌəp ˌtīm }

call sign *See* call letters. { 'kȯl ˌsīn }

call up [COMPUT SCI] To retrieve data from computer memory, especially for display and user interaction. { 'kȯl ˌəp }

CAM *See* computer-aided manufacturing. { ¦sē ¦ā'em *or* kam }

camera *See* television camera. { 'kam·rə }

camera chain [COMMUN] A television camera, associated amplifiers, a monitor, and the cable needed to bring the camera output signal to the control room. { 'kam·rə 'chān }

camera tube [ELECTR] An electron-beam tube used in a television camera to convert an optical image into a corresponding charge-density electric image and to scan the resulting electric image in a predetermined sequence to provide an equivalent electric signal. Also known as pickup tube; television camera tube. { 'kam·rə ˌtüb }

camp-on system [COMMUN] A circuit control feature whereby a user attempting to establish a telephone call and encountering a busy station will hold the connection for a preset time, to the exclusion of other callers, in case the original conversation should terminate. { 'kamp ¦ȯn ˌsis·təm }

canceler [ELECTR] A circuit used in providing moving-target indication in radar, in which small sets of successive pulses are compared such that invariant returns, presumed indicative of stationary objects, are cancelled and ignored; a primitive form of Doppler processing. Usually cited as a "two-pulse" or "threee-pulse canceler," for example. { 'kan·səl·ər }

canned cycle [COMPUT SCI] Any set of operations, either software or hardware, that is activated by a single command. { 'kand 'sī·kəl }

canned program [COMPUT SCI] A program which has been written to solve a particular problem, is available to users of a computer system, and is usually fixed in form and capable of little or no modification. { ¦kand 'prō·grəm }

canonical schema [COMPUT SCI] A model that represents the structure and interrelationships of data within a database. { kə'nän·ə·kəl 'skē·mə }

capability [COMPUT SCI] A permission that is given to a user of a computing system in advance to access a particular object in the system in a particular way, and that the user can later present to a reference monitor as a prevalidated ticket to gain access. { ˌkāp·ə'bil·ə·dē }

capability list [COMPUT SCI] A row of an access matrix that contains the access rights of a given user to various files and other resources of a computer system. { ˌkā·pə'bil·əd·ē ˌlist }

capacitance hat [ELECTROMAG] A network of wires that is placed at the top of an antenna either to increase its bandwidth or to lower its resonant frequency. { kə'pas·əd·əns ˌhat }

capacitive diaphragm [ELECTROMAG] A resonant window used in a waveguide to provide the equivalent of capacitive reactance at the frequency being transmitted. { kə'pas·əd·iv 'dī·ə ˌfram }

capacitive loading [ELECTROMAG] **1.** Raising the resonant frequency of an antenna by connecting a fixed capacitor or capacitors in series with it. **2.** Lowering the resonant frequency of an antenna by installing a capacitance hat. { kə'pas·əd·iv 'lōd·iŋ }

capacitive post [ELECTROMAG] Metal post or screw extending across a waveguide at right angles to the E field, to provide capacitive susceptance in parallel with the waveguide for tuning or matching purposes. { kə¦pas·əd·iv ¦pōst }

capacitive reactance [ELECTROMAG] Reactance due to the capacitance of a capacitor or circuit, equal to the inverse of the product of the capacitance and the angular frequency. { kə ¦pas·əd·iv rē'ak·təns }

capacitive tuning [ELECTR] Tuning involving use of a variable capacitor. { kə¦pas·əd·iv 'tün·iŋ }

capacitive window [ELECTROMAG] Conducting diaphragm extending into a waveguide from one or both sidewalls, producing the effect of a capacitive susceptance in parallel with the waveguide. { kə¦pas·əd·iv 'win·dō }

capacitor [ELEC] A device which consists essentially of two conductors (such as parallel metal plates) insulated from each other by a dielectric and which introduces capacitance into a circuit, stores electrical energy, blocks the flow of direct current, and permits the flow of alternating current to a degree dependent on the capacitor's capacitance and the current frequency.

Symbolized C. Also known as condenser; electric condenser. { kə'pas·əd·ər }

capacitor antenna [ELECTROMAG] Antenna consisting of two conductors or systems of conductors, the essential characteristic of which is its capacitance. Also known as condenser antenna. { kə'pas·əd·ər an'ten·ə }

capacity *See* storage capacity. { kə'pas·əd·ē }

capacity-rate product [COMMUN] The product of the capacity of a data-storage device in gigabytes and the data rate in megabits per second. { kə'pas·ə·dē ,rāt ,präd·əkt }

capture effect [ELECTR] The effect wherein a strong frequency-modulation signal in an FM receiver completely suppresses a weaker signal on the same or nearly the same frequency. { 'kap·chər i'fekt }

capture ratio [COMMUN] A measure of the ability of a frequency-modulation tuner to reject the weaker of two stations that are on the same frequency; the lower the ratio of desired to undesired signals, the better the performance of the tuner. { 'kap·chər ,rā·shō }

CAR *See* computer-assisted retrieval. { kär }

carbon transducer [ENG] A transducer consisting of carbon granules in contact with a fixed electrode and a movable electrode, so that motion of the movable electrode varies the resistance of the granules. { 'kär·bən tranz'dü·sər }

card [COMPUT SCI] *See* punch card. [ELECTR] A printed circuit board or other arrangement of miniaturized components that can be plugged into a computer or peripheral device. { kärd }

card cage [ELECTR] A rack built into a computer to hold printed circuit boards and allow them to be installed or removed easily. { 'kärd ,kāj }

card dialer [COMMUN] A telephone in which a number can be dialed automatically and almost instantly by inserting a coded card for that number in a slot on the dialer; now obsolete, having been replaced by automatic dialers using electronic memory. { 'kärd ,dī·lər }

card holder [ELECTR] A U-shaped slot designed to hold the edge of a printed circuit board securely in a card cage. { 'kärd ,hōl·dər }

cardioid pattern [ENG] Heart-shaped pattern obtained as the response or radiation characteristic of certain directional antennas, or as the response characteristic of certain types of microphones. { 'kärd·ē,ȯid ,pad·ərn }

card slot [ELECTR] A groove where a printed circuit board fits into a card cage or backplane. { 'kärd ,slät }

carriage return [COMPUT SCI] The operation that causes the next character to be printed at the extreme left margin, and usually advances to the next line at the same time. { 'kar·ij ri'tərn }

carrier [COMMUN] **1.** The radio wave produced by a transmitter when there is no modulating signal, or any other wave, recurring series of pulses, or direct current capable of being modulated. Also known as carrier wave; signal carrier. **2.** A wave generated locally at a receiver that, when combined with the sidebands of a suppressed-carrier transmission in a suitable detector, produces the modulating wave. **3.** *See* carrier system. { 'kar·ē·ər }

carrier amplifier [ELECTR] A direct-current amplifier in which the dc input signal is filtered by a low-pass filter, then used to modulate a carrier so it can be amplified conventionally as an alternating-current signal; the amplified dc output is obtained by rectifying and filtering the rectified carrier signal. { 'kar·ē·ər ,am·plə,fī·ər }

carrier amplitude regulation [COMMUN] Change in amplitude of the carrier wave in an amplitude-modulated transmitter when modulation is applied under conditions of symmetrical modulation. { 'kar·ē·ər 'am·plə,tüd reg·yə'lā·shən }

carrier beat [COMMUN] An undesirable heterodyne of facsimile signals, each synchronous with a different stable reference oscillator, causing a pattern in received copy. { 'kar·ē·ər ,bēt }

carrier channel [COMMUN] The equipment and lines that make up a complete carrier-current circuit between two or more points. { 'kar·ē·ər ,chan·əl }

carrier chrominance signal *See* chrominance signal. { 'kar·ē·ər 'krō·mə·nəns ,sig·nəl }

carrier current [COMMUN] A higher-frequency alternating current superimposed on ordinary telephone, telegraph, and power-line frequencies for communication and control purposes. { 'kar·ē·ər ,kər·ənt }

carrier detect [COMPUT SCI] A signal sent by a modem to a computer or a terminal to indicate that it is receiving a character. { 'kar·ē·ər di ,tekt }

carrier frequency [COMMUN] The frequency generated by an unmodulated radio, radar, carrier communication, or other transmitter, or the average frequency of the emitted wave when modulated by a symmetrical signal. Also known as center frequency; resting frequency. { 'kar·ē·ər ,frē·kwən·sē }

carrier leak [COMMUN] Carrier remaining after carrier suppression in a suppressed-carrier transmission system. { 'kar·ē·ər ,lēk }

carrier level [COMMUN] The strength or level of an unmodulated carrier signal at a particular point in a radio system, expressed in decibels in relation to some reference level. { 'kar·ē·ər ,lev·əl }

carrier loading [ELECTROMAG] The addition of lumped inductances to the cable section of a transmission line specifically designed for carrier transmission; it serves to minimize impedance mismatch between cable and open wire and to reduce the cable attenuation. { 'kar·ē·ər ,lōd·iŋ }

carrier noise [COMMUN] Noise produced by undesired variation of a radio-frequency signal in the absence of any intended modulation. Also known as residual modulation. { 'kar·ē·ər ,nȯiz }

carrier power output rating [COMMUN] Power available at the output terminals of a transmitter when the output terminals are connected to the normal-load circuit or to a circuit equivalent thereto. { 'kar·ē·ər ¦paủ·ər 'aủt,pủt ,rād·iŋ }

carrier repeater [ELECTR] Equipment designed to raise carrier signal levels to such a value that they may traverse a succeeding line section at such amplitude as to preserve an adequate signal-to-noise ratio; while the heart of a repeater is the amplifier, necessary adjuncts are filters, equalizers, level controls, and so on, depending upon the operating methods. { 'kar·ē·ər ri'pēd·ər }

carrier sense multiple access with collision detection *See* CSMA/CD. { 'kar·ē·ər ¦sens 'məl·tə·pəl 'ak,ses with kə'lizh·ən di,tek·shən }

carrier shift [COMMUN] **1.** Transmission of information by radio through shifting the carrier frequency in one direction for a mark signal and in the opposite direction for a spacing signal. **2.** Condition resulting from imperfect modulation whereby the positive and negative excursions of the envelope pattern are unequal, thus effecting a change in the power associated with the carrier. { 'kar·ē·ər ,shift }

carrier signaling [COMMUN] Method by which busy signals, ringing, or dial signaling relays are operated by the transmission of a carrier-frequency tone. { 'kar·ē·ər ,sig·nəl·iŋ }

carrier suppression [COMMUN] **1.** Suppression of the carrier frequency after conventional modulation at the transmitter, with reinsertion of the carrier at the receiving end before demodulation. **2.** Suppression of the carrier when there is no modulation signal to be transmitted; used on ships to reduce interference between transmitters. { 'kar·ē·ər sə'presh·ən }

carrier swing [COMMUN] The total deviation of a frequency-modulated or phase-modulated wave from the lowest instantaneous frequency to the highest instantaneous frequency. { 'kar·ē·ər ,swiŋ }

carrier system [COMMUN] A system permitting a number of simultaneous, independent communications over the same circuit. Also known as carrier. { 'kar·ē·ər ,sis·təm }

carrier telegraphy [COMMUN] Telegraphy in which a single-frequency carrier wave is modulated by the transmitting apparatus for transmission over wire lines. { 'kar·ē·ər tə'leg·rə·fē }

carrier telephony [COMMUN] Telephony in which a single-frequency carrier wave is modulated by a voice-frequency signal for transmission over wire lines. { 'kar·ē·ər tə'lef·ə·nē }

carrier terminal [ELECTR] Apparatus at one end of a carrier transmission system, whereby the processes of modulation, demodulation, filtering, amplification, and associated functions are effected. { 'kar·ē·ər ¦tərm·ən·əl }

carrier-to-noise ratio [COMMUN] The ratio of the magnitude of the carrier to that of the noise after specified band limiting and before any nonlinear process such as amplitude limiting and detection. { ¦kar·ē·ər tə ¦nȯiz ,rā·shō }

carrier transfer filters [ELECTR] Filters arranged as a carrier-frequency crossover or bridge between two transmission circuits. { 'kar·ē·ər ¦tranz·fər ,fil·tərz }

carrier transmission [COMMUN] Transmission in which the transmitted electric wave is a wave resulting from the modulation of a single-frequency wave by a modulating wave. { 'kar·ē·ər tranz'mish·ən }

carrier wave *See* carrier. { 'kar·ē·ər ,wāv }

carry [MATH] An arithmetic operation that occurs in the course of addition when the sum of the digits in a given position equals or exceeds the base of the number system; a multiple *m* of the base is subtracted from this sum so that the remainder is less than the base, and the number *m* is then added to the next-higher-order digit. { 'kar·ē }

carry-complete signal [COMPUT SCI] A signal generated by a digital parallel adder, indicating that all carries from an adding operation have been generated and propagated, and that the addition operation is completed. { ¦kar·ē kəm ¦plēt ,sig·nəl }

carry flag [COMPUT SCI] A flip-flop circuit which indicates overflow in arithmetic operations. { 'kar·ē ,flag }

carry lookahead [COMPUT SCI] A circuit which allows low-order carries to ripple through all the way to the highest-order bit to output a completed sum. { 'kar·ē 'lu̇k·ə,hed }

carry-save adder [COMPUT SCI] A device for the rapid addition of three operands; consists of a sequence of full adders, in which one of the operands is entered in the carry inputs, and the carry outputs, instead of feeding the carry inputs of the following full adders, form a second output word which is then added to the ordinary output in a two-operand adder to form the final sum. { ¦kar·ē ¦sāv 'ad·ər }

carry signal [COMPUT SCI] A signal produced in a computer when the sum of two digits in the same column equals or exceeds the base of the number system in use or when the difference between two digits is less than zero. { 'kar·ē ,sig·nəl }

carry time [COMPUT SCI] The time needed to transfer all carry digits to the next higher column. { 'kar·ē ,tīm }

cartridge [COMPUT SCI] A self-contained module that contains disks, magnetic tape, or integrated circuits for storing data. { 'kär·trij }

cartridge disk [COMPUT SCI] A type of disk storage device consisting of a single disk encased in a compact container which can be inserted in and removed from the disk drive unit; used extensively with computer systems. { 'kär·trij ,disk }

cartridge font [COMPUT SCI] A font for a computer printer that is stored on a read-only memory chip within a cartridge (a module that is inserted in a slot in the printer). { 'kär·trij ,fänt }

cartridge tape drive [COMPUT SCI] A tape drive which will automatically thread the tape on the takeup reels without human assistance. Formerly known as hypertape drive. { 'kär·trij ,tāp ,drīv }

cascade [COMPUT SCI] A series of actions that take place in the course of data processing, each triggered by the previous action in the series. { ka'skād }

cascade compensation [CONT SYS] Compensation in which the compensator is placed in series with the forward transfer function. Also known as series compensation; tandem compensation. { ka'skād käm·pən'sā·shən }

cascade control [CONT SYS] An automatic control system in which various control units are linked in sequence, each control unit regulating the operation of the next control unit in line. { ka'skād kən,trōl }

cascaded carry [COMPUT SCI] A carry process in which the addition of two numerals results in a sum numeral and a carry numeral that are in turn added together, this process being repeated until no new carries are generated. { ka'skād·əd 'kar·ē }

cascade noise [ELECTR] The noise in a communications receiver after an input signal has been subjected to two tandem stages of amplification. { ka'skād 'nȯiz }

cascading menu [COMPUT SCI] A menu that appears next to a pull-down menu as the result of selecting a choice on the latter. { ka,skād·iŋ 'men·yü }

cascading windows [COMPUT SCI] Two or more windows displayed so that they overlap but their title bars are still visible. { ka,skād·iŋ 'win,dōz }

cascode amplifier [ELECTR] An amplifier consisting of a grounded-emitter input stage that drives a grounded-base output stage; advantages include high gain and low noise; widely used in television tuners. { 'ka,skōd 'am·plə,fī·ər }

case [COMPUT SCI] **1.** In computers, a set of data to be used by a particular program. **2.** The metal box that houses a computer's circuit boards, disk drives, and power supply. Also known as system unit. { kās }

CASE *See* computer-aided software engineering. { kās }

case-sensitive language [COMPUT SCI] A programming language in which upper-case letters are distinguished from lower-case letters. { ¦kās ,sens·ə·tiv 'laŋ·gwij }

case structure [COMPUT SCI] A group of program statements in which a condition is tested and, according to the results of the test, one of at least three specific groups of program statements is executed, after which the program returns to the original location. { 'kās ,strək·chər }

Cassegrain antenna [ELECTROMAG] A microwave antenna in which the feed radiator is mounted at or near the surface of the main reflector and aimed at a mirror at the focus; energy from the feed first illuminates the mirror, then spreads outward to illuminate the main reflector. { kas·gran an'ten·ə }

cassette cartridge system [COMPUT SCI] An input system often used in computers; its low cost and ease in mounting often offset its slow access time. { kə'set ,kär·trij ,sis·təm }

cassette memory [COMPUT SCI] A removable magnetic tape cassette that stores computer programs and data. { kə'set 'mem·rē }

catalog [COMPUT SCI] **1.** All the indexes to data sets or files in a system. **2.** The index to all other indexes; the master index. **3.** To add an entry to an index or to build an entire new index. **4.** A list of items in a data storage device, usually arranged so that a particular kind of information can be located easily. { 'kad·əl,äg }

catalog-order device [ELECTR] A logic circuit element that is readily obtainable from a manufacturer, and can be combined with other such elements to provide a wide variety of logic circuits. { 'kad·əl,äg ¦ȯr·dər di'vīs }

catastrophic error [COMPUT SCI] A situation in which so many errors are detected in a computer program that its compilation or execution is automatically terminated. { ,kad·ə¦sträf·ik 'er·ər }

categorization [COMPUT SCI] Process of separating multiple addressed messages to form individual messages for singular addresses. { ,kad·ə·gə·rə'zā·shən }

catena [COMPUT SCI] A series of data items that appears in a chained list. { kə'tē·nə }

catenate [COMPUT SCI] To arrange a collection of items in a chained list or catena. { 'kat·ən ,āt }

cathode [ELECTR] **1.** The primary source of electrons in an electron tube; in directly heated tubes the filament is the cathode, and in indirectly heated tubes a coated metal cathode surrounds a heater. Designated K. Also known as negative electrode. **2.** The terminal of a semiconductor diode that is negative with respect to the other terminal when the diode is biased in the forward direction. { 'kath,ōd }

cathode modulation [ELECTR] Amplitude modulation accomplished by applying the modulating voltage to the cathode circuit of an electron tube in which the carrier is present. { 'kath,ōd ,mäj·ə'lā·shən }

cathode-ray tube [ELECTR] An electron tube in which a beam of electrons can be focused to a small area and varied in position and intensity on a surface. Abbreviated CRT. Originally known as Braun tube; also known as electron-ray tube. { 'kath,ōd ¦rā ,tüb }

CATT *See* controlled avalanche transit-time triode. { kat }

CATV *See* cable television.

Cauer filter *See* elliptic-integral filter. { 'kaü·ər ,fil·tər }

causal system [CONT SYS] A system whose response to an input does not depend on values of the input at later times. Also known as nonanticipatory system; physical system. { 'kȯ·zəl ,sis·təm }

cavity *See* cavity resonator. { 'kav·əd·ē }

cavity coupling [ELECTROMAG] The extraction of electromagnetic energy from a resonant cavity, either waveguide or coaxial, using loops, probes, or apertures. { 'kav·əd·ē ,kəp·liŋ }

cavity filter [ELECTROMAG] A microwave filter that uses quarter-wavelength-coupled cavities inserted in waveguides or coaxial lines to provide band-pass or other response characteristics at frequencies in the gigahertz range. { 'kav·əd·ē ˌfil·tər }

cavity magnetron [ELECTR] A magnetron having a number of resonant cavities forming the anode; used as a microwave oscillator. { 'kav·əd·ē 'mag·nəˌträn }

cavity resonance [ELECTROMAG] The resonant oscillation of the electromagnetic field in a cavity. { 'kav·əd·ē 'rez·ən·əns }

cavity resonator [ELECTROMAG] A space totally enclosed by a metallic conductor and excited in such a way that it becomes a source of electromagnetic oscillations. Also known as cavity; microwave cavity; microwave resonance cavity; resonant cavity; resonant chamber; resonant element; rhumbatron; tuned cavity; waveguide resonator. { 'kav·əd·ē 'rez·ənˌād·ər }

cavity tuning [ELECTROMAG] Use of an adjustable cavity resonator as a tuned circuit in an oscillator or amplifier, with tuning usually achieved by moving a metal plunger in or out of the cavity to change the volume, and hence the resonant frequency of the cavity. { 'kav·əd·ē ˌtün·iŋ }

CAW *See* channel address word.

C band [COMMUN] A band of radio frequencies extending from 4 to 8 gigahertz. { 'sē ˌband }

C-band fixed satellite service [COMMUN] Satellite communication at frequencies in and near the C band, with the uplink frequency in a band from 5.85 to 7.075 gigahertz and the downlink frequency in bands from 3.4 to 4.2 gigahertz and 4.5 to 4.8 gigahertz. { 'sē ˌband ¦fikst ¦sad·əˌlīt ˌsər·vəs }

C-band waveguide [ELECTROMAG] A rectangular waveguide, with dimensions 3.48 by 1.58 centimeters, which is used to excite only the dominant mode (TE_{01}) for wavelengths in the range 3.7–5.1 centimeters. { 'sē ˌband 'wāvˌgīd }

CBC *See* cipher block chaining.

CBX *See* computerized branch exchange.

CCD *See* charge-coupled device.

CCIS *See* common-channel interoffice signaling.

CCIT 2 code [COMMUN] A printing-telegraph code in which each character is represented by five binary digits. Also known as international telegraph alphabet; International Telegraphic Consultative Committee code 2. { ˌsēˌsēˌīˌtē 'tü ˌkōd }

CCTV *See* closed-circuit television.

CCU *See* communications control unit.

CCW *See* channel command word.

CD *See* compact disk.

C-display [ELECTR] A radar display format in which targets appear as spots with azimuth angle as the horizontal axis, and elevation angle as the vertical. Also known as C-indicator; C-scan; C-scope. { 'sē di'splā }

CDM *See* code-division multiplex.

CDMA *See* code-division multiple access.

CD-R [COMMUN] A compact-disk format that allows users to record audio or other digital data in such a way that the recording is permanent (nonerasable) and may be read indefinitely. Derived from compact-disk recordable. Also known as compact-disk write-once (CD-WO).

CD-ROM *See* compact-disk read-only memory. { ¦sē¦dē 'räm }

CD-RW [COMMUN] A compact-disk format that allows audio or other digital data to be written, read, erased, and rewritten. Derived from compact-disk rewritable. Also known as compact-disk erasable.

CDTV *See* conventional definition television.

CD-WO *See* CD-R.

cell [COMPUT SCI] **1.** An elementary unit of data storage. **2.** In a spreadsheet, the intersection of a row and a column. { sel }

cell address [COMPUT SCI] A combination of a letter and a number that specifies the column and row in which a cell is located on a spreadsheet. { 'sel əˌdres }

cellar *See* push-down storage. { 'sel·ər }

cell pointer [COMPUT SCI] A rectangular highlight that indicates the active cell in a spreadsheet program. { 'sel ˌpȯint·ər }

cell protection [COMPUT SCI] A format applied to a cell or range of cells in a spreadsheet, or to the entire spreadsheet, that prevents the contents of the cells in question from being altered. { 'sel prəˌtek·shən }

cell reference [COMPUT SCI] The address of a cell that contains a value that is needed to solve a formula in a spreadsheet program. { 'sel ˌref·rəns }

cellular automaton [COMPUT SCI] A theoretical model of a parallel computer which is subject to various restrictions to make practicable the formal investigation of its computing powers. [MATH] A mathematical construction consisting of a system of entities, called cells, whose temporal evolution is governed by a collection of rules, so that its behavior over time may appear highly complex or chaotic. { 'sel·yə·lər ȯ'täm·ə·tən }

cellular chain [COMPUT SCI] A chain which is not allowed to cross a cell boundary. { 'sel·yə·lər 'chān }

cellular horn *See* multicellular horn. { 'sel·yə·lər 'hȯrn }

cellular mobile radio [COMMUN] A system that serves portable and mobile radio receivers in which the service area is subdivided into multiple cells or zones, and unique radio channel frequencies are assigned to each cell. { 'sel·yə·lər 'mō·bəl 'rād·ē·ō }

cellular multilist [COMPUT SCI] A type of multilist organization composed of cellular chains. { 'sel·yə·lər 'məl·tiˌlist }

cellular splitting [COMPUT SCI] A method of adding records to a file in which the records are grouped into cells and each cell is divided into two when it becomes full. { 'sel·yə·lər 'splid·iŋ }

CELP coder *See* code-excited linear predictive coder. { ¦sē¦ē¦el'pē ˌkōd·ər or 'selp ˌkōd·ər }

center frequency *See* carrier frequency. { 'sen·tər 'frē·kwən·sē }

center line *See* stroke center line. { 'sen·tər ˌlīn }

center loading [ELECTROMAG] Alteration of the resonant frequency of a transmitting antenna by inserting an inductance or capacitance about halfway between the feed point and the end of the antenna. { 'sen·tər 'lōd·iŋ }

centimetric waves [COMMUN] Microwaves having wavelengths between 1 and 10 centimeters, corresponding to frequencies between 3 and 30 gigahertz. { ¦sent·ə¦me·trik 'wāvz }

central-battery system [COMMUN] A telephone or telegraph system which obtains all the energy for signaling (and for speaking, in the case of the telephone) from a single battery of secondary cells located at the main exchange. { ¦sen·trəl 'bad·ə·rē ,sis·təm }

central control [SYS ENG] Control exercised over an extensive and complicated system from a single center. { 'sen·trəl kən'trōl }

centralized configuration *See* star network. { 'sen·trə,līzd kən,fig·yə'rā·shən }

centralized data base [COMPUT SCI] A data base at a single physical location, usually employed in conjunction with centralized data processing. { 'sen·trə,līzd 'dad·ə ,bās }

centralized data processing [COMPUT SCI] The processing of all the data concerned with a given activity at one place, usually with fixed equipment within one building. { 'sen·trə,līzd 'dad·ə 'präs,əs·iŋ }

central office [COMMUN] A switching unit, installed in a telephone system serving the general public, having the necessary equipment and operating arrangements for terminating and interconnecting lines and trunks. Also known as telephone central office. { 'sen·trəl 'ȯ·fəs }

central processing unit [COMPUT SCI] The part of a computer containing the circuits required to interpret and execute the instructions. Abbreviated CPU. { 'sen·trəl 'präs,əs·iŋ ,yü·nət }

central-processing-unit time [COMPUT SCI] The time actually required to process a set of instructions in the logic unit of a computer. { 'sen·trəl 'präs,es·iŋ ,yü·nət ,tīm }

central terminal [COMPUT SCI] A communication device which queues tellers' requests for processing and which channels answers to the consoles originating the transactions. { 'sen·trəl 'tər·mən·əl }

centroid [NAV] In radar, the estimate of a contact's position as a single point, whereas the echoes may have occupied adjacent beam positions and-or range cells on successive pulses; the result of a centroiding algorithm in a radar contact generator. { 'sen,trȯid }

certificate [COMMUN] A data record containing an identification, a digital signature from a third party who is believed to be trustworthy, attesting to the authenticity of the identity, and an encryption key which provides a basis for two unknown entities to establish a shared encryption. { sər'tif·i·kət }

CFIA *See* component-failure-impact analysis.

CGI *See* common gateway interface.

CGI script [COMPUT SCI] A program, written in a language such as Perl, that is used for creating interactive Web pages; for example, it allows a Web server to process a request from a user, communicate with a database, and reply to the user by creating a Web page. { ¦sē¦jē'ī ,skript }

CGM *See* computer graphics metafile.

chad [COMPUT SCI] The piece of material removed when forming a hole or notch in a punched tape or punched card. Also known as chip. { chad }

chaff [ELECTROMAG] Reflective particulate matter, such as tiny strips of coated films or of metallic foil, that can be dispensed by aircraft in the airspace covered by an enemy radar, so as to create such an echo density that echoes of interest to that radar are obscured or the radar is distracted by the chaff return. { chaf }

chain [COMMUN] A network of radio, television, radar, navigation, or other similar stations connected by telephone lines, coaxial cables, or radio relay links so all can operate as a group for broadcast purposes, communication purposes, or determination of position. [COMPUT SCI] **1.** A series of data or other items linked together in some way. **2.** A sequence of binary digits used to construct a code. [ELECTR] A series of amplifiers in a transmitter, achieving a higher overall gain than any one amplifier could reasonably achieve. { chān }

chain code [COMPUT SCI] A binary code consisting of a cyclic sequence of some or all of the possible binary words at a given length such that each word is derived from the previous one by moving the binary digits one position to the left, dropping the leading bit, and inserting a new bit at the end, in such a way that no word recurs before the cycle is complete. { 'chān ,kōd }

chain command [COMPUT SCI] Any input/output command in a sequence of input/output commands such as WRITE, READ, SENSE. { 'chān kə'mand }

chain data flag [COMPUT SCI] A value of 1 given to a specific bit of a channel command word, commonly used with scatter read or scatter write operations. { 'chān 'dad·ə ,flag }

chained block encryption [COMMUN] The use of a block cipher in which the bits of a given output block depend not only on the bits in the corresponding input block and in the key, but also on any or all prior data bits, either inputted to or produced during the enciphering or deciphering process. Also known as block chaining. { ¦chānd 'bläk in'krip·shən }

chained list [COMPUT SCI] A collection of data items arranged in a sequence so that each item contains an address giving the location of the next item in a computer storage device. Also known as linked list. { ¦chānd 'list }

chained records [COMPUT SCI] A file of records arranged according to the chaining method. { ¦chānd 'rek·ərdz }

chaining [COMPUT SCI] A method of storing records which are not necessarily contiguous, in which the records are arranged in a sequence and each record contains means to identify its successor. { 'chān·iŋ }

chaining search [COMPUT SCI] A method of searching for a data item in a chained list in which an initial key is used to obtain the location of either the item sought or another item in the list, and the search then progresses through the chain until the required item is obtained or the chain is completed. { 'chān·iŋ ˌsərch }

chain pointer [COMPUT SCI] The part of a data item in a chained list that gives the address of the next data item. { 'chān 'pȯint·ər }

chain printer [COMPUT SCI] A high-speed printer in which the type slugs are carried by the links of a revolving chain. { 'chān ˌprint·ər }

chain printing [COMPUT SCI] The printing of a group of linked files by placing commands at the end of each file that direct the program to continue printing the next one. { ¦chān 'print·iŋ }

chain radar beacon [COMMUN] A beacon with a fast recovery time to permit simultaneous interrogation and tracking of the beacon by a number of radars. { ¦chān 'rāˌdär ˌbē·kən }

challenge [COMMUN] To cause an interrogator to transmit a signal which puts a transponder into operation. { 'chal·ənj }

challenger *See* interrogator. { 'chal·ən·jər }

challenge-response [COMPUT SCI] A method of identifying and authenticating persons seeking access to a computing system; each user is issued a device resembling a pocket calculator and is given a different problem to solve (the challenge), to which the calculator provides part of the answer, each time the person seeks authentication. { 'chal·ənj ri'späns }

challenging signal *See* interrogation. { 'chal·ən·jiŋ ˌsig·nəl }

chance-constrained programming [COMPUT SCI] Type of nonlinear programming wherein the deterministic constraints are replaced by their probabilistic counterparts. { ¦chans kən'strānd 'prōˌgram·iŋ }

changed memory routine [COMPUT SCI] A selective memory dump routine in which only those words that have been changed in the course of running a program are printed. { ¦chānjd 'mem·rē rüˌtēn }

change dump [COMPUT SCI] A type of dump in which only those locations in a computer memory whose contents have changed since some previous event are copied. { 'chānj ˌdəmp }

change file [COMPUT SCI] A transaction file that is used to update a master file. { 'chānj ˌfīl }

change of control [COMPUT SCI] **1.** A break in a series of records at which processing of the records may be interrupted and some predetermined action taken. **2.** *See* jump. { 'chānj əv kən'trōl }

change record [COMPUT SCI] A record that is used to alter information in a corresponding master record. Also known as amendment record; transaction record. { 'chānj ˌrek·ərd }

change tape [COMPUT SCI] A paper tape or magnetic tape carrying information that is to be used to update filed information; the latter is often on a master tape. Also known as transaction tape. { 'chānj ˌtāp }

channel [COMMUN] **1.** A band of radio frequencies allocated for a particular purpose; a standard broadcasting channel is 10 kilohertz wide, an FM channel is 200 kHz wide, and a television channel 6 megahertz wide. **2.** A path through which electrical transmission of information takes place. [COMPUT SCI] **1.** A path along which digital or other information may flow in a computer. **2.** A path for a signal, as an audio amplifier may have several input channels. **3.** The main current path between the source and drain electrodes in a field-effect transistor or other semiconductor device. { 'chan·əl }

channel adapter [COMPUT SCI] Equipment that allows devices operating at different rates of speed to be connected and data to be transferred at the slower data rate. { 'chan·əl əˌdap·tər }

channel address word [COMPUT SCI] A four-byte code containing the protection key and the main storage address of the first channel command word at the start of an input/output operation. Abbreviated CAW. { 'chan·əl 'adˌres ˌwərd }

channel-attached device [COMPUT SCI] Equipment that is directly connected to a computer by a channel. { 'chan·əl ə¦tacht diˌvīs }

channel bank [ELECTR] Part of a carrier-multiplex terminal that performs the first step of modulation of the transmitting voice frequencies into a higher-frequency band, and the final step in the demodulation of the received higher-frequency band into the received voice frequencies. { 'chan·əl ˌbaŋk }

channel capacity [COMMUN] The maximum number of bits or other information elements that can be handled in a particular channel per unit time. { 'chan·əl kə'pas·əd·ē }

channel command [COMPUT SCI] The step, equivalent to a program instruction, required to tell an input/output channel what operation is to be performed, and where the data are or should be located. { 'chan·əl kə'mand }

channel command word [COMPUT SCI] A code specifying an operation, one or more flags, a count, and a storage location. Abbreviated CCW. { 'chan·əl kə'mand ˌwərd }

channel configuration [COMPUT SCI] The types, number, and logical relationships of devices connected to a given computer channel. { 'chan·əl kənˌfig·yəˌrā·shən }

channel control command [COMPUT SCI] An order to a control unit to perform a nondata input/output operation. { 'chan·əl kən'trōl kə'mand }

channel design [COMPUT SCI] The type of channel, characterized by the tasks it can perform, available to a computer. { 'chan·əl di'zīn }

channel director [COMPUT SCI] A unit in some very large computers that controls the functioning of several channels. { 'chan·əl diˌrek·tər }

channel-end condition [COMPUT SCI] A signal indicating that the use of an input/output channel is no longer required. { 'chan·əl ˌend kən'dish·ən }

channeling [COMMUN] A type of multiplex transmission in which the separation between communication channels is accomplished through the use of carriers or subcarriers. { 'chan·əl·iŋ }

channelization [COMMUN] The division of a single wide-band (high-capacity) communications channel into many relatively narrow-band (lower-capacity) channels. { ˌchan·əl·ə'zā·shən }

channelizing [COMMUN] The process of subdividing a wide-band transmission facility so as to handle a number of different circuits requiring comparatively narrow bandwidths. { 'chan·əl ˌīz·iŋ }

channel mask [COMPUT SCI] A portion of a program status word indicating which channels may interrupt the task by their completion signals. { 'chan·əl ˌmask }

channel miles [COMMUN] The summation, in miles, of the electrical path of individual channels between two points; these points may be connected by wire or radio, or a combination of both. { 'chan·əl ˌmīlz }

channel program [COMPUT SCI] The set of steps, called channel commands, by means of which an input/output channel is controlled. { 'chan·əl ˌprō·grəm }

channel read-backward command [COMPUT SCI] A command to transfer data from tape device to main storage while the tape is moving backward. { 'chan·əl 'rēd ¦bak·wərd kəˌmand }

channel read command [COMPUT SCI] A command to transfer data from an input/output device to main storage. { ¦chan·əl 'rēd kə'mand }

channel reliability [COMMUN] The percent of time a channel was available for use in a specific direction during a specified period of time. { 'chan·əl riˌlī·ə'bil·əd·ē }

channel sense command [COMPUT SCI] A command commonly used to denote an unusual condition existing in an input/output device and requesting more information. { 'chan·əl 'sens kə'mand }

channel shifter [ELECTR] Radiotelephone carrier circuit that shifts one or two voice-frequency channels from normal channels to higher voice-frequency channels to reduce cross talk between channels; the channels are shifted back by a similar circuit at the receiving end. { 'chan·əl ˌshif·tər }

channel skip [COMPUT SCI] A control character that causes a printer to skip down to a specified line on a page or to the top of the next page. { 'chan·əl ˌskip }

channel spacing [COMMUN] The difference in frequency between successive radio or television channels. { 'chan·əl ˌspās·iŋ }

channel status table [COMPUT SCI] A table that is set up by an executive program to show the status of the various channels that connect the central processing unit with peripheral units, enabling the program to control input/output operations. { ¦chan·əl 'stad·əs ˌtā·bəl }

channel status word [COMPUT SCI] A storage register containing the status information of the input/output operation which caused an interrupt. Abbreviated CSW. { ¦chan·əl 'stad·əs ˌwərd }

channel synchronizer [ELECTR] An electronic device providing the proper interface between the central processing unit and the peripheral devices. { 'chan·əl 'siŋ·krəˌnīz·ər }

channel-to-channel adapter [COMPUT SCI] A device which provides two computer systems with interchannel communications. { ¦chan·əl tə ¦chan·əl ə'dap·tər }

channel write command [COMPUT SCI] A command which transfers data from main storage to an input/output device. { ¦chan·əl 'wrīt kə'mand }

character [COMPUT SCI] **1.** An elementary mark used to represent data, usually in the form of a graphic spatial arrangement of connected or adjacent strokes, such as a letter or a digit. **2.** A small collection of adjacent bits used to represent a piece of data, addressed and handled as a unit, often corresponding to a digit or letter. { 'kar·ik·tər }

character-addressable computer [COMPUT SCI] A computer that processes data as single characters, and is therefore able to handle words of varying length. { 'kar·ik·tər ə¦dres·ə·bəl kəm'pyüd·ər }

character adjustment [COMPUT SCI] An address modification affecting a specific number of characters of the address part of the instruction. { 'kar·ik·tər ə'jəs·mənt }

character boundary [COMPUT SCI] In character recognition, a real or imaginary rectangle which serves as the delimiter between consecutive characters or successive lines on a source document. { 'kar·ik·tər ˌbaün·drē }

character cell [COMPUT SCI] A matrix of dots that is used to form a single character on a printer or display screen. { 'kar·ik·tər ˌsel }

character code [COMMUN] A bit pattern assigned to a particular character in a coded character set. { 'kar·ik·tər ˌkōd }

character data type [COMPUT SCI] A scalar data type which provides an internal representation of printable characters. { 'kar·ik·tər 'dad·ə ˌtīp }

character density [COMPUT SCI] The number of characters recorded per unit of length or area. Also known as record density. { 'kar·ik·tər ˌden·səd·ē }

character display terminal [COMPUT SCI] A console that can display only alphanumeric characters, and cannot show arbitrary lines or curves. { 'kar·ik·tər di'splāˌtərm·ə·nəl }

character emitter [COMPUT SCI] In character recognition, an electromechanical device which conveys a specimen character in the form of a time pulse or group of pulses. { 'kar·ik·tər i'mid·ər }

character fill [COMPUT SCI] To fill one or more locations in a computer storage device by repeated insertion of some particular character, usually blanks or zeros. { 'kar·ik·tər ˌfil }

character generator [COMPUT SCI] A hard-wired subroutine which will display alphanumeric characters on a screen. { 'kar·ik·tər ˌjen·əˌrād·ər }

character graphics [COMPUT SCI] A collection of special symbols that can be strung together like letters of the alphabet to generate graphics. { 'kar·ik·tər ˌgraf·iks }

characteristic frequency [COMMUN] Frequency which can be easily identified and measured in a given emission. { ˌkar·ik·tə'ris·tik 'frē·kwən·sē }

characteristic impedance [COMMUN] The impedance that, when connected to the output terminals of a transmission line of any length, makes the line appear to be infinitely long, for there are then no standing waves on the line, and the ratio of voltage to current is the same for each point on the line. Also known as surge impedance. { ˌkar·ik·tə'ris·tik im'pēd·əns }

characteristic overflow [COMPUT SCI] An error condition encountered when the characteristic of a floating point number exceeds the limit imposed by the hardware manufacturer. { ˌkar·ik·tə'ris·tik 'ō·vərˌflō }

characteristic underflow [COMPUT SCI] An error condition encountered when the characteristic of a floating point number is smaller than the smallest limit imposed by the hardware manufacturer. { ˌkar·ik·tə'ris·tik 'ən·dərˌflō }

character mode [COMPUT SCI] A mode of computer operation in which only text is displayed. { 'kar·ik·tər ˌmōd }

character-oriented computer [COMPUT SCI] A computer in which the locations of individual characters, rather than words, can be addressed. { ¦kar·ik·tər ¦ȯr·ēˌen·təd kəmˌpyüd·ər }

character-oriented protocol *See* byte-oriented protocol. { 'kar·ik·tər ˌȯr·ēˌent·əd 'prōd·əˌkȯl }

character outline [COMPUT SCI] The graphic pattern formed by the stroke edges of a printed or handwritten character in character recognition. { 'kar·ik·tər 'au̇tˌlīn }

character reader [COMPUT SCI] In character recognition, any device capable of locating, identifying, and translating into machine code the handwritten or printed data appearing on a source document. { 'kar·ik·tər ˌrēd·ər }

character recognition [COMPUT SCI] The technology of using a machine to sense and encode into a machine language the characters which are originally written or printed by human beings. { 'kar·ik·tər ˌrek·ig'nish·ən }

character set [COMMUN] A set of unique representations called characters, for example, the 26 letters of the English alphabet, the Boolean 0 and 1, the set of signals in Morse code, and the 128 characters of the USASCII. { 'kar·ik·tər ˌset }

character skew [COMPUT SCI] In character recognition, an improper appearance of a character to be recognized, in which it appears in a tilted condition with respect to a real or imaginary horizontal base line. { 'kar·ik·tər ˌskyü }

character string [COMPUT SCI] A sequence of characters in a computer memory or other storage device. Also known as alphabetic string. { 'kar·ik·tər 'striŋ }

character string constant [COMPUT SCI] An arbitrary combination of letters, digits, and other symbols which, in the processing of nonnumeric data involving character strings, performs a function analogous to that of a numeric constant in the processing of numeric data. { 'kar·ik·tər ˌstriŋ ˌkän·stənt }

character stroke *See* stroke. { 'kar·ik·tər ˌstrōk }

character style [COMPUT SCI] In character recognition, a distinctive construction that is common to all members of a particular character set. { 'kar·ik·tər ˌstīl }

character terminal [COMPUT SCI] A screen that can display only text. { 'kar·ik·tər ˌtər·mə·nəl }

charge carrier [SOLID STATE] A mobile conduction electron or mobile hole in a semiconductor. Also known as carrier. { 'chärj ˌkar·ē·ər }

charge-coupled device [ELECTR] A semiconductor device wherein minority charge is stored in a spatially defined depletion region (potential well) at the surface of a semiconductor and is moved about the surface by transferring this charge to similar adjacent wells. Abbreviated CCD. { 'chärj ¦kəp·əld di'vīs }

charge-coupled memory [COMPUT SCI] A computer memory that uses a large number of charge-coupled devices for data storage and retrieval. { 'chärj ¦kəp·əld 'mem·rē }

charge coupling [COMPUT SCI] Transfer of all electric charges within a semiconductor storage element to a similar, nearby element by means of voltage manipulations. { 'chärj ˌkəp·liŋ }

chat mode [COMPUT SCI] A communications option that allows two or more computers to conduct a conversation by typing in turn. { 'chat ˌmōd }

chat room [COMPUT SCI] A Web site or server space on the Internet where live keyboard conversations (usually organized around a specific topic) with other people occur. { 'chat ˌrüm }

Chebyshev filter [ELECTR] A filter in which the transmission frequency curve has an equal-ripple shape, with very small peaks and valleys. { 'cheb·ə·shəf ˌfil·tər }

check [COMPUT SCI] A test which is necessary to detect a mistake in computer programming or a computer malfunction. [ENG] A device attached to something in order to limit the movement, such as a door check. { chek }

check bit [COMPUT SCI] A binary check digit. { 'chek ˌbit }

check box [COMPUT SCI] In a graphical user interface, a small box on which an x or check mark appears when the option indicated next to the box is turned on, and disappears when the option is turned off. { 'chek ˌbäks }

check character [COMPUT SCI] A redundant character used to perform a check. { 'chek ˌkar·ik·tər }

check digit [COMPUT SCI] A redundant digit used to perform a check. { 'chek ˌdij·ət }

check indicator [COMPUT SCI] A console device, usually a light, informing the operator that an error has occurred. { 'chek ˌin·dəˌkād·ər }

check indicator instruction [COMPUT SCI] A computer instruction which directs that a signal device is turned on to call the operator's attention to the fact that there is some discrepancy in the instruction now in use. { 'chek ˌin·dəˌkăd·ər in'strək·shən }

checking program [COMPUT SCI] A computer program which detects and determines the nature of errors in other programs, particularly those that involve incorrect coding or punching of wrong characters. Also known as checking routine. { 'chek·iŋ ˌprō·grəm }

checking routine *See* checking program. { 'chek·iŋ rü'tēn }

check number [COMPUT SCI] A number denoting a specific type of hardware malfunction. { 'chek ˌnəm·bər }

checkout [COMPUT SCI] A collection of routines that are built into a compiler to test and debug programs. { 'chekˌau̇t }

checkout compiler [COMPUT SCI] A special compiler designed specifically to test and debug programs by using checkout routines. { 'chek ˌau̇t kəmˌpī·lər }

checkpoint [COMPUT SCI] That place in a routine at which the entire state of the computer (memory, registers, and so on) is written out on auxiliary storage from which it may be read back into the computer if the program is to be restarted later. [NAV] Geographical location on land or water above which the position of an aircraft in flight may be determined by observation or by electronic means. { 'chekˌpȯint }

checkpoint/restart [COMPUT SCI] The procedures for resuming a processing run after it has been halted either accidentally or deliberately. { 'chekˌpȯint 'rēˌstärt }

check problem *See* check routine. { 'chek ˌpräb·ləm }

check protect symbol [COMPUT SCI] A character, usually an asterisk, that is printed in place of leading zeros in a number, such as a dollar amount on a check. { 'chek prə'tekt ˌsim·bəl }

check register [COMPUT SCI] A register in which transferred data are temporarily stored so that they may be compared with a second transfer of the same data, to verify the accuracy of the transfer. { 'chek ˌrej·ə·stər }

check routine [COMPUT SCI] A routine or problem designed primarily to indicate whether a fault exists in a computer, without giving detailed information on the location of the fault. Also known as check problem; test program; test routine. { 'chek rü'tēn }

check row [COMPUT SCI] A row (or one of two or more rows) on a paper tape which contains the cumulated sum of existing rows, column by column, resulting in either 1 or 0 by column, thus verifying that all rows have been properly read. { 'chek ˌrō }

check sum [COMPUT SCI] A sum of digits or numbers used in a summation check. { 'chek ˌsəm }

check symbol [COMPUT SCI] One or more digits generated by performing an arithmetic check or summation check on a data item which are then attached to the item and copied along with it through various stages of processing, allowing the check to be repeated to verify the accuracy of the copying processes. { 'chek ˌsim·bəl }

check word [COMPUT SCI] A computer word, containing data from a block of records, that is joined to the block and serves as a check symbol during transfers of the block between different locations. { 'chek ˌwərd }

cheese antenna [ELECTROMAG] An antenna having a parabolic reflector between two metal plates, dimensioned to permit propagation of more than one mode in the desired direction of polarization. { 'chēz an'ten·ə }

child [COMPUT SCI] **1.** An element that follows a given element in a data structure. **2.** In object-oriented programming, a subclass. { chīld }

child process [COMPUT SCI] One of the subsidiary processes that branches out from the root task in the fork-join model of programming on parallel machines. { 'chīld ˌpräs·es }

chip [COMPUT SCI] *See* chad. [ELECTR] **1.** The shaped and processed semiconductor die that is mounted on a substrate to form a transistor, diode, or other semiconductor device. **2.** An integrated microcircuit performing a significant number of functions and constituting a subsystem. Also known as microchip. { chip }

chip card *See* smart card. { 'chip ˌkärd }

chip circuit *See* large-scale integrated circuit. { 'chip ˌsər·kət }

chipset [COMPUT SCI] A number of integrated circuits, packaged as one unit, which perform one or more related functions. { 'chipˌset }

Chireix antenna [ELECTROMAG] A phased array composed of two or more coplanar square loops, connected in series. Also known as Chireix-Mesny antenna. { ki'rāks anˌten·ə }

Chireix-Mesny antenna *See* Chireix antenna. { ki'rāks ˌmezˌnē anˌten·ə }

chirp [COMMUN] **1.** An undesirable variation in the frequency of a continuous-wave carrier when it is keyed. **2.** The sound heard in a code receiver when the transmitted carrier frequency is increased linearly for the duration of a pulse code. { chərp }

chirp modulation [COMMUN] A modulation of the carrier frequency from a lover to a higher frequency, or vice versa, often linearly, used in radar pulse compression. { ¦chərp mäj·ə'lā·shən }

chirp radar [ENG] Radar in which a swept-frequency signal is transmitted, received from a target, then compressed in time to give a narrow pulse called the chirp signal. { 'chərp ˌrāˌdär }

choke [ELECTROMAG] A groove or other discontinuity in a waveguide surface so shaped and dimensioned as to impede the passage of guided waves within a limited frequency range. { chōk }

choke coupling [ELECTROMAG] Coupling between two parts of a waveguide system that are not in direct mechanical contact with each other. { 'chōk ˌkəp·liŋ }

choke flange [ELECTROMAG] A waveguide flange having in its mating surface a slot (choke) so shaped and dimensioned as to restrict leakage of microwave energy within a limited frequency range. { 'chōk ˌflanj }

choke joint [ELECTROMAG] A connection between two waveguides that uses two mating choke flanges to provide effective electrical continuity without metallic continuity at the inner walls of the waveguide. { 'chōk ˌjȯint }

choke piston [ELECTROMAG] A piston in which there is no metallic contact with the walls of the waveguide at the edges of the reflecting surface; the short circuit for high-frequency currents is achieved by a choke system. Also known as noncontacting piston; noncontacting plunger. { 'chōk ˌpis·tən }

chopper amplifier [ELECTR] A carrier amplifier in which the direct-current input is filtered by a low-pass filter, then converted into a square-wave alternating-current signal by either one or two choppers. { 'chäp·ər 'am·pləˌfī·ər }

chopper transistor [ELECTR] A bipolar or field-effect transistor operated as a repetitive "on/off" switch to produce square-wave modulation of an input signal. { 'chäp·ər tran'zis·tər }

chroma band-pass amplifier *See* burst amplifier. { 'krō·mə 'band ˌpas 'am·pləˌfī·ər }

chroma control [ELECTR] The control that adjusts the amplitude of the carrier chrominance signal fed to the chrominance demodulators in an analog color television receiver, so as to change the saturation or vividness of the hues in the color picture. Also known as color control; color-saturation control. { 'krō·mə kən'trōl }

chroma oscillator [ELECTR] A crystal oscillator used in analog color television receivers to generate a 3.579545-megahertz signal for comparison with the incoming 3.579545-megahertz chrominance subcarrier signal being transmitted. Also known as chrominance-subcarrier oscillator; color oscillator; color-subcarrier oscillator. { 'krō·mə 'äs·əˌlād·ər }

chromatron [ELECTR] A single-gun color picture tube having color phosphors deposited on the screen in strips instead of dots. Also known as Lawrence tube. { 'krō·mə'trän }

chrominance carrier *See* chrominance subcarrier. { 'krō·mə·nəns ˌkar·ē·ər }

chrominance-carrier reference [COMMUN] A continuous signal having the same frequency as the chrominance subcarrier in a color television system and having fixed phase with respect to the color burst; this signal is the reference with which the phase of a chrominance signal is compared for the purpose of modulation or demodulation. Also known as chrominance-subcarrier reference; color-carrier reference; color-subcarrier reference. { 'krō·mə·nəns ¦kar·ē·ər ˌref·rəns }

chrominance channel [COMMUN] Any path that is intended to carry the chrominance signal in an analog color television system. { 'krō·mə·nəns ˌchan·əl }

chrominance demodulator [ELECTR] A demodulator used in an analog color television receiver for deriving the I and Q components of the chrominance signal from the chrominance signal and the chrominance-subcarrier frequency. Also known as chrominance-subcarrier demodulator. { 'krō·mə·nəns dē'mäj·əˌlād·ər }

chrominance frequency [COMMUN] The frequency of the chrominance subcarrier, equal to 3.579545 megahertz. { 'krō·mə·nəns ˌfrē·kwən·sē }

chrominance gain control [ELECTR] Variable resistors in red, green, and blue matrix channels that individually adjust primary signal levels in an color television receiver. { 'krō·mə·nəns 'gān kən'trōl }

chrominance modulator [ELECTR] A modulator used in an analog color television transmitter to generate the chrominance signal from the video-frequency chrominance components and the chrominance subcarrier. Also known as chrominance-subcarrier modulator. { 'krō·mə·nəns 'mäj·əˌlād·ər }

chrominance signal [COMMUN] One of the two components, called the I signal and Q signal, that add together to produce the total chrominance signal in an analog color television system. Also known as carrier chrominance signal. { 'krō·mə·nəns ˌsig·nəl }

chrominance subcarrier [COMMUN] The 3.579545-megahertz carrier whose modulation sidebands are added to the monochrome signal to convey color information in an analog color television receiver. Also known as chrominance carrier; color carrier; color subcarrier; subcarrier. { 'krō·mə·nəns səb'kar·ē·ər }

chrominance-subcarrier demodulator *See* chrominance demodulator. { 'krō·mə·nəns səb 'kar·ē·ər dē'mäj·əˌlād·ər }

chrominance-subcarrier modulator *See* chrominance modulator. { 'krō·mə·nəns səb'kar·ē·ər 'mäj·əˌlād·ər }

chrominance-subcarrier oscillator *See* chroma oscillator. { 'krō·mə·nəns səb'kar·ē·ər 'äs·əˌlād·ər }

chrominance-subcarrier reference *See* chrominance-carrier reference. { 'krō·mə·nəns səb'kar·ē·ər 'ref·rəns }

chrominance video signal [ELECTR] Voltage output from the red, green, or blue section of a color television camera or receiver matrix. { 'krō·mə·nəns 'vid·ē·ō ˌsig·nəl }

chute blades [COMPUT SCI] Thin metal bands which form channels to the various pockets of a sorter. { 'shüt ˌblādz }

C³I *See* command, control, communications, and intelligence. { 'sē 'thrē'ī }

CIM *See* computer input from microfilm; computer-integrated manufacturing.

cinching [COMPUT SCI] Creases produced in magnetic tape when the supply reel is wound at low tension and suddenly stopped during playback. { 'sin·chiŋ }

C-indicator *See* C-display. { 'sē ˌin·dəˌkād·ər }

cipher [COMMUN] A transposition or substitution code for transmitting secret messages. { 'sī·fər }

cipher block chaining [COMMUN] A technique for block chaining in which each block of ciphertext is produced by adding, through the

EXCLUSIVE OR operation, the previous block of ciphertext to the current block of plaintext. Abbreviated CBC. { 'sī·fər ,bläk ,chān·iŋ }

cipher feedback [COMMUN] An implementation of ciphertext autokey cipher in which the leftmost *n* bits of the data encryption standard (DES) output are added by the EXCLUSIVE OR operation to N bits of plaintext to produce N bits of ciphertext (where N is the number of bits enciphered at one time), and these N bits of ciphertext are fed back into the algorithm by first shifting the current DES input N bits to the left, and then appending the N bits of ciphertext to the right-hand side of the shifted input to produce a new DES input used for the next iteration of the algorithm. { ¦sī·fər ¦fēd,bak }

cipher machine [COMMUN] Mechanical or electrical apparatus for enciphering and deciphering. { 'sī·fər mə'shēn }

ciphertext [COMMUN] A message which has been transformed by a cipher so that it can be read only by those privy to the secrets of the cipher. { 'sī·fər,tekst }

ciphertext autokey cipher [COMMUN] A stream cipher in which the cryptographic bit stream generated at a given time is determined by the ciphertext generated at earlier times. { 'sī·fər ,tekst 'òd·ō,kē ,si·fər }

ciphony [COMMUN] A technique by which security is accomplished by converting speech into a series of on-off pulses and mixing these with the pulses supplied by a key generator; to recover the original speech, the identical key must be subtracted and the resultant on-off pulses reconverted into the original speech pattern; unauthorized listeners are unable to reconstruct the plain text unless they have an identical key generator and the daily key setting. { 'sī·fə·nē }

ciphony equipment [ELECTR] Any equipment attached to a radio transmitter, radio receiver, or telephone for scrambling or unscrambling voice messages. { 'sī·fə·nē i,kwip·mənt }

circuit [ELECTROMAG] A complete wire, radio, or carrier communications channel. { 'sər·kət }

circuit board *See* printed circuit board. { 'sər·kət ,bȯrd }

circuit capacity [COMMUN] Number of communications channels which can be handled by a given circuit at the same time. { 'sər·kət kə'pas·əd·ē }

circuit grade [COMMUN] A circuit rating defining the ability to carry information; grades include telegraph, voice, and broad-band. { 'sər·kət ,grād }

circuit noise [COMMUN] In telephone practice, the noise which is brought to the receiver electrically from a telephone system, excluding noise picked up acoustically by telephone transmitters. { 'sər·kət ,nȯiz }

circuit noise level [COMMUN] Ratio of the circuit noise at that point to some arbitrary amount of circuit noise chosen as a reference; usually expressed in decibels above reference noise, signifying the reading of a circuit noise meter, or in adjusted decibels, signifying circuit noise meter reading adjusted to represent interfering effect under specified conditions. { 'sər·kət ,nȯiz ,lev·əl }

circuit reliability [COMMUN] The percent of time a circuit was available to the user during a specified period of time. { 'sər·kət ri,lī·ə'bil·əd·ē }

circuit shift *See* cyclic shift. { 'sər·kət ,shift }

circuit switching [COMMUN] **1.** The method of providing communication service through a switching facility, either from local users or from other switching facilities. **2.** A method of transmitting messages through a communications network in which a path from the sender to the receiver of fixed bandwidth or speed is set up for the entire duration of a communication or call. { 'sər·kət ,swich·iŋ }

circular antenna [ELECTROMAG] A folded dipole that is bent into a circle, so the transmission line and the abutting folded ends are at opposite ends of a diameter. { 'sər·kyə·lər an'ten·ə }

circular buffering [COMPUT SCI] A technique for receiving data in an input-output control system which uses a single buffer that appears to be organized in a circle, with data wrapping around it. { 'sər·kyə·lər 'bəf·ə·riŋ }

circular file [COMPUT SCI] An organized collection of records, generally with a high turnover, in which new records are inserted by replacing the oldest records. { 'sər·kyə·lər 'fīl }

circular horn [ELECTROMAG] A circular-waveguide section that flares outward into the shape of a horn, to serve as a feed for a microwave reflector or lens. { 'sər·kyə·lər 'hȯrn }

circular polarized loop vee [ELECTROMAG] Airborne communications antenna with an omnidirectional radiation pattern to provide optimum near-horizon communications coverage. { 'sər·kyə·lər 'pō·lə,rīzd 'lüp ,vē }

circular polling [COMMUN] A form of polling in which each terminal is interrogated exactly once in every pass, regardless of its level of activity. { 'sər·kyə·lər 'pōl·iŋ }

circular reference [COMPUT SCI] A situation created by a programming error in which two or more entities each refer to the other so that the execution of the program is carried on endlessly with no resolution. { 'sər·kyə·lər 'ref·rəns }

circular scanning [ENG] Radar scanning in which the direction of maximum radiation describes a right circular cone. { 'sər·kyə·lər 'skan·iŋ }

circular shift *See* cyclic shift. { 'sər·kyə·lər 'shift }

circular sweep generation [ELECTR] The use of electronic circuits to provide voltage or current which causes an electron beam in a device such as a cathode-ray tube to move in a circular deflection path at constant speed. { 'sər·kyə·lər 'swēp ,jen·ə,rā·shən }

circular wait *See* mutual deadlock. { 'sər·kyə·lər 'wāt }

circular waveguide [ELECTROMAG] A waveguide whose cross-sectional area is circular. { 'sər·kyə·lər 'wāv,gīd }

circulating register [COMPUT SCI] A shift register in which data move out of one end and reenter the other end, as in a closed loop. { 'sər·kyə ,lād·iŋ 'rej·ə·stər }

circulator [ELECTROMAG] A waveguide component having a number of terminals so arranged that energy entering one terminal is transmitted to the next adjacent terminal in a particular direction. Also known as microwave circulator. { ,sər·kyə·'lād·ər }

CISC *See* complex instruction set computer. { sisk }

citizens' band [COMMUN] A frequency band allocated for citizens' radio service (462.550–467.425, 72–76, or 26.965–27.405 megahertz). { 'sit·ə·zənz ,band }

citizens' radio service [COMMUN] A radio communication service intended for private or personal radio communication, including radio signaling and control of objects by radio. { 'sit·ə·zənz 'rād·ē·ō ,sər·vəs }

cladding [COMMUN] A plastic or glass sheath that is fused to and surrounds the core of an optical fiber. { 'klad·iŋ }

clamping circuit [ELECTR] A circuit that reestablishes the direct-current level of a waveform; used in the dc-restorer stage of an analog television receiver to restore the dc component to the video signal after its loss in capacitance-coupled alternating-current amplifiers, to reestablish the average light value of the reproduced image. Also known as clamp. { 'klamp·iŋ ,sər·kət }

clamp-on [COMMUN] A method of holding a call for a line that is in use and of signaling when it becomes free. { 'klamp ,ȯn }

class [COMPUT SCI] In object-oriented programming, a description of the structure and operations of an object. A new class is defined by stating how it differs from an existing class. The new (more specific) class is said to inherit from the original (general) class and is referred to as a subclass of the original class. The original class is referred to as the superclass of the new class. { klas }

class A amplifier [ELECTR] **1.** An amplifier in which the grid bias and alternating grid voltages are such that anode current in a specific tube flows at all times. **2.** A transistor amplifier in which each transistor is in its active region for the entire signal cycle. { ,klas 'ā 'am·plə,fī·ər }

class A modulator [ELECTR] A class A amplifier used to supply the necessary signal power to modulate a carrier. { ,klas 'ā 'mäj·ə,lād·ər }

class B amplifier [ELECTR] **1.** An amplifier in which the grid bias is approximately equal to the cutoff value, so that anode current is approximately zero when no exciting grid voltage is applied, and flows for approximately half of each cycle when an alternating grid voltage is applied. **2.** A transistor amplifier in which each transistor is in its active region for approximately half the signal cycle. { ,klas 'bē 'am·plə,fī·ər }

class B modulator [ELECTR] A class B amplifier used to supply the necessary signal power to modulate a carrier; usually connected in push-pull. { ,klas 'bē 'mäj·ə,lād·ər }

class NP problems [COMPUT SCI] Problems that cannot necessarily be solved in polynomial time on a sequential computer but can be solved in polynomial time on a nondeterministic computer which, roughly speaking, guesses in turn each of 2N possible values of some N-bit quantity. { 'klas ¦en¦pē ,präb·ləmz }

class P problems [COMPUT SCI] Problems that can be solved in polynomial time on a conventional sequential computer. { 'klas 'pē ,präb·ləmz }

class S modulator [ELECTR] A modulator that is based on pulse-width modulation with a switching frequency several times the highest output frequency, and in which the pulse-width modulated signal is boosted to the desired power level by switching amplifiers, after which the desired audio output is obtained by a low-pass filter. { ,klas 'es 'mäj·ə,lād·ər }

clause [COMPUT SCI] A part of a statement in the COBOL language which may describe the structure of an elementary item, give initial values to items in independent and group work areas, or redefine data previously defined by another clause. { klȯz }

clean and certify [COMPUT SCI] To prepare a magnetic tape for a computer system by running it through a machine that cleans it, writes a data test pattern on it, and checks it for errors. { 'klēn ən 'sərd·ə,fī }

clean compile [COMPUT SCI] Conversion of a computer program from source to object language with no detection of significant errors by the compiler; logic errors not identified by the compiler may exist. { 'klēn kəm'pīl }

clear [COMPUT SCI] **1.** To restore a storage device, memory device, or binary stage to a prescribed state, usually that denoting zero. Also known as reset. **2.** A function key on calculators, to delete an entire problem or just the last keyboard entry. { klir }

clear area [COMPUT SCI] In optical character recognition, any area designated to be kept free of printing or any other extraneous markings. { 'klir ,er·ē·ə }

clear band [COMPUT SCI] In character recognition, a continuous horizontal strip of blank paper which must be obtained between consecutive code lines on a source document. { 'klir ,band }

clear channel [COMMUN] A standard broadcast channel in which the dominant station or stations render service over wide areas; stations are cleared of objectionable interference within their primary service areas and over all or a substantial portion of their secondary service areas. { ¦klir 'chan·əl }

clear text [COMMUN] Text or language which conveys an intelligible meaning in the language in which it is written with no hidden meaning. { 'klir ,tekst }

clear-voice override [COMMUN] The ability of a speech scrambler to receive a clear message even when the scrambler is set for scrambler operation. { ¦klir ¦vȯis 'ō·və,rīd }

click [COMMUN] A short-duration electric disturbance, such as that sometimes produced by a code-sending key or a switch. [COMPUT SCI] To select an object when the pointer is touching it by pressing and quickly releasing a button on a mouse. { klik }

client [COMPUT SCI] A hardware or software entity that requests shared services from a server. { 'klī·ənt }

client-based application [COMPUT SCI] An application that runs on a work station or personal computer in a network and is not available to others in the network. { 'klī·ənt ,bāst ,ap·lə¦kā·shən }

client-server system [COMPUT SCI] A computing system composed of two logical parts: a server, which provides information or services, and a client, which requests them. On a network, for example, users can access server resources from their personal computers using client software. { ¦klī·ənt 'sər·vər ,sis·təm }

clip art [COMPUT SCI] A collection of graphic images that are stored on a computer disk for use in desktop publishing, word processing, and presentation graphics programs. { 'klip ,ärt }

clipboard [COMPUT SCI] An area in memory or a file where cut or copied material is held temporarily before being inserted elsewhere in the same document or in another document. { 'klip,bȯrd }

Clipper Chip [COMPUT SCI] A chip proposed by the United States government to be used in all devices that might use encryption, such as computers and communications devices, for which the government would have at least some access or control over the decryption key for purposes of surveillance. { 'klip·ər ,chip }

clipping [COMMUN] The perceptible mutilation of signals or speech syllables during transmission, often due to limiting. [COMPUT SCI] *See* scissoring. [ELECTR] *See* limiting. { 'klip·iŋ }

CLIST [COMPUT SCI] A file containing a series of commands that are processed in the order given when the file is entered. Acronym for command list. { 'sē,list }

clobber [COMPUT SCI] To write new data and thereby erase good data in a file, or to otherwise damage the file so that it becomes useless. { 'kläb·ər }

clock [ELECTR] A source of accurately timed pulses, used for synchronization in a digital computer or as a time base in a transmission system. { kläk }

clock-doubled [COMPUT SCI] Describing a microprocessor that operates at twice the clock speed of the bus or motherboard to which it is attached. { 'kläk ¦dəb·əld }

clocked flip-flop [ELECTR] A flip-flop circuit that is set and reset at specific times by adding clock pulses to the input so that the circuit is triggered only if both trigger and clock pulses are present simultaneously. { 'kläkt 'flip ,fläp }

clocked logic [ELECTR] A logic circuit in which the switching action is controlled by repetitive pulses from a clock. { ¦kläkt ¦läj·ik }

clock frequency [ELECTR] The master frequency of the periodic pulses that schedule the operation of a digital computer. Also known as clock rate; clock speed. { 'kläk ,frē·kwən·sē }

clock pulses [COMPUT SCI] Electronic pulses which are emitted periodically, usually by a crystal device, to synchronize the operation of circuits in a computer. Also known as clock signals. { 'kläk ,pəl·səz }

clock rate *See* clock frequency. { 'kläk ,rāt }

clock signals *See* clock pulses. { 'kläk ,sig·nəlz }

clock speed *See* clock frequency. { 'kläk ,spēd }

clock time *See* internal cycle time. { 'kläk ,tīm }

clock track [COMPUT SCI] A track on a magnetic recording medium that generates clock pulses for the synchronization of read and write operations. { 'kläk ,trak }

clock-tripled [COMPUT SCI] Describing a microprocessor that operates at three times the clock speed of the bus or motherboard to which it is attached. { 'kläk ¦trip·əld }

clone [COMPUT SCI] A hardware or software product that closely resembles another product created by a different manufacturer or developer, in operation, appearance, or both. { klōn }

close [COMPUT SCI] To make a file unavailable to a computer program which previously had access to it. { klōs }

closed architecture [COMPUT SCI] A computer architecture whose detailed, technical specifications are available only to those authorized by the manufacturer. { ¦klōzd 'ärk·ə,tek·chər }

closed-bus system [COMPUT SCI] A computer that lacks receptacles for expansion boards and is difficult to upgrade. { ¦klōd 'bəs ,sis·təm }

closed-caption television [COMMUN] A method of captioning or subtitling television programs by coding captions as a vertical-interval data signal in an analog television system or in the transport of a digital television system that is decoded at the receiver and superimposed on the normal television picture. { ¦klōzd ¦kap·shən 'tel·ə,vizh·ən }

closed circuit [COMMUN] Program source that is not broadcast for general consumption but is fed to remote monitoring units. { ¦klōzd 'sər·kət }

closed-circuit communications system [COMMUN] A communications systems which is entirely self-contained, and does not exchange intelligence with other facilities and systems. { ¦klōzd ¦sər·kət kə,myü·nə'kā·shənz ,sis·təm }

closed-circuit signaling [COMMUN] Signaling in which current flows in the idle condition, and a signal is initiated by increasing or decreasing the current. { ¦klōzd ¦sər·kət 'sig·nə·liŋ }

closed-circuit telegraph system [COMMUN] Telegraph system in which, when no station is transmitting, the circuit is closed and current flows through the circuit. { ¦klōzd ¦sər·kət 'tel·ə ,graf ,sis·təm }

closed-circuit television [COMMUN] Any application of television that does not involve broadcasting for public viewing; the programs can be seen only on specified receivers connected to the television camera by circuits, which include microwave relays and coaxial cables. Abbreviated CCTV. { ¦klōzd ¦sər·kət 'tel·ə,vizh·ən }

closed file [COMPUT SCI] A file that cannot be accessed for reading or writing. { ¦klōzd 'fīl }

closed loop [COMPUT SCI] A loop whose execution continues indefinitely in the absence of any external intervention. [CONT SYS] A family of automatic control units linked together with a process to form an endless chain; the effects of control action are constantly measured so that if the controlled quantity departs from the norm, the control units act to bring it back. { ¦klōzd 'lüp }

closed-loop control system *See* feedback control system. { ¦klōzd ¦lüp kən'trōl ,sis·təm }

closed-loop telemetry system [ENG] **1.** A telemetry system which is also used as the display portion of a remote-control system. **2.** A system used to check out test vehicle or telemetry performance without radiation of radio-frequency energy. { ¦klōzd ¦lüp tə'lem·ə·trē ,sis·təm }

closed shop [COMPUT SCI] A data-processing center so organized that only professional programmers and operators have access to the center to meet the needs of users. { ¦klōzd 'shäp }

closed subroutine [COMPUT SCI] A subroutine that can be stored outside the main routine and can be connected to it by linkages at one or more locations. { ¦klōzd 'səb·rü,tēn }

closefile [COMPUT SCI] A procedure call in time sharing which enables an ALGOL program to close a file no longer required. { 'klōz,fīl }

close-out file [COMPUT SCI] A file created at the end of a processing cycle, usually encompassing a specified period of time. { 'klōz ,au̇t ,fīl }

close routine [COMPUT SCI] A computer program that changes the state of a file from open to closed. { 'klōz rü'tēn }

cloud attenuation [ELECTROMAG] The attenuation of microwave radiation by clouds (for the centimeter-wavelength band, clouds produce Rayleigh scattering); due largely to scattering, rather than absorption, for both ice and water clouds. { 'klau̇d ə,ten·yə'wā·shən }

cloverleaf antenna [ELECTROMAG] Antenna having radiating units shaped like a four-leaf clover. { 'klō·vər,lēf an 'ten·ə }

cluster [COMPUT SCI] **1.** In a clustered file, one of the classes into which records with similar sets of content identifiers are grouped. **2.** A grouping of hardware devices in a distributed processing system. **3.** A group of disk sectors that is treated as a single entity by the operating system. { 'kləs·tər }

cluster controller [COMPUT SCI] A control unit to which several peripheral devices are assigned. { 'kləs·tər kən,trōl·ər }

clustered file [COMPUT SCI] A collection of records organized so that items which exhibit similar sets of content identifiers are automatically grouped into common classes. { 'kləs·tərd 'fīl }

clustering algorithm [COMPUT SCI] A computer program that attempts to detect and locate the presence of groups of vectors, in a high-dimensional multivariate space, that share some property of similarity. { ¦kləs·tə·riŋ ¦al·gə ,rith·əm }

clutter [ELECTROMAG] Unwanted echoes on a radar screen, such as those caused by the ground, sea, rain, stationary objects, chaff, enemy jamming transmissions, and grass. Also known as background returns; radar clutter. { 'kləd·ər }

clutter gating [ELECTR] A technique which provides switching between moving-target-indicator and normal videos; this results in normal video being displayed in regions with no clutter and moving-target-indicator video being switched in only for the clutter areas. { 'kləd·ər ,gad·iŋ }

clutter suppression [ELECTR] Technique of reducing, by various means integral to the radar system, the effects of echoes from scatterers such as rain and surface features among the received signals. { 'kləd·ər sə,presh·ən }

CMI *See* computer-managed instruction.

CML *See* current-mode logic.

CMOS device [ELECTR] A device formed by the combination of a PMOS (*p*-type-channel metal oxide semiconductor device) with an NMOS (*n*-type-channel metal oxide semiconductor device). Derived from complementary metal oxide semiconductor device. { 'se,mȯs di'vīs }

CMRR *See* common-mode rejection ratio.

CNC *See* computer numerical control.

coastal refraction [ELECTROMAG] An apparent change in the direction of travel of a radio wave when it crosses a shoreline obliquely. Also known as land effect. { 'kōs·təl ri'frak·shən }

coax *See* coaxial cable. { 'kō,aks }

coaxial antenna [ELECTROMAG] An antenna consisting of a quarter-wave extension of the inner conductor of a coaxial line and a radiating sleeve that is in effect formed by folding back the outer conductor of the coaxial line for a length of approximately a quarter wavelength. { kō 'ak·sē·əl an'ten·ə }

coaxial attenuator [ELECTROMAG] An attenuator that has a coaxial construction and terminations suitable for use with coaxial cable. { kō 'ak·sē·əl ə'ten·yə,wād·ər }

coaxial cable [ELECTROMAG] A transmission line in which one conductor is centered inside and insulated from an outer metal tube that serves as the second conductor. Also known as coax; coaxial line; coaxial transmission line; concentric cable; concentric line; concentric transmission line. { kō'ak·sē·əl 'kā·bəl }

coaxial cavity [ELECTROMAG] A cylindrical resonating cavity having a central conductor in contact with its pistons or other reflecting devices. { kō'ak·sē·əl 'kav·əd·ē }

coaxial cavity magnetron [ELECTR] A magnetron which achieves mode separation, high efficiency, stability, and ease of mechanical tuning by coupling a coaxial high Q cavity to a normal set of quarter-wavelength vane cavities. { kō'ak·sē·əl ,kav·əd·ē 'mag·nə,trän }

coaxial connector [ELECTROMAG] An electric connector between a coaxial cable and an equipment circuit, so constructed as to maintain the conductor configuration, through the separable connection, and the characteristic impedance of the coaxial cable. { kō'ak·sē·əl kə'nek·tər }

coaxial-cylinder magnetron [ELECTR] A magnetron in which the cathode and anode consist of coaxial cylinders. { kō'ak·sē·əl ,sil·ən·dər 'mag·nə,trän }

coaxial filter [ELECTROMAG] A section of coaxial line having reentrant elements that provide the inductance and capacitance of a filter section. { kō'ak·sē·əl 'fil·tər }

coaxial hybrid [ELECTROMAG] A hybrid junction of coaxial transmission lines. { kō'ak·sē·əl 'hī ,brəd }

coaxial isolator [ELECTROMAG] An isolator used in a coaxial cable to provide a higher loss for energy flow in one direction than in the opposite direction; all types use a permanent magnetic field in combination with ferrite and dielectric materials. { kō'ak·sē·əl 'ī·sə,lād·ər }

coaxial line *See* coaxial cable. { kō'ak·sē·əl 'līn }

coaxial-line resonator [ELECTROMAG] A resonator consisting of a length of coaxial line short-circuited at one or both ends. { kō'ak·sē·əl ,līn 'rez·ən,ād·ər }

coaxial stub [ELECTROMAG] A length of nondissipative cylindrical waveguide or coaxial cable branched from the side of a waveguide to produce some desired change in its characteristics. { kō 'ak·sē·əl 'stəb }

coaxial transmission line *See* coaxial cable. { kō'ak·sē·əl tranz'mish·ən ,līn }

COBOL [COMPUT SCI] A business data-processing language that can be given to a computer as a series of English statements describing a complete business operation. Derived from common business-oriented language. { 'kō,bȯl }

cochannel cells [COMMUN] Two cells in a cellular mobile radio system that use the same frequency. { ¦kō,chan·əl 'selz }

cochannel interference [COMMUN] Interference caused on one communication channel by a transmitter operating in the same channel. { 'kō ,chan·əl ,in·tər'fir·əns }

cochannel interference reduction factor [COMMUN] The ratio of the minimum separation between two cochannel cells without interference to the radius of a cell. { ¦kō,chan·əl ,in·tər,fir·əns ri'dək·shən ,fak·tər }

codan [ELECTR] A device that silences a receiver except when a modulated carrier signal is being received. { 'kō,dan }

code [COMMUN] A system of symbols and rules for expressing information, such as the Morse code, Electronic Industries Association color code, and the binary and other machine languages used in digital computers. { kōd }

code book [COMMUN] A book containing a large number of plaintext words, phrases, and sentences and their codetext equivalents. { 'kōd ,bu̇k }

codec [ELECTR] A device that converts analog signals to digital form for transmission and converts signals traveling in the opposite direction from digital to analog form. Derived from coder-decoder. { 'kō,dek }

code-check [COMPUT SCI] To remove mistakes from a coded routine or program. { 'kōd ,chek }

code checking time [COMPUT SCI] Time spent checking out a problem on the computer, making sure that the problem is set up correctly and that the code is correct. { 'kōd ,chek·iŋ ,tīm }

code converter [COMPUT SCI] A converter that changes coded information to a different code system. { 'kōd kən'vərd·ər }

coded character set [COMPUT SCI] A set of characters together with the code assigned to each character for computer use. { 'kōd·əd 'kar·ik·tər ,set }

coded decimal *See* decimal-coded digit. { 'kōd·əd 'des·məl }

coded interrogator [COMMUN] An interrogator whose output signal forms the code required to trigger a specific radio or radar beacon; part of an address-selective system. { 'kōd·əd in'ter·ə ,gād·ər }

code-division multiple access [COMMUN] The transmission of messages from a large number of transmitters over a single channel by assigning each transmitter a pseudorandom noise code (typically more than 2000 symbols long for each bit of information) so that the codes are mathematically independent of each other. Abbreviated CDMA. { 'kōd də¦vizh·ən 'məl·tə·pəl 'ak,ses }

code-division multiplex [COMMUN] Multiplex in which two or more communication links occupy the entire transmission channel simultaneously, with code signal structures designed so a given receiver responds only to its own signals and treats the other signals as noise. Abbreviated CDM. { 'kōd də'vizh·ən 'məlt·i,pleks }

coded passive reflector antenna [ELECTROMAG] An object intended to reflect Hertzian waves and having variable reflecting properties according to a predetermined code for the purpose of producing an indication on a radar receiver. { 'kōd·əd 'pas·iv ri'flek·tər an,ten·ə }

coded program [COMPUT SCI] A program expressed in the required code for a computer. { 'kōd·əd 'prō·grəm }

coded stop [COMPUT SCI] A stop instruction built into a computer routine. { 'kōd·əd 'stäp }

code element [COMMUN] One of the separate elements or events constituting a coded message, such as the presence or absence of a pulse, dot, dash, or space. { 'kōd ,el·ə·mənt }

code error [COMPUT SCI] A surplus or lack of a bit or bits in a machine instruction. { 'kōd ,er·ər }

code-excited linear predictive coder [COMMUN] A speech coder that uses both short-term and long-term predictors, vector quantization techniques, and an analysis-by-synthesis approach to search for the best combination of coder parameters. Abbreviated CELP coder. { ¦kōd i¦sīd·əd ¦lin·ē·ər prə¦dik·tiv 'kōd·ər }

code extension [COMPUT SCI] A method of increasing the number of characters that can be represented by a code by combining characters into groups. { 'kōd ik,sten·chən }

code group [COMMUN] A combination of letters or numerals or both, assigned to represent one or more words of plain text in a coded message. { 'kōd ,grüp }

code line [COMPUT SCI] In character recognition, the area reserved for the inscription of the printed or handwritten characters to be recognized. { 'kōd ,līn }

coder [COMMUN] A device that generates a code by producing pulses having varying lengths or spacings, as required for radio beacons and interrogators. Also known as moder; pulse coder; pulse-duration coder. [COMPUT SCI] A person who translates a sequence of computer instructions into codes acceptable to the machine. { 'kōd·ər }

coder-decoder *See* codec. { ¦kōd·ər dē¦kōd·ər }

code reader [COMPUT SCI] A scanning device used for automated identification of a two-dimensional pattern, one part after the other, and generation of either analog or digital signals that correspond to the pattern. Also known as code scanner. { 'kōd ,rēd·ər }

code ringing [COMMUN] In telephone switching, party-line ringing wherein the number or duration of rings indicates which station is being called. { 'kōd ,riŋ·iŋ }

code scanner *See* code reader. { 'kōd ,skan·ər }

code sensitivity [COMPUT SCI] Property of hardware or software that can handle only data presented in a particular code. { 'kōd ,sen·sə ,tiv·əd·ē }

code signal [COMMUN] A sequence of discrete conditions or events corresponding to a coded message. { 'kōd ,sig·nəl }

codetext [COMMUN] A message which has been transformed by a code into a form which can be read only by those privy to the secrets of the code. { 'kōd,tekst }

code translation [COMMUN] Conversion of a directory code or number into a predetermined code for controlling the selection of an outgoing trunk or line. { 'kōd tranz,lā·shən }

code transparency [COMPUT SCI] Property of hardware or software that can handle data regardless of what form it is in. { 'kōd tranz ,par·ən·sē }

coding [COMPUT SCI] **1.** The process of converting a program design into an accurate, detailed representation of that program in some suitable language. **2.** A list, in computer code, of the successive operations required to carry out a given routine or solve a given problem. { 'kōd·iŋ }

coding disk [COMMUN] Disk with small projections for operating contacts to give a certain predetermined code to a transmission. { 'kōd·iŋ ,disk }

coding form *See* coding sheet. { 'kōd·iŋ ,fȯrm }

coding line *See* instruction word. { 'kōd·iŋ ,līn }

coding sheet [COMPUT SCI] A sheet of paper printed with a form on which one can conveniently write a coded program. Also known as coding form. { 'kōd·iŋ ,shēt }

coercion [COMPUT SCI] A method employed by many programming languages to automatically convert one type of data to another. { kō'ər·shən }

COGO [COMPUT SCI] A higher-level computer language oriented toward civil engineering, enabling one to write a program in a technical vocabulary familiar to engineers and feed it to the computer; several versions have been implemented. Derived from coordinated geometry. { 'kō,gō }

coherent [ELECTR] Referring to radar signals and signal processing and related equipment wherein attention is given to both the amplitude and the phase of the signal; many valuable processes in radar operation are coherent in nature. { kō 'hir·ənt }

coherent carrier system [NAV] Transponder system in which the interrogating carrier is retransmitted at a definite multiple frequency for comparison. { kō'hir·ənt 'kar·ē·ər ,sis·təm }

coherent detector [ELECTR] A detector used in coherent radar giving an output-signal amplitude that depends on the phase of the echo signal (rather than only its amplitude) relative to the phase of that which was transmitted, as required for sensing the radial velocity of targets. Also known as phase detector. { kō'hir·ənt di'tek·tər }

coherent echo [ELECTR] A radar echo whose phase and amplitude at a given range remain relatively constant. { kō'hir·ənt 'ek·ō }

coherent integration [ELECTR] A radar signal processing technique in which the phase relationships among successive pulses being echoed from a target are interpreted, usually to estimate or to separate signals based on the apparent Doppler shift of the signals. { kō'hir·ənt ,int·ə'grā·shən }

coherent interrupted waves [COMMUN] Interrupted continuous waves occurring in wave trains in which the phase of the waves is maintained through successive wave trains. { kō'hir·ənt in·tə'rəp·təd 'wāvz }

coherent light communications [COMMUN] Communications using the optical band as a transmission medium by modulating a laser in amplitude or pulse frequency. { kō'hir·ənt 'līt kə,myü·nə'kā·shənz }

coherent moving-target indicator [ENG] A radar system in which the Doppler frequency of the target echo is compared to a local reference frequency generated by a coherent oscillator. { kō'hir·ənt ¦müv·iŋ ¦tär·gət ,in·də,kād·ər }

coherent oscillator [ELECTR] An oscillator locked in phase to the transmitted signal as used in coherent radar to provide a reference by which changes in the phase of successively received pulses may be recognized. Abbreviated coho. { kō'hir·ənt 'äs·ə,lād·ər }

coherent processing interval [ELECTR] That period of time over which radar return signals are coherently integrated, permitting a resolution in Doppler shift being sensed as great as the reciprocal of the interval. { kō'hir·ənt 'präs·əs·iŋ 'in·tər·vəl }

coherent-pulse radar [ELECTR] A radar in which the radio-frequency oscillations of recurrent pulses bear a constant phase relation to those of a continuous oscillation. { kō'hir·ənt ˌpəls 'rā ˌdär }

coherent pulses [ELECTR] Characterizing pulses in which the phase of the radio-frequency waves is maintained through successive pulses. { kō 'hir·ənt 'pəl·səz }

coherent radar [ELECTR] A radar capable of comparing the phase of received signals with the phase of the transmitted signal, generally with the object of sensing pulse-to-pulse phase changes, indicative of radial motion, and hence the Doppler shift, of the target. { kō'hir·ənt 'rā ˌdär }

coherent reference [ELECTR] A reference signal, usually of stable frequency, to which other signals are phase locked to establish coherence throughout a system. { kō'hir·ənt 'ref·rəns }

coherent side-lobe canceler [ELECTR] A radar feature in which interfering signals in the side lobes of the radar antenna are cancelled by adaptively adjusting the phase and amplitude of signals received in a number of auxiliary antennas and subtracting those from the signal in the main antenna. { kō'hir·ənt 'sīd ˌlōb 'kan·səl·ər }

coherent signal [ELECTR] In coherent radar, a signal having a known phase, often constant, as that produced by the coherent oscillator to be mixed in the coherent detector with the echo signal to detect pulse-to-pulse phase changes indicative of target radial motion. { kō'hir·ənt 'sig·nəl }

coherent system [NAV] A navigation system in which the signal output is obtained by demodulating the received signal after mixing with a local signal having a fixed phase relation to that of the transmitted signal, to permit use of the information carrier by the phase of the received signal. { kō'hir·ənt 'sis·təm }

coherent transponder [ELECTR] A transponder in which a fixed relation between frequency and phase of input and output signals is maintained. { kō'hir·ənt tranz'pänd·ər }

coherent video [ELECTR] The video signal produced in a coherent radar by combining in a coherent detector a radar echo signal with the output of the continuous wave coherent oscillator. Also called bipolar video. { kō'hir·ənt 'vid·ē·ō }

coho *See* coherent oscillator. { 'kōˌhō }

coil [CONT SYS] Any discrete and logical result that can be transmitted as output by a programmable controller. [ELECTROMAG] A number of turns of wire used to introduce inductance into an electric circuit, to produce magnetic flux, or to react mechanically to a changing magnetic flux; in high-frequency circuits a coil may be only a fraction of a turn. Also known as electric coil; inductance coil; inductor. { kȯil }

coil antenna [ELECTROMAG] An antenna that consists of one or more complete turns of wire. { 'kȯil an'ten·ə }

coil loading [COMMUN] Loading in which inductors, commonly called loading coils, are inserted in a line at intervals. { 'kȯil ˌlōd·iŋ }

coincidence amplifier [ELECTR] An electronic circuit that amplifies only that portion of a signal present when an enabling or controlling signal is simultaneously applied. { kō'in·sə·dəns ˌam·pləˌfī·ər }

coincidence circuit [ELECTR] A circuit that produces a specified output pulse only when a specified number or combination of two or more input terminals receives pulses within an assigned time interval. Also known as coincidence counter; coincidence gate. { kō'in·sə·dəns ˌsər·kət }

coincidence counter *See* coincidence circuit. { kō'in·sə·dəns ˌkau̇nt·ər }

coincidence gate *See* coincidence circuit. { kō 'in·sə·dəns ˌgāt }

cold boot [COMPUT SCI] To turn the power on and boot a computer. { ¦kōld 'büt }

cold emission *See* field emission. { 'kōld i'mish·ən }

cold link [COMPUT SCI] A linking of information in two documents in which updating the link requires recopying the information from the source document to the target document. { 'kōld 'liŋk }

cold start [COMPUT SCI] To start running a computer program from the very beginning, without being able to continue the processing that was occurring previously when the system was interrupted. { 'kōld 'stärt }

collate [COMPUT SCI] To combine two or more similarly ordered sets of values into one set that may or may not have the same order as the original sets. { 'käˌlāt }

collating sequence [COMPUT SCI] The ordering of a set of items such that sets in that assigned order can be collated. { 'käˌlād·iŋ ˌsē·kwəns }

collector [ELECTR] **1.** A semiconductive region through which a primary flow of charge carriers leaves the base of a transistor; the electrode or terminal connected to this region is also called the collector. **2.** An electrode that collects electrons or ions which have completed their functions within an electron tube; a collector receives electrons after they have done useful work, whereas an anode receives electrons whose useful work is to be done outside the tube. Also known as electron collector. { kə'lek·tər }

collector modulation [ELECTR] Amplitude modulation in which the modulator varies the collector voltage of a transistor. { kə'lek·tər ˌmäj·ə'lā·shən }

collimation error [ENG] **1.** Angular error in magnitude and direction between two nominally parallel lines of sight. **2.** Specifically, the angle by which the line of sight of a radar differs from what it should be. { ˌkäl·ə'mā·shən ˌer·ər }

collimation tower [ENG] Tower on which a visual and a radio target are mounted to check the electrical axis of an antenna. { ˌkäl·əˈmā·shən ˌtau̇·ər }

collinear array *See* linear array. { kəˈlin·ē·ər əˈrā }

collision-avoidance radar [ENG] Radar equipment utilized in a collision-avoidance system. { kəˈlizh·ən əˈvȯid·əns ˌrāˌdär }

collision-avoidance system [ENG] Electronic devices and equipment used by a pilot to perform the functions of conflict detection and avoidance. { kəˈlizh·ən əˈvȯid·əns ˌsis·təm }

collision detection [COMPUT SCI] A procedure in which a computer network senses a situation where two computer devices attempt to access the network at the same time and blocks the messages, requiring each device to resubmit its message at a randomly selected time. { kəˈlizh·ən diˌtek·shən }

color balance [ELECTR] Adjustment of the circuits feeding the three electron guns of a television color picture tube to compensate for differences in light-emitting efficiencies of the three color phosphors on the screen of the tube. { ˈkəl·ər ˌbal·əns }

color-bar generator [ELECTR] A signal generator that delivers to the input of a video system the signal needed to produce a color-bar test pattern on a device or system. { ˈkəl·ər ˌbär ˈjen·əˌrād·ər }

color-bar test pattern [COMMUN] A test pattern of different colors of vertical bars, used to check the performance of a video system. { ˈkəl·ər ˌbär ˈtest ˌpad·ərn }

color breakup [COMMUN] A transient or dynamic distortion of the color in an analog color television picture that can originate in videotape equipment, a television camera, or a receiver. { ˈkəl·ər ˌbrākˌəp }

color burst [ELECTR] The portion of an analog composite color television signal consisting of a few cycles of a sine wave of chrominance subcarrier frequency. Also known as burst; reference burst. { ˈkəl·ər ˌbərst }

color carrier *See* chrominance subcarrier. { ˈkəl·ər ˌkar·ē·ər }

color-carrier reference *See* chrominance-carrier reference. { ˈkəl·ər ˌkar·ē·ər ˌref·rəns }

color coder *See* matrix. { ˈkəl·ər ˌkōd·ər }

color contamination [ELECTR] An error in the color rendition of an analog color television picture that results from incomplete separation of the paths that carry different color components of a picture. { ˈkəl·ər kənˌtam·əˈnā·shən }

color control *See* chroma control. { ˈkəl·ər kən ˈtrōl }

color decoder *See* matrix. { ˈkəl·ər dēˈkōd·ər }

color-difference signal [ELECTR] A signal that is added to the monochrome signal in an analog color television receiver to obtain a signal representative of one of the three tristimulus values needed by the color picture tube. { ¦kəl·ər ¦dif·rəns ˌsig·nəl }

color encoder *See* matrix. { ˈkəl·ər enˈkōd·ər }

color facsimile [COMMUN] A facsimile system for transmission of color photographs, in which three separate facsimile transmissions are made from the original color print, using color-separation filters in the optical system of the facsimile transmitter. { ˈkəl·ər ˌfakˈsim·ə·lē }

color fringing [ELECTR] Spurious chromaticity at boundaries of objects in a television picture. { ˈkəl·ər ˈfrinj·iŋ }

color killer circuit [ELECTR] The circuit in an analog color television receiver that biases chrominance amplifier tubes to cutoff during reception of monochrome programs. Also known as killer stage. { ˈkəl·ər ˌkil·ər ˌsər·kət }

color kinescope *See* color picture tube. { ¦kəl·ər ˈkin·ə·skōp }

color oscillator *See* chroma oscillator. { ˈkəl·ər ˌäs·əˌlād·ər }

color phase [COMMUN] The difference in phase between components (I or Q) of a chrominance signal and the chrominance-carrier reference in an analog color television receiver. { ˈkəl·ər ˌfāz }

color-phase alternation [COMMUN] The periodic changing of the color phase of one or more components of the chrominance subcarrier between two sets of assigned values after every field in an analog color television system. Abbreviated CPA. { ˈkəl·ər ˌfāz ȯl·tərˈnā·shən }

color-phase detector [ELECTR] The analog color television receiver circuit that compares the frequency and phase of the incoming burst signal with those of the locally generated 3.579545-megahertz chroma oscillator and delivers a correction voltage to ensure that the color portions of the picture will be in exact register with the black-and-white portions on the screen. { ˈkəl·ər ˌfāz diˈtek·tər }

color picture signal [COMMUN] The electric signal that represents complete color picture information, excluding all synchronizing signals. { ˈkəl·ər ˌpik·chər ˌsig·nəl }

color picture tube [ELECTR] A cathode-ray tube having three different colors of phosphors, so that when these are appropriately scanned and excited, a color picture is obtained. Also known as color kinescope; color television picture tube; tricolor picture tube. { ˈkəl·ər ˌpik·chər ˌtüb }

color purity [ELECTR] Absence of undesired colors in the spot produced on the screen by each beam of a color picture tube. { ˈkəl·ər ˌpyür·əd·ē }

color-saturation control *See* chroma control. { ˈkəl·ər sach·əˈrā·shən kənˈtrōl }

color signal [COMMUN] Any signal that controls the chromaticity values of a color picture in a video system. { ˈkəl·ər ˌsig·nəl }

color subcarrier *See* chrominance subcarrier. { ˈkəl·ər səbˈkar·ē·ər }

color-subcarrier oscillator *See* chroma oscillator. { ˈkəl·ər səbˈkar·ē·ər ˈä·səˌlād·ər }

color-subcarrier reference *See* chrominance-carrier reference. { ˈkəl·ər səbˈkar·ē·ər ˈref·rəns }

color sync signal [COMMUN] A signal that is transmitted with each line of an analog color television broadcast to ensure that the color relationships in the transmitted signal are established and maintained in the receiver. { 'kəl·ər 'siŋk ˌsig·nəl }

color television [COMMUN] A television system that reproduces an image approximately in its original colors. { ¦kəl·ər ¦tel·əˌvizh·ən }

color television picture tube *See* color picture tube. { ¦kəl·ər ¦tel·əˌvizh·ən 'pik·chər ˌtüb }

color transmission [COMMUN] In television, the transmission of a signal waveform that represents both the brightness values and the chromaticity values in the picture. { 'kəl·ər tranz'mish·ən }

column [COMPUT SCI] A vertical arrangement of characters or other expressions, usually referring to a specific print position on a printer. { 'käl·əm }

column order [COMPUT SCI] The storage of a matrix $a(m,n)$ as $a(1,1)$, $a(2,1)$,. . . ,$a(m,1)$, $a(1,2)$,. . . . { 'kä l·əm ˌȯr·dər }

column printer [COMPUT SCI] A small line printer used with some calculators to provide hard-copy printout of input and output data; typically consists of 20 columns of numerals and a limited number of alphabetic or other identifying characters. { 'käl·əm ˌprint·ər }

COM *See* computer output on microfilm.

coma [ELECTR] A cathode-ray tube image defect that makes the spot on the screen appear comet-shaped when away from the center of the screen. { 'kō·mə }

coma lobe [ELECTROMAG] Side lobe that occurs in the radiation pattern of a microwave antenna when the reflector alone is tilted back and forth to sweep the beam through space because the feed is no longer always at the center of the reflector; used to eliminate the need for a rotary joint in the feed waveguide. { 'kō·mə ˌlōb }

comb antenna [ELECTROMAG] A broad-band antenna for vertically polarized signals, in which half of a fishbone antenna is erected vertically and fed against ground by a coaxial line. { 'kōm an ˌten·ə }

combat information center [COMMUN] A shipboard location at which tactical information from radar, sonar, and other equipment is received, displayed for rapid analysis, and evaluated. { 'kämˌbat in·fər'mā·shən ˌsen·tər }

combinational circuit [ELECTR] A switching circuit whose outputs are determined only by the concurrent inputs. { ˌkäm·bə'nā·shən·əl 'sər·kət }

combined head *See* read/write head. { kəm 'bīnd 'hed }

combining network [COMPUT SCI] A switching system for accessing memory modules in a multiprocessor, in which each switch remembers the memory addresses it has used, and can then satisfy several requests with a single memory access. { kəm'bīn·iŋ 'netˌwərk }

COMIT [COMPUT SCI] A user-oriented, general-purpose, symbol-manipulation programming language for computers. { 'kōˌmit }

command [COMPUT SCI] A signal that initiates a predetermined type of computer operation that is defined by an instruction. [CONT SYS] An independent signal in a feedback control system, from which the dependent signals are controlled in a predetermined manner. { kə'mand }

command and control [SYS ENG] The process of military commanders and civilian managers identifying, prioritizing, and achieving strategic and tactical objectives by exercising authority and direction over human and material resources by utilizing a variety of computer-based and computer-controlled systems, many driven by decision-theoretic methods, tools, and techniques. Abbreviated C^2. { kə'mand ən kən 'trōl }

command button [COMPUT SCI] A small rectangle on a graphical user interface with a command, such as open, close, OK, or print, that is immediately activated upon selection of the button. { kə'mand ˌbət·ən }

command code *See* operation code. { kə'mand ˌkōd }

command control *See* command guidance. { kə'mand kənˌtrōl }

command, control, and communications [SYS ENG] A version of command and control in which the role of communications equipment is emphasized. Abbreviated C^3. { kə'mand kən'trōl ən kəˌmyü·ne'kā·shənz }

command, control, communications, and intelligence [SYS ENG] A version of command and control in which the roles of communications equipment and intelligence are emphasized. Abbreviated C^3I. { kə'mand kən'trōl kəˌmyü·nə'kā·shənz ən in'tel·ə·jəns }

command control program [COMPUT SCI] The interface between a time-sharing computer and its users by means of which they can create, edit, save, delete, and execute their programs. { kə'mand kənˌtrōl ˌprō·grəm }

command-driven program [COMPUT SCI] A computer program that accepts command words and statements typed in by the user. { kə¦mand ˌdriv·ən 'prō·grəm }

command guidance [ENG] The guidance of a missile, rocket, or spacecraft by means of electronic signals sent to receiving devices in the vehicle. Also known as command control. { kə'mand ˌgīd·əns }

command interpreter [COMPUT SCI] A program that processes commands and other input and output from an active terminal in a time-sharing system. { kə'mand ˌin'tər·prə·tər }

command language [COMPUT SCI] The language of an operating system, through which the users of a data-processing system describe the requirements of their tasks to that system. Also known as job control language. { kə'mand ˌlaŋ·gwij }

command level [COMPUT SCI] The ability to control a computer's operating system through the use of commands, normally available only to computer operators. { kə'mand ˌlev·əl }

command line [COMPUT SCI] On a display screen, the space following a prompt (such as $) where a text instruction to a computer or device is typed. { kə'mand ˌlīn }

command list *See* CLIST. { kə'mand ˌlist }

command mode [COMPUT SCI] The status of a terminal in a time-sharing environment enabling the programmer to use the command control program. { kə'mand ˌmōd }

command processor [COMPUT SCI] A computer program that converts a limited number of user commands into the machine commands that direct the operating system. Also known as command shell. { kə¦mand 'präˌses·ər }

command pulses [ELECTR] The electrical representations of bit values of 1 or 0 which control input/output devices. { kə'mand ˌpəl·səs }

command set [COMMUN] A radio set used to receive or give commands, as between one aircraft and another or between an aircraft and the ground. { kə'mand ˌset }

command shell *See* command processor. { kə'mand ˌshel }

comment [COMPUT SCI] An expression identifying or explaining one or more steps in a routine, which has no effect on execution of the routine. { 'kämˌent }

comment code [COMPUT SCI] One or more characters identifying a comment. { 'kämˌent ˌkōd }

comment out [COMPUT SCI] To render a statement in a computer program inactive by making it a comment. { 'käˌment 'au̇t }

common area [COMPUT SCI] An area of storage which two or more routines share. { ¦käm·ən ¦er·ē·ə }

common-base connection *See* grounded-base connection. { ¦käm·ən 'bās kə'nek·shən }

common battery [COMMUN] System of current supply where all direct current energy for a unit of a telephone system is supplied by one source in a central office or exchange. { ¦käm·ən ¦bad·ə·rē }

common business-oriented language *See* COBOL. { ¦käm·ən ¦biz·nəs ¦ȯr·ēˌent·əd ˌlaŋ·gwij }

common-channel interoffice signaling [COMMUN] A method of signaling in a telecommunications switching system in which a network of separate data communication paths separate from the communications transmission is used for transmitting all signaling information between offices. Abbreviated CCIS. { ¦käm·ən ¦chan·əl ˌin·tərˌȯ·fəs 'sig·nəl·iŋ }

common-collector connection *See* grounded-collector connection. { ¦käm·ən kə'lek·tər kə 'nek·shən }

common control unit [COMPUT SCI] Control unit that is shared by more than one machine. { ¦käm·ən kən'trōl ˌyü·nət }

common declaration statement [COMPUT SCI] A nonexecutable statement in FORTRAN which allows specified arrays or variables to be stored in an area available to other programs. { ¦käm·ən ˌdek·lə'rā·shən ˌstāt·mənt }

common-drain amplifier [ELECTR] An amplifier using a field-effect transistor so that the input signal is injected between gate and drain, while the output is taken between the source and drain. Also known as source-follower amplifier. { ¦käm·ən 'drān 'am·pləˌfī·ər }

common-emitter connection *See* grounded-emitter connection. { ¦käm·ən i'mid·ər kə'nek·shən }

common-gate amplifier [ELECTR] An amplifier using a field-effect transistor in which the gate is common to both the input circuit and the output circuit. { ¦käm·ən 'gāt 'am·pləˌfī·ər }

common gateway interface [COMPUT SCI] A protocol that allows the secure data transfer to and from a server and a network user by means of a program which resides on the server and handles the transaction. For example, if an intranet user sent a request with a Web browser for database information, a CGI program would execute on the server, retrieve the information from the database, format it in HTML, and send it back to the user. Abbreviated CGI. { ˌkäm·ən ˌgātˌwā 'in·tərˌfās }

common language [COMPUT SCI] A machine-readable language that is common to a group of computers and associated equipment. { ¦käm·ən ¦laŋ·gwij }

common-mode rejection [ELECTR] The ability of an amplifier to cancel a common-mode signal while responding to an out-of-phase signal. Also known as in-phase rejection. { ¦käm·ən ˌmōd ri'jek·shən }

common-mode rejection ratio [ELECTR] The ratio of the gain of an amplifier for difference signals between the input terminals, to the gain for the average or common-mode signal component. Abbreviated CMRR. { 'käm·ən ˌmōd ri'jek·shən 'rā·shō }

common-mode signal [ELECTR] A signal applied equally to both ungrounded inputs of a balanced amplifier stage or other differential device. Also known as in-phase signal. { ¦käm·ən ˌmōd 'sig·nal }

common-mode voltage [ELECTR] A voltage that appears in common at both input terminals of a device with respect to the output reference (usually ground). { ¦käm·ən ˌmōd 'vōl·tij }

common object request broker [COMPUT SCI] A system that provides interoperability among objects in a heterogeneous, distributed, object-oriented environment in a way that is transparent to the programmer; its design is based on the OMG object model. Abbreviated CORBA. { ¦käm·ən ¦äb·jekt ri'kwest ˌbrō·kər }

common-source amplifier [ELECTR] An amplifier stage using a field-effect transistor in which the input signal is applied between gate and source and the output signal is taken between drain and source. { ¦käm·ən ˌsȯrs 'am·pləˌfī·ər }

common storage [COMPUT SCI] A section of memory in certain computers reserved for temporary storage of program outputs to be used as input for other programs. { ¦käm·ən 'stȯr·ij }

common-user channel [COMMUN] Any of the communications channels which are available to all authorized agencies for transmission of command, administrative, and logistic traffic. { ¦käm·ən ,yü·zər ,chan·əl }

communicating word processor [COMPUT SCI] A word processor that can be linked to other word processors to exchange information. { kə'myü·nə,kād·iŋ 'wórd ,prä,ses·ər }

communication [COMMUN] The transmission of intelligence between two or more points over wires or by radio; the terms telecommunication and communication are often used interchangeably, but telecommunication is usually the preferred term when long distances are involved. { kə,myü·nə'kā·shən }

communication band [COMMUN] The band of frequencies effectively occupied by a radio transmitter for the type of transmission and the speed of signaling used. { kə,myü·nə'kā·shən ,band }

communication bus [COMMUN] A device that transfers control, timing, and data signals between switching processor subsystems; designed to provide physical and electrical isolation, to provide for simple addition of units on an in-service basis, and to provide pluggable connection for efficient factory testing, installation, and maintenance. { kə,myü·nə'kā·shən ,bəs }

communication cable [COMMUN] A metallic wire or fiber-optic material used in the telephone industry to connect customers to their local switching centers and to interconnect local and long-distance switching centers. { kə ,myü·nə'kā·shən ,kā·bəl }

communication channel [COMMUN] The wire or radio channel that serves to convey intelligence between two or more terminals. { kə,myü·nə'kā·shən ,chan·əl }

communication countermeasure [COMMUN] Any electronic countermeasure against communications, such as jamming. { kə,myü·nə'kā·shən 'kaủnt·ər,mezh·ər }

communication engineering [COMMUN] The design, construction, and operation of all types of equipment used for radio, wire, or other types of communication. { kə,myü·nə'kā·shən en·jə'nir·iŋ }

communication link *See* data link. { kə,myü·nə'kā·shən ,liŋk }

communication protocol [COMPUT SCI] Procedures that enable devices within a computer network to exchange information. Also known as protocol. { kə,myü·nə'kā·shən 'prōd·ə,kól }

communication receiver [ELECTR] A receiver designed especially for reception of voice or code messages transmitted by radio communication systems. { kə,myü·nə'kā·shən ri'sē·vər }

communications [ENG] The science and technology by which information is collected from an originating source, transformed into electric currents or fields, transmitted over electrical networks or space to another point, and reconverted into a form suitable for interpretation by a receiver. { kə,myü·nə'kā·shənz }

communications control unit [COMMUN] A device that handles data transmission between components of a communications network, and performs related functions such as multiplexing, message switching, and code conversion. Abbreviated CCU. { kə,myü·nə'kā·shənz kən'trōl ,yü·nət }

communications intelligence [COMMUN] Technical and intelligence information derived from communications by other than the intended recipients. { kə,myü·nə'kā·shənz in'tel·ə·jəns }

communications language [COMMUN] A language structure complete with conventions, syntax, and character set, used primarily for conveying knowledge of processes between two participants. { kə,myü·nə'kā·shənz ,laŋ·gwij }

communications network [COMMUN] Organization of stations capable of intercommunications but not necessarily on the same channel. { kə ,myü·nə'kā·shənz ,net,wərk }

communications package [COMPUT SCI] A software product that specifies communications protocols for data transmission within a computer network or between a computer and its peripheral equipment. { kə,myü·nə'kā·shənz ,pak·ij }

communication speed [COMMUN] The rate at which information is transmitted over a communications channel, adjusted for redundancies. { kə,myü·nə'kā·shən ,spēd }

communications program [COMPUT SCI] A computer program that transmits data to and receives data from local and remote terminals and other computers. { kə,myü·nə'kā·shənz ,prō·grəm }

communications relay station [COMMUN] Facility for rapidly passing message traffic from one tributary to another by automatic, semiautomatic, or manual means, or by electrically connecting circuits (circuit switching) between two tributaries for direct transmission. { kə ,myü·nə'kā·shənz 'rē,la ,stā·shən }

communications satellite [AERO ENG] An orbiting, artificial earth satellite that relays radio, television, and other signals between ground terminal stations thousands of miles apart. Also known as radio relay satellite; relay satellite. { kə,myü·nə'kā·shənz 'sad·ə,līt }

communications traffic [COMMUN] All transmitted and received messages. { kə,myü·nə'kā·shənz ,traf·ik }

communication system [COMMUN] A telephone, radio, television, data transmission, or other system in which information-bearing signals originated at one place are reproduced at a distant point. { kə,myü·nə'kā·shən ,sis·təm }

communications zone indicator [ELECTR] Device to indicate whether or not long-distance high-frequency broadcasts are successfully reaching their destinations. { kə,myü·nə'kā·shənz ,zōn 'in·də,kād·ər }

communication theory [COMMUN] The mathematical theory of the communication of information from one point to another. { kə,myü·nə'kā·shən ,thē·ə·rē }

community antenna television *See* cable television. { kə'myü·nə·dē an'ten·ə 'tel·ə,vizh·ən }

community dial office [COMMUN] Small dial office with no employees located in the building serving an exchange area. { kə'myü·nə·dē 'dīl ,óf·əs }

commutation [COMMUN] The sampling of various quantities in a repetitive manner for transmission over a single channel in telemetering. { ˌkäm·yəˈtā·shən }

commutator pulse [COMPUT SCI] One of a series of pulses indicating the beginning or end of a signal representing a single binary digit in a computer word. Also known as position pulse; P pulse. { ˈkäm·yəˌtād·ər ˌpəls }

commutator switch [ELEC] A switch that performs a set of switching operations in repeated sequential order, such as is required for telemetering many quantities. Also known as sampling switch; scanning switch. { ˈkäm·yəˌtād·ər ˌswich }

compact disk [COMMUN] A nonmagnetic (optical) disk, usually 4¾ inches (12 centimeters) in diameter, used for audio or video recording or for data storage; information is recorded using a laser beam to burn microscopic pits into the surface and is accessed by means of a lower-power laser to sense the presence or absence of pits. Abbreviated CD. { ˈkämˌpak ˈdisk }

compact-disk erasable *See* CD-RW. { ¦kämˌpak ˌdisk iˈrās·ə·bəl }

compact-disk read-only memory [COMPUT SCI] A compact disk used for the permanent storage of up to approximately 500 megabytes of data. Abbreviated CD-ROM. { ˈkämˌpakt ¦disk ¦rēd ¦ōn·lē ˈmem·rē }

compact-disk recordable *See* CD-R. { ¦kämˌpak ˌdisk riˈkȯrd·ə·bəl }

compact-disk rewritable *See* CD-RW. { ¦kämˌpak ˌdisk ˌrēˈrīd·ə·bəl }

compact-disk write-once *See* CD-R. { ¦kämˌpak ˌdisk ¦rīt ˈwəns }

compacting garbage collection [COMPUT SCI] The physical rearrangement of data cells so that those cells whose contents are no longer useful (garbage) are compressed into a contiguous array. { ˌkămˈpak·tiŋ ˈgär·bij kəˈlek·shən }

compaction [COMPUT SCI] A technique for reducing the space required for data storage without losing any information content. Also known as squishing. { kəmˈpak·shən }

companded single-sideband system [COMMUN] A long-haul microwave telecommunications system that employs repeaters and single-sideband amplitude modulation and achieves subjective noise improvement by companding to reduce circuit noise between syllables and during pauses in speech. Abbreviated CSSB system. { kəmˈpan·dəd ¦siŋ·gəl ¦sīdˌband ˌsis·təm }

companding [ELECTR] A process in which compression is followed by expansion; often used for noise reduction in equipment, in which case compression is applied before noise exposure and expansion after exposure. { kəmˈpand·iŋ }

compandor [ELECTR] A system for improving the signal-to-noise ratio by compressing the volume range of the signal at a transmitter or recorder by means of a compressor and restoring the normal range at the receiving or reproducing apparatus with an expander. { kəmˈpand·ər }

comparator [COMPUT SCI] A device that compares two transcriptions of the same information to verify the accuracy of transcription, storage, arithmetical operation, or some other process in a computer, and delivers an output signal of some form to indicate whether or not the two sources are equal or in agreement. [CONT SYS] A device which detects the value of the quantity to be controlled by a feedback control system and compares it continuously with the desired value of that quantity. { kəmˈpar·əd·ər }

comparator probe [COMPUT SCI] A component of a hardware monitor that is used to sense the number of bits that appear in parallel, as in an address register. { kəmˈpar·əd·ər ˌprōb }

comparison [COMPUT SCI] A computer operation in which two numbers are compared as to identity, relative magnitude, or sign. { kəm ˈpar·ə·sən }

comparison indicators [COMPUT SCI] Registers, one of which is activated during the comparison of two quantities to indicate whether the first quantity is lower than, equal to, or greater than the second quantity. { kəmˈpar·ə·sən ˌin·dəˌkād·ərz }

compatibility [COMPUT SCI] The ability of one device to accept data handled by another device without conversion of the data or modification of the code. [SYS ENG] The ability of a new system to serve users of an old system. { kəm ˌpad·əˈbil·ə·dē }

compatibility mode [COMPUT SCI] A feature of a computer or operating system that enables it to run programs written for another system. { kəm ˌpad·əˈbil·əd·ē ˌmōd }

compatible color television system [COMMUN] A color television system that permits substantially normal monochrome reception of the transmitted color picture signal on a typical unaltered monochrome receiver. { kəm¦pad·ə·bəl ˈkəl·ər ˈtel·əˌvizh·ən ˌsis·təm }

compatible monolithic integrated circuit [ELECTR] Device in which passive components are deposited by thin-film techniques on top of a basic silicon-substrate circuit containing the active components and some passive parts. { kəmˈpad·ə·bəl ˌmän·əˈlith·ik ˈin·təˌgrād·əd ˈsər·kət }

compatible single-sideband system [COMMUN] A single-sideband system that can be received by an ordinary amplitude-modulation radio receiver without distortion. { kəmˈpad·ə·bəl ˌsiŋ·gəlˈsīd ˌband ˌsis·təm }

compensation [CONT SYS] Introduction of additional equipment into a control system in order to reshape its root locus so as to improve system performance. Also known as stabilization. [ELECTR] The modification of the amplitude-frequency response of an amplifier to broaden the bandwidth or to make the response more nearly uniform over the existing bandwidth. Also known as frequency compensation. { ˌkäm·pənˈsā·shən }

compensation signals [ENG] In telemetry, signals recorded on a tape, along with the data and in the same track as the data, used during the playback of data to correct electrically the effects of tape-speed errors. { ˌkäm·pən'sā·shən ˌsig·nəlz }

compensator [CONT SYS] A device introduced into a feedback control system to improve performance and achieve stability. Also known as filter. [ELECTR] A component that offsets an error or other undesired effect. { 'käm·pən ˌsād·ər }

compile [COMPUT SCI] To prepare a machine-language program automatically from a program written in a higher programming language, usually generating more than one machine instruction for each symbolic statement. { kəm'pīl }

compile-and-go [COMPUT SCI] A continuous sequence of steps that combine compilation, loading, and execution of a computer program. { kəm'pīl ən 'gō }

compiler [COMPUT SCI] A program to translate a higher programming language into machine language. Also known as compiling routine. { kəm'pīl·ər }

compiler-level language [COMPUT SCI] A higher-level language normally supplied by the computer manufacturer. { kəm'pīl·ər ˌlev·əl ˌlaŋ·gwij }

compiler listing [COMPUT SCI] A report that is produced by a compiler and contains an annotated printout of the source program together with other useful information. { kəm'pī·lər ˌlist·iŋ }

compiler system [COMPUT SCI] The set consisting of a higher-level language, such as FORTRAN, and its compiler which translates the program written in that language into machine-readable instructions. { kəm'pīl·ər ˌsis·təm }

compiler toggle [COMPUT SCI] A piece of information transmitted to a compiler to activate some special feature or otherwise control the way in which the compiler operates. { kəm'pī·lər ˌtäg·əl }

compiling routine *See* compiler. { kəm'pil·iŋ rü ˌtēn }

complementary [ELECTR] Having *pnp* and *npn* or *p*- and *n*- channel semiconductor elements on or within the same integrated-circuit substrate or working together in the same functional amplifier state. { ˌkäm·plə'men·trē }

complementary constant-current logic [ELECTR] A type of large-scale integration used in digital integrated circuits and characterized by high density and very fast switching times. Abbreviated CCCL; C^3L. { ˌkäm·plə¦men·trē ¦kän·stənt ¦kə·rənt 'läj·ik }

complementary logic switch [ELECTR] A complementary transistor pair which has a common input and interconnections such that one transistor is on when the other is off, and vice versa. { ˌkäm·plə'men·trē 'läj·ik ˌswich }

complementary metal oxide semiconductor device *See* CMOS device. { ˌkäm·plə¦men·trē ¦med·əl ¦äkˌsīd 'sem·i·kənˌdək·tər di'vīs }

complementary symmetry [ELECTR] A circuit using both *pnp* and *npn* transistors in a symmetrical arrangement that permits push-pull operation without an input transformer or other form of phase inverter. { ˌkäm·plə'men·trē 'sim·ə·trē }

complementary transistors [ELECTR] Two transistors of opposite conductivity (*pnp* and *npn*) in the same functional unit. { ˌkäm·plə'men·trē tran'zis·tərs }

complement number system [COMPUT SCI] System of number handling in which the complement of the actual number is operated upon; used in some computers to facilitate arithmetic operations. { 'käm·plə·mənt 'nəm·bər ˌsis·təm }

complete carry [COMPUT SCI] In parallel addition, an arrangement in which the carries that result from the addition of carry digits are allowed to propagate from place to place. { kəm'plēt 'kar·ē }

complete operation [COMPUT SCI] An operation which includes obtaining all operands from storage, performing the operation, returning resulting operands to storage, and obtaining the next instruction. { kəm'plēt äp·ə'rā·shən }

complete routine [COMPUT SCI] A routine, generally supplied by a computer manufacturer, which does not have to be modified by the user before being applied. { kəm'plēt rü'tēn }

complex data type [COMPUT SCI] A scalar data type which contains two real fields representing the real and imaginary components of a complex number. { 'kämˌpleks 'dad·ə ˌtīp }

complex declaration statement [COMPUT SCI] A nonexecutable statement in FORTRAN used to specify that the type of identifier appearing in the program is of the form $a + bi$, where i is the square root of -1. { 'kämˌpleks ˌdek·lə'rā·shən ˌstāt·mənt }

complex impedance *See* electrical impedance. { 'kämˌpleks im'pēd·əns }

complex instruction set computer [COMPUT SCI] A computer in which relatively high-level or complex hardware incorporating microcode is used to implement a relatively large number of instructions. Abbreviated CISC. { ¦kämˌpleks in'strək·shən ˌset kəmˌpyüd·ər }

complexity [COMPUT SCI] The number of elementary operations used by a program or algorithm to accomplish a given task. { kəm 'plek·səd·ē }

complex reflector [ENG] A structure or group of structures having many radar-reflecting surfaces facing in different directions. { 'kämˌpleks ri'flek·tər }

complex target [ENG] A radar target composed of a number of reflecting surfaces that, in the aggregate, are smaller in all dimensions than the resolution capabilities of the radar. { 'käm ˌpleks 'tär·gət }

component-failure-impact analysis [SYS ENG] A study that attempts to predict the consequences of failures of the major components of a system. Abbreviated CFIA. { kəm'pō·nənt ¦fāl·yər 'imˌpakt əˌnal·ə·səs }

component name *See* metavariable. { kəm'pō·nənt ,nām }

composite color signal [COMMUN] The analog color television picture signal plus all blanking and synchronizing signals. Also known as composite picture signal. { kəm'päz·ət 'kəl·ər ,sig·nəl }

composite color sync [COMMUN] The signal comprising all the synchronization signals necessary for proper operation of an analog color television receiver. { kəm'päz·ət 'kəl·ər ,siŋk }

composite picture signal *See* composite color signal. { kəm'päz·ət 'pik·chər ,sig·nəl }

composite video signal [COMMUN] The video-only portion of the analog color television signal used in the United States, in which red, green, and blue signals are encoded. { kəm'päz·ət 'vid·ē·ō ,sig·nəl }

composite wave filter [ELECTR] A combination of two or more low-pass, high-pass, band-pass, or band-elimination filters. { kəm'päz·ət 'wāv ,fil·tər }

compound document [COMPUT SCI] A document that contains two or more different data structures, such as text, graphics, and sound. { ,käm,paund 'däk·yə·mənt }

compound modulation *See* multiple modulation. { 'käm,paund ,mäj·ə'lā·shən }

compound statement [COMPUT SCI] A single program instruction that contains two or more instructions which could stand alone. { 'käm ,paund 'stāt·mənt }

compressed file *See* packed file. { kəm,prest 'fīl }

compression [COMPUT SCI] *See* data compression. [ELECTR] **1.** Reduction of the effective gain of a device at one level of signal with respect to the gain at a lower level of signal, so that weak signal components will not be lost in background and strong signals will not overload the system. **2.** *See* compression ratio. { kəm'presh·ən }

compression ratio [ELECTR] The ratio of the gain of a device at a low power level to the gain at some higher level, usually expressed in decibels. Also known as compression. { kəm'presh·ən ,rā·shō }

compressive intercept receiver [ELECTR] An electromagnetic surveillance receiver that instantaneously analyzes and sorts all signals within a broad radio-frequency spectrum by using pulse compression techniques which perform a complete analysis up to 10,000 times faster than a superheterodyne receiver or spectrum analyzer. { kəm'pres·iv 'in·tər,sept ri'sē·vər }

compressor [COMPUT SCI] A routine or program that reduces the number of binary digits needed to represent data or information. [ELECTR] The part of a compandor that is used to compress the intensity range of signals at the transmitting or recording end of a circuit. { kəm'pres·ər }

compromising emanations [COMMUN] Unintentional data-related or intelligence-bearing signals which, if intercepted and analyzed by any technique, could disclose the classified information transmitted, received, handled, or otherwise processed by equipments. { 'käm·prə,miz·iŋ ,em·ə'nā·shənz }

computational fluid dynamics [FL MECH] A field of study concerned with the use of high-speed digital computers to numerically solve the complete nonlinear partial differential equations governing viscous fluid flows. { ,käm·pyə'tā·shən·əl 'flü·əd dī'nam·iks }

computational numerical control *See* computer numerical control. { ,käm·pyə'tā·shən·əl nü'mer·ə·kəl kən'trōl }

compute-bound program *See* CPU-bound program. { kəm'pyüt ¦baund 'prō·grəm }

computed go to [COMPUT SCI] A control procedure in FORTRAN which allows the transfer of control to the *i*th label of a set of *n* labels used as statement numbers in the program. { kəm'pyüd·əd 'gō,tü }

compute mode [COMPUT SCI] The operation of an analog computer in which input signals are used by the computing units to calculate a solution, in contrast to hold mode and reset mode. { kəm'pyüt ,mōd }

computer [COMPUT SCI] A device that receives, processes, and presents data; the two types are analog and digital. Also known as computing machine. { kəm'pyüd·ər }

computer-aided design [CONT SYS] The use of computers in converting the initial idea for a product into a detailed engineering design. Computer models and graphics replace the sketches and engineering drawings traditionally used to visualize products and communicate design information. Abbreviated CAD. { kəm'pyüd·ər ,ād·əd də'zīn }

computer-aided design and drafting [COMPUT SCI] The carrying out of computer-aided design with a system that has additional features for the drafting function, such as dimensioning and text entry. Abbreviated CADD. { kəm'pyüd·ər ,ād·əd di'zīn ən 'draft·iŋ }

computer-aided engineering [ENG] The use of computer-based tools to assist in solution of engineering problems. { kəm'pyüd·ər ,ād·əd ,en·jə'nir·iŋ }

computer-aided instruction *See* computer-assisted instruction. { kəm'pyüd·ər ,ād·əd in 'strək·shən }

computer-aided management of instruction *See* computer-managed instruction. { kəm 'pyüd·ər ,ād·əd 'man·ij·mənt əv in'strək·shən }

computer-aided manufacturing [CONT SYS] The use of computers in converting engineering designs into finished products. Computers assist managers, manufacturing engineers, and production workers by automating many production tasks, such as developing process plans, ordering and tracking materials, and monitoring production schedules, as well as controlling the machines, industrial robots, test equipment, and systems that move and store materials in the factory. Abbreviated CAM. { kəm'pyüd·ər ,ād·əd ,man·ə'fak·chə·riŋ }

computer-aided software engineering [COMPUT SCI] The use of software packages to assist in all phases of the development of an information system, including analysis, design, and programming. Abbreviated CASE. { kəm'pyüd·ər ˌād·əd ˌsȯftˌwer en·jə'nir·iŋ }

computer algebra system *See* symbolic system. { kəm¦pyüd·ər 'al·jə·brə ˌsis·təm }

computer analyst [COMPUT SCI] A person who defines a problem, determines exactly what is required in the solution, and defines the outlines of the machine solution; generally, an expert in automatic data processing applications. { kəm'pyüd·ər 'an·əˌlist }

computer animation [COMPUT SCI] The use of a computer to present, either continuously or in rapid succession, pictures on a cathode-ray tube or other device, graphically representing a time developing system at successive times. { kəm'pyüd·ər an·ə'mā·shən }

computer architecture [COMPUT SCI] The art and science of assembling logical elements to form a computing device. { kəm'pyüd·ər 'är·kə ˌtek·chər }

computer-assisted instruction [COMPUT SCI] The use of computers to present drills, practice exercises, and tutorial sequences to the student, and sometimes to engage the student in a dialog about the substance of the instruction. Abbreviated CAI. Also known as computer-aided instruction; computer-assisted learning. { kəm'pyüd·ər ə'sis·təd in'strək·shən }

computer-assisted learning *See* computer-assisted instruction. { kəm'pyüd·ər ə'sis·təd 'lərn·iŋ }

computer-assisted retrieval [COMPUT SCI] The use of a computer to locate documents or records stored outside of the computer, on paper or microfilm. Abbreviated CAR. { kəm'pyüd·ər ə'sis·təd ri'trē·vəl }

computer center *See* electronic data-processing center. { kəm'pyüd·ər ˌsen·tər }

computer code [COMPUT SCI] The code representing the operations built into the hardware of a particular computer. { kəm'pyüd·ər ˌkōd }

computer conferencing *See* computer networking. { kəm'pyüd·ər 'kän·frəns·iŋ }

computer control [CONT SYS] Process control in which the process variables are fed into a computer and the output of the computer is used to control the process. { kəm'pyüd·ər kən'trōl }

computer control counter [COMPUT SCI] Counter which stores the next required address; any counter which furnishes information to the control unit. { kəm'pyüd·ər kən'trōl ˌkau̇nt·ər }

computer-controlled system [CONT SYS] A feedback control system in which a computer operates on both the input signal and the feedback signal to effect control. { kəm'pyüd·ər kən'trōld ˌsis·təm }

computer control register *See* program register. { kəm'pyüd·ər kən'trōl rej·ə·stər }

computer efficiency [COMPUT SCI] **1.** The ratio of actual operating time to scheduled operating time of a computer. **2.** In time-sharing, the ratio of user time to the sum of user time plus system time. { kəm'pyüd·ər i'fish·ən·sē }

computer graphics [COMPUT SCI] The process of pictorial communication between humans and computers, in which the computer input and output have the form of charts, drawings, or appropriate pictorial representation; such devices as cathode-ray tubes, mechanical plotting boards, curve tracers, coordinate digitizers, and light pens are employed. { kəm'pyüd·ər 'graf·iks }

computer graphics interface [COMPUT SCI] A standard format for writing graphics drivers. Abbreviated CGI. { kəm¦pyüd·ər ¦graf·iks 'in·tər ˌfās }

computer graphics metafile [COMPUT SCI] A standard device-independent graphics format that is used to transfer graphics images between computer programs and storage devices. Abbreviated CGM. { kəm¦pyüd·ər ¦graf·iks 'med·əˌfīl }

computer input from microfilm [COMPUT SCI] The technique of reading images on microfilm and transforming them into a form which is understandable to a computer. Abbreviated CIM. { kəm'pyüd·ər 'inˌpu̇t frəm 'mī·krəˌfilm }

computer-integrated manufacturing [IND ENG] A computer-automated system in which individual engineering, production, marketing, and support functions of a manufacturing enterprise are organized; functional areas such as design, analysis, planning, purchasing, cost accounting, inventory control, and distribution are linked through the computer with factory floor functions such as materials handling and management, providing direct control and monitoring of all process operations. Abbreviated CIM. { kəm'pyüd·ər ¦int·əˌgrād·əd ˌman·ə'fak·chər·iŋ }

computerized branch exchange [COMMUN] A computer-controlled telephone switching system that supports such services as conference calling, least-cost routing, direct inward dialing, and automatic reringing of a busy line. Abbreviated CBX. { kəm'pyüd·əˌrīzd 'branch iks'chānj }

computer-limited [COMPUT SCI] Pertaining to a situation in which the time required for computation exceeds the time required to read inputs and write outputs. { kəm'pyüd·ər ˌlim·əd·əd }

computer literacy [COMPUT SCI] Knowledge and understanding of computers and computer systems and how to apply them to the solution of problems. { kəm'pyüd·ər 'lit·rə·sē }

computer-managed instruction [COMPUT SCI] The use of computer assistance in testing, diagnosing, prescribing, grading, and record keeping. Abbreviated CMI. Also known as computer-aided management of instruction. { kəm'pyüd·ər ¦man·ijd in'strək·shən }

computer memory *See* memory. { kəm'pyüd·ər 'mem·rē }

computer modeling [COMPUT SCI] The use of a computer to develop a mathematical model of a complex system or process and to provide conditions for testing it. { kəm'pyüd·ər 'mäd·əl·iŋ }

computer network [COMPUT SCI] A system of two or more computers that are interconnected by communication channels. Also known as network. { kəm'pyüd·ər 'netˌwərk }

computer networking [COMMUN] The use of a network of computers and computer terminals by individuals at various locations to interact with each other by entering data into the computer system. Also known as computer conferencing. { kəm'pyüd·ər 'net,wərk·iŋ }

computer numerical control [CONT SYS] A control system in which numerical values corresponding to desired tool or control positions are generated by a computer. Abbreviated CNC. Also known as computational numerical control; soft-wired numerical control; stored-program numerical control. { kəm'pyüd·ər nü'mer·i·kəl kən'trōl }

computer operation [COMPUT SCI] The electronic action that is required in a computer to give a desired computation. { kəm'pyüd·ər äp·ə'rā·shən }

computer-oriented language [COMPUT SCI] A low-level programming language developed for use on a particular computer or line of computers produced by a specific manufacturer. Also known as machine-oriented language. { kəm'pyüd·ər ¦ȯr·ē,ent·əd 'laŋ·gwij }

computer output on microfilm [COMPUT SCI] The generation of microfilm which displays information developed by a computer. Abbreviated COM. { kəm'pyüd·ər 'au̇t,pu̇t ȯn 'mī·krə,film }

computer performance evaluation [COMPUT SCI] The measurement and evaluation of the performance of a computer system, aimed at ensuring that a minimum amount of effort, expense, and waste is incurred in the production of data-processing services, and encompassing such tools as canned programs, source program optimizers, software monitors, hardware monitors, simulation, and bench-mark problems. Abbreviated CPE. { kəm'pyüd·ər pər 'fȯr·məns i,val·yə'wā·shən }

computer programming *See* programming. { kəm'pyüd·ər 'prō,gram·iŋ }

computer science [COMPUT SCI] The study of computers and computing, including computer hardware, software, programming, networking, database systems, information technology, interactive systems, and security. { kəm'pyüd·ər 'sī·əns }

computer security [COMPUT SCI] Measures taken to protect computers and their contents from unauthorized use. { kəm'pyüd·ər sə 'kyu̇r·əd·ē }

computer storage device *See* storage device. { kəm'pyüd·ər 'stȯr·ij di'vīs }

computer system [COMPUT SCI] **1.** A set of related but unconnected components (hardware) of a computer or data-processing system. **2.** A set of hardware parts that are related and connected, and thus form a computer. { kəm'pyüd·ər ,sis·təm }

computer systems architecture [COMPUT SCI] The discipline that defines the conceptual structure and functional behavior of a computer system, determining the overall organization, the attributes of the component parts, and how these parts are combined. { kəm'pyüd·ər ¦sis·təmz 'ar·kə,tek·chər }

computer theory [COMPUT SCI] A discipline covering the study of circuitry, logic, microprogramming, compilers, programming languages, file structures, and system architectures. { kəm'pyüd·ər ,thē·ə·rē }

computer utility [COMPUT SCI] A computer that provides service on a time-sharing basis, generally over telephone lines, to subscribers who have appropriate terminals. { kəm'pyüd·ər yü'til·əd·ē }

computer vision [COMPUT SCI] The use of digital computer techniques to extract, characterize, and interpret information in visual images of a three-dimensional world. Also known as machine vision. { kəm'pyüd·ər 'vizh·ən }

computer word *See* word. { kəm'pyüd·ər ,wərd }

computing machine *See* computer. { kəm 'pyüd·iŋ mə'shēn }

computing power [COMPUT SCI] The number of operations that a computer can carry out in 1 second. { kəm'pyüd·iŋ ,pau̇·ər }

computing unit [COMPUT SCI] The section of a computer that carries out arithmetic, logical, and decision-making operations. { kəm'pyüd·iŋ ,yü·nət }

concatenate [COMPUT SCI] To unite in a sequence, link together, or link to a chain. { kən 'kat·ən,āt }

concatenation [COMPUT SCI] **1.** An operation in which a number of conceptually related components are linked together to form a larger, organizationally similar entity. **2.** In string processing, the synthesis of longer character strings from shorter ones. { kən,kat·ən'ā·shən }

concentrator [ELECTR] Buffer switch (analog or digital) which reduces the number of trunks required. [ENG] **1.** An apparatus used to concentrate materials. **2.** A plant where materials are concentrated. { 'kän·sən,trād·ər }

concentric cable *See* coaxial cable. { kən'sen·trik 'kā·bəl }

concentric line *See* coaxial cable. { kən'sen·trik 'līn }

concentric transmission line *See* coaxial cable. { kən'sen·trik tranz'mish·ən ,līn }

conceptual modeling [COMPUT SCI] Writing a program by means of which a given result will be obtained, although the result is incapable of proof. Also known as heuristic programming. { kən'sep·chə·wəl 'mäd·liŋ }

conceptual schema [COMPUT SCI] The logical structure of an entire data base. { kən'sep·chə·wəl 'skē·mə }

concurrency [COMPUT SCI] Referring to two or more tasks of a computer system which are in progress simultaneously. { kən'kər·ən·sē }

concurrent input/output [COMPUT SCI] The simultaneous reading from and writing on different media by a computer. { kən'kər·ənt ¦in,pu̇t ¦au̇t ,pu̇t }

concurrent operations control [COMPUT SCI] The supervisory capability required by a computer to handle more than one program at a time. { kən'kər·ənt äp·ə'rā·shənz kən'trōl }

concurrent processing [COMPUT SCI] The conceptually simultaneous execution of more than one sequential program on a computer or network of computers. { kən'kər·ənt 'präs,əs·iŋ }

concurrent real-time processing [COMPUT SCI] The capability of a computer to process simultaneously several programs, each of which requires responses within a time span related to its particular time frame. { kən'kər·ənt 'rēl ,tīm ,präs,əs·iŋ }

condenser See capacitor. { kən'den·sər }

condenser antenna See capacitor antenna. { kən'den·sər an'ten·ə }

conditional [COMPUT SCI] Subject to the result of a comparison made during computation in a computer, or subject to human intervention. { kən'dish·ən·əl }

conditional assembly [COMPUT SCI] A feature of some assemblers which suppresses certain sections of code if stated program conditions are not met at assembly time. { kən'dish·ən·əl ə'sem·blē }

conditional branch See conditional jump. { kən'dish·ən·əl 'branch }

conditional breakpoint [COMPUT SCI] A conditional jump that, if a specified switch is set, will cause a computer to stop; the routine may then be continued as coded or a jump may be forced. { kən'dish·ən·əl 'brāk,pȯint }

conditional expression [COMPUT SCI] A COBOL language expression which is either true or false, depending upon the status of the variables within the expression. { kən'dish·ən·əl ik'spresh·ən }

conditional jump [COMPUT SCI] A computer instruction that will cause the proper one of two or more addresses to be used in obtaining the next instruction, depending on some property of a numerical expression that may be the result of some previous instruction. Also known as conditional branch; conditional transfer; decision instruction; discrimination; IF statement. { kən'dish·ən·əl 'jəmp }

conditional replenishment [COMMUN] A form of differential pulse-code modulation in which the only information transmitted consists of addresses specifying the locations of picture samples in the moving area, and information by which the intensities of moving area picture samples can be reconstructed at the receiver. { kən'dish·ən·əl ri'plen·ish·mənt }

conditional statement [COMPUT SCI] A statement in a computer program that is executed only when a certain condition is satisfied. { kən'dish·ən·əl 'stāt·mənt }

conditional transfer See conditional jump. { kən'dish·ən·əl 'tranz·fər }

condition code [COMPUT SCI] Portion of a program status word indicating the outcome of the most recently executed arithmetic or boolean operation. { kən'dish·ən ,kōd }

conditioned line [COMPUT SCI] A communications channel, usually a telephone line, that has been adapted for data transmission. { kən'dish·ənd 'līn }

conditioned stop instruction [COMPUT SCI] A computer instruction which causes the execution of a program to stop if some given condition exists, such as the specific setting of a switch on a computer console. { kən'dish·ənd 'stäp in'strek·shən }

condition entries [COMPUT SCI] The upper-right-hand portion of a decision table, indicating, for each of the conditions, whether the condition satisfies various criteria listed in the condition stub, or the values of various parameters listed in the condition stub. { kən'dish·ən ,en,trēz }

conditioning [ELECTR] Equipment modifications or adjustments necessary to match transmission levels and impedances or to provide equalization between facilities. { kən 'dish·ən·iŋ }

condition portion [COMPUT SCI] The upper portion of a decision table, comprising the condition stub and condition entires. { kən'dish·ən ,pȯr·shən }

condition stub [COMPUT SCI] The upper-left-hand portion of a decision table, consisting of a single column listing various criteria or parameters which are used to specify the conditions. { kən'dish·ən ,stəb }

conducted interference [COMMUN] Interfering signals arriving by direct coupling such as on communications and power lines. { kən'dək·təd ,in·tər'fir·əns }

conductive interference [ELECTR] Interference to electronic equipment that orginates in power lines supplying the equipment, and is conducted to the equipment and coupled through the power supply transformer. { kən'dək·tiv ,in·tər'fir·əns }

cone antenna See conical antenna. { 'kōn an 'ten·ə }

conference communications [COMMUN] Communications facilities whereby direct speech conversation may be conducted between three or more locations simultaneously. { 'kän·frəns kə,myü·nə'kā·shənz }

configuration [COMPUT SCI] For a computer system, the relationship of hardware elements to each other, and the manner in which they are electronically connected. [SYS ENG] A group of machines interconnected and programmed to operate as a system. { kən,fig·yə'rā·shən }

confirmation message [COMPUT SCI] A message that appears on a computer screen asking the user to confirm an action that could have destructive effects, such as loss of data. { ,kän·fər'mā·shən ,mes·ij }

conformal array [ELECTR] An array-type antenna in which the radiating elements are mounted on a surface shaped for other purposes, such as aerodynamics, or on a surface more convenient of beneficial than a plane. Circular or cylindrical arrays provide an antenna-pattern consistency particularly valuable in TACAN, IFF, and secondary radar applications. { kən'fȯr·məl ə'rā }

confusion matrix [COMPUT SCI] In pattern recognition, a matrix used to represent errors in assigning classes to observed patterns in which the *ij*th element represents the number of samples from class *i* which were classified as class *j*. { kən'fyü·zhən ˌmā·triks }

congruential generator [COMPUT SCI] A method of generating a sequence of random numbers $x_0, x_1, x_2, \ldots$, in which each member is generated from the previous one by the formula $x_{i+1} \equiv ax_i + b$ modulus m, where a, b, and m are constants. { ¦känˌgrü¦en·chəl 'jen·əˌrād·ər }

conical antenna [ELECTROMAG] A wide-band antenna in which the driven element is conical in shape. Also known as cone antenna. { 'kän·ə·kəl an'ten·ə }

conical-horn antenna [ELECTROMAG] A horn antenna having a circular cross section and straight sides. { 'kän·ə·kəl ˌhȯrn an'ten·ə }

conical monopole antenna [ELECTROMAG] A variation of a biconical antenna in which the lower cone is replaced by a ground plane and the upper cone is usually bent inward at the top. { 'kän·ə·kəl 'män·əˌpōl an'ten·ə }

conical scanning [ELECTR] Scanning in radar in which the direction of maximum radiation generates a cone, the vertex angle of which is of the order of the beam width; may be either rotating or nutating, according to whether the direction of polarization rotates or remains unchanged. Done to effect accurate angle measurement in precision tracking radars. { 'kän·ə·kəl 'skan·iŋ }

conjunctive search [COMPUT SCI] A search to identify items having all of a certain set of characteristics. { kən'jəŋk·tiv 'sərch }

connect function [COMPUT SCI] A signal sent over a data line to a selected peripheral device to connect it with the central processing unit. { kə'nekt ˌfəŋk·shən }

connecting circuit [ELECTR] A functional switching circuit which directly couples other functional circuit units to each other to exchange information as dictated by the momentary needs of the switching system. { kə'nekt·iŋ ˌsər·kət }

connectionless transmission [COMMUN] Data transmission by packets that include addresses of the source and destination, so that a direct connection between these nodes is unnecessary. { kəˌnek·shən·ləs tranz'mish·ən }

connection-oriented transmission [COMMUN] Data transmission in which a physical path between the source and destination must be established and maintained for the duration of the transmission. { kə¦nek·shən ˌȯr·ēˌent·əd tranz'mish·ən }

connector [COMPUT SCI] In database management, a pointer or link between two data structures. [ELECTR] A switch, or relay group system in old electromechanical central offices, which sounds the telephone line being called as a result of digits being dialed; it also caused interrupted ringing voltage to be placed on the called line or returned a busy tone to the calling party if the line were busy. [ENG] **1.** A detachable device for connecting electrical conductors. **2.** A symbol on a flowchart indicating that the flow jumps to a different location on the chart. { kə'nek·tər }

connector block [ELECTR] A device for connecting two cables without using plugs, similar to a barrier strip but larger, in which wires from one cable are attached to lugs of screws on one side, and wires from the other cable are fastened to corresponding points on the opposite side. { kə'nek·tər ˌbläk }

connect time [COMPUT SCI] The time that a user at a terminal is signed on to a computer. { kə'nekt ˌtīm }

conode *See* tie line. { 'kōˌnōd }

consequence finding program [COMPUT SCI] A computer program that attempts to deduce mathematical consequences from a set of axioms and to select those consequences that will be significant. { 'kän·sə·kwəns ¦fīnd·iŋ ˌprō·grəm }

consistency routine [COMPUT SCI] A debugging routine which is used to determine whether the program being checked gives consistent results at specified check points; for example, consistent between runs or with values calculated by other means. { kən'sis·tən·sē rü'tēn }

console [COMPUT SCI] **1.** The section of a computer that is used to control the machine manually, correct errors, manually revise the contents of storage, and provide communication in other ways between the operator or service engineer and the central processing unit. Also known as master console. **2.** A display terminal together with its keyboard. [ENG] **1.** A main control desk for electronic equipment, as at a radar station, radio or television station, or airport control tower. Also known as control desk. **2.** A large cabinet for a radio or television receiver, standing on the floor rather than on a table. **3.** A grouping of controls, indicators, and similar items contained in a specially designed model cabinet for floor mounting; constitutes an operator's permanent working position. { 'känˌsōl }

console display [COMPUT SCI] The visible representation of information, whether in words, numbers, or drawings, on a console screen connected to a computer. { 'känˌsōl di'splā }

console file adapter [COMPUT SCI] A special input/output device which allows the operator to load reloadable control storage from the system console. { 'känˌsōl 'fīl ə'dap·tər }

console switch [COMPUT SCI] A switch on a computer console whose setting can be sensed by a computer, so that an instruction in the program can direct the computer to use this setting to determine which of various alternative courses of action should be followed. { 'känˌsōl ˌswich }

constant area [COMPUT SCI] A part of storage used for constants. { 'kän·stənt ¦er·ē·ə }

constant bit rate [COMMUN] A mode of operation in a digital system where the bit rate is constant from start to finish of the compressed bit stream. { 'kän·stənt 'bit ˌrāt }

constant-current modulation [COMMUN] System of amplitude modulation in which output circuits of the signal amplifier and the carrier-wave generator or amplifier are connected via a common coil to a constant-current source. Also known as Heising modulation. { ¦kän·stənt 'kər·ənt ,mäj·ə'lā·shən }

constant-false-alarm rate [ELECTR] Radar system devices used to prevent receiver saturation and overload so as to present clean video information to the display, and to present a constant noise level to an automatic detector. { ¦kän·stənt ,fȯls ə'lärm ,rāt }

constant-false-alarm-rate detection [ELECTR] Radar detection in which the sensitivity threshold is adjusted to adapt to a changing and uncertain background of clutter or interference. { 'kän·stənt fȯls ə'lärm rāt di,tek·shən }

constant instruction [COMPUT SCI] A nonexecutable instruction. { 'kän·stənt in¦strək·shən }

constant-k filter [ELECTR] A filter in which the product of the series and shunt impedances is a constant that is independent of frequency. { ¦kän·stənt ¦kā 'fil·tər }

constant-luminance transmission [COMMUN] Type of transmission in which the transmission primaries are a luminance primary and two chrominance primaries. { ¦kän·stənt ¦lü·mə·nəns tranz'mish·ən }

constant radio code [COMMUN] Code in which all characters are represented by combinations having a fixed ratio of ones to zeros. { 'kän·stənt 'rād·ē·ō ,kōd }

constraint matrix [COMPUT SCI] The set of equations and inequalities defining the set of admissible solutions in linear programming. { kən 'strānt ,mā·triks }

constraint programming language [COMPUT SCI] A programming language in which constraints (relationships that must hold among a number of variables) are directly usable as programming constructs. { kən¦strānt 'prō,gram·iŋ ,laŋ·gwij }

construction operator [COMPUT SCI] The part of a data structure which is used to construct composite objects from atoms. { kən'strək·shən 'äp·ə,rād·ər }

contact [ENG] A report of a target of interest in a radar's data processing; a detection. Also known as plot. { 'kän,takt }

contact head [COMPUT SCI] A read/write head that remains in contact with the recording surface of a hard disk, rather than hovering above it. { 'kän,takt ,hed }

contact-mask read-only memory *See* last-mask read-only memory. { 'kän,takt ,mask 'rēd ,ōn·lē 'mem·rē }

contact piston [ELECTROMAG] A waveguide piston that makes contact with the walls of the waveguide. Also known as contact plunger. { 'kän,takt ,pis·tən }

contact plunger *See* contact piston. { 'kän,takt ,plən·jər }

contamination [COMPUT SCI] Placement of data at incorrect locations in storage, where it generally overlays valid information or a program code and produces bizarre results. { kən,tam·ə'nā·shən }

content analysis [COMPUT SCI] A method of automatically assigning words that identify the content of information items or search requests in an information retrieval system. { 'kän,tent ə'nal·ə·səs }

content indicator [COMPUT SCI] Display unit that indicates the content in a computer, and the program or mode being used. { 'kän,tent ,in·də ,kād·ər }

contention [COMMUN] A method of operating a multiterminal communication channel in which any station may transmit if the channel is free; if the channel is in use, the queue of contention requests may be maintained in predetermined sequence. [COMPUT SCI] **1.** The condition arising when two or more units attempt to transmit over a time-division-multiplex channel at the same time. **2.** Competition for the same computer resources by two or more devices or programs, such as an attempt by several programs to use the same disk drive simultaneously, or by several users in a multiaccess system to use the system's resources. { kən'ten·chən }

contention resolver [COMPUT SCI] A device that enables a central processing unit, memory, or channel whose attention is being requested over several pathways to give its attention to one pathway and ignore all others. { kən'ten·chən ri'zäl·vər }

contents [COMPUT SCI] The information stored at any address or in any register of a computer. { 'kän,tens }

context-driven line editor [COMPUT SCI] A line editor in which the user need not know or keep track of line numbers but can call up text by line content; the computer will then search for the indicated pattern. { 'kän,tekst ,driv·ən 'līn ,ed·əd·ər }

context-free grammar [COMPUT SCI] A grammar in which any occurrence of a metavariable may be replaced by one of its alternatives. { 'kän,tekst ,frē 'gram·ər }

context-sensitive grammar [COMPUT SCI] A grammar in which the rules are applicable only when a metavariable occurs in a specified context. { 'kän,tekst ,sen·səd·iv 'gram·ər }

context-sensitive help [COMPUT SCI] A help screen that provides specific information about the current status or mode of a computer program or instructions for dealing with a particular error condition that has just occurred. { 'kän,tekst ,sen·səd·iv 'help }

context switch [COMPUT SCI] The action of a central processing unit that suspends work on one process to work on another. { 'kän,text ,swich }

context switching *See* task switching. { 'kän ,text ,swich·iŋ }

contextual analysis [COMPUT SCI] A phase of natural language processing, following semantic analysis, whose purpose is to elaborate the semantic representation of what has been made explicit in the utterance with what is implicit from context. { kən'teks·chə·wəl ə'nal·ə·səs }

contextual search [COMPUT SCI] A search for documents or records based upon the data they contain, rather than their file names or key fields. { kən'teks·chə·wəl 'sərch }

contiguous data [COMPUT SCI] Data that are stored in a collection of adjacent locations in a computer memory device. { kən'tig·yə·wəs 'dad·ə }

continental code [COMMUN] The code commonly used for manual telegraph communication, consisting of short (dot) and long (dash) symbols, but not the various-length spaces used in the original Morse code. Also known as international Morse code. { ¦känt·ən¦ent·əl 'kōd }

contingency interrupt [COMPUT SCI] A processing interruption due to an operator's action or due to an abnormal result from the system or from a program. { kən'tin·jən·sē 'in·tə,rəpt }

continue statement [COMPUT SCI] A nonexecutable statement in FORTRAN used principally as a target for transfers, particularly as the last statement in the range of a do statement. { kən'tin·yü ,stāt·mənt }

continuous carrier [COMMUN] A carrier signal that is transmitted at all times during maintenance of a communications link, whether or not data are being transmitted. { kən¦tin·yə·wəs 'kar·ē·ər }

continuous control [CONT SYS] Automatic control in which the controlled quantity is measured continuously and corrections are a continuous function of the deviation. { kən¦tin·yə·wəs kən'trōl }

continuous film scanner [ELECTR] A television film scanner in which the motion picture film moves continuously while being scanned by a flying-spot device. { kən¦tin·yə·wəs 'film ,skan·ər }

continuous forms [COMPUT SCI] **1.** In character recognition, any batch of source information that exists in reel form, such as tally rolls or cash-register receipts. **2.** Preprinted forms that repeat on each page, with the bottom of one page joined to the top of the next by a perforated attachment, so that they can be fed through a printer. { kən¦tin·yə·wəs 'fórmz }

continuous stationery [COMPUT SCI] A continuous ribbon of paper consisting of several hundred or more sheets separated by perforations and folded to form a pack, used to feed a computer printer and generally having sprocket holes along the margin for this purpose. { kən¦tin·yə·wəs 'stā·shə,ner·ē }

continuous stationery reader [COMPUT SCI] A type of character reader which processes only continuous forms of predefined dimensions. { kən¦tin·yə·wəs 'stā·shə,ner·ē 'rēd·ər }

continuous system [CONT SYS] A system whose inputs and outputs are capable of changing at any instant of time. Also known as continuous-time signal system. { kən¦tin·yə·wəs 'sis·təm }

continuous-time signal system *See* continuous system. { kən¦tin·yə·wəs ¦tīm 'sig·nəl ,sis·təm }

continuous-tone squelch [ELECTR] Squelch in which a continuous subaudible tone, generally below 200 hertz, is transmitted by frequency-modulation equipment along with a desired voice signal. { kən¦tin·yə·wəs ¦tōn 'skwelch }

continuous variable [COMPUT SCI] A variable that can take on any of a range of values. { kən ¦tin·yə·wəs 'ver·ē·ə·bəl }

continuous wave [ELECTROMAG] A radio or radar wave whose successive sinusoidal oscillations are identical under steady-state conditions. Abbreviated CW. Also known as type A wave. { kən¦tin·yə·wəs 'wāv }

continuous-wave Doppler radar *See* continuous-wave radar. { kən¦tin·yə·wəs ¦wāv 'däp·lər ,rā,där }

continuous-wave jammer [ELECTR] An electronic jammer that emits a single frequency continuously, giving the appearance of a picket or rail fence on an elementary radar display. Also known as rail-fence jammer. { kən¦tin·yə·wəs ¦wāv 'jam·ər }

continuous-wave modulation [COMMUN] Modulation of a continuous wave by modification of its amplitude, frequency, or phase, in contrast to pulse modulation. { kən¦tin·yə·wəs ¦wāv ,mäj·ə'lā·shən }

continuous-wave radar [ENG] A radar system in which a transmitter sends out a continuous flow of radio energy; the target reradiates a small fraction of this energy to a separate receiving antenna. Also known as continuous-wave Doppler radar. { kən¦tin·yə·wəs ¦wāv 'rā ,där }

contour analysis [COMPUT SCI] In optical character recognition, a reading technique that employs a roving spot of light which searches out the character's outline by bouncing around its outer edges. { 'kän,tu̇r ə'nal·ə·səs }

contouring control [COMPUT SCI] The guidance by a computer of a machine tool along a programmed path by interpolating many intermediate points between selected points. { 'kän ,tu̇r·iŋ kən'trōl }

contour model [COMPUT SCI] A model for describing the run-time execution of programs written in block-structured languages, consisting of a program component, the data component, and the control component. { 'kän,tu̇r ,mäd·əl }

contrast [COMMUN] The degree of difference in tone between the lightest and darkest areas in a video or facsimile picture. [COMPUT SCI] In optical character recognition, the difference in color, reflectance, or shading between two areas of a surface, for example, a character and its background. { 'kän,trast }

contrast control [ELECTR] A manual control that adjusts the range of brightness between highlights and shadows on the reproduced image of a display device. { 'kän,trast kən'trōl }

contrast ratio [ELECTR] The ratio of the maximum to the minimum luminance values in a video image. { 'kän,trast ,rā·shō }

control [COMPUT SCI] **1.** The section of a digital computer that carries out instructions in proper sequence, interprets each coded instruction, and applies the proper signals to the arithmetic unit and other parts in accordance with this interpretation. **2.** A mathematical check used with some computer operations. [CONT SYS] A means or device to direct and regulate a process or sequence of events. { kən'trōl }

control and read-only memory [COMPUT SCI] A read-only memory that also provides storage, sequencing, execution, and translation logic for various microinstructions. Abbreviated CROM. { kən'trōl ən ¦rēd ,ōn·lē 'mem·rē }

control bit [COMPUT SCI] A bit which marks either the beginning or the end of a character transmitted in asynchronous communication. { kən'trōl ,bit }

control block [COMPUT SCI] A storage area containing (in condensed, formalized form) the information required for the control of a task, function, operation, or quantity of information. { kən'trōl ,bläk }

control break [COMPUT SCI] **1.** A key change which takes place in a control data field, especially in the execution of a report program. **2.** A suspension of computer operation that is accomplished by simultaneously depressing the control key and the break key. { kən'trōl ,brāk }

control character [COMPUT SCI] A character whose occurrence in a particular context initiates, modifies, or stops a control operation in a computer or associated equipment. { kən'trōl ,kar·ik·tər }

control circuit [COMPUT SCI] One of the circuits that responds to the instructions in the program for a digital computer. [ELECTR] The circuit that feeds the control winding of a magnetic amplifier. { kən'trōl ,sər·kət }

control code [COMPUT SCI] A special code that is entered by a user to carry out a particular function, such as the moving or deleting of text in a word-processing program. { kən'trōl ,kōd }

control computer [COMPUT SCI] A computer which uses inputs from sensor devices and outputs connected to control mechanisms to control physical processes. { kən'trōl kəm 'pyüd·ər }

control counter [COMPUT SCI] A counter providing data used to control the execution of a computer program. { kən'trōl ,kau̇n·tər }

control data [COMPUT SCI] Data used for identifying, selecting, executing, or modifying another set of data, a routine, a record, or the like. { kən'trōl ,dad·ə }

control desk *See* console. { kən'trōl ,desk }

control diagram *See* flow chart. { kən'trōl ,dī·ə ,gram }

control element [CONT SYS] The portion of a feedback control system that acts on the process or machine being controlled. { kən'trōl ,el·ə·mənt }

control flow graph [COMPUT SCI] A graph describing the logic structure of a software module, in which the nodes represent computational statements or expressions, the edges represent transfer of control between nodes, and each possible execution path of the module has a corresponding path from the entry to the exit node of the graph. { kən¦trōl 'flō,graf }

control handle *See* handle. { kən'trōl ,hand·əl }

control head gap [COMPUT SCI] The distance maintained between the read/write head of a disk drive and the disk surface. { kən'trōl ¦hed ,gap }

control hierarchy *See* hierarchical control. { kən'trōl 'hī·ər,är·kē }

control instructions [COMPUT SCI] Those instructions in a computer program which ensure proper sequencing of instructions so that a programmed task can be performed correctly. { kən'trōl in'strək·shənz }

control key [COMPUT SCI] A special key on a computer keyboard which, when depressed together with another key, generates a different signal than would be produced by the second key alone. { kən'trōl ,kē }

controllability [CONT SYS] Property of a system for which, given any initial state and any desired state, there exists a time interval and an input signal which brings the system from the initial state to the desired state during the time interval. { kən,trōl·ə'bil· əd·ē }

control lead [COMPUT SCI] A character or sequence of characters indicating that the information following is a control code and not data. { kən'trōl ,lēd }

controlled avalanche transit-time triode [ELECTR] A solid-state microwave device that uses a combination of IMPATT diode and *npn* bipolar transistor technologies; avalanche and drift zones are located between the base and collector regions. Abbreviated CATT. { kən¦trōld 'av·ə ,lanch ¦tranz·ət ,tīm 'trī,ōd }

controlled carrier modulation [COMMUN] System of modulation wherein the carrier is amplitude-modulated by the signal frequencies and, in addition, the carrier is amplitude-modulated according to the envelope of the signal so that the modulation factor remains constant regardless of the amplitude of the signal. Also known as floating carrier modulation; variable carrier modulation. { kən¦trōld 'kar·ē·ər ,mäj·ə'lā·shən }

controlled variable [CONT SYS] In process automatic-control work, that quantity or condition of a controlled system that is directly measured or controlled. { kən¦trōld 'ver·ē·ə·bəl }

controller *See* automatic controller. { kən'trōl· ər }

control limits [ELECTR] In radar evaluation, upper and lower control limits are established at those performance figures within which it is expected that 95% of quality-control samples will fall when the radar is performing normally. { kən'trōl ,lim·əts }

control logic [COMPUT SCI] The sequence of steps required to perform a specific function. { kən'trōl ,läj·ik }

control mark *See* tape mark. { kən'trōl ,märk }

control-message display [COMPUT SCI] A device, such as a console typewriter, on which control information, such as information on the progress of a running computer program, is displayed in ordinary language. { kən'trōl ˌmes·ij di'splā }

control module [COMPUT SCI] The set of registers and circuitry required to carry out a specific function. { kən'trōl ˌmä·jül }

control operation [COMPUT SCI] Any action that affects data processing but is not directly included, such as managing input/output operations or determining job sequence. { kən'trōl ˌäp·əˌrā·shən }

control panel [COMPUT SCI] An array of jacks or sockets in which wires (or other elements) may be plugged to control the action of an electromechanical device in a data-processing system such as a printer. Also known as plugboard; wiring board. { kən'trōl ˌpan·əl }

control point [COMPUT SCI] **1.** The numerical value of the controlled variable (speed, temperature, and so on) which, under any fixed set of operating conditions, an automatic controller operates to maintain. **2.** One of the hardware locations at which the output of the instruction decoder of the processor activates the input to and output from specific registers as well as operational resources of the system. [NAV] A position marked by a buoy, boat, aircraft, electronic device, conspicuous terrain feature, or other identifiable object which is given a name or number and used as an aid for navigation or control of ships, boats, or aircraft. { kən'trōl ˌpȯint }

control program [COMPUT SCI] A program which carries on input/output operations, loading of programs, detection of errors, communication with the operator, and so forth. { kən'trōl ˌprō·grəm }

control record [COMPUT SCI] A special record added to the end of a file to provide information about the file and the records in it. { kən'trōl ˌrek·ərd }

control register [COMPUT SCI] Any one of the registers in a computer used to control the execution of a computer program. { kən'trōl ˌrej·ə·stər }

control room [COMMUN] A room from which engineers and production people control and direct a video or audio program or a recording session. [ENG] A room from which space flights are directed. { kən'trōl ˌrüm }

control section [COMPUT SCI] **1.** The smallest integral subsection of a program, that is, the smallest unit of code that can be separately relocated during loading. **2.** The part of a central processing unit that controls other sections of the unit. { kən'trōl ˌsek·shən }

control sequence [COMPUT SCI] The order in which a set of executions are carried to perform a specific function. { kən'trōl ˌsē·kwəns }

control signal [COMPUT SCI] A set of pulses used to identify the channels to be followed by transferred data. [CONT SYS] The signal applied to the device that makes corrective changes in a controlled process or machine. { kən'trōl ˌsig·nəl }

control state [COMPUT SCI] The operating mode of a system which permits it to override its normal sequence of operations. { kən'trōl ˌstāt }

control statement [COMPUT SCI] A statement in a computer program that controls program execution, such as a GOTO statement, conditional jump, or a loop. { kən'trōl ˌstāt·mənt }

control supervisor [COMPUT SCI] The computer software which controls the processing of the system. { kən'trōl ¦sü·pərˌvī·zər }

control switching point [COMMUN] A telephone office which is an important switching center in the routing of long-distance calls in the direct distance dialing system. Abbreviated CSP. { kən'trōl 'swich·iŋ ˌpȯint }

control symbol [COMPUT SCI] A symbol which, coded into the machine memory, controls certain steps in the mechanical translation process; since control symbols are not contextual symbols, they appear neither in the input nor in the output. { kən'trōl ˌsim·bəl }

control system [ENG] A system in which one or more outputs are forced to change in a desired manner as time progresses. { kən'trōl ˌsis·təm }

control systems equipment [COMPUT SCI] Computers which are an integral part of a total facility or larger complex of equipment and have the primary purpose of controlling, monitoring, analyzing, or measuring a process or other equipment. { kən'trōl ˌsis·təmz i'kwip·mənt }

control total [COMPUT SCI] The sum of the numbers in a specified record field of a batch of records, determined repetitiously during computer processing so that any discrepancy from the control indicates an error. { kən'trōl ˌtōd·əl }

control unit [COMPUT SCI] An electronic device containing data buffers and logical circuitry, situated between the computer channel and the input/output device, and controlling data transfers and such operations as tape rewind. { kən'trōl ˌyü·nət }

control unit terminal emulation [COMPUT SCI] A technique that enables a personal computer to imitate a terminal of a main frame. Abbreviated CUT emulation. { kən'trōl ˌyü·nət ¦tər·mə·nəl ˌem·yə'lā·shən }

control word [COMPUT SCI] A computer word specifying a certain action to be taken. { kən 'trōl ˌwərd }

conventional algorithm [COMMUN] A cryptographic algorithm in which the enciphering and deciphering keys are easily derivable from each other, or are identical, and both must be kept secret. { kən'ven·chən·əl 'al·gəˌrith·əm }

conventional definition television [COMMUN] The analog NTSC (National Television Standards Committee) television system. Abbreviated CDTV. { kən'ven·chən·əl 'def·əˌnish·ən 'tel·əˌvizh·ən }

conventional programming [COMPUT SCI] The use of standard programming languages, as opposed to application development languages, financial planning languages, query languages, and report programs. { kən'ven·chən·əl 'prō ˌgram·iŋ }

convergence [ELECTR] A condition in which the electron beams of a multibeam cathode-ray tube intersect at a specified point, such as at an opening in the shadow mask of a three-gun color television picture tube; both static convergence and dynamic convergence are required. { kən'vər·jəns }

convergence circuit [ELECTROMAG] An auxiliary deflection system in a color television receiver which maintains convergence, having separate convergence coils for electromagnetic controls of the positions of the three beams in a convergence yoke around the neck of the kinescope. { kən'vər·jəns ˌsər·kət }

convergence coil [ELECTR] One of the coils used to obtain convergence of electron beams in a three-gun color television picture tube. { kən'vər·jəns ˌkȯil }

convergence control [ELECTR] A control used in a color display device to adjust certain parameters of the three-gun color picture tube to achieve convergence. { kən'vər·jəns kən 'trōl }

convergence magnet [ELECTR] A magnet assembly whose magnetic field converges two or more electron beams; used in three-gun color picture tubes. Also known as beam magnet. { kən'vər·jəns ˌmag·nət }

Conversational Algebraic Language *See* CAL. { kän·vərˌsā·shən·əl al·jəˌbrā·ik 'laŋ·gwij }

conversational compiler [COMPUT SCI] A compiler which immediately checks the validity of each source language statement entered to the computer and informs the user if the next statement can be entered or if a mistake must be corrected. Also known as interpreter. { kän·vər'sā·shən·əl kəm'pīl·ər }

conversational mode [COMMUN] A computer operating mode that permits queries and responses between the computer and human operators at keyboard terminals. { kän·vər'sā·shən·əl ˌmōd }

conversational processing [COMPUT SCI] The operating mode of a computer system which enables a user to have each statement he keys into the system processed immediately. { kän·vər'sā·shən·əl 'präs·əs·iŋ }

conversational time-sharing [COMPUT SCI] The simultaneous utilization of a computer system by multiple users, each user being equipped with a remote terminal with which he communicates with the computer in conversational mode. { kän·vər'sā·shən·əl 'tīm ˌsher·iŋ }

conversion *See* data conversion. { kən'vər·zhən }

conversion gain [ELECTR] **1.** Ratio of the intermediate-frequency output voltage to the input signal voltage of the first detector of a superheterodyne receiver. **2.** Ratio of the available intermediate-frequency power output of a converter or mixer to the available radio-frequency power input. { kən'vər·zhən ˌgān }

conversion program [COMPUT SCI] A set of instructions which allows a program written for one system to be run on a different system. { kən'vər·zhən ˌprō·grəm }

conversion rate [COMPUT SCI] The number of complete conversions an analog-to-digital converter can perform per unit time, usually specified in cycles (or conversions) per second. { kən'vər·zhən ˌrāt }

conversion routine [COMPUT SCI] A flexible, self-contained, and generalized program used for data conversion, which only requires specifications about very few facts in order to be used by a programmer. { kən'vər·zhən rü'tēn }

conversion time [COMPUT SCI] The time required to read in data from one code into another code. { kən'vər·zhən ˌtīm }

convert [COMPUT SCI] To transform the representation of data. { kən'vərt }

converter [COMPUT SCI] A computer unit that changes numerical information from one form to another, as from decimal to binary or vice versa, from fixed-point to floating-point representation, from magnetic tape to disk storage, or from digital to analog signals and vice versa. Also known as data converter. [ELECTR] **1.** The section of a superheterodyne radio receiver that converts the desired incoming radio-frequency signal to an intermediate-frequency value; the converter section includes the oscillator and the mixer-first detector. Also known as heterodyne conversion transducer; oscillator-mixer-first-detector. **2.** An auxiliary unit used with a television or radio receiver to permit reception of channels or frequencies for which the receiver was not originally designed. **3.** In facsimile, a device that changes the type of modulation delivered by the scanner. **4.** Unit of a radar system in which the mixer of a heterodyne receiver and often first stages of intermediate-frequency amplification are located, often called a down-converter. Similar mixing circuitry is employed on transmit, the signal being up-converted from the synthesizer output, usually at the intermediate frequency, to the intended carrier frequency. **5.** *See* remodulator. { kən'vərd·ər }

converter tube [ELECTR] An electron tube that combines the mixer and local-oscillator functions of a heterodyne conversion transducer. { kən'vərd·ər ˌtüb }

convolutional code [COMMUN] An error-correcting code that processses incoming bits serially rather than in large blocks. { ˌkän·vəlü·shən·əl 'kōd }

convolver [ELECTR] A surface acoustic-wave device in which signal processing is performed by a nonlinear interaction between two waves traveling in opposite directions. Also known as acoustic convolver. { kən'väl·vər }

cookbook [COMPUT SCI] A document that describes how to install and use a software product or carry out other complex tasks in step-by-step fashion. { 'ku̇kˌbu̇k }

cookie [COMPUT SCI] A data file written to a hard drive by some Web sites, contains information the site can use to track such things as passwords, login, registration or identification, user preferences, online shopping cart information, and lists of pages visited. { 'ku̇k·ē }

cooperative multitasking [COMPUT SCI] A method of running more than one program on a computer at a time in which the program currently in control of the processor retains the control until it yields the control to another program voluntarily, which it can do only at certain points in the program. Also known as nonpreemptive multitasking. { kō,äp·rəd·iv 'məl·tə,task·iŋ }

coordinate addressing [COMPUT SCI] The use of cartesian coordinates to specify a location, such as the position of a character in an electronic display. { kō'ȯrd·ən·ət 'ad,res·iŋ }

coordinated geometry *See* COGO. { kō'ȯrd·ən ,ād·əd jē'äm·ə·trē }

coordinate indexing [COMPUT SCI] An indexing scheme in which equal-rank descriptors are used to describe a document, for information retrieval by a computer or other means. { kō'ȯrd·ən·ət 'in,deks·iŋ }

coordinate storage *See* matrix storage. { kō 'ȯrd·ən·ət 'stȯr·ij }

coprocessor [COMPUT SCI] A processing unit that works together with a primary central processing unit to speed a computer's execution of time-consuming operations. { kō'prä,ses·ər }

copy [COMMUN] **1.** To transcribe Morse code signals into written form. **2.** A string procedure in ALGOL by means of which a new byte string can be generated from an existing byte string. { 'käp·ē }

copying program [COMPUT SCI] A system program which copies a data or program file from one peripheral device onto another. { 'käp·ē·iŋ ,prō·grəm }

copy protection *See* software protection. { 'käp·ē prə,tek·shən }

CORBA *See* common object request broker. { 'kȯr·bə }

cord circuit [ELEC] Connecting circuit terminating in a plug at one or both ends and used at manually operated switchboard positions in establishing telephone connections. { 'kȯrd ,sər·kət }

cordless telephone [COMMUN] A telephone whose headset and base are equipped with small antennas and are linked by low-power radio instead of a wire. { 'kȯrd·ləs 'tel·ə,fōn }

core-dump [COMPUT SCI] To copy the contents of all or part of core storage, usually into an external storage device. { 'kȯr ,dəmp }

core image [COMPUT SCI] **1.** A computer program whose storage addresses have been assigned so that it can be loaded directly into main storage for processing. **2.** A visual representation of a computer's main storage. { 'kȯr ,im·ij }

core-image library [COMPUT SCI] A collection of computer programs residing on mass-storage device in ready-to-run form. { 'kȯr ¦im·ij ,lī ,brer·ē }

core memory *See* magnetic core storage. { 'kȯr ,mem·rē }

core memory resident [COMPUT SCI] A control program which is in the main memory of a computer at all times to supervise the processing of the computer. { 'kȯr ,mem·rē ,rez·ə·dənt }

core rope storage [COMPUT SCI] Direct-access storage consisting of a large number of doughnut-shaped ferrite cores arranged on a common axis, with sense, inhibit, and set wires threaded through or around individual cores in a predetermined manner to provide fixed storage of digital data; each core rope stores one or more complete words, rather than just a single bit. { 'kȯr ,rōp ,stȯr·ij }

coresident [COMPUT SCI] A computer program or program module that is stored in a computer memory along with other programs. { kō 'rez·ə·dənt }

core storage [COMPUT SCI] **1.** The main memory of a computer. **2.** *See* magnetic core storage. { 'kȯr ,stȯr·ij }

corner effect [ELECTR] The departure of the frequency-response curve of a band-pass filter from a perfect rectangular shape, so that the corners of the rectangle are rounded. [ENG] In ultrasonic testing, reflection of an ultrasonic beam directed perpendicular to the intersection of two surfaces 90° apart. { 'kȯr·nər i'fekt }

corner frequency *See* break frequency. { 'kȯr·nər ,frē·kwən·sē }

corner reflector [ELECTROMAG] An antenna consisting of two conducting surfaces intersecting at an angle that is usually 90°, with a dipole or other antenna located on the bisector of the angle. { 'kȯr·nər ri'flek·tər }

coroutine [COMPUT SCI] A program module for which the lifetime of a particular activation record is independent of the time when control enters or leaves the module, and in which the activation record maintains a local instruction counter so that, whenever control enters the module, execution begins at the point where it stopped when control last left that particular instance of execution. { 'kō·rü,tēn }

correction time [CONT SYS] The time required for the controlled variable to reach and stay within a predetermined band about the control point following any change of the independent variable or operating condition in a control system. Also known as settling time. { kə'rek·shən ,tīm }

corrective maintenance [COMPUT SCI] The maintenance performed as required, on an unscheduled basis, by the contractor following equipment failure. Also known as remedial maintenance. { kə'rek·tiv mānt·ən·əns }

correlation detection [ENG] A method of detection of aircraft or space vehicles in which a signal is compared, point to point, with an internally generated reference. Also known as cross-correlation detection. { ,kär·ə'lā·shən di'tek·shən }

correlation direction finder [ENG] Satellite station separated from a radar to receive jamming signals; by correlating the signals received from several such stations, range and azimuth of many jammers may be obtained. { ˌkär·əˈlā·shən dəˈrek·shən ˌfīnd·ər }

correlation distance [COMMUN] In tropospheric scatter propagation, the minimum spatial separation between antennas which will give rise to independent fading of the received signals. { ˌkär·əˈlā·shən ˌdis·təns }

correlation tracking system [ENG] A trajectory-measuring system utilizing correlation techniques where signals derived from the same source are correlated to derive the phase difference between the signals. { ˌkär·əˈlā·shən ˈtrak·iŋ ˌsis·təm }

correlation-type receiver *See* correlator. { ˌkär·əˈlā·shən ˌtīp riˈsē·vər }

correlator [ELECTR] A device that detects weak signals in noise by performing an electronic operation approximating the computation of a correlation function. Also known as correlation-type receiver. { ˈkär·əˌlād·ər }

correspondence *See* relation. { ˌkär·əˈspän·dəns }

correspondence printer *See* letter-quality printer. { ˌkär·əˈspän·dəns ˌprint·ər }

corrugated conical-horn antenna [ELECTROMAG] A horn antenna that has a circular cross section and a series of equally spaced ridges protruding from otherwise straight sides. { ¦kär·əˌgād·əd ˌkän·ə·kəl ˌhȯrn anˈten·ə }

corrupt [COMPUT SCI] To destroy or alter information so that it is no longer reliable. { kəˈrəpt }

cosecant antenna [ELECTROMAG] An antenna that gives a beam whose amplitude varies as the cosecant of the angle of depression below the horizontal; used in navigation radar. { kō ˈsēˌkant anˈten·ə }

cosecant-squared antenna [ELECTROMAG] An antenna that has a cosecant-squared pattern. { kōˈsēˌkant ¦skwerd anˈten·ə }

cosecant-squared pattern [ELECTROMAG] A ground radar-antenna radiation pattern that sends less power to nearby objects than to those farther away in the same sector; the field intensity varies as the square of the cosecant of the elevation angle. { kōˈsēˌkant ¦skwerd ˈpad·ərn }

cosmic noise [COMMUN] Radio static caused by a phenomenon outside the earth's atmosphere, such as sunspots. { ˈkäz·mik ˈnȯiz }

cost function [SYS ENG] In decision theory, a loss function which does not depend upon the decision rule. { ˈkȯst ˌfəŋk·shən }

count cycle [COMPUT SCI] An increase or decrease of the cycle index by unity or by an arbitrary integer. { ˈkau̇nt ˌsī·kəl }

countdown [COMMUN] The ratio of the number of interrogation pulses not answered by a transponder to the total number received. { ˈkau̇ntˌdau̇n }

counter [COMPUT SCI] **1.** A register or storage location used to represent the number of occurrences of an event. **2.** *See* accumulator; scaler. { ˈkau̇nt·ər }

counter circuit *See* counting circuit. { ˈkau̇nt·ər ˌsər·kət }

counter coupling [COMPUT SCI] The technique of combining two or more counters into one counter of larger capacity in electromechanical devices by means of control panel wiring. { ˈkau̇nt·ər ˌkəp·liŋ }

counter decade *See* decade scaler. { ˈkau̇nt·ər ˌdekˌād }

counter-free machine [COMPUT SCI] A sequential machine that cannot count modulo any integer greater than 1. { ˈkau̇nt·ər ˌfrē məˈshēn }

counting circuit [ELECTR] A circuit that counts pulses by frequency-dividing techniques, by charging a capacitor in such a way as to produce a voltage proportional to the pulse count, or by other means. Also known as counter circuit. { ˈkau̇nt·iŋ ˌsər·kət }

counting-down circuit *See* frequency divider. { ˈkau̇nt·iŋ ˌdau̇n ˌsər·kət }

coupled antenna [ELECTROMAG] An antenna electromagnetically coupled to another. { ˈkəp·əld anˈten·ə }

coupled systems [COMPUT SCI] Computer systems that share equipment and can exchange information. { ˈkəp·əld ˈsis·təmz }

coupler [ELECTROMAG] **1.** A passage which joins two cavities or waveguides, allowing them to exchange energy. **2.** A passage which joins the ends of two waveguides, whose cross section changes continuously from that of one to that of the other. [NAV] The portion of a navigation system that receives signals of one type from a sensor and transmits signals of a different type to an actuator. { ˈkəp·lər }

coupling aperture [ELECTROMAG] An aperture in the wall of a waveguide or cavity resonator, designed to transfer energy to or from an external circuit. Also known as coupling hole; coupling slot. { ˈkəp·liŋ ˌap·ə·chər }

coupling hole *See* coupling aperture. { ˈkəp·liŋ ˌhōl }

coupling loop [ELECTROMAG] A conducting loop projecting into a waveguide or cavity resonator, designed to transfer energy to or from an external circuit. { ˈkəp·liŋ ˌlüp }

coupling probe [ELECTROMAG] A probe projecting into a waveguide or cavity resonator, designed to transfer energy to or from an external circuit. { ˈkəp·liŋ ˌprōb }

coupling slot *See* coupling aperture. { ˈkəp·liŋ ˌslät }

courseware [COMPUT SCI] Computer programs designed to be used in computer-aided instruction or computer-managed instruction. { ˈkȯrs ˌwer }

coverage [COMMUN] *See* service area. [ELECTROMAG] A spatial account of the regions of useful sensitivity in a radar's surroundings that can be affected, for example, by multipath propagation or by obscuring terrain. { ˈkəv·rij }

COZI [COMMUN] An ionospheric sounding system for determining propagation characteristics of the ionosphere at various angles at any instant; used to determine how well long-distance, high-frequency broadcasts are reaching their intended destinations. Derived from communications zone indicator. { ¦kō¦zī }

CPA *See* color-phase alternation.

CPE *See* computer performance evaluation.

CPU *See* central processing unit.

CPU-bound program [COMPUT SCI] A computer program that involves a large amount of calculation and internal rearrangement of data, so that the speed of execution depends on the speed of the central processing unit (CPU) and memory. Also known as compute-bound program; cycle-bound program; process-bound program. { ,sē ,pe'yü ¦bau̇nd ,prō·grəm }

CPU fan [COMPUT SCI] A fan mounted directly over the integrated-circuit chip containing a computer's central processing unit to prevent overheating. { ¦sē¦pē¦yü 'fan }

crash [COMPUT SCI] **1.** A breakdown, hardware failure, or software problem that renders a computer system inoperative. **2.** *See* abend. { krash }

crash locator beacon [COMMUN] An automatic radio beacon carried in aircraft to guide searching forces in the event of a crash. { 'krash 'lō,kād·ər ,bē·kən }

CRC *See* cyclic redundancy check.

creation operator [COMPUT SCI] The part of a data structure which allows components to be created. { krē'ā·shən ,äp·ə,rād·ər }

credence [ELECTROMAG] In radar, a measure of confidence in a target detection, generally proportional to target return amplitude. { 'krēd·əns }

crippled leap-frog test [COMPUT SCI] A variation of the leap-frog test, modified so the computer tests are repeated from a single set of storage locations rather than a changing set of locations. { ¦krip·əld 'lēp ,fräg ,test }

crippled mode [COMPUT SCI] The operation of a computer at reduced capacity when certain parts are not working. { 'krip·əld ,mōd }

critical area *See* picture element. { 'krid·ə·kəl 'er·ē·ə }

critical frequency [ELECTR] *See* cutoff frequency. [ELECTROMAG] The limiting frequency below which a radio wave will be reflected by an ionospheric layer at vertical incidence at a given time. { 'krid·ə·kəl 'frē·kwən·sē }

critical wavelength [COMMUN] The free-space wavelength corresponding to the critical frequency. { 'krid·ə·kəl 'wāv,leŋkth }

CROM *See* control and read-only memory. { 'sē ,räm }

cross antenna [ELECTROMAG] An array of two or more horizontal antennas connected to a single feed line and arranged in the pattern of a cross. { 'krȯs an,ten·ə }

cross assembler [COMPUT SCI] An assembly program that allows a computer program written on one type of computer to be used on another type. { 'krȯs ə,sem·blər }

crossbar system [COMMUN] Automatic telephone switching system which is generally characterized by the following features: selecting mechanisms are crossbar switches, common circuits select and test the switching paths and control the operation of the selecting mechanisms, and method of operations is one in which the switching information is received and stored by controlling mechanisms that determine the operations necessary in establishing a telephone connection; largly replaced by electronic switching systems using digital switching techniques. { 'krȯs,bär ,sis·təm }

cross-color [ELECTR] In analog color television, the interference in the receiver chrominance channel caused by cross talk from monochrome signals. { 'krȯs ,kəl·ər }

cross compiler [COMPUT SCI] A compiler that allows a computer program written on one type of computer to be used on another type. { 'krȯs kəm,pī·lər }

cross-correlation detection *See* correlation detection. { 'krȯs kär·ə'lā·shən di'tek·shən }

cross-correlation function [COMMUN] A function, $\phi_{12}(\tau)$, where τ is a time-delay parameter, equal to the limit, as T approaches infinity, of the reciprocal of 2T times the integral over t from $-T$ to T of $f_1(t)f_2(t-\tau)$, where f_1 and f_2 are functions of time, such as the input and output of a communication system. { 'krȯs kär·ə'lā·shən ,fəŋk·shən }

cross-correlator [ELECTR] A correlator in which a locally generated reference signal is multiplied by the incoming signal and the result is smoothed in a low-pass filter to give an approximate computation of the cross-correlation function. Also known as synchronous detector. { ¦krȯs'kär·ə,lā d·ər }

cross-coupling [COMMUN] A measure of the undesired power transferred from one channel to another in a transmission medium. { ¦krȯs 'kəp·liŋ }

crossed-field tubes [ELECTR] Vacuum tubes often used in radar transmitters, either as oscillators or as amplifiers, in which the electrons leaving the cathode surface travel in a plasma to the anode in paths determined by the crossed electric and magnetic bias fields applied to the tube, so that the density of the plasma can be easily affected by the electromagnetic signal with which the electrons are interacting. { 'krȯst ,fēld ,tübz }

cross fire [COMMUN] Interfering current in one telegraph or signaling channel resulting from telegraph or signaling currents in another channel. { 'krȯs ,fīr }

crossfoot [COMPUT SCI] To add numbers in several different ways in a computer, for checking purposes. { 'krȯs,fu̇t }

crosshatch generator [ELECTR] A signal generator that generates a crosshatch pattern for adjusting a video display device. { 'krȯs,hach ,jen·ə,rād·ər }

cross modulation [COMMUN] A type of interference in which the carrier of a desired signal becomes modulated by the program of an undesired signal on a different carrier frequency; the program of the undesired station is then heard in the background of the desired program. { ¦krȯs ˌmäj·əˈlā·shən }

cross office switching time [COMMUN] Time required to connect any input through the switching center to any selected output. { ˈkrȯs ˌȯf·əs ˈswich·iŋ ˌtīm }

cross-platform computing [COMPUT SCI] The use of very similar user interfaces for versions of programs running on different operating systems and computer architectures. { ˌkrȯs ¦plat‚form kəmˈpyüd·iŋ }

cross-referencing program [COMPUT SCI] A computer program used in debugging that produces indexed lists of both the variable names and the statement numbers of the source program. { ¦krȯs ˈref·rəns·iŋ ˌprō·grəm }

crosstalk [COMMUN] **1.** The sound heard in a receiver along with a desired program because of cross modulation or other undesired coupling to another communication channel; it is also observed between adjacent pairs in a telephone cable. **2.** Interaction of audio and video signals in an analog television system, causing video modulation of the audio carrier or audio modulation of the video signal at some point. **3.** Interaction of the chrominance and luminance signals in an analog color television receiver. { ˈkrȯs‚tȯk }

crosstalk coupling [COMMUN] The cross coupling between speech communications channels or their component parts. Also known as crosstalk loss. { ˈkrȯs‚tȯk ˌkəp·liŋ }

crosstalk level [COMMUN] Volume of crosstalk energy, measured in decibels, referred to a reference level. { ˈkrȯs‚tȯk ˌlev·əl }

crosstalk loss *See* crosstalk coupling. { ˈkrȯs ˌtȯk ˌlȯs }

crosstalk unit [COMMUN] A measure of the coupling between two circuits; the number of crosstalk units is 1 million times the ratio of the current or voltage at the observing point to the current or voltage at the origin of the disturbing signal, the impedances at these points being equal. Abbreviated cu. { ˈkrȯs‚tȯk ˌyü·nət }

CRT *See* cathode-ray tube.

cryogenic film [COMPUT SCI] A storage element using superconducting thin films of lead at liquid-helium temperature. { ˌkrī·əˈjen·ik ˈfilm }

cryptanalysis [COMMUN] Steps and operations performed in converting encrypted messages into plain text without previous knowledge of the key employed. { ˌkrip·təˈnal·ə·səs }

cryptochannel [COMMUN] A complete system of communication that uses electronic encryption and decryption equipment and has two or more radio or wire terminals. { ¦krip·tōˈchan·əl }

cryptogram [COMMUN] Information written in code or cipher. { ˈkrip·tə‚gram }

cryptographic algorithm [COMMUN] An unchanging set of rules or steps for enciphering and deciphering messages in a cipher system. { ¦krip·tə¦graf·ik ˈal·gə‚rith·əm }

cryptographic bitstream [COMMUN] An unending sequence of digits which is combined with ciphertext to produce plaintext or with plaintext to recover ciphertext in a stream cipher system. { ¦krip·tə¦graf·ik ˈbit‚strēm }

cryptographic key [COMMUN] A sequence of numbers or characters selected by the user of a cipher system to implement a cryptographic algorithm for enciphering and deciphering messages. Also known as key. { ¦krip·tə¦graf·ik ˈkē }

cryptography [COMMUN] The science of preparing messages in a form which cannot be read by those not privy to the secrets of the form. { kripˈtäg·rə·fē }

cryptology [COMMUN] The science of preparing messages in forms which are intended to be unintelligible to those not privy to the secrets of the form, and of deciphering such messages. { kripˈtäl·ə·jē }

cryptopart [COMMUN] One of several portions of a cryptotext; each cryptopart bears a different message indicator. { ˈkrip·tō‚pärt }

cryptotext [COMMUN] In cryptology, a text of visible writing which conveys no intelligible meaning in any language, or which apparently conveys an intelligible meaning that is not the real meaning. { ˈkrip·tō‚tekst }

crystal-audio receiver [ELECTR] Similar to the crystal-video receiver, except for the path detection bandwidth which is audio rather than video. { ¦krist·əl ¦ȯd·ē·ōriˈsē·vər }

crystal-controlled transmitter [ELECTR] A transmitter whose carrier frequency is directly controlled by the electromechanical characteristics of a quartz crystal unit. { ¦krist·əl kən¦trōld ˈtranz ˌmid·ər }

crystal detector [ELECTR] **1.** A crystal used to rectify a modulated radio-frequency signal to obtain the audio or video signal directly. **2.** A crystal diode used in a microwave receiver to combine an incoming radio-frequency signal with a local oscillator signal to produce an intermediate-frequency signal. { ˈkrist·əl diˈtek·tər }

crystal diode *See* semiconductor diode. { ¦krist·əl ˈdī‚ōd }

crystal filter [ELECTR] A highly selective tuned circuit employing one or more quartz crystals; sometimes used in intermediate-frequency amplifiers of communication receivers to improve the selectivity. { ¦krist·əl ˈfil·tər }

crystal mixer [ELECTR] A mixer that uses the nonlinear characteristic of a crystal diode to mix two frequencies; widely used in radar receivers to convert the received radar signal to a lower intermediate-frequency value by mixing it with a local oscillator signal. { ¦krist·əl ˈmik·sər }

crystal oscillator [ELECTR] An oscillator in which the frequency of the alternating-current output is determined by the mechanical properties of a piezoelectric crystal. Also known as piezoelectric oscillator. { ¦krist·əl ˈäs·ə‚lād·ər }

crystal rectifier *See* semiconductor diode. { ¦krist·əl ˈrek·tə‚fī·ər }

crystal-stabilized transmitter [ELECTR] A transmitter employing automatic frequency control, in which the reference frequency is that of a crystal oscillator. { ¦krist·əl ¦stā·bə‚līzd ˈtranz‚mid·ər }

crystal video receiver [ELECTR] A broad-tuning radar or other microwave receiver consisting only of a crystal detector and a video or audio amplifier. { ¦krist·əl ¦vid·ē·ō ri'sē·vər }

crystal video rectifier [ELECTR] A crystal rectifier transforming a high-frequency signal directly into a video-frequency signal. { ¦krist·əl ¦vid·ē·ō 'rek·tə,fī·ər }

C-scan *See* C-display. { 'sē ,skan }

C-scope *See* C-display. { 'sē ,skōp }

CSMA/CD [COMPUT SCI] A method of controlling multiaccess computer networks in which each station on the network senses traffic and waits for it to clear before sending a message, and two devices that try to send concurrent messages must both step back and try again. Abbreviation for carrier-sense multiple access with collision detection.

CSP *See* control switching point.

CSSB system *See* companded single-sideband system. { ¦sē,es,es¦bē ,sis·təm }

CSW *See* channel status word.

cu *See* crosstalk unit.

cubical antenna [ELECTROMAG] An antenna array, the elements of which are positioned to form a cube. { 'kyü·bə·kəl an'ten·ə }

cue circuit [ELECTR] A one-way communication circuit used to convey program control information. { 'kyü ,sər·kət }

current awareness system [COMPUT SCI] A system for notifying users on a periodic basis of the acquisition, by a central file or library, of information (usually literature) which should be of interest to the user. { 'kər·ənt ə'wer·nəs ,sis·təm }

current cell *See* active cell. { ,kər·ənt 'sel }

current feed [ELECTR] Feed to a point where current is a maximum, as at the center of a half-wave antenna. { 'kər·ənt ,fēd }

current gain [ELECTR] The fraction of the current flowing into the emitter of a transistor which flows through the base region and out the collector. { 'kər·ənt ,gān }

current-instruction register *See* instruction register. { 'kər·ənt in'strək·shən ,rej·ə·stər }

current location reference [COMPUT SCI] A symbolic expression, such as a star, which indicates the current location reached by the program; a transfer to * + 2 would bring control to the second statement after the current statement. { 'kər·ənt lō'kā·shən ,ref·rəns }

current margin [COMMUN] Difference between the steady-state currents flowing through a telegraph receiving instrument corresponding respectively to the two positions of the telegraph transmitter. { 'kər·ənt ,mär·jən }

current-mode filter [ELECTR] An integrated-circuit filter in which the signals are represented by current levels rather than voltage levels. { 'kər·ənt,mōd ,fil·tər }

current-mode logic [ELECTR] Integrated-circuit logic in which transistors are paralleled so as to eliminate current hogging. Abbreviated CML. { 'kər·ənt ,mōd 'läj·ik }

current-type telemeter [COMMUN] A telemeter in which the magnitude of a single current is the translating means. { 'kər·ənt ,tīp tə'lem·əd·ər }

cursor [COMPUT SCI] A movable spot of light that appears on the screen of a visual display terminal and can be positioned horizontally and vertically through keyboard controls to instruct the computer at what point a change is to be made. { 'kər·sər }

cursor arrows [COMPUT SCI] Arrows marked on keys of a computer keyboard that control the movement of the cursor. { 'kər·sər ,ar·ōz }

curtain array [ELECTROMAG] An antenna array consisting of vertical wire elements stretched between two suspension cables. { 'kərt·ən ə'rā }

curtain rhombic antenna [ELECTROMAG] A multiple-wire rhombic antenna having a constant input impedance over a wide frequency range; two or more conductors join at the feed and terminating ends but are spaced apart vertically from 1 to 5 feet (30 to 150 centimeters) at the side poles. { 'kərt·ən 'räm·bik an'ten·ə }

curve follower [COMPUT SCI] A device in which a photoelectric, capacitive or inductive pick-off guided by a servomechanism reads data in the form of a graph, such as a curve drawn on paper with suitable ink. Also known as graph follower. { 'kərv ,fäl·ə·wər }

cut and paste [COMPUT SCI] An editing function of a word processing system in which a portion of text is marked with a particular character at the beginning and at the end and is then copied to another location within the text. Also known as block move. { ¦kət ən 'pāst }

cut constraint [SYS ENG] A condition sometimes imposed in an integer programming problem which excludes parts of the feasible solution space without excluding any integer points. { 'kət kən'strānt }

CUT emulation *See* control unit terminal emulation. { 'kət ,em·yə,lā·shən }

cut form [COMPUT SCI] In optical character recognition, any document form, receipt, or such, of standard dimensions which must be issued a separate read command in order to be recognized. { ¦kət ¦fȯrm }

Cutler feed [ELECTROMAG] A resonant cavity that transfers radio-frequency energy from the end of a waveguide to the reflector of a radar spinner assembly. { 'kət·lər ,fēd }

cut methods [SYS ENG] Methods of solving integer programming problems that employ cut constraints derived from the original problem. { 'kət ,meth·əds }

cutoff [ELECTR] **1.** The minimum value of bias voltage, for a given combination of supply voltages, that just stops output current in an electron tube, transistor, or other active device. **2.** *See* cutoff frequency. { 'kət,ȯf }

cutoff frequency [ELECTR] A frequency at which the attenuation of a device begins to increase sharply, such as the limiting frequency below which a traveling wave in a given mode cannot be maintained in a waveguide, or the frequency above which an electron tube loses efficiency rapidly. Also known as critical frequency; cutoff. { 'kət,ȯf ,frē·kwən·sē }

cutoff wavelength [ELECTROMAG] **1.** The ratio of the velocity of electromagnetic waves in free space to the cutoff frequency in a uniconductor waveguide. **2.** The wavelength corresponding to the cutoff frequency. { 'kət,ȯf 'wāv,leŋkth }

cut-sheet printer [COMPUT SCI] A printer designed to print on separate sheets of paper. { 'kət ,shēt ¦print·ər }

cut-signal-branch operation [ELECTR] In systems where radio reception continues without cutting off the carrier, the cut-signal-branch operation technique disables a signal branch in one direction when it is enabled in the other to preclude unwanted signal reflections. { ¦kət ¦sig·nəl ¦branch ,äp·ə,rā·shən }

CW *See* continuous wave.

cyberspace [COMPUT SCI] The digital realms, including Web sites and virtual worlds. { 'sī·bər ,spās }

cycle-bound program *See* CPU-bound program. { 'sī·kəl ¦bau̇nd 'prō·grəm }

cycle count [COMPUT SCI] The operation of keeping track of the number of cycles a computer system goes through during processing time. { 'sī·kəl ,kau̇nt }

cycle criterion [COMPUT SCI] Total number of times a cycle in a computer program is to be repeated. { 'sī·kəl krī'tir·ē·ən }

cycle index [COMPUT SCI] **1.** The number of times a cycle has been carried out by a computer. **2.** The difference, or its negative, between the number of executions of a cycle which are desired and the number which have actually been carried out. { 'sī·kəl ,in,deks }

cycle index counter [COMPUT SCI] A device that counts the number of times a given cycle of instructions in a computer program has been carried out. { 'sī·kəl ,in,deks ,kau̇nt·ər }

cycle-matching loran *See* low-frequency loran. { 'sī·kəl ,mach·iŋ ,lȯ'ran }

cycle reset [COMPUT SCI] The resetting of a cycle index to its initial or other specified value. { 'sī·kəl 'rē,set }

cycle stealing [COMPUT SCI] A technique for memory sharing whereby a memory may serve two autonomous masters, commonly a central processing unit and an input-output channel or device controller, and in effect provide service to each simultaneously. { 'sī·kəl ,stēl·iŋ }

cycle time [COMPUT SCI] The shortest time elapsed between one store (or fetch) and the next store (or fetch) in the same memory unit. Also known as memory cycle. { 'sī·kəl ,tīm }

cycle timing diagram [COMPUT SCI] A diagram showing the activity that occurs in each clock cycle of a computer during the execution of a machine-language instruction. { 'sī·kəl ¦tīm·iŋ ,dī·ə,gram }

cyclic code [COMPUT SCI] A code, such as a binary code, that changes only in one digit when going from one number to the number immediately following, and in that digit by only one unit. { 'sīk·lik 'kōd }

cyclic feeding [COMPUT SCI] In character recognition, a system employed by character readers in which each input document is issued to the document transport in a predetermined and constant period of time. { 'sīk·lik 'fēd·iŋ }

cyclic redundancy check [COMPUT SCI] A block check character in which each bit is calculated by adding the first bit of a specified byte to the second bit of the next byte, and so forth, spiraling through the block; used to verify the correctness of data. Abbreviated CRC. { 'sīk·lik ri'dən·dən·sē ,chek }

cyclic shift [COMPUT SCI] A computer shift in which the digits dropped off at one end of a word are returned at the other end of the word. Also known as circuit shift; circular shift; end-around shift; nonarithmetic shift; ring shift. { 'sīk·lik 'shift }

cyclic storage [COMPUT SCI] A computer storage device, such as a magnetic drum, whose storage medium is arranged in such a way that information can be read into or extracted from individual locations at only certain fixed times in a basic cycle. { 'sīk·lik 'stȯr·ij }

cyclic transfer [COMPUT SCI] The automatic transfer of data from some medium to memory or from memory to some medium until all the data are read. { 'sīk·lik 'tranz·fər }

cyclomatic complexity [COMPUT SCI] A measure of the complexity of a software module, equal to $e - n + 2$, where e is the number of edges in the control flow graph and n is the number of nodes in this graph (that is, the cyclomatic number of the graph plus one). { ¦sī·klə,mad·ik kəm'plek·səd·ē }

cylinder [COMPUT SCI] **1.** The virtual cylinder represented by the tracks of equal radius of a set of disks on a disk drive. **2.** *See* seek area. { 'sil·ən·dər }

cylindrical antenna [ELECTROMAG] An antenna in which hollow cylinders serve as radiating elements. { sə'lin·drə·kəl an'ten·ə }

cylindrical array [ELECTR] An antenna, generally using electronic scanning, in which columns of radiating elements are arranged in a circle; used in some secondary radars. { sə'lin·drə·kəl ə'rā }

D

DAB *See* digital audio broadcasting.

DABS *See* Mode S. { dabz *or* ˌdēˌāˌbē'es }

DAC *See* digital-to-analog converter.

daemon [COMPUT SCI] In Unix, a program that runs in the background, such as a server. { 'dē·mən }

daily keying element [COMMUN] Part of a specific cipher key that changes at predetermined intervals, usually daily. { ¦dā·lē ˌkē·iŋ ˌel·ə·mənt }

daisy chain [COMPUT SCI] A means of connecting devices (readers, printers, and so on) to a central processor by party-line input/output buses which join these devices by male and female connectors, the last female connector being shorted by a suitable line termination. { 'dāz·ē ˌchān }

daisy wheel printer [COMPUT SCI] A serial printer in which the printing element is a plastic hub that has a large number of flexible radial spokes, each spoke having one or more different raised printing characters; the wheel is rotated as it is moved horizontally step by step under computer control, and stops when a desired character is in a desired print position so a hammer can drive that character against an inked ribbon. { 'dāz·ē ˌwēl ˌprint·ər }

damaged pack [COMPUT SCI] A disk drive whose use is impaired by physical damage such as a scratch on the recording surface or by a serious software error that renders control information on the disk unreadable. { 'dam·ijd 'pak }

damper [ELECTR] A diode used in the horizontal deflection circuit of a CRT display device to make the sawtooth deflection current decrease smoothly to zero instead of oscillating at zero; the diode conducts each time the polarity is reversed by a current swing below zero. { 'dam·pər }

dance-hall machine [COMPUT SCI] A multiprocessor in which the memory is spread over several modules, and a switch is used to make connections between memory modules and processors, so that several processors can use the memory simultaneously. { 'dans ˌhȯl məˌshēn }

dangling ELSE [COMPUT SCI] A situation in which it is not clear to which part of a compound conditional statement an ELSE instruction belongs. { ¦daŋ·gliŋ 'els }

dark current *See* electrode dark current. { 'därk ˌkər·ənt }

DARS *See* direct audio radio service. { ¦dē¦ä¦är'es *or* därz }

DASD *See* direct-access storage device. { 'daz ˌdē }

DAT *See* digital audio tape.

data [COMPUT SCI] **1.** General term for numbers, letters, symbols, and analog quantities that serve as input for computer processing. **2.** Any representations of characters or analog quantities to which meaning, if not information, may be assigned. { 'dad·ə, 'dād·ə, *or* 'däd·ə }

data acquisition [COMMUN] The phase of data handling that begins with the sensing of variables and ends with a magnetic recording or other record of raw data; may include a complete radio telemetering link. { 'dad·ə ˌak·wəˌzish·ən }

data acquisition computer [COMPUT SCI] A computer that is used to acquire and analyze data generated by instruments. { 'dad·ə ˌak·wə ˌzish·ən kəm'pyüd·ər }

data aggregate [COMPUT SCI] The set of data items within a record. { 'dad·ə ˌag·rə·gət }

data analysis [COMPUT SCI] The evaluation of digital data. { 'dad·ə əˌnal·ə·səs }

data attribute [COMPUT SCI] A characteristic of a block of data, such as the type of representation used or the length in characters. { 'dad·ə ¦a·trə'byüt }

data automation [COMPUT SCI] The use of electronic, electromechanical, or mechanical equipment and associated techniques to automatically record, communicate, and process data and to present the resultant information. { ¦dad·ə ȯd·ə'mā·shən }

data bank [COMPUT SCI] A complete collection of information such as contained in automated files, a library, or a set of computer disks. { 'dad·ə ˌbaŋk }

database [COMPUT SCI] A nonredundant collection of interrelated data items that can be shared and used by several different subsystems. { 'dad·əˌbās }

database/data communication [COMPUT SCI] An advanced software product that combines a database management system with data communications procedures. Abbreviated DB/DC. { 'dad·əˌbās 'dad·ə kəˌmyü·nə'kā·shən }

database machine [COMPUT SCI] A computer that handles the storage and retrieval of data into and out of a database. { 'dad·əˌbās məˌshēn }

database management system [COMPUT SCI] A special data processing system, or part of a data processing system, which aids in the storage, manipulation, reporting, management, and control of data. Abbreviated DBMS. { 'dad·ə,bās 'man·ij·mənt ,sis·təm }

database server [COMPUT SCI] An independently functioning computer in a local-area network that holds and manages the database. { 'dad·ə,bās ,sər·vər }

data break [COMPUT SCI] A facility which permits input/output transfers to occur without disturbing program execution in a computer. { 'dad·ə ,brāk }

data buffering [COMPUT SCI] The temporary collection and storage of data awaiting further processing in physical storage devices, allowing a computer and its peripheral devices to operate at different speeds. { 'dad·ə ,bəf·ə·riŋ }

data bus [ELECTR] An internal channel that carries data between a computer's central processing unit and its random-access memory. { 'dad·ə ,bəs }

data capture [COMPUT SCI] The acquisition of data to be entered into a computer. { 'dad·ə ,kap·chər }

data carrier [COMPUT SCI] A medium on which data can be recorded, and which is usually easily transportable, such as disks or tape. { 'dad·ə ,kar·ē·ər }

data carrier storage [COMPUT SCI] Any type of storage in which the storage medium is outside the computer, such as disks and tape, in contrast to inherent storage. { 'dad·ə ,kar·ē·ər ,stör·ij }

data cartridge [COMPUT SCI] A tape cartridge used for nonvolatile and removable data storage in small digital systems. { 'dad·ə ,kar·trij }

data cell drive [COMPUT SCI] A large-capacity storage device consisting of strips of magnetic tape which can be individually transferred to the read-write head. { 'dad·ə ,sel ,drīv }

data center [COMPUT SCI] An organization established primarily to acquire, analyze, process, store, retrieve, and disseminate one or more types of data. { 'dad·ə ,sen·tər }

data chain [COMPUT SCI] Any combination of two or more data elements, data items, data codes, and data abbreviations in a prescribed sequence to yield meaningful information; for example, "date" consists of data elements year, month, and day. { 'dad·ə ,chān }

data chaining [COMPUT SCI] A technique used in scatter reading or scatter writing in which new storage areas are defined for use as soon as the current data transfer is completed. { 'dad·ə ,chān·iŋ }

data channel [COMPUT SCI] A bidirectional data path between input/output devices and the main memory of a digital computer permitting one or more input/output operations to proceed concurrently with computation. { 'dad·ə ,chan·əl }

data circuit [ELECTR] A telephone facility that allows transmission of digital data pulses with minimum distortion. { 'dad·ə ,sər·kət }

data code [COMPUT SCI] A number, letter, character, symbol, or any combination thereof, used to represent a data item. { 'dad·ə ,kōd }

data collection [COMPUT SCI] The process of sending data to a central point from one or more locations. { 'dad·ə kə,lek·shən }

data communication network [COMPUT SCI] A set of nodes, consisting of computers, terminals, or some type of communication control units in various locations, connected by links consisting of communication channels providing a data path between the nodes. { 'dad·ə kə,myü·nə ,kā·shən 'net,wərk }

data communications [COMMUN] The conveying from one location to another of information that originates or is recorded in alphabetic, numeric, or pictorial form, or as a signal that represents a measurement; includes telemetering and facsimile but not voice or television. Also known as data transmission. { 'dad·ə kə ,myü·nə'kā·shənz }

data communications processor [COMPUT SCI] A small computer used to control the flow of data between machines and terminals over communications channels. { 'dad·ə kə,myü·nə ¦kā·shənz 'präs,es·ər }

data compression [COMPUT SCI] Reduction in the number of bits used to represent an item of data. Also known as compression. { 'dad·ə kəm ,presh·ən }

data concentrator [ELECTR] A device, such as a microprocessor, that takes data from several different teletypewriter or other slow-speed lines and feeds them to a single higher-speed line. { 'dad·ə kän·sən,trād·ər }

data conversion [COMPUT SCI] The changing of the representation of data from one form to another, as from binary to decimal, or from one physical recording medium to another (as from tape to disk), or from one file format to another, or from one programming language to another. Also known as conversion. { 'dad·ə kən,vər·zhən }

data conversion line [COMPUT SCI] The channel, electronic or manual, through which data elements are transferred between data banks. { 'dad·ə kən,vər·zhən ,līn }

data converter *See* converter. { 'dad·ə kən ,vərd·ər }

data definition [COMPUT SCI] The statements in a computer program that specify the physical attributes of the data to be processed, such as location and quantity of data. { 'dad·ə ,def· ə'nish·ən }

data dependence graph [COMPUT SCI] A chart that represents a program in a data flow language, in which each node is a function and each arc carries a value. { 'dad·ə di,pen·dəns ,graf }

data description language [COMPUT SCI] A programming language used to specify the arrangement of data items within a database. { 'dad· ə di¦skrip·shən ,laŋ·gwij }

data descriptor [COMPUT SCI] A pointer indicating the memory location of a data item. { 'dad· ə di'skrip·tər }

data dictionary [COMPUT SCI] A catalog which contains the names and structures of all data types. { 'dad·ə ,dik·shə,ner·ē }

data display [COMPUT SCI] Visual presentation of processed data by specially designed electronic or electromechanical devices, such as video monitors, through interconnection (either on- or off-line) with digital computers or component equipments. { 'dad·ə di,splā }

data distribution [COMPUT SCI] Data transmission to one or more locations from a central point. { 'dad·ə ,dis·trə,byü·shən }

data division [COMPUT SCI] The section of a program (written in the COBOL language) which describes each data item used for input, output, and storage. { 'dad·ə di,vizh·ən }

data-driven execution [COMPUT SCI] A mode of carrying out a program in a data flow system, in which an instruction is carried out whenever all its input values are present. { 'dad·ə ,driv·ən ,ek·sə'kyü·shən }

data element [COMPUT SCI] A set of data items pertaining to information of one kind, such as months of a year. [COMMUN] An item of data as represented before encoding and after decoding. { 'dad·ə ,el·ə·mənt }

data encryption standard [COMMUN] A cryptographic algorithm of validated strength which is in the public domain and is accepted as a standard. Abbreviated DES. { 'dad·ə en,krip·shən 'stan·dərd }

data entry [COMPUT SCI] The procedures for placing data in a computer system. { 'dad·ə ,en·trē }

data entry program [COMPUT SCI] An application program that receives data from a keyboard or other input device and stores it in a computer system. Also known as input program. { 'dad·ə ¦en·trē ,prō·grəm }

data entry terminal [COMPUT SCI] A portable keyboard and small numeric display designed for interactive communication with a computer. { 'dad·ə ¦en·trē ,tər·mən·əl }

data error [COMPUT SCI] A deviation from correctness in data, usually an error, which occurred prior to processing the data. { 'dad·ə ,er·ər }

data exchange system [COMPUT SCI] A combination of hardware and software designed to accept data from various sources, sort the data according to its destination and priority, carry out any necessary code conversions, and transmit the data to its destination. { 'dad·ə iks¦chānj ,sis·təm }

data expansion [COMPUT SCI] The reproduction in its original form of information that has undergone data compression. { 'dad·ə ik,span·chən }

data field [COMPUT SCI] An area in the main memory of the computer in which a data record is contained. { 'dad·ə ,fēld }

data flow [COMMUN] The route followed by a data message from its origination to its destination, including all the nodes through which it travels. [COMPUT SCI] The transfer of data from an external storage device, through the processing unit and memory, and out to an external storage device. { 'dad·ə ,flō }

data flow analysis [COMPUT SCI] The development of models for the movement of information within an organization, indicating the sources and destinations of information and where and how information is transmitted, processed, and stored. { 'dad·ə ¦flō ə,nal·ə·səs }

data flow diagram [COMPUT SCI] A chart that traces the movement of data in a computer system and shows how the data is to be processed, using circles to represent data. Also known as bubble chart; system flowchart. { 'dad·ə ¦flō ,dī·ə,gram }

data flow language [COMPUT SCI] A programming language used in a data flow system. { 'dad·ə ¦flō ,laŋ·gwij }

data flow system [COMPUT SCI] An alternative to conventional programming languages and architectures which is able to achieve a high degree of parallel computation, in which values rather than value containers are dealt with, and in which all processing is achieved by applying functions to values to produce new values. { 'dad·ə ¦flō ,sis·təm }

data flow technique [COMPUT SCI] A method of computer system design in which diagrams and charts that show how data is to be handled by the system are used to prepare detailed specifications from which actual programs can be written. { 'dad·ə ¦flō tek,nēk }

data formatting [COMPUT SCI] Structuring the presentation of data as numerical or alphabetic and specifying the size and type of each datum. { 'dad·ə fȯr'mad·iŋ }

data fusion [ELECTR] The combining of data as from several radars or other sensors with common fields of view, in order to improve the accuracy of the estimations being made about features of interest. { 'dad·ə ,fyü·zhən }

data generator [COMPUT SCI] A specialized word generator in which the programming is designed to test a particular class of device, the pulse parameters and timing are adjustable, and selected words may be repeated, reinserted later in the sequence, omitted, and so forth. { 'dad·ə ,jen·ə,rād·ər }

datagram [COMPUT SCI] A unit of information in the Internet Protocol (IP) containing both data and address information. In TCP/IP networks, datagrams are referred to as packets. { 'dad·ə ,gram }

data-handling system [COMPUT SCI] Automatically operated equipment used to interpret data gathered by instrument installations. Also known as data reduction system. { 'dad·ə ,hand·liŋ ,sis·təm }

data independence [COMPUT SCI] Separation of data from processing, either so that changes in the size or format of the data elements require no change in the computer programs processing them or so that these changes can be made automatically by the database management system. { 'dad·ə in·də'pen·dəns }

data-initiated control [COMPUT SCI] The automatic handling of a program dependent only upon the value of input data fed into the computer. { 'dad·ə i,nish·ē,ād·əd kən'trōl }

data-intense application [COMPUT SCI] A program or computer system that handles large quantities of data and extremely repetitive tasks. { 'dad·ə in¦tens ,ap·lə'kā·shən }

data interchange [COMPUT SCI] Switching of data in and out of storage units. { 'dad·ə 'in·tər ,chānj }

data item [COMPUT SCI] A single member of a data element. Also known as datum. { 'dad·ə ,ī·dəm }

data level [COMPUT SCI] The rank of a data element in a source language with respect to other elements in the same record. { 'dad·ə ,lev·əl }

data library [COMPUT SCI] A center for the storage of data not in current use by the computer. { 'dad·ə lī,brer·ē }

data line [COMMUN] An individual circuit that transmits data within a communications or computer channel. { 'dad·ə ,līn }

data line monitor [COMMUN] A test instrument that analyzes the signals transmitted over a communications line and provides a visual display or stores the results for further analysis, or both. { ¦dad·ə ,līn 'män·əd·ər }

data link [COMMUN] The physical equipment for automatic transmission and reception of information. Also known as communication link; information link; tie line; tie-link. { 'dad·ə ,liŋk }

data logging [COMPUT SCI] Conversion of electrical impulses from process instruments into digital data to be recorded, stored, and periodically tabulated. { 'dad·ə ,läg·iŋ }

data management [COMPUT SCI] The collection of functions of a control program that provide access to data sets, enforce data storage conventions, and regulate the use of input/output devices. { 'dad·ə ,man·ij·mənt }

data management program [COMPUT SCI] A computer program that keeps track of what is in a computer system and where it is located, and of the various means to store and access the data efficiently. { 'dad·ə ,man·ij·mənt ,prō·grəm }

data manipulation [COMPUT SCI] The standard operations of sorting, merging, input/output, and report generation. { 'dad·ə mə,nip·yə,lā·shən }

data manipulation language [COMPUT SCI] The interface between a data base and an applications program, which is embedded in the language of the applications program and provides the programmer with procedures for accessing data in the data base. { 'dad·ə mə,nip·yə¦lā·shən ,laŋ·gwij }

data mining [COMPUT SCI] **1.** The identification or extraction of relationships and patterns from data using computational algorithms to reduce, model, understand, or analyze data. **2.** The automated process of turning raw data into useful information by which intelligent computer systems sift and sort through data, with little or no help from humans, to look for patterns or to predict trends. { 'dad·ə ,mīn·iŋ }

data module [COMPUT SCI] A sealed disk drive unit that includes mechanical and electronic components for handling data stored on the disk. { 'dad·ə ,mäj·yül }

data move instruction [COMPUT SCI] An instruction in a computer program to transfer data between memory locations and registers or between the central processor and peripheral devices. { 'dad·ə ,müv in'strək·shən }

data name [COMPUT SCI] A symbolic name used to represent an item of data in a source program, in place of the address of the data item. { 'dad·ə ,nām }

data organization [COMPUT SCI] Any one of the data management conventions for physical and spatial arrangement of the physical records of a data set. Also known as data set organization. { 'dad·ə ,ȯr·gə·nə,zā·shən }

data origination [COMPUT SCI] The process of putting data in a form that can be read by a machine. { 'dad·ə ə,rij·ə'nā·shən }

data patch panel [COMMUN] A plugboard used to rearrange communications lines and modems by connecting them with double-ended cables, or to attach monitoring devices to analyze circuit signals. { 'dad·ə 'pach ,pan·əl }

data plotter [COMPUT SCI] A device which plots digital information in a continuous fashion. { 'dad·ə ,pläd·ər }

data processing [COMPUT SCI] Any operation or combination of operations on data, including everything that happens to data from the time they are observed or collected to the time they are destroyed. Also known as information processing. { 'dad·ə 'präs,es·iŋ }

data processing center [COMPUT SCI] A computer installation providing data processing service for others, sometimes called customers, on a reimbursable or nonreimbursable basis. { 'dad·ə ¦präs,es·iŋ ,sent·ər }

data processing inventory [COMPUT SCI] An identification of all major data processing areas in an agency for the purpose of selecting and focusing upon those in which the use of automatic data processing (ADP) techniques appears to be potentially advantageous, establishing relative priorities and schedules for embarking on ADP studies, and identifying significant relationships among areas to pinpoint possibilities for the integration of systems. { 'dad·ə¦präs,es·iŋ ,in·vən ,tȯr·ē }

data processor [COMPUT SCI] **1.** Any device capable of performing operations on data, for instance, a desk calculator, an analog computer, or a digital computer. **2.** Person engaged in processing data. { 'dad·ə 'präs,es·ər }

data protection [COMPUT SCI] The safeguarding of data against unauthorized access or accidental or deliberate loss or damage. { 'dad·ə prə,tek·shən }

data purification [COMPUT SCI] The process of removing as many inaccurate or incorrect items as possible from a mass of data before automatic data processing is begun. { 'dad·ə pyu̇r·ə·fə'kā·shən }

data rate [COMMUN] The number of digital bits per second that are recorded or retrieved from a data storage device during the transfer of a large data block. { 'dad·ə ˌrāt }

data record [COMPUT SCI] A collection of data items related in some fashion and usually contiguous in location. { 'dad·ə ˌrek·ərd }

data recorder [COMPUT SCI] A keyboard device for entering data onto magnetic tape. { 'dad·ə riˌkȯr·dər }

data reduction [COMPUT SCI] The transformation of raw data into a more useful form. { 'dad·ə riˌdək·shən }

data reduction system *See* data-handling system. { ˌdad·ə riˌdək·shən ˌsis·təm }

data redundancy [COMPUT SCI] The occurrence of values for data elements more than once within a file or database. { 'dad·ə riˌdən·dən·sē }

data register [COMPUT SCI] A register used in microcomputers to temporarily store data being transmitted to or from a peripheral device. { 'dad·ə ˌrej·ə·stər }

data representation [COMPUT SCI] **1.** The way that the physical properties of a medium are used to represent data. **2.** The manner in which data is expressed symbolically by binary digits in a computer. { 'dad·ə ˌrep·ri·zen'tā·shən }

data retrieval [COMPUT SCI] The searching, selecting, and retrieving of actual data from a personnel file, data bank, or other file. { 'dad·ə ri'trē·vəl }

data rules [COMPUT SCI] Conditions which must be met by data to be processed by a computer program. { 'dad·ə ˌrülz }

data scope [ELECTR] An electronic display that shows the content of the information being transmitted over a communications channel. { 'dad·ə ˌskōp }

data security [COMPUT SCI] The protection of data against the deliberate or accidental access of unauthorized persons. Also known as file security. { 'dad·ə səˌkyu̇r·əd·ē }

data set [COMPUT SCI] **1.** A named collection of similar and related data records recorded upon some computer-readable medium. **2.** A data file in IBM 360 terminology. { 'dad·ə ˌset }

data set coupler [COMPUT SCI] The interface between a parallel computer input/output bus and the serial input/output of a modem. { 'dad·ə ˌset ˌkəp·lər }

data set label [COMPUT SCI] A data element that describes a data set, and usually includes the name of the data set, its boundaries in physical storage, and certain characteristics of data items within the set. { 'dad·ə ˌset ˌlā·bəl }

data set migration [COMPUT SCI] The process of moving inactive data sets from on-line storage to back up storage in a time-sharing environment. { 'dad·ə ˌset mīˌgrā·shən }

data set organization *See* data organization. { 'dad·ə ˌset ˌȯr·gə·nəˌzā·shən }

data sink [COMPUT SCI] A memory or recording device capable of accepting data signals from a data transmission device and storing data for future use. { 'dad·ə ˌsiŋk }

data source [COMPUT SCI] A device capable of originating data signals for a data transmission device. { 'dad·ə ˌsȯrs }

data stabilization [ELECTR] Stabilization of the display of radar signals with respect to a selected reference, regardless of changes in radar-carrying vehicle attitude, as in azimuth-stabilized plan-position indicator. { 'dad·ə ˌstā·bə·ləˌzā·shən }

data statement [COMPUT SCI] An instruction in a source program that identifies an item of data in the program and specifies its format. { 'dad·ə ˌstāt·mənt }

data station [COMPUT SCI] A remote input/output device which handles a variety of transmissions to and from certain centralized computers. { 'dad·ə ˌstā·shən }

data station control [COMPUT SCI] The supervision of a data station by means of a program resident in the central computer. { 'dad·ə ˌstā·shən kənˌtrōl }

data stream [COMMUN] The continuous transmission of data from one location to another. { 'dad·ə ˌstrēm }

data striping *See* disk striping. { 'dad·ə ˌstrīp·iŋ }

data structure [COMPUT SCI] A collection of data components that are constructed in a regular and characteristic way. { 'dad·ə ˌstrək·chər }

data switch [COMPUT SCI] A manual or automatic device that connects data-processing machines to one another. { 'dad·ə ˌswich }

data system [COMPUT SCI] The means, either manual or automatic, of converting data into action or decision information, including the forms, procedures, and processes which together provide an organized and interrelated means of recording, communicating, processing, and presenting information relative to a definable function or activity. { 'dad·ə ˌsis·təm }

data system interface [COMPUT SCI] **1.** A common aspect of two or more data systems involving the capability of intersystem communications. **2.** A common boundary between automatic data-processing systems or parts of a single system. { 'dad·ə ¦sis·təm 'in·tərˌfās }

data systems integration [COMPUT SCI] Achievement through systems design of an improved or broader capability by functionally or technically relating two or more data systems, or by incorporating a portion of the functional or technical elements of one data system into another. { 'dad·ə ˌsis·təmz ˌin·təˌgrā·shən }

data system specifications [COMPUT SCI] **1.** The delineation of the objectives which a data system is intended to accomplish. **2.** The data processing requirements underlying that accomplishment; includes a description of the data output, the data files and record content, the volume of data, the processing frequencies, training, and such other facts as may be necessary to provide a full description of the system. { 'dad·ə ˌsis·təm ˌspes·ə·fəˌkā·shən }

data table [COMPUT SCI] An on-screen display of the information in a database management system, presented in columnar format, with field names at the top. { 'dad·ə ˌtā·bəl }

data tablet See electronic tablet. { 'dad·ə ˌtab·lət }

data tracks [COMPUT SCI] Information storage positions on drum storage devices; information is stored on the drum surface in the form of magnetized or nonmagnetized areas. { 'dad·ə ˌtraks }

data transcription equipment [COMPUT SCI] Those devices or equipment designed to convey data from its original state to a data processing media. { ¦dad·ə tranzˌkrip·shən iˌkwip·mənt }

data transfer [COMPUT SCI] The technique used by the hardware manufacturer to transmit data from computer to storage device or from storage device to computer; usually under specialized program control. { 'dad·ə 'tranz·fər }

data transmission See data communications. { 'dad·ə tranz'mish·ən }

data transmission equipment [COMPUT SCI] The communications equipment used in direct support of data processing equipment. { 'dad·ə tranz'mish·ən iˌkwip·mənt }

data transmission-utilization measure [COMMUN] The ratio of useful data output to the sum total of data input. { ¦dad·ə tranz'mish·ən yüd·əl·ə'zā·shən ˌmezh·ər }

data type [COMPUT SCI] The manner in which a sequence of bits represents data in a computer program. { 'dad·ə ˌtīp }

data under voice [COMMUN] A telephone digital data service that allows digital signals to travel on the lower portion of the frequency spectrum of existing microwave radio systems; digital channels initially available handled speeds of 2.4, 4.8, 9.6, and 56 kilobits per second. Abbreviated DUV. { 'dad·ə ˌən·dər 'vȯis }

data unit [COMPUT SCI] A set of digits or characters treated as a whole. { 'dad·ə ˌyü·nət }

data validation [COMPUT SCI] The checking of data for correctness, or the determination of compliance with applicable standards, rules, and conventions. { 'dad·ə val·ə'dā·shən }

data warehouse [COMPUT SCI] **1.** A large specialized database, holding perhaps hundreds of terabytes of data. **2.** A database specifically structured for information access and reporting. { ¦dad·ə 'werˌhau̇s }

data word [COMPUT SCI] A computer word that is part of the data which the computer is manipulating, in contrast with an instruction word. Also known as information word. { 'dad·ə ˌwərd }

date time group [COMMUN] The date and time, expressed in digits and zone suffix, at which the message was prepared for transmission (expressed as six digits followed by the zone suffix; first pair of digits denoting the date, second pair the hours, third pair the minutes). { ¦dāt ¦tīm ˌgrüp }

datum See data item. { 'dad·əm, 'dād·əm, *or* 'däd·əm }

daughter board [COMPUT SCI] A small printed circuit board that is attached to another printed circuit board. { 'dȯd·ər ˌbȯrd }

day clock [COMPUT SCI] An internal binary counter, with a resolution usually of a microsecond and a cycle measured in years, providing an accurate measure of elapsed time independent of system activity. { 'dā ˌkläk }

dBa See adjusted decibel.

DB/DC See database/data communication.

DBMS See database management system.

DBRT diode See double-barrier resonant tunneling diode. { ¦dē¦bē¦är¦tē 'dīˌōd }

DB server [COMPUT SCI] The database portion of a Web server, which serves as a repository of data and content. { ¦dē¦bē ˌsər·vər }

DBS system See direct broadcasting satellite system. { ¦dē¦bē 'es ˌsis·təm }

DCFL See direct-coupled FET logic.

DCT See discrete cosine transform.

DCTL See direct-coupled transistor logic.

DDA See digital differential analyzer.

DDBS See Digital Data Broadcast System.

D-display [ELECTR] A radar display format in which the coordinates are the same as in the C-display, with target spots extended vertically to indicate range. Also known as D-indicator; D-scan; D-scope. { 'dē diˌsplā }

DDR See double data rate.

DDS See digital data service.

deaccentuator [ELECTR] A circuit used in a frequency-modulation receiver to offset the preemphasis of higher audio frequencies introduced at the transmitter. { ˌdē·ak'sen·chəˌwād·ər }

dead code [COMPUT SCI] Statements in a computer program that are not executed, usually as the result of modification of a large program. { 'ded 'kōd }

dead halt See drop-dead halt. { ¦ded 'hȯlt }

dead letter box [COMMUN] A file for storing undeliverable messages in a data communications system, particularly a message switching system. { ¦ded 'led·ər ˌbäks }

deadlock [COMPUT SCI] A situation in which a task in a multiprogramming system cannot proceed because it is waiting for an event that will never occur. Also known as deadly embrace; interlock; knot. { 'dedˌläk }

deadly embrace See deadlock. { ¦ded·lē im 'brās }

dead spot [COMMUN] A geographic location in which signals from a radio or television transmitter are received poorly or not at all. { 'ded ˌspät }

dead time [CONT SYS] The time interval between a change in the input signal to a process control system and the response to the signal. [ENG] The time interval, after a response to one signal or event, during which a system is unable to respond to another. Also known as insensitive time. { 'ded ˌtīm }

dead-time compensation [CONT SYS] The modification of a controller to allow for time delays between the input to a control system and the response to the signal. { 'ded ,tīm kăm·pən'sā·shən }

dead zone unit [COMPUT SCI] An analog computer device that maintains an output signal at a constant value over a certain range of values of the input signal. { 'ded ,zōn ,yü·nət }

deallocation [COMPUT SCI] The release of a portion of computer storage or a peripheral unit from control by a computer program when it is no longer needed. { dē,al·ə'kā·shən }

debatable time [COMPUT SCI] In the keeping of computer usage statistics, time that cannot be attributed with certainty to any one of various categories of computer use. { di'bād·ə·bəl 'tīm }

deblocking [COMPUT SCI] Breaking up a block of records into individual records. { dē'bläk·iŋ }

debug [COMPUT SCI] To test for, locate, and remove mistakes from a program or malfunctions from a computer. [ELECTR] To detect and remove secretly installed listening devices popularly known as bugs. { dē'bəg }

debugging routine [COMPUT SCI] A routine to aid programmers in the debugging of their routiness; some typical routines are storage printout, tape printout, and drum printout routines. { dē'bəg·iŋ rü,tēn }

debugging statement [COMPUT SCI] Temporary instructions inserted into a program being tested so as to pinpoint problem areas. { dē'bəg·iŋ ,stāt·mənt }

debug on-line [COMPUT SCI] **1.** To detect and correct errors in a computer program by using only certain parts of the hardware of a computer, while other routines are being processed simultaneouly. **2.** To detect and correct errors in a program from a console distant from a computer in a multiaccess system. { dē'bəg·iŋ ȯn 'līn }

decade counter *See* decade scaler. { de'kād ,kaủnt·ər }

decade scaler [ELECTR] A scaler that produces one output pulse for every 10 input pulses. Also known as counter decade; decade counter; scale-of-ten circuit. { de'kād ,skāl·ər }

Decca [NAV] A hyperbolic navigation system which establishes a line of position from measurement of the phase difference between two continuous-wave signals; the intersection of the two lines of position from two pairs of transmitting stations establishes a navigational fix, or location. { 'dek·ə }

deceleration time [COMPUT SCI] For a storage medium, such as magnetic tape that must be physically moved in order for reading or writing to take place, the minimum time that must elapse between the completion of a reading or writing operation and the moment that motion ceases. Also known as stop time. { dē,sel·ə'rā·shən ,tīm }

decentralized data processing [COMPUT SCI] An arrangement comprising a data-processing center for each division or location of a single organization. { dē'sen·trə,lizd 'dad·ə 'präs,es·iŋ }

decibel adjusted *See* adjusted decibel. { 'des·ə ,bel ə'jəs·təd }

decibel loss [COMMUN] Signal attenuation over a transmission path or a conductor expressed in decibels. { 'des·ə,bel ,lȯs }

decimal code [COMPUT SCI] A code in which each allowable position has one of 10 possible states; the conventional decimal number system is a decimal code. { ¦des·məl ¦kōd }

decimal-coded digit [COMPUT SCI] One of 10 arbitrarily selected patterns of 1 and 0 used to represent the decimal digits. Also known as coded decimal. { ¦des·məl ¦kōd·əd 'dij·ət }

decimal processor [COMPUT SCI] A digital computer organized to calculate by decimal arithmetic. { ¦des·məl 'präs,es·ər }

decimal-to-binary conversion [COMPUT SCI] The mathematical process of converting a number written in the scale of 10 into the same number written in the scale of 2. { ¦des·məl tə ¦bin·ə·re kən'vər·zhən }

decision [COMPUT SCI] The computer operation of determining if a certain relationship exists between words in storage or registers, and taking alternative courses of action; this is effected by conditional jumps or equivalent techniques. { di'sizh·ən }

decision box [COMPUT SCI] A flow-chart symbol indicating a decision instruction; usually diamond-shaped. { di'sizh·ən ,bäks }

decision calculus [SYS ENG] A guide to the process of decision-making, often outlined in the following steps: analysis of the decision area to discover applicable elements; location or creation of criteria for evaluation; appraisal of the known information pertinent to the applicable elements and correction for bias; isolation of the unknown factors; weighting of the pertinent elements, known and unknown, as to relative importance; and projection of the relative impacts on the objective, and synthesis into a course of action. { di'sizh·ən 'kal·kyə·ləs }

decision element [ELECTR] A circuit that performs a logical operation such as "and," "or," "not," or "except" on one or more binary digits of input information representing "yes" or "no" and that expresses the result in its output. Also known as decision gate. { di'sizh·ən ,el·ə·mənt }

decision gate [ELECTR] *See* decision element. [NAV] In an instrument landing, that point along the path at which the pilot must decide to land or to execute a missed-approach procedure. { di'sizh·ən ,gāt }

decision instruction *See* conditional jump. { di'sizh·ən in'strək·shən }

decision mechanism [COMPUT SCI] In character recognition, that component part of a character reader which accepts the finalized version of the input character and makes an assessment as to its most probable identity. { di'sizh·ən ,mek·ə ,niz·əm }

decision rule [SYS ENG] In decision theory, the mathematical representation of a physical system which operates upon the observed data to produce a decision. { di'sizh·ən ,rül }

decision support [COMPUT SCI] The process of filtering, optimizing, and organizing mined information to support decision making. { di'sizh·ən sə,pȯrt }

decision support system [COMPUT SCI] A computer-based system that enables management to interrogate the computer system on an ad hoc basis for various kinds of information on the organization and to predict the effect of potential decisions beforehand. Abbreviated DSS. { di'sizh·ən sə'pȯrt ,sis·təm }

decision table [COMPUT SCI] **1.** A table of contingencies to be considered in the definition of a problem, together with the actions to be taken; sometimes used in place of a flow chart for program documentation. **2.** *See* DETAB. { di'sizh·ən ,tā·bəl }

decision theory [SYS ENG] A broad spectrum of concepts and techniques which have been developed to both describe and rationalize the process of decision making, that is, making a choice among several possible alternatives. { di'sizh·ən ,thē·ə·rē }

deck [ENG] A magnetic-tape transport mechanism. { dek }

declaration *See* declarative statement. { ,dek·lə'rā·shən }

declarative language [COMPUT SCI] A nonprocedural programming language that allows the programmer to state the task to be accomplished without specifying the procedures needed to carry it out. { di,klar·əd·iv 'laŋ·gwij }

declarative macroinstruction [COMPUT SCI] An instruction in an assembly language which directs the compiler to take some action or take note of some condition and which does not generate any instruction in the object program. { di¦klar·əd·iv ¦mak·rō·in¦strək·shən }

declarative markup language [COMPUT SCI] A system of codes for identifying the subdivisions of a text-processing document, without carrying out the actual formatting. { di,klar·əd·iv 'mär·kəp ,laŋ·gwij }

declarative statement [COMPUT SCI] Any program statement describing the data which will be used or identifying the memory locations which will be required. Also known as declaration. { di¦klar·əd·iv 'stāt·mənt }

decode [COMMUN] **1.** To translate coded characters into a more understandable form. **2.** *See* demodulate. { dē'kōd }

decoded stream [COMMUN] The decoded reconstruction of a compressed bit stream. { dē'kōd·əd 'strēm }

decoder [ELECTR] **1.** A matrix of logic elements that selects one or more output channels, depending on the combination of input signals present. **2.** *See* decoder circuit; matrix; tree. { dē'kōd·ər }

decoder circuit [ELECTR] A circuit that responds to a particular coded signal while rejecting others. Also known as decoder. { dē'kōd·ər ,sər·kət }

decoding gate [COMPUT SCI] The use of combinatorial logic in circuitry to select a device identified by a binary address code. Also known as recognition gate. { dē'kōd·iŋ ,gāt }

decollator [COMPUT SCI] A device which separates the sheets of continuous stationery that form the output of a computer printer into separate stacks. { dē'kō,lād·ər }

decommutation [ELECTR] The process of recovering a signal from the composite signal previously created by a commutation process. { dē ,käm·yə'tā·shən }

decommutator [ELECTR] The section of a telemetering system that extracts analog data from a time-serial train of samples representing a multiplicity of data sources transmitted over a single radio-frequency link. { dē'käm·yə ,tād·ər }

decrement [COMPUT SCI] **1.** A specific part of an instruction word in some binary computers, thus a set of digits. **2.** For a counter, to subtract 1 or some other number from the current value. { 'dek·rə·mənt }

decrement field [COMPUT SCI] That part of an instruction word which is used to modify the contents of a storage location or register. { 'dek·rə·mənt ,fēld }

decrypt [ELECTR] To convert a crypotogram or series of electronic pulses into plain text by electronic means. { dē'kript }

dedicated file server [COMPUT SCI] A computer that operates solely to provide services to other computers in a particular local-area network and to manage the network operating system. Also known as dedicated server. { ,ded·ə,kād·əd 'fīl ,sər·vər }

dedicated line [COMPUT SCI] A permanent communications link that is used solely to transmit information between a computer and a data-processing system. { 'ded·ə,kād·əd 'līn }

dedicated server *See* dedicated file server. { ,ded·ə,kād·əd ,sər·vər }

dedicated terminal [COMPUT SCI] A computer terminal that is permanently connected to a data-processing system by a communications link that is used only to transmit information between the two. { 'ded·ə,kād·əd 'tərm·ən·əl }

Deep Space Network [AERO ENG] A spacecraft network operated by NASA which tracks, commands, and receives telemetry for all types of spacecraft sent to explore deep space, the moon, and solar system planets. Abbreviated DSN. { ¦dēp ¦spās 'net,wərk }

deerhorn antenna [ELECTROMAG] A dipole antenna whose ends are swept back to reduce wind resistance when mounted on an airplane. { 'dir ,hȯrn an'ten·ə }

de facto standard [COMPUT SCI] A set of criteria for software, hardware, or communications procedures that is widely accepted because of the dominance of a particular technology over others rather than the action of a recognized standards organization. { dē 'fak·tō 'stan·dərd }

default [COMPUT SCI] A value automatically used or an action automatically carried out unless another is specified. { di'fȯlt }

default printer [COMPUT SCI] The printer that is automatically used by a program unless another printer is specifically designated. { di'fölt ˌprint·ər }

defective track [COMPUT SCI] Any circular path on the surface of a magnetic disk which is detected by the system as unable to accept one or more bits of data. { di'fek·tiv 'trak }

deferred addressing [COMPUT SCI] A type of indirect addressing in which the address part of an instruction specifies a location containing an address, the latter in turn specifies another location containing an address, and so forth, the number of iterations being controlled by a preset counter. { di'fərd ə'dres·iŋ }

deferred data item [COMPUT SCI] A quantity or attribute that is assigned a value only at the time it is actually processed. { di'fərd 'dad·ə ˌīd·əm }

deferred entry [COMPUT SCI] The passing of control of the central processing unit to a subroutine or to an entry point as the result of an asynchronous event. { di'fərd 'en·trē }

deferred mount [COMPUT SCI] Postponement of the placement of a tape on a tape drive until it is actually needed, rather than when the program starts to run. { di'fərd 'maůnt }

deferred processing [COMPUT SCI] The making of computer runs which are postponed until nonpeak periods. { di'fərd 'präsˌes·iŋ }

definite network [COMPUT SCI] A sequential network in which no feedback loops exist. { ¦def·ə·nət 'netˌwərk }

definition [COMMUN] The fidelity with which an imaging system conveys and reproduces an image. [ELECTR] The extent to which the fine-line details of a printed circuit correspond to the master drawing. { ˌdef·ə'nish·ən }

deflection [COMPUT SCI] Encouraging a potential attacker of a computer system to direct the attack elsewhere. [ELECTR] The displacement of an electron beam from its straight-line path by an electrostatic or electromagnetic field. { di'flek·shən }

deflection circuit [ELECTR] A circuit which controls the deflection of an electron beam in a cathode-ray tube. { di'flek·shən ˌsər·kət }

deflection coil [ELECTR] One of the coils in a deflection yoke. { di'flek·shən ˌkȯil }

deflection electrode [ELECTR] An electrode whose potential provides an electric field that deflects an electron beam. Also known as deflection plate. { di'flek·shən iˌlekˌtrōd }

deflection-modulated indicator *See* amplitude-modulated indicator. { di'flek·shən ¦mäj·əˌlād·əd 'in·dəˌkād·ər }

deflection plate *See* deflection electrode. { di'flek·shən ˌplāt }

deflection voltage [ELECTR] The voltage applied between a pair of deflection electrodes to produce an electric field. { di'flek·shən ˌvōl·tij }

deflection yoke [ELECTR] An assembly of one or more electromagnets that is placed around the neck of an electron-beam tube to produce a magnetic field for deflection of one or more electron beams. Also known as scanning yoke; yoke. { di'flek·shən ˌyōk }

defragmentation [COMPUT SCI] A procedure in which portions of files on a computer disk are moved until all parts of each file occupy continuous sectors, resulting in a substantial improvement in disk access times. { ˌdēˌfrag·mən'tā·shən }

defragmenter [COMPUT SCI] A program that analyzes storage locations of files on a computer disk and then carries out defragmentation. { ˌdēˌfrag'men·tər }

defruit [ELECTR] To remove random asynchronous replies from the video input of a display unit in a secondary (beacon) radar system by such means as comparing the video signals on successive sweeps. { dē'früt }

degauss [ELECTROMAG] To neutralize (demagnetize) a magnetic field of, for example, a ship hull or television tube; a direct current of the correct value is sent through a cable around the ship hu

degaussing coil [ELECTROMAG] A plastic-encased coil, about 1 foot (0.3 meter) in diameter, that can be plugged into a 120-volt alternating-current wall outlet and moved slowly toward and away from a color television picture tube to demagnetize adjacent parts. { dē'gaůs·iŋ ˌkȯil }

deglitcher [ELECTR] A nonlinear filter or other special circuit used to limit the duration of switching transients in digital converters. { dē'glich·ər }

degradation [COMPUT SCI] Condition under which a computer operates when some area of memory or some units of peripheral equipment are not available to the user. { ˌdeg·rə'dā·shən }

delay [COMMUN] **1.** Time required for a signal to pass through a device or a conducting medium. **2.** Time which elapses between the instant at which any designated point of a transmitted wave passes any two designated points of a transmission circuit; such delay is primarily determined by the constants of the circuit. { di'lā }

delay counter [COMPUT SCI] A counter which inserts a time delay in a sequence of events. { di'lā ˌkaůnt·ər }

delay distortion [ELECTR] Phase distortion in which the rate of change of phase shift with frequency of a circuit or system is not constant over the frequency range required for transmission. Also called envelope delay distortion. { di'lā di'stȯrˌshən }

delayed automatic gain control [ELECTR] An automatic gain control system that does not operate until the signal exceeds a predetermined magnitude; weaker signals thus receive maximum amplification. Also known as biased automatic gain control; delayed automatic volume control; quiet automatic volume control. { di'lād ˌȯd·ə¦mad·ik 'gān kənˌtrōl }

delayed automatic volume control *See* delayed automatic gain control. { di'lād ˌȯd·ə¦mad·ik 'väl·yəm kənˌtrōl }

delay/frequency distortion [COMMUN] That form of distortion which occurs when the delay of a circuit or system is not constant over the frequency range required for transmissions. { di¦lā ¦frē·kwən·sē di'stȯr·shən }

delay line [ELECTR] **1.** A transmission line (as dissipationless as possible), or an electric network approximation of it, which, if terminated in its characteristic impedance, will reproduce at its output a waveform applied to its input terminals with little distortion, but at a time delayed by an amount dependent upon the electrical length of the line. Also known as artificial delay line. **2.** A circuit component, analog or digital, in a radar system by which pulses may be delayed a controllable amount; used typically for pulse comparisons as in canceler circuits. { di'lā,līn }

delay unit *See* transport delay unit. { di'lā ,yü·nət }

deleted representation [COMPUT SCI] In paper tape codes, the superposition of a pattern of holes upon another pattern of holes representing a character, to effectively remove or obliterate the latter. { di'lēd·əd ,rep·rə,zen'tā·shən }

deletion operator [COMPUT SCI] The part of a data structure which allows components to be deleted. { di'lē·shən ,äp·ə,rād·ər }

deletion record [COMPUT SCI] A record which removes and replaces an existing record when it is added to a file. { di'lē·shən ,rek·ərd }

delimiter [COMPUT SCI] A character that separates items of data. { də'lim·əd·ər }

Dellinger fadeout [COMMUN] Type of fadeout that occurs during shortwave reception, believed to be caused by rapid shifting of ionosphere layers during solar eruptions. { 'del·ən·jər 'fād,au̇t }

delta-gun tube [ELECTR] A color television picture tube in which three electron guns, arranged in a triangle, provide electron beams that fall on phosphor dots on the screen, causing them to emit light in three primary colors; a shadow mask located just behind the screen ensures that each beam excites only dots of one color. { 'del·tə ,gən ,tüb }

delta modulation [ELECTR] A pulse-modulation technique in which a continuous signal is converted into a binary pulse pattern, for transmission through low-quality channels. { 'del·tə ,mäj·ə'lā·shən }

delta pulse code modulation [ELECTR] A modulation system that converts audio signals into corresponding trains of digital pulses to give greater freedom from interference during transmission over wire or radio channels. { 'del·tə ¦pəls ,kōd ,mäj,ə'lā·shən }

delta-sigma converter *See* sigma-delta converter. { ¦del·tə ¦sig·mə kən'vərd·ər }

delta-sigma modulator *See* sigma-delta modulator. { ¦del·tə¦sig·mə 'mä·jə,lād·ər }

deltic method [ELECTR] A method of sampling incoming radar, sonar, seismic, speech, or other waveforms along with reference signals, compressing the samples in time, and comparing them by autocorrelation. { 'del·tik ,meth·əd }

demand assignment multiple access [COMMUN] The allocation of bandwidth in a communications system among multiple users based on demand, such as by multiplexing. Abbreviated DAMA. { di¦mand ə,sīn·mənt ¦məl·tə·pəl 'ak ,ses }

demand-driven execution [COMPUT SCI] A mode of carrying out a program in a data flow system in which no calculation is carried out until its results are demanded as input to another calculation. Also known as lazy evaluation. { də'mand ,driv·ən ,ek·sə'kyü·shən }

demand paging [COMPUT SCI] The characteristic of a virtual memory system which retrieves only that part of a user's program which is required during execution. { də'mand ,pā·jiŋ }

demand processing [COMPUT SCI] The processing of data by a computer system as soon as it is received, so that it is not necessary to store large amounts of raw data. Also known as immediate processing. { də'mand ,präs,es·iŋ }

demand reading [COMPUT SCI] A method of carrying out input operations in which blocks of data are transmitted to the central processing unit as needed for processing. { də'mand ,rēd·iŋ }

demand staging [COMPUT SCI] Moving blocks of data from one storage device to another when programs request them. { də'mand ,stā·jiŋ }

demand writing [COMPUT SCI] A method of carrying out output operations in which blocks of data are transmitted from the central processing unit as they are needed by the user. { də'mand ,rīd·iŋ }

demodifier [COMPUT SCI] A data element used to restore part of an instruction which has been modified to its original value. { dē'mäd·ə,fī·ər }

demodulate [COMMUN] To recover the modulating wave from a modulated carrier. Also known as decode; detect. { dē'mäj·ə,lāt }

demodulation [COMMUN] The recovery, from a modulated carrier, of a signal having substantially the same characteristics as the original signal. { dē,mäj·ə'lā·shən }

demodulator *See* detector. { dē'mäj·ə,lad·ər }

demount [COMPUT SCI] To take out a magnetic storage medium from a device that reads or writes on it. { dē'mau̇nt }

demountable pack [COMPUT SCI] A disk pack that can be taken out and replaced by another. { dē'mau̇nt·ə·bəl 'pak }

DEMS *See* Digital Electronic Message Service.

demultiplexer [ELECTR] A device used to separate two or more signals that were previously combined by a compatible multiplexer and transmitted over a single channel. { dē,məl·tə ,plek·sər }

demultiplexing [COMMUN] The separation of two or more channels previously multiplexed. { dē'məl·tə,pleks·iŋ }

demultiplexing circuit [ELECTR] A circuit used to separate the signals that were combined for transmission by multiplex. { dē'məl·tə,plek·siŋ ,sər·kət }

dense binary code [COMPUT SCI] A code in which all possible states of the binary pattern are used. { 'dens ¦bī·nə·rē 'kōd }

dense list [COMPUT SCI] A list in which all the cells contain records of the file. { ¦dens ¦list }

density packing [COMPUT SCI] In computers, the number of binary digit magnetic pulses stored on tape or drum per linear inch on a single track by a single head. { 'den·səd·ē ,pak·iŋ }

density step tablet [COMMUN] Facsimile test chart consisting of a series of areas; density of the areas increases from a low value to a maximum value in steps. Also known as step tablet. { 'den·səd·ē 'step ,tab·lət }

dependency [COMPUT SCI] The necessity for a computer to complete work on some job before execution of another can begin. { di'pen·dən·sē }

dependent segment [COMPUT SCI] In a database management system, a block of data that depends on data at a higher level for its full meaning. { di'pen·dənt 'seg·mənt }

depletion layer [ELECTR] An electric double layer formed at the surface of contact between a metal and a semiconductor having different work functions, because the mobile carrier charge density is insufficient to neutralize the fixed charge density of donors and acceptors. Also known as barrier layer (deprecated); blocking layer (deprecated); space-charge layer. { də'plē·shən ,lā·ər }

depletion mode [ELECTR] Operation of a field-effect transistor in which current flows when the gate-source voltage is zero, and is increased or decreased by altering the gate-source voltage. { də'plē·shən ,mōd }

depletion-mode HEMT [ELECTR] A high-electron mobility transistor (HEMT) in which application of negative bias to the gate electrode cuts off the current between source and drain. Abbreviated D-HEMT. { də'plē·shən ,mōd ,āch ,ē,em'tē }

depletion region [ELECTR] The portion of the channel in a metal oxide field-effect transistor in which there are no charge carriers. { də'plē·shən ,rē·jən }

deposit [COMPUT SCI] To preserve the contents of a portion of a computer memory by copying it in a backing storage. { də'päz·ət }

derivative action [CONT SYS] Control action in which the speed at which a correction is made depends on how fast the system error is increasing. Also known as derivative compensation; rate action. { də'riv·əd·iv ,ak·shən }

derivative compensation *See* derivative action. { də'riv·əd·iv ,käm·pən'sā·shən }

derivative network [CONT SYS] A compensating network whose output is proportional to the sum of the input signal and its derivative. Also known as lead network. { də'riv·əd·iv 'net,wərk }

DES *See* data encryption standard.

descending sort [COMPUT SCI] The arranging of data records from high to low sequence (9 to 0, and Z to A). { di'send·iŋ 'sȯrt }

describing function [CONT SYS] A function used to represent a nonlinear transfer function by an approximately equivalent linear transfer function; it is the ratio of the phasor representing the fundamental component of the output of the nonlinearity, determined by Fourier analysis, to the phasor representing a sinusoidal input signal. { di'skrīb·iŋ ,fəŋk·shən }

descriptor [COMPUT SCI] A word or phrase used to identify a document in a computer-based information storage and retrieval system. { di'skrip·tər }

desensitization [COMMUN] Reduction in receiver sensitivity due to the presence of a high-level off-channel signal overloading the radio-frequency amplifier or mixer stages, or causing automatic gain control action. { dē,sen·sə·tə'zā·shən }

deserialize [COMMUN] To convert a data stream from a serial stream of bits to parallel streams of bits. { dē'sir·ē·ə,līz }

designation [COMPUT SCI] An item of data forming part of a computer record that indicates the type of record and thus determines how it is to be processed. [ENG] In radar, the act of reporting a target just detected or being tracked routinely by one radar to cause another radar to acquire the target for a different purpose, such as precision tracking for weapon control. { ,dez·əg'nā·shən }

design-oriented system [COMPUT SCI] A computer system developed primarily to maximize performance of hardware and software, rather than ease of use. { di'zīn ¦ȯr·ē,ent·əd ,sis·təm }

desk calculator [COMPUT SCI] A device that is used to perform arithmetic operations and is small enough to be conveniently placed on a desk. { ¦desk 'kal·kyə,lād·ər }

desk check *See* dry run. { 'desk ,chek }

desktop [COMPUT SCI] In a graphical user interface, a screen on which frequently used software resources are represented by icons. { 'desk ,täp }

desktop accessory software [COMPUT SCI] A set of computer programs providing functions that simulate the office accessories normally found on a desktop, such as a notepad, appointment calendar, and calculator. Also known as desktop application; desktop organizer. { ¦desk ,täp ik¦ses·ə·rē 'sȯf,wer }

desktop application *See* desktop accessory software. { ¦desk,täp ,ap·lə'kā·shən }

desktop organizer *See* desktop accessory software. { ¦desk,täp 'ȯr·gə,nīz·ər }

desktop publishing [COMPUT SCI] The use of a personal computer to produce printed output of high quality that is camera-ready for a printing facility. { ¦desk,täp 'pəb·lish·iŋ }

despooler [COMPUT SCI] Software that reads computer output information from a buffer and routes it to a printer. { dē'spül·ər }

despun antenna [ELECTROMAG] Satellite directional antenna pointed continuously at earth by electrically or mechanically despinning the antenna at the same rate that the satellite is spinning for stabilization. { dē'spən an'ten·ə }

destination [COMPUT SCI] The location (record, file, document, program, device, or disk) to which information is moved or copied. { ,des·tə'nā·shən }

destination address [COMPUT SCI] The location to which a jump instruction passes control in a program. { ˌdes·tə'nā·shən ə'dres }

destination time [COMPUT SCI] The time involved in a memory access plus the time required for indirect addressing. { ˌdes·tə'nā·shən ˌtīm }

destination warning mark *See* tape mark. { ˌdes·tə'nā·shən ˌwȯrn·iŋ ˌmärk }

destructive memory *See* destructive readout memory. { di¦strək·tiv 'mem·rē }

destructive read [COMPUT SCI] Reading that partially or completely erases the stored information as it is being read. { di'strək·tiv 'rēd }

destructive readout memory [COMPUT SCI] A memory type in which reading the contents of a storage location destroys the contents of that location. Also known as destructive memory. { di'strək·tiv 'rēdˌau̇t ˌmem·rē }

DETAB [COMPUT SCI] A programming language based on COBOL in which problems can be specified in the form of decision tables. Acronym for decision table. { 'dēˌtab }

detachable plugboard [COMPUT SCI] A control panel that can be removed from the computer or other system and exchanged for another without altering the positions of the plugs and cords. Also known as removable plugboard. { di'tach·ə·bəl 'pləgˌbȯrd }

detail chart [COMPUT SCI] A flow chart representing every single step of a program. { 'dē ˌtāl ˌchärt }

detail file [COMPUT SCI] A file containing current or transient data used to update a master file or processed with the master file to obtain a specific result. Also known as transaction file. { 'dēˌtāl ˌfīl }

detect *See* demodulate. { di'tekt }

detection [COMMUN] The recovery of information from an electrical or electromagnetic signal. { di'tek·shən }

detector [ELECTR] The stage in a receiver at which demodulation takes place; in a superheterodyne receiver this is called the second detector. Also known as demodulator; envelope detector. { di'tek·tər }

deterministic algorithm *See* static algorithm. { dəˌtər·mə'nis·tik 'al·gəˌrith·əm }

deterrence [COMPUT SCI] Making an attack on a computer sufficiently difficult to discourage potential attackers. { di'tər·əns }

detune [ELECTR] To change the inductance or capacitance of a tuned circuit so its resonant frequency is different from the incoming signal frequency. { dē'tün }

detuning stub [ELECTROMAG] Quarter-wave stub used to match a coaxial line to a sleeve-stub antenna; the stub detunes the outside of the coaxial feed line while tuning the antenna itself. { dē'tün·iŋ 'stəb }

developer's toolkit [COMPUT SCI] A collection of program subroutines that are used to help write an application program in a particular programming language or with a particular operating system. { di¦vel·əp·ərz 'tülˌkit }

development system [COMPUT SCI] The computer and software that are used to create a computer program. { di'vel·əp·mənt ˌsis·təm }

development tool [COMPUT SCI] A piece of hardware or software that is used to help design a computer or write a computer program. { di'vel·əp·mənt ˌtül }

deviation absorption [COMMUN] Distortion in a frequency-modulated receiver due to inadequate bandwidth, inadequate amplitude-modulation rejection, or inadequate discriminator linearity. { ˌdēv·ē'ā·shən əbˌsȯrp·shən }

deviation ratio [COMMUN] Ratio of the maximum frequency deviation to the maximum modulating frequency of a frequency-modulated system under specified conditions. { ˌdēv·ē'ā·shən ˌrā·shō }

device [COMPUT SCI] A general-purpose term used, often indiscriminately, to refer to a computer component or the computer itself. [ELECTR] An electronic element that cannot be divided without destroying its stated function; commonly applied to active elements such as transistors and transducers. [ENG] A mechanism, tool, or other piece of equipment designed for specific uses. { di'vīs }

device address [COMPUT SCI] The binary code which corresponds to a unique device, referred to when selecting this specific device. { di'vīs ə'dres }

device assignment [COMPUT SCI] The use of a logical device number used in conjunction with an input/output instruction, and made to refer to a specific device. { di'vīs ə'sīn·mənt }

device cluster [COMPUT SCI] A collection of peripheral devices (usually terminals) that have a common control unit. { di'vīs ˌkləs·tər }

device control character [COMPUT SCI] A special character used to direct a peripheral or communications device to perform a specific function. { di'vīs kən'trōl ˌkar·ik·tər }

device dependence [COMPUT SCI] Property of a computer program that will operate only with specified hardware. { di 'vīs deˌpen·dəns }

device driver [COMPUT SCI] A subroutine which handles a complete input/output operation. { di'vīs ˌdrīv·ər }

device-end condition [COMPUT SCI] The completion of an input/output operation, such as the transfer of a complete data block, recognized by the hardware in the absence of a byte count. { di'vīs ˌend kən'dish·ən }

device end pending [COMPUT SCI] A hardware error in which a peripheral device does not respond when addressed by the central processing unit, usually because the device has become inoperative. { di'vīs 'end ˌpend·iŋ }

device flag [COMPUT SCI] A flip-flop output which indicates the ready status of an input/output device. { di'vīs ˌflag }

device independence [COMPUT SCI] Property of a computer program whose successful execution (without recompilation) does not depend on the type of physical unit associated with a given logical unit employed by the program. { di'vīs ˌin·də'pen·dəns }

device-independent colors [COMPUT SCI] Colors produced by printers, monitors, and other output devices that have been modified to conform with a standard method of color description. { di¦vīs ˌin·dəˌpen·dənt 'kəl·ərz }

device-name assignment [COMPUT SCI] The designation of a peripheral device by a symbolic name rather than an address. { di'vīs ¦nām ə ˌsīn·mənt }

device number [COMPUT SCI] The physical or logical number which refers to a specific input/output device. { di'vīs ˌnəm·bər }

device selector [COMPUT SCI] A circuit which gates data-transfer or command pulses to a specific input/output device. { di'vīs si'lek·tər }

D-frame [COMMUN] A frame coded according to an MPEG-1 mode that uses dc (direct-current or zero-frequency) coefficients only. { 'dē ˌfrām }

DGPS *See* differential GPS.

D-HEMT *See* depletion-mode HEMT.

diagnosis [COMPUT SCI] The process of locating and explaining detectable errors in a computer routine or hardware component. { ˌdī·əg'nō·səs }

diagnostic check *See* diagnostic routine. { ˌdī·əg'näs·tik 'chek }

diagnostic message [COMPUT SCI] A statement produced automatically during some computer processing activity, such as program compilation, that provides information on the status of the computer or its software, particularly errors or potential problems. { ¦dī·əg¦näs·tik 'mes·ij }

diagnostic routine [COMPUT SCI] A routine designed to locate a computer malfunction or a mistake in coding. Also known as diagnostic check; diagnostic subroutine; diagnostic test; error detection routine. { ˌdī·əg'näs·tik rü'tēn }

diagnostic subroutine *See* diagnostic routine. { ˌdī·əg'näs·tik 'səb·rüˌtēn }

diagnostic test *See* diagnostic routine. { ˌdī·əg'näs·tik 'test }

diagnotor [COMPUT SCI] A combination diagnostic and edit routine which questions unusual situations and notes the implied results. { ˌdī·əg'nōd·ər }

diagonal horn antenna [ELECTROMAG] Horn antenna in which all cross sections are square and the electric vector is parallel to one of the diagonals; the radiation pattern in the far field has almost perfect circular symmetry. { dī'ag·ən·əl 'hȯrn an'ten·ə }

diagram [COMPUT SCI] A schematic representation of a sequence of subroutines designed to solve a problem; it is a coarser and less symbolic representation than a flow chart, frequently including descriptions in English words. { 'dī·ə ˌgram }

dial [COMMUN] In automatic telephone switching, either a type of calling device that, when wound up and released, generates pulses required for establishing connections or a pushbutton array that, with associated electronics, generates dual-tone multifrequency (DTMF) signals. { dīl }

dial backup [COMMUN] A dial telephone line that can be used in case a point-to-point line fails, so that data transmission can continue. { 'dīl 'bakˌəp }

dial central office [COMMUN] Telephone or teletypewriter office where necessary automatic equipment is located for connecting two or more users together by wires for communications purposes. { ¦dīl ¦sen·trəl 'ȯf·əs }

dialect [COMPUT SCI] A version of a programming language that differs from other versions in some respects but generally resembles them. { 'dī·əˌlekt }

dial exchange [COMMUN] A telephone exchange area in which all subscribers originate their calls by dialing. { 'dīl iksˌchānj }

dialing key [COMMUN] Method of dialing in which a set of numerical keys is used to originate dial pulses instead of a dial; generally used in connection with voice-frequency dialing. { 'dī·liŋ ˌkē }

dial office [COMMUN] Central office operating on dial signals. { 'dīl ˌȯf·əs }

dialog [COMPUT SCI] A form of data processing involving an interaction between a computer system and a terminal operator who uses a keyboard and electronic display to enter data which the computer edits and may respond to. { 'dī·əˌläg }

dialog box [COMPUT SCI] On a computer screen, a small window that is used to emphasize the importance of some action or to request an answer to a question. { 'dī·əˌläg ˌbäks }

dial pulse interpreter [ELECTR] A device that converts the signaling pulses of a dial telephone to a form suitable for data entry to a computer. { 'dīl ˌpəls in'tər·prəd·ər }

dial pulsing *See* loop pulsing. { 'dīl ˌpəls·iŋ }

dial telephone system [COMMUN] A telephone system in which telephone connections between customers are ordinarily established by electronic and mechanical apparatus, controlled by manipulations of dials operated by calling parties. { 'dīl 'tel·əˌfōn ˌsist·əm }

dial tone [COMMUN] A tone employed in a dial telephone system to indicate that the equipment is ready for dialing operation. { 'dīl ˌtōn }

dial-up [COMMUN] **1.** The service whereby a dial telephone can be used to initiate and effect station-to-station telephone calls. **2.** In computer networks, pertaining to terminals which must dial up to receive service, as contrasted with those hand-wired or permanently connected into the network. { 'dīl ˌəp }

dial-up telephone system [COMMUN] The switched telephone network that is regulated by national governments; operated in the United States by various carriers. { ¦dīl ˌəp 'tel·əˌfōn ˌsis·təm }

diamond antenna *See* rhombic antenna. { 'dī ˌmənd an'ten·ə }

diaphragm *See* iris. { 'dī·əˌfram }

diathermy interference [COMMUN] Television interference caused by diathermy equipment; produces a herringbone pattern in a dark horizontal band across the picture. { 'dī·ə ˌthər·mē ˌin·tər'fir·əns }

dibit [COMPUT SCI] A pair of binary digits, used to specify one of four values. { 'dīˌbit }

dichotomizing search [COMPUT SCI] A procedure for searching an item in a set, in which, at each step, the set is divided into two parts, one part being then discarded if it can be logically shown that the item could not be in that part. { dī'käd·əˌmīz·iŋ ˌsərch }

dichotomy [COMPUT SCI] A division into two subordinate classes; for example, all white and all nonwhite, or all zero and all nonzero. { dī'käd·ə·mē }

dictionary [COMPUT SCI] A table establishing the correspondence between specific words and their code representations. { 'dik·shəˌner·ē }

dictionary code [COMPUT SCI] An alphabetical arrangement of English words and terms, associated with their code representations. { 'dik·shə ˌner·ē ˌkōd }

dictionary encoding [COMPUT SCI] A method of data compression in which each word is replaced by a number which is the position of that word in a dictionary. { 'dik·shəˌner·ē in'kōd·iŋ }

dictionary sort [COMPUT SCI] A sort algorithm that ignores capitalization, punctuation, and spaces, and treats numbers as if they were spelled out alphabetically. { 'dik·shəˌner·ē ˌsȯrt }

dielectric antenna [ELECTROMAG] An antenna in which a dielectric is the major component used to produce a desired radiation pattern. { ˌdī·ə'lek·trik an'ten·ə }

dielectric lens [ELECTROMAG] A lens made of dielectric material so that it refracts radio waves in the same manner that an optical lens refracts light waves; used with microwave antennas. { ˌdī·ə'lek·trik 'lenz }

dielectric-lens antenna [ELECTROMAG] An aperture antenna in which the beam width is determined by the dimensions of a dielectric lens through which the beam passes. { ˌdī·ə¦lek·trik ¦lenz an'ten·ə }

dielectric-rod antenna [ELECTROMAG] A surface-wave antenna in which an end-fire radiation pattern is produced by propagation of a surface wave on a tapered dielectric rod. { ˌdī·ə¦lek·trik ¦räd an'ten·ə }

dielectric wedge [ELECTROMAG] A wedge-shaped piece of dielectric used in a waveguide to match its impedance to that of another waveguide. { dī·ə'lek·trik 'wej }

difference amplifier *See* differential amplifier. { 'dif·rəns ˌam·pləˌfī·ər }

difference detector [ELECTR] A detector circuit in which the output is a function of the difference between the amplitudes of the two input waveforms. { 'dif·rəns diˌtek·tər }

difference encoding [COMPUT SCI] A method of data compression that takes advantage of a sequence of data that differs little from one value to the next by encoding each value as the difference from the previous value. { 'dif·rəns inˌkōd·iŋ }

difference equation [MATH] An equation expressing a functional relationship of one or more independent variables, one or more functions dependent on these variables, and successive differences of these functions. { 'dif·rəns i'kwā·zhən }

difference in depth modulation [COMMUN] In directive systems employing overlapping lobes with modulated signals, a ratio obtained by subtracting from the percentage of modulation of the larger signal the percentage of modulation of the smaller signal and dividing by 100. { 'dif·rəns 'in ¦depth ˌmäj·ə'lā·shən }

difference mapping [COMMUN] A method of coding information in which a sample value is presented as an error term formed by the difference between the sample and the previous sample. { 'dif·rəns ˌmap·iŋ }

differential amplifier [ELECTR] An amplifier whose output is proportional to the difference between the voltages applied to its two inputs. Also called difference amplifier. { ˌdif·ə'ren·chəl 'am·pləˌfī·ər }

differential analyzer [COMPUT SCI] A mechanical or electromechanical device designed primarily to solve differential equations. { ˌdif·ə'ren·chəl 'an·əˌlīz·ər }

differential backup [COMPUT SCI] Backup of only files that have been changed or added since the last backup. { ˌdif·əˌren·chəl 'bakˌəp }

differential comparator [ELECTR] A comparator having at least two high-gain differential-amplifier stages, followed by level-shifting and buffering stages, as required for converting a differential input to single-ended output for digital logic applications. { ˌdif·ə'ren·chəl kəm'par·əd·ər }

differential delay [COMMUN] The difference between the maximum and minimum frequency delays occurring across a band. { ˌdif·ə'ren·chəl di'lā }

differential duplex system [ELECTR] System in which the sent currents divide through two mutually inductive sections of a receiving apparatus, connected respectively to the line and to a balancing artificial line in opposite directions, so that there is substantially no net effect on the receiving apparatus; the received currents pass mainly through one section, or through the two sections in the same direction, and operate the apparatus. { ˌdif·ə'ren·chəl 'düˌpleks ˌsis·təm }

differential encoding [COMMUN] A method of compressing television signals by transmitting only differences between pixels in neighboring lines and successive frames. { ˌdif·əˌren·chəl in'kōd·iŋ }

differential gain control [ELECTR] Device for altering the gain of a radio receiver according to expected change of signal level, to reduce the amplitude differential between the signals at the output of the receiver. Also known as gain sensitivity control. { ˌdif·ə'ren·chəl ˌgān kən ˌtrōl }

differential GPS [NAV] A technique for improving the accuracy of the Global Positioning System

(GPS) in which error corrections are transmitted to users based on measurements of GPS signals by one or more reference receivers situated at known locations. Abbreviated DGPS. { ˌdif·ə¦ren·chəl ¦jē¦pē'es }

differentially coherent phase-shift keying *See* differential phase-shift keying. { ˌdif·ə'ren·chə·lē kō'hir·ənt 'fāz ˌshift ˌkē·iŋ }

differential modulation [COMMUN] Modulation in which the choice of the significant condition for any signal element is dependent on the choice for the previous signal element. { ˌdif·ə'ren·chəl ˌmäj·ə'lā·shən }

differential operational amplifier [ELECTR] An amplifier that has two input terminals, used with additional circuit elements to perform mathematical functions on the difference in voltage between the two input signals. { ˌdif·ə'ren·chəl äp·ə'rā·shən·əl 'am·pləˌfī·ər }

differential phase-shift keying [COMMUN] Form of phase-shift keying in which the reference phase for a given keying interval is the phase of the signal during the preceding keying interval. Also known as differentially coherent phase-shift keying. { ˌdif·ə'ren·chəl 'fāz ˌshift ˌkē·iŋ }

differential pulse-code modulation [COMMUN] A type of pulse-code modulation in which an analog signal is sampled and the difference between its actual value and its predicted value, based on a previous sample or samples, is quantized; for example, in television transmission, only the differences between the continuous picture elements on the scanning lines are transmitted, enabling the bandwidth of the signal to be reduced. Abbreviated DPCM. { dif·ə'ren·chəl 'pəls ˌkōd ˌmäj·ə'lā·shən }

digicom [COMMUN] A wire communication system that transmits speech signals in the form of corresponding trains of pulses and transmits digital information directly from computers, radar, tape readers, teleprinters, and telemetering equipment. { 'dij·əˌkäm }

digit [COMPUT SCI] In a decimal digital computer, the space reserved for storage of one digit of information. { 'dij·ət }

digital [COMPUT SCI] Pertaining to data in the form of digits. { 'dij·əd·əl }

digital audio broadcasting [COMMUN] The radio broadcasting of audio signals encoded in digital form. Abbreviated DAB. { ¦dij·əd·əl ¦ȯd·ē·ō 'brȯdˌkast·iŋ }

digital audio tape [COMPUT SCI] A magnetic tape on which sound is recorded and played back in digital form. Abbreviated DAT. { ¦dij·əd·əl 'ȯd·ē·ō ˌtāp }

digital camera [ELECTR] A television camera that breaks up a picture into a fixed number of pixels and converts the light intensity (or the intensities of each of the primary colors) in each pixel to one of a finite set of numbers. { 'dij·əd·əl 'kam·rə }

digital channel [COMMUN] A transmission path that carries only digital signals. { 'dij·əd·əl 'chan·əl }

digital circuit [ELECTR] A circuit designed to respond at input voltages at one of a finite number of levels and, similarly, to produce output voltages at one of a finite number of levels. { 'dij·əd·əl 'sər·kət }

digital circuit multiplication equipment [COMMUN] Equipment that uses digital compression techniques to increase the capacity of digital satellite and cable links carrying voice, facsimile, and voice-frequency modem traffic. { ˌdij·əd·əl ˌsər·kət ˌməl·tə·plə'kā·shən iˌkwip·mənt }

digital communications [COMMUN] System of telecommunications employing a nominally discontinuous signal that changes in frequency, amplitude, time, or polarity. { 'dij·əd·əl kə ˌmyü·nə'kā·shənz }

digital comparator [ELECTR] A comparator circuit operating on input signals at discrete levels. Also known as discrete comparator. { 'dij·əd·əl kəm'par·əd·ər }

digital computer [COMPUT SCI] A computer operating on discrete data by performing arithmetic and logic processes on these data. { 'dij·əd·əl kəm'pyüd·ər }

digital control [CONT SYS] The use of digital or discrete technology to maintain conditions in operating systems as close as possible to desired values despite changes in the operating environment. { 'dij·əd·əl kən'trōl }

digital converter [ELECTR] A device that converts voltages to digital form; examples include analog-to-digital converters, pulse-code modulators, encoders, and quantizing encoders. { 'dij·əd·əl kən'vərd·ər }

digital counter [ELECTR] A discrete-state device (one with only a finite number of output conditions) that responds by advancing to its next output condition. { 'dij·əd·əl 'kau̇nt·ər }

digital data [COMPUT SCI] Data that are electromagnetically stored in the form of discrete digits. { 'dij·əd·əl 'dad·ə }

Digital Data Broadcast System [NAV] A system that will provide information aiding air-traffic control; digital data to aircraft over vortac channels will carry information on the geographic location, elevation, magnetic variation, and related data of the vortac station being received. Abbreviated DDBS. { ˌdij·əd·əl 'dad·ə 'brȯdˌkast ˌsis·təm }

digital data modulation system [COMMUN] A digital communications system in which the information source consists of a finite number of discrete messages which are coded into a sequence of waveforms or symbols, each one selected from a specified and finite set. { 'dij·əd·əl 'dad·ə ˌmäj·ə'lā·shən ˌsis·təm }

digital data recorder [COMPUT SCI] Electronic device that converts continuous electrical analog signals into number (digital) values and records these values onto a data log via a high-speed typewriter. { 'dij·əd·əl ¦dad·ə riˌkȯrd·ər }

digital data service [COMMUN] A telephone communication system developed specifically for digital data, using existing local digital lines combined with data-under-voice microwave transmission facilities. Abbreviated DDS. { 'dij·əd·əl 'dad·ə ˌsər·vəs }

digital differential analyzer [COMPUT SCI] A differential analyzer which uses numbers to represent analog quantities. Abbreviated DDA. { 'dij·əd·əl ˌdif·əˌren·chəl 'an·əˌlīz·ər }

digital display [COMPUT SCI] A display in which the result is indicated in directly readable numerals. { 'dij·əd·əl di'splā }

Digital Electronic Message Service [COMMUN] A communication system whose purpose is to provide efficient means for two-way high-speed data communications, transfer of graphic images (fascimile), and teleconferencing between cities and within a city environment. Abbreviated DEMS. { 'dij·əd·əl iˌlek'trän·ik 'mes·ij ˌsər·vəs }

digital filter [ELECTR] An electrical filter that responds to an input which has been quantified, usually as pulses. { 'dij·əd·əl 'fil·tər }

digital format [COMPUT SCI] Use of discrete integral numbers in a given base to represent all the quantities that occur in a problem or calculation. { 'dij·əd·əl 'fȯr·mat }

digital incremental plotter [COMPUT SCI] A device for converting digital signals in the output of a computer into graphical form, in which the digital signals control the motion of a plotting pen and of a drum that carries the paper on which the graph is drawn. { 'dij·əd·əl ˌiŋ·krəˌment·əl 'pläd·ər }

digital integrator [COMPUT SCI] A device for computing definite integrals in which increments in the input variables and output variable are represented by digital signals. { 'dij·əd·əl 'in·tə ˌgrād·ər }

digital loop carrier [COMMUN] A technology for providing 24 or more telephone circuits on many fewer pairs of wires, in which analog input signals are first sampled and digitized, and the binary digital signals from each user is then time-multiplexed into a single bit stream. { 'dij·əd·əl 'lüp ˌkar·ē·ər }

digital microwave radio [COMMUN] Transmission of voice and data signals in digital form on microwave links, as in the 2-gigahertz common-carrier bands; pulse-code modulation is used. { ¦dij·əd·əl ¦mī·krōˌwāv 'rād·ē·ō }

digital modulation [COMMUN] A method of placing digital traffic on a microwave system without use of modems, by transmitting the information in the form of discrete phase or frequency states determined by the digital signal. { 'dij·əd·əl ˌmäj·ə'lā·shən }

digital monitor [ELECTR] A display unit that accepts digital signals and converts them to analog signals internally in order to illuminate the screen. { 'dij·əd·əl 'män·əd·ər }

Digital Multiplexed Interface [COMPUT SCI] A cost-effective, high-speed interconnection between terminals and host computers in a private branch exchange environment. { 'dij·əd·əl 'məl·təˌplekst 'in·tərˌfās }

digital multiplier [ELECTR] A multiplier that accepts two numbers in digital form and gives their product in the same digital form, usually by making repeated additions; the multiplying process is simpler if the numbers are in binary form wherein digits are represented by a 0 or 1. { 'dij·əd·əl 'məl·təˌplī·ər }

digital object identifier [COMPUT SCI] A system for identifying and exchanging intellectual properties (including, for example, physical objects as well as digital files) in the digital environment. { ¦dij·əd·əl ¦äbˌjekt ī'den·təˌfī·ər }

digital output [ELECTR] An output signal consisting of a sequence of discrete quantities coded in an appropriate manner for driving a printer or digital display. { 'dij·əd·əl 'au̇tˌpu̇t }

digital plotter [ELECTR] A recorder that produces permanent hard copy in the form of a graph from digital input data. { 'dij·əd·əl 'pläd·ər }

digital printer [COMPUT SCI] A printer that provides a permanent readable record of binary-coded decimal or other coded data in a digital form that may include some or all alphanumeric characters and special symbols along with numerals. Also known as digital recorder. { 'dij·əd·əl 'print·ər }

digital private automatic branch exchange [COMMUN] A central communications switching system for a local-area network, which employs existing telephone wires in a building for the connection of telephones and computer terminals and systems. { 'dij·əd·əl ¦prīv·ət ¦ȯd·ə¦mad·ik 'branch iksˌchānj }

digital radio [COMMUN] The microwave transmission of digital signals through space or the atmosphere. { ¦dij·əl·əl 'rād·ē·ō }

digital recorder *See* digital printer. { 'dij·əd·əl ri'kȯrd·ər }

digital recording [ELECTR] Magnetic recording in which the information is first coded in a digital form, generally with a binary code that uses two discrete values of residual flux. { 'dij·əd·əl ri'kȯrd·iŋ }

digital representation [COMPUT SCI] The use of discrete impulses or quantities arranged in coded patterns to represent variables or other data in the form of numbers or characters. { 'dij·əd·əl ˌrep·rəˌzen'tā·shən }

digital resolution [COMPUT SCI] The ability of a digital computer to approach a truly correct answer, generally established by the number of places expressed, and the value of the least significant digit in a digitally coded representation. { 'dij·əd·əl ˌrez·ə'lü·shən }

digital set-top box [COMMUN] A device that is attached to a television receiver and can collect, store, and output digitally compressed television signals. { ˌdij·əd·əl 'setˌtäp ˌbäks }

digital signal analyzer [ELECTR] A signal analyzer in which one or more analog inputs are sampled at regular intervals, converted to digital form, and fed to a memory. { 'dij·əd·əl 'sig·nəl ˌan·əˌlīz·ər }

digital signal processing *See* signal processing. { ˌdij·əd·əl ˌsig·nəl 'prä·səs·iŋ }

digital signal processing chip [COMPUT SCI] A digital device for executing algorithms for the transformation or extraction of information from signals originally in analog form, such as audio or images. Abbreviated DSP chip. Also known as digital signal processor. { ˌdij·əd·əl ˌsig·nəl ˈprä·səs·iŋ ˌchip }

digital signal processor *See* digital signal processing chip. { ˌdij·əd·əl ˈsig·nəl ˌpräˌses·ər }

digital signature [COMMUN] A set of alphabetic or numeric characters used to authenticate a cryptographic message by ensuring that the sender cannot later disavow the message, the receiver cannot forge the message or signature, and the receiver can prove to others that the contents of the message are genuine and originated with the sender. { ˈdij·əd·əl ˈsig·nə·chər }

digital simulation [COMPUT SCI] The representation of a system in a form acceptable to a digital computer as opposed to an analog computer. { ˈdij·əd·əl ˌsim·yəˈlā·shən }

digital speech communications [COMMUN] Transmission of voice in digitized or binary form via landline or radio. { ˈdij·əd·əl ˈspēch kə ˌmyün·əˌkā·shənz }

digital speech interpolation [COMMUN] In digital speech communications, the use of periods of inactivity or constant signal level to increase the transmission efficiency by insertion of additional signals. Abbreviated DSI. { ˌdij·əd·əl ˈspēch ˌin·tər·pəˌlā·shən }

digital subscriber line [COMMUN] A system that provides subscribers with continuous, uninterrupted connections to the Internet over existing telephone lines, offering a choice of speeds ranging from 32 kilobits per second to more than 50 megabits per second. Abbreviated DSL. { ¦dij·əd·əl səbˈskrīb·ər ˌlīn }

digital system [COMPUT SCI] Any of the levels of operation for a digital computer, including the wires and mechanical parts, the logical elements, and the functional units for reading, writing, storing, and manipulating information. { ˈdij·əd·əl ˈsis·təm }

digital telemetering [COMPUT SCI] Conversion of a continuous electrical analog signal into a digital (number system) code prior to transmitting the signal to a receiver. { ˈdij·əd·əl ¦tel·ə¦mēd·ər·iŋ }

digital television [COMMUN] Television in which picture information is encoded into digital signals on the transmitter, and decoded at the receiver. Abbreviated DTV. { ˈdij·əd·əl ˈtel·ə ˌvizh·ən }

digital television converter [ELECTR] A converter used to convert television programs from one system to another, such as for converting 525-line 60-field United States broadcasts to 625-line 50-field European PAL (phase-alternation line) or SECAM (sequential couleur á memoire) standards; the video signal is digitized before conversion. { ˈdij·əd·əl ¦tel·əˌvizh·ən kənˈvərd·ər }

digital-to-analog converter [ELECTR] A converter in which digital input signals are changed to essentially proportional analog signals. Abbreviated DAC. { ˈdij·əd·əl tü ¦an·əˌläg kən ˈvərd·ər }

digital versatile disk *See* DVD. { ¦dij·əd·əl ˈvər·səd·əl ˌdisk }

digital video disk *See* DVD. { ¦dij·əd·əl ˈvid·ē·ō ˌdisk }

digital watermark [COMPUT SCI] Invisible or inaudible data (a random pattern of bits or noise) permanently embedded in a graphic, video, or audio file for protecting copyright or authenticating data. { ˌdij·əd·əl ˈwȯd·ərˌmärk }

digit-coded voice [COMPUT SCI] A limited, spoken vocabulary, each word of which corresponds to a code and which, upon keyed inquiry, can be strung in meaningful sequence and can be outputted as audio response to the inquiry. { ˈdij·ət ˌkōd·əd ˈvȯis }

digit compression [COMPUT SCI] Any process which increases the number of digits stored at a given location. { ˈdij·ət kəmˈpresh·ən }

digit delay element [ELECTR] A logic element that introduces a delay of one digit period in a series of signals or pulses. { ˈdij·ət diˈlā ˌel·ə·mənt }

digitize [COMPUT SCI] To convert an analog measurement of a quantity into a numerical value. { ˈdij·əˌtīz }

digitizer [COMPUT SCI] A large drawing table connected to a computer video display and equipped with a penlike or pucklike instrument whose motions are reproduced on the screen. Also known as digitizer tablet. { ˈdij·əˌtīz·ər }

digitizer tablet *See* digitizer. { ˈdij·əˌtīz·ər ˌtab·lət }

digit period [ELECTR] The time interval between successive pulses, usually representing binary digits, in a computer or in pulse modulation, determined by the pulse-repetition frequency. Also known as digit time. { ˈdij·ət ˌpir·ē·əd }

digit plane [COMPUT SCI] In a computer memory consisting of magnetic cores arranged in a three-dimensional array, a plane containing elements for a particular digit position in various words. { ˈdij·ət ˌplān }

digit rearrangement [COMPUT SCI] A method of hashing which consists of selecting and shifting digits of the original key. { ˈdij·ət ˌrē·əˈrānj·mənt }

digit time *See* digit period. { ˈdij·ət ˌtīm }

digram encoding [COMPUT SCI] A method of data compression that relies on the fact that there are unused characters in the alphabet and uses these characters to represent common pairs of characters. { ˈdīˌgram inˌkōd·iŋ }

diheptal base [ELECTR] A tube base having 14 pins or 14 possible pin positions; used chiefly on television cathode-ray tubes. { dīˈhept·əl ˈbās }

dimension [COMPUT SCI] A declarative statement that specifies the width and height of an array of data items. { dəˈmen·chən }

dimension declaration statement [COMPUT SCI] A FORTRAN statement identifying arrays and specifying the number and bounds of the subscripts. { də'men·chən·əl dek·lə'rā·shən ˌstāt·mənt }

diminution [COMPUT SCI] Limiting the negative effect of an attack on a computer system. { ˌdim·ə'nü·shən }

DIMM [COMPUT SCI] A small circuit board that holds semiconductor memory chips with two independent rows of input/output contacts. Derived from dual in-line memory module.

D-indicator *See* D-display. { 'dē ˌin·dəˌkād·ər }

diode [ELECTR] **1.** A two-electrode electron tube containing an anode and a cathode. **2.** *See* semiconductor diode. { 'dīˌōd }

diode amplifier [ELECTR] A microwave amplifier using an IMPATT, TRAPATT, or transferred-electron diode in a cavity, with a microwave circulator providing the input/output isolation required for amplification; center frequencies are in the gigahertz range, from about 1 to 100 gigahertz, and power outputs are up to 20 watts continuous-wave or more than 200 watts pulsed, depending on the diode used. { 'dīˌōd 'am·plə ˌfī·ər }

diode-capacitor transistor logic [ELECTR] A circuit that uses diodes, capacitors, and transistors to provide logic functions. { ¦dīˌōd kə¦pas·əd·ər tran'zis·tər ˌläj·ik }

diode demodulator [ELECTR] A demodulator using one or more diodes to provide a rectified output whose average value is proportional to the original modulation. Also known as diode detector. { 'dīˌōd dē'mäj·əˌlād·ər }

diode detector *See* diode demodulator. { 'dī ˌōd di'tek·tər }

diode gate [ELECTR] An AND gate that uses diodes as switching elements. { 'dīˌōd ˌgāt }

diode laser *See* semiconductor laser. { 'dīˌōd ˌlāz·ər }

diode logic [ELECTR] An electronic circuit using current-steering diodes, such that the relations between input and output voltages correspond to AND or OR logic functions. { 'dīˌōd ˌläj·ik }

diode modulator [ELECTR] A modulator using one or more diodes to combine a modulating signal with a carrier signal; used chiefly for low-level signaling because of inherently poor efficiency. { 'dīˌōd 'mäj·əˌlād·ər }

diode transistor logic [ELECTR] A circuit that uses diodes, transistors, and resistors to provide logic functions. Abbreviated DTL. { ¦dī ˌōd tran'zis·tər ˌläj·ik }

DIP *See* dual in-line package. { dip }

diplexer [ELECTR] A coupling system that allows two different transmitters to operate simultaneously or separately from the same antenna. { 'dī ˌplek·sər }

diplex operation [COMMUN] Simultaneous transmission or reception of two signals using a specified common element, such as a single antenna or a single carrier. { 'dīˌpleks ˌäp·əˌrā·shən }

diplex radio transmission [COMMUN] The simultaneous transmission of two signals by using a common carrier wave. { ¦dīˌpleks 'rād·ē·ō tranz ˌmish·ən }

dipole antenna [ELECTROMAG] An antenna approximately one-half wavelength long, split at its electrical center for connection to a transmission line whose radiation pattern has a maximum at right angles to the antenna. Also known as doublet antenna; half-wave dipole. { 'dīˌpōl an 'ten·ə }

dipole disk feed [ELECTROMAG] Antenna, consisting of a dipole near a disk, used to reflect energy to the disk. { 'dīˌpōl 'disk ˌfēd }

DIP switch [COMPUT SCI] A unit with several small rocker-type switches that plugs into a dual in-line package (DIP) on a printed circuit board. { 'dip ˌswich }

dipulse [COMMUN] Transmission of a binary code in which the presence of one cycle of a sine-wave tone represents a binary "1" and the absence of one cycle represents a binary "0." { 'dīˌpəls }

direct access *See* random access. { də'rekt 'ak·ses }

direct-access library [COMPUT SCI] A disk-stored set of programs, each of which is directly accessible without sequential search. { də¦rekt ¦ak·ses 'līˌbrer·ē }

direct-access memory *See* random-access memory. { də¦rekt ¦ak·ses 'mem·rē }

direct-access method [COMPUT SCI] A technique for directly determining the location of data on a disk (track and sector address) from an identifying key in the record. { də¦rekt 'akˌses ˌmeth·əd }

direct-access storage *See* random-access memory. { də¦rekt ¦ak·ses 'stȯr·ij }

direct-access storage device [COMPUT SCI] Any peripheral storage device, such as a disk or drum, that can be directly addressed by a computer. Abbreviated DASD. { də¦rekt ¦akˌses 'stȯr·ij diˌvīs }

direct address [COMPUT SCI] Any address specifying the location of an operand. { də¦rekt 'a ˌdres }

direct-address processing [COMPUT SCI] Any computer operation during which data are accessed by means of addresses rather than contents. { də¦rekt ¦aˌdres 'präsˌes·iŋ }

direct allocation [COMPUT SCI] A system in which the storage locations and peripheral units to be assigned to use by a computer program are specified when the program is written, in contrast to dynamic allocation. { də¦rekt ˌal·əˌkā·shən }

direct-aperture antenna [ELECTROMAG] An antenna whose conductor or dielectric is a surface or solid, such as a horn, mirror, or lens. { də ¦rekt ¦ap·ə·chər an'ten·ə }

direct audio radio service [COMMUN] Radio broadcasting from satellites directly to receivers on the ground. Abbreviated DARS. { dəˌrekt ¦ȯd·ē·ō'rād·ē·ōˌ sər·vəs }

direct broadcasting satellite system [COMMUN] A television broadcasting system in which program signals are transmitted from ground stations to satellite repeater stations in geostationary orbit, and from there directly to home consumer terminals. Abbreviated DBS. { də'rekt 'brȯd,kast·iŋ 'sad·əl,īt ,sis·təm }

direct broadcast radio satellite [COMMUN] A satellite in geosynchronous orbit that broadcasts radio programming directly to inexpensive home, car-mounted, and portable radio receivers. { di ¦rekt ¦brȯd,kast 'rād·ē·ō,sad·əl,īt }

direct code [COMPUT SCI] A code in which instructions are written in the basic machine language. { də¦rekt 'kōd }

direct connect modem [COMMUN] A device that transforms binary signals into electronic pulses (as opposed to sound modulations) that can be carried over a communications channel. { də'rekt kə¦nekt 'mō,dem }

direct control [COMPUT SCI] The control of one machine in a data-processing system by another, without human intervention. { də¦rekt kən'trōl }

direct control function *See* regulatory control function. { də¦rekt kən'trōl ,fəŋk·shən }

direct-coupled FET logic [ELECTR] A logic gate configuration used with gallium arsenide field-effect transistors operating in the enhancement mode, whose low power consumption and circuit simplicity lead to high packing density and potential use in very large-scale integrated circuits. Abbreviated DCFL. { də'rekt ¦kəp·əld ¦ef¦ē¦tē 'läj·ik }

direct-coupled transistor logic [ELECTR] Integrated-circuit logic using only resistors and transistors, with direct conductive coupling between the transistors; speed can be up to 1 megahertz. Abbreviated DCTL. { də¦rekt ¦kəp·əld tran¦zis·tər 'läj·ik }

direct-current component [COMMUN] The average value of a signal; in television, it represents the average lumininance of the picture being transmitted; in radar, the level from which the transmitted and received pulses rise. { də¦rekt ¦kə·rənt kəm'pō·nənt }

direct-current dump [ELECTR] Removal of all direct-current power from a computer system or component intentionally, accidentally, or conditionally; in some types of storage, this results in loss of stored information. { də¦rekt ¦kə·rənt 'dəmp }

direct-current inserter [ELECTR] An analog television transmitter stage that adds to the video signal a dc component known as the pedestal level. { də¦rekt ¦kə·rənt in'sərd·ər }

direct-current picture transmission [COMMUN] Television transmission in which the signal contains a dc component that represents the average illumination of the entire scene. Also known as direct-current transmission. { də¦rekt ¦kə·rənt 'pik·chər tranz,mish·ən }

direct-current quadruplex system [COMMUN] Direct-current telegraph system which affords simultaneous transmission of two messages in each direction over the same line, achieved by superimposing neutral telegraph upon polar telegraph. { də¦rekt ¦kə·rənt 'kwä·drə,pleks ,sis·təm }

direct-current telegraphy [COMMUN] Telegraphy in which direct current controlled by the transmitting apparatus is supplied to the line to form the transmitted signal. { də¦rekt ¦kə·rənt tə'leg·rə·fē }

direct-current transmission *See* direct-current picture transmission. { də¦rekt ¦kə·rənt tranz 'mish·ən }

direct digital control [CONT SYS] The use of a digital computer generally on a time-sharing or multiplexing basis, for process control in petroleum, chemical, and other industries. { də ¦rekt ¦dij·əd·əl kən'trōl }

direct distance dialing [COMMUN] A telephone exchange service that allows a telephone user to dial subscribers outside the local area using a standard routing pattern from the local or end office. { də¦rekt ¦dis·təns 'dīl·iŋ }

direct-entry terminal [COMPUT SCI] A device from which data are received into a computer immediately, and which edits data at the time of receipt, allowing computer files to be accessed to validate the information entered, and allowing the terminal operator to be notified immediately of any errors. { də¦rekt ¦en·trē 'term·ən·əl }

direct expert control system [CONT SYS] An expert control system that contains rules that directly associate controller output values with different values of the controller measurements and set points. Also known as rule-based control system. { də¦rekt ,eks·pərt kən'trōl ,sis·təm }

direct hierarchy control [COMPUT SCI] A method of manipulating data in a computer storage hierarchy in which data transfer is completely under the control of built-in algorithms and the user or programmer is not concerned with the various storage subsystems. { də¦rekt 'hī·ər,är·kē kən,trōl }

direct input/output [COMPUT SCI] The transfer of data to and from a computer's main storage by passing it through the central processing unit. { də'rekt 'in,pu̇t 'au̇t,pu̇t }

direct-insert subroutine [COMPUT SCI] A body of coding or a group of instructions inserted directly into the logic of a program, often in multiple copies, whenever required. { də¦rekt ¦in·sərt 'səb·rü,tēn }

direct instruction [COMPUT SCI] An instruction containing the address of the operand on which the operation specified in the instruction is to be performed. { də'rekt in'strək·shən }

direct inward dialing [COMMUN] The capability for dialing individual telephone extensions in a large organization directly from outside, without going through a central switchboard. { də'rekt ¦in·wərd 'dīl·iŋ }

directional antenna [ELECTROMAG] An antenna that radiates or receives radio waves more effectively in some directions than others. { də'rek·shən·əl an'ten·ə }

directional beam [ELECTROMAG] A radio or radar wave that is concentrated in a given direction. { də'rek·shən·əl 'bēm }

directional coupler [ELECTR] A device that couples a secondary system only to a wave traveling in a particular direction in a primary transmission system, while completely ignoring a wave traveling in the opposite direction. Also known as directive feed. { də'rek·shən·əl 'kəp·lər }

directional filter [ELECTR] A low-pass, band-pass, or high-pass filter that separates the bands of frequencies used for transmission in opposite directions in a carrier system. Also known as directional separation filter. { də'rek·shən·əl 'fil·tər }

directional pattern *See* radiation pattern. { də'rek·shən·əl 'pad·ərn }

directional separation filter *See* directional filter. { də'rek·shən·əl sep·ə'rā·shən ˌfil·tər }

direction finder *See* radio direction finder. { də'rek·shən ˌfīnd·ər }

directive [COMPUT SCI] An instruction in a source program that guides the compiler in making the translation to machine language, and is usually not translated into instructions in the object program. { də'rek·tiv }

directive feed *See* directional coupler. { də 'rek·tiv ˌfēd }

directive gain [ELECTROMAG] Of an antenna in a given direction, 4π times the ratio of the radiation intensity in that direction to the total power radiated by the antenna. { də'rek·tiv ˌgān }

directivity [ELECTR] The ability of a logic circuit to ensure that the input signal is not affected by the output signal. [ELECTROMAG] **1.** The value of the directive gain of an antenna in the direction of its maximum value. **2.** The ratio of the power measured at the forward-wave sampling terminals of a directional coupler, with only a forward wave present in the transmission line, to the power measured at the same terminals when the direction of the forward wave in the line is reversed; the ratio is usually expressed in decibels. { dəˌrek'tiv·əd·ē }

direct keying device [COMPUT SCI] A computer input device which enables direct entry of information by means of a keyboard. { də'rekt 'kē·iŋ diˌvīs }

direct-map cache [COMPUT SCI] A cache memory that is organized by linking it to locations in random-access memory. { dəˌrekt ˌmap 'kash }

direct memory access [COMPUT SCI] The use of special hardware for direct transfer of data to or from memory to minimize the interruptions caused by program-controlled data transfers. Abbreviated dma. { də¦rekt ¦mem·rē 'akˌses }

direct numerical control [COMPUT SCI] The use of a computer to program, service, and log a process such as a machine-tool cutting operation. { də¦rekt nü¦mer·i·kəl kən'trōl }

director [ELECTR] Telephone switch which translates the digits dialed into the directing digits actually used to switch the call. [ELECTROMAG] A parasitic element placed a fraction of a wavelength ahead of a dipole receiving antenna to increase the gain of the array in the direction of the major lobe. { də'rek·tər }

direct organization [COMPUT SCI] A type of processing in which records within data sets stored on direct-access devices may be fetched directly if their physical locations are known. { də'rekt ȯr·gə·nə'zā·shən }

director-type computer [COMPUT SCI] A gun-sight computer used in the director-sight system which, in response to the gunner's action of tracking, computes the angle at which a gun must be fired in order to hit a target. { də'rek·tər ˌtīp kəmˌpyüd·ər }

directory [COMPUT SCI] The listing and description of all the fields of the records making up a file. { də'rek·trē }

directory service [COMPUT SCI] **1.** A directory of the names and addresses of all the mail recipients on a particular network, which provides electronic mail addresses. **2.** A provider of online directories of Web sites and search engines. { də'rek·trē ˌsər·vəs }

directory tree [COMPUT SCI] A graphic representation of the hierarchical branching structure in which files are organized in a hard disk or other storage device. { də'rek·trē ˌtrē }

direct outward dialing [COMMUN] A private automatic branch telephone exchange that permits all local stations to dial outside numbers. Abbreviated DOD. { də¦rekt ¦au̇t·wərd 'dīl·iŋ }

direct point repeater [ELECTR] Telegraph repeater in which the receiving relay controlled by the signals received over a line repeats corresponding signals directly into another line or lines without the interposition of any other repeating or transmitting apparatus. { də¦rekt ¦pȯint ri'pēd·ər }

direct read after write [COMPUT SCI] The reading of data immediately after the data have been written in order to check for errors in the recoding process. Abbreviated DRAW. { də¦rekt ¦rēd ˌaf·tər 'rīt }

direct realization [ELECTR] An active filter configuration that is derived by systematically replacing the elements of a passive RLC prototype filter (a filter that consists entirely of resistors, inductors, and capacitors) according to some rule. { di¦rekt ˌrē·ə·lə'zā·shən }

direct sequence system [COMMUN] A system for generating spread spectrum transmissions by phase-modulating a sine wave pseudorandomly by an unending string of pseudonoise code symbols, each of duration much smaller than a bit. { də'rekt 'sē·kwəns ˌsis·təm }

direct symbol recognition [COMPUT SCI] Recognition by sensing the unique geometrical properties of symbols. { də¦rekt 'sim·bəl ˌrek·ig ˌnish·ən }

direct wave [COMMUN] A radio wave that is propagated directly through space from transmitter to receiver without being refracted by the ionosphere. { də¦rekt 'wāv }

disability glare *See* glare. { dis·ə'bil·əd·ē ˌglār }

disable [COMPUT SCI] **1.** To prevent some action from being carried out. **2.** To turn off a computer system or a piece of equipment. { dis'ā·bəl }

disassemble [COMPUT SCI] To translate a program from machine language to assembly language to aid in its understanding. { ˌdis·ə ˈsem·bəl }

disassembler [COMPUT SCI] A program that translates machine language into assembly language.. { ˌdis·əˈsem·blər }

disaster dump [COMPUT SCI] A listing of the contents of a computer's central processing unit that is created when the computer detects an error that it cannot handle in the course of processing. { diˈzas·tər ˌdəmp }

discomfort glare *See* glare. { disˈkəm·fərt ˌgler }

discone antenna [ELECTROMAG] A biconical antenna in which one of the cones is spread out to 180° to form a disk; the center conductor of the coaxial line terminates at the center of the disk, and the cable shield terminates at the vertex of the cone. { ˈdisˌkōn anˈten·ə }

discrete address beacon system *See* Mode S. { diˌskrēt ˈad·res ˈbē·kən ˌsis·təm }

discrete comparator *See* digital comparator. { diˈskrēt kəmˈpar·əd·ər }

discrete cosine transform [COMMUN] A mathematical transform, used in bit rate reduction applications, in which the reconstructed bit stream is identical to the bit stream input to the system; in this regard, the transform is a mathematical process that can be perfectly undone. Abbreviated DCT. { diˈskrēt ˈkōˌsīn ˈtranzˌfȯrm }

discrete system [CONT SYS] A control system in which signals at one or more points may change only at discrete values of time. Also known as discrete-time system. { diˈskrēt ˈsis·təm }

discrete-time system *See* discrete system. { diˈskrēt ˌtīm ˈsis·təm }

discrete transfer function *See* pulsed transfer function. { di¦skrēt ˈtranz·fər ˌfəŋk·shən }

discrete-word intelligibility [COMMUN] The percent of intelligibility obtained when the speech units under consideration are words, usually presented so as to minimize the contextual relation between them. { di¦skrēt ˌwərd in ˌtel·ə·jəˈbil·əd·ē }

discrimination [COMMUN] **1.** In frequency-modulated systems, the detection or demodulation of the imposed variations in the frequency of the carriers. **2.** In a tuned circuit, the degree of rejection of unwanted signals. **3.** Of any system or transducer, the difference between the losses at specified frequencies with the system or transducer terminated in specified impedances. [COMPUT SCI] *See* conditional jump. { diˌskrim·əˈnā·shən }

discriminator transformer [ELECTR] A transformer designed to be used in a stage where frequency-modulated signals are converted directly to audio-frequency signals or in a stage where frequency changes are converted to corresponding voltage changes. { diˈskrim·əˌnād·ər tranzˈfȯr·mər }

dish *See* parabolic reflector. { dish }

disjunctive search [COMPUT SCI] A search to find items that have at least one of a given set of characteristics. { disˈjənk·tiv ˈsərch }

disk [COMPUT SCI] A rotating circular plate having a surface on which information may be stored as a pattern of magnetically polarized spots (on a magnetic disk) or holes (on an optical disk) on concentric recording tracks. Also known as magnetic disk. Also spelled disc. { disk }

disk cache [COMPUT SCI] A portion of random-access memory that contains the data most recently read from or written to the disk, allowing rapid access by the central-processing unit. { ˈdisk ˌkash }

disk cartridge [COMPUT SCI] A removable module that contains a single magnetic disk platter which remains attached to the housing when placed into the disk drive. { ˈdisk ˌkär·trij }

disk crash *See* head crash. { ˈdisk ˌkrash }

disk drive [COMPUT SCI] The physical unit that holds, spins, reads, and writes the magnetic disks. Also known as disk unit. { ˈdisk ˌdrīv }

disk drive controller [COMPUT SCI] A device that enables a microcomputer to control the functioning of a disk drive. { ˈdisk ¦drīv kənˈtrō·lər }

diskette *See* floppy disk. { diˈsket }

disk file [COMPUT SCI] An organized collection of records held on a magnetic disk. { ˈdisk ˌfīl }

diskless work station [COMPUT SCI] A computer in a network that has no disk storage of its own. { ¦disk·ləs ˈwərk ˌstā·shən }

disk memory *See* disk storage. { ˈdisk ˌmem·rē }

disk operating system [COMPUT SCI] An operating system which uses magnetic disks as its primary on-line storage. Abbreviated DOS. { ¦disk ¦äp·əˌrād·iŋ ˌsis·təm }

disk pack [COMPUT SCI] A set of magnetic disks that can be removed from a disk drive as a unit. { ˈdisk ˌpak }

disk storage [ELECTR] An external computer storage device consisting of one or more disks spaced on a common shaft, and magnetic heads mounted on arms that reach between the disks to read and record information on them. Also known as disk memory; magnetic disk storage. { ¦disk ¦stȯr·ij }

disk striping [COMPUT SCI] The distribution of a unit of data over two or more hard disks, enabling the data to be read more quickly. Also known as data striping. { ˈdisk ˌstrīp·iŋ }

disk unit *See* disk drive. { ˈdisk ˌyü·nət }

dispatching [COMPUT SCI] The control of priorities in a queue of requests in a multiprogramming or multitasking environment. { disˈpach·iŋ }

dispatching priority [COMPUT SCI] In a multiprogramming or multitasking environment, the priority assigned to an active (non-real time, nonforeground) task. { disˈpach·iŋ prīˌär·əd·ē }

disperse [COMPUT SCI] A data-processing operation in which grouped input items are distributed among a larger number of groups in the output. { dəˈspərs }

dispersion [COMMUN] The entropy of the output of a communications channel when the input is known. [ELECTROMAG] Scattering of microwave radiation by an obstruction. { də ˈspər·zhən }

displacement [COMPUT SCI] The number of character positions or memory locations from some point of reference to a specified character or data item. Also known as offset. { dis'plās·mənt }

display [ELECTR] **1.** A visible representation of information, in words, numbers, or drawings, as on the cathode-ray tube screen of a radar set, navigation system, or computer console. **2.** The device on which the information is projected. Also known as display device. **3.** The image of the information. { di'splā }

display adapter See video display board. { di'splā ə,dap·tər }

display console [COMPUT SCI] A cathode-ray tube or other display unit on which data being processed or stored in a computer can be presented in graphical or character form; sometimes equipped with a light pen with which the user can alter the information displayed. { di'splā ,kän,sōl }

display control [COMPUT SCI] A unit in a computer system consisting of channels and associated control circuitry that connect a number of visual display units with a central processor. { di'splā kən,trōl }

display cycle [COMPUT SCI] In computer graphics, the sequence of operations carried out to display an image. { di,splā ,sī·kəl }

display device See display. { di'splā di,vīs }

display element [COMPUT SCI] In computer graphics, a basic component of a display, such as a circle, line, or dot. { di'splā,el·ə·mənt }

display entity [COMPUT SCI] In computer graphics, a group of display elements that can be manipulated as a unit. { di'splā,en·təd·ē }

display formats See radar display formats. { di ,splā ,fȯr·matz }

display frame [COMPUT SCI] In computer graphics, one of a sequence of frames making up a computer-generated animation. { di'splā ,frām }

display information processor [COMPUT SCI] Computer used to generate situation displays in a combat operations center. { di'splā in·fər 'mā·shən ,präs,es·ər }

display list [COMPUT SCI] In computer graphics, a set of vectors that form an image stored in vectors graphics format. { di'splā ,list }

display packing [COMPUT SCI] An efficient means of transmitting the *x* and *y* coordinates of a point packed in a single word to halve the time required to freshen the spot on a cathode-ray tube display. { di'splā ,pak·iŋ }

display power management signaling [COMPUT SCI] Signaling whereby a video adapter can instruct a monitor to reduce its power level to conserve electricity. Abbreviated DPMS. { di ¦splā 'pau̇·ər ,man·ij·mənt ,sig·nəl·iŋ }

display primary [COMMUN] One of the primary colors produced in a video system that, when mixed in proper proportions, serve to produce the other desired colors. { di'splā 'prī,mer·ē }

display processor [COMPUT SCI] A section of a computer which handles the routines required to display an output on a cathode-ray tube. { di'splā ,präs,es·ər }

display screen See video monitor. { di'splā ,skrēn }

display system [COMPUT SCI] The total system, combining hardware and software, needed to achieve a visible representation of information in a data-processing system. { di'splā ,sis·təm }

display terminal [COMPUT SCI] A computer output device in which characters and sometimes graphic information appear on the screen of a cathode-ray tube; now largely replaced by monitors using bit-mapped displays. Also known as display unit; video display terminal (VDT). { di'splā ,tər·mən·əl }

display tube [ELECTR] A cathode-ray tube used to provide a visual display. Also known as visual display unit. { di'splā ,tüb }

display unit See display terminal. { di'splā ,yü·nət }

display window [COMMUN] Width of the portion of the frequency spectrum presented on panoramic presentation; expressed in frequency units, usually megahertz. { di'splā ,win,dō }

disposition [COMPUT SCI] The status of a file after it has been closed by a computer program, for example, retained or deleted. { ,dis·pə'zish·ən }

dissector tube [ELECTR] Camera tube having a continuous photo cathode on which is formed a photoelectric emission pattern which is scanned by moving its electron-optical image over an aperture. { də'sek·tər ,tüb }

dissipation line [ELECTROMAG] A length of stainless steel or Nichrome wire used as a noninductive terminating impedance for a rhombic transmitting antenna when several kilowatts of power must be dissipated. { ,dis·ə'pā·shən ,līn }

distance marker [ENG] One of a series of concentric circles, painted or otherwise fixed on the screen of a plan position indicator, from which the distance of a target from the radar antenna can be read directly; used for surveillance and navigation where the relative distances between a number of targets are required simultaneously. Also known as radar range marker; range marker. { 'dis·təns ,märk·ər }

distance-measuring equipment [NAV] A radio aid to navigation that provides distance information by measuring total round-trip time of transmission from an airborne interrogator to a ground-based transponder and return. Abbreviated DME. { 'dis·təns ,mezh·ər·iŋ i'kwip·mənt }

distance reception [COMMUN] Reception of messages from, or communication with, distant radio stations. Abbreviated DX. { 'dis·təns ri'sep·shən }

distance resolution [ENG] The minimum radial distance by which targets must be separated to be separately distinguishable by a particular radar. Also known as range discrimination; range resolution. { 'dis·təns ,rez·ə,lü·shən }

distant field [ELECTROMAG] The electromagnetic field at a distance of five wavelengths or more from a transmitter, where the radial electric field becomes negligible. { ¦dis·tənt ¦fēld }

distortion [ELECTR] Any undesired change in the waveform of an electric signal passing through a circuit or other transmission medium. [ENG] In general, the extent to which a system fails to accurately reproduce the characteristics of an input signal at its output. { di'stȯr·shən }

distortion factor [COMMUN] Ratio of the effective value of the residue of a wave after elimination of the fundamental to the effective value of the original wave. { di'stȯr·shən ˌfak·tər }

distress frequency [COMMUN] A frequency allotted to distress calls, generally by international agreement; for ships at sea and aircraft over the sea, it is 500 kilohertz. { də'stres ˌfrē·kwən·sē }

distress signal [COMMUN] An international signal used when a ship, aircraft, or other vehicle is threatened by grave and imminent danger and requests immediate assistance; examples are special radiotelegraph and radiotelephone signals or special signal flags or flares. { də'stres ˌsig·nəl }

distributed bulletin board [COMPUT SCI] A collection of newsgroups on a wide-area network, whose postings are available to every user. { di ˌstrib·yəd·əd 'bu̇l·ət·ən ˌbȯrd }

distributed communications [COMMUN] Information transfer beyond the local level that may involve the originating source to transmit information to all communications centers on any one network, and may also cause an interchange of communications among several whole networks. { di'strib·yəd·əd kə'myü·nə'kā·shənz }

distributed computing [COMPUT SCI] The use of multiple network-connected computers for solving a problem or for information processing. { diˌstrib·yəd·əd kəm'pyüd·iŋ }

distributed control system [CONT SYS] A collection of modules, each with its own specific function, interconnected tightly to carry out an integrated data acquisition and control application. { di'strib·yəd·əd kən'trōl ˌsis·təm }

distributed database [COMPUT SCI] A database maintained in physically separated locations and supported by a computer network so that it is possible to access all parts of the database from various points in the network. { di'strib·yəd·əd 'dad·ə ˌbās }

distributed free space [COMPUT SCI] Empty spaces in a data layout to allow new data to be inserted at a future time. { di'strib·yəd·əd ¦frē ¦spās }

distributed intelligence [COMPUT SCI] The existence of processing capability in terminals and other peripheral devices of a computer system. Also known as distributed logic. { di'strib·yəd·əd in'tel·ə·jəns }

distributed logic *See* distributed intelligence. { di¦strib·yəd·əd 'läj·ik }

distributed logic cluster word processor [COMPUT SCI] A system of word processors each of which can operate independently, although printers are generally shared by a number of terminals. { di'strib·yəd·əd 'läj·ikˌkləs·tər 'wərd ˌpräsˌes·ər }

distributed network [COMMUN] A communications network in which there exist alternative routings between the various nodes. [COMPUT SCI] A computer network in which at least some of the processing is done at individual work stations and information is shared by and often stored at the work stations. { di'strib·yəd·əd 'netˌwərk }

distributed-parameter system *See* distributed system. { di'strib·yəd·əd pə'ram·əd·ər ˌsis·təm }

distributed processing system [COMPUT SCI] An information processing system consisting of two or more programmable devices, connected so that information can be exchanged. { di'strib·yəd·əd 'präsˌes·iŋ ˌsis·təm }

distributed system [COMPUT SCI] A computer system consisting of a collection of autonomous computers linked by a network and equipped with software that enables the computers to coordinate their activities and to share the resources of system hardware, software, and data, so that users perceive a single, integrated computing facility. [CONT SYS] A collection of modules, each with its own specific function, interconnected to carry out integrated data acquisition and control in a critical environment. [SYS ENG] A system whose behavior is governed by partial differential equations, and not merely ordinary differential equations. Also known as distributed-parameter system. { di'strib·yəd·əd 'sis·təm }

distributing frame [ELECTR] Structure for terminating permanent wires of a central office, private branch exchange, or private exchange, and for permitting the easy change of connections between them by means of cross-connecting wires. { di'strib·yəd·iŋ ˌfrām }

distribution amplifier [ELECTR] A radio-frequency power amplifier used to feed television or radio signals to a number of receivers, as in an apartment house or a hotel. { ˌdis·trə'byü·shən 'am·pləˌfī·ər }

distribution frame [COMMUN] A place where a number of cables converge and signals are redistributed among them. { ˌdis·trə'byü·shən ˌfrām }

disturbance [COMMUN] An undesired interference or noise signal affecting radio, television, or data reception. [CONT SYS] An undesired command signal in a control system. { də'stər·bəns }

disturbed-one output [ELECTR] One output of a magnetic cell to which partial-read pulses have been applied since that cell was last selected for writing. { də¦stərbd ¦wən 'au̇tˌpu̇t }

dither [COMMUN] A technique for representing the entire gray scale of a picture by picture elements with only one of two levels ("white" and "black"), in which a multilevel input image signal is compared with a position-dependent set of thresholds, and picture elements are set to "white" only where the image input signal exceeds the threshold. [CONT SYS] A force having a controlled amplitude and frequency, applied continuously to a device driven by a servomotor so that the device is constantly in small-amplitude motion and cannot stick at its null position. Also known as buzz. { 'dith·ər }

dither matrix [COMMUN] A square matrix of threshold values that is repeated as a regular array to provide a threshold pattern for an entire image in the dither method of image representation. { 'dith·ər ,mā·triks }

diversity [COMMUN] Method of signal extraction by which an optimum resultant signal is derived from a combination of, or selection from, a plurality of transmission paths, channels, techniques, or physical arrangements; the system may employ space diversity, polarization diversity, frequency diversity, or any other arrangement by which a choice can be made between signals. { də'vər·səd·ē }

diversity gain [COMMUN] Gain in reception as a result of the use of two or more receiving antennas. { də'vər·səd·ē ,gān }

diversity radar [ENG] A radar that uses two or more transmitters and receivers, each pair operating at a slightly different frequency but sharing a common antenna and video display, to obtain greater effective range and reduce susceptibility to jamming. { də'vər·səd·ē 'rā ,där }

diversity receiver [ELECTR] A radio receiver designed for space or frequency diversity reception. { də'vər·səd·ē ri'sē·vər }

diversity reception [COMMUN] Radio reception in which the effects of fading are minimized by combining two or more sources of signal energy carrying the same modulation. { də'vər·səd·ē ri'sep·shən }

divide check [COMPUT SCI] An error signal indicating that an illegal division (such as dividing by zero) was attempted. { də'vīd ,chek }

divided slit scan [COMPUT SCI] In optical character recognition, a device consisting of a narrow column of photoelectric cells which scans an input character at given intervals for the purpose of obtaining its horizontal and vertical components. { də'vīd·əd 'slit ,skan }

division [COMPUT SCI] One of four required parts of a COBOL program, labeled identification, environment, data, and procedure, each with a set of rules governing the contents. { də'vizh·ən }

division subroutine [COMPUT SCI] A built-in program which achieves division by methods such as repetitive subtraction. { də'vizh·ən 'səb·rü,tēn }

D layer [GEOPHYS] The lowest layer of ionized air above the earth, occurring in the D region only in the daytime hemisphere; reflects frequencies below about 50 kilohertz and partially absorbs higher-frequency waves. { 'dē ,lā·ər }

dma *See* direct memory access.

DME *See* distance-measuring equipment.

DNS *See* domain name system.

docking station [COMPUT SCI] A device that connects a portable computer with peripherals such as an external monitor, keyboard, and so on, allowing a portable computer to function as a desktop computer. { 'däk·iŋ ,stā·shən }

document [COMPUT SCI] **1.** Any record, printed or otherwise, that can be read by a human or a machine. **2.** To prepare a written text and charts describing the purpose, nature, usage, and operation of a program or a system of programs. { 'däk·yə·mənt }

document alignment [COMPUT SCI] The phase of the reading process in which a transverse force is applied to a document to line up its reference edge with that of the reading station. { 'däk·yə·mənt ə,līn·mənt }

documentation [COMPUT SCI] The collection, organized and stored, of records that describe the purpose, use, structure, details, and operational requirements of a program, for the purpose of making this information easily accessible to the user. { ,däk·yə·mən'tā·shən }

document comparison utility [COMPUT SCI] A program that compares two documents created by word-processing programs and provides a display of the differences between them. { ,däk·yə·mənt kəm'par·ə·sən yü,til·əd·ē }

document flow [COMPUT SCI] The path taken by documents as they are processed through a record handling system. { 'däk·yə·mənt ,flō }

document handling [COMPUT SCI] In character recognition, the process of loading, feeding, transporting, and unloading a cut-form document that has been submitted for character recognition. { 'däk·yə·mənt ,hand·liŋ }

document image processing [COMPUT SCI] The scanning of paper documents followed by the storage, retrieval, display, and management of the resulting electronic images. Also known as document imaging. { ¦däk·yə·mənt 'im·ij ,prä ,ses·iŋ }

document imaging *See* document image processing. { 'däk·yə·mənt ,im·ij·iŋ }

document leading edge [COMPUT SCI] In character recognition, that edge which is the foremost one encountered during the reading process and whose relative position defines the document's direction of travel. { 'däk·yə·mənt ,lēd·iŋ 'ej }

document misregistration [COMPUT SCI] In character recognition, the improper state of appearance of a document, on site in a character reader, with respect to real or imaginary horizontal baselines. { 'däk·yə·mənt ,mis·rej·ə'strā·shən }

document number [COMPUT SCI] The number given to a document by its originators to be used as a means for retrieval; it will follow any one of various systems, such as chronological, subject area, or accession. { 'däk·yə·mənt ,nəm·bər }

document processing [COMPUT SCI] The creation, handling, labeling, and modification of text documents, such as in word processing and in the indexing of documents for retrieval based on their content. { ¦däk·yə·mənt 'prä,ses·iŋ }

document reader [COMPUT SCI] An optical character reader which reads a limited amount of information (one to five lines) and generally operates from a predetermined format. { 'däk·yə·mənt ,rēd·ər }

document reference edge [COMPUT SCI] In character recognition, that edge of a source document which provides the basis of all subsequent reading processes, insofar as it indicates the relative position of registration marks, and the impending text. { 'däk·yə·mənt 'ref·rəns ,ej }

Document Type Definition [COMPUT SCI] In Standard Generalized Markup Language, a file that specifies the tags in a particular document and the relationships among the fields that they represent. Abbreviated DTD. { 'däk·yə·mənt ,tīp ,def·ə,nish·ən }

docuterm [COMPUT SCI] A word or phrase descriptive of the subject matter or concept of an item of information and considered important for later retrieval of information. { 'däk·yə,tərm }

DOD *See* direct outward dialing.

dog [COMPUT SCI] A name for the hexadecimal digit whose decimal equivalent is 13. { dȯg }

do loop [COMPUT SCI] A FORTRAN iterative technique which enables any number of instructions to be executed repeatedly. { 'dü ,lüp }

domain [COMPUT SCI] **1.** The set of all possible values contained in a particular field for every record of a file. **2.** The protected resources that are surrounded by the security perimeter of a distributed computer system. Also known as enclave; protected subnetwork. **3.** The final two or three letters of an Internet address, which specifies the highest subdivision; in the United States this is the type of organization, such as commercial, educational, or governmental, while outside the United States it is usually a country. { dō'mān }

domain name [COMPUT SCI] An alphanumeric string which identifies a particular computer or a network on the Internet. { dō'mān ,nām }

domain name system [COMPUT SCI] Abbreviated DNS. **1.** A system used on the Internet to map the easily remembered names of host computers (domain names) to their respective Internet Protocol (IP) numbers. **2.** A software database program that converts domain names to Internet Protocol addresses, and vice versa. { dō,mān 'nām ,sis·təm }

domain tip memory [COMPUT SCI] A computer memory in which the presence or absence of a magnetic domain in a localized region of a thin magnetic film designates a 1 or 0. Abbreviated DOT memory. Also known as magnetic domain memory. { dō'mān ,tip 'mem·rē }

domestic public-frequency bands [COMMUN] Radio-frequency bands reserved for public service within the United States. { də'mes·tik ¦pəb·lik 'frē·kwən·sē ,banz }

domestic satellite [AERO ENG] A satellite in stationary orbit 22,300 miles (35,680 kilometers) above the equator for handling 12 or more separate color television programs, thousands of private-line telephone calls, or an equivalent number of channels for other communication services within the United States. Abbreviated DOMSAT. { də'mes·tik 'sad·əl,īt }

dominant mode *See* fundamental mode. { 'däm·ə·nənt 'mōd }

DOMSAT *See* domestic satellite. { 'däm,sat }

dongle [COMPUT SCI] A hardware device that plugs into a computer or printer port and serves as a copy-protection device for certain software, which must verify its presence in order to run properly. Also known as hardware key. { 'daŋ·gəl }

do-nothing instruction *See* NO OP. { 'dü ,nəth·iŋ in,strək·shən }

dopant *See* doping agent. { 'dō·pənt }

dope *See* doping agent. { dōp }

doping [ELECTR] The addition of impurities to a semiconductor to achieve a desired characteristic, as in producing an *n*-type or *p*-type material. Also known as semiconductor doping. { 'dōp·iŋ }

doping agent [ELECTR] An impurity element added to semiconductor materials used in crystal diodes and transistors. Also known as dopant; dope. { 'dōp·iŋ ,ā·jənt }

Doppler filtering [ELECTR] A form of coherent signal processing in a Doppler radar involving, in a pulsed radar, multiple pulses in a coherent processing interval so that one Doppler shift, indicative of the target radial velocity, may be distinguished from another; similar Doppler-sensitive processing in a continuous-wave radar. { 'däp·lər ,fil·tər·iŋ }

Doppler radar [ENG] Coherent radar, either continuous wave or pulsed, capable of sensing the radial motion of targets by sensing the Doppler shift of the echoes. { 'däp·lər 'rā,där }

Doppler sonar [ENG] Sonar based on Doppler shift measurement technique. Abbreviated DS. { 'däp·lər 'sō,när }

Doppler tracking [ENG] Tracking of a target by using Doppler radar. { 'däp·lər ,trak·iŋ }

Doppler VOR [NAV] A ground-based navigational aid operating at very high frequency and using a wide-aperture radiation system to reduce azimuth errors caused by reflection from terrain and other obstacles; makes use of the Doppler principle to solve the problem of ambiguity that arises from the use of a radiation system with apertures that exceed one-half wavelength; the system is so designed that its signals may be received on the equipment used for the narrow-aperture VOR (very-high-frequency omnidirectional radio range). { 'däp·lər ¦vē¦ō'är }

DOS *See* disk operating system. { däs }

dot-addressable [COMPUT SCI] The ability of an electronic display or a dot-matrix printer to specify the individual dots that form images of characters. { ¦dät ə'dres·ə·bəl }

dot character printer *See* dot matrix printer. { 'dät 'kar·ik·tər ,print·ər }

dot cycle [COMMUN] In teletypewriter systems, an on-off or mark-space cycle in which both mark and space have the same length as the unit pulse. { 'dät ,sī·kəl }

dot generator [ELECTR] A signal generator that produces a dot pattern on the screen of a color display device for use in convergence adjustments. { 'dät ˌjen·əˌrād·ər }

dot matrix [COMPUT SCI] An array of dots that forms a character or graphic symbol. { ¦dät 'mā·triks }

dot matrix printer [COMPUT SCI] A type of printer that forms each character as a group of small dots, using a group of wires located in the printing element. Also known as dot character printer. { 'dät ¦mā·triks 'prin·tər }

dot-sequential color television [ELECTR] An analog color television system in which the red, blue, and green primary-color dots are formed in rapid succession along each scanning line. { ¦dät sə¦kwen·chəl 'kəl·ər 'tel·əˌvizh·ən }

double-barrier resonant tunneling diode [ELECTR] A variant of the tunnel diode with thin layers of aluminum gallium arsenide and gallium arsenide that have sharp interfaces and have widths comparable to the Schrödinger wavelengths of the electrons, permitting resonant behavior. Abbreviated DBRT diode. { ¦dəb·əl ˌbar·ē·ər ¦rez·ən·ənt ˌtən·əl·iŋ 'dīˌōd }

double-buffered data transfer [COMPUT SCI] The transmission of data into the buffer register and from there into the device register proper. { ¦dəb·əl ¦bəf·ərd'dad·ə ˌtrans·fər }

double-channel duplex [COMMUN] A method that provides for simultaneous communication between two stations through use of two radio-frequency channels, one in each direction. { ¦dəb·əl ¦chan·əl 'düˌpleks }

double-channel simplex [COMMUN] A method that provides for nonsimultaneous communication between two stations through use of two radio-frequency channels, one in each direction. { ¦dəb·əl ¦chan·əl 'simˌpleks }

double-click [COMPUT SCI] To depress and release a mouse button twice in quick succession; often used to initiate an action such as opening a file, and to extend actions that result from a single click. { ¦dəb·əl 'klik }

double-current cable code [COMMUN] A cable code in which characters are determined by bipolar characters of equal length. { ¦dəb·əl ¦kə·rənt 'kā·bəl ˌkōd }

double-current signaling [COMMUN] A system of telegraph signaling that uses both positive and negative currents. { ¦dəb·əl ¦kə·rənt 'sig·nəl·iŋ }

double data rate [COMPUT SCI] A clocking technique that increases the transfer speeds of synchronous memories by using both the leading and trailing edges of the clock signal to transfer data, effectively doubling the transfer rate or bandwidth. { ¦dəb·əl 'dad·ə ˌrāt }

double density [COMPUT SCI] Property of a computer storage medium that holds twice as much data per unit of storage space as the standard; applied particularly to floppy disks. { 'dəb·əl 'den·səd·ē }

double-doublet antenna [ELECTROMAG] Two half-wave doublet antennas criss-crossed at their center, one being shorter than the other to give broader frequency coverage. { ¦dəb·əl ¦dəb·lət an'ten·ə }

double frequency shift keying [COMMUN] Multiplex system in which two telegraph signals are combined and transmitted simultaneously by a method of frequency shifting between four radio frequencies. { ¦dəb·əl 'frē·kwən·sē ¦shift 'kē·iŋ }

double image [ELECTR] A television picture consisting of two overlapping images due to reception of the analog signal over two paths of different length so that signals arrive at slightly different times. { ¦dəb·əl 'im·ij }

double-length number [COMPUT SCI] A number having twice as many digits as are ordinarily used in a given computer. Also known as double-precision number. { ¦dəb·əl ¦leŋkth 'nəm·bər }

double-list sorting [COMPUT SCI] A method of internal sorting in which the entire unsorted list is first placed in one portion of main memory and sorting action then takes place, creating a sorted list, generally in another area of memory. { ¦dəb·əl ˌlist 'sȯrd·iŋ }

double modulation [COMMUN] A method of modulation in which a subcarrier is first modulated with the desired intelligence, and the modulated subcarrier is then used to modulate a second carrier having a higher frequency. { ¦dəb·əl ˌmäj·ə'lā·shən }

double-polarity pulse-amplitude modulation [COMMUN] Pulse-amplitude modulation employing pulses of positive and negative polarity, the average value being equal to zero. Also known as bidirectional pulse-amplitude modulation. { ¦dəb·əl pə'lar·əd·ē 'pəls ¦am·pləˌtüd ˌmäj·ə'lā·shən }

double precision [COMPUT SCI] The use of two computer words to represent a double-length number. { ¦dəb·əl prə'sizh·ən }

double-precision hardware [COMPUT SCI] Special arithmetic units in a computer designed to handle double-length numbers, employed in operations in which greater accuracy than normal is desired. { ¦dəb·əl prə¦sizh·ən 'härdˌwer }

double-precision number *See* double-length number. { ¦dəb·əl prə¦sizh·ən 'nəm·bər }

double-pulse recording [COMPUT SCI] A technique for recording binary digits in magnetic cells in which each cell consists of two regions that can be magnetized in opposite directions and the value of each bit (0 or 1) is determined by the order in which the regions occur. { ¦dəb·əl ¦pəls ri'kȯrd·iŋ }

doubler *See* frequency doubler. { 'dəb·lər }

double refraction *See* birefringence. { ¦dəb·əl ri'frak·shən }

double-sideband modulation [COMMUN] Amplitude modulation in which the modulated wave is composed of a carrier, an upper sideband whose frequency is the sum of the carrier and modulation frequencies, and a lower sideband whose frequency is the difference between the carrier and modulation frequencies. Abbreviated DSB. Also known as double-sideband transmitted-carrier modulation (DSB-TC modulation; DSTC modulation). { ¦dəb·əl ¦sīdˌband ˌmäj·ə'lā·shən }

double-sideband reduced-carrier modulation [COMMUN] A form of amplitude modulation in which both the upper and lower sidebands are transmitted but the power contained in the unmodulated carrier is reduced to a fixed level below that provided to the modulator. Abbreviated DSB-RC modulation. { ,dəb·əl ¦sīd,band ri,düst ¦kar·ē·ər ,mä·jə,lā·shən }

double-sideband suppressed-carrier modulation [COMMUN] A form of amplitude modulation in which both the upper and lower sidebands are transmitted but the power contained in the unmodulated carrier is reduced to a fixed level below that provided to the modulator. Abbreviated DSB-SC modulation. { ,dəb·əl ¦sīd ,band sə,prest ¦kar·ē·ər ,mäj·ə,lā·shən }

double-sideband transmission [COMMUN] The transmission of a modulated carrier wave accompanied by both of the sidebands resulting from modulation; the upper sideband corresponds to the sum of the carrier and modulation frequencies, whereas the lower sideband corresponds to the difference between the carrier and modulation frequencies. { ¦dəb·əl ¦sīd,band tranz'mish·ən }

double-sideband transmitted-carrier modulation *See* double-sideband modulation. { ¦dəb·əl ¦sīd ,band tranz¦mid·əd ¦kar·ē·ər ,mäj·ə'lā·shən }

double-sided board [ELECTR] A printed wiring board that contains circuitry on both external layers. { ¦dəb·əl ,sīd·əd 'bȯrd }

double-sided disk [COMPUT SCI] A diskette that can be written on both of its sides. { ¦dəb·əl ¦sīd·əd 'disk }

double-stub tuner [ELECTROMAG] Impedance-matching device, consisting of two stubs, usually fixed three-eighths of a wavelength apart, in parallel with the main transmission lines. { ¦dəb·əl ,stəb 'tün·ər }

double-superheterodyne reception [COMMUN] Method of reception in which two frequency converters are employed before final detection. Also known as triple detection. { ¦dəb·əl ,sü·pər ¦het·rə,dīn ri'sep·shən }

doublet antenna *See* dipole antenna. { 'dəb·lət an'ten·ə }

doublet trigger [ELECTR] A trigger signal consisting of two pulses spaced a predetermined amount for coding purposes. { 'dəb·lət ,trig·ər }

double-tuned detector [ELECTR] A type of frequency-modulation discriminator in which the limiter output transformer has two secondaries, one tuned above the resting frequency and the other tuned an equal amount below. { ¦dəb·əl ,tünd di'tek·tər }

double word [COMPUT SCI] A unit containing twice as many bits as a word. { ¦dəb·əl 'wərd }

double-word addressing [COMPUT SCI] An addressing mode in computers with short words (less than 16 bits) in which the second of two consecutive instruction words contains the address of a location. { ¦dəb·əl ,wərd 'a,dres·iŋ }

doubly linked ring [COMPUT SCI] A cycle arrangement of data elements in which searches are possible in both directions. { ¦dəb·lē ¦liŋkt 'riŋ }

do-until structure [COMPUT SCI] A set of program statements that is executed once, and may then be executed repeatedly, depending on the results of a test specified in the first statement. { 'dü ən'til ,strək·chər }

do-while structure [COMPUT SCI] A set of program statements that is executed repeatedly, as long as some condition, specified in the first statement, remains in effect. { 'dü 'wīl ,strək·chər }

downlink [COMMUN] The radio or optical transmission path downward from a communications satellite to the earth or an aircraft, or from an aircraft to the earth. { 'daün,liŋk }

download [COMPUT SCI] To transfer a program or data file from a central computer to a remote computer or to the memory of an intelligent terminal. { 'daün,lōd }

downward compatibility [COMPUT SCI] The ability of an older or smaller computer to accept programs from a newer or larger one. Also known as backward compatibility. { 'daün·wərd kəm ,pad·ə'bil·əd·ē }

DPCM *See* differential pulse-code modulation.

DPMS *See* display power management signaling.

drag [COMPUT SCI] To move an object across a screen by moving a pointing device while holding down the control button. { drag }

drag and drop [COMPUT SCI] A feature whereby operations are performed on objects, such as icons or blocks of text, by dragging them across the screen with a mouse. { ¦drag ən 'dräp }

drain [ELECTR] The region into which majority carriers flow in a field-effect transistor; it is comparable to the collector of a bipolar transistor and the anode of an electron tube. { drān }

DRAM *See* dynamic random-access memory. { 'dē,ram }

DRAW *See* direct read after write. { drȯ }

drawing program [COMPUT SCI] A graphics program that maintains images in vector graphics format, allowing the user to design and illustrate objects on the display screen. Also known as illustration program. { 'drȯ·iŋ ,prō·grəm }

drift-corrected amplifier [ELECTR] A type of amplifier that includes circuits designed to reduce gradual changes in output, used in analog computers. { ¦drift kə¦rek·təd 'am·plə,fī·ər }

drift error [COMPUT SCI] An error arising in the use of an analog computer due to gradual changes in the output of circuits (such as amplifiers) in the computer. { 'drift ,er·ər }

drill circuit [COMMUN] A telegraph circuit used only to practice sending and receiving. { 'dril ,sər·kət }

drill down [COMPUT SCI] In data mining, viewing data at a greater level of detail; for example, viewing individual sales as opposed to viewing total sales. { ¦dril 'daün }

drill up [COMPUT SCI] In data mining, viewing data in less detail; for example, viewing total sales as opposed to individual sales. { ¦dril 'əp }

drive array [COMPUT SCI] A collection of hard disks organized to increase speed and improve reliability, often with the help of data stripping. { 'drīv ə,rā }

drive bay [COMPUT SCI] A space in the cabinet of a personal computer where disk drives, tape drives, and CD-ROM drives can be installed. Also known as bay. { 'drīv ,bā }

driveless work station [COMPUT SCI] A computer or terminal in a local area network that does not have its own disk drives and relies on a central mass storage facility for information storage. { 'driv·ləs 'wərk ,stā·shən }

drive light [COMPUT SCI] A lamp on the front of a disk drive that lights to indicate when the unit is reading or writing data. { 'drīv ,līt }

driven array [ELECTROMAG] An antenna array consisting of a number of driven elements, usually half-wave dipoles, fed in phase or out of phase from a common source. { ¦driv·ən ə'rā }

driven element [ELECTROMAG] An antenna element that is directly connected to the transmission line. { ¦driv·ən 'el·ə·mənt }

drive pattern [COMMUN] In a facsimile system, undesired pattern of density variations caused by periodic errors in the position of the recording spot. { 'drīv ,pad·ərn }

driver [COMPUT SCI] A sequence of program instructions that controls an input/output device such as a tape drive or disk drive. [ELECTR] The amplifier stage preceding the output stage in a receiver or transmitter. { 'drī·vər }

driver element [ELECTROMAG] Antenna array element that receives power directly from the transmitter. { 'drī·vər ,el·ə·mənt }

driver transformer [ELECTR] A transformer in the input circuit of an amplifier, especially in the transmitter. { 'drī·vər tranz'fȯr·mər }

driving-point function [CONT SYS] A special type of transfer function in which the input and output variables are voltages or currents measured between the same pair of terminals in an electrical network. { 'drīv·iŋ ,pȯint, fəŋk·shən }

driving signal [ELECTR] Television signal that times the scanning at the pickup point. { 'drīv·iŋ ,sig·nəl }

drop-dead halt [COMPUT SCI] A machine halt from which there is no recovery; such a halt may occur through a logical error in programming; examples in which a drop-dead halt could occur are division by zero and transfer to a nonexistent instruction word. Also known as dead halt. { ¦dräp ¦ded 'hȯlt }

drop-in [COMPUT SCI] The accidental appearance of an unwanted bit, digit, or character on a magnetic recording surface or during reading from or writing to a magnetic storage device. { 'dräp ,in }

dropout [COMPUT SCI] The accidental disappearance of a valid bit, digit, or character from a storage medium or during reading from or writing to a storage device. [ELECTR] A reduction in output signal level during reproduction of recorded data, sufficient to cause a processing error. { 'dräp,au̇t }

dropout error [ELECTR] Loss of a recorded bit or any other error occurring in recorded magnetic tape due to foreign particles on or in the magnetic coating or to defects in the backing. { 'dräp,au̇t ,er·ər }

drop repeater [ELECTR] Microwave repeater that is provided with the necessary equipment for local termination of one or more circuits. { 'dräp ri,pēd·ər }

drum [ELECTR] A computer storage device consisting of a rapidly rotating cylinder with a magnetizable external surface on which data can be read or written by many read/write heads floating a few millionths of an inch off the surface; once used as a primary storage device but now used as an auxiliary device. { drəm }

drum mark [COMPUT SCI] A character indicating the termination of a record on a magnetic drum. { 'drəm ,märk }

drum memory *See* drum. { ¦drəm 'mem·rē }

drum parity error [COMPUT SCI] Parity error occurring during transfer of information onto or from drums. { ¦drəm 'par·əd·ē ,er·ər }

drum printer [COMPUT SCI] An impact printer in which a complete set of characters for each print position on a line is on a continuously rotating drum behind an inked ribbon, with paper in front of the ribbon; identical characters are printed simultaneously at all required positions on a line, on the fly, by signal-controlled hammers. { 'drəm ,print·ər }

drum recorder [ELECTR] A facsimile recorder in which the record sheet is mounted on a rotating drum or cylinder. { 'drəm ri,kȯrd·ər }

drum storage *See* drum. { 'drəm ,stȯr·ij }

drum transmitter [ELECTR] A facsimile transmitter in which the subject copy is mounted on a rotating drum or cylinder. { ¦drəm tranz'mid·ər }

drunk mouse [COMPUT SCI] A mouse whose pointer jumps irrationally, usually as a result of dirt or grease on the rollers. { ¦drəŋk 'mau̇s }

dry run [COMPUT SCI] A check of the logic and coding of a computer program in which the program's operations are followed from a flow chart and written instructions, and the results of each step are written down, before the program is run on a computer. Also known as desk check. [ENG] Any practice test or session. { ¦drī 'rən }

DS *See* Doppler sonar.

DSB *See* double-sideband modulation.

DSB-RC modulation *See* double-sideband reduced-carrier modulation. { ¦dē¦es'bē ¦är'sē ,mäj·ə,lā·shən }

DSB-SC modulation *See* double-sideband suppressed-carrier modulation. { ¦dē,es,bē ¦es'sē ,mäj·ə,lā·shən }

DSB-TC modulation *See* double-sideband modulation. { ¦dē¦es¦bē ¦tē'sē ,mäj·ə,lā·shən }

D-scan *See* D-display. { 'dē ,skan }

D-scope *See* D-display. { 'dē ,skōp }

DSECT *See* dummy section. { ¦dē'sekt }

D-shell connector [COMPUT SCI] The connector at the end of the cable between a video adapter and a monitor that is plugged into the video adapter. { 'dē,shel kə,nek·tər }

DSI See digital speech interpolation.

DSL See digital subscriber line.

DSN See Deep Space Network.

DSP chip See digital signal processing chip. { ¦de¦es'pē ,chip }

DSS See decision support system.

DSTC modulation See double-sideband modulation. { ¦dē¦es¦tē'sē ,mäj·ə,lā·shən }

DTD See Document Type Definition.

DTL See diode transistor logic.

DTMF See dual-tone mulitfrequency.

DTMF dialing See push-button dialing. { ¦dē¦tē ¦em'ef ,dī·liŋ }

DTV See digital television.

D/U [COMMUN] Ratio of desired to undesired signals, usually expressed in decibels.

dual-actuator hard disk [COMPUT SCI] A hard disk that is equipped with two read/write heads. { ¦dül 'ak·chə,wād·ər ¦härd ,disk }

dual control [CONT SYS] An optimal control law for a stochastic adaptive control system that gives a balance between keeping the control errors and the estimation errors small. { ¦dü·əl kən'trōl }

dual diversity receiver [ELECTR] A diversity radio receiver in which the two antennas feed separate radio-frequency systems, with mixing occurring after the converter. { ¦dü·əl də'vər·səd·ē ri,sē·vər }

dual in-line package [ELECTR] Microcircuit package with two rows of seven vertical leads that are easily inserted into an etched circuit board. Abbreviated DIP. { ¦dü·əl ¦in ,līn 'pak·ij }

dual-mode control [CONT SYS] A type of control law which consists of two distinct types of operation; in linear systems, these modes usually consist of a linear feedback mode and a bang-bang-type mode. { 'dü·əl ,mōd kən'trōl }

dual modulation [COMMUN] The process of modulating a common carrier wave or subcarrier with two different types of modulation, each conveying separate information. { 'dü·əl ,mäj·ə'lā·shən }

dual-scanned liquid-crystal display [ELECTR] A passive matrix liquid-crystal display that is improved by being refreshed twice as frequently as standard displays of this type. { ¦dül ,skand lik·wəd 'krist·əl di,splā }

dual-stripe magnetoresistive head [COMPUT SCI] A type of read/write head for hard disks that has separate areas for reading and writing, reduced vulnerability to outside interference, and the ability to pack data densely on disks { ¦dül ¦strīp mag,ned·ō·ri,zis·div 'hed }

dual-tone multifrequency [COMMUN] Signaling method employing set combinations of two specific frequencies used by subscribers and telephone private branch exchange attendants, if their switchboard positions are so equipped, to indicate telephone address digits, precedence ranks, and end of signaling. Abbreviated DTMF. { 'dü·əl ,tōn ,məl·tē'frē·kwən·sē }

dual-tone multifrequency dialing See push-button dialing. { 'dü·əl ,tōn ,məl·tē'frē·kwən·sē 'dī·liŋ }

dual-use line [COMMUN] Communications link normally used for more than one mode of transmission, such as voice and data. { 'dü·əl ,yüs ,līn }

dual-use radar [ENG] Radar designed to perform both as surveillance radar and weather radar, of particular value in air traffic management where both the monitoring of aircraft and estimation of the weather environment are important. { 'dü·əl ,yüs 'rā,där }

duct [COMMUN] An enclosed runway for cables. { dəkt }

dumb terminal [COMPUT SCI] A computer input/output device that lacks the capability to process or format data, and is thus entirely dependent on the main computer for these activities. { ¦dəm 'term·ən·əl }

dummy [COMMUN] Telegraphy network simulating a customer's loop for adjusting a telegraph repeater; the dummy side of the repeater is that toward the customer. [COMPUT SCI] An artificial address, instruction, or other unit of information inserted in a digital computer solely to fulfill prescribed conditions (such as word length or block length) without affecting operations. { 'dəm·ē }

dummy antenna [ELECTR] A device that has the impedance characteristic and power-handling capacity of an antenna but does not radiate or receive radio waves; used chiefly for testing a transmitter. Also known as artificial antenna. { ¦dəm·ē an'ten·ə }

dummy argument [COMPUT SCI] The variable appearing in the definition of a macro or function which will be replaced by an address at call time. { ¦dəm·ē 'är·gyə·mənt }

dummy file [COMPUT SCI] A nonexistent file which is treated by a computer program as if it were receiving its output data, when in fact the data are being ignored; used to suppress the creation of files that are needed only occasionally. { 'dəm·ē 'fīl }

dummy instruction [COMPUT SCI] An artificial instruction or address inserted in a list to serve a purpose other than the execution as an instruction. { ¦dəm·ē in'strək·shən }

dummy load [ELECTR] A dissipative device used at the end of a transmission line or waveguide to convert transmitted energy into heat, so that essentially no energy is radiated outward or reflected back to its source. { 'dəm·ē ,lōd }

dummy message [COMMUN] A message sent for some purpose other than its content, which may consist of dummy groups or may have a meaningless text. { ¦dəm·ē 'mes·ij }

dummy parameter [COMPUT SCI] A parameter whose value has no significance but which is included in an instruction or command to satisfy the requirements of the system. { 'dəm·ē pə'ram·əd·ər }

dummy record [COMPUT SCI] Meaningless information that is stored for some purpose such as fulfillment of a length requirement. { ˈdəm·ē ˈrek·ərd }

dummy section [COMPUT SCI] The part of an assembly language program in which the arrangement of the data in memory is specified. Abbreviated DSECT. { ˈdəm·ē ˈsek·shən }

dump [COMPUT SCI] To copy the contents of all or part of a storage, usually from an internal storage device into an external storage device. [ELECTR] To withdraw all power from a system or component accidentally or intentionally. { dəmp }

dump check [COMPUT SCI] A computer check that usually consists of adding all the digits during dumping, and verifying the sum when retransferring. { ˈdəmp ˌchek }

dump routine [COMPUT SCI] A program within a computer's operating system that handles the processing of dumps. { ˈdəmp rüˌtēn }

duplex channel [COMMUN] A communication channel providing simultaneous transmission in both directions. { ¦düˌpleks ˈchan·əl }

duplex computer [COMPUT SCI] Two identical computers, either one of which can ensure continuous operation of the system when the other is shut down. { ¦düˌpleks kəmˈpyüd·ər }

duplexer [ELECTR] A switching device used in radar to permit alternate use of the same antenna for both transmitting and receiving; other forms of duplexers serve for two-way radio communication using a single antenna at lower frequencies. Also known as duplexing assembly. { ˈdüˌplek·sər }

duplexing [COMMUN] *See* duplex operation. [COMPUT SCI] The provision of redundant hardware or excess capacity which can pick up the work load in the event of failure of one part of a computer system. { ˈdüˌpleks·iŋ }

duplexing assembly *See* duplexer. { ˈdüˌpleks·iŋ əˌsem·blē }

duplex operation [COMMUN] The operation of associated transmitting and receiving apparatus concurrently, as in ordinary telephones, without manual switching between talking and listening periods. Also known as duplexing; duplex transmission. [ENG] In radar, operation in which two identical and interchangeable equipments are provided, generally to enhance system reliability, one in an active state and the other immediately available for operation. { ¦düˌpleks äp·əˈrā·shən }

duplex transmission *See* duplex operation. { ¦düˌpleks tranzˈmish·ən }

duplicate record [COMPUT SCI] An unwanted record that has the same key as another record in the same file. { ˈdüp·lə·kət ˈrek·ərd }

duplication check [COMPUT SCI] A check based on the identity in results of two independent performances of the same task. { ˌdüp·ləˈkā·shən ˌchek }

duty cycle [COMMUN] The product of the pulse duration and pulse repetition frequency of a pulse carrier, equal to the time per second that pulse power is applied. Also known as duty factor. { ˈdüd·ē ˌsī·kəl }

duty factor [COMMUN] In a pulse radar or similar system, the ratio of average to pulse power; basically, the product of the pulse width (for square pulses) and the pulse repetition frequency. Also known as duty ratio. { ˈdüd·ē ˌfak·tər }

duty ratio *See* duty factor. { ˈdüd·ē ˌrā·shō }

DUV *See* data under voice.

DVD [COMMUN] An optical disk that has formats for audio, video, and computer storage applications, and that uses the same basic structure as the compact disk (CD) to store data, but achieves a greater storage capability by using a track pitch less than half that of the CD, pits and lands as little as half as long as the shortest on a CD, and two substrates, bonded together. Derived from digital versatile disk; digital video disk.

DVD-audio [COMMUN] A DVD format for digital storage of audio information. Also known as Book C. { ¦dē¦vē¦dē ˈȯd·ē·ō }

DVD-RAM *See* DVD-rewritable. { ¦dē¦vē¦dē ˈram }

DVD-read-only [COMMUN] A DVD format in which data written on the disk at the time of its manufacture are permanent, and the disk cannot be written or erased after that. Also known as Book A; DVD-ROM. { ¦dē¦vē¦dē ˌrēd ˈōn·lē }

DVD-rewritable [COMMUN] A DVD format that allows audio or other digital data to be written, read, erased, and rewritten. Also known as Book E; DVD-RAM. { ¦dē¦vē¦dē rēˈrīd·ə·bəl }

DVD-ROM *See* DVD-read-only. { ¦dē¦vē¦dē ˈräm }

DVD-video [COMMUN] A DVD format for digital storage of video information. Also known as Book B. { ¦dē¦vē¦dē ˈvid·ē·ō }

DVD-write once [COMMUN] A DVD format that allows users to record audio or other digital data in such a way that the recording is permanent and may be read indefinitely but cannot be erased. Also known as Book D. { ¦dē¦vē¦dē ˌrīt ˈwəns }

dwell time [ELECTR] The length of time a radar examines a single target in making a single estimate about it; it is limited by the antenna rotation rate and beam width in simple radars, while in more flexible radars it is established by the computer-generated scheduling of operations. Also known as look time. { ˈdwel ˌtīm }

DX *See* distance reception.

dyadic processor [COMPUT SCI] A type of multiprocessor that includes two processors which operate under control of the same copy of the operating system. { dīˈad·ik ˈpräsˌes·ər }

dye polymer recording [COMPUT SCI] An optical recording technique in which dyed plastic layers are used as the recording medium. { ¦dī ˈpäl·ə·mər riˈkȯrd·iŋ }

dynamic acceleration *See* dynamic resolution. { dī¦nam·ik ikˌsel·əˈrā·shən }

dynamic address translator [COMPUT SCI] A hardware device used in a virtual memory system to automatically identify a virtual address inquiry in terms of segment number, page number within the segment, and position of the record with reference to the beginning of the page. { dī¦nam·ik ˈaˌdres ˌtranzˌlād·ər }

dynamic algorithm [COMPUT SCI] An algorithm whose operation is, to some extent, unpredictable in advance, generally because it contains logical decisions that are made on the basis of quantities computed during the course of the algorithm. Also known as heuristic algorithm. { dī¦nam·ik 'al·gə,rith·əm }

dynamic convergence [ELECTR] The process whereby the locus of the point of convergence of electron beams in a multibeam cathode-ray tube is made to fall on a specified surface during scanning. { dī¦nam·ik kən'vər·jəns }

dynamic debugging routine [COMPUT SCI] A debugging routine which operates in conjunction with the program being checked and interacts with it while the program is running. { dī¦nam·ik dē'bəg·iŋ rü,tēn }

dynamic dump [COMPUT SCI] A dump performed during the execution of a program. { dī¦nam·ik 'dəmp }

dynamic focusing [ELECTR] The process of varying the focusing electrode voltage for a color picture tube automatically so the electron-beam spots remain in focus as they sweep over the flat surface of the screen. { dī¦nam·ik 'fō·kəs·iŋ }

dynamicizer [COMPUT SCI] A device that converts a collection of data represented by a spatial arrangement of bits in a computer storage device into a series of signals occurring in time. { dī'nam·ə,sīz·ər }

dynamic link [COMPUT SCI] A linking of data in two different programs, whereby modification in either program causes a similar change of the data in the other. { dī¦nam·ik 'liŋk }

dynamic memory *See* dynamic storage. { dī¦nam·ik 'mem·rē }

dynamic memory allocation *See* dynamic storage allocation. { dī¦nam·ik 'mem·rē al·ə,kā·shən }

dynamic printout [COMPUT SCI] A printout of data which occurs during the machine run as one of the sequential operations. { dī¦nam·ik 'print ,au̇t }

dynamic problem check [COMPUT SCI] Any dynamic check used to ascertain that the computer solution satisfies the given system of equations in an analog computer operation. { dī¦nam·ik 'präb·ləm ,chek }

dynamic programming [MATH] A mathematical technique, more sophisticated than linear programming, for solving a multidimensional optimization problem, which transforms the problem into a sequence of single-stage problems having only one variable each. { dī¦nam·ik 'prō·grə·miŋ }

dynamic program relocation [COMPUT SCI] The act of moving a partially executed program to another location in main memory, without hindering its ability to finish processing normally. { dī¦nam·ik 'prō·grəm ,rē·lō,kā·shən }

dynamic random-access memory [COMPUT SCI] A read-write random-access memory whose storage cells are based on transistor-capacitor combinations, in which the digitalinformation is represented by charges that are stored on the capacitors and must be repeatedly replenished in order to retain the information. Abbreviated DRAM. { dī¦nam·ik ¦ran·dəm 'ak·ses ,mem·rē }

dynamic regulator [ELECTR] Transmission regulator in which the adjusting mechanism is in self-equilibrium at only one or a few settings and requires control power to maintain it at any other setting. { dī¦nam·ik 'reg·yə,lād·ər }

dynamic relocation [COMPUT SCI] The ability to move computer programs or data from auxiliary memory into main memory at any convenient location. { dī¦nam·ik ,rē·lō'kā·shən }

dynamic resolution [COMPUT SCI] A feature of some mice whereby the pointer moves a larger distance in proportion to the mouse's actual displacement when the mouse is moved quickly and a smaller distance when it is moved slowly. Also known as automatic acceleration; ballistic tracking; dynamic acceleration; variable acceleration. { dī¦nam·ik ,rez·ə'lü·shən }

dynamic sequential control [COMPUT SCI] Method of operation of a digital computer through which it can alter instructions as the computation proceeds, or the sequence in which instructions are executed, or both. { dī¦nam·ik sə¦kwen·chəl kən'trōl }

dynamic shift register [COMPUT SCI] A shift register that stores information by using temporary charge storage techniques. { dī¦nam·ik 'shift ,rej·ə·stər }

dynamic stop [COMPUT SCI] A loop in a computer program which is created by a branch instruction in the presence of an error condition, and which signifies the existence of this condition. { dī¦nam·ik 'stäp }

dynamic storage [COMPUT SCI] **1.** Computer storage in which information at a certain position is not always available instantly because it is moving, as in an acoustic delay line or magnetic drum. Also known as dynamic memory. **2.** Computer storage consisting of capacitively charged circuit elements which must be continually refreshed or recharged at regular intervals. { dī¦nam·ik 'stȯr·ij }

dynamic storage allocation [COMPUT SCI] A computer system in which memory capacity is made available to a program on the basis of actual, momentary need during program execution, and areas of storage may be reassigned at any time. Also known as dynamic allocation; dynamic memory allocation. { dī¦nam·ik ¦stȯr·ij ,al·ə'kā·shən }

dynamic subroutine [COMPUT SCI] Subroutine that involves parameters, such as decimal point position or item size, from which a relatively coded subroutine is derived by the computer itself. { dī¦nam·ik 'səb·rü,tēn }

dynamo *See* generator. { 'dī·nə,mō }

dynode [ELECTR] An electrode whose primary function is secondary emission of electrons; used in multiplier phototubes and some types of television camera tubes. Also known as electron mirror. { 'dī,nōd }

E

EA *See* electronic attack.

EADI *See* electronic attitude directional indicator.

E and M lead signaling [COMMUN] Communications between a trunk circuit and a separate signaling unit over two leads: an M lead that transmits battery or ground signals to the signaling equipment, and an E lead which receives open or ground signals from the signaling unit. { ¦ē ən ¦em ˈlēd ˌsig·nəl·iŋ }

early binding [COMPUT SCI] The assignment of data types (such as integer or string) to variables during the compilation of a computer program rather than at run time. { ˈər·lē ¦bīnd·iŋ }

EAROM *See* electrically alterable read-only memory. { ˈēˌräm }

earth station [COMMUN] A facility with a land-based antenna used to transmit and receive information to and from a communications satellite. { ˈərth ˌstā·shən }

easy [COMPUT SCI] A name for the hexadecimal digit whose decimal equivalent is 14. { ˈē·zē }

EBCDIC *See* extended binary-coded decimal interchange code. { ˈeb·səˌdik }

E bend [ELECTROMAG] A smooth change in the direction of the axis of a waveguide, throughout which the axis remains in a plane parallel to the direction of polarization. Also known as E-plane bend. { ˈē ˌbend }

e-business *See* electronic commerce. { ˈē ˌbiz·nəs }

ECB *See* block encryption.

Eccles-Jordan circuit *See* bistable multivibrator. { ¦ek·əlz ˈjȯrd·ən ˌsər·kət }

Eccles-Jordan multivibrator *See* bistable multivibrator. { ¦ek·əlz ˈjȯrd·ən ˌməl·tiˈvīˌbrād·ər }

ECDIS *See* electronic chart display and information system. { ˈekˌdis or ¦ē¦sē¦dē¦ī'es }

echo [ELECTR] **1.** The signal reflected, or backscattered, by a radar target, or that scattered in the receiver's direction in a bistatic radar; also, the indication of this signal on the radar display. Also known as echo pulse; radar echo; return. **2.** *See* ghost signal. { ˈek·ō }

echo area [ELECTROMAG] In radar, the area of a fictitious perfect reflector of electromagnetic waves that would reflect the same amount of energy back to the radar as the actual target. Also known as target cross section. { ˈekˌō ˌer·ē·ə }

echo box [ELECTR] A calibrated high-Q resonant cavity that stores part of the transmitted radar pulse power and gradually feeds this energy into the receiving system after completion of the pulse transmission; used to provide an artificial target signal for test and tuning purposes; being replace in design by other forms of built-in test equipment (BITE). { ˈekˌō ˌbäks }

echo check [COMPUT SCI] A method of ascertaining the accuracy of transmission of data in which the transmitted data are returned to the sending end for comparison with original data. Also known as loopback check; loop check; read-back check. { ˈek·ō ˌchek }

echo contour [ELECTR] A trace of equal signal intensity of the radar echo displayed on a range height indicator or plan position indicator. { ¦ek·ō ˈkänˌtu̇r }

echo frequency [ELECTR] The number of fluctuations, per unit time, in the power or amplitude of a radar target signal, often in reference to a moving target's echo going through cycles of constructive and destructive interference with coincident stationary clutter echo. { ˈek·ō ˌfrē·kwən·sē }

echo intensity [ELECTR] The brightness or brilliance of a radar echo as displayed on an intensity-modulated indicator; echo intensity is, within certain limits, proportional to the voltage of the target signal or to the square root of its power. { ¦ek·ō inˈten·səd·ē }

echo matching [ENG] Rotating an antenna to a position in which the pulse indications of an echo-splitting radar are equal. { ˈek·ōˌmach·iŋ }

echoplex technique [COMPUT SCI] A technique for detecting errors in a data communication system with full duplex lines, in which the signal generated when a character is typed on a keyboard is transmitted to a receiver and retransmitted to a display terminal, enabling the operator to check if the character displayed is the same as the character typed. { ˈek·ō ˌpleks tekˌnēk }

echo power [ELECTR] The electrical strength, or power, of a radar target signal, normally measured in watts or dBm (decibels referred to 1 milliwatt). { ˈek·ō ˌpau̇·ər }

echo pulse *See* echo. { ˈek·ō ˌpəls }

echo-ranging sonar [ENG] Active sonar, in which underwater sound equipment generates bursts of ultrasonic sound and picks up echoes reflected from submarines, fish, and other objects within range, to determine both direction and distance to each target. { 'ek·ō ˌrānj·iŋ 'sōˌnär }

echo recognition [ENG] Identification of a sonar reflection from a target, as distinct from energy returned by other reflectors. { 'ek·ō ˌrek·ig ˌnish·ən }

echo signal *See* target signal. { 'ek·ō ˌsig·nəl }

echo-splitting radar [ENG] Radar in which the echo is split by special circuits associated with the antenna lobe-switching mechanism, to give two echo indications on the radarscope screen; when the two echo indications are equal in height, the target bearing is read from a calibrated scale. { ¦ek·ō ˌsplid·iŋ 'rāˌdär }

echo suppressor [ELECTR] **1.** A circuit that desensitizes radar navigation equipment for a fixed period after the reception of one pulse, for the purpose of rejecting delayed pulses arriving from longer, indirect reflection paths. **2.** A relay or other device used on a transmission line to prevent a reflected wave from returning to the sending end of the line. { 'ek·ō səˌpres·ər }

echo talker [COMPUT SCI] The interference created by the retransmission of a message back to its source while the source is still transmitting. { 'ek·ō ˌtȯk·ər }

ECL *See* emitter-coupled logic.

ECM *See* embrittlement control message.

e-commerce *See* electronic commerce. { 'ē ˌkäm·ərs }

economy [COMPUT SCI] The ratio of the number of characters to be coded to the maximum number available with the code; for example, binary-coded decimal using 4 bits provides 16 possible characters but uses only 10 of them. { ē'kän·ə·mē }

ECSW *See* extended channel status word.

EDFA *See* erbium-doped fiber amplifier. { 'edˌfä *or* ¦ē¦dē¦ef'ā }

edge connector [ELECTR] A row of etched lines on the edge of a printed circuit board that is inserted into a slot to establish a connection with another printed circuit board. { 'ej kəˌnek·tər }

E-display [ELECTR] A radar display format in which the horizontal coordinate indicates range, the vertical indicates elevation, and the intensity of the target spot is proportional to signal strength. Also known as E-indicator; E-scan; E-scope. { 'ē diˌsplā }

edit [COMPUT SCI] **1.** To modify the form or format of an output or input by inserting or deleting characters such as page numbers or decimal points. **2.** A computer instruction directing that this step be performed. { 'ed·ət }

edit capability [COMPUT SCI] The degree of sophistication available to the programmer to modify his or her statements while in the time-sharing mode. { 'ed·ət ˌkāp·əˌbil·əd·ē }

edit check [COMPUT SCI] A program instruction or subroutine that tests the validity of input in a data entry program. Also known as edit test. { 'ed·ət ˌchek }

edit mask [COMPUT SCI] The receiving word through which a source word is filtered, allowing for the suppression of leading zeroes, the insertion of floating dollar signs and decimal points, and other such formatting. { 'ed·ət ˌmask }

edit mode [COMPUT SCI] A software mode of operation in which previously entered text or data can be modified or replaced. { 'ed·ət ˌmōd }

editor program [COMPUT SCI] A special program by means of which a user can easily perform corrections, insertions, modifications, or deletions in an existing program or data file. { 'ed·ə·tər ˌprō·grəm }

edit test *See* edit check. { 'ed·ət ˌtest }

EDO RAM *See* extended data out random-access memory. { ˌā·dō'ram *or* ¦ē¦dē¦ō }

EDP *See* electronic data processing.

EDP center *See* electronic data-processing center. { ¦ē¦dē'pē ˌsen·tər }

edulcorate [COMPUT SCI] To eliminate irrelevant data from a data file. { ē'dəl·kəˌrāt }

EDVAC [COMPUT SCI] The first stored program computer, built in 1952. Derived from electron discrete variable automatic compiler. { 'ed ˌvak }

EEPROM *See* electrically erasable programmable read-only memory. { ¦ē'ēˌpräm }

EER *See* equal error rate.

effective address [COMPUT SCI] The address that is obtained by applying any specified indexing or indirect addressing rules to the specified address; the effective address is then used to identify the current operand. { ə¦fek·tiv 'aˌdres }

effective bandwidth [ELECTR] The bandwidth of an assumed rectangular band-pass having the same transfer ratio at a reference frequency as a given actual band-pass filter, and passing the same mean-square value of a hypothetical current having even distribution of energy throughout that bandwidth. { ə¦fek·tiv 'band ˌwidth }

effective earth radius [COMMUN] A radius value used in place of the geometric radius to correct for atmospheric refraction in estimating ranges of antennas when the index of refraction in the atmosphere changes linearly with height; under conditions of standard refraction it is $\frac{4}{3}$ the geometric radius. Also known as effective radius of the earth. { ə¦fek·tiv 'ərth ˌrād·ē·əs }

effective facsimile band [COMMUN] Frequency band of a facsimile signal wave equal in width to that between zero frequency and maximum keying frequency. { ə¦fek·tiv fak'sim·ə·lē ˌband }

effective horizon [COMMUN] A horizon whose distance at a given height above sea level is the distance to the horizon of a fictitious earth, having a radius $\frac{4}{3}$ times the earth's true radius; used to estimate ranges of antennas, taking atmospheric refraction into account. { ə¦fek·tiv hə'rīz·ən }

effective instruction [COMPUT SCI] The computer instruction that results from changing a basic instruction during program modification. Also known as actual instruction. { ə¦fek·tiv in'strək·shən }

effective isotropic radiated power [COMMUN] A measure of the strength of the signal leaving a satellite antenna in a particular direction, equal to the product of the power supplied to the satellite transmit antenna and its gain in that direction. Abbreviated eirp. { i,fek·tiv ,ī·sə ,träp·ik ,rād·ē,ād·əd 'pau̇·ər }

effectiveness level [COMPUT SCI] A measure of the effectiveness of data-processing equipment, equal to the ratio of the operational use time to the total performance period, expressed as a percentage. Also known as average effectiveness level. { ə'fek·tiv·nəs ,lev·əl }

effective percentage modulation [COMMUN] For a single sinusoidal input component, the ratio of the peak value of the fundamental component of the envelope to the average amplitude of the modulated wave expressed in percent. { ə¦fek·tiv pər¦sent·ij ,mäj·ə'lā·shən }

effective radiated power [ELECTROMAG] The product of antenna input power and antenna power gain, expressed in kilowatts. Abbreviated ERP. { ə¦fek·tiv ,rād·ē,ād·əd 'pau̇·ər }

effective radius of the earth See effective earth radius. { ə¦fek·tiv 'rād·ē·əs əv thē 'ərth }

effective speed [COMPUT SCI] The actual speed that a computer system can sustain over a period of time when the time devoted to various control, error-detection, and other overhead activities is taken into account. { ə¦fek·tiv 'spēd }

effective time [COMPUT SCI] The time during which computer equipment is in actual use and produces useful results. { ə¦fek·tiv 'tīm }

effective value See root-mean-square value. { ə¦fek·tiv 'val·yü }

EFL See error frequency limit.

e format [COMPUT SCI] A decimal, normalized form of a floating point number in FORTRAN in which a number such as 18.756 appears as .18756E + 02, which stands for $.18756 \times 10^2$. { 'ē ,fȯr,mat }

EGNOS See European Geostationary Navigation Overlay System. { 'eg,nōs }

E-HEMT See enhancement-mode high-electron-mobility transistor.

EHF See extremely high frequency.

EHSI See electronic horizontal-situation indicator.

E-H T junction [ELECTROMAG] In microwave waveguides, a combination of E- and H-plane T junctions forming a junction at a common point of intersection with the main waveguide. { ¦ē ¦āch 'tē ,jəŋk·shən }

E-H tuner [ELECTROMAG] Tunable E-H T junction having two arms terminated in adjustable plungers used for impedance transformation. { ¦ē ¦āch 'tün·ər }

eight-level code [COMMUN] A teletypewriter code that uses eight impulses, in addition to the start and stop impulses, to define a character. { ¦āt ¦lev·əl 'kōd }

8 VSB [COMMUN] Vestigial sideband modulation with 8 discrete amplitude levels.

E-indicator See E-display. { 'ē ,in·də,kād·ər }

eirp See effective isotropic radiated power.

eject [COMPUT SCI] To move the printing mechanism to the top of the following page, skipping the remainder of the current page. { ē'jekt }

E-JFET See enhancement-mode junction field-effect transistor.

elaboration [COMPUT SCI] A technique, used chiefly in the Ada programming language, of setting up a hierarchy of calculated constants so that the values of one or more of them determine others further down in the hierarchy. { i,lab·ə'rā·shən }

E layer [GEOPHYS] A layer of ionized air occurring at altitudes between 60 and 72 miles (100 and 120 kilometers) in the E region of the ionosphere, capable of bending radio waves back to earth. Also known as Heaviside layer; Kennelly-Heaviside layer. { 'ē ,lā·ər }

elbow [ELECTROMAG] In a waveguide, a bend of comparatively short radius, normally 90°, and sometimes for acute angles down to 15°. { 'el ,bō }

electrical impedance [ELEC] **1.** The total opposition that a circuit presents to an alternating current, equal to the complex ratio of the voltage to the current in complex notation. Also known as complex impedance. **2.** The ratio of the maximum voltage in an alternating-current circuit to the maximum current; equal to the magnitude of the quantity in the first definition. Also known as impedance. { i'lek·trə·kəl im'pēd·əns }

electrical interference See interference. { i'lek·trə·kəl ,in·tər'fir·əns }

electrical length [ELECTROMAG] The length of a conductor expressed in wavelengths, radians, or degrees. { i'lek·trə·kəl 'leŋkth }

electrically alterable read-only memory [COMPUT SCI] A read-only memory that can be reprogrammed electrically in the field a limited number of times, after the entire memory is erased by applying an appropriate electric field. Abbreviated EAROM. { i'lek·trə·klē 'ȯl·trə·bəl 'rēd ¦ōn·lē 'mem·rē }

electrically erasable programmable read-only memory [COMPUT SCI] An integrated-circuit memory chip that has an internal switch to permit a user to erase the contents of the chip and write new contents into it by means of electrical signals. Abbreviated EEPROM. { i'lek·trə·klē i'rās·ə·bəl prō'gram·ə·bəl 'rēd ¦ōn·lē 'mem·rē }

electrical resonator See tank circuit. { i'lek·trə·kəl 'rez·ən,ād·ər }

electric circuit [ELEC] **1.** A path or group of interconnected paths capable of carrying electric currents. **2.** An arrangement of one or more complete, closed paths for electron flow. Also known as circuit. { i¦lek·trik 'sər·kət }

electric coil See coil. { i¦lek·trik 'kȯil }

electric condenser See capacitor. { i¦lek·trik kən'den·sər }

electric eye See photocell; phototube. { i¦lek·trik 'ī }

electric filter [ELECTR] **1.** A network that transmits alternating currents of desired frequencies while substantially attenuating all other frequencies. Also known as frequency-selective device. **2.** *See* filter. { i¦lek·trik 'fil·tər }

electric network *See* network. { i¦lek·trik 'net ˌwərk }

electric relay *See* relay. { i¦lek·trik 'rēˌlā }

electric shielding [ELECTROMAG] Any means of avoiding pickup of undesired signals or noise, suppressing radiation of undesired signals, or confining wanted signals to desired paths or regions, such as electrostatic shielding or electromagnetic shielding. Also known as screening; shielding. { i¦lek·trik 'shēld·iŋ }

electric solenoid *See* solenoid. { i¦lek·trik 'sō·ləˌnȯid }

electric switchboard *See* switchboard. { i¦lek·trik 'swichˌbȯrd }

electric telemetering [COMMUN] System to transmit electric impulses from the primary detector to a remote receiving station, with or without wire interconnections. { i¦lek·trik ˌtel·ə'mēd·ə·riŋ }

electric tuning [ELECTR] Tuning a receiver to a desired station by switching a set of preadjusted trimmer capacitors or coils into the tuning circuits. { i¦lek·trik 'tün·iŋ }

electric-wave filter *See* filter. { i¦lek·trik ¦wāv 'fil·tər }

electrochromic display [ELECTR] A solid-state passive display that uses organic or inorganic insulating solids which change color when injected with positive or negative charges. { i¦lek·trō¦krō·mik di'splā }

electrode current [ELECTR] Current passing to or from an electrode, through the interelectrode space within a vacuum tube. { i'lekˌtrōd ˌkə·rənt }

electrode dark current [ELECTR] The electrode current that flows when there is no radiant flux incident on the photocathode in a phototube or camera tube. Also known as dark current. { i'lek ˌtrōd ¦därk 'kə·rənt }

electromagnetic compatibility [ELECTR] The capability of electronic equipment or systems to be operated in the intended electromagnetic environment at design levels of efficiency. { i¦lek·trō·mag'ned·ik kəmˌpat·ə'bil·əd·ē }

electromagnetic constant *See* speed of light. { i¦lek·trō·mag'ned·ik 'kän·stənt }

electromagnetic deflection [ELECTR] Deflection of an electron stream by means of a magnetic field. { i¦lek·trō·mag'ned·ik di'flek·shən }

electromagnetic energy [ELECTROMAG] The energy associated with electric or magnetic fields. { i¦lek·trō·mag'ned·ik 'en·ər·jē }

electromagnetic environment [COMMUN] The radio-frequency fields existing in a given area. { i¦lek·trō·mag'ned·ik en'vi·rən·mənt }

electromagnetic field [ELECTROMAG] An electric or magnetic field, or a combination of the two, as in an electromagnetic wave. { i¦lek·trō·mag'ned·ik 'fēld }

electromagnetic field equations *See* Maxwell field equations. { i¦lek·trō·mag'ned·ik 'fēld iˌkwā·zhənz }

electromagnetic focusing [ELECTR] Focusing the electron beam in a video display device by means of a magnetic field parallel to the beam; the field is produced by an adjustable value of direct current through a focusing coil mounted on the neck of the tube. { i¦lek·trō·mag'ned·ik 'fō·kəs·iŋ }

electromagnetic horn *See* horn antenna. { i¦lek·trō·mag'ned·ik 'hȯrn }

electromagnetic induction [ELECTROMAG] The production of an electromotive force either by motion of a conductor through a magnetic field so as to cut across the magnetic flux or by a change in the magnetic flux that threads a conductor. Also known as induction. { i¦lek·trō·mag'ned·ik in'dək·shən }

electromagnetic pulse [ELECTROMAG] The pulse of electromagnetic radiation generated by a large thermonuclear explosion; althought not a direct threat to human health, it is a threat to electronic communications systems. { i¦lek·trō·mag'ned·ik 'pəls }

electromagnetic radiation [ELECTROMAG] Electromagnetic waves and, especially, the associated electromagnetic energy. { i¦lek·trō·mag'ned·ik ˌrād·ē'ā·shən }

electromagnetic susceptibility [ELECTR] The tolerance of circuits and components to all sources of interfering electromagnetic energy. { i¦lek·trō·mag'ned·ik səˌsep·tə'bil·əd·ē }

electromagnetic wave [ELECTROMAG] A disturbance which propagates outward from any electric charge which oscillates or is accelerated; far from the charge it consists of vibrating electric and magnetic fields which move at the speed of light and are at right angles to each other and to the direction of motion. { i¦lek·trō·mag'ned·ik 'wāv }

electromechanical dialer [ELECTR] Telephone dialer which activates one of a set of desired numbers, precoded into it, when the user selects and presses a start button. { i¦lek·trō·mi'kan·ə·kəl 'dī·lər }

electromechanical plotter [COMPUT SCI] An automatic device used in conjunction with a digital computer to produce a graphic or pictorial representation of computer data on hard copy. { i¦lek·trō·mi'kan·ə·kəl 'pläd·ər }

electron beam [ELECTR] A narrow stream of electrons moving in the same direction, all having about the same velocity. { i'lekˌträn ˌbēm }

electron-beam lithography [ELECTR] Lithography in which the radiation-sensitive film or resist is placed in the vacuum chamber of a scanning-beam electron microscope and exposed by an electron beam under digital computer control. { i'lekˌträn ˌbēm li'thäg·rə·fē }

electron-beam parametric amplifier [ELECTR] A parametric amplifier in which energy is pumped from an electrostatic field into a beam of electrons traveling down the length of the tube, and electron couplers impress the input signal at one end of the tube and translate spiraling electron motion into electric output at the other. { i'lekˌträn ˌbēm ˌpar·ə¦me·trik 'am·pləˌfī·ər }

electron collector *See* collector. { i'lek,trän kə ,lek·tər }

electron gun [ELECTR] An electrode structure that produces and may control, focus, deflect, and converge one or more electron beams in an electron tube. { i'lek,trän ,gən }

electron hole *See* hole. { i'lek,trän ¦hōl }

electronically agile radar [ENG] An airborne radar that uses a phased-array antenna which changes radar beam shapes and beam positions at electronic speeds. { i,lek'trän·ik·lē ,a·jəl 'rā ,där }

electronic attack [ELECTR] A term embracing all means in electronic warfare both to counter the enemy's electronic or electromagnetic sensing and communications and also to effect offense with high-power electromagnetic weaponry. Abbreviated EA. { i,lək'trän·ik ə'tak }

electronic attitude directional indicator [NAV] A multicolor cathode-ray-tube display of attitude information (roll and pitch) showing the aircraft's position in relation to the instrument landing system or a very high-frequency omnirange station. Abbreviated EADI. { i,lek'trän·ik 'ad·ə ,tüd də'rek·shən·əl 'in·də,kād·ər }

electronic calculator [ELECTR] A calculator in which integrated circuits perform calculations and show results on a digital display; the displays usually use either seven-segment light-emitting diodes or liquid crystals. { i,lek'trän·ik 'kal·kyə ,lād·ər }

electronic chart display and information system [ENG] A navigation information system with an electronic chart database, as well as navigational and piloting information (typically, vessel-route-monitoring, track-keeping, and track-planning information). Abbreviated ECDIS. { i·lek¦trän·ik 'chärt di¦splāən ,in·fər'mā·shən ,sis·təm }

electronic chart reader [COMPUT SCI] A device which scans curves by a graphical recorder on a continuous paper form and converts them into digital form. { i,lek'trän·ik 'chärt ,rēd·ər }

electronic circuit [ELECTR] An electric circuit in which the equilibrium of electrons in some of the components (such as electron tubes, transistors, or magnetic amplifiers) is upset by means other than an applied voltage. { i,lek'trän·ik 'sər·kət }

electronic codebook mode *See* block encryption. { i,lek'trän·ik 'kōd,bu̇k ,mōd }

electronic commerce [COMPUT SCI] Business done on the Internet. Also known as e-business; e-commerce. { i·lek¦trän·ik 'kä·mərs }

electronic countermeasure [ELECTR] An offensive or defensive tactic or device using electronic, electromagnetic, and reflecting apparatus to reduce the military effectiveness of enemy equipment involving electromagnetic radiation, such as radar, communication, guidance, or other radio-wave devices. Abbreviated ECM. { i,lek'trän·ik 'kau̇nt·ər,mezh·ər }

electronic data processing [COMPUT SCI] Processing data by using equipment that is predominantly electronic in nature, such as an electronic digital computer. Abbreviated EDP. { i,lek'trän·ik 'dad·ə ,prä·səs·iŋ }

electronic data-processing center [COMPUT SCI] The complex formed by the computer, its peripheral equipment, the personnel related to the operation of the center and control functions, and, usually, the office space housing hardware and personnel. Abbreviated EDP center. Also known as computer center. { i,lek'trän·ik 'dad·ə ,präs·əs·iŋ ,sen·tər }

electronic data-processing management science [COMPUT SCI] The field consisting of a class of management problems capable of being handled by computer programs. { i,lek'trän·ik 'dad·ə ,prä·səs·iŋ 'man·ij·mənt ,sī·əns }

electronic data-processing system [COMPUT SCI] A system for data processing by means of machines using electronic circuitry at electronic speed, as opposed to electromechanical equipment. { i,lek'trän·ik 'dad·ə ,prä·səs·iŋ ,sis·təm }

electronic differential analyzer [COMPUT SCI] A form of analog computer using interconnected electronic integrators to solve differential equations. { i,lek'trän·ik ,dif·ə'ren·chəl 'an·ə,liz·ər }

electronic display [ELECTR] An electronic component used to convert electric signals into visual imagery in real time suitable for direct interpretation by a human operator. { i,lek'trän·ik di 'splā }

electronic distance-measuring equipment [NAV] A navigation system consisting of airborne devices that transmit microsecond pulses to special ground beacons, which retransmit the signals to the aircraft; the length of expired time between transmission and reception is measured, converted to kilometers or miles, and presented to the pilot. { i,lek'trän·ik 'dis·təns ,mezh·ə·riŋ i,kwip·mənt }

electronic horizontal-situation indicator [NAV] An integrated multicolor map display of an airplane's position combined with a color weather radar display, with a scale selected by the pilot, together with information on wind direction and velocity, horizontal situation, and deviation from the planned vertical path. Abbreviated EHSI. { i,lek'trän·ik ,här·ə'zänt·əl ,sich·ə¦wā·shən ,in·də ,kād·ər }

electronic interference [ELECTR] Any electrical or electromagnetic disturbance that causes undesirable response in electronic equipment. { i,lek'trän·ik ,int·ər·'fir·əns }

electronic jammer *See* jammer. { i,lek'trän·ik 'jam·ər }

electronic jamming *See* jamming. { i,lek'trän·ik 'jam·iŋ }

electronic line scanning [ELECTR] Method which provides motion of the scanning spot along the scanning line by electronic means. { i,lek'trän·ik 'līn ,skan·iŋ }

electronic mail [COMMUN] The electronic transmission of letters, messages, and memos through a communications network. Also known as e-mail. { i,lek'trän·ik 'māl }

Electronic Numerical Integrator and Calculator *See* ENIAC. { i,lek'trän·ik nü'mer·ə·kəl 'int·ə,grād·ər ən 'kal·kyə,lād·ər }

electronic packaging [ENG] The technology of packaging electronic equipment; in current usage it refers to inserting discrete components, integrated circuits, and MSI and LSI chips (usually attached to a lead frame by beam leads) into plates through holes on multilayer circuit boards (also called cards), where they are soldered in place. { i,lek'trän·ik 'pak·ij·iŋ }

electronic protection [ELECTR] Measures taken to counteract the effects of electronic attack. Abbreviated EP. { i,lək'trän·ik prə'tek·shən }

electronic publishing [COMMUN] The provision of information with high editorial and value-added content in electronic form, allowing the user some degree of control and interactivity. { i,lek|trän·ik 'pəb·lish·iŋ }

electronic scanning [ELECTR] Scanning in which the radar beam direction is determined by control of the relative phases of the signals fed to the elements of an otherwise stationary antenna array. { i,lek'trän·ik 'skan·iŋ }

electronic sculpturing [COMPUT SCI] Procedure for constructing a model of a system by using an analog computer, in which the model is devised at the console by interconnecting components on the basis of analogous configuration with real system elements; then, by adjusting circuit gains and reference voltages, dynamic behavior can be generated that corresponds to the desired response, or is recognizable in the real system. { i,lek'trän·ik 'skəlp·chə·riŋ }

electronic spreadsheet [COMPUT SCI] A type of computer software for performing mathematical computations on numbers arranged in rows and columns, in which the numbers can depend on the values in other rows and columns, allowing large numbers of calculations to be carried out simultaneously. { i,lek'trän·ik 'spred,shēt }

electronic support [ELECTR] Means employed in electronic warfare to intercept and interpret the enemy's electromagnetic radiations to one's own advantage. Abbreviated ES. Also known as electronic support measures (ESM). { i,lək 'trän·ik sə,pȯrt }

electronic support measures *See* electronic support. { i,lek'trän·ik sə'pȯrt ,mezh·ərz }

electronic switching [COMMUN] Telephone switching using a computer with a storage containing program switching logic, whose output actuates switches that set up telephone connections automatically. [ELECTR] The use of electronic circuits to perform the functions of a high-speed switch. { i,lek'trän·ik 'swich·iŋ }

electronic tablet [COMPUT SCI] A data-entry device consisting of stylus, writing surface, and circuitry that produces a pair of digital coordinate values corresponding continuously to the position of the stylus upon the surface. Also known as data tablet. { i,lek'trän·ik 'tab·lət }

electronic tuning [ELECTR] Tuning of a transmitter, receiver, or other tuned equipment by changing a control voltage rather than by adjusting or switching components by hand. { i,lek'trän·ik 'tün·iŋ }

electronic typewriter [COMPUT SCI] A typewriter whose operation is enhanced through the use of microprocessor technology to provide many of the functions of a word-processing system but which has at most a partial-line visual display. Also known as memory typewriter. { i'lek,trän·ik 'tīp,rīd·ər }

electronic warfare [ELECTR] The entire realm of military capability, strategies, and equipment in using electronic attack, electronic protection, and electronic support. { i,lek'trän·ik 'wȯr,fer }

electron image tube *See* image tube. { i'lek ,trän 'im·ij ,tüb }

electron mirror *See* dynode. { i'lek,trän mir·ər }

electron multiplier [ELECTR] An electron-tube structure which produces current amplification; an electron beam containing the desired signal is reflected in turn from the surfaces of each of a series of dynodes, and at each reflection an impinging electron releases two or more secondary electrons, so that the beam builds up in strength. Also known as multiplier. { i'lek ,trän 'məl·tə,plī·ər }

electron-ray tube *See* cathode-ray tube. { i'lek ,trän ,rā,tüb }

electron tube [ELECTR] An electron device in which conduction of electricity is provided by electrons moving through a vacuum or gaseous medium within a gastight envelope. Also known as radio tube; tube; valve (British usage). { i'lek ,trän ,tüb }

electrooptical birefringence *See* electrooptical Kerr effect. { i,lek·trō'äp·tə·kəl bī·ri'frin·jəns }

electrooptical character recognition *See* optical character recognition. { i,lek·trō'äp·tə·kəl 'kar·ik·tər ,rek·ig,nish·ən }

electrooptical Kerr effect [OPTICS] Birefringence induced by an electric field. Also known as electrooptical birefringence; Kerr effect. { i,lek·trō'äp·tə·kəl 'kər i,fekt }

electrooptical modulator [COMMUN] An optical modulator in which a Kerr cell, an electrooptical crystal, or other signal-controlled electrooptical device is used to modulate the amplitude, phase, frequency, or direction of a light beam. { i,lek·trō'äp·tə·kəl 'mäj·ə,lād·ər }

electrooptic material [OPTICS] A material in which the indices of refraction are changed by an applied electric field. { i,lek·trō'äp·tik mə'tir·ē·əl }

electrooptic radar [ENG] Radar system using electrooptic techniques and equipment instead of microwave to perform the acquisition and tracking operation. { i,lek·trō'äp·tik 'rā,där }

electrooptics [OPTICS] The study of the influence of an electric field on optical phenomena, as in the electrooptical Kerr effect and the Stark effect. Also known as optoelectronics. { i,lek·trō 'äp·tiks }

electrostatic deflection [ELECTR] The deflection of an electron beam by means of an electrostatic field produced by electrodes on opposite sides of the beam; used chiefly in cathode-ray tubes for oscilloscopes. { i,lek·trə'stad·ik di'flek·shən }

electrostatic error *See* antenna effect. { i¦lek·trə'stad·ik 'er·ər }

element [COMPUT SCI] A circuit or device performing some specific elementary data-processing function. [ELECTROMAG] Radiator, active or parasitic, that is a part of an antenna. { 'el·ə·mənt }

elemental area *See* picture element. { ¦el·ə'ment·əl 'er·ē·ə }

elementary item [COMPUT SCI] An item considered to have no subordinate item in the COBOL language. { ¦el·ə'men·trē ¦īd·əm }

elementary stream [COMMUN] A generic term for one of the coded video, coded audio, or other coded bit streams in a digital television system. { ¦el·ə'mən·trē 'strēm }

elevation angle [ELECTROMAG] The angle that a radio, radar, or other such beam makes with the horizontal. { ¦el·ə'vā·shən ¦aŋ·gəl }

elevation-angle error [ELECTROMAG] In radar, the error in the measurement of the elevation angle of a target resulting from the vertical bending or refraction of radio energy in traveling through the atmosphere. Also known as elevation error. { ¦el·ə'vā·shən ¦aŋ·gəl ¦er·ər }

elevation error *See* elevation-angle error. { ¦el·ə'vā·shən ¦er·ər }

ELF *See* extremely low frequency.

elimination factor [COMPUT SCI] In information retrieval, the ratio obtained in dividing the number of documents that have not been retrieved by the total number of documents in the file. { ə¦lim·ə'nā·shən ¦fak·tər }

elliptic-integral filter [ELECTR] An electronic filter whose gain characteristic has both an equal-ripple shape in the pass-band and equal minima of attenuation in the stop-band. Also known as Cauer filter. { ə¦lip·tik ¦int·ə·grəl 'fil·tər }

ellipticity *See* axial ratio. { ē¦lip'tis·əd·ē }

elongation [COMMUN] The extension of the envelope of a signal due to delayed arrival of multipath components. { ē¦loŋ'gā·shən }

ELSE instruction [COMPUT SCI] An instruction in a programming language which tells a program what actions to take if previously specified conditions are not met. { 'els in¦strək·shən }

ELSE rule [COMPUT SCI] A convention in decision tables which spells out which action to take in the case specified conditions are not met. { 'els ¦rül }

e-mail *See* electronic mail. { 'ē¦māl }

embedded command [COMPUT SCI] In word processing, a code inserted in a text document that instructs the printer to change its print attributes. { em¦bed·əd kə'mand }

embedded pointer [COMPUT SCI] A pointer set in a data record instead of in a directory. { em'bed·əd 'pȯint·ər }

embedded system [COMPUT SCI] A computer system that cannot be programmed by the user because it is preprogrammed for a specific task and embedded within the equipment which it serves. { em'bed·əd 'sis·təm }

embossed plate printer [COMPUT SCI] In character recognition, a data preparation device which accomplishes printing by allowing a raised character behind the paper to push the paper against the printing ribbon in front of the paper. { em¦bäst ¦plāt 'print·ər }

emergency alert system [COMMUN] A system of radio, television, and cable networks and wire services for communicating with the general public in emergency situations. { ə¦mər·jən·sē ə'lərt ¦sis·təm }

emergency broadcast system [COMMUN] A system of broadcast stations and interconnecting facilities authorized by the U.S. Federal Communications Commission to operate in a controlled manner during a war, threat of war, state of public peril or disaster, or other national emergency. { ə'mər·jən·sē 'brȯd¦kast ¦sis·təm }

emergency radio channel [COMMUN] Any radio frequency reserved for emergency use, particularly for distress signals. { ə'mər·jən·sē 'rād·ē·ō ¦chan·əl }

emergency receiver [COMMUN] Receiver immediately available in a station for emergency communications and capable of being energized by self-contained or emergency power supply. { ə'mər·jən·sē ri'sē·vər }

emission [ELECTROMAG] Any radiation of energy by means of electromagnetic waves, as from a radio transmitter. { i'mish·ən }

emission security [ELECTR] That component of communications security which results from all measures taken to protect any unintentional emissions of a telecommunications system from any form of exploitation other than cryptanalysis. { i'mish·ən sə'kyu̇r·əd·ē }

emitter [ELECTR] A transistor region from which charge carriers that are minority carriers in the base are injected into the base, thus controlling the current flowing through the collector; corresponds to the cathode of an electron tube. Symbolized E. Also known as emitter region. { i'mid·ər }

emitter-coupled logic [ELECTR] A form of current-mode logic in which the emitters of two transistors are connected to a single current-carrying resistor in such a way that only one transistor conducts at a time. Abbreviated ECL. { i'mid·ər ¦kəp·əld 'läj·ik }

emitter region *See* emitter. { i'mid·ər ¦rē·jən }

EMM *See* entitlement management message.

E mode *See* transverse magnetic mode. { 'ē ¦mōd }

emoticon [COMPUT SCI] A combination of keyboard characters that depicts a sideways face whose expression conveys an emotional response. Also known as smiley. { i'mōd·ə¦kän }

emphasizer *See* preemphasis network. { 'em·fə¦sīz·ər }

empty medium [COMPUT SCI] A material which has been prepared to have data recorded on it by the entry of some preliminary data, such as feed holes punched in a paper tape or header labels written on a magnetic tape; in contrast to a virgin medium. { 'em·tē 'mēd·ē·əm }

empty shell [COMPUT SCI] A room that has been fully prepared for the installation of computer and data-processing equipment. { 'em·tē 'shel }

emulation [COMPUT SCI] Imitation of one computer system by another so that the latter functions in exactly the same way and runs the same programs. { ˌem·yə'lā·shən }

emulation mode [COMPUT SCI] A method of operation in which a computer actually executes the instructions of a different (simpler) computer, in contrast to normal mode. { ˌem·yə'lā·shən ˌmōd }

emulator [COMPUT SCI] The microprogram-assisted macroprogram which allows a computer to run programs written for another computer. { 'em·yəˌlād·ər }

emulator circuit [COMPUT SCI] A circuit built into a computer's control section to enable it to process instructions that were written for another computer. { 'em·yəˌlād·ər ˌsər·kət }

enable [COMPUT SCI] **1.** To authorize an activity which would otherwise be suppressed, such as to write on a tape. **2.** To turn on a computer system or a piece of equipment. [ELECTR] To initiate the operation of a device or circuit by applying a trigger signal or pulse. { ə'nā·bəl }

enabled instruction [COMPUT SCI] An instruction in a program in data flow language, all of whose input values are present, so that the instruction may be carried out. { ə'nā·bəld in'strək·shən }

enabling pulse [ELECTR] A pulse that prepares a circuit for some subsequent action. { ə'nāb·liŋ ˌpəls }

encipher [COMMUN] To convert a plain-text message into unintelligible language by means of a cryptosystem. Also known as encrypt. { en'sī·fər }

enciphered facsimile communications [COMMUN] Communications in which security is accomplished by mixing pulses produced by a key generator with the output of the facsimile converter; plain text is recovered by subtracting the identical key at the receiving terminal; unauthorized listeners are unable to reconstruct the plain text unless they have an identical key generator and the daily key setting. { en'sī·fərd fak'sim·ə·lē kəˌmyün·ə'kā·shənz }

enclave See domain. { 'änˌklāv }

encode [COMMUN] To express given information by means of a code. [COMPUT SCI] To prepare a routine in machine language for a specific computer. { en'kōd }

encoded abstract [COMPUT SCI] An abstract prepared to be scanned by automatic electronic machines. { en'kōd·əd 'abˌstrakt }

encoded question [COMPUT SCI] A question set up and encoded in the form appropriate for operating, programming, or conditioning a searching device. { en'kōd·əd 'kwes·chən }

encoder [COMMUN] An embodiment of an encoding process. [COMPUT SCI] In character recognition, that class of printer which is usually designed for the specific purpose of printing a particular type font in predetermined positions on certain size forms. [ELECTR] In an electronic computer, a network or system in which only one input is excited at a time and each input produces a combination of outputs. { en'kōd·ər }

encoding strip [COMPUT SCI] In character recognition, the area reserved for the inscription of magnetic-ink characters, as in bank checks. { en'kōd·iŋ 'strip }

encrypt See encipher. { en'kript }

encryption [COMPUT SCI] The coding of a clear text message by a transmitting unit so as to prevent unauthorized eavesdropping along the transmission line; the receiving unit uses the same algorithm as the transmitting unit to decode the incoming message. { en'krip·shən }

end-around carry [COMPUT SCI] A carry from the most significant digit place to the least significant digit place. { ¦end ə¦raủnd 'kar·ē }

end-around shift See cyclic shift. { ¦end ə¦raủnd 'shift }

end distortion [COMMUN] The displacement of trailing edges of marking pulses transmitted over a teletypewriter circuit relative to the leading edge of the start pulse. { 'end diˌstȯr·shən }

end effect [ELECTROMAG] The effect of capacitance at the ends of an antenna; it requires that the actual length of a half-wave antenna be about 5% less than a half wavelength. { 'end iˌfekt }

end effector [CONT SYS] The component of a robot that comes into contact with the workpiece and does the actual work on it. Also known as hand. { 'end iˌfek·tər }

end-fire antenna See end-fire array. { 'end ˌfīr an'ten·ə }

end-fire array [ELECTROMAG] A linear array whose direction of maximum radiation is along the axis of the array; it may be either unidirectional or bidirectional; the elements of the array are parallel and in the same plane, as in a fishbone antenna. Also known as end-fire antenna. { 'end ˌfīr ə'rā }

endless loop [COMPUT SCI] A sequence of instructions in a computer program that is repeated over and over without end, due to a mistake in the programming. { 'end·ləs 'lüp }

end loss [ELECTROMAG] The difference between the actual and the effective lengths of a radiating antenna element. { 'end ˌlȯs }

end mark [COMPUT SCI] A mark which signals the end of a unit of information. { 'end ˌmärk }

end-of-block character [COMPUT SCI] A character that indicates the completion of a block of code. { ¦end əv ¦bläk 'kar·ik·tər }

end-of-data mark [COMPUT SCI] A character or word signaling the end of all data held in a particular storage unit. { ¦end əv 'dad·ə ˌmärk }

end-of-field mark [COMPUT SCI] A data item signaling the end of a field of data, generally a variable-length field. { ¦end əv 'fēld ˌmärk }

end of file [COMPUT SCI] **1.** Termination or point of completion of a quantity of data; end of file marks are used to indicate this point. **2.** Automatic procedures to handle tapes when the end of an input or output tape is reached; a reflective spot, called a record mark, is placed on the physical end of the tape to signal the end. { ¦end əv 'fīl }

end-of-file gap [COMPUT SCI] A gap of precise dimension to indicate the end of a file on tape. Abbreviated EOF gap. { ¦end əv 'fīl ˌgap }

end-of-file indicator [COMPUT SCI] **1.** A device that indicates the end of a file on tape. **2.** *See* end-of-file mark. { ¦end əv 'fīl 'in·dəˌkād·ər }

end-of-file mark [COMPUT SCI] A control character which signifies that the last record of a file has been read. Also known as end-of-file indicator. { ¦end əv 'fīl ˌmärk }

end-of-file routine [COMPUT SCI] A program which checks that the contents of a file read into the computer were correctly read; may also start the rewind procedure. { ¦end əv 'fīl rüˌtēn }

end-of-file spot [COMPUT SCI] A reflective piece of tape indicating the end of the tape. { ¦end əv 'fīl ˌspät }

end-of-message [COMMUN] A character or series of characters signifying the end of a message or record, such as a message sent by teletypewriter. { ¦end əv 'mes·ij }

end-of-record gap [COMPUT SCI] A gap of precise dimension (shorter than the end-of-file gap) which indicates the physical end of a record on a magnetic tape. Abbreviated EOR gap. { ¦end əv 're·kərd ˌgap }

end-of-record word [COMPUT SCI] The last word in a record, usually written in a special format that enables identification of the end of the record. { ¦end əv 're·kərd ˌwərd }

end-of-run routine [COMPUT SCI] A routine that carries out various housekeeping operations such as rewinding tapes and printing control totals before a run is completed. { ¦end əv 'rən rüˌtēn }

end-of-tape routine [COMPUT SCI] A program which is brought into play when the end of a tape is reached; may involve a series of validity checks and initiate the tape rewind. { ¦end əv 'tāp rü ˌtēn }

end-of-transmission card [COMMUN] Last card of each message; used to signal the end of a transmission and contains the same information as the header card, plus additional data for traffic analysis. { ¦end əv tranz'mish·ən ˌkärd }

end-of-transmission recognition [COMPUT SCI] The capability of a computer to recognize the end of transmission of a data string even if the buffer area is not filled. { ¦end əv tranz'mish·ən rek·ig ˌnish·ən }

endorser [COMPUT SCI] A special feature available on most magnetic-ink character-recognition readers that imprints a bank's endorsement on successful document reading. { en'dȯr·sər }

end point [COMPUT SCI] In vector graphics, one of the two ends of a line or vector. [CONT SYS] The point at which a robot stops along its path of motion. { 'end ˌpȯint }

end section [COMMUN] Additional portion of switchboard added to each end of a large multiple switchboard and used to extend some of the trunks or locals to these end positions to place all jacks within easy reach of the first and last operator. Also known as head section. { 'end ˌsek·shən }

end sentinel [COMPUT SCI] A character that indicates the end of a message or record. { 'end ˌsent·nəl }

end-to-end encryption [COMMUN] Encryption of a message at its point of origination so that it travels in encrypted form all the way to its destination. { ¦end·tü¦end in'krip·shən }

end user [COMPUT SCI] The person for whom the output of a computer is ultimately intended. { 'end ˌyüz·ər }

engineering channel circuit [COMMUN] Auxiliary circuit or channel (radio or wire) for use by operating or maintenance personnel for communications incident to the establishment, operation, maintenance, and control of communications facilities. { ˌen·jə'nir·iŋ 'chan·əl ˌsər·kət }

engineering time [COMPUT SCI] The nonproductive time of a computer, reserved for maintenance and servicing. { ˌen·jə'nir·iŋ ˌtīm }

enhanceable language [COMPUT SCI] A computer language that has a modest degree of semantic extensibility. { en¦han·sə·bəl 'laŋ·gwij }

enhanced carrier demodulation [COMMUN] Amplitude demodulation system in which a synchronized local carrier of proper phase is fed into the demodulator to reduce demodulation distortion. { en¦hanst 'kar·ē·ər dēˌmäj·ə'lā·shən }

enhanced small device interface [COMPUT SCI] A standard method of connecting disk and tape drives to computers which allows for the transfer of 1–3 megabytes per second from disk drives holding up to 1 gigabyte of storage. Abbreviated ESDI. { en¦hanst ¦smȯl di¦vīs 'in·tər ˌfās }

enhancement [COMPUT SCI] A substantial increase in the capabilities of hardware or software. [ELECTR] An increase in the density of charged carriers in a particular region of a semiconductor. { en'hans·mənt }

enhancement mode [ELECTR] Operation of a field-effect transistor in which no current flows when zero gate voltage is applied, and increasing the gate voltage increases the current. { en'hans·mənt ˌmōd }

enhancement-mode high-electron-mobility transistor [ELECTR] A high-electron-mobility transistor in which application of a positive bias to the gate electrode is required for current to flow between the source and drain electrodes. Abbreviated E-HEMT. { en'hans·mənt ¦mōd 'hī i¦lekˌträn mō¦bil·əd·ē tran'zis·tər }

enhancement-mode junction field-effect transistor [ELECTR] A type of gallium arsenide field-effect transistor in which the gate consists of the junction between the *n*-type gallium arsenide forming the conducting channel and *p*-type material implanted under a metal electrode. Abbrevate E-JFET. { en'hans·mənt ¦mōd 'jəŋk·shən 'fēld iˌfekt tran'zis·tər }

ENIAC [COMPUT SCI] The first digital computer in the modern sense of the word, built 1942–1945. Derived from Electronic Numerical Integrator and Calculator. { 'ē·nē·ak }

E notation [COMPUT SCI] A type of scientific notation in which the phrase "times 10 to the power of" is replaced by the letter E; for example, 3.1×10^7 is written 3.1E+7 and 5.1×10^{-9} is written 5.1E−9. { 'ē nō,tā·shən }

enquiry character [COMPUT SCI] A control character used to request a response from receiving equipment. { in'kwīr·ē ,kar·ik·tər }

enter key [COMPUT SCI] A key on a computer keyboard that corresponds to the return key on a typewriter and usually signals the computer to act on the information just entered on the keyboard. { 'en·tər ,kē }

entitlement control message [COMMUN] Private conditional access information which specifies control words and possibly other stream-specific, scrambling, or control parameters. { in'tī·təl·mənt kən'trōl ,məs·ij }

entitlement management message [COMMUN] Private conditional access information which specifies the authorization level or the services of specific decoders; addrressed to single decoders or groups of decoders. { in'tī·təl·mənt 'man·ij·mənt ,məs·ij }

entity *See* record. { 'ent·ə·tē }

entity type [COMPUT SCI] A particular kind of file in a database, such as an employee, customer, or product file. { 'ent·ə·tē ,tīp }

entrance [COMPUT SCI] The location of a program or subroutine at which execution is to start. Also known as entry point. [ENG] A place of physical entering, such as a door or passage. { 'en·trəns }

entropy [COMMUN] A measure of the absence of information about a situation, or, equivalently, the uncertainty associated with the nature of a situation. { 'en·trə·pē }

entropy coding [COMMUN] Variable-length lossless coding of the digital representation of a signal to reduce redundancy. { 'en·trə·pē ,kōd·iŋ }

entry [COMPUT SCI] Input data fed during the execution of a program by means of a terminal. { 'en·trē }

entry block [COMPUT SCI] The area of main memory reserved for the data which will be introduced at execution time. { 'en·trē ,bläk }

entry condition [COMPUT SCI] A requirement that must be met before a program or routine can be entered by a computer program. Also known as initial condition. { 'en·trē kən,dish·ən }

entry instruction [COMPUT SCI] The first instruction to be executed in a subroutine. { 'en·trē in ,strək·shən }

entry point [COMMUN] A point in a coded bit stream after which a decoder can become properly initialized and commence syntactically correct decoding. The first transmitted picture after an entry point is either an I-picture or a P-picture. If the first transmitted picture is not an I-picture, the decoder may produce one or more pictures during acquisition. [COMPUT SCI] *See* entrance. { 'en·trē ,pȯint }

entry portion [COMPUT SCI] The right-hand portion of a decision table, which comprises the condition entries and action entries, and whose columns are the decision rules. { 'en·trē ,pȯr·shən }

entry sorting [COMPUT SCI] A method of internal sorting in which records or blocks of records are placed, one at a time, in a buffer area and then integrated into the sorted list before the next record is placed in the buffer. { 'en·trē ,sȯrd·iŋ }

envelope [COMMUN] A curve drawn to pass through the peaks of a graph, such as that of a moduated radio-frequency carrier signal. { 'en·və,lōp }

envelope delay [COMMUN] The time required for the envelope of a modulated signal to travel between two points in a system. { 'en·və,lōp di ,lā }

envelope delay distortion *See* delay distortion. { 'en·və,lōp di,lādi'stȯr·shən }

envelope detector *See* detector. { 'en·və,lōp di ,tek·tər }

environment [COMPUT SCI] The computer system in which an applications program is running, including the hardware and system software. { in'vī·ərn·mənt *or* in'vī·rən·ment }

environment division [COMPUT SCI] The section of a program written in COBOL which defines the hardware and files to be used by the program. { in¦vī·ərn¦mənt di'vizh·ən }

environment pointer [COMPUT SCI] **1.** A component of a task descriptor that designates where the instructions and data code for the task are located. **2.** A control component element belonging to the stack model of block structure execution that points to the current environment. { in¦vī·ərn¦mənt ,pȯint·ər }

EOF gap *See* end-of-file gap. { ¦ē¦ō'ef ,gap }

EOR gap *See* end-of-record gap. { ¦ē¦ō'är ,gap }

EP *See* electronic protection.

E-plane antenna [ELECTROMAG] An antenna which lies in a plane parallel to the electric field vector of the radiation that it emits. { 'ē ,plān an,ten·ə }

E-plane bend *See* E bend. { 'ē ,plān ,bend }

E-plane T junction [ELECTROMAG] Waveguide T junction in which the change in structure occurs in the plane of the electric field. Also known as series T junction. { 'ē ,plān 'tē ,jəŋk·shən }

EPROM *See* erasable programmable read-only memory. { 'ē,präm }

equal error rate [COMMUN] The error rate of a verification system when the operating threshold for the accept/reject decision is adjusted such that the probability of false acceptance and that of false rejection become equal. Abbreviated EER. { ¦ē·kwəl 'er·ər ,rāt }

equality gate *See* equivalence gate. { ē'kwal·əd·ē ,gāt }

equalization [ELECTR] The effect of all frequency-discriminating means employed in transmitting, recording, amplifying, or other signal-handling systems to obtain a desired overall frequency response. Also known as frequency-response equalization. { ˌē·kwə·lə ˈzā·shən }

equalizer [ELECTR] A network designed to compensate for an undesired amplitude-frequency or phase-frequency response of a system or component; usually a combination of coils, capacitors, and resistors. Also known as equalizing circuit. { ˈē·kwəˌlīz·ər }

equalizing circuit *See* equalizer. { ˈēkwəˌlīz·iŋ ˌsər·kət }

equalizing pulses [ELECTR] In analog television, pulses at twice the line frequency, occurring just before and after the vertical synchronizing pulses, which minimize the effect of line frequency pulses on the interlace. { ˈē·kwəˌlīz·iŋ ˌpəl·səs }

equal ripple [ELECTR] Property of an amplitude or phase characteristic whose local maxima all have the same value, and whose local minima all have the same value, within a specified frequency range. { ˌē·kwəl ˈrip·əl }

equal-zero indicator [COMPUT SCI] A circuit component which is on when the result of an operation is zero. { ¦ē·kwəl ¦zir·ōˈin·dəˌkād·ər }

equation solver [COMPUT SCI] A machine, usually analog, for solving systems of simultaneous equations, which may be linear, nonlinear, or differential, and for finding roots of polynomials. { iˈkwā·zhən ˌsälv·ər }

equiangular spiral antenna [ELECTROMAG] A frequency-independent broad-band antenna, cut from sheet metal, that radiates a very broad, circularly polarized beam on both sides of its surface; this bidirectional radiation pattern is its chief limitation. { ¦ē·kwē¦aŋ·gyə·lər ¦spī·rəl anˈten·ə }

equipment augmentation [COMPUT SCI] **1.** Procuring additional automatic data-processing equipment capability to accommodate increased work load within an established data system. **2.** Obtaining additional sites or locations. { əˈkwip·mənt ˌȯg·mənˈtā·shən }

equipment characteristic distortion [COMMUN] Teletypewriter transmission repetitive display or disruption peculiar to specific portions of a signal, normally caused by maladjusted or dirty contacts of the sending or receiving equipment. { əˈkwip·mənt ˌkar·ik·tə¦ris·tik diˈstȯr·shən }

equipment compatibility [COMPUT SCI] The ability of a device to handle data prepared or handled by other equipment, without alteration of the code or of the form of the data. { əˈkwip·mənt kəmˌpad·əˈbil·əd·ē }

equipment failure [COMPUT SCI] A fault in equipment that results in its improper behavior or prevents the execution of a job as scheduled. { əˈkwip·mənt ˌfāl·yər }

equisignal [COMMUN] **1.** Pertaining to two signals of equal intensity, used particularly with reference to the signals of a radio range station. **2.** Referring to a radio system in which two identifiable separate radio signals are received with the same intensity. { ¦e·kwə¦sig·nəl }

equisignal surface [ELECTROMAG] Surface around an antenna formed by all points at which, for transmission, the field strength (usually measured in volts per meter) is constant. { ¦e·kwə¦sig·nəl ˌsər·fəs }

equivalence element *See* equivalence gate. { iˈkwiv·ə·ləns ˌel·ə·mənt }

equivalence gate [COMPUT SCI] A logic circuit that produces a binary output signal of 1 if its two binary input signals are the same, and an output signal of 0 if the input signals differ. Also known as biconditional gate; equality gate; equivalence element; exclusive-NOR gate; match gate. { iˈkwiv·ə·ləns ˌgāt }

equivalent binary digits [COMPUT SCI] The number of binary positions required to enumerate the elements of a given set. { iˈkwiv·ə·lənt ˈbīˌner·ē ˈdij·əts }

equivalent four-wire system [COMMUN] A transmission system in which multiplex techniques are used to carry on duplex operation over a single pair of wires. { iˈkwiv·ə·lənt ¦fȯr ¦wīr ˈsis·təm }

erasable programmable read-only memory [COMPUT SCI] A read-only memory in which stored data can be erased by ultraviolet light or other means and reprogrammed bit by bit with appropriate voltage pulses. Abbreviated EPROM. { i¦rās·ə·bəl prō¦gram·ə·bəl ¦rēd ˌōn·lē ˈmem·rē }

erasable storage [COMPUT SCI] Any storage medium which permits new data to be written in place of the old, such as magnetic disk or tape. { i¦rās·ə·bəl ˈstȯr·ij }

erase [COMPUT SCI] To change all the binary digits in a digital computer storage device to binary zeros. [ELECTR] **1.** To remove recorded material from magnetic tape by passing the tape through a strong, constant magnetic field (dc erase) or through a high-frequency alternating magnetic field (ac erase). **2.** To eliminate previously stored information in a charge-storage tube by charging or discharging all storage elements. { iˈrās }

erase character *See* ignore character. { i¦rās ˈkar·ik·tər }

erase oscillator [ELECTR] The oscillator used in a magnetic recorder to provide the high-frequency signal needed to erase a recording on magnetic tape; the bias oscillator usually serves also as the erase oscillator. { iˈrās ˌäs·əˌlād·ər }

erasing head [ELECTR] A magnetic head used to obliterate material previously recorded on magnetic tape. { iˈrās·iŋ ˌhed }

erbium-doped fiber amplifier [COMMUN] An optical-fiber amplifier whose fiber core is lightly doped with trivalent erbium ions which absorb light at pump wavelengths of 0.98 and 1.48 micrometers and emit it at a signal wavelength around 1.5 micrometers through stimulated emission. Abbreviated EDFA. { ¦ər·bē·əm ˌdōpt ˌfī·bər ˈam·pləˌfī·ər }

erlang [COMMUN] A unit of communication traffic load, equal to the traffic load whose calls, if placed end to end, will keep one path continuously occupied. { ˈerˌläŋ }

ERP *See* effective radiated power.

error [COMPUT SCI] An incorrect result arising from approximations used in numerical methods, rather than from a human mistake or computer malfunction. { 'er·ər }

error analysis [COMPUT SCI] In the solution of a problem on a digital computer, the estimation of the cumulative effect of rounding or truncation errors associated with basic arithmetic operations. { 'er·ər ə,nal·ə·səs }

error burst [COMPUT SCI] The condition when more than one bit is in error in a given number of bits. { 'er·ər ,bərst }

error character [COMPUT SCI] A character that indicates the existence of an error in the data being processed or transmitted, and usually specifies that a certain amount of preceding or following data is to be ignored. { 'er·ər ,kar·ik·tər }

error checking and recovery [COMPUT SCI] An automatic procedure which checks for parity and will proceed with the execution after error correction. { 'er·ər ,chek·iŋ ən ri'kəv·ə·rē }

error-checking code *See* self-checking code. { 'er·ər ,chek·iŋ ,kōd }

error coefficient [CONT SYS] The steady-state value of the output of a control system, or of some derivative of the output, divided by the steady-state actuating signal. Also known as error constant. { 'er·ər ,kō·i'fish·ənt }

error constant *See* error coefficient. { 'er·ər ,kän·stənt }

error-control procedures [COMMUN] Methods of detecting errors and correcting or recovering from those that occur in data transmission. { 'er·ər kən,trōl prə,sē·jərz }

error-correcting code [COMPUT SCI] Data representation that allows for error detection and error correction if the error is of a specific kind. Also known as error-correction code. Abbreviated ECC. { 'er·ər kə¦rek·tiŋ 'kōd }

error-correcting telegraph system [COMMUN] System employing an error-detecting code, and so conceived that any false signal initiates a repetition of the transmission of the character incorrectly received. { 'er·ər kə¦rek·tiŋ 'tel·ə,graf ,sis·təm }

error correction [COMMUN] Any system for reducing errors in an incoming message, such as sending redundant signals as a check. [COMPUT SCI] Computer device for automatically locating and correcting a machine error of dropping a bit or picking up an extraneous bit, without stopping the machine or having it go to a programmed recovery routine. { 'er·ər kə,rek·shən }

error-correction code *See* error-correcting code. { 'er·ər kə,rek·shən 'kōd }

error correction routine [COMPUT SCI] A program which corrects specific error conditions in another program, routine, or subroutine. { 'er·ər kə,rek·shən rü,tēn }

error-detecting code *See* self-checking code. { 'er·ər di,tek·tiŋ ,kōd }

error-detecting system [COMPUT SCI] An automatic system which detects an error due to a lack of data, or erroneous data during transmission. { 'er·ər di,tek·tiŋ,sis·təm }

error detection and feedback system [COMPUT SCI] An automatic system which retransmits a piece of data detected by the computer as being in error. { 'er·ər di,tek·shən ən 'fēd,bak ,sis·təm }

error detection routine *See* diagnostic routine. { 'er·ər di,tek·shən rü,tēn }

error diagnostic [COMPUT SCI] A computer printout of an instruction or data statement, pinpointing an error in the instruction or statement and spelling out the type of error involved. { 'er·ər ,dī·əg'näs·tik }

error frequency limit [COMPUT SCI] The maximum number of single bit errors per unit of time that a computer will accept before a machine check interrupt is initiated. Abbreviated EFL. { 'er·ər ,frē·kwən·sē ,lim·ət }

error handling [COMPUT SCI] The ability of a computer program to deal with errors automatically. { 'er·ər ,hand·liŋ }

error-indicating system [COMPUT SCI] Built-in circuits designed to indicate automatically that certain computational errors have occurred. { 'er·ər ,in·də,kād·iŋ ,sis·təm }

error interrupt [COMPUT SCI] The halt in execution of a program because of errors which the computer is not capable of correcting. { 'er·ər 'int·ə,rəpt }

error list [COMPUT SCI] A list generated by a compiler showing invalid or erroneous instructions in a source program. { 'er·ər ,list }

error log [COMPUT SCI] A file that is created during data processing to hold data known to contain errors, and that is usually printed after completion of processing so that the errors can be corrected. { 'er·ər ,läg }

error message [COMPUT SCI] A message indicating detection of an error. { 'er·ər ,mes·ij }

error range [COMPUT SCI] A range of values such that an error condition will result if a specified data item falls within it. { 'er·ər ,rānj }

error rate [COMMUN] The number of erroneous bits or characters received for some fixed number of bits transmitted. { 'er·ər ,rāt }

error ratio [COMPUT SCI] The ratio of the number of erroneous items to the total number of bits or characters transmitted. { 'er·ər ,rā·shō }

error recovery routine [COMPUT SCI] A part of a computer program that attempts to handle errors without terminating the program. { 'er·ər ri¦kəv·ə·rē rü,tēn }

error report [COMPUT SCI] A list produced by a computer showing the error conditions, such as overflows and errors resulting from incorrect or unmatched data, that are generated during program execution. { 'er·ər ri,pȯrt }

error routine [COMPUT SCI] A routine which takes control of a program and initiates corrective actions when an error is detected. { 'er·ər rü ,tēn }

error signal [CONT SYS] In an automatic control device, a signal whose magnitude and sign are used to correct the alignment between the controlling and the controlled elements. [ELECTR] A voltage that depends on the signal received from the target in a tracking system, having a polarity and magnitude dependent on the angle between the target and the center of the scanning beam. { 'er·ər ˌsig·nəl }

error tape [COMPUT SCI] The magnetic tape on which erroneous records are stored during processing. { 'er·ər ˌtāp }

ES *See* electronic support; elementary stream.

Esaki tunnel diode *See* tunnel diode. { e'sä·kē ¦tən·əl 'dīˌōd }

E-scan *See* E-display. { 'ē ˌskan }

escape [COMPUT SCI] To exit from a program, routine, or mode. { i'skāp }

escape character [COMPUT SCI] A character used to indicate that the succeeding character or characters are expressed in a code different from the code currently in use. { ə'skāp ˌkar·ik·tər }

E-scope *See* E-display. { 'ē ˌskōp }

ESD *See* external symbol dictionary.

ESDI *See* enhanced small device interface. { 'ez ˌdē }

ESM *See* electronic support measures.

esoteric name [COMPUT SCI] A symbolic name that is chosen in a computer program to designate a collection of devices. { es·ə'ter·ik 'nām }

estimation theory [STAT] A branch of probability and statistics concerned with deriving information about properties of random variables, stochastic processes, and systems based on observed samples. { ˌes·tə'mā·shən ˌthē·ə·rē }

Ethernet [COMPUT SCI] A protocol for interconnecting computers and peripheral devices in a local area network. { 'ē·thərˌnet }

EU *See* expected value.

European Geostationary Navigation Overlay System [NAV] A satellite-based augmentation system developed jointly by the European Union, European Space Agency, and EUCONTROL. Abbreviated EGNOS. { ˌyu̇r·ə¦pē·ən ˌjē·ō¦stā·shə·ner·ē ˌnav·ə¦gā·shən 'ō·vərˌlay ˌsis·təm }

EV *See* expected value.

even parity check [COMPUT SCI] A parity check in which the number of 0's or 1's in each word is expected to be even. { ¦ē·vən 'par·əd·ē ˌchek }

event [COMMUN] A collection of elementary streams with a common time base, an associated start time, and an associated end time. [COMPUT SCI] The moment of time at which a specified change of state occurs; usually marks the completion of an asynchronous input/output operation. { i'vent }

event-driven monitor [COMPUT SCI] A computer program that measures the performance of a computer system by counting the tasks performed by the system. { i'vent ¦driv·ən 'män·əd·ər }

even-word boundary [COMPUT SCI] A storage address that is an integral multiple of the computer's word length. { 'ē·vən ¦wərd 'bau̇n·drē }

evolutionary computation *See* evolutionary programming. { ˌev·ə¦lü·shəˌner·ē ˌkam·pyə'tā·shən }

evolutionary programming [COMPUT SCI] Computer programming with genetic algorithms. Also known as evolutionary computation; genetic programming. { ˌev·ə¦lü·shəˌner·ē 'prōˌgram·iŋ }

evolutionary strategy *See* genetic algorithm. { ˌev·ə¦lü·shəˌner·ē 'strad·ə·jē }

E wave *See* transverse magnetic wave. { 'ē ˌwāv }

exalted-carrier receiver [ELECTR] Receiver that counteracts selective fading by maintaining the carrier at a high level at all times; this minimizes the second harmonic distortion that would otherwise occur when the carrier drops out while leaving most of the sidebands at their normal amplitudes. { ig¦zȯl·təd 'kar·ē·ər riˌsēv·ər }

except gate [ELECTR] A gate that produces an output pulse only for a pulse on one or more input lines and the absence of a pulse on one or more other lines. { ek'sept ˌgāt }

exception handling [COMPUT SCI] Programming techniques for dealing with error conditions, generally without terminating execution of the program. [CONT SYS] The actions taken by a control system when unpredictable conditions or situations arise in which the controller must respond quickly. { ek'sep·shən ˌhand·liŋ }

exception-item encoding [COMPUT SCI] A technique which allows the uninterrupted flow of a process by the automatic shunting of erroneous records to an error tape for later corrections. { ek'sep·shən ˌīd·əm en'kōd·iŋ }

exception-principle system [COMPUT SCI] A technique which assumes no printouts except when an error is encountered. { ek'sep·shən ˌprin·sə·pəl ˌsis·təm }

exception reporting [COMPUT SCI] A form of programming in which only values that are outside predetermined limits, representing significant changes, are selected for printout at the output of a computer. { ek'sep·shən riˌpȯrd·iŋ }

excess-fifty code [COMPUT SCI] A number code in which the number n is represented by the binary equivalent of $n + 50$. { 'ekˌses 'fif·tē ˌkōd }

excess-three code [COMPUT SCI] A number code in which the decimal digit n is represented by the four-bit binary equivalent of $n + 3$. Also known as XS-3 code. { ¦ekˌses 'thrē ˌkōd }

exchange [COMMUN] **1.** A unit established by a telephone company for the administration of telephone service in a specified area, usually a town, a city, or a village and its environs, and consisting of one or more central offices together with the associated plant used in furnishing telephone service in that area. Also known as local exchange. **2.** Room or building equipped so telephone lines terminating there may be interconnected as required; equipment may include a switchboard or automatic switching apparatus. [COMPUT SCI] The interchange of contents between two locations. { iksˌchānj }

exchangeable disk storage [COMPUT SCI] A type of disk storage, used as a backing storage, in which the disks come in capsules, each containing several disks; the capsules can be replaced during operation of the computer and can be stored until needed. { iks¦chānj·ə·bəl 'disk,stȯr·ij }

exchange buffering [COMPUT SCI] An input/output buffering technique that avoids the internal moving of data. { iks'chānj ,bəf·ə·riŋ }

exchange message [COMPUT SCI] A device, placed between a communication line and a computer, in order to take care of certain communication functions and thereby free the computer for other work. { iks'chānj ,mes·ij }

exchange plant [COMMUN] Plant used to serve subscriber's local needs as distinguished from that used for long-distance communication. { iks'chānj ,plant }

exchange sort [COMPUT SCI] A method of arranging records or other types of data into a specified order, in which adjacent pairs of records are exchanged until the correct order is achieved. { iks'chānj ,sȯrt }

exciter [ELECTR] A crystal oscillator or self-excited oscillator used to generate the carrier frequency of a transmitter. [ELECTROMAG] **1.** The portion of a directional transmitting antenna system that is directly connected to the transmitter. **2.** A loop or probe extending into a resonant cavity or waveguide. { ek'sīd·ər }

exclusive-NOR gate *See* equivalence gate. { ik ¦sklü·siv 'nȯr ,gāt }

exclusive or [COMPUT SCI] An instruction which performs the "exclusive or" operation on a bit-by-bit basis for its two operand words, usually storing the result in one of the operand locations. Abbreviated XOR. { ik¦sklü·siv 'ȯr }

exclusive segments [COMPUT SCI] Parts of an overlay program structure that cannot be resident in main memory simultaneously. { ik'sklü·siv 'seg·mənts }

executable module [COMPUT SCI] A file holding a computer program written in machine language so that it is ready to run. { ,ek·sə'kyüd·ə·bəl 'mäj·yül }

executable program [COMPUT SCI] A program that is ready to run on a computer. { ,ek·sə ¦kyüd·ə·bəl 'prō·grəm }

executable statement [COMPUT SCI] A program statement that causes the computer to carry out some operation, in contrast to a declarative statement. { ,ek·sə'kyüd·ə·bəl 'stāt·mənt }

execute [COMPUT SCI] Usually, to run a compiled or assembled program on the computer; by extension, to compile or assemble and to run a source program. { 'ek·sə,kyüt }

execute statement [COMPUT SCI] A program statement that indicates the beginning of a job statement in a job control language. { 'ek·sə ,kyüt ,stāt·mənt }

execution control program [COMPUT SCI] The program delivered by the manufacturer which permits the computer to handle the programs fed to it. { ,ek·sə¦kyü·shən kən'trōl ,prō·grəm }

execution cycle [COMPUT SCI] The time during which an elementary operation takes place. { ,ek·sə'kyü·shən ,sī·kəl }

execution error detection [COMPUT SCI] The detection of errors which become apparent only during execution time. { ,ek·sə¦kyü·shən 'er·ər di,tek·shən }

execution time [COMPUT SCI] The time during which actual work, such as addition or multiplication, is carried out in the execution of a computer instruction. { ,ek·sə'kyü·shən ,tīm }

executive communications [COMPUT SCI] The routine information transmitted to the operator on the status of programs being executed and of the requirements made by these programs of the various components of the system. { ig ¦zek·yəd·iv kə·,myü·nə'kā·shənz }

executive control language [COMPUT SCI] The generic term for a finite set of instructions which enables the programmer to run a program more efficiently. { ig¦zek·yəd·iv kən'trōl ,laŋ·gwij }

executive file-control system [COMPUT SCI] The assignment of intermediate storage devices performed by the computer, and over which the programmer has no control. { ig¦zek·yəd·iv 'fīl kən,trōl ,sis·təm }

executive guard mode [COMPUT SCI] A protective technique which prevents the programmer from accessing, or using, the executive instructions. { ig¦zek·yəd·iv 'gärd ,mōd }

executive instruction [COMPUT SCI] Instruction to determine how a specially written computer program is to operate. { ig¦zek·yəd·iv in ¦strək·shən }

executive logging [COMPUT SCI] The automatic bookkeeping of time utilization by programs of the various components of a computer system. { ig¦zek·yəd·iv 'läg·iŋ }

executive routine [COMPUT SCI] A digital computer routine designed to process and control other routines. Also known as master routine; monitor routine. { ig'zek·yəd·iv rü,tēn }

executive schedule maintenance [COMPUT SCI] The scheduling of jobs to be run according to priorities as established and maintained by a computer's executive supervisor. { ig¦zek·yəd·iv 'sked·jəl ,mān·tə·nəns }

executive supervisor [COMPUT SCI] The component of the computer system which controls the sequencing, setup, and execution of the jobs presented to it. { ig¦zek·yəd·iv 'sü·pər,viz·ər }

executive system concurrency [COMPUT SCI] The capability of a computer system's executive supervisor to handle more than one job at the same time if these jobs do not require the same components at the same time. { ig'zek·yəd·iv ,sis·təm kən'kər·ən·sē }

executive system utilities [COMPUT SCI] The set of programs, such as diagnostic programs or file utility programs, which enables the executive supervisor to handle the jobs efficiently and completely. { ig'zek·yəd·iv ,sis·təm yü'til·əd·ēz }

exit [COMPUT SCI] **1.** A way of terminating a repeated cycle of operations in a computer program. **2.** A place at which such a cycle can be stopped. { 'eg·zət }

expanded batch [COMPUT SCI] A level of computer processing more complex than basic batch, in which computer programs perform complex computations and produce reports that analyze performance in addition to reporting it. { ik'spand·əd 'bach }

expandor [ELECTR] The part of a compandor that is used at the receiving end of a circuit to return the compressed signal to its original form; attentuates weak signals and amplifies strong signals. { ik'spand·dər }

expansion [ELECTR] A process in which the effective gain of an amplifier is varied as a function of signal magnitude, the effective gain being greater for large signals than for small signals; the result is greater volume range in an audio amplifier and greater contrast range in facsimile. { ik'span·shən }

expansion board [COMPUT SCI] A printed circuit board that can be plugged into a computer to provide it with additional peripherals or enhancements, such as increased memory or communications facilities. { ik'span·shən ˌbȯrd }

expansion bus [COMPUT SCI] The wiring and protocols that connect a computer's motherboard with the peripheral devices. { ik'span·shən ˌbəs }

expansion slot [COMPUT SCI] A location in a computer system where additional facilities, especially circuit boards, can be plugged in to extend the computer's capability. { ik'span·shən ˌslät }

expected utility *See* expected value. { ek'spek·təd yü'til·əd·ē }

expected value [SYS ENG] In decision theory, a measure of the value or utility expected to result from a given strategy, equal to the sum over states of nature of the product of the probability of the state times the consequence or outcome of the strategy in terms of some value or utility parameter. Abbreviated EV. Also known as expected utility (EU). { ek'spek·təd 'val·yü }

expert control system [CONT SYS] A control system that uses expert systems to solve control problems. { ¦ekˌspərt kən'trōl ˌsis·təm }

expert system [COMPUT SCI] A computer system composed of algorithms that perform a specialized, usually difficult professional task at the level of (or sometimes beyond the level of) a human expert. { 'ekˌspərt ˌsis·təm }

explicit programming [CONT SYS] Robotic programming that employs detailed and exact descriptions of the tasks to be performed. { ik'splis·ət 'prōˌgram·iŋ }

exploded file [COMPUT SCI] A file in which more data have been added to each record in order to adapt it to a new application. { ik'splōd·əd 'fīl }

expression [COMPUT SCI] A mathematical or logical statement written in a source language, consisting of a collection of operands connected by operations in a logical manner. { ik'spresh·ən }

extended-area service [COMMUN] Telephone exchange service, without toll charges, that extends over an area where there is a community of interest, often in return for a somewhat higher exchange service rate. { ik¦stend·əd ¦er·ē·ə 'sər·vəs }

extended ASCII [COMMUN] An addition to the standard American Standard Code for Information Interchange, namely, characters 128 through 255; includes letters with diacritics, Greek letters, and special symbols. { ik¦sten·dəd 'asˌkē }

extended binary - coded decimal interchange code [COMPUT SCI] A computer code that uses eight binary positions to represent a single character, giving a possible maximum of 256 characters. Abbreviated EBCDIC. { ik¦stend·əd 'bīˌner·ē ¦kōd·əd ¦des·məl 'int·ərˌchānj ˌkōd }

extended channel status word [COMPUT SCI] Stored information which follows an input/output interrupt. Abbreviated ECSW. { ik¦stend·əd 'chan·əl 'stad·əs ˌwərd }

extended data out random-access memory [COMPUT SCI] A type of dynamic random-access memory that was optimized for the 66-megahertz bus but largely has been replaced by faster systems. Abbreviated EDO RAM. { ik¦stend·əd ˌdad·ə ¦aůt ˌran·dəm 'akˌses ˌmem·rē }

extended-entry decision table [COMPUT SCI] A decision table in which the condition stub cites the identification of the condition but not the particular values, which are entered directly into the condition entries. { ik¦stend·əd 'en·trē di'sizh·ən ˌtā·bəl }

extended-hybrid FM IBOC [COMMUN] The second of three modes in the FM IBOC system approved by the Federal Communications Commission for use in the United States that increases data capacity by adding additional carriers closer to the analog host signal. The extended -hybrid IBOC mode adds two frequency partitions around the analog carrier, where digital audio date rate can range from 64 to 96 kbits/s, and the corresponding ancillary data rate will range from 83 kbits/s for 64-kbits/s audio to 51 kbits/s for 96-kbits/s audio. { ik'stend·əd 'hī·brəd 'efˌem 'Ībäk }

extended-precision word [COMPUT SCI] A piece of data of 16 bytes in floating-point arithmetic when additional precision is required. { ik ¦stend·əd prə'sizh·ən ˌwərd }

extended time scale *See* slow time scale. { ik ¦stend·əd 'tīm ˌskāl }

extend flip-flop [COMPUT SCI] A special flag set when there is a carry-out of the most significant bit in the register after an addition or a subtraction. { ik'stend 'flipˌfläp }

extensible language [COMPUT SCI] A programming language which can be modified by adding new features or changing existing ones. { ik'sten·sə·bəl 'laŋ·gwij }

Extensible Markup Language [COMMUN] A set of rules for writing markup languages which provides a robust, machine-readable information protocol that can handle complex objects. Abbreviated XML. { ik¦sten·sə·bəl 'märkˌəp ˌlaŋ·gwij }

extensible system [COMPUT SCI] A computer system in which users may extend the basic system by implementing their own languages and subsystems and making them available for others to use. { ik'sten·sə·bəl 'sis·təm }

extension mechanism [COMPUT SCI] One of the components of an extensible language which allows the definition of new language features in terms of the primitive facilities of the base language. { ik'sten·chən ˌmek·əˌniz·əm }

extension register [COMPUT SCI] A register that is combined with an accumulator register for calculations involving multiple precision arithmetic. { ik'sten·chən ˌrej·ə·stər }

extent [COMPUT SCI] The physical locations in a mass-storage device or volume allocated for use by a particular data set. { ik'stent }

extern [COMPUT SCI] A pseudoinstruction found in several assembly languages which explicitly tells an assembler that a symbol is external, that is, not defined in the program module. { ek'stərn }

external buffer [COMPUT SCI] A buffer storage located outside the computer's main storage, often within a control unit or other peripheral device. { ek'stərn·əl 'bəf·ər }

external declaration [COMPUT SCI] A declarative statement in a computer program that specifies that a symbolic name used in the program is defined in another program. { ek'stərn·əl ˌdek·lə'rā·shən }

external delay [COMPUT SCI] Time during which a computer cannot be operated due to circumstances beyond the reasonable control of the operators and maintenance engineers, such as a failure of the public power supply. { ek'stərn·əl di'lā }

external-device address [COMPUT SCI] The address of a component such as a tape drive. { ek¦stərn·əl di¦vīs 'aˌdres }

external-device control [COMPUT SCI] The capability of an external device to create an interrupt during the execution of a job. { ek¦stərn·əl di¦vīs kənˌtrōl }

external-device operands [COMPUT SCI] The part of an instruction referring to an external device such as a tape drive. { ek¦stərn·əl di¦vīs 'äp·əˌranz }

external-device response [COMPUT SCI] The signal from an external device, such as a tape drive, that it is not busy. { ek¦stərn·əl di¦vīs ri ˌspäns }

external error [COMPUT SCI] An error sensed by the computer when this error occurs in a device such as a disk drive. { ek¦stərn·əl 'er·ər }

external interrupt [COMPUT SCI] Any interrupt caused by the operator or by some external device such as a tape drive. { ek¦stərn·əl 'int·əˌrəpt }

external-interrupt status word [COMPUT SCI] The content of a special register which indicates, among other things, the source of the interrupt. { ek¦stərn·əl 'int·əˌrəpt 'stad·əs ˌwərd }

external label [COMPUT SCI] A reference to a variable not defined in a program segment. { ek ¦stərn·əl 'lā·bəl }

externally stored program [COMPUT SCI] A program achieved by wiring plugboards, as in some tabulating equipment. { ek¦stərn·əl·ē ¦stȯrd 'prō·grəm }

external memory [COMPUT SCI] Any storage device not an integral part of a computer system, such as a magnetic tape or disk. { ek¦stərn·əl 'mem·rē }

external Q [ELECTR] The inverse of the difference between the loaded and unloaded Q values of a microwave tube. { ek¦stərn·əl 'kyü }

external reference [COMPUT SCI] In a computer program, a branch or call to a separate independent program or routine. { ek¦stərn·əl 'ref·rəns }

external signal [COMPUT SCI] Any message to an operator for which no printout is required but which is self-explanatory, such as a light condition indicating whether the equipment is on or off. { ek¦stərn·əl 'sig·nəl }

external sorting [COMPUT SCI] The sorting of a list of items by a computer in which the list is too large to be brought into the memory at one time, and instead is brought into the memory a piece at a time so as to produce a collection of ordered sublists which are subsequently reordered by the computer to produce a single list. { ek¦stərn·əl 'sȯrd·iŋ }

external storage [COMPUT SCI] Large-capacity, slow-access data storage attached to a digital computer and used to store information that exceeds the capacity of main storage. { ek ¦stərn·əl 'stȯr·ij }

external symbol dictionary [COMPUT SCI] A list of external symbols and their relocatable addresses which allows the linkage editor to resolve interprogram references. Abbreviated ESD. { ek ¦stərn·əl 'sim·bəl ˌdik·shəˌner·ē }

external table [COMPUT SCI] A table whose data are located outside a computer program, usually in a separate file. { ek'stərn·əl 'tā·bəl }

extract [COMPUT SCI] **1.** To form a new computer word by extracting and putting together selected segments of given words. **2.** To remove from a computer register or memory all items that meet a specified condition. { ik'strakt }

extract instruction [COMPUT SCI] An instruction that requests the formation of a new expression from selected parts of given expressions. { ik'strakt inˌstrək·shən }

extraneous emission [ELECTR] Any emission of a transmitter or transponder, other than the output carrier fundamental, plus only those sidebands intentionally employed for the transmission of intelligence. { ik'strān·ē·əs ə'mish·ən }

extraneous response [ELECTR] Any undersired response of a receiver, recorder, or other susceptible device, due to the desired signals, undersired signals, or any combination or interaction among them. { ik'strān·ē·əs ri'späns }

extranet [COMPUT SCI] A secure, Internet-based private network that allows organizations to share information with vendors, partners, customers, and so on; access requires either a password or digital encryption. { 'ek·strə ˌnet }

extrapolation [MATH] Estimating a function at a

point which is larger than (or smaller than) all the points at which the value of the function is known. { ik,strap·ə'lā·shən }

extraterrestrial noise [ELECTROMAG] Cosmic and solar noise; radio disturbances from sources other than those related to the earth. { ¦ek·strə·tə'res·trē·əl 'nȯiz }

extremely high frequency [COMMUN] The frequency band from 30,000 to 300,000 megahertz in the radio spectrum. Abbreviated EHF. { ek'strēm·lē 'hī 'frē·kwən·sē }

extremely low frequency [COMMUN] A frequency below 300 hertz in the radio spectrum. Appreviated ELF. { ek'strēm·lē 'lō 'frē·kwən·sē }

extrinsic semiconductor [ELECTR] A semiconductor whose electrical properties are dependent on impurities added to the semiconductor crystal, in contrast to an intrinsic semiconductor, whose properties are characteristic of an ideal pure crystal. { ek¦strinz·ik 'sem·i·kən,dək·tər }

e-zine [COMPUT SCI] A Web-published magazine. { 'ē ,zēn }

F

face *See* faceplate. { fās }

face-bonding [ELECTR] Method of assembling hybrid microcircuits wherein semiconductor chips are provided with small mounting pads, turned facedown, and bonded directly to the ends of the thin-film conductors on the passive substrate. { 'fās ˌbänd·iŋ }

faceplate [ELECTR] The transparent or semitransparent glass front of a cathode-ray tube, through which the image is viewed or projected; the inner surface of the face is coated with fluorescent chemicals that emit light when hit by an electron beam. Also known as face. { 'fās ˌplāt }

facility assignment [COMPUT SCI] The allocation of core memory and external devices by the executive as required by the program being executed. { fə'sil·əd·ē əˌsīn·mənt }

facility dispersion [COMMUN] The distribution of circuits between two points over more than one physical or geographic route to reduce the likelihood of a trunk group being put completely out of service by facility damage or other circuit failure. { fə'sil·əd·ē di'spər·zhən }

facsimile [COMMUN] **1.** A system of communication in which a transmitter scans a photograph, map, or other fixed graphic material and converts the information into signal waves for transmission by wire or radio to a facsimile receiver at a remote point. Also known as fax; phototelegraphy; radiophoto; telephoto; telephotography; wirephoto. **2.** A photograph transmitted by radio to a facsimile receiver. Also known as radiophoto. { fak'sim·ə·lē }

facsimile modulation [COMMUN] Process in which the amplitude, frequency, or phase of a transmitted wave is varied with time in accordance with a facsimile transmission signal. { fak'sim·ə·lē ˌmäj·ə·'lā·shən }

facsimile posting [COMPUT SCI] The process of transferring by a duplicating process a printed line of information from a report, such as a listing of transactions prepared on an accounting machine, to a ledger or other recorded sheet. { fak'sim·ə· lē 'pōst·iŋ }

facsimile receiver [ELECTR] The receiver used to translate the facsimile signal from a wire or radio communication channel into a facsimile record of the subject copy. { fak'sim·ə·lē ri'sē·vər }

facsimile recorder [ELECTR] The section of a facsimile receiver that performs the final conversion of electric signals to an image of the subject copy on the record medium. { fak'sim·ə·lē ri'kȯrd·ər }

facsimile signal [COMMUN] The picture signal produced by scanning the subject copy in a facsimile transmitter. { fak'sim· ə·lē ˌsig·nəl }

facsimile signal level [ELECTR] Maximum facsimile signal power or voltage (root mean square or direct current) measured at any point in a facsimile system. { fak'sim·ə·lē 'sig·nəl ˌlev·əl }

facsimile synchronizing [ELECTR] Maintenance of predetermined speed relations between the scanning spot and the recording spot within each scanning line. { fak'sim·ə·lē 'siŋ· krəˌniz·iŋ }

facsimile telegraph [COMMUN] A telegraph system designed to transmit pictures. { fak'sim·ə·lē 'tel·əˌgraf }

facsimile transmitter [ELECTR] The apparatus used to translate the subject copy into facsimile signals suitable for delivery over a communication system. { fak'sim·ə·lē tranz'mid·ər }

fade-out [COMMUN] A gradual and temporary loss of a received radio or television signal due to magnetic storms, atmospheric disturbances, or other conditions along the transmission path. { 'fādˌau̇t }

fader [ELECTR] A multiple-unit level control used for gradual changeover from one audio video source { 'fād·ər }

fading [COMMUN] Variations in the field strength of a radio signal that are caused by changes in the transmission medium. { 'fād·iŋ }

fading margin [COMMUN] **1.** Number of decibels of attenuation which may be added to a specified radio-frequency propagation path before the signal-to-noise ratio of a specified channel falls below a specified minimum in order to avoid disruption of service. **2.** Allowance made in radio system planning to accommodate estimated fading. { 'fād·iŋ ˌmär·jən }

failsafe tape *See* incremental dump tape. { 'fāl ¦sāf ˌtāp }

fail-soft system [COMPUT SCI] A computer system with automatic controls that allow function to continue after a malfunction and, if necessary, permit the shutdown of the system without loss of data. { 'fāl ¦sȯft ˌsis·təm }

failure logging [COMPUT SCI] The automatic recording of the state of various components of a computer system following detection of a machine fault; used to initiate corrective procedures, such as repeating attempts to read or write a magnetic tape, and to aid customer engineers in diagnosing errors. { 'fāl·yər ˌläg·iŋ }

fallback [COMPUT SCI] The system, electronic or manual, which is substituted for the computer system in case of breakdown. { 'fȯlˌbak }

fallback switch [COMMUN] A mechanical switch to transfer a communications path from a primary device to an identical standby device in the event of a primary device failure. { 'fȯlˌbak ˌswich }

false alarm [ELECTR] In radar, an indication of a detected target even though one does not exist, due to noise or interference levels exceeding the set threshold of detection. { ¦fȯls ə'lärm }

false drop *See* false retrieval. { ¦fȯls 'dräp }

false retrieval [COMPUT SCI] An item retrieved in an automatic library search which is unrelated or vaguely related to the subject of the search. Also known as false drop. { ¦fȯls ri'trē·vəl }

false sorts [COMPUT SCI] Entries irrelevant to the subject sought which are retrieved in a search. { ¦fȯls 'sȯrts }

false target [ELECTR] In radar, a contact (target) estimated to be where none exists, generally as the result of ambiguity in the data processing. { ¦fȯls 'tär·gət }

false-target generator [ELECTR] An electronic countermeasure device that generates a delayed return signal on an enemy radar frequency to give erroneous position information. { ¦fȯls ¦tär·gət 'jen·əˌrād·ər }

FAMOS device *See* floating-gate avalanche-injection metal-oxide semiconductor device. { 'fāˌmȯs di'vīs }

fan antenna [ELECTROMAG] An array of folded dipoles of different length forming a wide-band ultra-high-frequency or very-high-frequency antenna. { 'fan anˌten·ə }

fan beam [ELECTROMAG] **1.** A radio beam having an elliptically shaped cross section in which the ratio of the major to the minor axis usually exceeds 3 to 1; the beam is broad in the vertical plane and narrow in the horizontal plane. **2.** A radar beam having the shape of a fan. { 'fan ˌbēm }

fanfold [COMPUT SCI] Continuous paper that is perforated at page boundaries and can be folded back and forth at the perforations to form a stack. { 'fanˌfōld }

fan marker *See* fan-marker beacon. { 'fan ˌmärk·ər }

fan-marker beacon [NAV] A very-high frequency radio facility having a vertically directed fan beam interesecting an airway to provide a fix. Also known as fan marker; radio fan-marker beacon. { 'fan ˌmärk·ər ˌbē·kən }

fanned-beam antenna [ELECTROMAG] Unidirectional antenna so designed that transverse cross sections of the major lobe are approximately elliptical. { ¦fand ¦bēm anˌten·ə }

fanning beam [ELECTROMAG] Narrow antenna beam which is repeatedly scanned over a limited arc. { 'fan·iŋ ˌbēm }

FAQ *See* Frequently Asked Questions.

Faraday birefringence [OPTICS] Difference in the indices of refraction of left and right circularly polarized light passing through matter parallel to an applied magnetic field; it is responsible for the Faraday effect. { 'far·əˌdā ˌbī·ri'frin·jəns }

Faraday effect [OPTICS] Rotation of polarization of a beam of linearly polarized light when it passes through matter in the direction of an applied magnetic field; it is the result of Faraday birefringence. Also known as Faraday rotation; Kundt effect; magnetic rotation. { 'far·əˌdā i'fekt }

Faraday rotation *See* Faraday effect. { 'far·əˌdā rō'tā·shən }

Faraday rotation isolator *See* ferrite isolator. { 'far·əˌdā rō'tā·shən 'ī·səˌlād·ər }

far-end crosstalk [COMMUN] Crosstalk that travels along the disturbed circuit in the same direction as desired signals in that circuit. { ¦fär ¦end 'krȯsˌtȯk }

far field *See* Fraunhofer region. { ¦fär ¦fēld }

far-infrared radiation [ELECTROMAG] Infrared radiation the wavelengths of which are the longest of those in the infrared region, about 50–1000 micrometers; requires diffraction gratings for spectroscopic analysis. { ¦fär in·frə'red ˌrād·ē'ā·shən }

far region *See* Fraunhofer region. { ¦fär ¦rē·jən }

far zone *See* Fraunhofer region. { ¦fär ¦zōn }

fast-access storage [COMPUT SCI] The section of a computer storage from which data can be obtained most rapidly. { ¦fast ¦ak·ses 'stȯr·ij }

fast automatic gain control [ELECTR] Radar automatic gain control method characterized by a response time that is long with respect to a pulse width, and short with respect to the time on target. { 'fast ˌȯd·əˌmad·ik 'gān kənˌtrōl }

fast time constant [ELECTR] Circuit with short time constant used to emphasize signals of short duration to produce discrimination against low-frequency components of clutter in radar. { 'fast 'tīm ˌkän·stənt }

fast time scale [COMPUT SCI] In simulation by an analog computer, a scale in which the time duration of a simulated event is less than the actual time duration of the event in the physical system under study. { 'fast 'tīm ˌskāl }

FAT *See* file allocation table. { fat *or* ¦ef¦ā'tē }

fatal error [COMPUT SCI] An error in a computer program which causes running of the program to be terminated. { ¦fād·əl 'er·ər }

father file [COMPUT SCI] A copy of the master file from the cycle or generation that precedes the one being updated. { 'fäth·ər ˌfīl }

fatware [COMPUT SCI] Software that is overly laden with features or is inefficiently designed, so that it occupies inordinate space in disk storage and random-access memory, and requires an inappropriate share of microprocessor power. Also known as bloatware. { 'fatˌwer }

fault [ELECTR] Any physical condition that causes a component of a data-processing system to fail in performance. { fȯlt }

fault detection and exclusion [NAV] The capability of a user of the Global Positioning System, once the presence of a fault in the system has been detected, to identify and exclude the malfunctioning satellite in order to continue navigating using the remaining satellites. Abbreviated FDE. { ¦fȯlt di‚tek·shən ən ik'sklü·zhən }

fault masking [COMPUT SCI] Any type of hardware redundancy in which faults are corrected immediately and the operations of fault detection, location, and correction are indistinguishable. { 'fȯlt ‚mask·iŋ }

fault monitoring [SYS ENG] A procedure for systematically checking for errors and malfunctions in the software and hardware of a computer or control system. { 'fȯlt ‚män·ə·triŋ }

fault tolerance [SYS ENG] The capability of a system to perform in accordance with design specifications even when undesired changes in the internal structure or external environment occur. { 'fȯlt ‚täl·ə·rəns }

fax *See* facsimile. { faks }

FB data set [COMPUT SCI] A data set which has F-format logical records and whose physical records are all some multiple of the size of the logical record, except possibly for a few truncated blocks. Also known as blocked F-format data set. { ¦ef¦bē ‚dad·ə ‚set }

FBM data set [COMPUT SCI] An FB data set which has a machine-control (M) character in its first byte of information. { ¦ef¦bē¦em ‚dad·ə ‚set }

FBSA data set [COMPUT SCI] An FBS data set which has an ASCII (American Standard Code for Information Interchange) control (A) character in its first byte of information. { ¦ef¦bē¦es¦ā ‚dad·ə ‚set }

FBS data set [COMPUT SCI] An FB data set which has at most one truncated block, which must be the last one in the data set. Also known as standard blocked F-format data set. { ¦ef¦bē¦es ‚dad·ə ‚set }

F connector [ELECTR] A plug and socket for interconnecting coaxial cables; commonly used to interconnect television receivers and cable or antenna sources. { 'ef kə‚nek·tər }

FDDI *See* fiber-optic data distribution interface.

FDE *See* fault detection and exclusion.

F-display [ELECTR] A radar display format in which the target appears as a spot in the center when the antenna of a tracking radar is aimed directly at it, with any displacement indicating pointing error. Also known as F-indicator; F-scan; F-scope. { 'ef di‚splā }

FDM *See* frequency-division multiplexing.

FDMA *See* frequency-division multiple access.

feasible solution [COMPUT SCI] In linear programming, any set of values for the variables x_j, $j = 1, 2, \ldots, n$, that (1) satisfy the set of restrictions

$$\sum_{j=1}^{n} a_{ij}x_j \le b_i, i = 1, 2, \ldots, m$$

$$\left(\text{alternatively, } \sum_{j=1}^{n} a_{ij}x_j \le b_i, \text{ or } \sum_{j=1}^{n} a_{ij}x_j \le b_i\right)$$

where the b_i are numerical constants known collectively as the right-hand side and the a_{ij} are coefficients of the variables x_j, and (2) satisfy the restrictions $x_j \ge 0$. { 'fēz·ə·bəl sə'lü·shən }

feature [COMPUT SCI] In automatic pattern recognition, a property of an image that is useful for its interpretation. { 'fē·chər }

feature extraction-classification model [COMPUT SCI] A method of automatic pattern recognition in which recognition is achieved by making measurements on the patterns to be recognized, and then deriving features from these measurements. { 'fē·chər ik¦strak·shən ‚klas·ə·fə¦kā·shən ‚mäd·əl }

Federal Telecommunications System [COMMUN] System of commercial telephone lines, leased by the government, for use between major government installations for official telecommunications. { 'fed·rəl ‚tel·ə·kə‚myü·nə'kā·shənz ‚sis·təm }

fedsim star [COMPUT SCI] The starlike shape that is characteristic of the Kiviat graph of a well-balanced computer system. { 'fed‚sim ‚stär }

feed [COMPUT SCI] **1.** To supply the material to be operated upon to a machine. **2.** A device capable of so feeding. [ELECTR] To supply a signal to the input of a circuit, transmission line, or antenna. [ELECTROMAG] The part of a radar antenna that is connected to or mounted on the end of the transmission line and serves to radiate radio-frequency electromagnetic energy to the reflector or receive energy therefrom; in multiple-element (array) antennas, the constrained network, radiation means or digital means for distributing the energy to the radiating elements and collecting the energy received by them. { fēd }

feedback [ELECTR] The return of a portion of the output of a circuit or device to its input. { 'fēd ‚bak }

feedback circuit [ELECTR] A circuit that returns a portion of the output signal of an electronic circuit or control system to the input of the circuit or system. { 'fēd‚bak ‚sər·kət }

feedback compensation [CONT SYS] Improvement of the response of a feedback control system by placing a compensator in the feedback path, in contrast to cascade compensation. Also known as parallel compensation. { 'fēd‚bak ‚käm·pən‚sā·shən }

feedback control loop *See* feedback loop. { 'fēd‚bak kən'trōl ‚lüp }

feedback control signal [CONT SYS] The portion of an output signal which is retransmitted as an input signal. { 'fēd‚bak kən'trōl ‚sig·nəl }

feedback control system [CONT SYS] A system in which the value of some output quantity is controlled by feeding back the value of the controlled quantity and using it to manipulate an input quantity so as to bring the value of the controlled quantity closer to a desired value. Also known as closed-loop control system. { ¦fēd ˌbak kənˈtrōl ˌsis·təm }

feedback loop [CONT SYS] A closed transmission path or loop that includes an active transducer and consists of a forward path, a feedback path, and one or more mixing points arranged to maintain a prescribed relationship between the loop input signal and the loop output signal. Also known as feedback control loop. { ˈfēdˌbak ˌlüp }

feedback oscillator [ELECTR] An oscillating circuit, including an amplifier, in which the output is fed back in phase with the input; oscillation is maintained at a frequency determined by the values of the components in the amplifier and the feedback circuits. { ˈfēdˌbak ˌäs·əˌlād·ər }

feedback regulator [CONT SYS] A feedback control system that tends to maintain a prescribed relationship between certain system signals and other predetermined quantities. { ˈfēdˌbak ˌreg·yəˌlād·ər }

feedback transfer function [CONT SYS] In a feedback control loop, the transfer function of the feedback path. { ˈfēdˌbak ˈtranz·fər ˌfəŋk·shən }

feeder cable [COMMUN] In communications practice, a cable extending from the central office along a primary route (main feeder cable) or from a main feeder cable along a secondary route (branch feeder cable) and providing connections to one or more distribution cables. { ˈfēd·ər ¦kā·bəl }

feeder distribution center [COMMUN] Distribution center at which feeders or subfeeders are connected. { ˈfēd·ər dis·trəˈbyü·shən ˌsen·tər }

feedforward control [CONT SYS] Process control in which changes are detected at the process input and an anticipating correction signal is applied before process output is affected. { ¦fēd ¦fȯr·wərd kənˌtrōl }

feed holes [COMPUT SCI] Holes along the edges of continuous-feed computer paper that are engaged by sprockets to move the paper and maintain alignment during printing. { ˈfēd ˌhōlz }

feed horn [ELECTROMAG] A device located at the focus of a receiving paraboloidal antenna that acts as a receiver of radio waves which the antenna collects, focuses, and couples to transmission lines to the amplifier. { ˈfēd ˌhȯrn }

feed shelf [COMPUT SCI] **1.** A device for supporting documents for manual sensing. **2.** The first few feet of a tape reel, used to prime the tape drive. { ˈfēd ˌshelf }

feed-tape [COMPUT SCI] A mechanism which will feed tape to be read or sensed. { ˈfēdˌtāp }

fence [ENG] **1.** A line of data-acquisition or tracking stations used to monitor orbiting satellites. **2.** A line of radar or radio stations for detection of satellites or other objects in orbit. **3.** A line or network of early-warning radar stations. **4.** A concentric steel fence erected around a ground radar transmitting antenna to serve as an artificial horizon and suppress ground clutter that would otherwise drown out weak signals returning at a low angle from a target. { fens }

fence cell [COMPUT SCI] A criterion for dividing a list into two equal or nearly equal parts in the course of a binary search. { ˈfens ˌsel }

ferrimagnetic amplifier [ELECTR] A microwave amplifier using ferrites. { ˌfe·ri·magˈned·ik ˈam·pləˌfī·ər }

ferrite attenuator *See* ferrite limiter. { ˈfeˌrīt əˈten·yəˌwād·ər }

ferrite circulator [ELECTROMAG] A combination of two dual-mode transducers and a 45° ferrite rotator, used with rectangular waveguides to control and switch microwave energy. Also known as ferrite phase-differential circulator. { ˈfeˌrīt ˈsər·kyəˌlād·ər }

ferrite isolator [ELECTROMAG] A device consisting of a ferrite rod, centered on the axis of a short length of circular waveguide, located between rectangular-waveguide sections displaced 45° with respect to each other, which passes energy traveling through the waveguide in one direction while absorbing energy from the opposite direction. Also known as Faraday rotation isolator. { ˈfeˌrīt ˈī·səˌlād·ər }

ferrite limiter [ELECTROMAG] A passive, low-power microwave limiter having an insertion loss of less than 1 decibel when operating in its linear range, with minimum phase distortion; the input signal is coupled to a single-crystal sample of either yttrium iron garnet or lithium ferrite, which is biased to resonance by a magnetic field. Also known as ferrite attenuator. { ˈfeˌrīt ˈlim·əd·ər }

ferrite phase-differential circulator *See* ferrite circulator. { ˈfeˌrīt ¦fāz dif·ə¦ren·chəl ˈsər·kyə ˌlād·ər }

ferrite-rod antenna [ELECTROMAG] An antenna consisting of a coil wound on a rod of ferrite; used in place of a loop antenna in radio receivers. Also known as ferrod; loopstick antenna. { ˈfeˌrīt ¦räd anˈten·ə }

ferrite rotator [ELECTROMAG] A gyrator consisting of a ferrite cylinder surrounded by a ring-type permanent magnet, inserted in a waveguide to rotate the plane of polarization of the electromagnetic wave passing through the waveguide. { ˈfeˌrīt ˈrōˌtād·ər }

ferrite switch [ELECTROMAG] A ferrite device that blocks the flow of energy through a waveguide by rotating the electric field vector 90°; the switch is energized by sending direct current through its magnetizing coil; the rotated electromagnetic wave is then reflected from a reactive mismatch or absorbed in a resistive card. { ˈfe ˌrīt ˈswich }

ferrite-tuned oscillator [ELECTR] An oscillator in which the resonant characteristic of a ferrite-loaded cavity is changed by varying the ambient magnetic field, to give electronic tuning. { ˈfe ˌrīt ¦tünd ˈäs·əˌlād·ər }

ferrod *See* ferrite-rod antenna. { ˈfeˌräd }

ferroelectric liquid-crystal display [ELECTR] An electronic display that employs a liquid crystal that is ferroelectric, such as smectic C*, which has two different stable molecular configurations; polarizers are positioned such that one state is optically transmissive while the other is dark. { ˌfer·ō·i¦lek·trik ¦lik·wəd ¦kris·təl dis¦plā }

ferromagnetic amplifier [ELECTR] A parametric amplifier based on the nonlinear behavior of ferromagnetic resonance at high radio-frequency power levels; incorrectly known as garnet maser. { ¦fe·rō·mag¦ned·ik 'am·pləˌfī·ər }

FET *See* field-effect transistor.

fetch [COMPUT SCI] To locate and load into main memory a requested load module, relocating it as necessary and leaving it in a ready-to-execute condition. { fech }

fetch ahead *See* instruction lookahead. { ¦fech ə'hed }

fetch bit [COMPUT SCI] The fifth bit in a storage key; the value of the fetch bit can protect a stored block from destruction or from being accessed by unauthorized programs. { 'fech ˌbit }

fetch cycle [COMPUT SCI] The period during which a machine language instruction is read from memory into the control section of the central processing unit. { 'fech ˌsī kəl }

F format [COMPUT SCI] **1.** In data management, a fixed-length logical record format. **2.** In FORTRAN, a real variable formatted as Fμ.*d*, where μ is the width of the field and *d* represents the number of digits to appear after the decimal point. { 'ef ˌfȯr·mat }

fiber [OPTICS] A transparent threadlike object made of glass or clear plastic, used to conduct light along selected paths. { 'fī·bər }

fiber bundle [OPTICS] A flexible bundle of glass or other transparent fibers, parallel to each other, used in fiber optics to transmit a complete image from one end of the bundle to the other. { 'fī·bər ˌbən·dəl }

fiber-optic circuit [COMMUN] A path for data transmission in which light acts as the information carrier and is transmitted by total internal reflection through a transparent optical waveguide. { ¦fī·bər ¦äp·tik 'sər·kət }

fiber-optic data distribution interface [COMMUN] A set of standards for high-speed fiber-optic local-area networks. Abbreviated FDDI. { ¦fī·bər ¦äp·tik ¦dad·ə ˌdis·trə¦byü·shən 'in·tər ˌfās }

fiber optics [OPTICS] The technique of transmitting light through long, thin, flexible fibers of glass, plastic, or other transparent materials; bundles of parallel fibers can be used to transmit complete images. { 'fī·bər ˌäp·tiks }

fiber waveguide *See* optical waveguide. { 'fī·bər 'wāvˌgīd }

fidelity [COMMUN] The degree to which a system accurately reproduces at its output the essential characteristics of the signal impressed on its input. { fə'del·əd·ē }

field [COMPUT SCI] **1.** A location in a record in a database that contains a specific piece of information **2.** A specified area on a geographical user interface for the input of a particular category of data. [ELECTR] One of the equal parts into which a frame is divided in interlaced scanning for television; includes one complete scanning operation from top to bottom of the picture and back again. { fēld }

fieldata code [COMMUN] A standardized military data transmission code, seven data bits plus one parity bit. { 'fēlˌdad·ə ˌkōd }

field delimiter [COMPUT SCI] Any symbol, such as a slash, colon, tab, or space, which enables an assembler to recognize the end of a field. { 'fēld dəˌlim·əd·ər }

field designator [COMPUT SCI] A character generally placed at the beginning of a field to specify the nature of the data contained in it. { 'fēld ¦dez·igˌnād·ər }

field effect [ELECTR] The local change from the normal value that an electric field produces in the charge-carrier concentration of a semiconductor. { 'fēld iˌfekt }

field-effect device [ELECTR] A semiconductor device whose properties are determined largely by the effect of an electric field on a region within the semiconductor. { 'fēld iˌfekt diˌvīs }

field-effect phototransistor [ELECTR] A field-effect transistor that responds to modulated light as the input signal. { 'fēld iˌfekt ¦fōd·ō·tran'zis·tər }

field-effect transistor [ELECTR] A transistor in which the resistance of the current path from source to drain is modulated by applying a transverse electric field between grid or gate electrodes; the electric field varies the thickness of the depletion layer between the gates, thereby reducing the conductance. Abbreviated FET. { 'fēld iˌfekt tran'zis·tər }

field emission [ELECTR] The emission of electrons from the surface of a metallic conductor into a vacuum (or into an insulator) under influence of a strong electric field; electrons penetrate through the surface potential barrier by virtue of the quantum-mechanical tunnel effect. Also known as cold emission. { 'fēld əˌmish·ən }

field-emission display [ELECTR] A flat-panel electronic display in which electrons are extracted from an array of cold-cathode emitters by applying a voltage between the cathode and a control electrode, and the electrons are then accelerated without deflection over a distance of less than 1 millimeter before colliding with a phosphor-coated flat faceplate. { 'fēld iˌmish·ən diˌsplā }

field-emitter array [ELECTR] An array of pyramidal silicon structures, with spacing on the order of 10 micrometers, designed for field emission of electrons into a vacuum. { ¦fēld i¦mid·ər ə'rā }

field engineer [COMPUT SCI] A professional who installs computer hardware on customers' premises, performs routine preventive maintenance, and repairs equipment when it is out of order. Also known as field service representative. { 'fēld en·jəˌnir }

field frequency [ELECTR] The number of fields transmitted per second in a video system; equal to the frame frequency multiplied by the number of fields that make up one frame. Also known as field repetition rate. { 'fēld ˌfrē·kwən·sē }

field intensity [COMMUN] In Federal Communications Commission regulations, the electric field intensity in the horizontal direction. { 'fēld inˌten·səd·ē }

field length [COMPUT SCI] The number of columns, characters, or bits in a specified field. { 'fēld ˌleŋkth }

field of search [ELECTR] The space that a radar set or installation can cover effectively. { 'fēld əv 'sərch }

field of view [ELECTR] The space in which a radar can operate effectively. { 'fēld əv 'vyü }

field pattern *See* radiation pattern. { 'fēld ˌpad·ərn }

field-programmable gate array [ELECTR] A gate-array device that can be configured and reconfigured by the system manufacturer and sometimes by the end user of the system. { ¦fēld prōˌgram·ə·bəl 'gāt əˌrā }

field-programmable logic array [ELECTR] A programmed logic array in which the internal connections of the logic gates can be programmed once in the field by passing high current through fusible links, by using avalanche-induced migration to short base-emitter junctions at desired interconnections, or by other means. Abbreviated FPLA. Also known as programmable logic array. { 'fēld prō¦gram·ə·bəl 'läj·ik ə'rā }

field repetition rate *See* field frequency. { ¦fēld rep·ə'tish·ən ˌrāt }

field scan [ELECTR] Television term denoting the vertical excursion of an electron beam downward across a cathode-ray tube face, the excursion being made in order to scan alternate lines. { 'fēld ˌskan }

field section [COMPUT SCI] A portion of a field, such as the section formed by the second and third character of a 10-character field. { 'fēld ˌsek·shən }

field separator [COMPUT SCI] A character that is used to mark the boundary between fields in a record. { 'fēld ˌsep·əˌrād·ər }

field-sequential color television [COMMUN] An analog color television system in which the individual red, green, and blue primary colors are associated with successive fields. { ¦fēld sə ¦kwen·chəl ¦kəl·ər 'tel·əˌvizh·ən }

field service representative *See* field engineer. { 'fēld ˌsər·vəs ˌrep·rəˌzent·əd·iv }

field squeeze [COMPUT SCI] In a mail merge operation, the elimination of extra blank spaces in a data field so that the data field is correctly printed within the text of the letter. { 'fēld ˌskwēz }

field telephone [COMMUN] A portable telephone designed for field or combat use. { 'fēld ˌtel·ə ˌfōn }

field waveguide [ELECTROMAG] A single wire, threaded or coated with dielectric, which guides an electromagnetic field. Also known as G string. { 'fēld 'wāvˌgīd }

fifth-generation computer [COMPUT SCI] A computer that would use artificial intelligence techniques to learn, reason, and converse in natural languages resembling human languages. { 'fifth ˌjen·ə¦rā·shən kəm'pyüd·ər }

figurative constant [COMPUT SCI] A predefined constant in COBOL which does not require a description in data division, such as ZERO which stands for 0. { 'fig·yə·rəd·iv 'kän·stənt }

figures shift [COMMUN] A physical movement that permits a teletypewriter to print uppercase characters, numbers, symbols, and the like. { 'fig·yərz ˌshift }

file [COMPUT SCI] A collection of related records treated as a unit. { fīl }

file allocation table [COMPUT SCI] A table stored on hard or removable disks used to locate files or sections of files if scattered about the disk. Abbreviated FAT. { ˌfīl ˌal·ə'kā·shən ˌtā·bəl }

file compression program *See* file compression utility. { 'fīl kəmˌpresh·ən ˌprō·grəm }

file compression utility [COMPUT SCI] A utility program that encodes files so that they take up less space in storage. Also known as file compression program. { 'fīl kəmˌpresh·ən yüˌtil·əd·ē }

file control system [COMPUT SCI] Software package which handles the transfer of data from any device into any device. { 'fīl kənˌtrōl ˌsis·təm }

file event [COMPUT SCI] A single access to any storage device for either input or output. { 'fīl iˌvent }

file format [COMPUT SCI] The rules that determine the organization of data in a file. { 'fīl ˌfȯr·mat }

file gap [COMPUT SCI] An area in a data storage medium which is used mainly to indicate the end of a file and sometimes the beginning of another. { 'fīl ˌgap }

file-handling routine [COMPUT SCI] A part of a computer program that deals with reading and writing of data from and to a file. { 'fīl ˌhand·liŋ rüˌtēn }

file header [COMPUT SCI] A set of words comprising the file name and various characteristics of the file, found at the beginning of a file stored on magnetic tape or disk. { 'fīl ˌhed·ər }

file identification [COMPUT SCI] A device, such as a label or tag, used to identify, describe, or name a physical medium, such as a disk or reel of magnetic tape, which contains data. { ¦fīl īˌdent·ə·fə'kā·shən }

file layout [COMPUT SCI] A description of the arrangement of the data in a file. { 'fīl ˌlāˌau̇t }

file locking [COMPUT SCI] A technique that prevents processing of a file by more than one program or user at a time, ensuring that a file in use by one user is made unavailable to others. { 'fil ˌläk·iŋ }

file maintenance [COMPUT SCI] Data-processing operation in which a master file is updated on the basis of one or more transaction files. { 'fīl ˌmānt·ən·əns }

file management system [COMPUT SCI] Computer programs that control the space used for file storage and provide such services as input/output control and indexing. { 'fīl ¦man·ij·mənt ˌsis·təm }

file manager [COMPUT SCI] Software for managing data that works only with single files and lacks relational capability. { 'fīl ˌman·ə·jər }

file name [COMPUT SCI] The name given by the programmer to a specific set of data. { 'fīl ˌnām }

file opening [COMPUT SCI] The process, carried out by computer software, of identifying a file and comparing the file header with specifications in the program being run to ensure that the file corresponds. { 'fīl ˌōp·ə·niŋ }

file organization [COMPUT SCI] The structure of a file meeting two requirements: to minimize the running time of the program, and to simplify the work involved in modifying the contents of the file. { ¦fīl ˌȯrg·ə·nə'zā·shən }

file organization routine [COMPUT SCI] A program which allocates data files into random-access storage devices. { ¦fīl ˌȯrg·ə·nə'zā·shən rüˌtēn }

file-oriented system [COMPUT SCI] A computer configuration which considers a heavy, or exclusive, usage of data files. { 'fīl ˌȯr·ēˌent·əd ˌsis·təm }

file printout [COMPUT SCI] Output from a computer printer consisting of a copy of the contents of a file held in some storage device, usually to assist in debugging a program. { 'fīl ˌprinˌtau̇t }

file processing [COMPUT SCI] The job of updating, sorting, or validating a data file. { 'fīl ˌpräs ˌes·iŋ }

file protection [COMPUT SCI] A mechanical device or a computer command which prevents erasing of or writing upon a magnetic tape but allows a program to read the data from the tape. { 'fīl prəˌtek·shən }

file protection ring [COMPUT SCI] A ring that can be attached to, or detached from, the hub of a reel of magnetic tape, used to identify the reel's status and, in some computer systems, to prevent writing upon the tape when the ring is attached or detached. { 'fīl prəˌtek·shən ˌriŋ }

file reference [COMPUT SCI] An operation involving looking up and retrieving the information on file for a specified item or items. { 'fīl ˌref·rəns }

file reorganization [COMPUT SCI] An activity performed periodically on files and data bases, involving such operations as deletion of unneeded records, in order to minimize space requirements of files and improve efficiency of processing. { 'fīl rēˌȯr·gə·nə'zā·shən }

file search [COMPUT SCI] An operation involving looking through the file for information on all items falling in a specified category, extracting the information for any item where the information recorded meets certain criteria, and determining whether or not there exists a specified pattern of information anywhere in the file. { 'fīl ˌsərch }

file security *See* data security. { 'fīl səˌkyür·əd·ē }

file server [COMPUT SCI] A mass storage device that holds programs and data that can be accessed and shared by the workstations connected to a local-area network. Also known as network server. { 'fīl ˌsər·vər }

file sharing [COMPUT SCI] The common use, by two or more users, of data and program files, usually located in a file server. { 'fīl ˌsher·iŋ }

FileSize metric [COMPUT SCI] A measure of computer program size, equal to the total number of characters in the source file of the program. { ¦fīl'sīz ¦me·trik }

file specification [COMPUT SCI] A designation that enables a file to be located on a disk and includes the disk drive, name of the directory/subdirectory, and name of file. { 'fīl ˌspes·ə·fəˌkā·shən }

file storage unit [COMPUT SCI] The component of a computer system that stores information required for reference. { 'fīl ˌstȯr·ij ˌyü·nət }

file transfer [COMPUT SCI] The movement, under program control, of a file from one storage device to another. { 'fīl ˌtranz·fər }

file transfer access and management [COMPUT SCI] A standard communications protocol for transferring files between systems of different vendors. Abbreviated FTAM. { ¦fīl ¦tranz·fər ¦ak ˌses ən 'man·ij·mənt }

file transfer protocol [COMPUT SCI] A set of standards that allows the user of any computer on the Internet to receive files from another computer, or to transmit files to another computer, after the user has specified a name and password for the other computer. Abbreviated FTP. { 'fīl ¦tranz·fər ˌprōd·ə·kȯl }

file transfer utility [COMPUT SCI] A computer program specifically designed to handle file transfers. { 'fīl ¦tranz·fər yüˌtil·əd·ē }

file virus [COMPUT SCI] A computer virus that infects application files such as spreadsheets, computer games, or accounting software. { 'fīl ˌvī·rəs }

fill characters [COMPUT SCI] Nondata characters or bits which are used to fill out a field on the left if data are right-justified or on the right if data are left-justified. { 'fil ˌkar·ik·tərz }

filler [COMPUT SCI] Storage space that does not contain significant data but is needed to comply with length requirements or is reserved to fulfill some future need. { 'fil·ər }

filler metal *See* filler. { 'fil·ər ˌmed·əl }

film optical-sensing device [COMPUT SCI] A device capable of digitizing the information stored on a film. { 'film ¦äp·tə·kəl ¦sens·iŋ diˌvīs }

film reader [ELECTR] A device for converting a pattern of transparent or opaque spots on a photographic film into a series of electric pulses. [OPTICS] A device for projecting or displaying microfilm so that an operator can read the data on the film; usually provided with equipment for moving or holding the film. { 'film ˌrēd·ər }

film recorder [ELECTR] A device which places data, usually in the form of transparent and opaque spots or light and dark spots, on photographic film. { 'film ri,kȯrd·ər }

film scanning [ELECTR] The process of converting motion picture film into corresponding electric signals that can be transmitted by a video system. { 'film ,skan·iŋ }

filter [COMPUT SCI] A device or program that separates data or signals in accordance with specified criteria. [CONT SYS] *See* compensator. [ELECTR] Any transmission network used in electrical systems for the selective enhancement of a given class of input signals. Also known as electric filter; electric-wave filter. [ENG ACOUS] A device employed to reject sound in a particular range of frequencies while passing sound in another range of frequencies. Also known as acoustic filter. [OPTICS] An optical element that partially absorbs incident electromagnetic radiation in the visible, ultraviolet, or infrared spectra, consisting of a pane of glass or other partially transparent material, or of films separated by narrow layers; the absorption may be either selective or nonselective with respect to wavelength. Also known as optical filter. { 'fil·tər }

filter discrimination [ELECTR] Difference between the minimum insertion loss at any frequency in a filter attenuation band and the maximum insertion loss at any frequency in the operating range of a filter transmission band. { 'fil·tər di,skrim·ə'nā·shən }

filtered radar data [ELECTR] Radar data from which unwanted returns have been removed by mapping. { ¦fil·tərd 'rā,där ,dad·ə }

filter impedance compensator [ELECTR] Impedance compensator which is connected across the common terminals of electric wave filters when the latter are used in parallel to compensate for the effects of the filters on each other. { ¦fil·tər im'ped·əns ,käm·pən'sād·ər }

filter pass band *See* filter transmission band. { ¦fil·tər 'pas ,band }

filter slot [ELECTROMAG] Choke in the form of a slot designed to suppress unwanted modes in a waveguide. { 'fil·tər ,slät }

filter transmission band [ELECTR] Frequency band of free transmission; that is, frequency band in which, if dissipation is neglected, the attenuation constant is zero. Also known as filter pass band. { ¦fil·tər tranz'mish·ən ,band }

final amplifier [ELECTR] The transmitter stage that feeds the antenna. { ¦fīn·əl 'am·plə,fī·ər }

financial planning system [COMPUT SCI] A decision-support system that allows the financial planner or manager to examine and evaluate many alternatives before making final decisions, and which employs the use of a model, usually a matrix of data elements which is constructed as a series of equations. { fī¦nan·chəl 'plan·iŋ ,sis·təm }

finder [COMMUN] Switch or relay group in telephone switching systems that selects the path which the call is to take through the system; operates under the instruction of the calling station's dial. { 'fīnd·ər }

finder beam [COMPUT SCI] A beam of light projected by a light pen on the spot on the display screen where the light pen photodetector is focused, in order to aid the user in positioning the light pen. { 'fīnd·ər ,bēm }

F-indicator *See* F-display. { 'ef ,in·də,kād·ər }

finding circuit *See* lockout circuit. { 'fīnd·iŋ ,sər·kət }

fine index [COMPUT SCI] The more specific of two indices consulted to gain access to a record. { ¦fīn 'in,deks }

finite difference [MATH] The difference between the values of a function at two discrete points, used to approximate the derivative of the function. { ¦fī,nīt 'dif·rəns }

finite-difference equations [MATH] Equations arising from differential equations by substituting difference quotients for derivatives, and then using these equations to approximate a solution. { ¦fī,nīt ¦dif·rəns i,kwā·zhənz }

finite element method [ENG] A numerical analysis technique for obtaining approximate descriptions of continuous physical systems, used in structural mechanics, electrical field theory, and fluid mechanics; the system is broken into discrete elements interconnected at discrete node points, and the values of various physical quantities for the elements or node points are calculated numerically. { ¦fī,nīt 'el·ə·mənt ,meth·əd }

finite impulse response filter [ELECTR] An electric filter that will settle to a steady state within a finite amount of time after being exposed to a change in input. Abbreviated FIR filter. { ¦fī,nīt ,im,pəls ri'späns ,fil·tər }

finite precision number [COMPUT SCI] A number that can be represented by a finite set of symbols in a given numeration system. { ¦fī,nīt prə¦sizh·ən 'nəm·bər }

finite-state machine [COMPUT SCI] An automaton that has a finite number of distinguishable internal configurations. { 'fī,nīt ,stāt mə,shēn }

fin waveguide [ELECTROMAG] Waveguide containing a thin longitudinal metal fin that serves to increase the wavelength range over which the waveguide will transmit signals efficiently; usually used with circular waveguides. { ¦fin 'wāv ,gīd }

firewall [COMPUT SCI] Hardware and software programs that protect the resources of a private network from users in other networks, controlling all traffic according to a predefined access policy. { 'fī,wȯl }

firewire *See* IEEE 1394. { 'fir,wīr }

FIR filter *See* finite impulse response filter. { 'fər *or* ¦ef¦ī'är ,fil·ter }

firmware [COMPUT SCI] A computer program or instruction, such as a microprogram, used so often that it is stored in a read-only memory instead of being included in software; often used in computers that monitor production processes. { 'fərm,wer }

first detector *See* mixer. { ¦fərst di'tek·tər }

first Fresnel zone [ELECTROMAG] Circular portion of a wavefront transverse to the line between an emitter and a more distant point, where the resultant disturbance is being observed, whose center is the intersection of the front with the direct ray, and whose radius is such that the shortest path from the emitter through the periphery to the receiving point is one-half wavelength longer than the direct ray. { ¦fərst frə'nel ˌzōn }

first-generation [COMPUT SCI] Denoting electronic hardware, logical organization, and software characteristic of a first-generation computer. { ¦fərst jen·ə'rā·shən }

first-generation computer [COMPUT SCI] A computer from the earliest stage of computer development, ending in the early 1960s, characterized by the use of vacuum tubes, the performance of one operation at a time in strictly sequential fashion, and elementary software, usually including a program loader, simple utility routines, and an assembler to assist in program writing. { ¦fərst jen·ə¦rā·shən kəm'pyüd·ər }

first-item list [COMPUT SCI] A series of records that is printed with descriptive information from only the first record of each group. { ¦fərst 'ī·dəm ˌlist }

first-level address [COMPUT SCI] The location of a referenced operand. { ¦fərst ¦lev·əl ə'dres }

first-level controller [CONT SYS] A controller that is associated with one of the subsystems into which a large-scale control system is partitioned by plant decomposition, and acts to satisfy local objectives and constraints. Also known as local controller. { ¦fərst ¦lev·əl kən'trōl·ər }

first-level interrupt handler [COMPUT SCI] A software or hardware routine that is activated by interrupt signals sent by peripheral devices and decides, based on the relative importance of the interrupts, how they should be handled. Abbreviated FLIH. { 'fərst ¦lev·əl 'int·əˌrəpt ˌhand·lər }

first-level packaging [ELECTR] Electronic packaging which provides interconnection directly to the integrated circuit chip. { ˌfərst ˌlev·əl 'pak·ij·iŋ }

first-order subroutine [COMPUT SCI] A subroutine which is entered directly from a main routine or program and which leads back to that program. Also known as first-remove subroutine. { ¦fərst ˌȯrd·ər 'səb·rüˌtēn }

first-remove subroutine *See* first-order subroutine. { 'fərst 'rəˌmüv 'səb·rüˌtēn }

fish-bone antenna [ELECTROMAG] **1.** Antenna consisting of a series of coplanar elements arranged in collinear pairs, loosely coupled to a balanced transmission line. **2.** Directional antenna in the form of a plane array of doublets arranged transversely along both sides of a transmission line. { 'fish ˌbōn anˌten·ə }

fishpole antenna *See* whip antenna. { 'fishˌpōl anˌten·ə }

five-level code [COMPUT SCI] A code which uses five bits to specify each character. { 'fīv ˌlev·əl 'kōd }

five-level start-stop operation [COMMUN] Simplex mode of operation used in teletypewriter circuits; each code character is divided into five electrical units; the machine distributor unit makes a positive start and stop for the transmission of each character. { 'fīv ˌlev·əl ¦stärt ¦stäp ˌäp·əˌrā·shən }

fix [COMPUT SCI] A piece of coding that is inserted in a computer program to correct an error. { fiks }

fixed area [COMPUT SCI] That portion of the main storage occupied by the resident portion of the control program. { ¦fikst 'er·ē·ə }

fixed attenuator *See* pad. { ¦fikst ə'ten·yə ˌwād·ər }

fixed-block [COMPUT SCI] Pertaining to an arrangement of data in which all the blocks of data have the same number of words or characters, as determined by either the hardware requirements of the computer or the programmer. { ¦fikst 'bläk }

fixed-cycle operation [COMPUT SCI] An operation completed in a specified number of regularly timed execution cycles. { ¦fikst 'sī·kəl ˌäp·ə'rā·shən }

fixed disk [COMPUT SCI] A disk drive that permanently holds the disk platters. { ¦fikst 'disk }

fixed echo [ELECTR] A persistent echo indication that remains stationary on the radar display, indicating the presence of a fixed target. Also known as permanent echo. { ¦fikst 'ek·ō }

fixed-field method [COMPUT SCI] A method of data storage in which the same type of data is always placed in the same relative position. { 'fikst ˌfēld 'meth·əd }

fixed form coding [COMPUT SCI] Any method of coding a source language in which each part of the instruction appears in a fixed field. { 'fikst ˌfȯrm 'kōd·iŋ }

fixed-head disk [COMPUT SCI] A disk storage device in which the read-write heads are fixed in position, one to a track, and the arms to which they are attached are immovable. { 'fikst ˌhed 'disk }

fixed-length field [COMPUT SCI] A field that always has the same number of characters, regardless of its content. { ¦fikst ˌleŋkth 'fēld }

fixed-length operation [COMPUT SCI] A computer operation whose operands always have the same number of bits or characters. { ¦fikst ˌleŋkth ˌäp·ə'rā·shən }

fixed-length record [COMPUT SCI] One of a file of records, each of which must have the same specified number of data units, such as blocks, words, characters, or digits. { ¦fikst ˌleŋkth 'rek·ərd }

fixed logic [COMPUT SCI] Circuit logic of computers or peripheral devices that cannot be changed by external controls; connections must be physically broken to arrange the logic. { ¦fikst 'läj·ik }

fixed medium [COMPUT SCI] A data storage device in which the reading and writing of data do not involve mechanical motion. { 'fikst 'mē·dē·əm }

fixed memory [COMPUT SCI] Of a computer, a nondestructive readout memory that is only mechanically alterable. { ¦fikst 'mem·rē }

fixed-point arithmetic [COMPUT SCI] **1.** A method of calculation in which the computer does not consider the location of the decimal or radix point because the point is given a fixed position. **2.** A type of arithmetic in which the operands and results of all arithmetic operations must be properly scaled so as to have a magnitude between certain fixed values. { ¦fikst ˌpȯint əˈrith·mə·tik }

fixed-point calculation [COMPUT SCI] A calculation made with fixed-point arithmetic. { ¦fikst ˌpȯint ˌkal·kyəˈlā·shən }

fixed-point computer [COMPUT SCI] A computer in which numbers in all registers and storage locations must have an arithmetic point which remains in the same fixed location. { ¦fikst ˌpȯint kəmˈpyüd·ər }

fixed-point part *See* mantissa. { ¦fikst ˌpȯint ˈpärt }

fixed-point representation [COMPUT SCI] Any method of representing a number in which a fixed-point convention is used. { ¦fikst ˌpȯint ˌrep·rə·zenˈtā·shən }

fixed-point system [COMPUT SCI] A number system in which the location of the point is fixed with respect to one end of the numerals, according to some convention. { ¦fikst ˌpȯint ˈsis·təm }

fixed-position addressing [COMPUT SCI] Direct access to an item in a data file on disk or drum, as opposed to a sequential search for this item starting with the first item in the file. { ¦fikst pə ˌzish·ən əˈdres·iŋ }

fixed-product area [COMPUT SCI] The area in core memory where multiplication takes place for certain types of computers. { ¦fikst ˌpräd·əkt ˈer·ē·ə }

fixed-program computer [COMPUT SCI] A special-purpose computer having a program permanently wired in. { ¦fikst ˌprō·grəm kəm ˈpyüd·ər }

fixed-satellite service [COMMUN] A radiocommunication service between earth stations at given positions that uses one or more satellites. Abbreviated FSS. { ¦fikst ˈsad·əlˌīt ˌsər·vis }

fixed-sequence robot *See* fixed-stop robot. { ˈfikst ¦sē·kwəns ˈrōˌbät }

fixed service [COMMUN] Service providing radio communications between fixed points. { ¦fikst ˈsər·vəs }

fixed-stop robot [CONT SYS] A robot in which the motion along each axis has a fixed limit, but the motion between these limits is not controlled and the robot cannot stop except at these limits. Also known as fixed-sequence robot; limited-sequence robot; nonservo robot. { ˈfikst ¦stäp ˈrōˌbät }

fixed storage [COMPUT SCI] A storage for data not alterable by computer instructions, such as magnetic-core storage with a lockout feature. { ¦fikst ˈstȯr·ij }

fixed transmitter [ELECTR] Transmitter that is operated in a fixed or permanent location. { ¦fikst ˈtranzˈmid·ər }

fixed word length [COMPUT SCI] The length of a computer machine word that always contains the same number of characters or digits. { ¦fikst ˈwərd ˌleŋkth }

flag [COMPUT SCI] Any of various types of indicators used for identification, such as a work mark, or a character that signals the occurrence of some condition, such as the end of a word. { flag }

flag flip-flop [COMPUT SCI] A one-bit register which indicates overflow, carry, or sign bit from past or current operations. { ˈflag ˈflip ˌfläp }

flag operand [COMPUT SCI] A part of the instruction of some assembly languages denoting which elements of the object instruction will be flagged. { ˈflag ˈäp·əˌrand }

flange isolator *See* short waveguide isolator. { ¦flanj ˈī·səˌlād·ər }

flap attenuator [ELECTROMAG] A waveguide attenuator in which a contoured sheet of dissipative material is moved into the guide through a nonradiating slot to provide a desired amount of power absorption. Also known as vane attenuator. { ¦flap əˈten·yəˌwād·ər }

flare *See* horn antenna. { fler }

flash memory [COMPUT SCI] A type of electrically erasable programmable read-only memory (EEPROM). While EPROM is reprogrammed bit-by-bit, flash memory is reprogrammed in blocks, making it faster. It is nonvolatile. { ¦flash ˈmem·rē }

flash message [COMMUN] A category of precedence reserved for initial enemy contact messages or operational combat messages of extreme urgency; brevity is mandatory. { ¦flash ˈmes·ij }

flat fading [COMMUN] Type of fading in which all components of the received radio signal fluctuate in the same proportion simultaneously. { ¦flat ˈfād·iŋ }

flat file [COMPUT SCI] A two-dimensional array. { ˈflat ˌfīl }

flat line [ELECTROMAG] A radio-frequency transmission line, or part thereof, having essentially 1-to-1 standing wave ratio. { ˈflat ˌlīn }

flat-panel display *See* panel display. { ˈflat ¦pan·əl diˈsplā }

flat-top antenna [ELECTROMAG] An antenna having two or more lengths of wire parallel to each other and in a plane parallel to the ground, each fed at or near its midpoint. { ˈflat ˌtäp an ˌten·ə }

flat-top response *See* band-pass response. { ˈflat ¦täp riˈspäns }

flat tuning [ELECTR] Tuning of a radio receiver in which a change in frequency of the received waves produces only a small change in the current in the tuning apparatus. { ¦flat ˈtün·iŋ }

F layer [GEOPHYS] An ionized layer in the F region of the ionosphere which consists of the F_1 and F_2 layers in the day hemisphere, and the F_2 layer alone in the night hemisphere; it is capable of reflecting radio waves to earth at frequencies up to about 50 megahertz. { ˈef ˌlā·ər }

F_1 layer [GEOPHYS] The ionosphere layer beneath the F_2 layer during the day, at a virtual height of 120–180 miles (200–300 kilometers), being closest to earth around noon; characterized by a distinct maximum of free-electron density, except at high latitudes during winter, when the layer is not detectable. { ¦ef 'wən ˌlā·ər }

F_2 layer [GEOPHYS] The highest constantly observable ionosphere layer, characterized by a distinct maximum of free-electron density at a virtual height from about 135 miles (225 kilometers) in the polar winter to more than 240 miles (400 kilometers) in daytime near the magnetic equator. Also known as Appleton layer. { ¦ef 'tü ˌlā·ər }

fleet broadcast [COMMUN] The radio broadcast (in addition to the general broadcast) to all U.S. Navy ships and merchant ships in which storm warnings are given. { 'flēt ˌbrȯdˌkast }

flexible circuit [ELECTR] A printed circuit made on a flexible plastic sheet that is usually die-cut to fit between large components. { ˌflek·sə·bəl 'sər·kət }

flexible waveguide [ELECTROMAG] A waveguide that can be bent or twisted without appreciably changing its electrical properties. { ˌflek·sə·bəl 'wāvˌgīd }

flexional symbols [COMPUT SCI] Symbols in which the meaning of each component digit is dependent on those which precede it. { ¦flek·shən·əl ¦sim·bəlz }

flexowriter [COMPUT SCI] A typewriterlike device to read in manually or to read out information of a computer to which it is connected; it can also be used to punch paper tape. { 'flek·səˌwrīd·ər }

flight-path computer [COMPUT SCI] A computer that includes all of the functions of a course-line computer and also provides means for controlling the altitude of an aircraft in accordance with a desired plan of flight. { 'flīt ˌpath kəm'pyüd·ər }

FLIH *See* first-level interrupt handler.

flip chip [ELECTR] A tiny semiconductor die having terminations all on one side in the form of solder pads or bump contacts; after the surface of the chip has been passivated or otherwise treated, it is flipped over for attaching to a matching substrate. Also known as solder-ball flip chip. { 'flip ˌchip }

flip-flop circuit *See* bistable multivibrator. { 'flipˌfläp ˌsər·kət }

floating address [COMPUT SCI] The symbolic address used prior to its conversion to a machine address. { ¦flōd·iŋ ə'dres }

floating carrier modulation *See* controlled carrier modulation. { ¦flōd·iŋ ˌkar·ē·ər ˌmäj·ə'lā·shən }

floating dollar sign [COMPUT SCI] A dollar sign used with an edit mask, allowing the sign to be inserted before the nonzero leading digit of a dollar amount. { ¦flōd·iŋ 'däl·ər ˌsīn }

floating-gate avalanche-injection metal-oxide semiconductor device [ELECTR] An erasable programmable read-only memory chip that holds its contents until they are erased by ultraviolet light. Abbreviated FAMOS device. { ¦flōd·iŋ ¦gāt ¦av·əˌlanch in¦jek·shən ¦med·əl ¦äkˌsīd ˌsem·i·kən'dək·tər diˌvīs }

floating graphic [COMPUT SCI] A picture or graph that moves up or down on a page of a document as text is deleted or inserted above it. { ˌflōd·iŋ 'graf·ik }

floating-point calculation [COMPUT SCI] A calculation made with floating-point arithmetic. { ¦flōd·iŋ ¦pȯint ˌkal·kyə'lā·shən }

floating-point coefficient *See* mantissa. { ¦flōd·iŋ ¦pȯint ˌkō·i'fish·ənt }

floating-point package [COMPUT SCI] A program which enables a computer to perform arithmetic operations when such capabilities are not wired into the computer. Also known as floating-point routine. { ¦flōd·iŋ ¦pȯint 'pak·ij }

floating-point processor [COMPUT SCI] A separate processor or a special section of a computer's main storage that is for the efficient handling of floating-point operations. { 'flōd·iŋ ¦pȯint 'präs ˌes·ər }

floating-point routine *See* floating-point package. { ¦flōd·iŋ ¦pȯint rü'tēn }

floating-point system [COMPUT SCI] A number system in which the location of the point does not remain fixed with respect to one end of the numerals. { ¦flōd·iŋ ¦pȯint 'sis·təm }

floppy disk [COMPUT SCI] A flexible plastic disk coated with magnetic oxide and used for data storage and data entry to a computer; a slot in its protective envelope or housing, which remains stationary while the disk rotates, exposes the track positions for the magnetic read/write head of the drive unit. Also known as diskette. { ¦fläp·ē 'disk }

flops [COMPUT SCI] A unit of computer speed, equal to one floating-point arithmetic operation per second. { fläps }

flow [COMPUT SCI] The sequence in which events take place or operations are carried out. [ENG] A forward movement in a continuous stream or sequence of fluids or discrete objects or materials, as in a continuous chemical process or solids-conveying or production-line operations. { flō }

flow chart [ENG] A graphical representation of the progress of a system for the definition, analysis, or solution of a data-processing or manufacturing problem in which symbols are used to represent operations, data or material flow, and equipment, and lines and arrows represent interrelationships among the components. Also known as control diagram; flow diagram; flow sheet. { 'flō ˌchärt }

flow-chart symbol [ENG] Any of the existing symbols normally used to represent operations, data or materials flow, or equipment in a data-processing problem or manufacturing-process description. { 'flō ˌchärt ˌsim·bəl }

flow diagram *See* flow chart. { 'flō ˌdī·əˌgram }

flow graph [COMPUT SCI] A directed graph that represents a computer program, wherein a node in the graph corresponds to a block of sequential code and branches correspond to decisions taken in the program. [SYS ENG] *See* signal-flow graph. { 'flō ˌgraf }

flow sheet *See* flow chart. { 'flō ˌshēt }

fluctuation [ELECTROMAG] The change in amplitude of a radar echo due to a target of some complexity changing its attitude or structural features. Fluctuations can be rapid (pulse-to-pulse) or somewhat slow (scan-to-scan). { ˌflək·chə ˈwā·shən }

fluid computer [COMPUT SCI] A digital computer constructed entirely from air-powered fluid logic elements; it contains no moving parts and no electronic circuits; all logic functions are carried out by interaction between jets of air. { ¦flü·əd kəm¦pyüd·ər }

fluidics [ENG] A control technology that employs fluid dynamic phenomena to perform sensing, control, information processing, and actuation functions without the use of moving mechanical parts. { flüˈid·iks }

fluorescent screen [ENG] A sheet of material coated with a fluorescent substance so as to emit visible light when struck by ionizing radiation such as x-rays or electron beams. { flu̇¦res·ənt ˈskrēn }

flush left See left-justify. { ¦fləsh ˈleft }

flush right See right-justify. { ¦fləsh ˈrīt }

flutter [ELECTROMAG] A fast-changing variation in received signal strength, such as may be caused by antenna movements in a high wind or interaction with a signal or another frequency. { ˈfləd·ər }

flutter echo [ELECTROMAG] A radar echo consisting of a rapid succession of reflected pulses resulting from a single transmitted pulse. { ˈfləd·ər ˌek·ō }

flyback [ELECTR] The time interval in which the electron beam of a cathode-ray tube returns to its starting point after scanning one line or one field of a video. Also known as retrace; return trace. { ˈflīˌbak }

flyback power supply [ELECTR] A high-voltage power supply used to produce the direct-current voltage of about 10,000–25,000 volts required for the second anode of a cathode-ray tube in a video display. { ˈflīˌbak ˈpau̇r səˌplī }

flyback transformer See horizontal output transformer. { ˈflīˌbak tranzˌfȯr·mər }

flying-aperture scanner [ELECTR] An optical scanner, used in character recognition, in which a document is flooded with light, and light is collected sequentially spot by spot from the illuminated image. { ¦flī·iŋ ¦ap·ər·chər ˌskan·ər }

flying head [ELECTR] A read/write head used on magnetic disks and drums, so designed that it flies a microscopic distance off the moving magnetic surface and is supported by a film of air. { ¦flī·iŋ ˈhed }

flying spot [ELECTR] A small point of light, controlled mechanically or electrically, which moves rapidly in a rectangular scanning pattern in a flying-spot scanner. { ¦flī·iŋ ˈspät }

flying-spot scanner [ELECTR] A scanner used for video film and slide transmission, electronic writing, and character recognition, in which a moving spot of light, controlled mechanically or electrically, scans the image field, and the light reflected from or transmitted by the image field is picked up by a device that generates a corresponding electric signal output. Also known as optical scanner. { ¦flī·iŋ ˌspät ˈskan·ər }

flywheel synchronization [ELECTR] Automatic frequency control of a scanning system by using the average timing of the incoming sync signals, rather than by making each pulse trigger the scanning circuit; used in analog television receivers designed for fringe-area reception, when noise pulses might otherwise trigger the sweep circuit prematurely. { ˈflīˌwēl ˌsiŋ·krə·nəˈzā·shən }

FM See frequency modulation.

FM/AM multiplier [ELECTR] Multiplier in which the frequency deviation from the central frequency of a carrier is proportional to one variable, and its amplitude is proportional to the other variable; the frequency-amplitude-modulated carrier is then consecutively demodulated for frequency modulation (FM) and for amplitude modulation (AM); the final output is proportional to the product of the two variables. { ˈefˌem ˈāˌem ˈməl·təˌplī·ər }

focus [ELECTR] To control convergence or divergence of the electron paths within one or more beams, usually by adjusting a voltage or current in a circuit that controls the electric or magnetic fields through which the beams pass, in order to obtain a desired image or a desired current density within the beam. { ˈfō·kəs }

focus control [ELECTR] A control that adjusts spot size at the screen of a cathode-ray tube to give the sharpest possible image; it may vary the current through a focusing coil or change the position of a permanent magnet. { ˈfō·kəs kən ˌtrōl }

focusing coil [ELECTR] A coil that produces a magnetic field parallel to an electron beam for the purpose of focusing the beam. { ˈfō·kəs·iŋ ˌkȯil }

focus projection and scanning [ELECTR] Method of magnetic focusing and electrostatic deflection of the electron beam of a hybrid vidicon; a transverse electrostatic field is used for beam deflection; this field is immersed with an axial magnetic field that focuses the electron beam. { ¦fō·kəs prəˌjek·shən ən ˈskan·iŋ }

folded cavity [ELECTR] Arrangement used in a klystron repeater to make the incoming wave act on the electron stream from the cathode at several places and produce a cumulative effect. { ¦fōld·əd ˈkav·əd·ē }

folded dipole See folded-dipole antenna. { ¦fōld·əd ˈdīˌpōl }

folded-dipole antenna [ELECTROMAG] A dipole antenna whose outer ends are folded back and joined together at the center; the impedance is about 300 ohms, as compared to 70 ohms for a single-wire dipole; widely used with television and frequency-modulation receivers. Also known as folded dipole. { ¦fōld·əd ˈdīˌpōl anˈten·ə }

folding [COMPUT SCI] A method of hashing which consists of splitting the original key into two or more parts and then adding the parts together. { ˈfōld·iŋ }

font cartridge [COMPUT SCI] A removable module that can be plugged into a slot in a printer and has one or more fonts stored in a read-only memory chip. { 'fänt ˌkär·trij }

font compiler *See* font generator. { 'fänt kəm ˌpīl·ər }

font generator [COMPUT SCI] A computer program that converts an outline font into the patterns of dots required for a particular size of font. Also known as font compiler. { 'fänt ˌjen·ə ˌrād·ər }

footprint [COMMUN] The area of the earth's surface that can be covered by a communications satellite at any given time. [COMPUT SCI] The amount and shape of the area occupied by equipment, such as a terminal or microcomputer, on desktop, floor, or other surface area. { 'fu̇t ˌprint }

forbidden-character code [COMPUT SCI] A bit code which exists only when an error occurs in the binary coding of characters. { fər¦bid·ən 'kar·ik·tər ˌkōd }

forbidden-combination check [COMPUT SCI] A test for the occurrence of a nonpermissible code expression in a computer; used to detect computer errors. { fər¦bid·ən ˌkäm·bə'nā·shən ˌchek }

force [COMPUT SCI] To intervene manually in a computer routine and cause the computer to execute a jump instruction. { fȯrs }

force-controlled motion commands [CONT SYS] Robot control in which motion information is provided by computer software but sensing of forces or feedback is used by the robot to adapt this information to the environment. { 'fȯrs kən ¦trōld 'mō·shən kəˌmanz }

forced programming *See* minimum-access programming. { ¦fȯrst 'prōˌgram·iŋ }

force feedback [CONT SYS] A method of error detection in which the force exerted on the effector is sensed and fed back to the control, usually by mechanical, hydraulic, or electric transducers. { ¦fȯrs ¦fēdˌbak }

forecasting [COMMUN] The prediction of conditions of radio propagation for a period extending anywhere from a few hours to a few months. { 'fȯrˌkast·iŋ }

foreground [COMPUT SCI] A program or process of high priority that utilizes machine facilities as needed, with less critical, background work performed in otherwise unused time. { 'fȯr ˌgrau̇nd }

fork-join model [COMPUT SCI] A method of programming on parallel machines in which one or more child processes branch out from the root task when it is time to do work in parallel, and end when the parallel work is done. { 'fȯrk ˌjȯin 'mäd·əl }

formal language [COMPUT SCI] An abstract mathematical object used to model the syntax of a programming or natural language. { ¦fȯr·məl 'laŋ·gwij }

format [COMPUT SCI] **1.** The specific arrangement of data on a printed page, display screen, or such, or in a record, data file, or storage device. **2.** To prepare a disk to store information by using a special program that divides the disk into storage units such as tracks and sectors. { 'fȯr ˌmat }

format effector *See* layout character. { 'fȯrˌmat iˌfek·tər }

formatted tape [COMPUT SCI] A magnetic tape which employs a prerecorded timing track by means of which blocks of data can be found after reference to a directory table. { fȯr¦mad·əd 'tāp }

formatting [COMPUT SCI] The preparation of a magnetic storage device to receive data structures; for example, the recording of track and sector information on a floppy disk. { 'fȯr ˌmad·iŋ }

form feed character [COMPUT SCI] A control character that determines when a printer or display device moves to the next page, form, or equivalent unit of data. { 'fȯrm ¦fēd ˌkar·ik·tər }

form feeding [COMPUT SCI] The positioning of documents in order to move them past printing or sensing devices, either singly or in continuous rolls. { 'fȯrm ˌfēd·iŋ }

form feed printer [COMPUT SCI] A computer printer that accepts continuous forms or continuous sheets of paper. { 'fȯrm ¦fēd ˌprint·ər }

forms [COMPUT SCI] Web pages that allow users to fill in and submit information, they are written in HTML and processed by CGI scripts. { fȯrmz }

forms control buffer [COMPUT SCI] A reserved storage containing coordinates for a page position on the printer; earlier printers utilized a carriage control tape, allowing the page to be set at a specific position. { 'fȯrmz kənˌtrōl 'bəf·ər }

form stop [COMPUT SCI] A device which stops a machine when its supply of paper has run out. { 'fȯrm ˌstäp }

formula translation *See* FORTRAN. { ¦fȯr·myə·lə tranz¦lā·shən }

for-next loop [COMPUT SCI] In computer programming, a high-level logic statement which defines a part of a computer program that will be repeated a certain number of times. { ¦fȯr ¦nekst ˌlüp }

FOR statement [COMPUT SCI] A statement in a computer program that is repeatedly executed a specified number of times, generally while a control variable takes on successive values over a specified range. { 'fȯr ˌstāt·mənt }

Forth [COMPUT SCI] A high-level programming language developed primarily for microcomputers and characterized by a number of features that make it highly adaptable and readily extensible, such as the ability to be used as an interpreter or an operating system. { fȯrth }

FORTRAN [COMPUT SCI] A family of procedure-oriented languages used mostly for scientific or algebraic applications; derived from formula translation. { 'fȯrˌtran }

forty-four-type repeater [ELECTR] Type of telephone repeater employing two amplifiers and no hybrid arrangements; used in a four-wire system. { ¦fȯrd·ēˌfȯr ¦tīp ri'pēd·ər }

forum *See* newsgroup. { 'fȯr·əm }

forward-acting regulator [ELECTR] Transmission regulator in which the adjustment made by the regulator does not affect the quantity which caused the adjustment. { 'fȯr·wərd ˌak·tiŋ 'reg·yəˌlād·ər }

forward-backward counter [COMPUT SCI] A counter that has both an add and a subtract input so as to count in either an increasing or a decreasing direction. Also known as bidirectional counter. { ¦fȯr·wərd ¦bak·wərd 'kau̇nt·ər }

forward chaining [COMPUT SCI] In artificial intelligence, a method of reasoning which begins with a statement of all the relevant data and works toward the solution using the system's rules of inference. { ¦fȯr·wərd 'chān·iŋ }

forward compatibility *See* upward compatibility. { ¦fȯr·wərd kəmˌpad·ə'bil·əd·ē }

forward error analysis [COMPUT SCI] A method of error analysis based on the assumption that small changes in the input data lead to small changes in the results, so that bounds for the errors in the results caused by rounding or truncation errors in the input can be calculated. { ¦fȯr·wərd 'er·ər əˌnal·ə·səs }

forward error correction [COMMUN] The location and correction of errors occurring in data communications by the receiver without retransmission of data. { 'fȯr·wərd 'er·ər kəˌrek·shən }

forward path [CONT SYS] The transmission path from the loop actuating signal to the loop output signal in a feedback control loop. { 'fȯr·wərd ˌpath }

forward propagation by ionospheric scatter [COMMUN] Radio communications technique using the scattering phenomenon exhibited by electromagnetic waves in the 30–100-megahertz region when passing through the ionosphere at an elevation of about 50 miles (85 kilometers). { ¦fȯr·wərd ˌpräp·əˌgā·shən bī ī'än·əˌsfir·ik ˌskad·ər }

forward propagation by tropospheric scatter [COMMUN] Radio communications technique using high transmitting power levels, large antenna arrays, and the scattering phenomenon of the troposphere to permit communications far beyond line-of-sight distances. { ¦fȯr·wərd ˌpräp·əˌgā·shən bī 'träp·əˌsfir·ik ˌskad·ər }

forward reference [COMPUT SCI] Reference to a data element that has not yet been defined in the program being compiled. { 'fȯr·wərd 'ref·rəns }

forward scatter [COMMUN] **1.** Propagation of electromagnetic waves at frequencies above the maximum usable high frequency through use of the scattering of a small portion of the transmitted energy when the signal passes from an unionized medium into a layer of the ionosphere. **2.** Collectively, the very-high-frequency forward propagation by ionospheric scatter and ultra-high-frequency forward propagation by tropospheric scatter communications techniques. { ¦fȯr·wərd 'skad·ər }

forward-scatter propagation *See* scatter propagation. { ¦fȯr·wərd ¦skad·ər präp·ə'gā·shən }

forward transfer function [CONT SYS] In a feedback control loop, the transfer function of the forward path. { ¦fȯr·wərd 'tranz·fər ˌfəŋk·shən }

forward wave [ELECTR] Wave whose group velocity is the same direction as the electron stream motion. { ¦fȯr·wərd ¦wāv }

Foster-Seely discriminator *See* phase-shift discriminator. { ¦fȯs·tər ¦sē·lē di'skrim·əˌnād·ər }

four-address [COMPUT SCI] Pertaining to an instruction address which contains four address parts. { 'fȯr əˌdres }

four-frequency diplex telegraphy [COMMUN] Frequency-shift telegraphy in which each of the four possible signal combinations corresponding to two telegraph channels is represented by a separate frequency. { ¦fȯr ¦frē·kwən·sē ¦dīˌpleks tə'leg·rə·fē }

Fourier analyzer [ENG] A digital spectrum analyzer that provides push-button or other switch selection of averaging, coherence function, correlation, power spectrum, and other mathematical operations involved in calculating Fourier transforms of time-varying signal voltages for such applications as identification of underwater sounds, vibration analysis, oil prospecting, and brain-wave analysis. { ˌfu̇r·ēˌā 'an·əˌlīz·ər }

four-phase modulation [COMMUN] Modulation in which data are encoded on a carrier frequency as a succession of phase shifts that will be 45, 135, 225, or 315°; each phase shift contains 2 bits of information called dibits, as follows: 225° represents 00, 315° is 01, 45° is 11, and 135° is 10. { ¦fȯr ¦fāz ˌmäj·ə'lā·shən }

four-plus-one address [COMPUT SCI] An instruction that contains four operand addresses and a control address. { ¦fȯr ˌpləs ¦wən ə'dres }

four-quadrant multiplier [COMPUT SCI] A multiplier in an analog computer in which both the reference signal and the number represented by the input may be bipolar, and the multiplication rules for algebraic sign are obeyed. Also known as quarter-square multiplier. { ¦fȯr ¦kwäd·rənt 'məl·təˌplī·ər }

four-tape [COMPUT SCI] To sort input data, supplied on two tapes, into incomplete sequences alternately on two output tapes; the output tapes are used for input on the succeeding pass, resulting in longer and longer sequences after each pass, until the data are all in one sequence on one output tape. { ¦fȯr ¦tāp }

fourth-generation computer [COMPUT SCI] A type of general-purpose digital computer used in the 1970s and 1980s that is characterized by increasingly advanced very large-scale integrated circuits and increasing use of a hierarchy of memory devices. { 'fȯrth ˌjen·ə¦rā·shən kəm'pyüd·ər }

fourth-generation language [COMPUT SCI] A higher-level programming language that automates many of the basic functions that must be spelled out in conventional languages, and can obtain results with an order-of-magnitude less coding because of its richer content of commands. { 'fȯrth ˌjen·ə¦rā·shən 'laŋ·gwij }

four-wire circuit [COMMUN] A two-way circuit using two paths so arranged that communication currents are transmitted in one direction only on one path, and in the opposite direction on the other path; the transmission path may or may not employ four wires. { 'fór ,wīr 'sər·kət }

four-wire repeater [ELECTR] Telephone repeater for use in a four-wire circuit and in which there are two amplifiers, one serving to amplify the telephone currents in one side of the four-wire circuit, and the other serving to amplify the telephone currents in the other side of the four-wire circuit. { 'fór ,wīr ri'pēd·ər }

four-wire subscriber line [COMMUN] Four-wire circuit connecting a subscriber directly to a switching center. { 'fór ,wīr səb'skrīb·ər ,līn }

four-wire terminating set [ELECTR] Hybrid arrangement by which four-wire circuits are terminated on a two-wire basis for interconnection with two-wire circuits. { 'fór ,wīr 'ter·mə,nād·iŋ ,set }

fox [COMPUT SCI] A name for the hexadecimal digit whose decimal equivalent is 15. { fäks }

Fox broadcast [COMMUN] Radio broadcast of messages for which receiving stations make no acknowledgment. { 'fäks ,bród,kast }

FPLA *See* field-programmable logic array.

fragmentation [COMPUT SCI] The tendency of files in disk storage to be divided up into many small areas scattered around the disk. { ,frag·mən'tā·shən }

fragmenting [COMPUT SCI] The breaking up of a document into its various components. { 'frag ,ment·iŋ }

frame [COMMUN] **1.** One cycle of a regularly recurring series of pulses. **2.** An elementary block of data for transmission over a network or communications system. [COMPUT SCI] **1.** Subdivision of a browser window, with each section containing a separate Web page. **2.** *See* main frame. [ELECTR] One complete representation of a video image. { frām }

frame buffer [COMPUT SCI] A device that stores a television picture or frame for processing. { 'frām ,bəf·ər }

frame frequency [ELECTR] The number of times per second that the frame is completely scanned in a video system. Also known as picture frequency. { 'frām ,frē·kwən·sē }

frame grabber [COMPUT SCI] An external device that digitizes standard television video images for storage or processing in a computer. { 'frām ,grab·ər }

frame period [ELECTR] A time interval equal to the reciprocal of the frame frequency. { 'frām ,pir·ē·əd }

framer [ELECTR] Device for adjusting facsimile equipment so the start and end of a recorded line are the same as on the corresponding line of the subject copy. { 'frām·ər }

framing [ELECTR] Adjusting a facsimile picture to a desired position in the direction of line progression. Also known as phasing. { 'frām·iŋ }

framing control [ELECTR] **1.** A control that adjusts the centering, width, or height of the image on a video display device. **2.** A control that shifts a received facsimile picture horizontally. { 'frām·iŋ kən,trō }

Fraunhofer region [ELECTROMAG] The region far from an antenna compared to the dimensions of the antenna and the wavelength of the radiation. Also known as far field; far region; far zone; radiation zone. { 'fraún,hōf·ər ,rē·jən }

free field [COMPUT SCI] A property of information retrieval devices which permits recording of information in the search medium without regard to preassigned fixed fields. { 'frē ,fēld }

free-field storage [COMPUT SCI] Data storage that allows recording of the data without regard for fixed or preassigned fields. { 'frē ,fēld ,stór·ij }

freeform language [COMPUT SCI] A programming or command language that does not require rigid formatting. { ¦frē,fórm 'laŋ·gwij }

freeform text [COMPUT SCI] A record, or a variable-length portion of a record, that stores plain, unformatted English. { 'frē,fórm 'tekst }

freenet [COMPUT SCI] A bulletin board system, based in a public library or other community or government organization, that provides access to useful resources. { 'frē,net }

free-running multivibrator *See* astable multivibrator. { 'frē ,rən·iŋ məl·tə'vī,brād·ər }

free-space field intensity [ELECTROMAG] Radio field intensity that would exist at a point in a uniform medium in the absence of waves reflected from the earth or other objects. { 'frē ,spās 'fēld in,ten·səd·ē }

free-space loss [ELECTROMAG] The theoretical radiation loss, depending only on frequency and distance, that would occur if all variable factors were disregarded when transmitting energy between two antennas. { 'frē ,spās ,lós }

free-space propagation [ELECTROMAG] Propagation of electromagnetic radiation over a straight-line path in a vacuum or ideal atmosphere, sufficiently removed from all objects that affect the wave in any way. { 'frē ,spās ,präp·ə'gā·shən }

free-space radiation pattern [ELECTROMAG] Radiation pattern that an antenna would have if it were in free space where there is nothing to reflect, refract, or absorb the radiated waves. { 'frē ,spās rād·ē'ā·shən ,pad·ərn }

free symbol [COMPUT SCI] A contextual symbol preceded and followed by a space; it is always meaningful and always used to symbolize both grammatical and nongrammatical meaning; an example is the English "I." { ¦frē 'sim·bəl }

free symbol sequence [COMPUT SCI] A symbol sequence not preceded, or not followed, or neither preceded nor followed by space. { ¦frē 'sim·bəl 'sēk·wəns }

freeware [COMPUT SCI] Copyrighted software that is downloaded from the Internet for which there is no charge. { 'frē,wer }

frequency agility [ELECTR] A feature of modern radar permitting rapid changes of the carrier frequency within the band of operating frequencies for which the radar is designed; electronic rather than mechanical tuning permits pulse-to-pulse agility. { 'frē·kwən·sē ə,jil·əd·ē }

frequency allocation [COMMUN] Assignment of available frequencies in the radio spectrum to specific stations and for specific purposes, to give maximum utilization of frequencies with minimum interference between stations. { 'frē·kwən·sē ˌal·ə'kā·shən }

frequency analysis [COMPUT SCI] A determination of the number of times certain parts of an algorithm are executed, indicating which parts of the algorithm consume large quantities of time and hence where efforts should be directed toward improving the algorithm. { 'frē·kwən·sē əˌnal·ə·səs }

frequency analyzer [ELECTR] A device which measures the intensity of many different frequency components in some oscillation, as in a radio band; used to identify transmitting sources. { 'frē·kwən·sē 'an·əˌlīz·ər }

frequency carrier system [COMMUN] A form of frequency division multiplex in which intelligence is carried on subcarriers. { 'frē·kwən·sē ˌkar·ē·ər ˌsis·təm }

frequency characteristic *See* frequency-response curve. { ¦frē·kwən·sē ˌkar·ik·tə'ris·tik }

frequency compensation *See* compensation. { ¦frē·kwən·sē ˌkäm·pən'sā·shən }

frequency conversion [ELECTR] Converting the carrier frequency of a received signal from its original value to the intermediate frequency value in a superheterodyne receiver. { 'frē·kwən·sē kən ˌvər·zhən }

frequency counter [ELECTR] An electronic counter used to measure frequency by counting the number of cycles in an electric signal during a preselected time interval. { 'frē·kwən·sē ˌkaünt·ər }

frequency deviation [COMMUN] The peak difference between the instantaneous frequency of a frequency-modulated wave and the carrier frequency. { ¦frē·kwən·sē ˌdē·vē'ā·shən }

frequency discriminator [ELECTR] A discriminator circuit that delivers an output voltage which is proportional to the deviations of a signal from a predetermined frequency value. { ¦frē·kwən·sē di'skrim·əˌnād·ər }

frequency distortion [ELECTR] Distortion in which the relative magnitudes of the different frequency components of a wave are changed during transmission or amplification. Also known as amplitude distortion; amplitude-frequency distortion; waveform-amplitude distortion. { ¦frē·kwən·sē di'stȯr·shən }

frequency diversity [COMMUN] Diversity reception involving the use of carrier frequencies separated 500 hertz or more and having the same modulation, to take advantage of the fact that fading does not occur simultaneously on different frequencies. { ¦frē·kwən·sē də'vər·səd·ē }

frequency divider [ELECTR] A harmonic conversion transducer in which the frequency of the output signal is an integral submultiple of the input frequency. Also known as counting-down circuit. { 'frē·kwən·sē diˌvīd·ər }

frequency-division data link [COMMUN] Data link using frequency division techniques for channel spacing. { 'frē·kwən·sē diˌvizh·ən 'dad·ə ˌliŋk }

frequency-division multiple access [COMMUN] A technique by which multiple users who are geographically dispersed gain access to a communications channel to which they are assigned distinct and nonoverlapping sections of the electromagnetic spectrum. Abbreviated FDMA. { ¦frē·kwən·sē di¦vizh·ən ˌməl·tə·pəl 'akˌses }

frequency-division multiplexing [COMMUN] A multiplex system for transmitting two or more signals over a common path by using a different frequency band for each signal. Abbreviated fdm; FDM. Also known as frequency multiplexing. { 'frē·kwən·sē diˌvizh·ən 'məl·təˌplek·siŋ }

frequency domain [COMMUN] A plane on which signal strength can be represented graphically as a function of frequency, instead of a function of time. { 'frē·kwən·sē dəˌmān }

frequency-domain optical storage [COMPUT SCI] A technique whereby up to 1000 bits of information would be stored at each spatial location in an optical storage medium by using persistent spectral holeburning. { 'frē·kwən·sē də¦mān 'äp·tə·kəl 'stȯr·ij }

frequency doubler [ELECTR] An amplifier stage whose resonant anode circuit is tuned to the second harmonic of the input frequency; the output frequency is then twice the input frequency. Also known as doubler. { 'frē·kwən·sē ˌdəb·lər }

frequency drift [ELECTR] A gradual change in the frequency of an oscillator or transmitter due to temperature or other changes in the circuit components that determine frequency. { 'frē·kwən·sē ˌdrift }

frequency frogging [COMMUN] Interchanging of frequency allocations for carrier channels to prevent singing, reduce crosstalk, and reduce the need for equalization; modulators in each repeater translate a low-frequency group to a high-frequency group, and vice versa. { 'frē·kwən·sē ˌfräg·iŋ }

frequency hopping [COMMUN] A spread-spectrum technique in which the frequency of the carrier changes pseudorandomly according to a pseudonoise code, with a consecutive group of code symbols defining a particular frequency. { 'frē·kwən·sē ˌhäp·iŋ }

frequency interlace [COMMUN] Carrier chrominance signal frequency chosen so I and J sidebands are interwoven with luminance sidebands in the same bandwidth and in a manner that causes no mutual interference. { ¦frē·kwən·sē 'in·tərˌlās }

frequency locus [CONT SYS] The path followed by the frequency transfer function or its inverse, either in the complex plane or on a graph of amplitude against phase angle; used in determining zeros of the describing function. { 'frē·kwən·sē ˌlō·kəs }

frequency-modulated carrier current telephony [COMMUN] Telephony involving the use of a frequency-modulated carrier signal transmitted over power-line wires or other wires. { 'frē·kwən·sē ,mäj·ə,lād·əd 'kar·ē·ər 'kə·rənt tə 'lef·ə·nē }

frequency-modulated jamming [ELECTR] Jamming technique consisting of a constant amplitude radio-frequency signal that is varied in frequency about a center frequency to produce a signal over a band of frequencies. {'frē·kwən·sē ,mäj·ə,lād·əd 'jam·iŋ }

frequency-modulated radar [ENG] Form of radar in which the radiated wave is frequency modulated, and the returning echo beats with the wave being radiated, thus enabling range to be measured. { 'frē·kwən·sē ,mäj·ə,lād·əd 'rā ,där }

frequency modulation [COMMUN] Modulation in which the instantaneous frequency of the modulated wave differs from the carrier frequency by an amount proportional to the instantaneous value of the modulating wave. Abbreviated FM. { 'frē·kwən·sē ,mäj·ə,lā·shən }

frequency-modulation broadcast band [COMMUN] The band of frequencies extending from 88 to 108 megahertz; used for frequency-modulation radio broadcasting in the United States. { 'frē·kwən·sē ,mäj·ə,lā·shən 'bròd,kast ,band }

frequency-modulation detector [ELECTR] A device, such as a Foster-Seely discriminator, for the detection or demodulation of a frequency-modulated wave. { 'frē·kwən·sē ,mäj·ə,lā·shən di'tek·tər }

frequency-modulation Doppler [ENG] Type of radar involving frequency modulation of both carrier and modulation on radial sweep. { 'frē·kwən·sē ,mäj·ə,lā·shən 'däp·lər }

frequency modulation-frequency modulation [COMMUN] System in which frequency-modulated subcarriers are used to frequency-modulate a second carrier. { 'frē·kwən·sē ,mäj·ə,lā·shən 'frē·kwən·sē ,mäj·ə,lā·shən }

frequency-modulation noise level on carrier [COMMUN] Residual frequency modulation resulting from disturbance produced in an aural transmitter operating within the band of 50 to 15,000 hertz. { 'frē·kwən·sē ,mäj·ə,lā·shən 'nòiz ,lev·əl òn 'kar·ē·ər }

frequency modulation-phase modulation [COMMUN] System in which the several frequency-modulated subcarriers are used to phase modulate a second carrier. { 'frē·kwən·sē ,mäj·ə,lā·shən 'fāz ,mäj·ə,lā·shən }

frequency-modulation receiver [ELECTR] A radio receiver that receives frequency-modulated waves and delivers corresponding sound waves. { 'frē·kwən·sē ,mäj·ə,lā·shən ri'sē·vər }

frequency-modulation receiver deviation sensitivity [ELECTR] Least frequency deviation that produces a specified output power. { 'frē·kwən·sē ,mäj·ə,lā·shən ri'sē·vər dē·vē¦ā·shən sen·sə'tiv·əd·ē }

frequency-modulation transmitter [ELECTR] A radio transmitter that transmits a frequency-modulated wave. { 'frē·kwən·sē ,mäj·ə,lā·shən tranz'mid·ər }

frequency-modulation tuner [ELECTR] A tuner containing a radio-frequency amplifier, converter, intermediate-frequency amplifier, and demodulator for frequency-modulated signals, used to feed a low-level audio-frequency signal to a separate audio-frequency amplifier and loudspeaker. { 'frē·kwən·sē ,mäj·ə,lā·shən 'tün·ər }

frequency modulator [ELECTR] A circuit or device for producing frequency modulation. { ¦frē·kwən·sē 'mäj·ə,lād·ər }

frequency monitor [ELECTR] An instrument for indicating the amount of deviation of the carrier frequency of a transmitter from its assigned value. { 'frē·kwən·sē ,män·əd·ər }

frequency multiplexing *See* frequency-division multiplexing. { ¦frē·kwən·sē 'məl·tə,plek·siŋ }

frequency multiplier [ELECTR] A harmonic conversion transducer in which the frequency of the output signal is an exact integral multiple of the input frequency. Also known as multiplier. { ¦frē·kwən·sē 'məl·tə,plī·ər }

frequency offset [COMMUN] A small difference in the carrier frequencies of television stations in adjacent cities operating on the same channel. { ¦frē·kwən·sē 'òf,set }

frequency-offset transponder [ELECTR] Transponder that changes the signal frequency by a fixed amount before retransmission. { ¦frē·kwən·sē ¦òf·set tran'spän·dər }

frequency optimum traffic *See* optimum working frequency. { 'frē·kwən·sē ,äp·tə·məm 'traf·ik }

frequency prediction chart [COMMUN] Graph showing curve for the maximum usable frequency, frequency optimum traffic, and lowest usable frequency between two specific points for various times throughout a 24-hour period. { 'frē·kwən·sē prə,dik·shən ,chärt }

frequency pulling [ELECTR] A change in the frequency of an oscillator due to a change in load impedance. { 'frē·kwən·sē ,pu̇l·iŋ }

frequency response [ENG] A measure of the effectiveness with which a circuit, device, or system transmits the different frequencies applied to it; it is a phasor whose magnitude is the ratio of the magnitude of the output signal to that of a sine-wave input, and whose phase is that of the output with respect to the input. Also known as amplitude-frequency response; sine-wave response. { 'frē·kwən·sē ri,späns }

frequency-response curve [ENG] A graph showing the magnitude or the phase of the freqency response of a device or system as a function of frequency. Also known as frequency characteristic. { 'frē·kwən·sē ri,späns ,kərv }

frequency-response equalization *See* equalization. { 'frē·kwən·sē ri,späns ,ē·kwə·lə'zā·shən }

frequency-response trajectory [CONT SYS] The path followed in the complex plane by the phasor that represents the frequency response as the frequency is varied. { 'frē·kwən·sē ri,späns trə'jek·trē }

frequency scan antenna [ELECTROMAG] A radar antenna similar to a phased array antenna in which one dimensional scanning is accomplished through frequency variation. { 'frē·kwən·sē ,skan an'ten·ə }

frequency scanning [ELECTR] Type of system in which output frequency is made to vary at a mechanical rate over a desired frequency band. { 'frē·kwən·sē ,skan·iŋ }

frequency-selective device *See* electric filter. { 'frē·kwən·sē si,lek·tiv di,vīs }

frequency separation multiplier [ELECTR] Multiplier in which each of the variables is split into a low-frequency part and a high-frequency part that are multiplied separately, and the results added to give the required product; this system makes it possible to get high accuracy and broad bandwidth. { 'frē·kwən·sē ,sep·ə,rā·shən 'məl·tə,plī·ər }

frequency separator [ELECTR] The circuit that separates the horizontal and vertical synchronizing pulses in an analog monochrome or color television receiver. { 'frē·kwən·sē ,sep·ə,rād·ər }

frequency shift [ELECTR] A change in the frequency of a radio transmitter or oscillator. Also known as radio-frequency shift. { 'frē·kwən·sē ,shift }

frequency-shift converter [ELECTR] A device that converts a received frequency-shift signal to an amplitude-modulated signal or a direct-current signal. { 'frē·kwən·sē ,shift kən 'vərd·ər }

frequency-shift keying [COMMUN] A form of frequency modulation used especially in telegraph, data, and facsimile transmission, in which the modulating wave shifts the output frequency between predetermined values corresponding to the frequencies of correlated sources. Abbreviated FSK. Also known as frequency-shift modulation; frequency-shift transmission. { 'frē·kwən·sē ,shift 'kē·iŋ }

frequency-shift modulation *See* frequency-shift keying. { 'frē·kwən·sē ,shift ,mäj·ə'lā·shən }

frequency-shift transmission *See* frequency-shift keying. { 'frē·kwən·sē ,shift tranz'mish·ən }

frequency-slope modulation [COMMUN] Type of modulation in which the carrier signal is swept periodically over the entire width of the band, much as in chirp radar; modulation of the carrier with a voice or other communication signal changes the bandwidth of the system without affecting the uniform distribution of energy over the band. { 'frē·kwən·sē ,slōp ,mäj·ə'lā·shən }

frequency spectrum [SYS ENG] In the analysis of a random function of time, such as the amplitude of noise in a system, the limit as T approaches infinity of $1/(2\pi T)$ times the ensemble average of the squared magnitude of the amplitude of the Fourier transform of the function from $-T$ to T. Also known as power-density spectrum; power spectrum; spectral density. { 'frē·kwən·sē ,spek·trəm }

frequency stability [ELECTR] The ability of an oscillator to maintain a desired frequency; usually expressed as percent deviation from the assigned frequency value. { 'frē·kwən·sē stə ,bil·əd·ē }

frequency stabilization [COMMUN] Process of controlling the center or carrier frequency so that it differs from that of a reference source by not more than a prescribed amount. { 'frē·kwən·sē ,stā·bə·lə'zā·shən }

frequency swing [COMMUN] **1.** Peak difference between the maximum and the minimum values of the instantaneous frequency. **2.** In frequency modulation, a term used to describe the change in frequency resulting from the modulation. { 'frē·kwən·sē ,swiŋ }

frequency telemetering [COMMUN] The transmittal of an alternating-current signal from a primary element by variations in the signal frequency, instead of amplitude. { ¦frē·kwən·sē ¦tel·ə¦mēd·ə·riŋ }

frequency tolerance [ELECTR] Of a radio transmitter, extent to which the carrier frequency of the transmitter may be permitted to depart from the frequency assigned. { 'frē·kwən·sē ,täl·ə·rəns }

frequency transformation [CONT SYS] A transformation used in synthesizing a band-pass network from a low-pass prototype, in which the frequency variable of the transfer function is replaced by a function of the frequency. Also known as low-pass band-pass transformation. { ¦frē·kwən·sē ,tranz·fər'mā·shən }

frequency translation [COMMUN] Moving a modulated radio-frequency carrier signal to a new location in the frequency spectrum. { ¦frē·kwən·sē tranz'lā·shən }

frequency-type telemeter [ELECTR] Telemeter that employs frequency of an alternating current or voltage as the translating means. { 'frē·kwən·sē ,tīp 'tel·ə,mēd·ər }

frequency variation [ELECTR] The change over time of the deviation from assigned frequency of a radio-frequency carrier (or power supply system); usually tightly controlled because of national or industry standards. { ¦frē·kwən·sē ,ver·ē¦ā·shən }

Frequently Asked Questions [COMPUT SCI] Abbreviated FAQ. **1.** A document containing answers to common questions about the subjects of other documents to which it is linked. **2.** In particular, a document associated with a Web site that contains answers to common questions about the site. { ,frē·kwənt·lē ,askt 'kwes·chənz }

Fresnel region [ELECTROMAG] The region between the near field of an antenna (close to the antenna compared to a wavelength) and the Fraunhofer region. { frā'nel ,rē·jən }

Fresnel zones [ELECTROMAG] Circular portions of a wavefront transverse to a line between an emitter and a point where the disturbance is being observed; the nth zone includes all paths whose lengths are between $n - 1$ and n half-wavelengths longer than the line-of-sight path. Also known as half-period zones. { frā'nel ,zōnz }

friction-feed printer [COMPUT SCI] A computer printer in which a roller is used to hold and advance the paper, much as in an ordinary typewriter. { 'frik·shən ¦fēd ,print·ər }

fringe area [COMMUN] An area just beyond the limits of the reliable service area of a television or radio transmitter, in which signals are weak and the reception is erratic. { 'frinj ˌer·ē·ə }

front-end [COMPUT SCI] Of a computer, under programmed instructions, performing data transfers and control operations to relieve a larger computer of these routines. { ¦frənt 'end }

front-end edit [COMPUT SCI] The process of checking and correcting data at the time it is entered into a computer system. { 'frənt ¦end 'ed·it }

front-end processor [COMPUT SCI] A computer which connects to the main computer at one end and communications channels at the other, and which directs the transmitting and receiving of messages, detects and corrects transmission errors, assembles and disassembles messages, and performs other processing functions so that the main computer receives pure information. { ¦frənt ¦end ˌpräsˌes·ər }

front porch [COMMUN] Portion of a composite picture signal which lies between the leading edge of the horizontal blanking pulse and the leading edge of the corresponding synchronizing pulse. { ¦frənt 'pȯrch }

front-to-back ratio [ELECTROMAG] Ratio of the effectiveness of a directional antenna, loudspeaker, or microphone toward the front and toward the rear. { ¦frənt tə ¦bak 'rā·shō }

fruit [ELECTR] Undesired signals received by a secondary radar from transponders responding to other radars. [NAV] Radar-beacon-system video display of a synchronous beacon return which results when several interrogator stations are located within the same general area; each interrogator receives its own interrogated reply as well as many synchronous replies resulting from interrogation of the airborne transponders by other ground stations. { früt }

F-scan *See* F-display. { 'ef ˌskan }

F-scope *See* F-display. { 'ef ˌskōp }

FSK *See* frequency-shift keying.

FSS *See* fixed-satellite service.

FTAM *See* file transfer access and management. { 'efˌtam }

FTP *See* file transfer protocol.

full adder [ELECTR] A logic element which operates on two binary digits and a carry digit from a preceding stage, producing as output a sum digit and a new carry digit. Also known as three-input adder. { ¦fu̇l 'ad·ər }

full duplex [COMMUN] Telegraph or other data channel able to operate in both directions simultaneously. [COMPUT SCI] The complete duplication of any data-processing facility. { ¦fu̇l 'dü ˌpleks }

full-duplex operation [COMMUN] Simultaneous communications in both directions between two points. { ¦fu̇l ¦düˌpleks ˌäp·ə'rā·shən }

full-featured software [COMPUT SCI] Software with the most advanced available functionality. { ¦fu̇l ¦fē·chərd 'sȯftˌwer }

full-motion video adapter [COMPUT SCI] A video adapter capable of displaying moving video images from a video cassette recorder, laser disk player, or camcorder on a computer screen. { ˌfu̇l ¦mō·shən 'vid·ē·ō əˌdap·tər }

full-screen editor [COMPUT SCI] A computer program that allows the user to work with the computer in an interactive manner by using all or most of the area of a cathode-ray tube or similar electronic display. { 'fu̇l ¦skrēn 'ed·əd·ər }

full section filter [ELECTR] A filter network whose graphical representation has the shape of the Greek letter pi, connoting capacitance in the upright legs and inductance or reactance in the horizontal member. { 'fu̇l ˌsek·shən 'fil·tər }

full subtracter [ELECTR] A logic element which operates on three binary input signals representing a minuend, subtrahend, and borrow digit, producing as output a different digit and a new borrow digit. Also known as three-input subtracter. { ¦fu̇l səb'trak·tər }

full-word boundary [COMPUT SCI] In the IBM 360 system, any address which ends in 00, and is therefore a natural boundary for a four-byte machine word. { 'fu̇l ˌwərd 'bau̇n·drē }

fully populated board [COMPUT SCI] A printed circuit board on which no room remains to install additional chips or other electronic components that would provide additional capabilities. { 'fu̇l·ē ¦päp·yəˌlād·əd 'bȯrd }

function [COMPUT SCI] In FORTRAN, a subroutine of a particular kind which returns a computational value whenever it is called. [MATH] A mathematical rule between two sets that assigns to each member of the first, exactly one member of the second. { 'fəŋk·shən }

functional [COMPUT SCI] In a linear programming problem involving a set of variables $x_j, j = 1, 2, \ldots, n$, a function of the form $c_1x_1 + c_2x_2 + \cdots + c_nx_n$ (where the c_j are constants) which one wishes to optimize (maximize or minimize, depending on the problem) subject to a set of restrictions. { 'fəŋk·shən·əl }

functional analysis [SYS ENG] A part of the design process that addresses the activities that a system, software, or organization must perform to achieve its desired outputs, that is, the transformations necessary to turn available inputs into the desired outputs. { ¦'fəŋk·shən·əl ə'nal·ə·səs }

functional analysis diagram [SYS ENG] A representation of functional analysis and, in particular, the transformations necessary to turn available inputs into the desired outputs, the flow of data or items between functions, the processing instructions that are available to guide the transformation, and the control logic that dictates the activation and termination of functions. { ¦fəŋk·shən·əl ə'nal·ə·səs ˌdī·ə ˌgram }

functional application [COMPUT SCI] A program or computer system, particularly a real-time system, that deals with the primary, ongoing operations of a business enterprise. { 'fəŋk·shən·əl ˌap·lə'kā·shən }

functional decomposition [CONT SYS] The partitioning of a large-scale control system into a nested set of generic control functions, namely the regulatory or direct control function, the optimizing control function, the adaptive control function, and the self-organizing function. { 'fəŋk·shən·əl dē,käm·pə'zish·ən }

functional design [COMPUT SCI] A level of the design process in which subtasks are specified and the relationships among them defined, so that the total collection of subsystems performs the entire task of the system. [SYS ENG] The aspect of system design concerned with the system's objectives and functions, rather than its specific components. { 'fəŋk·shən·əl di'zīn }

functional diagram [COMPUT SCI] A diagram that indicates the functions of the principal parts of a total system and also shows the important relationships and interactions among these parts. { 'fəŋk·shən·əl 'dī·ə,gram }

functional error recovery [COMPUT SCI] A procedure whereby the operating system intervenes in certain common errors and attempts actions to allow execution of the computer program to continue. { 'fəŋk·shən·əl 'er·ər ri,kəv·ə·rē }

functional failure [COMPUT SCI] Failure of a computer system to generate the correct results for a set of inputs. { ,fəŋk·shən·əl 'fāl·yər }

functional generator *See* function generator. { 'fəŋk·shən·əl 'jen·ə,rād·ər }

functional interleaving [COMPUT SCI] Alternating the parts of a number of sequences in a cyclic fashion, such as a number of accesses to memory followed by an access to a data channel. { ¦fəŋk·shən·əl 'in·tər,lēv·iŋ }

functional multiplier *See* function multiplier. { ¦fəŋk·shən·əl ¦məl·tə,plī·ər }

functional programming [COMPUT SCI] A type of computer programming in which functions are used to control the processing of logic. { 'fəŋk·shən·əl 'prō,gram·iŋ }

functional requirement [COMPUT SCI] The documentation which accompanies a program and states in detail what is to be performed by the system. { ¦fəŋk·shən·əl ri'kwīr·mənt }

functional specifications [COMPUT SCI] The documentation for the design of an information system, including the data base; the human and machine procedures; and the inputs, outputs, and processes for each data entry, query, update, and report program in the system. { ¦fəŋk·shən·əl ,spes·ə·fə'kā·shənz }

functional switching circuit [ELECTR] One of a relatively small number of types of circuits which implements a Boolean function and constitutes a basic building block of a switching system; examples are the AND, OR, NOT, NAND, and NOR circuits. { ¦fəŋk·shən·əl 'swich·iŋ ,sər·kət }

functional unit [COMPUT SCI] The part of the computer required to perform an elementary process such as an addition or a pulse generation. { ¦fəŋk·shən·əl 'yü·nət }

function code [COMPUT SCI] Special code which appears on a medium such as a paper tape and which controls machine functions such as a carriage return. { 'fəŋk·shən ,kōd }

function-evaluation routine [COMPUT SCI] A canned routine such as a log function or a sine function. { ¦fəŋk·shən i,val·yə'wā·shən rü,tēn }

function generator Also known as functional generator. [ELECTR] **1.** An analog computer device that indicates the value of a given function as the independent variable is increased. **2.** A signal generator that delivers a choice of a number of different waveforms, with provisions for varying the frequency over a wide range. { 'fəŋk·shən ,jen·ə,rād·ər }

function key [COMPUT SCI] A special key on a keyboard to control a mechanical function, initiate a specific computer operation, or transmit a signal that would otherwise require multiple key strokes. { 'fəŋk·shən ,kē }

function multiplier [ELECTR] An analog computer device that takes in the changing values of two functions and puts out the changing value of their product as the independent variable is changed. Also known as functional multiplier. { 'fəŋk·shən ¦məl·tə,plī·ər }

function switch [ELECTR] A network having a number of inputs and outputs so connected that input signals expressed in a certain code will produce output signals that are a function of the input information but in a different code. { 'fəŋk·shən ,swich }

function table [COMPUT SCI] **1.** Sets of computer information arranged so an entry in one set selects one or more entries in the other sets. **2.** A computer device that converts multiple inputs into a single output or encodes a single input into multiple outputs. { 'fəŋk·shən ,tā·bəl }

function unit [COMPUT SCI] In computer systems, a device which can store a functional relationship and release it continuously or in increments. { 'fəŋk·shən ,yü·nət }

functor *See* logic element. { 'fəŋk·tər }

fundamental group [COMMUN] In wire communications, a group of trunks that connect each local or trunk switching center to a trunk switching center of higher rank on which it homes; the term also applies to groups that interconnect zone centers. { ¦fən·də¦ment·əl 'grüp }

fundamental mode [ELECTROMAG] The waveguide mode having the lowest critical frequency. Also known as dominant mode; principal mode. { ¦fən·də¦ment·əl 'mōd }

fuse PROM [COMPUT SCI] A programmable read-only memory in which the programming is carried out either by blowing open microscopic fuse links to define a logic one or zero for each cell in the memory array, or by causing metal to short out base-emitter transistor junctions to program the ones or zeros into the memory. { 'fyüz ,präm }

future address patch [COMPUT SCI] A computer output containing the address of a symbol and the address of the last reference to that symbol. { ¦fyü·chər a'dres ,pach }

future label [COMPUT SCI] An address referenced in the operand field of an instruction, but which has not been previously defined. { 'fyü·chər ,lā·bəl }

fuzzy [MATH] Property of objects or processes that are not amenable to precise definition or precise measurement. { ¦fəz·ē }

fuzzy algorithm [COMPUT SCI] An ordered set of instructions, comprising fuzzy assignment statements, fuzzy conditional statements, and fuzzy unconditional action statements, that, upon execution, yield an approximate solution to a specified problem. { ¦fəz·ē 'al·gə,rith·əm }

fuzzy assignment statement [COMPUT SCI] An instruction in a fuzzy algorithm that assigns a possibly fuzzy value to a variable. { ¦fəz·ē ə'sīn·mənt ,stāt·mənt }

fuzzy conditional statement [COMPUT SCI] An instruction in a fuzzy algorithm that assigns a possibly fuzzy value to a variable or causes an action to be executed, provided that a fuzzy condition holds. { ¦fəz·ē kən'dish·ən·əl ,stāt·mənt }

fuzzy controller [CONT SYS] An automatic controller in which the relation between the state variables of the process under control and the action variables, whose values are computed from observations of the state variables, is given as a set of fuzzy implications or as a fuzzy relation. { ¦fəz·ē kən'trōl·ər }

fuzzy logic [MATH] The logic of approximate reasoning, bearing the same relation to approximate reasoning that two-valued logic does to precise reasoning. { ¦fəz·ē 'läj·ik }

fuzzy mathematics [MATH] A methodology for systematically handling concepts that embody imprecision and vagueness. { ¦fəz·ē ,math·ə'mad·iks }

fuzzy model [MATH] A finite set of fuzzy relations that form an algorithm for determining the outputs of a process from some finite number of past inputs and outputs. { ¦fəz·ē 'mäd·əl }

fuzzy relation [MATH] A fuzzy subset of the cartesian product $X \times Y$, denoted as a relation from a set X to a set Y. { ¦fəz·ē ri'lā·shən }

fuzzy relational equation [MATH] An equation of the form $A \cdot R = B$, where A and B are fuzzy sets, R is a fuzzy relation, and $A \cdot R$ stands for the composition of A with R. { ¦fəz·ē ri¦lā·shən·əl i'kwā·zhən }

fuzzy set [MATH] An extension of the concept of a set, in which the characteristic function which determines membership of an object in the set is not limited to the two values 1 (for membership in the set) and 0 (for nonmembership), but can take on any value between 0 and 1 as well. { 'fəz·ē 'set }

fuzzy system [SYS ENG] A process that is too complex to be modeled by using conventional mathematical methods, and that gives rise to data that are, in general, soft, with no precise boundaries; examples are large-scale engineering complex systems, social systems, economic systems, management systems, medical diagnostic processes, and human perception. { ¦fəz·ē 'sis·təm }

fuzzy unconditional action statement [COMPUT SCI] An instruction in a fuzzy algorithm that specifies a possibly fuzzy mathematical operation or an action to be executed. { ¦fəz·ē ən·kən¦dish·ən·əl 'ak·shən ,stāt·mənt }

G

GaAs FET *See* gallium arsenide field-effect transistor. { 'gas,fet }

gain [ELECTR] **1.** The increase in signal power that is produced by an amplifier; usually given as the ratio of output to input voltage, current, or power, expressed in decibels. Also known as transmission gain. **2.** *See* antenna gain. { gān }

gain-bandwidth product [ELECTR] The midband gain of an amplifier stage multiplied by the bandwidth in megacycles. { ¦gān ¦band,width ,präd·əkt }

gain control [ELECTR] A device for adjusting the gain of a system or component. { 'gān kən ,trōl }

gain-crossover frequency [CONT SYS] The frequency at which the magnitude of the loop ratio is unity. { ¦gān ¦krȯs,ō·vər ,frē·kwən·sē }

gain margin [CONT SYS] The reciprocal of the magnitude of the loop ratio at the phase crossover frequency, frequently expressed in decibels. { 'gān ,mär·jən }

gain sensitivity control *See* differential gain control. { ¦gān ,sen·sə'tiv·əd·ē kən,trōl }

gallium arsenide field-effect transistor [ELECTR] A field-effect transistor in which current between the ohmic source and drain contacts is carried by free electrons in a channel consisting of *n*-type gallium arsenide, and this current is modulated by a Schottky-barrier rectifying contact called the gate that varies the cross-sectional area of the channel. Abbreviated GaAs FET. { 'gal·ē·əm 'ärs·ən,īd 'fēld i¦fekt tran'zis·tər }

game theory [MATH] The mathematical study of games or abstract models of conflict situations from the viewpoint of determining an optimal policy or strategy. Also known as theory of games. { 'gām ,thē·ə·rē }

game tree [MATH] A tree graph used in the analysis of strategies for a game, in which the vertices of the graph represent positions in the game, and a given vertex has as its successors all vertices that can be reached in one move from the given position. Also known as lookahead tree. { 'gām ,trē }

gap [COMMUN] A region not adequately covered by the main lobes of a radar antenna, or in a larger area, not well covered by the fields of view of the radars of a network. [COMPUT SCI] A uniformly magnetized area in a magnetic storage device (tape, disk), used to indicate the end of an area containing information. { gap }

gap coding [COMMUN] A process for conveying information by inserting gaps or periods of nontransmission in a system that normally transmits continuously. { 'gap ,kōd·iŋ }

gap digit [COMPUT SCI] A digit in a machine word that does not represent data or instructions, such as a parity bit or a digit included for engineering purposes. { 'gap ,dij·it }

gap-filler radar [ENG] Radar used to fill gaps in radar coverage of other radar. { 'gap ,fil·ər 'rā ,där }

gapless tape [COMPUT SCI] A magnetic tape upon which raw data is recorded in a continuous manner; the data are streamed onto the tape without the word gaps; the data still may contain signs and end-of-record marks in the gapless form. { ¦gap·ləs 'tāp }

gapped tape [COMPUT SCI] A magnetic tape upon which blocked data has been recorded; it contains all of the flag bits and format to be read directly into a computer for immediate use. { ¦gapt 'tāp }

gap scatter [COMPUT SCI] The deviation from the exact distance required between read/write heads and the magnetized surface. { 'gap ,skad·ər }

garbage *See* hash. { 'gär·bij }

garbage collection [COMPUT SCI] In a computer program with dynamic storage allocation, the automatic process of identifying those memory cells whose contents are no longer useful for the computation in progress and then making them available for some other use. { 'gär·bij kə ,lek·shən }

garbage in, garbage out [COMPUT SCI] A phrase often stressed during introductory courses in computer utilization as a reminder that, regardless of the correctness of the logic built into the program, no answer can be valid if the input is erroneous. Abbreviated GIGO. { ¦gär·bij 'in ¦gär·bij 'au̇t }

garble [COMMUN] To alter a message intentionally or unintentionally so that it is difficult to understand. { 'gär·bəl }

garbling [ELECTR] Confusion resulting from a secondary radar receiving overlapping coded responses from transponders in a dense target environment. { 'gär·bliŋ }

gate [ELECTR] **1.** A circuit having an output and a multiplicity of inputs and so designed that the output is energized only when a certain combination of pulses is present at the inputs. **2.** A circuit in which one signal, generally a square wave, serves to switch another signal on and off. **3.** One of the electrodes in a field-effect transistor. **4.** To control the passage of a pulse or signal. **5.** In radar, an electric waveform which is applied to the control point of a circuit to alter the mode of operation of the circuit at the time when the waveform is applied. Also known as gating waveform. **6.** In radar, an electronic waveform applied to a circuit or a timing cue applied to logic to alter the operation of the circuit or logic at the appropriate time; generally used in anticipation of an input of particular interest. { gāt }

gate-array device [ELECTR] An integrated logic circuit that is manufactured by first fabricating a two-dimensional array of logic cells, each of which is equivalent to one or a few logic gates, and then adding final layers of metallization that determine the exact function of each cell and interconnect the cells to form a specific network when the customer orders the device. { 'gāt ə,rā di,vīs }

gate-controlled switch [ELECTR] A semiconductor device that can be switched from its nonconducting or "off" state to its conducting or "on" state by applying a negative pulse to its gate terminal and that can be turned off at any time by applying reverse drive to the gate. Abbreviated GCS. { 'gāt kən,trōld 'swich }

gate equivalent circuit [ELECTR] A unit of measure for specifying relative complexity of digital circuits, equal to the number of individual logic gates that would have to be interconnected to perform the same function as the digital circuit under evaluation. { 'gāt i,kwiv·ə·lənt ,sər·kət }

gate generator [ELECTR] A circuit used to generate gate pulses; in one form it consists of a multivibrator having one stable and one unstable position. { 'gāt ,jen·ə·rād·ər }

gate pulse [ELECTR] A pulse that triggers a gate circuit so it will pass a signal. { 'gāt ,pəls }

gateway [COMMUN] A point of entry and exit to another system, such as the connection point between a local-area network and an external-communications network. { 'gāt,wā }

gather write [COMPUT SCI] An operation that creates a single output record from data items gathered from nonconsecutive locations in main memory. { ¦gath·ər 'wrīt }

gating [ELECTR] The process of selecting those portions of a wave that exist during one or more selected time intervals or that have magnitudes between selected limits. { 'gād·iŋ }

gating waveform *See* gate. { ¦gād·iŋ 'wāv,fȯrm }

Gaussian beam [ELECTROMAG] A beam of electromagnetic radiation whose wave front is approximately spherical at any point along the beam and whose transverse field intensity over any wave front is a Gaussian function of the distance from the axis of the beam. { 'gau̇s·ē·ən 'bēm }

Gaussian noise [COMMUN] Random electromagnetic signals inherent in nature, both in the surroundings of a receiver and produced in the receiver itself; typically produced by the thermal agitation of molecular structures, and having Gaussian statistics in its components. Also known as thermal noise. { ¦gau̇·sē·ən 'nȯiz }

gc *See* gigahertz.

GCA radar *See* ground-controlled approach radar. { ¦jē¦sē¦ā 'rā,där }

GCS *See* gate-controlled switch.

G-display [ELECTR] A radar display format in which the target of a tracking radar appears as a spot, as in an F-display, with "wings" (horizontal extensions of the plot) that increase in length as the range decreases. Also known as G-indicator; G-scan; G-scope. { 'jē di,splā }

gen [COMPUT SCI] To install an operating system or a systems software package for a particular configuration of computer equipment. Abbreviation for generate. { jen }

general address [COMMUN] Group of characters included in the heading of a message that causes the message to be routed to all addresses included in the general address category. { ¦jen·rəl ə'dres }

generalized routine [COMPUT SCI] A routine which can process a wide variety of jobs; for example, a generalized sort routine which will sort in ascending or descending order on any number of fields whether alphabetic or numeric, or both, and whether binary coded decimals or pure binaries. { 'jen·rə,līzd rü'tēn }

generalized system [COMPUT SCI] A computer system developed for a broad range of users. { 'jen·rə,līzd 'sis·təm }

general program [COMPUT SCI] A computer program designed to solve a specific type of problem when values of appropriate parameters are supplied. { ¦jen·rəl 'prō·grəm }

general-purpose computer [COMPUT SCI] A device that manipulates data without detailed, step-by step control by human hand and is designed to be used for many different types of problems. { ¦jen·rəl ¦pər·pəs kəm'pyüd·ər }

general-purpose function generator [COMPUT SCI] A function generator which can be adjusted to generate many different functions, rather than being designed for a particular function. Also known as arbitrary function generator. { ¦jen·rəl ¦pər·pəs 'fəŋk·shən ,jen·ə,rād·ər }

general-purpose language [COMPUT SCI] A computer programming language whose use is not restricted to a particular type of computer or a specialized application. { 'jen·rəl ¦pər·pəs 'laŋ·gwij }

general-purpose systems simulation *See* GPSS. { 'jen·rəl ¦pər·pəs 'sis·təmz ,sim·yə,lā·shən }

general register *See* local register. { ¦jen·rəl 'rej·ə·stər }

general routine [COMPUT SCI] In computers, a routine, or program, applicable to a class of problems; it provides instructions for solving a specific problem when appropriate parameters are supplied. { ¦jen·rəl rü'tēn }

generate [COMPUT SCI] **1.** To create a particular program by selecting parts of a general-program skeleton (or outline) and specializing these parts into a cohesive entity. **2.** *See* gen. { 'jen·ə,rāt }

generate and test [COMPUT SCI] A computer problem-solving method in which a sequence of candidate solutions is generated, and each is tested to determine if it is an appropriate solution. { 'jen·ə,rāt ən 'test }

generated address [COMPUT SCI] An address calculated or determined by instructions contained in a computer program for subsequent use by that program. Also known as calculated address; synthetic address. { ¦jen·ə,rād·əd ə'dres }

generating area *See* fetch. { 'jen·ə,rād·iŋ ,er·ē·ə }

generating routine *See* generator. { 'jen·ə,rād·iŋ rü,tēn }

generation [COMPUT SCI] **1.** Any one of three groups used to historically classify computers according to their electronic hardware components, logical organization and software, or programming techniques; computers are thus known as first-, second-, or third-generation; a particular computer may possess characteristics of all generations simultaneously. **2.** One of a family of data sets, related to one another in that each is a modification of the next most recent data set. { ,jen·ə'rā·shən }

generation data group [COMPUT SCI] A collection of files, each a modification of the previous one, with the newest numbered 0, the next −1, and so forth, and organized so that each time a new file is added the oldest is deleted. Abbreviated GDG. { ,jen·ə'rā·shən 'dad·ə ,grüp }

generation number [COMPUT SCI] A number contained in the file label of a reel of magnetic tape that indicates the generation of the data set of the tape. { ,jen·ə'rā·shən ,nəm·bər }

generative grammar [COMPUT SCI] A set of rules that describes the valid expressions in a formal language on the basis of a set of the parts of speech (formally called the set of metavariables or phrase names) and the alphabet or character set of the language. { 'jen·rəd·iv 'gram·ər }

generator [COMPUT SCI] A program that produces specific programs as directed by input parameters. Also known as generating routine. [ELECTR] **1.** A vacuum-tube oscillator or any other nonrotating device that generates an alternating voltage at a desired frequency when energized with direct-current power or low-frequency alternating-current power. **2.** A circuit that generates a desired repetitive or nonrepetitive waveform, such as a pulse generator. { 'jen·ə,rād·ər }

generator lock [ELECTR] Circuitry that synchronizes two video signals so that they can be mixed. Abbreviated genlock. { 'jen·ə,rād·ər ,läk }

genetic algorithm [COMPUT SCI] A search procedure based on the mechanics of natural selection and genetics. Also known as evolutionary strategy. { jə,ned·ik 'al·gə,rith·əm }

genetic programming *See* evolutionary programming. { jə,ned·ik 'prō,gram·iŋ }

genlock *See* generator lock. { 'gen,läk }

GEO *See* geosynchronous orbit. { ¦jē¦ē'ō *or* 'jē·ō }

geomagnetic noise [COMMUN] Interference in radio communications arising from terrestrial magnetism. { ¦jē·ō·mag¦ned·ik 'nȯiz }

geometrical distortion [COMPUT SCI] A discrepancy between the horizontal and vertical dimensions of the picture elements on an electronic display, causing, for example, circles to appear as ovals unless corrected for in software. { ¦jē·ə ¦me·trə·kəl di'stȯr·shən }

geometric programming [SYS ENG] A nonlinear programming technique in which the relative contribution of each of the component costs is first determined; only then are the variables in the component costs determined. { ¦jē·ə ¦me·trik 'prō,gram·iŋ }

geostationary satellite [AERO ENG] A satellite that follows a circular orbit in the plane of the earth's equator from west to east at such a speed as to remain fixed over a given place on the equator at an altitude of 22,280 miles (35,860 kilometers). { ¦jē·ō¦stā·shə,ner·ē 'sad·əl,īt }

geosynchronous orbit [AERO ENG] A satellite orbit that has a period of one sidereal day (23 hours, 56 minutes, 4 seconds). Abbreviated GEO. { ,jē·ō¦siŋ·krə·nəs 'ȯr·bət }

geosynchronous satellite [AERO ENG] An earth satellite that makes one revolution in one sidereal day (23 hours, 56 minutes, 4 seconds), synchronous with the earth's rotation; the orbit can have arbitrary eccentricity and arbitrary inclination to the earth's equator. Also known as synchronous satellite. { ,jē·ō¦siŋ·krə·nəs 'sad·əl,īt }

get [COMPUT SCI] An instruction in a computer program to read data from a file. { get }

getmain [COMPUT SCI] An instruction used in some programming languages to request dynamic allocation of additional storage space to the program. { 'get,mān }

ghost [COMPUT SCI] To display a menu option in a dimmed, fuzzy typeface to indicate that this option is no longer available. [ELECTR] In radar, a contact generated where in fact no target exists, resulting from measurement ambiguity or attempts to resolve ambiguities with multiple observations in a multiple-target situation. { gōst }

ghost algebraic manipulation language [COMPUT SCI] An algebraic manipulation language which externally gives the appearance of manipulating quite general mathematical expressions, although internally it is functioning with canonically represented data, much like the simpler seminumerical languages. { ¦gōst al·jə'brā·ik mə,nip·yə'lā·shən ,laŋ·gwij }

ghost image [ELECTR] An undesired duplicate image offset from the desired image on a video display device. { 'gōst ,im·ij }

ghost mode [ELECTROMAG] Waveguide mode having a trapped field associated with an imperfection in the wall of the waveguide; a ghost mode can cause trouble in a waveguide operating close to the cutoff frequency of a propagation mode. { 'gōst ˌmōd }

ghost signal [ELECTR] The reflection-path signal that produces a ghost image on an analog television receiver. Also known as echo. { 'gōst ˌsig·nəl }

GHz *See* gigahertz.

gibberish *See* hash. { 'jib·rish }

GIF *See* graphics interchange format. { gif }

gigabit [COMMUN] One billion bits, or 1,000,000,000 bits. { 'gig·əˌbit }

gigacycle *See* gigahertz. { 'gig·əˌsī·kəl }

gigaflops [COMPUT SCI] A unit of computer speed, equal to 10^9 flops. { 'gig·əˌfläps }

gigahertz [COMMUN] Unit of frequency equal to 10^9 hertz. Abbreviated GHz. Also known as gigacycle (gc); kilomegacycle; kilomegahertz. { 'gig·əˌhərts }

GIGO *See* garbage in, garbage out. { 'gīˌgō }

Gilbert-Varshamov bound [COMPUT SCI] In the theory of quantum computation, a sufficient condition for an algorithm that encodes N logical qubits into N' carrier qubits (with N' larger than N) to correct any error on any M carrier qubits; namely, that N/N' be smaller than

$$1 - 2[-x \log_{2x} - (1 - x) \log_2 (1 - x)],$$

where $x = 2M/N'$.

{ ¦gil·bərt ˌvär¦sha·məv ¦bau̇nd }

G-indicator *See* G-display. { 'jē ˌin·dəˌkād·ər }

GKS *See* graphical kernel system.

glare [COMMUN] The interference that arises when an attempt is made to place a telephone call just as an incoming call is arriving; in the case of data transmission under the control of a computer, this can render the line or even the computer temporarily inoperative. { gler }

glint [ELECTR] **1.** Pulse-to-pulse variation in the apparent angular center of a target, due to target scattering complexity and dynamics; causes angle errors in tracking radars using either conical scan or monopulse techniques. **2.** The use of this effect to degrade tracking or seeking functions of an enemy weapons system. { glint }

global format [COMPUT SCI] A choice of label alignment or numeric format in a spreadsheet program that applies to all the cells of the spreadsheet. { ¦glō·bəl 'fȯrˌmat }

global memory [COMPUT SCI] Computer storage that can be used by a number of processors connected together in a multiprocessor system. { 'glō·bəl 'mem·rē }

global orbiting navigation satellite system *See* GLONASS. { ¦glō·bəl ¦ȯrb·əd·iŋ ˌnav·ə¦gā·shən 'sad·əˌlīt ˌsis·təm }

Global Positioning System [NAV] A positioning or navigation system designed to use 24 satellites, each carrying atomic clocks, to provide a receiver anywhere on earth with extremely accurate measurements of its three-dimensional position, velocity, and time. Abbreviated GPS. { 'glō·bəl pə'zish·niŋ ˌsis·təm }

global resource sharing [COMPUT SCI] The ability of all of the users of a local-area network to share any of the resources (storage devices, input/output devices, and so forth) connected to the network. { 'glō·bəl ri'sȯrs ˌsher·iŋ }

global search and replace [COMPUT SCI] A text-editing function of a word-processing system in which text is scanned for a given combination of characters, and each such combination is replaced by another set of characters. { ¦glō·bəl ¦sərch ən ri'plās }

global system for mobile communications *See* GSM. { ¦glō·bəl ¦sis·təm fər ˌmō·bəl kəˌmyü·nə'kā·shənz }

global variable [COMPUT SCI] A variable which can be accessed (used or changed) throughout a computer program and is not confined to a single block. { ¦glō·bəl 'ver·ē·ə·bəl }

GLONASS [NAV] A worldwide Russian navigation system designed to use 24 satellites in three uniformly spaced orbital planes to provide three-dimensional position and velocity data to equipped users on or above the earth's surface. Acronym for global orbiting navigation satellite system. { 'glōˌnas }

glossary [COMPUT SCI] A file of commonly used phrases that can be retrieved in a word-processing program, usually through use of a command and a keyword. { 'gläs·ə·rē }

GNU [COMPUT SCI] Freely distributed software for producing and distributing nonproprietary software that is compatible with Unix, but is not Unix. { gə'nü }

Golay code [COMMUN] A linear, block-based error-correcting code that is particularly suited to applications where short code word length and low latency are important. { gə'lā ˌkōd }

golden-section search [COMPUT SCI] A dichotomizing search in which, in each step, the remaining items are divided as closely as possible according to the golden section. { 'gōl·dən 'sek·shən ˌsərch }

goovoo [COMPUT SCI] A file within a generation data group, so called because of the notation used in some systems in which, for example, G003 V001 is volume 1, generation −3 of a generation data group. { 'güˌvü }

GOP *See* group of pictures.

Gopher [COMPUT SCI] A menu-based program for browsing the Internet and finding and gaining access to files, programs, definitions, and other Internet resources. { 'gō·fər }

GOTO-less programming [COMPUT SCI] The writing of computer programs without the use of GOTO statements. { ¦gō·tüˌles 'prōˌgram·iŋ }

GOTO statement [COMPUT SCI] A statement in a computer program that provides for the direct transfer of control to another statement with the identifier that is the argument of the GOTO statement. { 'gōˌtü ˌstāt·mənt }

government frequency bands [COMMUN] Radio-frequency bands which are allotted to various departments and services of the federal government. { 'gəv·ər·mənt ,fre·kwən·se ,banz }

GPS *See* Global Positioning System.

GPSS [COMPUT SCI] A problem-oriented programming language designed to assist the user in developing models. Acronym for general-purpose systems simulation.

graceful degradation [COMPUT SCI] A programming technique to prevent catastrophic system failure by allowing the machine to operate, though in a degraded mode, despite failure or malfunction of several integral units or subsystems. { 'grās·fu̇l ,deg·rə'dā·shən }

graceful exit [COMPUT SCI] The ability to escape from a problem situation in a computer program without having to reboot the computer. { ¦grās·fəl 'eg·zət }

grade [COMMUN] One of two types of television service, designated grade A and grade B, each having a specified signal strength, that of grade A being several timeshigher than B. { grād }

graded periodicity technique [ELECTR] A technique for modifying the response of a surface acoustic wave filter by varying the spacing between successive electrodes of the interdigital transducer. { ¦grād·əd ,pir·ē·ə'dis·əd·ē tek,nēk }

grain direction [COMPUT SCI] In character recognition, the arrangement of paper fibers in relation to a document's travel through a character reader. { 'grān də,rek·shən }

grandfather [COMPUT SCI] A data set that is two generations earlier than the data set under consideration. { 'gran,fat͟h·ər }

grandfather cycle [COMPUT SCI] The period during which records are kept but not used except to reconstruct other records which are accidentally lost. { 'gran,fat͟h·ər ,sī·kəl }

granularity [SYS ENG] The degree to which a system can be broken down into separate components, making it customizable and flexible. { ,gran·yə'lar·əd·ē }

graph follower *See* curve follower. { 'graf ,fäl·ə·wər }

graphical kernel system [COMPUT SCI] A standard system and language for creating two- and three-dimensional master graphics images on many types of display devices. Abbreviated GKS. { ¦graf·i·kəl 'kər·nəl ,sis·təm }

graphical user interface [COMPUT SCI] A user interface in which program features are represented by icons that the user can access and manipulate with a pointing device. Abbreviated GUI. { ¦graf·ə·kəl ,yü·zər 'in·tər,fās }

graphical visual display device [COMPUT SCI] A computer input-output device which enables the user to manipulate graphic material in a visible two-way, real-time communication with the computer, and which consists of a light pen, keyboard, or other data entry devices, and a visual display unit monitored by a controller. Also known as graphoscope. { ¦graf·ə·kəl ¦vizh·ə·wəl di'spla di,vīs }

graphics driver [COMPUT SCI] A series of instructions that activates a graphics device, such as a display screen or plotter. { 'graf·iks ,drīv·ər }

graphics engine [COMPUT SCI] A specialized processor that carries out graphics processing independently of the main central processing unit. Also known as graphics processor. { 'graf·iks ¦en·jən }

graphics interchange format [COMPUT SCI] Common file format for compressed graphic images on the World Wide Web that is limited to 256 colors. Abbreviated GIF. { ¦graf·iks 'in·tər ,chānj ,fȯr,mat }

graphics interface [COMPUT SCI] A user interface that displays icons to represent objects. { 'graf·iks ¦in·tər,fās }

graphics primitive [COMPUT SCI] A basic building block for graphic images, such as a dot, line, or curve. { 'graf·iks ¦prim·əd·iv }

graphics processor *See* graphics engine. { 'graf·iks ¦prä·ses·ər }

graphics program [COMPUT SCI] A program for the generation of images, ranging in complexity from simple line drawings to realistically shaded pictures that resemble photographs. { 'graf·iks ,prō·grəm }

graphics tablet [COMPUT SCI] A padlike peripheral device which is designed so that shapes appear on the monitor's screen when the tablet is drawn upon with a pointed device. { 'graf·iks ,tab·lət }

graphics terminal [COMPUT SCI] An input/output device that can accept and display picture images. { 'graf·iks ¦tər·mən·əl }

graphoscope *See* graphical visual display device. { 'graf·ə,skōp }

graph theory [MATH] **1.** The mathematical study of the structure of graphs and networks. **2.** The body of techniques used in graphing functions in the plane. { 'graf ,thē·ə·rē }

grass [ELECTR] Clutter due to circuit noise in a radar receiver, seen on an A scope as a pattern resembling a cross section of turf. Also known as hash. { gras }

Gray code [COMMUN] A modified binary code in which sequential numbers are represented by expressions that differ only in one bit, to minimize errors. Also known as reflective binary code. { 'grā,kōd }

greeking [COMPUT SCI] The display of the format of a document without displaying the characters. { 'grēk·iŋ }

grid [COMPUT SCI] In optical character recognition, a system of two groups of parallel lines, perpendicular to each other, used to measure or specify character images. [ELECTR] An electrode located between the cathode and anode of an electron tube, which has one or more openings through which electrons or ions can pass, and serves to control the flow of electrons from cathode to anode. { grid }

Grosh's law [COMPUT SCI] The law that the processing power of a computer is proportional to the square of its cost. { 'grōsh·əz ,lȯ }

gross index [COMPUT SCI] The first of two indexes consulted to gain access to a record. { ¦grōs 'in,deks }

gross information content [COMMUN] Measure of the total information, redundant or otherwise, contained in a message; expressed as the number of bits, nits, or Hartleys required to transmit the message with specified accuracy over a noiseless medium without coding. { ¦grōs ˌin·fər'mā·shən ˌkän·tent }

ground absorption [ELECTROMAG] Loss of energy in transmission of radio waves, due to dissipation in the ground. { 'grau̇nd əbˌsȯrp·shən }

ground clutter [ELECTROMAG] Clutter on a ground or airborne radar due to reflection of signals from the ground or objects on the ground. Also known as ground flutter; ground return; land return; terrain echoes. { 'grau̇nd ˌkləd·ər }

ground-controlled approach radar [ENG] A ground radar system providing information by which aircraft approaches may be directed by radio communications. Abbreviated GCA radar. { 'grau̇nd kənˌtrōld əˌprōch 'rāˌdär }

ground-controlled intercept radar [ENG] A radar system by means of which a controller may direct an aircraft to make an interception of another aircraft. Abbreviated GCI radar. { 'grau̇nd kənˌtrōld 'in·tərˌsept ˌrāˌdär }

grounded-base connection [ELECTR] A transistor circuit in which the base electrode is common to both the input and output circuits; the base need not be directly connected to circuit ground. Also known as common-base connection. { ¦grau̇nd·əd 'bās kəˌnek·shən }

grounded-collector connection [ELECTR] A transistor circuit in which the collector electrode is common to both the input and output circuits; the collector need not be directly connected to circuit ground. Also known as common-collector connection. { ¦grau̇nd·əd kə'lek·tər kə ˌnek·shən }

grounded-emitter connection [ELECTR] A transistor circuit in which the emitter electrode is common to both the input and output circuits; the emitter need not be directly connected to circuit ground. Also known as common-emitter connection. { ¦grau̇nd·əd i'mid·ər kəˌnek·shən }

ground effect [COMMUN] The effect of ground conditions on radio communications. { 'grau̇nd iˌfekt }

ground equalizer inductors [ELECTROMAG] Coils, having relatively low inductance, inserted in the circuit to one or more of the grounding points of an antenna to distribute the current to the various points in any desired manner. { ¦grau̇nd 'ē·kwəˌlīz·ər inˌdək·tərz }

ground flutter *See* ground clutter. { 'grau̇nd ˌfləd·ər }

ground instrumentation *See* spacecraft ground instrumentation. { 'grau̇nd ˌin·strə·mən'tā·shən }

ground loop [COMMUN] Return currents or magnetic fields from relatively high-powered circuits or components which generate unwanted noisy signals in the common return of relatively low-level signal circuits. { 'grau̇nd ˌlüp }

ground-penetrating radar *See* ground-probing radar. { ¦grau̇nd ˌpen·ə¦trād·iŋ 'rāˌdär }

ground-plane antenna [ELECTROMAG] Vertical antenna combined with a grounded horizontal disk, turnstile element, or similar ground-plane simulation; such antennas may be mounted several wavelengths above the ground, and provide a low radiation angle. { 'grau̇nd ˌplān an'ten·ə }

ground-probing radar [ENG] A nondestructive technique using electromagnetic waves to locate objects or interfaces buried beneath the earth's surface or located within a visually opaque structure. Also known as ground-penetrating radar; subsurface radar; surface-penetrating radar. { ¦grau̇nd ¦prōb·iŋ 'rāˌdär }

ground-reflected wave [ELECTROMAG] Component of the ground wave that is reflected from the ground. { 'grau̇nd ri¦flek·təd 'wāv }

ground return [ELECTROMAG] **1.** An echo received from the ground by an airborne radar set. **2.** *See* ground clutter. { 'grau̇nd riˌtərn }

groundscatter propagation [COMMUN] Multi-hop ionospheric radio propagation along other than the great-circle path between transmitting and receiving stations; radiation from the transmitter is first reflected back to earth from the ionosphere, then scattered in many directions from the earth's surface. { 'grau̇nd ˌskad·ər ˌpräp·ə'gā·shən }

ground surveillance radar [ENG] **1.** A surveillance radar operated at a fixed point on the earth's surface for observation and control of the position of aircraft or other vehicles in the vicinity. **2.** A radar system capable of detecting objects on the ground from points on the ground. { 'grau̇nd sərˌvā·ləns ˌrāˌdär }

ground system [ELECTROMAG] The portion of an antenna that is closely associated with an extensive conducting surface, which may be the earth itself. { 'grau̇nd ˌsis·təm }

ground-up read-only memory [COMPUT SCI] A read-only memory which is designed from the bottom up, and for which all fabrication masks used in the multiple mask process are custom-generated. { 'grau̇nd ¦əp 'rēd ¦ōn·lē 'mem·rē }

ground wave [COMMUN] A radio wave that is propagated along the earth and is ordinarily affected by the presence of the ground and the troposphere; includes all components of a radio wave over the earth except ionospheric and tropospheric waves. Also known as surface wave. { 'grau̇nd ˌwāv }

group [COMMUN] A communication transmission subdivision containing a number of voice channels, either within a supergroup or separately, normally comprised of up to 12 voice channels occupying the frequency band 60–108 kilohertz; each voice channel may be multiplexed for teletypewriter operation, if required. { grüp }

group busy tone [COMMUN] High tone connected to the jack sleeves of an outgoing trunk group as an indication that all trunks in the group are busy. { 'grüp ¦biz·ē ˌtōn }

group code See systematic error-checking code. { 'grüp ˌkōd }

group-coded record [COMPUT SCI] A method of recording data on magnetic tape with eight tracks of data and one parity track, in which every eighth byte in conjunction with the parity track is used for detection and correction of all single-bit errors. { 'grüp ˌkōd·əd 'rek·ərd }

group communications software See groupware. { ˌgrüp kəˌmyü·nəˌkā·shənz 'sȯfˌwer }

grouped-frequency operation [COMMUN] Use of different frequency bands for channels in opposite directions in a two-wire carrier system. { 'grüpt ¦frē·kwən·sē ˌäp·əˌrā·shən }

grouped records [COMPUT SCI] Two or more records placed together and identified by a single key, to save storage space or reduce access time. { 'grüpt ¦rek·ərdz }

group-indicate [COMPUT SCI] To print indicative information from only the first record of a group. { 'grüp ¦in·dəˌkāt }

grouping [COMMUN] Periodic error in the spacing of recorded lines in a facsimile system. { 'grüp·iŋ }

grouping circuits [COMMUN] Circuits used to interconnect two or more switchboard positions together, so that one operator may handle the several switchboard positions from one operator's set. { 'grüp·iŋ ˌsər·kəts }

grouping of records [COMPUT SCI] Placing records together in a group to either conserve storage space or reduce access time. { 'grüp·iŋ əv 'rek·ərdz }

group mark [COMPUT SCI] A character signaling the beginning or end of a group of data. { 'grüp ¦märk }

group modulation [COMMUN] Process by which a number of channels, already separately modulated to a specific frequency range, are again modulated to shift the group to another range. { 'grüp ˌmäj·ə'lā·shən }

group of pictures [COMMUN] In MPEG-2, a group of pictures consists of one or more pictures in sequence. Abbreviated GOP. { 'grüp əv 'pik·chərz }

group printing [COMPUT SCI] The printing of information summarizing the data on a group of cards or other records when a key change occurs. { 'grüp ¦print·iŋ }

groupware [COMPUT SCI] Multiuser software that supports information sharing through digital media, such as electronic mail and messaging, electronic meeting systems and audio conferencing, group calendaring and scheduling, workflow process diagramming and analysis tools, and group document handling including group editing. { 'grüpˌwer }

Grover's algorithm [COMPUT SCI] An algorithm for finding an item in a database of 2N items, using a quantum computer, in a time of order $2^{N/2}$ steps instead of order 2^N steps. { ¦grō·vərz 'al·gəˌrith·əm }

G-scan See G-display. { 'jē ˌskan }

G-scope See G-display. { 'jē ˌskōp }

GSM [COMMUN] A digital cellular telephone technology that is based on time-division multiple access; it operates on the 900-megahertz and 1.8-gigahertz bands in Europe, where it is the predominant cellular system, and on the 1.9-gigahertz band in the United States. Derived from global system for mobile communications.

G string See field waveguide. { 'jē ˌstriŋ }

guard band [ELECTR] A narrow frequency band provided between adjacent channels in certain portions of the radio spectrum to prevent interference between stations. { 'gärd ˌband }

guarded command [COMPUT SCI] A program statement within a group of such statements that determines whether the other statements will be executed by the computer. { 'gärd·əd kə'mand }

guard signal [COMPUT SCI] A signal used in digital-to-analog converters, analog-to-digital converters, or other converters which permits values to be read or converted only when the values are not changing, usually to avoid ambiguity error. { 'gärd ˌsig·nəl }

guest computer [COMPUT SCI] A computer that operates under the control of another computer (the host). { 'gest kəmˌpyüd·ər }

GUI See graphical user interface. { 'güˌē *or* ¦jē ¦yü'ī }

guided propagation [COMMUN] Type of radio-wave propagation in which radiated rays are bent excessively by refraction in the lower layers of the atmosphere; this bending creates an effect much as if a duct or waveguide has been formed in the atmosphere to guide part of the radiated energy over distances far beyond the normal range. Also known as trapping. { 'gīd·əd präp·ə'gā·shən }

guided wave [ELECTROMAG] A wave whose energy is concentrated near a boundary or between substantially parallel boundaries separating materials of different properties and whose direction of propagation is effectively parallel to these boundaries; waveguides transmit guided waves. { 'gīd·əd 'wāv }

guide wavelength [ELECTROMAG] Wavelength of electromagnetic energy conducted in a waveguide; guide wavelength for all air-filled guides is always longer than the corresponding free-space wavelength. { 'gīd ¦wāvˌleŋkth }

gulp [COMPUT SCI] A series of bytes considered as a unit. { gəlp }

Gunn amplifier [ELECTR] A microwave amplifier in which a Gunn oscillator functions as a negative-resistance amplifier when placed across the terminals of a microwave source. { 'gən ¦am·pləˌfī·ər }

Gunn diode See Gunn oscillator. { 'gən ¦dīˌōd }

Gunn effect [ELECTR] Development of a rapidly fluctuating current in a small block of a semiconductor (perhaps *n*-type gallium arsenide) when a constant voltage above a critical value is applied to contacts on opposite faces. { 'gən iˌfekt }

Gunn oscillator [ELECTR] A microwave oscillator utilizing the Gunn effect. Also known as Gunn diode. { 'gən ¦äs·əˌlād·ər }

gyrator [ELECTROMAG] A waveguide component that uses a ferrite section to give zero phase shift for one direction of propagation and 180° phase shift for the other direction; in other words, it causes a reversal of signal polarity for one direction of propagation but not for the other direction. Also known as microwave gyrator. { 'jī,rād·ər }

gyrator filter [ELECTR] A highly selective active filter that uses a gyrator which isterminated in a capacitor so as to have an inductive input impedance. { 'jī,rād·ər ,fil·tər }

gyromagnetic coupler [ELECTR] A coupler in which a single-crystal yig (yttrium iron garnet) resonator provides coupling at the required low signal levels between two crossed strip-line resonant circuits. { ¦jī·rō·mag'ned·ik 'kəp·lər }

H

hacker [COMPUT SCI] A person who uses a computer system without a specific, constructive purpose or without proper authorization. { 'hak·ər }

hacking [COMPUT SCI] Use of a computer system without a specific, constructive purpose, or without proper authorization. { 'hak·iŋ }

half-adder [ELECTR] A logic element which operates on two binary digits (but no carry digits) from a preceding stage, producing as output a sum digit and a carry digit. { ¦haf ¦ad·ər }

half-adjust [COMPUT SCI] A rounding process in which the least significant digit is dropped and, if the least significant digit is one-half or more of the number base, one is added to the next more significant digit and all carries are propagated. { ¦haf ə¦jəst }

half block [COMPUT SCI] The unit of transfer between main storage and the buffer control unit; it consists of a column of 128 elements, each element 16 bytes long. { 'haf ¦bläk }

half carry [COMPUT SCI] A flag used in the central processing unit of some computers to indicate that a carry has occurred from the low-order N bits of a 2N-bit number to the high-order N bits. { 'haf ˌkar·ē }

half-cycle transmission [COMMUN] Data transmission and control system that uses synchronized sources of 60-hertz power at the transmitting and receiving ends; either of two receiver relays can be actuated by choosing the appropriate half-cycle polarity of the 60-hertz transmitter power supply. { 'haf ¦sī·kəl tranz'mish·ən }

half-duplex circuit [COMMUN] A circuit designed for half-duplex operation. Abbreviated HDX. { 'haf ¦dü,pleks ˌsər·kət }

half-duplex operation [COMMUN] Operation of a telegraph system in either direction over a single channel, but not in both directions simultaneously. { 'haf ¦dü,pleks äp·ə'rā·shən }

half-height drive [COMPUT SCI] A personal-computer disk drive whose height is half that of earlier disk drives. { ¦haf ˌhīt 'drīv }

half-period zones *See* Fresnel zones. { 'haf ˌpir·ē·əd ˌzōnz }

half-power beamwidth [ELECTROMAG] The angle across the main lobe of an antenna pattern between the two directions at which the antenna's sensitivity is half its maximum value at the center of the lobe. Abbreviated HPBW. { 'haf ¦paù·ər 'bēm,width }

half-power frequency [ELECTR] One of the two values of frequency, on the sides of an amplifier response curve, at which the voltage is $1/\sqrt{2}$ (70.7%) of a midband or other reference value. Also known as half-power point. { 'haf ¦paù·ər 'frē·kwən·sē }

half-power point [ELECTR] **1.** A point on the graph of some quantity in an antenna, network, or control system, versus frequency, distance, or some other variable at which the power is half that of a nearby point at which power is a maximum. **2.** *See* half-power frequency. { 'haf ¦paù·ər ˌpȯint }

half-pulse-repetition-rate delay [ELECTR] In the loran navigation system, an interval of time equal to half the pulse repetition rate of a pair of loran transmitting stations, introduced as a delay between transmission of the master and slave signals, to place the slave station signal on the B trace when the master station signal is mounted on the A trace pedestal. { 'haf ˌpəls ˌrep·ə'tish·ən ˌrāt di,lā }

half-shift register [ELECTR] Logic circuit consisting of a gated input storage element, with or without an inverter. { 'haf ˌshift ˌrej·ə·stər }

half-subtracter [ELECTR] A logic element which operates on two digits from a preceding stage, producing as output a difference digit and a borrow digit. Also known as one-digit subtracter; two-input subtracter. { 'haf səb'trak·tər }

half-wave antenna [ELECTROMAG] An antenna whose electrical length is half the wavelength being transmitted or received. { 'haf ¦wāv an'ten·ə }

half-wave dipole *See* dipole antenna. { 'haf ¦wāv 'dī,pōl }

half-wavelength [ELECTROMAG] The distance corresponding to an electrical length of half a wavelength at the operating frequency of a transmission line, antenna element, or other device. { 'haf ¦wāv,leŋkth }

half-wave transmission line [ELECTROMAG] Transmission line which has an electrical length equal to one-half the wavelength of the signal being transmitted or received. { 'haf ¦wāv tranz'mish·ən ˌlīn }

half-word I/O buffer [COMPUT SCI] A buffer, the upper half being used to store the upper half of a word for both input and output characters, the lower half of the buffer being used for purposes such as the storage of constants. { 'haf ˌwərd ¦ī ¦ō ˌbəf·ər }

Hall-effect modulator [ELECTR] A Hall-effect multiplier used as a modulator to give an output voltage that is proportional to the product of two input voltages or currents. { 'hȯl iˌfekt 'mäj·ə ˌlād·ər }

Hall-effect multiplier [ELECTR] A multiplier based on the Hall effect, used in analog computers to solve such problems as finding the square root of the sum of the squares of three independent variables. { 'hȯl iˌfekt 'məl·təˌplī·ər }

halo [ELECTR] An undesirable bright or dark ring surrounding an image on the fluorescent screen of a television cathode-ray tube; generally due to overloading or maladjustment of the camera tube. { 'hā·lō }

halt [COMPUT SCI] The cessation of the execution of the sequence of operations in a computer program resulting from a halt instruction, hang-up, or interrupt. { hȯlt }

Hamming code [COMMUN] An error-correcting code used in data transmission. { 'ham·iŋ ˌkōd }

hamming distance *See* signal distance. { 'ham·iŋ ˌdis·təns }

ham radio *See* amateur radio. { ¦ham 'rād·ē·ō }

hand *See* end effector. { hand }

hand-held computer [COMPUT SCI] A small, battery-powered mobile computer for personal or business use. Also known as palmtop, personal digital assistant (PDA). { ¦handˌheld kəm'pyüd·ər }

handle [COMPUT SCI] **1.** One of several small squares that appear around a selected object in an object-oriented computer-graphics program, and can be dragged with a mouse to move, enlarge, reduce, or change the shape of the object. **2.** In particular, one of the two interior points on a Bézier curve that can be dragged to alter its shape. Also known as control handle. { 'han·dəl }

handler [COMPUT SCI] A computer program developed to perform one particular function, such as control of input from, and output to, a specific peripheral device. { ˌhand·lər }

handset [ENG] A combination of a telephone-type receiver and transmitter, and sometimes also other components, designed for holding in one hand. { 'handˌset }

handshaking [COMMUN] The establishment of synchronization between sending and receiving equipment by means of exchange of specific character configurations. { 'handˌshak·iŋ }

hang-up [COMPUT SCI] A nonprogrammed stop in a computer routine caused by a human mistake or a computer malfunction. { 'haŋˌəp }

haptic interface [COMPUT SCI] A device that allows a user to interact with a computer by receiving tactile feedback; for example, glove or pen devices that allow users to touch and manipulate three-dimensional virtual objects. { ˌhap·tik 'in·tərˌfās }

haptics [COMPUT SCI] The study of the use of touch in order to produce computer interfaces that will allow users to interact with digital objects by means of force feedback and tactile feedback. { 'hap·tiks }

hard code [COMPUT SCI] Program statements that are written into the computer program itself, in contrast to external tables and files to hold values and parameters used by the program. { 'härd 'kōd }

hard-coded program [COMPUT SCI] A software program or program subroutine that is designed to perform a specific task and is not easily modified. { ¦härd ˌkōd·əd 'prō·grəm }

hard copy [COMPUT SCI] Human-readable typewritten or printed characters produced on paper at the same time that information is being keyboarded in a coded machine language, as when punching cards or paper tape. { 'härd ¦käp·ē }

hard crash [COMPUT SCI] An abrupt halting of operations by a computer due to a malfunction, allowing the users or operators of the computer little or no time to minimize its effects. { 'härd 'krash }

hard disk [COMPUT SCI] A magnetic disk made of rigid material, providing high-capacity random-access storage. { 'härd ¦disk }

hard disk drive [COMPUT SCI] A high-capacity magnetic storage device that holds one or more hard disks and controls their positioning, reading, and writing; used to store programs and data, and to transfer instructions or information to the computer's working memory for use or further processing. Also known as hard drive. { 'härd ¦disk ¦drīv }

hard drive *See* hard disk drive. { 'härd ˌdrīv }

hard edit [COMPUT SCI] The process of checking and correction that causes data containing errors to be rejected by a computer system. { 'härd 'ed·it }

hardened links [COMMUN] Transmission links that require special construction or installation to assure a high probability of survival under nuclear attack. { 'härd·ənd ¦liŋks }

hard error [COMPUT SCI] Any error that results from malfunctioning of hardware, including storage devices and data transmission equipment. { 'härd 'er·ər }

hard failure [COMPUT SCI] Equipment failure that requires repair by a person with specialized knowledge before the equipment can be put back into operation. { 'härd 'fāl·yər }

hard-limiting [COMMUN] Limiting condition for which there is little variation in the output signal over the input signal range where the input is subject to limiting. { 'härd ˌlim·əd·iŋ }

hard page [COMPUT SCI] A page break that is inserted in a document by the user, and whose location is not changed by the addition, deletion, or reformatting of text. { 'härd ˌpāj }

hard patch [COMPUT SCI] A modification of a computer program, generally to repair a software error, which is applied to a stored copy of the program in machine language, so that recompilation of the source program is unnecessary and the change is permanent. { 'härd 'pach }

hard return [COMPUT SCI] A control code that is entered into a document by pressing the enter key. { 'härd ri,tərn }

hard-sectored disk [COMPUT SCI] A disk whose sectors are set up during manufacture. { 'härd ¦sek·tərd 'disk }

hardware [COMPUT SCI] The physical, tangible, and permanent components of a computer or a data-processing system. { 'härd,wer }

hardware check *See* machine check. { 'härd ,wer ,chek }

hardware compatibility [COMPUT SCI] Property of two computers such that the object code from one machine can be loaded and executed on the other to produce exactly the same results. { 'härd,wer kəm,pad·ə'bil·əd·ē }

hardware control [COMPUT SCI] The control of, and communications between, the various parts of a computer system. { 'härd,wer kən,trōl }

hardware description language [COMPUT SCI] A computer language that facilitates the documentation, design, and manufacturing of digital systems, particularly very large-scale integrated circuits, and combines program verification techniques with expert system design methodologies. { 'här,dwer di'skrip·shən ,laŋ·gwij }

hardware diagnostic [COMPUT SCI] A computer program designed to determine whether the components of a computer are operating properly. { 'härd,wer dī·əg'näs·tik }

hardware division [COMPUT SCI] Mathematical division performed by electronic circuitry on a large computer as a result of a single machine instruction. { 'här,dwer di,vizh·ən }

hardware floating point [COMPUT SCI] Complex circuitry within a central processing unit that carries out floating-point arithmetic. { 'här ,dwer 'flōd·iŋ 'pòint }

hardware key *See* dongle. { 'här,dwer ,kē }

hardware monitor [COMPUT SCI] A system used to evaluate the performance of computer hardware; it collects information such as central processing unit usage from voltage level sensors that are attached to the circuitry and measure the length of time or the number of times various signals occur, and displays this information or stores it on a medium that is then fed into a special data-reduction program. { 'härd,wer ,män·əd·ər }

hardware multiplexing [COMPUT SCI] A procedure in which a servicing unit interleaves its attention among a family of serviced units in such a way that the serviced units appear to be receiving constant attention. { 'härd,wer 'məl·tə,plek·siŋ }

hardware multiplication [COMPUT SCI] Multiplication performed by electronic circuitry on a large computer as a result of a single machine instruction. { 'här,dwer ,məl·tə·plə,kā·shən }

hard-wired [COMPUT SCI] Having a fixed wired program or control system built in by the manufacturer and not subject to change by programming. { 'härd ¦wīrd }

hard-wire telemetry *See* wire-link telemetry. { 'härd ,wīr tə'lem·ə·trē }

harmful interference [COMMUN] Radiation, emission, or induction which endangers the functioning of a radionavigation broadcasting service or of a safety broadcasting service, or obstructs or repeatedly interrupts a radio service operating in accordance with the appropriate regulations. { 'härm·fül ,int·ə'fir·əns }

harmonica bug [ELECTR] A surreptitious interception technique applied to telephone lines; the target instrument is modified so that a tuned relay bypasses the switch hook and ringing circuit when a 500-hertz tone is received; this tone was originally generated by use of a harmonica. { här'män·ə·kə ,bəg }

harmonic analyzer [ELECTR] An instrument that measures the strength of each harmonic in a complex wave. Also known as harmonic wave analyzer. { här'man·ik 'an·ə,līz·ər }

harmonic antenna [ELECTROMAG] An antenna whose electrical length is an integral multiple of a half-wavelength at the operating frequency of the transmitter or receiver. { här'män·ik an'ten·ə }

harmonic attenuation [ELECTR] Attenuation of an undesired harmonic component in the output of a transmitter. { här'män·ik ə,ten·yə'wā·shən }

harmonic distortion [ELECTR] Nonlinear distortion in which undesired harmonics of a sinusoidal input signal are generated because of circuit nonlinearity. { här'män·ik di'stór·shən }

harmonic filter [ELECTR] A filter that is tuned to suppress an undesired harmonic in a circuit. { här'män·ik 'fil·tər }

harmonic generator [ELECTR] A generator operated under conditions such that it generates strong harmonics along with the fundamental frequency. { här'män·ik 'jen·ə,rād·ər }

harmonic interference [COMMUN] Interference due to the presence of harmonics in the output of a radio transmission. { här'män·ik ,in·tər'fir·əns }

harmonic oscillator *See* sinusoidal oscillator. { här'män·ik 'äs·ə,lād·ər }

harmonic selective ringing [COMMUN] Selective ringing which employs currents of several frequencies and ringers, each tuned mechanically or electrically to the frequency of one of the ringing currents, so that only the desired ringer responds. { här'män·ik si¦lek·tiv 'riŋ·iŋ }

harmonic telephone ringer [ELECTR] Telephone ringer which responds only to alternating current within a very narrow frequency band. { här'män·ik 'tel·ə,fōn ,riŋ·ər }

harmonic wave analyzer *See* harmonic analyzer. { här'män·ik 'wāv ,an·ə,līz·ər }

hartley [COMMUN] A unit of information content, equal to the designation of 1 of 10 possible and equally likely values or states of anything used to store or convey information. { 'härt·lē }

Hartley principle [COMMUN] The principle that the total number of bits of information that can be transmitted over a channel in a given time is proportional to the product of channel bandwidth and transmission time. { 'härt·lē ˌprin·sə·pəl }

hash [COMPUT SCI] Data which are obviously meaningless, caused by human mistakes or computer malfunction. Also known as garbage; gibberish. [ELECTR] *See* grass. { hash }

hash coding *See* hashing. { 'hash ˌkōd·iŋ }

hashing [COMPUT SCI] **1.** A method for converting representations of values within fields, usually keys, to a more compact form. **2.** An addressing technique that uses keys to store and retrieve data in a file. { 'hash·iŋ }

hash total [COMPUT SCI] A sum obtained by adding together numbers having different meanings; the sole purpose is to ensure that the correct number of data have been read by the computer. { 'hash ¦tōd·əl }

HASP [COMPUT SCI] A technique used on some types of larger computers to control input and output between a computer and its peripheral devices by utilizing mass-storage devices to temporarily store data. Acronym for Houston Automatic Spooling Processor. { hasp }

hatted code [COMMUN] Randomized code consisting of an encoding section; the plain text groups are arranged in alphabetical or other significant order, accompanied by their code groups arranged in a nonalphabetical or random order. { 'had·əd ¦kōd }

H bend *See* H-plane bend. { 'āch ˌbend }

HBT *See* heterojunction bipolar transistor.

HDA *See* head/disk assembly.

H-display [NAV] A radar display format in which a short cursor is added to the target spot in a B-display format, the slope of which is proportional to the sine of the elevation angle. Also known as H-indicator; H-scan; H-scope. { 'āch diˌsplā }

HDLC *See* high-level data-link control.

HDTV *See* high-definition television.

HDX *See* half-duplex circuit.

head [COMPUT SCI] A device that reads, records, or erases data on a storage medium such as a drum or tape; examples are a small electromagnet or a sensing or punching device. [ELECTR] The photoelectric unit that converts the sound track on motion picture film into corresponding audio signals in a motion picture projector. { hed }

head crash [COMPUT SCI] The collision of the read-write head and the magnetic recording surface of a hard disk. Also known as disk crash. { 'hed ˌkrash }

head/disk assembly [COMPUT SCI] An airtight assembly including a disk pack and read/write heads. Abbreviated HDA. { 'hed 'disk ə'sem·blē }

header [COMMUN] The first section of a message, which contains information such as the addressee, routing, data, and origination time. *See* header label. { 'hed·ər }

header label [COMPUT SCI] A block of data at the beginning of a magnetic tape file containing descriptive information to identify the file. Also known as header. { 'hed·ər ˌlā·bəl }

header record [COMPUT SCI] Computer input record containing common, constant, or identifying information for records that follow. { 'hed·ər ˌrek·ərd }

head gap [COMPUT SCI] The space between the read/write head and the recording medium, such as a disk in a computer. { 'hed ˌgap }

heading-upward plan position indicator [ELECTR] A plan position indicator in which the heading of the craft appears at the top of the indicator at all times. { ¦hed·iŋ 'əp·wərd ¦plan pə¦zish·ən 'ind·əˌkād·ər }

head-mounted display [COMPUT SCI] A tracking device incorporating liquid-crystal displays or miniature cathode-ray tubes worn on a user's head to simulate a virtual environment (a three-dimensional sensation of depth) and to provide information on head movements for updating visual images. { ¦hed ˌmau̇nt·əd di'splā }

head parking [COMPUT SCI] The positioning of the read/write head of a hard disk over the landing zone to ensure against head crashes. { 'hed ˌpärk·iŋ }

head-per-track [COMPUT SCI] An arrangement having one read/write head for each magnetized track on a disk or drum to eliminate the need to move a single head from track to track. { 'hed pər 'trak }

head section *See* end section. { 'hed ˌsek·shən }

head stepping rate [COMPUT SCI] The rate at which the read/write head of a disk drive moves from one track to another on the disk surface. { 'hed ¦step·iŋ ˌrāt }

Heaviside layer *See* E layer. { 'hev·ēˌsīd ˌlā·ər }

hectometric wave [COMMUN] A radio wave between the wavelength limits of 100 and 1000 meters, corresponding to the frequency range of 3000 to 300 kilohertz. { ˌhek·tə'me·trik 'wāv }

height control [ELECTR] The video display control that adjusts picture height. { 'hīt kənˌtrōl }

height finder [ENG] A radar equipment, used to determine height of aerial targets. { 'hīt ˌfīn·dər }

height-finding radar [ENG] A radar set that measures and determines the height of an airborne object. { 'hīt ¦fīnd·iŋ 'rāˌdär }

height gain [ELECTR] A radio-wave interference phenomenon which results in a more or less periodic signal strength variation with height; this specifically refers to interference between direct and surface-reflected waves; maxima or minima in these height-gain curves occur at those elevations at which the direct and reflected waves are exactly in phase or out of phase respectively. { 'hīt ˌgān }

height input [ELECTR] Radar height information on target received by a computer from height finders and relayed via ground-to-ground data link or telephone. { 'hīt 'inˌpu̇t }

height-range indicator display *See* range-height indicator display. { 'hīt ¦rānj 'in·dəˌkād·ər }

Heising modulation *See* constant-current modulation. { 'hī·ziŋ ˌmäj·əˌlā·shən }

helical antenna [ELECTROMAG] An antenna having the form of a helix. Also known as helix antenna. { 'hel·ə·kəl an'ten·ə }

helical line [ELECTROMAG] A transmission line with a helical inner conductor. { 'hel·ə·kəl 'līn }

helical resonator [ELECTROMAG] A cavity resonator with a helical inner conductor. { 'hel·ə·kəl 'rez·ən,ād·ər }

helical scanning [COMMUN] A method of facsimile scanning in which a single-turn helix rotates against a stationary bar to give horizontal movement of an elemental area. [ELECTR] A method of recording on videotape and digital audio tape in which the tracks are recorded diagonally from top to bottom by wrapping the tape around the rotating-head drum in a helical path. [ENG] A method of radar scanning in which the antenna beam rotates continuously about the vertical axis while the elevation angle changes slowly from horizontal to vertical, so that a point on the radar beam describes a distorted helix. { 'hel·ə·kəl 'skan·iŋ }

helical traveling-wave tube *See* helix tube. { 'hel·ə·kəl |trav·ə·liŋ 'wāv ,tüb }

heliogram [COMMUN] A message transmitted on a heliograph. { 'hē·lē·ə,gram }

heliograph [COMMUN] An instrument for sending telegraphic messages by reflecting the sun's rays from a mirror. { 'hē·lē·ə,graf }

helix antenna *See* helical antenna. { 'hē,liks an'ten·ə }

helix tube [ELECTR] A traveling-wave tube in which the electromagnetic wave travels along a wire wound in a spiral about the path of the beam, so that the wave travels along the tube at a velocity approximately equal to the beam velocity. Also known as helical traveling-wave tube. { 'hē,liks ,tüb }

help screen [COMPUT SCI] Instructions that explain how to use the software of a computer system and that can be presented on the screen of a video display terminal at any time. { 'help ,skrēn }

HEMT *See* high-electron-mobility transistor.

herringbone pattern [ELECTR] An interference pattern sometimes seen on television receiver screens, consisting of a horizontal band of closely spaced V- or S-shaped lines. { 'her·iŋ ,bōn ,pad·ərn }

Hertz antenna [ELECTROMAG] An ungrounded half-wave antenna. { 'hərts an¦ten·ə }

hesitation [COMPUT SCI] A brief automatic suspension of the operations of a main program in order to perform all or part of another operation, such as rapid transmission of data to or from a peripheral unit. { ,hez·ə'tā·shən }

Hesser's variation [COMPUT SCI] A variation of a Kiviat graph in which all variables are arranged so that their plots approach the circumference of the graph as the system being evaluated approaches saturation, and the scales on the various axes may not cover the full 0–100% range, or may be in units other than percent. { 'hes·ərz ,ver·ē,ā·shən }

heterodyne [ELECTR] To mix two alternating-current signals of different frequencies in a nonlinear device for the purpose of producing two new frequencies, the sum of and difference between the two original frequencies. { 'hed·ə·rə ,dīn }

heterodyne conversion transducer *See* converter. { 'hed·ə·rə,dīn kən¦vər·zhən tranz ,dü·sər }

heterodyne detector [ELECTR] A detector in which an unmodulated carrier frequency is combined with the signal of a local oscillator having a slightly different frequency, to provide an audio-frequency beat signal that can be heard with a loudspeaker or headphones; used chiefly for code reception. { 'hed·ə·rə,dīn di'tek·tər }

heterodyne frequency [COMMUN] Either of the two new frequencies resulting from heterodyne action between the two input frequencies of a heterodyne detector. { 'hed·ə·rə,dīn 'frē·kwən·sē }

heterodyne frequency meter [ELECTR] A frequency meter in which a known frequency, which may be adjusted or fixed, is heterodyned with an unknown frequency to produce a zero beat or an audio-frequency signal whose value is measured by other means. Also known as heterodyne wavemeter. { 'hed·ə·rə,dīn 'frē·kwən·sē ,mēd·ər }

heterodyne interference *See* heterodyne whistle. { 'hed·ə·rə,dīn ,in·tər'fir·əns }

heterodyne modulator *See* mixer. { 'hed·ə·rə ,dīn 'mäj·ə,lād·ər }

heterodyne oscillator [ELECTR] **1.** A separate variable-frequency oscillator used to produce the second frequency required in a heterodyne detector for code reception. **2.** *See* beat-frequency oscillator. { 'hed·ə·rə,dīn 'äs·ə,lād·ər }

heterodyne reception [ELECTR] **1.** Radio reception in which the incoming radio-frequency signal is combined with a locally generated RF signal of different frequency, followed by detection. Also known as beat reception. **2.** In radar, use of a receiver that is tuned by adjusting a local oscillator signal within the receiver to a frequency differing from the frequency desired to be received by a fixed amount. When received signals are mixed with the reference signal, the difference frequency, called the intermediate frequency, is produced, permitting further signal processing at that convenient fixed frequency. { 'hed·ə·rə,dīn ri'sep·shən }

heterodyne wavemeter *See* heterodyne frequency meter. { 'hed·ə·rə,dīn 'wāv,mēd·ər }

heterodyne whistle [COMMUN] A steady, high-pitched audio tone heard in an ordinary amplitude-modulation radio receiver under certain conditions when two signals that differ slightly in carrier frequency enter the receiver and heterodyne to produce an audio beat. Also known as heterodyne interference. { 'hed·ə·rə ,dīn 'wis·əl }

heterojunction [ELECTR] The boundary between two different semiconductor materials, usually with a negligible discontinuity in the crystal structure. { ¦hed·ə·rō'jəŋk·shən }

heterojunction bipolar transistor [ELECTR] A bipolar transistor that has two or more materials making up the emitter, base, and collector regions, giving it a much higher maximum frequency than a silicon bipolar transistor. Abbreviated HBT. { ¦hed·ə·rə,jəŋk·shən 'bī,pōl·ər tran,zis·tər }

heterojunction field-effect transistor *See* high-electron-mobility transistor. { ¦hed·ə·rə ,jəŋk·shən 'fēld i,fekt tran,zis·tər }

heuristic algorithm *See* dynamic algorithm. { hyü'ris·tik 'al·gə,rith·əm }

heuristic program [COMPUT SCI] A program in which a computer tries each of several methods of solving a problem and judges whether the program is closer to solution after each attempt. Also known as heuristic routine. { hyü'ris·tik 'prō·grəm }

heuristic routine *See* heuristic program. { hyü 'ris·tik rü'tēn }

hexadecimal notation [COMPUT SCI] A notation in the scale of 16, using decimal digits 0 to 9 and six more digits that are sometimes represented by A, B, C, D, E, and F. { ,hek·sə'des·məl nō'tā·shən }

HF *See* high frequency.

HFET *See* high-electron-mobility transistor. { 'āch,fet }

HH beacon [NAV] Nondirectional radio homing beacon which has a power output of 2000 watts or greater. { ,āch'āch ,bē·kən }

hidden file [COMPUT SCI] A disk file that does not appear in a directory listing and cannot be displayed, changed, or deleted. { ¦hid·ən ¦fīl }

hierarchical control [CONT SYS] The organization of controllers in a large-scale system into two or more levels so that controllers in each level send control signals to controllers in the level below and feedback or sensing signals to controllers in the level above. Also known as control hierarchy. { ¦hī·ər¦är·kə·kəl kən'trōl }

hierarchical distributed processing system [COMPUT SCI] A type of distributed processing system in which processing functions are distributed outward from a central computer to intelligent terminal controllers or satellite information processors. Also known as host-centered system; host/satellite system. { ¦hī·ər¦är·kə·kəl di¦strib·yəd·əd 'prä,ses·iŋ ,sis·təm }

hierarchical file [COMPUT SCI] A file with a grandfather-father-son structure. { ¦hī·ər¦är·kə·kəl 'fīl }

hierarchical storage management [COMPUT SCI] A method of managing large amounts of data in which files are assigned to various storage media based on how soon or how frequently they will be needed. { ¦hī·ər¦är·kə·kəl 'stór·ij ,man·ij·mənt }

high-altitude radio altimeter *See* radar altimeter. { 'hī ¦al·tə,tüd 'rād·ē·ōal'tim·əd·ər }

high boost *See* high-frequency compensation. { 'hī ¦büst }

high core [COMPUT SCI] The locations with higher addresses in a computer's main storage, usually occupied by the operating system. { 'hī 'kór }

high definition [COMMUN] Television or facsimile equivalent of high fidelity, in which the reproduced image contains such a large number of accurately reproduced elements that picture details approximate those of the original scene. { 'hī ,def·ə'nish·ən }

high-definition television [COMMUN] A television system with a resolution of more than 1000 scan lines, as compared to 525–625 scan lines in conventional systems. Abbreviated HDTV. { ¦hī def·ə¦nish·ən 'tel·ə,vizh·ən }

high-density disk [COMPUT SCI] A diskette that holds two or more times as much data per unit area as a double-density disk of the same size. { ¦hī ¦den·səd·ē 'disk }

high-density drive [COMPUT SCI] A disk drive that accepts both high-density and double-density disks. { ¦hī ¦den·səd·ē 'drīv }

high-electron-mobility transistor [ELECTR] A type of field-effect transistor consisting of gallium arsenide and gallium aluminum arsenide, with a Schottky metal contact on the gallium aluminum arsenide layer and two ohmic contacts penetrating into the gallium arsenide layer, serving as the gate, source, and drain respectively. Abbreviated HEMT. Also known as heterojunction field-effect transistor (HFET); modulation-doped field-effect transistor (MODFET); selectively doped heterojunction transistor (SDHT); two-dimensional electron gas field-effect transistor (TEGFET). { 'hī i'lek,trän mō ¦bil·əd·ē tran,zis·tər }

higher-level language *See* high-level language. { 'hī·ər ,lev·əl ,laŋ·gwij }

higher-order language *See* high-level language. { 'hī·ər ,ór·dər ,laŋ·gwij }

higher-order software [COMPUT SCI] Software for designing and documenting an information system by decomposing the system into elementary components that are mathematically correct and error-free. Abbreviated HOS. { ¦hī·ər ,ór·dər 'sóft,wer }

higher than high-level language [COMPUT SCI] A programming language, such as an application development language, report program, or financial planning language, that is oriented toward a particular application and is much easier to use for that application than a conventional programming language. { 'hī·ər than 'hī ,lev·əl ,laŋ·gwij }

high frequency [COMMUN] Federal Communications Commission designation for the band from 3 to 30 megahertz in the radio spectrum. Abbreviated HF. { 'hī ¦frē·kwən·sē }

high-frequency carrier telegraphy [COMMUN] Form of carrier telegraphy in which the carrier currents have their frequencies above the range transmitted over a voice-frequency telephone channel. { 'hī ¦frē·kwən·sē 'kar·ē·ər tə'leg·rə·fē }

high-frequency compensation [ELECTR] Increasing the amplification at high frequencies with respect to that at low and middle frequencies in a given band, such as in a video band or an audio band. Also known as high boost. { 'hī ¦frē·kwən·sē ˌkäm·pən'sā·shən }

high-frequency propagation [COMMUN] Propagation of radio waves in the high-frequency band, which depends entirely on reflection from the ionosphere. { 'hī ¦frē·kwən·sē ˌpräp·ə'gā·shən }

high-frequency transformer [ELECTR] A transformer which matches impedances and transmits a frequency band in the carrier (or higher) frequency ranges. { 'hī ¦frē·kwən·sē tranz'fór·mər }

high-information-content display [ELECTR] An electronic display that has a sufficient number of pixels (75,000 to 2,000,000) to show standard or high-definition television images or comparable computer images. { ¦hī ˌin·fər'mā·shən ˌkän ˌtent diˌsplā }

high level [COMMUN] A range of allowed picture parameters defined by the MPEG-2 video coding specifications which corresponds to high-definition television. [ELECTR] The more positive of the two logic levels or states in a binary digital logic system. { 'hī ¦lev·əl }

high-level data-link control Abbreviated HDLC. [COMMUN] A bit-oriented protocol for managing information flow in a data communications channel that supports both full-duplex and half-duplex transmission, and both point-to-point and multipoint communications using synchronous data transmission. [COMPUT SCI] A communications protocol that allows devices from different manufacturers to interface with each other and standardizes the transmission of packets of information between them. { 'hī ˌlev·əl 'dad·ə ˌliŋk kənˌtrōl }

high-level index [COMPUT SCI] The first part of a file name, which frequently specifies the category of data to which it belongs. { 'hīˌlev·əl 'inˌdeks }

high-level language [COMPUT SCI] A computer language whose instructions or statements each correspond to several machine language instructions, designed to make coding easier. Also known as higher-level language; higher-order language. { 'hī ˌlev·əl 'laŋ·gwij }

high-level modulation [COMMUN] Modulation produced at a point in a system where the power level approximates that at the output of the system. { 'hī ˌlev·əl ˌmäj·ə'lā·shən }

highlights [ELECTR] Bright areas occurring in a video image. { 'hīˌlīts }

high-order [COMPUT SCI] Pertaining to a digit location in a numeral, the leftmost digit being the highest-order digit. { 'hī ¦ór·dər }

high-pass filter [ELECTR] A filter that transmits all frequencies above a given cutoff frequency and substantially attenuates all others. { 'hī ˌpas 'fil·tər }

high-positive indicator [COMPUT SCI] A component in some computers whose status is "on" if the number tested is positive and nonzero. { 'hī ¦päz·əd·iv 'in·dəˌkād·ər }

high Q [ELECTR] A characteristic wherein a component has a high ratio of reactance to effective resistance, so that its Q factor is high. { ¦hī 'kyü }

high-Q cavity [ELECTROMAG] A cavity resonator which has a large Q factor, and thus has a small energy loss. Also known as high-Q resonator. { 'hī ˌkyü 'kav·əd·ē }

high-Q resonator *See* high-Q cavity. { 'hī ˌkyü 'rez·ənˌād·ər }

high-resolution radar [ENG] A radar system which can discriminate between two close targets. { 'hī ˌrez·əˌlü·shən 'rāˌdär }

high side [COMPUT SCI] The part of a remote device that communicates with a computer. { 'hī ˌsīd }

high-speed carry [COMPUT SCI] A technique in parallel addition to speed up the propagation of carries. { 'hī ˌspēd'kar·ē }

high-speed data acquisition system [COMPUT SCI] A system which collects and transmits data rapidly to a monitoring and controlling center. { 'hī ˌspēd 'dad·ə ¦ak·wəˌzish·ən ˌsis·təm }

high-speed printer [COMPUT SCI] A printer which can function at a high rate, relative to the state of the art; 600 lines per minute is considered high speed. Abbreviated HSP. { 'hī ˌspēd 'print·ər }

high-speed reader [COMPUT SCI] The fastest input device existing at a particular time in the state of the technology. { 'hī ˌspēd 'rēd·ər }

high-speed storage *See* rapid storage. { 'hī ˌspēd 'stór·ij }

high-tier system [COMMUN] A wireless telephone system that supports base stations with large coverage areas and low traffic densities, but provides low-quality voice service and has limited data-service capabilities with high delays. { ˌhī 'tir ˌsis·təm }

high-water mark [COMPUT SCI] The maximum number of jobs that are in a queue awaiting execution by a large computer system during a specified period of observation. { ¦hī 'wód·ər ˌmärk }

Hilbert transformer [ELECTR] An electric filter whose gain is $-j$ for positive frequencies and j for negative frequencies, where j is the square root of -1. { 'hil·bərt tranzˌfór·mər }

hill bandwidth [ELECTR] The difference between the upper and lower frequencies at which the gain of an amplifier is 3 decibels less than its maximum value. { ¦hil 'bandˌwidth }

H-indicator *See* H-display. { 'āch ¦in·dəˌkād·ər }

hiss [COMMUN] Random noise in the audio-frequency range, similar to prolonged sibilant sounds. { his }

historical data [COMPUT SCI] Any data that is not actively maintained by a computer system and cannot be readily revised or updated. { hi'stär·ə·kəl 'dad·ə }

hit [COMPUT SCI] **1.** The obtaining of a correct answer in an information-retrieval system. **2.** An attempt to access a specified piece of information on a website; a count of the number of such attempts is an indicator or the usage or popularity of the site. { hit }

hit-on-the-fly system [COMPUT SCI] A printer in a computer system where either the print roller or the paper is in continuous motion. { ¦hid ȯn thə ′flī ‚sis·təm }

hit rate [COMPUT SCI] The ratio of the number of records found and processed during a particular processing run, to the total number of records available. { ′hit ‚rāt }

H mode *See* transverse electric mode. { ′āch ‚mōd }

hog [COMPUT SCI] A computer program that uses excessive computer resources, such as memory or processing power, or requires excessive time to execute. { häg }

hoghorn antenna *See* horn antenna. { ′häg ‚hȯrn an′ten·ə }

hold [COMPUT SCI] To retain information in a computer storage device for further use after it has been initially utilized. { hōld }

hold control [ELECTR] A manual control that changes the frequency of the horizontal or vertical sweep oscillator in an analog television receiver, so that the frequency more nearly corresponds to that of the incoming synchronizing pulses. { ′hōld kən‚trōl }

hold facility [COMPUT SCI] The ability of a computer to operate in a hold mode. { ′hōld fə ‚sil·əd·ē }

holding time [COMMUN] Period of time a trunk or circuit is in use on a call, including operator's time in connecting and subscriber's or user's conversation time. { ′hōl·diŋ ‚tīm }

hold mode [COMPUT SCI] The state of an analog computer in which its operation is interrupted without altering the values of the variables it is handling, so that computation can continue when the interruption is over. Also known as interrupt mode. { ′hōld ‚mōd }

hold queue [COMPUT SCI] A queue consisting of jobs that have been submitted for execution by a large computer system and are waiting to be run. { ′hōld ‚kyü }

hole [SOLID STATE] A vacant electron energy state near the top of an energy band in a solid; behaves as though it were a positively charged particle. Also known as electron hole. { hōl }

hole conduction [ELECTR] Conduction occurring in a semiconductor when electrons move into holes under the influence of an applied voltage and thereby create new holes. { ′hōl kən ¦dək·shən }

hole mobility [ELECTR] A measure of the ability of a hole to travel readily through a semiconductor, equal to the average drift velocity of holes divided by the electric field. { ′hōl mō‚bil·əd·ē }

holistic masks [COMPUT SCI] In character recognition, that set of characters which resides within a character reader and theoretically represents the exact replicas of all possible input characters. { hō′lis·tik ′masks }

Hollerith string [COMPUT SCI] A sequence of characters preceded by an H and a character count in FORTRAN, as 4HSTOP. { ′häl·ə·rəth ‚striŋ }

hollow-pipe waveguide [ELECTROMAG] A waveguide consisting of a hollow metal pipe; electromagnetic waves are transmitted through the interior and electric currents flow on the inner surfaces. { ′häl·ō‚pīp ′wāv‚gīd }

holographic memory [COMPUT SCI] A memory in which information is stored in the form of holographic images on thermoplastic or other recording films. { ‚häl·ə′graf·ik ′mem·rē }

holographic storage [COMMUN] A form of data storage in which bits of information are distributed throughout the storage volume and recorded interferometrically, rather than being stored at discrete locations in the medium. { ¦häl·ə‚graf·ik ′stȯr·ij }

home [COMPUT SCI] The location at the upper left-hand corner of an electronic display. [NAV] To navigate toward a point by maintaining constant some navigational parameter other than altitude. { hōm }

home address [COMPUT SCI] A technique used to identify each disk track uniquely by means of a 9-byte record immediately following the index marker; the record contains a flag (good or defective track), cylinder number, head number, cyclic check, and bit count appendage. { ¦hōm ′ad‚res }

home-on-jam [ELECTR] A feature that permits radar or a passive seeker to track a jamming source in angle, to guide a weapon to the jammer. { ¦hōm ‚ȯn ¦jam }

home page [COMPUT SCI] A document in a hypertext system that serves as the point of entry to a web of related documents, and generally contains introductory information and hyperlinks to other documents in the web. Also known as welcome page. { ‚hōm ′pāj }

home record [COMPUT SCI] The first record in the chaining method of file organization. { ′hōm ¦rek·ərd }

hometaxial-base transistor [ELECTR] Transistor manufactured by a single-diffusion process to form both emitter and collector junctions in a uniformly doped silicon slice; the resulting homogeneously doped base region is free from accelerating fields in the axial (collector-to-emitter) direction, which could cause undesirable high current flow and destroy the transistor. { ′häm·ə‚tak·sē·əl ‚bās tran′zis·tər }

homing antenna [ELECTROMAG] A directional antenna array used in flying directly to a target that is emitting or reflecting radio or radar waves. { ′hōm·iŋ an‚ten·ə }

homing beacon [NAV] A radio beacon, either airborne or on the ground, toward which an aircraft can fly if equipped with a radio compass or homing adapter. Also known as radio homing beacon. { ′hōm·iŋ ‚bē·kən }

homing device [ENG] A device incorporated in a guided missile or the like to home it on a target. [NAV] A transmitter, receiver, or adapter used for homing aircraft or used by aircraft for homing purposes. { ′hōm·iŋ di‚vīs }

homodyne reception [ELECTR] **1.** A system of radio reception for suppressed-carrier systems of radiotelephony, in which the receiver generates a voltage having the original carrier frequency and combines it with the incoming signal. Also known as zero-beat reception. **2.** Referring to a radio or radar receiver in which received signals are mixed with a reference signal at the same frequency as the signal intended to be received; the mixing produces a voltage output dependent only on the phase difference of the two inputs, hence a voltage at the "beat" frequency if there is a slight difference. { 'hä·mə,dīn ri'sep·shən }

homogeneous network [COMPUT SCI] A computer network consisting of fairly similar computers from a single manufacturer. { ¦hō·mə ¦jē·nē·əs 'net,wərk }

homojunction bipolar transistor [ELECTR] Any bipolar transistor that is composed entirely of one type of semiconductor. { ¦hō·mō,jəŋk·shən bī,pō·lər tran'zis·tər }

hook [COMPUT SCI] A modification of a computer program to add instructions to an existing part of the program. [ELECTR] A circuit phenomenon occurring in four-zone transistors, wherein hole or electron conduction can occur in opposite directions to produce voltage drops that encourage other types of conduction. { hu̇k }

hoot stop [COMPUT SCI] A closed loop that generates an audible signal; usually employed to signal an error or for operating convenience. { 'hüt ,stäp }

hop [COMMUN] A single reflection of a radio wave from the ionosphere back to the earth in travelling from one point to another. { häp }

horizontal blanking [ELECTR] Blanking of a video picture tube during the horizontal retrace. { ,här·ə'zänt·əl 'blaŋk·iŋ }

horizontal blanking pulse [ELECTR] The rectangular pulse that forms the pedestal of the composite video signal between active horizontal lines and causes the display device to be cut off during retrace. Also known as line-frequency blanking pulse. { ,här·ə'zänt·əl 'blaŋk·iŋ ,pəls }

horizontal centering control [ELECTR] The centering control provided in a video display to shift the position of the entire image horizontally in either direction on the screen. { ,här·ə'zänt·əl 'sen·tə·riŋ kən,trōl }

horizontal convergence control [ELECTR] The control that adjusts the amplitude of the horizontal dynamic convergence voltage in a video display device. { ,här·ə'zänt·əl kən'vər·jəns kən ,trōl }

horizontal definition *See* horizontal resolution. { ,här·ə'zänt·əl ,def·ə'nish·ən }

horizontal deflection oscillator [ELECTR] The oscillator that produces, under control of the horizontal synchronizing signals, the sawtooth voltage waveform that is amplified to feed the horizontal deflection coils on the picture tube of a video display. Also known as horizontal oscillator. { ,här·ə'zänt·əl di'flek·shən 'äs·ə,lād·ər }

horizontal distributed processing system [COMPUT SCI] A type of distributed system in which two or more computers which are logically equivalent are connected together, with no hierarchy or master/slave relationship. { ,här·ə'zänt·əl di¦strib·yəd·əd 'prä,ses·iŋ ,sis·təm }

horizontal flyback [ELECTR] Flyback in which the electron beam of a picture tube returns from the end of one scanning line to the beginning of the next line. Also known as horizontal retrace. { ,här·ə'zänt·əl 'flī,bak }

horizontal frequency *See* line frequency. { ,här·ə'zänt·əl 'frē·kwən·sē }

horizontal hold control [ELECTR] The hold control that changes the free-running period of the horizontal deflection oscillator in an analog television receiver, so that the picture remains steady in the horizontal direction. { ,här·ə'zänt·əl 'hōld kən,trōl }

horizontal instruction [COMPUT SCI] An instruction in machine language to carry out independent operations on various operands in parallel or in a well-defined time sequence. { ,här·ə'zänt·əl in'strək·shən }

horizontal line frequency *See* line frequency. { ,här·ə'zänt·əl 'līn ,frē·kwən·sē }

horizontal oscillator *See* horizontal deflection oscillator. { ,här·ə'zänt·əl 'äs·ə,lād·ər }

horizontal output stage [ELECTR] The television receiver stage that feeds the horizontal deflection coils of the picture tube through the horizontal output transformer; may also include a part of the second-anode power supply for the picture tube. { ,här·ə'zänt·əl 'au̇t,pu̇t ,stāj }

horizontal output transformer [ELECTR] A transformer used in a television receiver to provide the horizontal deflection voltage, the high voltage for the second-anode power supply of the picture tube, and the filament voltage for the high-voltage rectifier tube. Also known as flyback transformer; horizontal sweep transformer. { ,här·ə'zänt·əl 'au̇t,pu̇t tranz ,fȯr·mər }

horizontal parity check *See* longitudinal parity check. { ,här·ə'zänt·əl 'par·əd·ē ,chek }

horizontal polarization [COMMUN] Transmission of linear polarized radio waves whose electric field vector is parallel to the earth's surface. { ,här·ə'zänt·əl ,pō·lə·rə'zā·shən }

horizontal resolution [ELECTR] The number of individual picture elements or dots that can be distinguished in a horizontal scanning line of a video display. Also known as horizontal definition. { ,här·ə'zänt·əl ,rez·ə'lü·shən }

horizontal retrace *See* horizontal flyback. { ,här·ə'zänt·əl 'rē,trās }

horizontal scanning frequency [ELECTR] The number of horizontal lines scanned by the electron beam in a video system in 1 second. { ,här·ə'zänt·əl 'skan·iŋ ,frē·kwən·sē }

horizontal sweep [ELECTR] The sweep of the electron beam from left to right across the screen of a cathode-ray tube. { ,här·ə'zänt·əl 'swēp }

horizontal sweep transformer *See* horizontal output transformer. { ,här·ə'zänt·əl 'swēp tranz ,fȯr·mər }

horizontal synchronizing pulse [ELECTR] The rectangular pulse transmitted at the end of each line in an analog television system, to keep the receiver in line-by-line synchronism with the transmitter. Also known as line synchronizing pulse. { ˌhär·əˈzänt·əl ˈsiŋ·krəˌniz·iŋ ˌpəls }

horizontal system [COMPUT SCI] A programming system in which instructions are written horizontally, that is, across the page. { ˌhär·əˈzänt·əl ˈsis·təm }

horizontal vee [ELECTROMAG] An antenna consisting of two linear radiators in the form of the letter V, lying in a horizontal plane. { ˌhär·əˈzänt·əl ˈvē }

horn *See* horn antenna. { hȯrn }

horn antenna [ELECTROMAG] A microwave antenna produced by flaring out the end of a circular or rectangular waveguide into the shape of a horn, for radiating radio waves directly into space. Also known as electromagnetic horn; flare (British usage); hoghorn antenna (British usage); horn; horn radiator. { ˈhȯrn anˈten·ə }

horn radiator *See* horn antenna. { ˈhȯrn ˈrād·ēˌād·ər }

HOS *See* higher-order software.

hospital information system [COMPUT SCI] The collection, evaluation or verification, storage, and retrieval of information about a patient. { ˈhäsˌpid·əl ˌin·fərˈmā·shən ˌsis·təm }

host-based system [COMMUN] A communications system that is controlled by a central computer system. { ˈhōst ˌbāst ˌsis·təm }

host-centered system *See* hierarchical distributed processing system. { ˈhōst ˌsen·tərd ˌsis·təm }

host computer [COMPUT SCI] **1.** The central or controlling computer in a time-sharing or distributed-processing system. **2.** The computer upon which depends a specialized computer handling the input/output functions in a real-time system. **3.** A computer that can function as the source or recipient of data transfers on a network. { ˈhōst kəm¦pyüd·ər }

host language database management system [COMPUT SCI] A database management system that, from a programmer's point of view, represents an extension of an existing programming language. { ˈhōst ˌlaŋ·gwij ˈdad·əˌbās ˈman·ij·mənt ˌsis·təm }

host processor [COMPUT SCI] The central computer in a hierarchical distributed processing system, which is typically located at some central site where it serves as a focal point for the collection of data, and often for the provision of services which cannot economically be distributed. { ˈhōst ¦präˌses·ər }

host/satellite system *See* hierarchical distributed processing system. { ˈhōst ˈsad·əlˌīt ˌsis·təm }

hot carrier [ELECTR] A charge carrier, which may be either an electron or a hole, that has relatively high energy with respect to the carriers normally found in majority-carrier devices such as thin-film transistors. { ˈhät ¦kar·ē·ər }

hot-carrier diode *See* Schottky barrier diode. { ˈhät ˌkar·ē·ər ˈdīˌōd }

hot editing [CONT SYS] A method for detecting errors in the programming of a robot in which as many errors as possible are identified and resolved during testing, without setting the robotic program to its starting condition. { ˈhät ˈed·əd·iŋ }

hot electron [ELECTR] An electron that is in excess of the thermal equilibrium number and, for metals, has an energy greater than the Fermi level; for semiconductors, the energy must be a definite amount above that of the edge of the conduction band. { ˈhät iˈlekˌträn }

hot-electron transistor [ELECTR] A transistor in which electrons tunnel through a thin emitter-base barrier ballistically (that is, without scattering), traverse a very narrow base region, and cross a barrier at the base-collector interface whose height, controlled by the collector voltage, determines the fraction of electrons coming to the collector. { ¦hät iˈlekˌträn ˌtranˈzis·tər }

hot key [COMPUT SCI] A computer key or key combination that causes a specified action to occur, regardless of what else the computer is currently doing. { ˈhät ˌkē }

hot line [COMMUN] Direct circuit between two points, available for immediate use without patching or switching. { ˈhät ˈlīn }

hot link [COMPUT SCI] A linking of information in two documents so that modification of the information in the source document results in the same change in the destination document. { ¦hät ˈliŋk }

hot spot [COMPUT SCI] A word in a multiprocessor memory that several processors attempt to access simultaneously, creating a conflict or bottleneck. { ˈhät ˌspät }

housekeeping [COMPUT SCI] Those operations or routines which do not contribute directly to the solution of a computer program, but rather to the organization of the program. { ˈhau̇sˌkēp·iŋ }

housekeeping run [COMPUT SCI] The performance of a program or routine to maintain the structure of files, such as sorting, merging, addition of new records, or deletion or modification of existing records. { ˈhau̇sˌkēp·iŋ ˌrən }

Houston Automatic Spooling Processor *See* HASP. { hyüs·tən ¦ȯd·ə¦mad·ik ˈspül·iŋ ˌpräˌses·ər }

howler [COMMUN] In telephone practice, an associated unit by which the test desk operator may connect a high tone of varying loudness to a subscriber's line to call the subscriber's attention to the fact that the phone receiver is off the hook. { ˈhau̇l·ər }

howl repeater [COMMUN] Condition in telephone repeater operation where more energy is returned than sent, resulting in an oscillation being set up on the circuit. { ˈhau̇l riˈpēd·ər }

h parameter [ELECTR] One of a set of four transistor equivalent-circuit parameters that conveniently specify transistor performance for small voltages and currents in a particular circuit. Also known as hybrid parameter. { ¦āch pəˌram·əd·ər }

HPBW *See* half-power beamwidth.

H plane [ELECTROMAG] The plane of an antenna in which lies the magnetic field vector of linearly polarized radiation. { ¦āch ˌplān }

H-plane bend [ELECTROMAG] A rectangular waveguide bend in which the longitudinal axis of the waveguide remains in a plane parallel to the plane of the magnetic field vector throughout the bend. Also known as H bend. { ¦āch ˌplān ˌbend }

H-plane T junction [ELECTROMAG] Waveguide T junction in which the change in structure occurs in the plane of the magnetic field. Also known as shunt T junction. { ¦āch ˌplān 'tē ˌjəŋk·shən }

H-scan *See* H-display. { 'āch ˌskan }

H-scope *See* H-display. { 'āch ˌskōp }

HSP *See* high-speed printer.

HTML *See* Hypertext Markup Language.

HTTP *See* Hypertext Transfer Protocol.

hub [COMPUT SCI] An electric socket in a plugboard into which one may insert or connect leads or may plug wires. { həb }

hub ring [COMPUT SCI] A thin plastic ring placed around the center hole of a floppy disk to prevent the disk from warping and damaging its contents if it is improperly inserted in a disk drive. { 'həb ˌriŋ }

hue control [ELECTR] A control that varies the phase of the chrominance signals with respect to that of the burst signal in an analog color television receiver, in order to change the hues in the image. Also known as phase control. { 'hyü kənˌtrōl }

Huffman method [COMPUT SCI] A data compression technique in which a bit representation for each character is determined that is as close as possible to the character's predicted information content, based on its frequency of occurrence. { 'həf·mən ˌmeth·əd }

hum [ELECTR] An electrical disturbance occurring at the power supply frequency or its harmonics, usually 60 or 120 hertz in the United States. { həm }

human-computer interaction [COMPUT SCI] The processes through which human users work with interactive computer systems. { ¦yü·mən kəm¦pyüd·ər ˌin·tər'ak·shən }

hum bar [ELECTR] A dark horizontal band extending across a television picture due to excessive hum in the video signal applied to the input of the picture tube. { 'həm ˌbär }

hum modulation [ELECTR] Modulation of a radio-frequency signal or detected audio-frequency signal by hum; heard in a radio receiver only when a station is tuned in. { 'həm ˌmäj·ə ˌlā·shən }

hunting [CONT SYS] Undesirable oscillation of an automatic control system, wherein the controlled variable swings on both sides of the desired value. { 'hənt·iŋ }

hunting circuit *See* lockout circuit. { 'hənt·iŋ ˌsər·kət }

H wave *See* transverse electric wave. { 'āch ˌwāv }

hybrid algebraic manipulation language [COMPUT SCI] The most ambitious type of algebraic manipulation language, which accepts the broadest spectrum of mathematical expressions but possesses, in addition, special representations and special algorithms for particular special classes of expressions. { 'hī·brəd ˌal·jə'brā·ik məˌnip·yə'lā·shən ˌlaŋ·gwij }

hybrid AM IBOC [COMMUN] The initial mode of the AM IBOC system approved by the Federal Communications Commission for use in the United States that adds digital audio capacity to an AM signal by inserting digital sidebands in the spectrum above, below, and within the analog AM signal. The digital audio data rate can range from 36 to 56 kbits/s, and the corresponding ancillary data rate is 0.4 kbits/s in both cases. { 'hī·brəd 'āˌem 'īˌbäk }

hybrid computer [COMPUT SCI] A computer designed to handle both analog and digital data. Also known as analog-digital computer; hybrid system. { 'hī·brəd kəm'pyüd·ər }

hybrid distributed processing system [COMPUT SCI] A distributed processing system that includes both horizontal and hierarchical distribution. { 'hī·brəd di¦strib·yəd·əd 'präˌses·iŋ ˌsis·təm }

hybrid electromagnetic wave [ELECTROMAG] Wave which has both transverse and longitudinal components of displacement. { 'hī·brəd i¦lek·trō·mag¦ned·ik 'wāv }

hybrid FM IBOC [COMMUN] The first of three modes in the FM IBOC system approved by the Federal Communications Commission for use in the United States that increases data capacity by a adding additional carriers closer to the analog host signal. The hybrid IBOC mode adds one frequency partition around the analog carrier and is characterized by the highest possible digital and analog audio quality with a limited amount of ancillary data available to the broadcaster. Digital audio data rates can range from 64 to 96 kbits/s, and the corresponding ancillary data rate can range from 33 kbits/s for 64-kbits/s audio and 1 kbit/s for 96-kbits/s audio. { 'hī·brəd 'efˌem 'īˌbäk }

hybrid hardware control [COMPUT SCI] The control of and communication between the various parts of a hybrid computer. { 'hī·brəd 'härdˌwer kənˌtrōl }

hybrid input/output [COMPUT SCI] The routines required to handle inputs to and outputs from a computer system comprising digital and analog computers. { 'hī·brəd ¦inˌpu̇t ¦au̇tˌpu̇t }

hybrid integrated circuit [ELECTR] A circuit in which one or more discrete components are used in combination with integrated-circuit construction. { 'hī·brəd ¦int·əˌgrād·əd 'sər·kət }

hybrid interface [COMPUT SCI] A device that joins a digital to an analog computer, converting digital signals transmitted serially by the digital computer into analog signals that are transmitted simultaneously to the various units of the analog computer, and vice versa. { 'hī·brəd 'in·tərˌfās }

hybrid junction [ELECTR] A transformer, resistor, or waveguide circuit or device that has four pairs of terminals so arranged that a signal entering at one terminal pair divides and emerges from the two adjacent terminal pairs, but is unable to reach the opposite terminal pair. Also known as bridge hybrid. { ˈhī·brəd ˈjəŋk·shən }

hybrid network [COMMUN] Nonhomogeneous communications network required to operate with signals of dissimilar characteristics (such as analog and digital modes). [ELECTR] A four-port circuit, useful in radar and other microwave applications as a power switch or signal comparator, in which two inputs add constructively at one output and destructively at the other, with good isolation between the two iinputs and between the two outputs; the waveguide magic tee and the rat race are types of hybrid networks. { ˈhī·brəd ˈnet,wərk }

hybrid parameter See h parameter. { ˈhī·brəd pəˈram·əd·ər }

hybrid problem analysis [COMPUT SCI] The determination of the parts of a problem best suited for the digital computer. { ˈhī·brəd ¦präb·ləm əˈnal·ə·səs }

hybrid programming [COMPUT SCI] Hybrid system routines that handle timing, function generation, and simulation. { ˈhī·brəd ˈprō,gram·iŋ }

hybrid redundancy [COMPUT SCI] A synthesis of triple modular redundancy and standby replacement redundancy, consisting of a triple modular redundancy system (or, in general, an N-modular redundancy system) with a bank of spares so that when one of the units in the triple modular redundancy system fails it is replaced by a spare unit. { ˈhī·brəd riˈdən·dən·sē }

hybrid simulation [COMPUT SCI] The use of a hybrid computer for purposes of simulation. { ˈhī·brəd ,sim·yəˈlā·shən }

hybrid system [COMPUT SCI] **1.** A computer system that performs two or more functions, such as data processing and word processing. **2.** See hybrid computer. { ˈhī,brid ,sis·təm }

hybrid system checkout [COMPUT SCI] The static check of a hybrid system and of the digital program and analog wiring required to solve a problem. { ˈhī·brəd ¦sis·təm ˈchek,au̇t }

hybrid tee [ELECTROMAG] A microwave hybrid junction composed of an E-H tee with internal matching elements; it is reflectionless for a wave propagating into the junction from any arm when the other three arms are match-terminated. Also known as magic tee. { ˈhī·brəd ˈtē }

HYCATS See Hydrofoil Collision Avoidance and Tracking System. { ˈhī,kats }

Hydrofoil Collision Avoidance and Tracking System [NAV] A computer-based system designed to automate target tracking and navigation functions in order to increase the safety of high-speed ships; it makes use of the superposition of chart data over radar data, allowing precise positioning to be accomplished. Abbreviated HYCATS. { ˈhī·drə,fȯil kəˈlish·ən ə,vȯid·əns ən ˈtrak·iŋ ,sis·təm }

hydrophone [ENG] A device which receives underwater sound waves and converts them to electric waves. { ˈhī·drə,fōn }

hydrophone array [COMMUN] A group of two or more hydrophones which feed into a common receiver. { ˈhī·drə,fōn ə,rā }

hyperbolic amplitude [COMMUN] Excursion of a signal measured along hyperbolic rather than Cartesian coordinates. { ¦hī·pər¦bäl·ik ˈam·plə ,tüd }

hyperbolic antenna [ELECTROMAG] A radiator whose reflector in cross section describes a half hyperbola. { ¦hī·pər¦bäl·ik anˈten·ə }

hyperbolic line of position [NAV] A line of position in the shape of a hyperbola, determined by measuring the difference in the phase or time of transit of radiations from fixed points; measurement may be made through the use of radio, sound, or light. { ¦hī·pər¦bäl·ik ¦līn əv pəˈzish·ən }

hyperbolic navigation [NAV] Navigation by maintaining constant the indication of two parameters; the parameters can have any reasonable ratio to each other. { ¦hī·pər¦bäl·ik ,nav·əˈgā·shən }

hypercube [COMPUT SCI] A configuration of parallel processors in which the locations of the processors correspond to the vertices of a mathematical hypercube and the links between them correspond to its edges. { ˈhī·pər ,kyüb }

hyperdisk [COMPUT SCI] A mass-storage technique which uses a large-capacity storage and a disk for overflow. { ˈhī·pər,disk }

hyperlink [COMPUT SCI] A highlighted word, phrase, or image in the display of a computer document which, when chosen, connects the user to another part of the same document or to different document (text, image, audio, video, or animation). In electronic documents, these cross references can be followed by a mouse click, and the target of the hyperlink may be on a physically distant computer connected by a network or the Internet. { ˈhī·pər,liŋk }

hypermedia [COMPUT SCI] Hypertext-based systems that combine data, text, graphics, video, and sound. { ˈhī·pər,mē·dē·ə }

hypertape control unit See tape control unit. { ˈhī·pər,tāp kənˈtrōl ,yü·nət }

hypertape drive See cartridge tape drive. { ˈhī·pər,tāp ,drīv }

hypertext [COMPUT SCI] A data structure in which there are links between words, phrases, graphics, or other elements and associated information so that selection of a key object can activate a linkage and reveal the information. { ˈhī·pər,tekst }

Hypertext Markup Language [COMPUT SCI] The language used to specifically encode the content and format of a document and to link documents on the World Wide Web. Abbreviated HTML. { ¦hī·pər,tekst ˈmärk,əp ,laŋ·gwij }

Hypertext Transfer Protocol [COMPUT SCI] The communication protocol for transmitting linked documents between computers; it is the basis for the World Wide Web and follows the TCP/IP protocol for the client-server model of computing. Abbreviated HTTP. { ˈhī·pər,tekst ˈtranz·fər ,prōd·ə,kȯl }

hypervisor [COMPUT SCI] A control program enabling two operating systems to share a common computing system. { 'hī·pər,vīz·ər }

hyphenation zone [COMPUT SCI] In word processing, the area adjacent to the right margin consisting of those positions at which words may be hyphenated. { 'hī·fə,nā·shən ,zōn }

hyposensitization *See* desensitization. { ¦hī·pō,sen·səd·ə'zā·shən }

IBOC *See* in-band/on-channel. { ¦ī¦bē¦ō'sē *or* 'ī ˌbäk }

IC *See* integrated circuit.

ICNI *See* integrated communications - navigation - identification.

icon [COMPUT SCI] A symbolic representation of a computer function that appears on an electronic display and makes it possible to command this function by selecting the symbol. { 'īˌkän }

ICS system *See* intercarrier sound system. { ¦ī ¦sē'es ˌsis·təm }

ICW *See* interrupted continuous wave.

I demodulator [ELECTR] Stage of an analog color television receiver which combines the chrominance signal with the color oscillator output to restore the I signal. { 'ī dē'mäj·əˌlād·ər }

identification [CONT SYS] The procedures for deducing a system's transfer function from its response to a step-function input or to an impulse. { īˌdent·ə·fə'kā·shən }

identification and authentication [COMPUT SCI] The process of determining with high assurance the identity of a person who is seeking access to a computing system. { īˌden·tə·fə¦kā·shən ən əˌthen·tə'kā·shən }

identification division [COMPUT SCI] The section of a program, written in the COBOL language, which contains the name of the program and the name of the programmer. { īˌdent·ə·fə'kā·shən di¦vizh·ən }

identification, friend or foe [ENG] A system using pulsed radio transmissions to which equipment carried by friendly forces automatically responds, by emitting a pulse code, thereby identifying themselves from enemy forces; a method of determining the friendly or unfriendly character of aircraft, ships, and army units by other aircraft, ships, or ground force units. Abbreviated IFF. { īˌdent·ə·fə'kā·shən 'frend ər 'fō }

identifier [COMPUT SCI] A symbol whose purpose is to specify a body of data. { ī'dent·əˌfī·ər }

identifier word [COMPUT SCI] A full-length computer word associated with a search function. { ī'dent·əˌfī·ər ˌwərd }

identity gate *See* identity unit. { ī'den·əˌdē ˌgāt }

identity unit [COMPUT SCI] A logic element with several binary input signals and a single binary output signal whose value is 1 if all the input signals have the same value and 0 if they do not. Also known as identity gate. { ī'den·əˌdē ˌyü·nət }

I-display [ELECTR] A radar display format in which the target appears as a circle, of radius proportional to range, when a tracking radar antenna is pointed at it exactly and as a segment of the circle when there is a pointing error. Also known as I-indicator; I-scan; I-scope. { 'ī di ˌsplā }

idle time [COMPUT SCI] The time during which a piece of hardware in good operating condition is unused. { 'īd·əl ˌtīm }

IDP *See* integrated data processing.

IEEE 1394 [COMPUT SCI] The standard for connecting storage, digital audio and video, and other peripheral devices to personal computers at data transfer rates up to 400 million bits per second. Also known as firewire.

i-f *See* intermediate frequency.

IF *See* intermediate frequency.

i-f amplifier *See* intermediate-frequency amplifier. { ¦ī'ef 'am·pləˌfī·ər }

IF canceler [ELECTR] In radar, a moving-target indicator canceler operating at the intermediate frequency using an internal phase reference as in a coherent radar, as opposed to a video canceler. { ¦ī'ef 'kans·lər }

IFF *See* identification, friend or foe.

I-frame *See* intra-coded picture. { 'ī ˌfrām }

IF statement *See* conditional jump. { 'if ˌstāt·mənt }

if then else [COMPUT SCI] A logic statement in a high-level programming language that defines the data to be compared and the actions to be taken as the result of a comparison. { ¦if then 'els }

i-f transformer *See* intermediate-frequency transformer. { ¦ī ¦ef tranz'fȯr·mər }

IGBT *See* insulated-gate bipolar transistor.

IGES *See* initial graphics exchange specification.

IGFET *See* metal oxide semiconductor field-effect transistor. { 'igˌfet }

ignition interference [COMMUN] Radio interference due to the spark discharges in an automotive or other ignition system. { ig'nish·ən ˌint·ə'fir·əns }

ignore character [COMPUT SCI] Also known as erase character. **1.** A character indicating that no action whatever is to be taken, that is, a character to be ignored; often used to obliterate an erroneous character. **2.** A character indicating that the preceding or following character is to be ignored, as specified. **3.** A character indicating that some specified action is not to be taken. { ig'nȯr ˌkar·ik·tər }

I-indicator *See* I-display. { 'ī ¦in·də,kād·ər }

IIR filter *See* infinite impulse response filter. { ¦ī¦ī'är ˌfil·tər }

I²L *See* integrated injection logic.

ILF *See* infralow frequency.

ill-conditioned problem [COMPUT SCI] A problem in which a small error in the data or in subsequent calculation results in much larger errors in the answers. { ¦il kən¦dish·ənd 'präb·ləm }

illegal character [COMPUT SCI] A character or combination of bits that is not accepted as a valid representation by a computer or by a specific routine; commonly detected and used as an indication of a machine malfunction. { i'lē·gəl 'kar·ik·tər }

illegal operation [COMPUT SCI] An operation specified by a program instruction that cannot be carried out by the computer. { i'lē·gəl äp·ə'rā·shən }

illustration program *See* drawing program. { ˌil·ə'strā·shən ˌprō·grəm }

ILS *See* instrument landing system.

image [COMMUN] **1.** One of two groups of side bands generated in the process of modulation; the unused group is referred to as the unwanted image. **2.** The scene reproduced by a video display. [COMPUT SCI] A copy of the information contained in one medium recorded on a different data medium. [ELECTROMAG] The input reflection coefficient corresponding to the reflection coefficient of a specified load when the load is placed on one side of a waveguide junction and a slotted line is placed on the other. { 'im·ij }

image antenna [ELECTROMAG] A fictitious electrical counterpart of an actual antenna, acting mathematically as if it existed in the ground directly under the real antenna and served as the direct source of the wave that is reflected from the ground by the actual antenna. { 'im·ij an ¦ten·ə }

image converter [ELECTR] *See* image tube. [OPTICS] A converter that uses a fiber optic bundle to change the form of an image, for more convenient recording and display or for the coding of secret messages. { 'im·ij kən,vərd·ər }

image converter camera [ELECTR] A camera consisting of an image tube and an optical system which focuses the image produced on the phosphorescent screen of the tube onto photographic film. { 'im·ij ˌkən,vərd·ər ˌkam·rə }

image dissector [COMPUT SCI] In optical character recognition, a device that optically examines an input character for the purpose of breaking it down into its prescribed elements. { 'im·ij di,sek·tər }

image effect [ELECTROMAG] Effect produced on the field of an antenna due to the presence of the earth; electromagnetic waves are reflected from the earth's surface, and these reflections often are accounted for by an image antenna at an equal distance below the earth's surface. { 'im·ij i,fekt }

image enhancement [COMPUT SCI] Improvement of the quality of a picture, with the aid of a computer, by giving it higher contrast or making it less blurred or less noisy. { 'im·ij in'hans·mənt }

image frequency [ELECTR] An undesired carrier frequency that differs from the frequency to which a superheterodyne receiver is tuned by twice the intermediate frequency. { 'im·ij ˌfrē·kwən·sē }

image iconoscope [ELECTR] A camera tube in which an optical image is projected on a semitransparent photocathode, and the resulting electron image emitted from the other side of the photocathode is focused on a separate storage target; the target is scanned on the same side by a high-velocity electron beam, neutralizing the elemental charges in sequence to produce the camera output signal at the target. Also known as superemitron camera (British usage). { 'im·ij ī'kän·ə,skōp }

image interference [COMMUN] Interference occurring in a superheterodyne receiver when a station broadcasting on the image frequency is received along with the desired station. { 'im·ij ˌin·tər'fir·əns }

image processing [COMPUT SCI] A technique in which the data from an image are digitized and various mathematical operations are applied to the data, generally with a digital computer, in order to create an enhanced image that is more useful or pleasing to a human observer, or to perform some of the interpretation and recognition tasks usually performed by humans. Also known as picture processing. { 'im·ij ˌprä·ses·iŋ }

image ratio [ELECTR] In a heterodyne receiver, the ratio of the image frequency signal input at the antenna to the desired signal input for identical outputs. { 'im·ij ˌrā·shō }

image reject mixer [ELECTR] Combination of two balanced mixers and associated hybrid circuits designed to separate the image channel from the signal channels normally present in a conventional mixer; the arrangement gives image rejection up to 30 decibels without the use of filters. { 'im·ij 'rē,jekt ˌmik·sər }

image response [ELECTR] The response of a superheterodyne receiver to an undesired signal at its image frequency. { 'im·ij ri,späns }

image restoration [COMPUT SCI] Operation on a picture with a digital computer to make it more closely resemble the original object. { 'im·ij ˌres·tə'rā·shən }

image-storage array [ELECTR] A solid-state panel or chip in which the image-sensing elements may be a metal oxide semiconductor or a charge-coupled or other light-sensitive device that can be manufactured in a high-density configuration. { 'im·ij ˌstȯr·ij ə,rā }

image tube [ELECTR] An electron tube that reproduces on its fluorescent screen an image of the optical image or other irradiation pattern incident on its photosensitive surface. Also known as electron image tube; image converter. { 'im·ij ˌtüb }

imaging radar [ENG] A radar of such high resolution in one or several dimensions that a visual likeness of a target or scene of interest is produced; for example, radar carried on aircraft that forms images of the terrain. { 'im·i·jiŋ 'rā ˌdär }

IMAP *See* Internet Mail Access Protocol. { 'ī,map }

immediate-access [COMPUT SCI] **1.** Pertaining to an access time which is relatively brief, or to a relatively fast transfer of information. **2.** Pertaining to a device which is directly connected with another device. { i'mē·dē·ət 'ak·ses }

immediate address [COMPUT SCI] The value of an operand contained in the address part of an instruction and used as data by this instruction. { i'mē·dē·ət 'a,dres }

immediate data [COMPUT SCI] Data that appears in an instruction exactly as it is to be processed. { i'mēd·ē·ət 'dad·ə }

immediate instruction [COMPUT SCI] A computer program instruction, part of which contains the actual data to be operated upon, rather than the address of that data. { i'mēd·ē·ət in'strək·shən }

immediate operand [COMPUT SCI] An operand contained in the instruction which specifies the operation. { i'mē·dē·ət 'äp·ə,rand }

immediate processing *See* demand processing. { i'mē·dē·ət 'präs,es·iŋ }

immersive simulation *See* virtual reality. { i¦mər·siv sim·yə'lā·shən }

impact avalanche and transit time diode *See* IMPATT diode. { 'im,pakt ¦av·ə,lanch ən 'tran·zit ˌtīm 'dī,ōd }

impact ionization [ELECTR] Ionization produced by the impact of a high-energy charge carrier on an atom of semiconductor material; the effect is an increase in the number of charge carriers. { 'im,pakt ˌī·ə·nə'zā·shən }

IMPATT amplifier [ELECTR] A diode amplifier that uses an IMPATT diode; operating frequency range is from about 5 to 100 gigahertz, primarily in the C and X bands, with power output up to about 20 watts continuous-wave or 100 watts pulsed. { 'im,pat ˌam·plə,fī·ər }

IMPATT diode [ELECTR] A *pn* junction diode that has a depletion region adjacent to the junction, through which electrons and holes can drift, and is biased beyond the avalanche breakdown voltage. Derived from impact avalanche and transit time diode. { 'im,pat ˌdī,ōd }

impedance *See* electrical impedance. { im'pēd·əns }

imperative language [COMPUT SCI] A programming language in which programs largely consist of a series of commands to assign values to objects. { im'per·əd·iv ˌlaŋ·gwij }

imperative statement [COMPUT SCI] A statement in a symbolic program which is translated into actual machine-language instructions by the assembly routine. { im'per·əd·iv ˌstāt·mənt }

implementation [COMPUT SCI] **1.** The installation of a computer system or an information system. **2.** The use of software on a particular computer system. { ˌim·plə,men'tā·shən }

implicit programming [CONT SYS] Robotic programming that uses descriptions of the tasks at hand which are less exact than in explicit programming. { im'plis·ət 'prō,gram·iŋ }

improvement factor [COMMUN] *See* noise improvement factor. [ELECTR] In radar, a measure of the effectiveness of Doppler-sensitive processes, given by the ratio of the signal-to-clutter power ratios with and without the use of the processing, averaged over all target velocities. { im 'prüv·mənt ˌfak·tər }

improvement threshold [COMMUN] The condition of unity for the ratio of peak carrier voltage to peak noise voltage after selection and before any nonlinear process such as amplitude limiting. { im'prüv·mənt ˌthresh,hōld }

impulse modulation [CONT SYS] Modulation of a signal in which it is replaced by a series of impulses, equally spaced in time, whose strengths (integrals over time) are proportional to the amplitude of the signal at the time of the impulse. { 'im,pəls ˌmäj·ə,lā·shən }

impulse period *See* pulse period. { 'im,pəls ˌpir·ē·əd }

impulse response [CONT SYS] The response of a system to an impulse which differs from zero for an infinitesimal time, but whose integral over time is unity; this impulse may be represented mathematically by a Dirac delta function. { 'im ˌpəls ri,späns }

impulse separator [ELECTR] In an analog television receiver, the circuit that separates the horizontal synchronizing impulses in the received signal from the vertical synchronizing impulses. { 'im,pəls ˌsep·ə,rād·ər }

impulse signaling [COMMUN] Conveying information by means of on-off conditions transmitted down a line or over free space. { 'im,pəls ¦sig·nə·liŋ }

impulse train [CONT SYS] An input consisting of an infinite series of unit impulses, equally separated in time. { 'im,pəls ˌtrān }

impulse transmission [COMMUN] Form of signaling which employs impulses of either or both polarities for transmission to indicate the occurrence of transitions in the signals; used principally to reduce the effects of low-frequency interference; the impulses are generally formed by suppressing the low-frequency components, including direct current, of the signals. { 'im ˌpəls tranz'mish·ən }

impulse-type telemeter [COMMUN] A telemeter that employs electric impulses as the translating means. { 'im,pəls ˌtīp tə'lem·əd·ər }

in-band/on-channel [COMMUN] A system of digital radio where the digital signals are placed within the current AM and FM bands and within the FCC-assigned channel of a radio station. Abbreviated IBOC. { ¦in ¦band ¦ȯn 'chan·əl }

incident wave [ELECTR] A current or voltage wave that is traveling through a transmission line in the direction from source to load. { 'in·sə·dənt ¦wāv }

inclusive or *See* or. { in'klü·siv 'ȯr }

incorporate [COMPUT SCI] To place in storage. { in'kȯr·pə,rāt }

incremental compiler [COMPUT SCI] A compiler that generates code for a statement, or group of statements, which is independent of the code generated for other statements. { ,iŋ·krə 'ment·əl kəm'pīl·ər }

incremental computer [COMPUT SCI] A special-purpose computer designed to process changes in variables as well as absolute values; for instance, a digital differential analyzer. { ,iŋ·krə 'ment·əl kəm'pyüd·ər }

incremental digital recorder [COMPUT SCI] Magnetic tape recorder in which the tape advances across the recording head step by step, as in a punched-paper-tape recorder; used for recording an irregular flow of data economically and reliably. { ,iŋ·krə'ment·əl ¦dij·əd·əl ri'kȯrd·ər }

incremental dump tape [COMPUT SCI] A safety technique used in time-sharing which consists in copying all files (created or modified by a user during a day) on a magnetic tape; in case of system failure, the file storage can then be reconstructed. Also known as failsafe tape. { ,iŋ·krə'ment·əl 'dəmp ,tāp }

incremental frequency shift [COMMUN] Method of superimposing incremental intelligence on another intelligence by shifting the center frequency of an oscillator a predetermined amount. { ,iŋ·krə'ment·əl 'frē·kwən·sē ,shift }

incremental mode [COMPUT SCI] The plotting of a curve on a cathode-ray tube by illuminating a fixed number of points at a time. { ,iŋ·krə 'ment·əl ¦mōd }

incremental representation [COMPUT SCI] A way of representing variables used in incremental computers, in which changes in the variables are represented instead of the values of the variables themselves. { ,iŋ·krə'ment·əl ,rep·rə·sən'tā·shən }

independent-sideband modulation [COMMUN] Modulation in which the radio-frequency carrier is reduced or eliminated and two channels of information are transmitted, one on an upper and one on a lower sideband. Abbreviated ISB modulation. { ,in·də'pen·dənt ¦sīd,band ,mäj·ə'lā·shən }

independent-sideband receiver [ELECTR] A radio receiver designed for the reception of independent-sideband modulation, having provisions for restoring the carrier. { ,in·də'pen·dənt ¦sīd,band ri'sē·vər }

independent-sideband transmitter [ELECTR] A transmitter which produces independent-sideband modulated signals. { ,in·də'pen·dənt ¦sīd,band tranz'mid·ər }

index [COMPUT SCI] **1.** A list of record surrogates arranged in order of some attribute expressible in machine-orderable form. **2.** To produce a machine-orderable set of record surrogates, as in indexing a book. **3.** To compute a machine location by indirection, as is done by index registers. **4.** The portion of a computer instruction which indicates what index register (if any) is to be used to modify the address of an instruction. { 'in,deks }

index arithmetic unit [COMPUT SCI] A section of some computers that performs addition or subtraction operations on address parts of instructions for the purpose of indexing, boundary tests for memory protection, and so forth. { 'in ,deks ə'rith·mə·tik ,yü·nət }

indexed address [COMPUT SCI] An address which is modified, generally by means of index registers, before or during execution of a computer instruction. { 'in,dekst ə'dres }

indexed array [COMPUT SCI] An array of data items in which the individual items can be accessed by specifying their position through use of a subscript. { 'in,dekst ə'rā }

indexed sequential data set [COMPUT SCI] A collection of related data items that are stored sequentially on a key, but are also accessible through index tables maintained by the system. { 'in,dekst si¦kwen·chəl 'dad·ə ,set }

indexed sequential organization [COMPUT SCI] A sequence of records arranged in collating sequence used with direct-access devices. { 'in ,dekst si¦kwen·chəl ,ȯr·gə·nə'zā·shən }

index marker [COMPUT SCI] The beginning (and end) of each track in a disk, which is recognized by a special sensing device within the disk mechanism. { 'in,deks ,märk·ər }

index of cooperation [COMMUN] In rectilinear scanning or recording, the product of the total length of a scanning or recording line by the number of scanning or recording lines per unit length divided by pi. { 'in,deks əv kō,äp·ə'rā·shən }

index of modulation *See* modulation factor. { 'in,deks əv ,mäj·ə'lā·shən }

index of refraction [OPTICS] The ratio of the phase velocity of light in a vacuum to that in a specified medium. Also known as absolute index of refraction; absolute refractive constant; refractive constant; refractive index. { 'in,deks əv ri'frak·shən }

index point [COMPUT SCI] A hardware reference mark on a disk or drum for use in timing. { 'in ,deks ,pȯint }

index register [COMPUT SCI] A hardware element which holds a number that can be added to (or, in some cases, subtracted from) the address portion of a computer instruction to form an effective address. Also known as base register; B box; B line; B register; B store; modifier register. { 'in,deks ,rej·ə·stər }

index word *See* modifier. { 'in,deks ,wərd }

indicative data [COMPUT SCI] Data which describe a specific item. { in'dik·əd·iv 'dad·ə }

indicator [COMPUT SCI] A device announcing an error or failure. [ELECTR] A cathode-ray tube or other device that presents information transmitted or relayed from some other source, as from a radar receiver. { 'in·də,kād·ər }

indifferent stability *See* neutral stability. { in 'dif·ərnt stə'bil·əd·ē }

indirect address [COMPUT SCI] An address in a computer instruction that indicates a location where the address of the referenced operand is to be found. Also known as multilevel address. { ˌin·də'rekt ə'dres }

indirect addressing [COMPUT SCI] A programming device whereby the address part of an instruction is not the address of the operand but rather the location in storage where the address of the operand may be found. { ˌin·də'rekt ə'dres·iŋ }

indirect control [COMPUT SCI] The control of one peripheral unit by another through some sequence of events that involves human intervention. { ˌin·də'rekt kən'trōl }

indirect-path echo [ELECTROMAG] An echo resulting from radar transmission and reception, not via the direct path to the target but rather via reflections, for example, from a large building very near the radar, resulting in incorrect bearing estimation. { ˌin·də'rekt ¦path 'ek·ō }

individual line [COMMUN] Subscriber line arranged to serve only one main station, although additional stations may be connected to the line as extensions; an individual line is not arranged for discriminatory ringing with respect to the stations on that line. { ¦in·də¦vij·ə·wəl 'līn }

inductance [ELECTROMAG] **1.** That property of an electric circuit or of two neighboring circuits whereby an electromotive force is generated (by the process of electromagnetic induction) in one circuit by a change of current in itself or in the other. **2.** Quantitatively, the ratio of the emf (electromotive force) to the rate of change of the current. { in'dək·təns }

inductance coil *See* coil. { in'dək·təns ˌkȯil }

induction *See* electromagnetic induction. { in 'dək·shən }

induction field [ELECTROMAG] A component of an electromagnetic field associated with an alternating current in a loop, coil, or antenna which carries energy alternately away from and back into the source, with no net loss, and which is responsible for self-inductance in a coil or mutual inductance with neighboring coils. { in'dək·shən ˌfēld }

induction problem [ELECTROMAG] An effect of potentials and currents induced in conductors of a telephone system by paralleling power facilities or power lines. { in'dək·shən ˌpräb·ləm }

inductive coordination [ELECTROMAG] Measures to reduce induction problems. { in'dək·tiv kōˌȯrd·ən'ā·shən }

inductive interference [COMMUN] Effect arising from the characteristics and inductive relations of electric supply and communications systems of such character and magnitude as would prevent the communications circuits from rendering service satisfactorily and economically if methods of inductive coordination were not applied. { in'dək·tiv ˌin·tər'fir·əns }

inductive line pair [COMMUN] A telephone line displaying induction whose effects are of consequence, as in crosstalk; opposed to twisted pair. { in'dək·tiv 'līn ˌper }

inductive tuning [ELECTR] Tuning involving the use of a variable inductance. { in'dək·tiv 'tün·iŋ }

inductor *See* coil. { in'dək·tər }

industrial frequency bands [COMMUN] The radio-frequency bands allocated in the United States for land mobile communications of private industries other than transportation. { in'dəs·trē·əl 'frē·kwən·sē ˌbanz }

industrial television [COMMUN] Closed-circuit video system used for remote viewing of industrial processes and operations; may also be used for training purposes. Abbreviated ITV. { in'dəs·trē·əl 'tel·əˌvizh·ən }

ineffective time [COMPUT SCI] Time during which a computer can operate normally but which is not used effectively because of mistakes or inefficiency in operating the installation or for other reasons. { ˌin·i'fek·tiv 'tīm }

inertial guidance [NAV] **1.** Guidance by means of accelerations measured and integrated within the craft. **2.** Guidance by the use of an inertial navigation system. { i'nər·shəl 'gīd·əns }

inertial navigation system [NAV] A self-contained system that can automatically determine the position, velocity, and attitude of a moving vehicle by means of the double integration of the outputs of accelerometers that are either strapped to the vehicle or stabilized with respect to inertial space. Also known as inertial navigator. { in'ər·shəl ˌnav·ə'gā·shən ˌsis·təm }

inertial navigator *See* inertial navigation system. { i'nər·shəl 'nav·əˌgād·ər }

inference control [COMPUT SCI] A method of preventing data about specific individuals from being inferred from statistical information in a data base about groups of people. { 'in·frəns kənˌtrōl }

inference program [COMPUT SCI] A computer program that uses certain facts provided as input to reach conclusions. { 'in·frəns ˌprō·grəm }

infinite impulse response filter [ELECTR] An electronic filter that will continue oscillating in a decaying manner forever after being exposed to a change in input. Abbreviated IIR filter. { ˌin·fə·nət ¦imˌpəls ri¦späns ˌfil·tər }

infinite sequence *See* sequence. { 'in·fə·nət 'sē·kwəns }

infinity [COMPUT SCI] Any number larger than the maximum number that a computer is able to store in any register. { in'fin·əd·ē }

infinity transmitter [ELECTR] A device used to tap a telephone; the telephone instrument is so modified that an interception device can be actuated from a distant source without the caller's becoming aware. { in'fin·əd·ē tranz'mid·ər }

infix operation [COMPUT SCI] An operation carried out within an operation, as the addition of a and b prior to the multiplication by c or division by d in the operation $(a+b)c/d$. { 'inˌfiks ˌäp·əˌrā·shən }

influence diagram [SYS ENG] A graph-theoretic representation of a decision, which may include four types of nodes (decision, chance, value, and deterministic), directed arcs between the nodes (which identify dependencies between them), a marginal or conditional probability distribution defined at each chance node, and a mathematical function associated with each of the other types of node. { 'in,flü·əns ,dī·ə,gram }

influence factor *See* telephone influence factor. { 'in,flü·əns ,fak·tər }

information [COMMUN] Data which has been recorded, classified, organized, related, or interpreted within a framework so that meaning emerges. { ,in·fər'mā·shən }

information architecture [COMPUT SCI] The organization of large bodies of content, as well as the organization and labeling (tagging) of content at the document level to make information easy to search, navigate, and manage. { ,in·fər,mā·shən 'är·kə,tek·chər }

information bit [COMMUN] Bit that is generated by the data source but is not used by the data-transmission system. { ,in·fər'mā·shən ,bit }

information center [COMMUN] Center designed specifically for storing, processing, and retrieving information for dissemination at regular intervals, on demand or selectively, according to express needs of users. { ,in·fər'mā·shən ,sen·tər }

information channel [COMMUN] A facility used to transmit information between data-processing terminals separated by large distances. { ,in·fər'mā·shən ¦chan·əl }

information content [COMMUN] A numerical measure of the information generated in selecting a specific symbol (or message), equal to the negative of the logarithm of the probability of the symbol (or message) selected. Also known as negentropy. { ,in·fər'mā·shən ¦kän,tent }

information engineering [COMPUT SCI] The process of networking, collecting, analyzing, and reporting information, as well as controlling business, manufacturing, or service operations. { ,in·fər'mā·shən ,en·jə,nir·iŋ }

information feedback system [COMMUN] An information transmission system in which a return transmission is used to verify the accuracy of the sent transmission. { ,in·fər'mā·shən 'fēd ,bak ,sis·təm }

information float [COMPUT SCI] Information that is not located in a file or data base but is traveling between systems or is not assigned to a particular computer system. { ,in·fər'mā·shən ,flōt }

information flow [COMPUT SCI] The graphic representation of data collection, data processing, and report distribution throughout an organization. { ,in·fər'mā·shən ,flō }

information flow control [COMPUT SCI] A restriction on the use of information generated by a computer system that is consistent with the access controls on the resources of the system itself. { ,in·fər'mā·shən 'flō kən,trōl }

information interchange [COMMUN] The exchange of information between machines. { ,in·fər'mā·shən 'in·tər,chānj }

information link *See* data link. { ,in·fər'mā·shən ,liŋk }

information management [COMMUN] The science that deals with definitions, uses, value and distribution of information that is processed by an organization, whether or not it is handled by a computer. { ,in·fər'mā·shən 'man·ij·mənt }

information network [COMPUT SCI] A service that provides a variety of information services to subscribers on a dial-up basis. Also known as subscription database. { ,in·fər'mā·shən 'net ,wərk }

information precedence relation [COMPUT SCI] A statement that some specified piece of data is required for the production of another piece of data. { ,in·fər'mā·shən 'pres·ə·dəns ri,lā·shən }

information processing [COMPUT SCI] **1.** The manipulation of data so that new data (implicit in the original) appear in a useful form. **2.** *See* data processing. { ,in·fər'mā·shən 'prä·ses·iŋ }

Information Processing Language *See* IPL. { ,in·fər'mā·shən 'prä·ses·iŋ 'laŋ·gwij }

information rate [COMMUN] The information content generated per symbol or per second by an information source. { ,in·fər'mā·shən ,rāt }

information redundancy [COMPUT SCI] The use of more information than is absolutely necessary, such as the application of error-detection and error-correction codes, in order to increase the reliability of a computer system. { ,in·fər'mā·shən rə'dan·dən·sē }

information requirements [COMPUT SCI] Actual or anticipated questions which may be posed to an information retrieval system. { ,in·fər'mā·shən rə'kwīr·məns }

information resources management [COMPUT SCI] A concept for processing information that focuses on the information and places data-processing technology (software and hardware) in a secondary role. { ,in·fər'mā·shən ri'sór·səz ,man·ij·mənt }

information retrieval [COMPUT SCI] The technique and process of searching, recovering, and interpreting information from large amounts of stored data. { ,in·fər'mā·shən ri,trē·vəl }

information selection systems [COMPUT SCI] A class of information processing systems which carry out a sequence of operations necessary to locate in storage one or more items assumed to have certain specified characteristics and to retrieve such items directly or indirectly, in whole or in part. { ,in·fər'mā·shən si'lek·shən ,sis·təmz }

information separator [COMPUT SCI] A character that separates items or fields of information in a record, especially a variable-length record. { ,in·fər'mā·shən 'sep·ə,rād·ər }

information source [COMMUN] A system which produces messages by making successive selections from a group of symbols. { ,in·fər'mā·shən ,sōrs }

information system [COMMUN] Any means for communicating knowledge from one person to another, ranging from simple verbal communication to completely computerized methods of storing, searching, and retrieving of information. { ˌin·fərˈmā·shən ˌsis·təm }

information system architecture [COMPUT SCI] The study of the structure of both computer systems and the organizations that use them, in order to develop computer systems that support the objectives of the organizations more effectively. { ˌin·fər¦mā·shən ¦sis·təm ˈär·kəˌtek·chər }

information systems engineering [ENG] The discipline concerned with the design, development, testing, and maintenance of information systems. { ˌin·fər¦mā·shən ¦sis·təmz ˌen·jəˈnir·iŋ }

information technology [COMPUT SCI] The collection of technologies that deal specifically with processing, storing, and communicating information, including all types of computer and communications systems as well as reprographics methodologies. { ˌin·fərˈmā·shən tek¦näl·ə·jē }

information theory [COMMUN] A branch of theory which is devoted to problems in communications, and which provides criteria for comparing different communications systems on the basis of signaling rate, using a numerical measure of the amount of information gained when the content of a message is learned. { ˌin·fərˈmā·shən ˌthē·ə·rē }

information unit [COMMUN] A unit of information content, equal to a bit, nit, or hartley, according to whether logarithms are taken to base 2, *e*, or 10. { ˌin·fərˈmā·shən ˌyü·nət }

information utility [COMPUT SCI] An information network that specializes in supplying information to businesses and other organizations. { ˌin·fərˈmā·shən yüˌtil·əd·ē }

information word *See* data word. { ˌin·fərˈmā·shən ˌwərd }

infradyne receiver [ELECTR] A superheterodyne receiver in which the intermediate frequency is higher than the signal frequency, so as to obtain high selectivity. { ˈin·frəˌdīn riˈsē·vər }

infralow frequency [COMMUN] A designation for the band from 0.3 to 3 kilohertz in the radio spectrum. Abbreviated ILF. { ˈin·frəˌlōˈfrē·kwən·sē }

infrared-emitting diode [ELECTR] A light-emitting diode that has maximum emission in the near-infrared region, typically at 0.9 micrometer for *pn* gallium arsenide. { ¦in·frə¦red i¦mid·iŋ ˈdīˌōd }

infrared heterodyne detector [ELECTR] A heterodyne detector in which both the incoming signal and the local oscillator signal frequencies are in the infrared range and are combined in a photodetector to give an intermediate frequency in the kilohertz or megahertz range for conventional amplification. { ¦in·frə¦red ¦hed·ə·rə ˌdīn diˈtek·tər }

infrared radiation [ELECTROMAG] Electromagnetic radiation whose wavelengths lie in the range from 0.75 or 0.8 micrometer (the long-wavelength limit of visible red light) to 1000 micrometers (the shortest microwaves). { ¦in·frə ¦red ˌrād·ēˈā·shən }

infrared receiver [ELECTR] A device that intercepts or demodulates infrared radiation that may carry intelligence. Also known as nancy receiver. { ¦in·frə¦red riˈsē·vər }

infrared transmitter [ELECTR] A transmitter that emits energy in the infrared spectrum; may be modulated with intelligence signals. { ¦in·frə ¦red tranzˈmid·ər }

infrared vidicon [ELECTR] A vidicon whose photoconductor surface is sensitive to infrared radiation. { ¦in·frə¦red ˈvid·əˌkän }

inherent storage [COMPUT SCI] Any type of storage in which the storage medium is part of the hardware of the computer medium. { inˈhir·ənt ˈstȯr·ij }

inheritance [COMPUT SCI] A feature of object-oriented programming that allows a new class to be defined simply by stating how it differs from an existing class. { inˈher·əd·əns }

inherited error [COMPUT SCI] The error existing in the data supplied at the beginning of a step in a step-by-step calculation as executed by a program. { inˈher·əd·əd ˈer·ər }

inhibit-gate [ELECTR] Gate circuit whose output is energized only when certain signals are present and other signals are not present at the inputs. { inˈhib·ətˌgāt }

inhibiting input [ELECTR] A gate input which, if in its prescribed state, prevents any output which might otherwise occur. { inˈhib·əd·iŋ ˈinˌpu̇t }

inhibiting signal [ELECTR] A signal, which when entered into a specific circuit will prevent the circuit from exercising its normal function; for example, an inhibit signal fed into an AND gate will prevent the gate from yielding an output when all normal input signals are present. { inˈhib·əd·iŋ ˌsig·nəl }

initial condition *See* entry condition. { iˈnish·əl kənˈdish·ən }

initial condition mode *See* reset mode. { iˈnish·əl kənˈdish·ən ˌmōd }

initial graphics exchange specification [COMPUT SCI] A standard graphics file format for three-dimensional wire-frame models. Abbreviated IGES. { i¦nish·əl ˈgraf·iks iksˌchānj ˌspes·ə·fəˈkā·shən }

initial instructions [COMPUT SCI] A routine stored in a computer to aid in placing a program in memory. Also known as initial orders. { iˈnish·əl inˈstrək·shənz }

initialize [COMPUT SCI] **1.** To set counters, switches, and addresses to zero or other starting values at the beginning of, or at prescribed points in, a computer routine. **2.** To begin an operation, and more specifically, to adjust the environment to the required starting configuration. { iˈnish·əˌlīz }

initial orders *See* initial instructions. { iˈnish·əl ˈȯr·dərz }

initial program load [COMPUT SCI] A routine, used in starting up a computer, that loads the operating system from a direct-access storage device, usually a disk or diskette, into the computer's main storage. Abbreviated IPL. { i'nish·əl 'prō·grəm ,lōd }

initial program load button *See* bootstrap button. { i'nish·əl 'prō·grəm ,lōd ¦bət·ən }

initiate *See* trigger. { i'nish·ē,āt }

initiator [COMPUT SCI] A part of an operating system of a large computer that runs several jobs at the same time, setting up the job, monitoring its progress, and performing any necessary cleanup after the job's completion. { i'nish·ē,ād·ər }

injection [ELECTR] **1.** The method of applying a signal to an electronic circuit or device. **2.** The process of introducing electrons or holes into a semiconductor so that their total number exceeds the number present at thermal equilibrium. { in'jek·shən }

injection locking [ELECTR] The capture or synchronization of a free-running oscillator by a weak injected signal at a frequency close to the natural oscillator frequency or to one of its subharmonics; used for frequency stabilization in IMPATT or magnetron microwave oscillators, gas-laser oscillators, and many other types of oscillators. { in'jek·shən ¦läk·iŋ }

ink bleed [COMPUT SCI] In character recognition, the capillary extension of ink beyond the original edges of a printed or handwritten character. { 'iŋk ,blēd }

ink smudge [COMPUT SCI] In character recognition, the overflow of ink beyond the original edges of a printed or handwritten character. { 'iŋk ,sməj }

ink squeezeout [COMPUT SCI] In character recognition, the overflow of ink from the stroke centerline to the edges of a printed or handwritten character. { 'iŋk 'skwē,zaůt }

in-line coding [COMPUT SCI] Any group of instructions within the main body of a program. { 'in ¦līn 'kōd·iŋ }

in-line guns [ELECTR] An arrangement of three electron guns in a horizontal line; used in color picture tubes that have a slot mask in front of vertical color phosphor stripes. { 'in ¦līn 'gənz }

in-line procedure [COMPUT SCI] A short body of coding or instruction which accomplishes some purpose. { 'in ¦līn prə'sē·jər }

in-line processing [COMPUT SCI] The processing of data in random order, not subject to preliminary editing or sorting. { 'in ¦līn 'prä·ses·iŋ }

in-line subroutine [COMPUT SCI] A subroutine which is an integral part of a program. { 'in ¦līn 'səb·rü,tēn }

in-line tuning [ELECTR] Method of tuning the intermediate-frequency strip of a superheterodyne receiver in which all the intermediate-frequency amplifier stages are made resonant to the same frequency. { 'in ¦līn 'tün·iŋ }

in-phase and quadrature video [ELECTR] The pair of video signals produced in a radar receiver using two homodyne reception channels in which the reference signal iin one has shifted by ninety degrees of phase from the reference signal in the other; the process overcomes certain limiting conditions in the use of just one homodyne channel. { 'in ,fāz ənd 'kwä·drə·chər 'vid·ē·ō }

in-phase rejection *See* common-mode rejection. { 'in ,fāz ri'jek·shən }

in-phase signal *See* common-mode signal. { 'in ,fāz 'sig·nəl }

input [COMPUT SCI] The information that is delivered to a data-processing device from the external world, the process of delivering this data, or the equipment that performs this process. [ELECTR] **1.** The power or signal fed into an electrical or electronic device. **2.** The terminals to which the power or signal is applied. { 'in ,půt }

input area [COMPUT SCI] A section of internal storage reserved for storage of data or instructions received from an input unit such as cards or tape. Also known as input block; input storage. { 'in,půt ,er·ē·ə }

input data [COMPUT SCI] Data employed as input. { 'in,půt ,dad·ə }

input equipment [COMPUT SCI] **1.** The equipment used for transferring data and instructions into an automatic data-processing system. **2.** The equipment by which an operator transcribes original data and instructions to a medium that may be used in an automatic data-processing system. { 'in,půt i,kwip·mənt }

input-limited [COMPUT SCI] Pertaining to a system or operation whose speed or efficiency depends mainly on the speed of input into the machine rather than the speed of the machine itself. { 'in,půt ¦lim·əd·əd }

input magazine [COMPUT SCI] A part of a card-handling device which supplies the cards to the processing portion of the machine. Also known as magazine. { 'in,půt ,mag·ə,zēn }

input/output [COMPUT SCI] Pertaining to all equipment and activity that transfers information into or out of a computer. Abbreviated I/O. { 'in ,půt 'aůt,půt }

input/output adapter [COMPUT SCI] A circuitry which allows input/output devices to be attached directly to the central processing unit. { 'in,půt 'aůt,půt ə,dap·tər }

input/output area [COMPUT SCI] A portion of computer memory that is reserved for accepting data from input devices and holding data for transfer to output devices. { 'in,půt 'aůt,půt ,er·ē·ə }

input/output bound [COMPUT SCI] Pertaining to a system or condition in which the time for input and output operation exceeds other operations. Also known as input/output limited. { 'in,půt 'aůt,půt ,baůnd }

input/output buffer [COMPUT SCI] An area of a computer memory used to temporarily store data and instructions transferred into and out of a computer, permitting several such transfers to take place simultaneously with processing of data. { 'in,půt 'aůt,půt ,bəf·ər }

input/output channel [COMPUT SCI] The physical link connecting the computer to an input device or to an output device. { 'in,pu̇t 'au̇t,pu̇t ,chan·əl }

input/output controller [COMPUT SCI] An independent processor which provides the data paths between input and output devices and main memory. { 'in,pu̇t 'au̇t,pu̇t kən,trōl·ər }

input/output control system [COMPUT SCI] A set of flexible routines that supervise the input and output operations of a computer at the detailed machine-language level. Abbreviated IOCS. { 'in,pu̇t 'au̇t,pu̇t kən,trōl ,sis·təm }

input/output control unit [COMPUT SCI] The piece of hardware which controls the operation of one or more of a type of devices such as tape drives or disk drives; this unit is frequently an integral part of the input/output device itself. { 'in,pu̇t 'au̇t,pu̇t kən'trōl ,yü·nət }

input/output device *See* peripheral device. { 'in,pu̇t 'au̇t,pu̇t di,vīs }

input/output generation [COMPUT SCI] A procedure involved in installing an operating system on a large computer in which addresses and attributes of peripheral equipment under the computer's control are described in a language that can be read by the operating system. Abbreviated IOGEN. { 'in,pu̇t 'au̇t,pu̇t ,jen·ə,rā·shən }

input/output instruction [COMPUT SCI] An instruction in a computer program that causes transfer of data between peripheral devices and main memory, and enables the central processing unit to control the peripheral devices connected to it. { 'in,pu̇t 'au̇t,pu̇t in,strək·shən }

input/output interrupt [COMPUT SCI] A technique by which the central processor needs only initiate an input/output operation and then handle other matters, while other units within the system carry out the rest of the operation. { 'in,pu̇t 'au̇t,pu̇t 'int·ə,rəpt }

input/output interrupt identification [COMPUT SCI] The ascertainment of the device and channel taking part in the transfer of information into or out of a computer that causes a particular input/output interrupt, and of the status of the device and channel. { 'in,pu̇t 'au̇t,pu̇t 'int·ə ,rəpt ī,dent·ə·fə,kā·shən }

input/output interrupt indicator [COMPUT SCI] A device which registers an input/output interrupt associated with a particular input/output channel; it can be used in input/output interrupt identification. { 'in,pu̇t 'au̇t,pu̇t 'int·ə,rəpt ,in·də,kād·ər }

input/output library [COMPUT SCI] A set of programs which take over the job from the programmer of creating the required instructions to access the various peripheral devices. Also known as input/output routines. { 'in,pu̇t 'au̇t ,pu̇t ,lī,brer·ē }

input/output limited *See* input/output bound. { 'in,pu̇t 'au̇t,pu̇t ,lim·əd·əd }

input/output order [COMPUT SCI] A procedure of transferring data between main memory and peripheral devices which is assigned to and performed by an input/output controller. { 'in ,pu̇t 'au̇t,pu̇t ,ȯr·dər }

input/output processor [COMPUT SCI] A hardware device or software processor whose sole function is to handle input and output operations. { 'in,pu̇t 'au̇t,pu̇t ,prä,ses·ər }

input/output referencing [COMPUT SCI] The use of symbolic names in a computer program to indicate data on input/output devices, the actual devices allocated to the program being determined when the program is executed. { 'in ,pu̇t 'au̇t,pu̇t ,ref·rən·siŋ }

input/output register [COMPUT SCI] Computer register that provides the transfer of information from inputs to the central computer, or from it to output equipment. { 'in,pu̇t 'au̇t,pu̇t ,rej·ə·stər }

input/output relation [SYS ENG] The relation between two vectors whose components are the inputs (excitations, stimuli) of a system and the outputs (responses) respectively. { 'in,pu̇t 'au̇t ,pu̇t ri,lā·shən }

input/output routines *See* input/output library. { 'in,pu̇t 'au̇t,pu̇t rü,tēnz }

input/output statement [COMPUT SCI] A statement in a computer language that summons data or stores data in a peripheral device. { 'in,pu̇t 'au̇t,pu̇t ,stāt·mənt }

input/output switching [COMPUT SCI] A technique in which a number of channels can connect input and output devices to a central processing unit; each device may be assigned to any available channel, so that several different channels may service a particular device during the execution of a program. { 'in,pu̇t 'au̇t,pu̇t ,swich·iŋ }

input/output traffic control [COMPUT SCI] The coordination, by both hardware and software facilities, of the actions of a central processing unit and the input, output, and storage devices under its control, in order to permit several input/output devices to operate simultaneously while the central processing unit is processing data. { 'in,pu̇t 'au̇t,pu̇t 'traf·ik kən,trōl }

input/output wedge [COMPUT SCI] The characteristic shape of a Kiviat graph of a system which is approaching complete input/output boundedness. { 'in,pu̇t 'au̇t,pu̇t ,wej }

input program *See* data entry program. { 'in ,pu̇t ,prō·grəm }

input record [COMPUT SCI] **1.** A record that is read from an input device into a computer memory during the performance of a program or routine. **2.** A record that has been stored in an input area and is ready to be processed. { 'in ,pu̇t ,rek·ərd }

input register [COMPUT SCI] A register that accepts input information from a computer at one speed and supplies the information to the central processing unit at another speed, usually much greater. { 'in,pu̇t ,rej·ə·stər }

input routine [COMPUT SCI] A routine which controls the loading and reading of programs, data, and other routines into a computer for storage or immediate use. Also known as loading routine. { 'in,pu̇t rü,tēn }

input section [COMPUT SCI] The part of a program which controls the reading of data into a computer memory from external devices. { ˈin ˌpu̇t ˌsek·shən }

input station [COMPUT SCI] A terminal in an in-plant communications system at which data can be entered into the system directly as events take place, enabling files to be immediately updated. { ˈinˌpu̇t ˌstā·shən }

input storage *See* input area. { ˈinˌpu̇t ˌstȯr·ij }

inquiry [COMPUT SCI] A request for the retrieval of a particular item or set of items from storage. { inˈkwī·ə·rē }

inquiry and communications system [COMPUT SCI] A computer system in which centralized records are maintained with data transmitted to and from terminals at remote locations or in an in-plant system, and which immediately responds to inquiries from remote terminals. { inˈkwī·ə·rē ən kəˌmyü·nəˈkā·shənz ˌsis·təm }

inquiry and subscriber display [COMPUT SCI] An inquiry display unit that is distant from its computer and communicates with it over wire lines. { inˈkwī·ə·rē ən səbˈskrīb·ər diˌsplā }

inquiry display terminal [COMPUT SCI] A cathode-ray-tube terminal which allows the user to query the computer through a keyboard, the answer appearing on the screen. { inˈkwī·ə·rē diˈsplā ˌtər·mən·əl }

inquiry station [COMPUT SCI] A remote terminal from which an inquiry may be sent to a computer over wire lines. { inˈkwī·ə·rē ˌstā·shən }

inquiry unit [COMPUT SCI] Any terminal which enables a user to query a computer and get a hard-copy answer. { inˈkwī·ə·rē ˌyü·nət }

inscribe [COMPUT SCI] To rewrite data on a document in a form which can be read by an optical or magnetic ink character recognition machine. { inˈskrīb }

insensitive time *See* dead time. { inˈsen·sə·tiv ˌtīm }

insertion gain [ELECTR] The ratio of the power delivered to a part of the system following insertion of an amplifier, to the power delivered to that same part before insertion of the amplifier; usually expressed in decibels. { inˈsər·shən ˌgān }

insertion loss [ELECTR] The loss in load power due to the insertion of a component or device at some point in a transmission system; generally expressed as the ratio in decibels of the power received at the load before insertion of the apparatus, to the power received at the load after insertion. { inˈsər·shən ˌlȯs }

insertion switch [COMPUT SCI] Process by which information is inserted into the computer by an operator who manually operates switches. { inˈsər·shən ˌswich }

instability [CONT SYS] A condition of a control system in which excessive positive feedback causes persistent, unwanted oscillations in the output of the system. { ˌin·stəˈbil·əd·ē }

installation processing control [COMPUT SCI] A system that automatically schedules the processing of jobs by a computer installation, in order to minimize waiting time and time taken to prepare equipment for operation. { ˌin·stəˈlā·shən ˈpräs ˌes·iŋ kənˌtrōl }

installation specification [COMPUT SCI] The criteria defined by a computer manufacturer for specifying correct physical installation. { ˌin·stəˈlā·shən ˌspes·ə·fəˌkā·shən }

installation tape number [COMPUT SCI] A number that is permanently assigned to a reel of magnetic tape to identify it. { ˌin·stəˈlā·shən ¦tāp ˌnəm·bər }

install program [COMPUT SCI] A computer program that adapts a software package for use on a particular computer system. { inˈstȯl ˌprō·grəm }

instance variable [COMPUT SCI] The data in an object of an object-oriented program. { ˈin·stəns ˌver·ē·ə·bəl }

instantaneous automatic gain control [ELECTR] Portion of a radar system that automatically adjusts the gain of an amplifier for each pulse to obtain a substantially constant output-pulse peak amplitude with different input-pulse peak amplitudes; the circuit is fast enough to act during the time a pulse is passing through the amplifier. { ¦in·stən¦tā·nē·əs ˌȯd·ə¦mad·ik ˈgān kənˌtrōl }

instantaneous bandwidth [COMMUN] A term in radar for the modulation bandwidth of the signal being used, so as not to be confused with the radar's agility bandwidth, or its approved operating bandwidth. { ¦in·stən¦tā·nē·əs ˈband ˌwidth }

instantaneous companding [ELECTR] Companding in which the effective gain variations are made in response to instantaneous values of the signal wave. { ¦in·stən¦tā·nē·əs kəmˈpan·diŋ }

instantaneous description [COMPUT SCI] For a Turing machine, the set of machine conditions at a given point in the computation, including the contents of the tape, the position of the read-write head on the tape, and the internal state of the machine. { ¦in·stən¦tā·nē·əs diˈskrip·shən }

instantaneous effects [COMMUN] Impairment of telephone or telegraph transmission caused by instantaneous changes in phase or amplitude of the wave in a transmission line. { ¦in·stən ¦tā·nē·əs iˈfeks }

instantaneous frequency [COMMUN] The time rate of change of the angle of an angle-modulated wave. { ¦in·stən¦tā·nē·əs ˈfrē·kwən·sē }

instantaneous sample [COMMUN] One of a sequence of instantaneous values of a wave taken at regular intervals. { ¦in·stən¦tā·nē·əs ˈsam·pəl }

instantiation [COMPUT SCI] **1.** An external declaration or a reference to another program or subprogram in the Ada programming language. **2.** The deduction of omitted values in a set of data from the known values. **3.** The creation of an object of a specific class in an object-oriented program. { inˌstan·chēˈā·shən }

instruction [COMPUT SCI] A pattern of digits which signifies to a computer that a particular operation is to be performed and which may also indicate the operands (or the locations of operands) to be operated on. { inˈstrək·shən }

instruction address [COMPUT SCI] The address of the storage location in which a given instruction is stored. { in'strək·shən ə'dres }

instruction address register [COMPUT SCI] A special storage location, forming part of the program controller, in which addresses of instructions are stored in order to control their sequential retrieval from memory during the execution of a program. { in'strək·shən ə¦dres 'rej·ə·stər }

instruction area [COMPUT SCI] A section of storage used for storing program instructions. { in'strək·shən ˌer·ē·ə }

instruction code [COMPUT SCI] That part of an instruction which distinguishes it from all other instructions and specifies the action to be performed. { in'strək·shən ˌkōd }

instruction constant [COMPUT SCI] A dummy instruction of the type K = I, where K is irrelevant to the program. { in'strək·shən ˌkän·stənt }

instruction counter [COMPUT SCI] A counter that indicates the location of the next computer instruction to be interpreted. Also known as location counter; program counter; sequence counter. { in'strək·shən ˌkaủnt·ər }

instruction cycle [COMPUT SCI] The steps involved in carrying out an instruction. { in 'strək·shən ˌsī·kəl }

instruction format [COMPUT SCI] Any rule which assigns various functions to the various digits of an instruction. { in'strək·shən ˌfȯrˌmat }

instruction length [COMPUT SCI] The number of bits or bytes (eight bits per byte) which defines an instruction. { in'strək·shən ˌleŋkth }

instruction lookahead [COMPUT SCI] A technique for speeding up the process of fetching and decoding instructions in a computer program, and of computing addresses of required operands and fetching them, in which the control unit fetches any unexecuted instructions on hand, to the extent this is feasible. Also known as fetch ahead. { in'strək·shən 'lu̇k·əˌhed }

instruction mix [COMPUT SCI] The proportion of various types of instructions that appear in a particular computer program, or in a benchmark representing a class of programs. { in'strək·shən ˌmiks }

instruction modification [COMPUT SCI] A change, carried out by the program, in an instruction so that, upon being repeated, this instruction will perform a different operation. { in'strək·shən ˌmäd·ə·fə'kā·shən }

instruction pointer [COMPUT SCI] **1.** A component of a task descriptor that designates the next instruction to be executed by the task. **2.** An element of the control component of the stack model of block structure execution, which points to the current instruction. { in'strək·shən ˌpȯint·ər }

instruction register [COMPUT SCI] A hardware element that receives and holds an instruction as it is extracted from memory; the register either contains or is connected to circuits that interpret the instruction (or discover its meaning). Also known as current-instruction register. { in'strək·shən ˌrej·ə·stər }

instruction repertory *See* instruction set. { in'strək·shən ˌrep·əˌtȯr·ē }

instruction set [COMPUT SCI] **1.** The set of instructions which a computing or data-processing system is capable of performing. **2.** The set of instructions which an automatic coding system assembles. Also known as instruction repertory. { in'strək·shən ˌset }

instruction time [COMPUT SCI] The time required to carry out an instruction having a specified number of addresses in a particular computer. { in'strək·shən ˌtīm }

instruction transfer [COMPUT SCI] An instruction which transfers control to one or another subprogram, depending upon the value of some operation. { in'strək·shən ˌtranz·fər }

instruction word [COMPUT SCI] A computer word containing an instruction rather than data. Also known as coding line. { in'strək·shən ˌwərd }

instrumented range [ELECTR] In radar, the distance from which an echo might be received just before the subsequent pulse is transmitted; sometimes called unambiguous range. Echoes from targets at greater ranges are liable to be associated with the subsequent pulse and thought to be coming from a target at a much shorter range. { 'in·strə¦mən·təd ˌrānj }

instrument landing system [NAV] A system of radio navigation which provides lateral and vertical guidance, as well as other navigational parameters required by a pilot in a low approach or a landing. Abbreviated ILS. { 'in·strə·mənt ¦lan·diŋ ˌsis·təm }

instrument landing system localizer [NAV] System of horizontal guidance embodied in the instrument landing system which indicates the horizontal deviation of the aircraft from its optimum path of descent along the axis of the runway. { 'in·strə·mənt ¦lan·diŋ ˌsis·təm ˌlō·kə ˌliz·ər }

insulated-gate bipolar transistor [ELECTR] A power semiconductor device that combines low forward voltage drop, gate-controlled turnoff, and high switching speed. It structurally resembles a vertically diffused MOSFET, featuring a double diffusion of a *p*-type region and an *n*-type region, but differs from the MOSFET in the use of a *p*+ substrate layer (in the case of an *n*-channel device) for the drain. The effect is to change the transistor into a bipolar device, as this *p*-type region injects holes into the *n*-type drift region. Abbreviated IGBT. { ¦in·səˌlād·ədˌgāt bīˌpō·lər tran'zis·tər }

insulated-gate field-effect transistor *See* metal oxide semiconductor field-effect transistor. { 'in·səˌlād·əd ¦gāt ¦fēld iˌfekt tran'zis·tər }

integer constant [COMPUT SCI] A constant that uses the values 0, 1, . . . , 9 with no decimal point in FORTRAN. { 'int·ə·jər ¦kän·stənt }

integer data type [COMPUT SCI] A scalar date type which is used to represent whole numbers, that is, values without fractional parts. { 'int·ə·jər 'dad·ə ˌtāp }

integer programming [SYS ENG] A series of procedures used in operations research to find maxima or minima of a function subject to one or more constraints, including one which requires that the values of some or all of the variables be whole numbers. { 'int·ə·jər 'prō,gram·iŋ }

integer variable [COMPUT SCI] A variable in FORTRAN whose first character is normally I, J, K, L, M, or N. { 'int·ə·jər 'ver·ē·ə·bəl }

integral action [CONT SYS] A control action in which the rate of change of the correcting force is proportional to the deviation. { 'int·ə·grəl ,ak·shən }

integral compensation [CONT SYS] Use of a compensator whose output changes at a rate proportional to its input. { 'int·ə·grəl ,käm·pən'sā·shən }

integral control [CONT SYS] Use of a control system in which the control signal changes at a rate proportional to the error signal. { 'int·ə·grəl kən,trōl }

integral-mode controller [CONT SYS] A controller which produces a control signal proportional to the integral of the error signal. { 'int·ə·grəl ¦mōd kən,trōl·ər }

integral modem [COMMUN] A modem built directly into a machine to enable it to communicate over a telephone line. Also known as internal modem. { 'int·ə·grəl ¦mō,dem }

integral network [CONT SYS] A compensating network which produces high gain at low input frequencies and low gain at high frequencies, and is therefore useful in achieving low steady-state errors. Also known as lagging network; lag network. { 'int·ə·grəl 'net,wərk }

integral square error [CONT SYS] A measure of system performance formed by integrating the square of the system error over a fixed interval of time; this performance measure and its generalizations are frequently used in linear optimal control and estimation theory. { 'int·ə·grəl ¦skwer ,er·ər }

integrated circuit [ELECTR] An interconnected array of active and passive elements integrated with a single semiconductor substrate or deposited on the substrate by a continuous series of compatible processes, and capable of performing at least one complete electronic circuit function. Abbreviated IC. Also known as integrated semiconductor. { 'int·ə,grād·əd 'sər·kət }

integrated-circuit filter [ELECTR] An electronic filter implemented as an integrated circuit, rather than by interconnecting discrete electrical components. { ¦int·i,grād·əd 'sər·kət ,fil·tər }

integrated-circuit memory *See* semiconductor memory. { 'int·ə,grād·əd ¦sər·kət 'mem·rē }

integrated communications-navigation-identification [NAV] The concept of coordinating the electronic units of a system to improve the efficiency of providing communications, navigation, and identification for civilian and military aircraft. Abbreviated ICNI. { 'int·ə,grād·əd kə,myü·nə'kā·shənz ,nav·ə'gā·shən ī,dent·ə·fə'kā·shən }

integrated communications system [COMMUN] Communications system on either a unilateral or joint basis in which a message can be filed at any communications center in that system and be delivered to the addressee by any other appropriate communications center in that system without reprocessing enroute. { 'int·ə ,grād·əd kə,myü·nə'kā·shənz ,sis·təm }

integrated console [COMMUN] Computer control console that is capable of controlling the operation of the switching center equipment of an integrated communications system. { 'int·ə ,grād·əd 'kän,sōl }

integrated data dictionary [COMPUT SCI] An index or catalog of information about a data base that is physically and logically integrated into the data base. { 'in·tə,grād·əd 'dad·ə 'dik·shə ,ner·ē }

integrated data processing [COMPUT SCI] Data processing that has been organized and carried out as a whole, so that intermediate outputs may serve as inputs for subsequent processing with no human copying required. Abbreviated IDP. { 'in·tə,grād·əd ¦dad·ə 'prä·ses·iŋ }

integrated data retrieval system [COMPUT SCI] A section of a data-processing system that provides facilities for simultaneous operation of several video-data interrogations in a single line and performs required communications with the rest of the system; it provides storage and retrieval of both data subsystems and files and standard formats for data representation. { 'in·tə,grād·əd ¦dad·ə ri'trē·vəl ,sis·təm }

integrated electronics [ELECTR] A generic term for that portion of electronic art and technology in which the interdependence of material, device, circuit, and system-design consideration is especially significant; more specifically, that portion of the art dealing with integrated circuits. { 'in·tə,grād·əd i,lek'trän·iks }

integrated inertial navigation system [NAV] An air navigation system in which an inertial navigation or inertial reference subsystem is interconnected with all the other subsystems associated with avionics, so that data from all subsystems are automatically blended and interpreted to provide the control data needed to operate aircraft efficiently and safely. { 'in·tə,grād·əd in'ər·shəl ,nav·ə'gā·shən ,sis·təm }

integrated information processing [COMPUT SCI] System of computers and peripheral systems arranged and coordinated to work concurrently or independently on different problems at the same time. { 'in·tə,grād·əd ,in·fər'mā·shən ,prä·ses·iŋ }

integrated information system [COMMUN] An expansion of a basic information system achieved through system design of an improved or broader capability by functionally or technically relating two or more information systems, or by incorporating a portion of the functional or technical elements of one information system into another. { 'in·tə,grād·əd ,in·fər'mā·shən ,sis·təm }

integrated injection logic [ELECTR] Integrated-circuit logic that uses a simple and compact bipolar transistor gate structure which makes possible large-scale integration on silicon for logic arrays, memories, watch circuits, and various other analog and digital applications. Abbreviated I^2L. Also known as merged-transistor logic. { 'in·tə ˌgrād·əd in'jek·shən 'läj·ik }

integrated optics [OPTICS] A thin-film device containing tiny lenses, prisms, and switches to transmit very thin laser beams, and serving the same purposes as the manipulation of electrons in thin-film devices of integrated electronics. { 'in·təˌgrād·əd 'äp·tiks }

integrated semiconductor *See* integrated circuit. { 'in·təˌgrād·əd ¦sem·i·kən¦dək·tər }

integrated services digital network [COMMUN] A public end-to-end digital communications network which has capabilities of signaling, switching, and transport over facilities such as wire pairs, coaxial cables, optical fibers, microwave radio, and satellites, and which supports a wide range of services, such as voice, data, video, facsimile, and music, over standard interfaces. Abbreviated ISDN. { 'in·təˌgrād·əd ¦sər·vəs·əz 'dij·əd·əl 'netˌwərk }

integrated software [COMPUT SCI] **1.** A collection of computer programs designed to work together to handle an application, either by passing data from one to another or as components of a single system. **2.** A collection of computer programs that work as a unit with a unified command structure to handle several applications, such as word processing, spread sheets, data-base management, graphics, and data communications. { 'in·təˌgrād·əd 'sȯft ˌwer }

integrating amplifier [ELECTR] An operational amplifier with a shunt capacitor such that mathematically the waveform at the output is the integral (usually over time) of the input. { 'int·ə ˌgrād·iŋ 'am·pləˌfī·ər }

integrating detector [ELECTR] A frequency-modulation detector in which a frequency-modulated wave is converted to an intermediate-frequency pulse-rate modulated wave, from which the original modulating signal can be recovered by use of an integrator. { 'int·ə ˌgrād·iŋ di'tek·tər }

integrating network [ELECTR] A circuit or network whose output waveform is the time integral of its input waveform. Also known as integrator. { 'int·əˌgrād·iŋ 'netˌwərk }

integration [SYS ENG] The arrangement of components in a system so that they function together in an efficient and logical way. { ˌint·ə'grā·shən }

integration test [COMPUT SCI] A stage in testing a computer system in which a collection of modules in the system is tested as a group. { ˌint·ə'grā·shən ˌtest }

integrator [ELECTR] **1.** A computer device that approximates the mathematical process of integration. **2.** *See* integrating network. { 'int·ə ˌgrād·ər }

integrity [COMPUT SCI] Property of data which can be recovered in the event of its destruction through failure of the recording medium, user carelessness, program malfunction, or other mishap. [NAV] The ability of a navigation system to inform the user in a timely manner of a latent failure that may cause a hazardous condition. { in'teg·rəd·ē }

intelligence [COMMUN] Data, information, or messages that are to be transmitted. { in'tel·ə·jəns }

intelligent agent *See* knowbot. { in¦tel·ə·jənt 'ā·jənt }

intelligent cable [COMMUN] A multiline communications cable that is equipped with a microprocessor to analyze or convert signals. { in'tel·ə·jənt ¦kā·bəl }

intelligent controller [COMPUT SCI] A peripheral control unit whose operation is controlled by a built-in microprocessor. { in'tel·ə·jənt kən 'trō·lər }

intelligent database [COMPUT SCI] **1.** A database that can respond to queries in a high-level, interactive language. **2.** A database that can store validation criteria with each item of data, so that all programs entering or updating the data must conform to these criteria. { in'tel·ə·jənt 'dad·əˌbās }

intelligent machine [COMPUT SCI] A machine that uses sensors to monitor the environment and thereby adjust its actions to accomplish specific tasks in the face of uncertainty and variability. { in'tel·ə·jənt mə'shēn }

intelligent robot [CONT SYS] A robot that functions as an intelligent machine, that is, it can be programmed to take actions or make choices based on input from sensors. { in'tel·ə·jənt 'rō ˌbät }

intelligent sensor *See* smart sensor. { in¦tel·ə·jənt 'sen·sər }

intelligent terminal [COMPUT SCI] A computer input/output device with its own memory and logic circuits which can perform certain operations normally carried out by the computer. Also known as smart terminal. { in'tel·ə·jənt 'ter·mən·əl }

intelligent work station [COMPUT SCI] A work station that has an intelligent terminal to carry out a variety of functions independently. { in'tel·ə·jənt 'wərk ˌstā·shən }

intelligibility [COMMUN] The percentage of speech units understood correctly by a listener in a communications system; customarily used for regular messages where the context aids the listener, in distinction to articulation. Also known as speech intelligibility. { inˌtel·ə·jə'bil·əd·ē }

intelligible crosstalk [COMMUN] Crosstalk which is sufficiently understandable under pertinent circuit and room noise conditions that meaningful information can be obtained by more sensitive listeners. { in'tel·ə·jə·bəl 'krȯsˌtȯk }

Intelsat [COMMUN] A satellite network, formerly under international control, used for global communication by more than 100 countries;

the system uses geostationary satellites over the Atlantic, Pacific, and Indian oceans and highly directional antennas at earth stations. Derived from International Telecommunications Satellite. { in'tel,sat }

intensity control *See* brightness control. { in'ten·səd·ē kən,trōl }

intensity modulation [ELECTR] Modulation of electron beam intensity in a cathode-ray tube in accordance with the magnitude of the received signal. { in'ten·səd·ē ,mäj·ə'lā·shən }

interactive graphical input [COMPUT SCI] Information which is delivered to a computer by using hand-held devices, such as writing styli used with electronic tablets and light-pens used with cathode-ray tube displays, to sketch a problem description in an on-line interactive mode in which the computer acts as a drafting assistant with unusual powers, such as converting rough freehand motions of a pen or stylus to accurate picture elements. { ¦in·tə¦rak·tiv ¦graf·ə·kəl 'in ,pu̇t }

interactive information system [COMPUT SCI] An information system in which the user communicates with the computing facility through a terminal and receives rapid responses which can be used to prepare the next input. { ¦in·tə¦rak·tiv ,in·fər'mā·shən ,sis·təm }

interactive language [COMPUT SCI] A programming language designed to operate in an environment in which the user and computer communicate as transactions are being processed. { ¦in·tə¦rak·tiv 'laŋ·gwij }

interactive processing [COMPUT SCI] Computer processing in which the user can modify the operation appropriately while observing results at critical steps. { ¦in·tə¦rak·tiv 'prä·ses·iŋ }

interactive television [COMMUN] A form of television in which the content is personalized and the viewer can control its various parameters. { ¦in·tə¦rak·tiv 'tel·ə,vizh·ən }

interactive terminal [COMPUT SCI] A computer terminal designed for two-way communication between operator and computer. { ¦in·tə¦rak·tiv 'ter·mən·əl }

interblock [COMPUT SCI] A device or system that prevents one part of a computing system from interfering with another. { 'in·tər,bläk }

interblock gap [COMPUT SCI] A space separating two blocks of data on a magnetic tape. { 'in·tər ,bläk ,gap }

intercarrier channel [COMMUN] A carrier telegraph channel in the available frequency spectrum between carrier telephone channels. { ¦in·tər'kar·ē·ər ,chan·əl }

intercarrier noise suppression [ELECTR] Means of suppressing the noise resulting from increased gain when a high-gain receiver with automatic volume control is tuned between stations; the suppression circuit automatically blocks the audio-frequency input of the receiver when no signal exists at the second detector. Also known as interstation noise suppression. { ¦in·tər'kar·ē·ər 'nȯiz sə,presh·ən }

intercarrier sound system [ELECTR] An analog television receiver arrangement in which the television picture carrier and the associated sound carrier are amplified together by the video intermediate-frequency amplifier and passed through the second detector, to give the conventional video signal plus a frequency-modulated sound signal whose center frequency is the 4.5 megahertz difference between the two carrier frequencies. Abbreviated ICS system. { ¦in·tər'kar·ē·ər 'sau̇nd ,sis·təm }

intercept call [COMMUN] In telephone practice, routing of a call placed to a disconnected or nonexisting telephone number, to an operator, or to a machine answering device, or to a tone. { ¦in·tər·sept ,ȯl }

interception [COMMUN] Tapping or tuning in to a telephone or radio message not intended for the listener. { ,in·tər'sep·shən }

intercept station [COMMUN] Provides service for subscribers whereby calls to disconnected stations or dead lines are either routed to an intercept operator for explanation, or the calling party receives a distinctive tone that informs the party that he has made such a call has been made. { 'in·tər,sept ,stā·shən }

intercept tape [COMMUN] A tape used for temporary storage of messages for trunk channels and tributary stations that are having equipment or circuit trouble. { 'in·tər,sept ,tāp }

intercept trunk [COMMUN] Trunk to which a call for a vacant number, a changed number, or a line out of order is connected for action by an operator. { 'in·tər,sept ,trəŋk }

interchannel crosstalk [COMMUN] Crosstalk between channels in a multiplex system. { ¦in·tər ¦chan·əl 'krȯs,tȯk }

intercom *See* intercommunicating system. { 'in·tər,käm }

intercommunicating system [COMMUN] **1.** A telephone system providing direct communication between telephones on the same premises. **2.** A two-way communication system having a microphone and loudspeaker at each station and providing communication within a limited area. Also known as intercom. { ¦in·tər·kə'myü·nə ,kād·iŋ ,sis·təm }

interconnected multiple processor [COMPUT SCI] A collection of computers that are physically separated but linked by communication channels to handle distributed data processing. { ,in·tər·kə'nek·təd 'məl·tə·pəl 'präs,es·ər }

interconversion [COMMUN] Changing the representation of information from one code to another, as from six-bit to ASCII. { ¦in·tər·kən'vər·zhən }

interdigital magnetron [ELECTR] Magnetron having axial anode segments around the cathode, alternate segments being connected together at one end, remaining segments connected together at the opposite end. { ,in·tər'dij·əd·əl 'mag·nə,trän }

interdigital transducer [ELECTR] Two interlocking comb-shaped metallic patterns applied to a piezoelectric substrate such as quartz or lithium niobate, used for converting microwave voltages to surface acoustic waves, or vice versa. { ˌin·tərˈdij·əd·əl tranzˈdü·sər }

interface [COMPUT SCI] **1.** Some form of electronic device that enables one piece of gear to communicate with or control another. **2.** A device linking two otherwise incompatible devices, such as an editing terminal of one manufacturer to typesetter of another. { ˈin·tər ˌfās }

interface adapter [COMMUN] A device that connects a terminal or computer to a network. { ˈin·tərˌfās əˌdap·tər }

interface card [COMPUT SCI] A card containing circuits that allow a device to interface with other devices. { ˈin·tərˌfās ˌkärd }

interface control module [COMPUT SCI] Relocatable modularized compiler allowing for efficient operation and easy maintenance. { ˈin·tər ˌfās kənˈtrōl ˌmäj·ül }

interfacility transfer trunk [COMMUN] Trunk interconnecting switching centers of two different facilities. { ˌin·tər·fəˈsil·əd·ē ˈtranz·fər ˌtrəŋk }

interference [COMMUN] Any undesired energy that tends to interfere with the reception of desired signals. Also known as electrical interference; radio interference. { ˌin·tərˈfir·əns }

interference blanker [ELECTR] Device that permits simultaneous operation of two or more pieces of radio or radar equipment without confusion of intelligence, or that suppresses undesired signals when used with a single receiver. { ˌin·tərˈfir·əns ¦blaŋ·kər }

interference fading [COMMUN] Fading of the signal produced by different wave components traveling slightly different paths in arriving at the receiver (often termed multipath). { ˌin·tərˈfir·əns ˌfād·iŋ }

interference filter [ELECTR] **1.** A filter used to attenuate artificial interference signals entering a receiver through its power line. **2.** A filter used to attenuate unwanted carrier-frequency signals in the tuned circuits of a receiver. [OPTICS] An optical filter in which the wavelengths that are not transmitted are removed by interference phenomena rather then by absorption or scattering. { ˌin·tərˈfir·əns ¦fil·tər }

interference pattern [ELECTR] Pattern produced on a radar display by undesired signals. Also, the vertical coverage pattern of a radar antenna resulting from the interference of the direct-path and earth-reflected path signals. { ˌin·tərˈfir·əns ¦pad·ərn }

interference prediction [ELECTR] Process of estimating the interference level of a particular equipment as a function of its future electromagnetic environment. { ˌin·tərˈfir·əns prəˈdik·shən }

interference reduction [ELECTR] Reduction of interference from such causes as power lines and equipment, radio transmitters, and lightning, usually through the use of electric filters. Also known as interference suppression. { ˌin·tərˈfir·əns riˈdək·shən }

interference region [COMMUN] That region in space in which interference between wave trains occurs; in microwave propagation, it refers to the region bounded by the ray path and the surface of the earth which is above the radio horizon. { ˌin·tərˈfir·əns ˌrē·jən }

interference rejection [ELECTR] Use of a filter to reject (to bypass to ground) unwanted input. { ˌin·tərˈfir·əns riˈjek·shən }

interference source suppression [ELECTR] Techniques applied at or near the source to reduce its emission of undesired signals. { ˌin·tərˈfir·əns ˈsōrs səˌpresh·ən }

interference spectrum [ELECTR] Frequency distribution of the jamming interference in the propagation medium external to the receiver. { ˌin·tərˈfir·əns ¦spek·trəm }

interference suppression *See* interference reduction. { ˌin·tərˈfir·əns səˈpresh·ən }

interference wave [COMMUN] A radio wave reflected by the lower atmosphere which produces an interference pattern when combined with the direct wave. { ˌin·terˈfir·əns ˌwāv }

interfix [COMPUT SCI] A technique for describing relationships of key words in an item or document in a way which prevents crosstalk from causing false retrievals when very specific entries are made. { ˈin·tərˌfiks }

interior label [COMPUT SCI] A label attached to the data that it identifies. { inˈtir·ē·ər ˈlā·bəl }

interlace [COMPUT SCI] To assign successive memory location numbers to physically separated locations on a storage tape or magnetic drum of a computer, usually to reduce access time. { ¦in·tər¦lās }

interlaced scanning [ELECTR] A scanning process in which the distance from center to center of successively scanned lines is two or more times the nominal line width, so that adjacent lines belong to different fields. Also known as line interlace. { ¦in·tər¦lāst ˈskan·iŋ }

interlace operation [COMPUT SCI] System of computer operation where data can be read out or copied into memory without interfering with the other activities of the computer. { ¦in·tər¦lās ˌäp·əˈrā·shən }

interleave [COMPUT SCI] **1.** To alternate parts of one sequence with parts of one or more other sequences in a cyclic fashion such that each sequence retains its identity. **2.** To arrange the members of a sequence of memory addresses in different memory modules of a computer system, in order to reduce the time taken to access the sequence. { ˌin·tərˈlēv }

interlock [COMPUT SCI] **1.** A mechanism, implemented in hardware or software, to coordinate the activity of two or more processes within a computing system, and to ensure that one process has reached a suitable state such that the other may proceed. **2.** *See* deadlock. { ˈin·tər ˌläk }

interlude [COMPUT SCI] A small routine or program which is designed to carry out minor preliminary calculations or housekeeping operations before the main routine begins to operate, and which can usually be overwritten after it has performed its function. { 'in·tər,lüd }

intermediate control data [COMPUT SCI] Control data at a level which is neither the most nor the least significant, or which is used to sort records into groups that are neither the largest nor the smallest used; for example, if control data are used to specify state, town, and street, then the data specifying town would be intermediate control data. { ,in·tər'mēd·ē·ət kən'trōl ,dad·ə }

intermediate frequency [ELECTR] In radio or radar heterodyne receivers, that frequency produced by mixing the received signal, presumably at the intended carrier frequency, with a local-oscillator signal offset from the carrier frequency, the intermediate frequency being the difference of the two. Abbreviated IF. { ,in·tər'mēd·ē·ət 'frē·kwən·sē }

intermediate-frequency amplifier [ELECTR] The section of a superheterodyne receiver that amplifies signals after they have been converted to the fixed intermediate-frequency value by the frequency converter. Abbreviated i-f amplifier. { ,in·tər'mēd·ē·ət ¦frē·kwən·sē 'am·plə,fī·ər }

intermediate-frequency jamming [ELECTR] Form of continuous wave jamming that is accomplished by transmitting two continuous wave signals separated by a frequency equal to the center frequency of the radar receiver intermediate-frequency amplifier, expecting the radar's own mixer to produce the obscuring intermediate-frequency signal. { ,in·tər 'mēd·ē·ət ¦frē·kwən·sē 'jam·iŋ }

intermediate-frequency response ratio [ELECTR] In a superheterodyne receiver, the ratio of the intermediate-frequency signal input at the antenna to the desired signal input for identical outputs. Also known as intermediate-interference ratio. { ,in·tər'mēd·ē·ət ¦frē·kwən·sē ri'späns ,rā·shō }

intermediate-frequency signal [ELECTR] A modulated or continuous-wave signal whose frequency is the intermediate-frequency value of a superheterodyne receiver and is produced by frequency conversion before demodulation. { ,in·tər'mēd·ē·ət ¦frē·kwən·sē ,sig·nəl }

intermediate-frequency stage [ELECTR] One of the stages in the intermediate-frequency amplifier of a superheterodyne receiver. { ,in·tər 'mēd·ē·ət ¦frē·kwən·sē ,stāj }

intermediate-frequency strip [ELECTR] A receiver subassembly consisting of the intermediate-frequency amplifier stages, installed or replaced as a unit. { ,in·tər'mēd·ē·ət ¦frē·kwən·sē ,strip }

intermediate-frequency transformer [ELECTR] The transformer used at the input and output of each intermediate-frequency amplifier stage in a superheterodyne receiver for coupling purposes and to provide selectivity. Abbreviated i-f transformer. { ,in·tər'mēd·ē·ət ¦frē·kwən·sē tranz 'fōr·mər }

intermediate-infrared radiation [ELECTROMAG] Infrared radiation having a wavelength between about 2.5 micrometers and about 50 micrometers; this range includes most molecular vibrations. Also known as mid-infrared radiation. { ,in·tər'mēd·ē·ət ¦in·frə¦red ,rād·ē'ā·shən }

intermediate-interference ratio *See* intermediate-frequency response ratio. { ,in·tər 'mēd·ē·ət ,in·tər'fir·əns ,rā·shō }

intermediate language level [COMPUT SCI] A computer program that has been converted by a compiler into a form that does not resemble the original program but that still requires further processing by an interpreter at run time before it can be executed. { ,in·tər'mēd·ē·ət 'laŋ·gwij ,lev·əl }

intermediate memory storage [COMPUT SCI] An electronic device for holding working figures temporarily until needed and for releasing final figures to the output. { ,in·tər'mēd·ē·ət 'mem·rē ,stór·ij }

intermediate repeater [ELECTR] Repeater for use in a trunk or line at a point other than an end. { ,in·tər'mēd·ē·ət ri'pēd·ər }

intermediate result [COMPUT SCI] A quantity or value derived from an operation performed in the course of a program or subroutine which is itself used as an operand in further operations. { ,in·tər'mēd·ē·ət ri'zəlt }

intermediate storage [COMPUT SCI] The portion of the computer storage facilities that usually stores information in the processing stage. { ,in·tər'mēd·ē·ət 'stór·ij }

intermediate total [COMPUT SCI] A sum that is produced when there is a change in the value of control data at a level that is neither the most nor the least significant. { ,in·tər'mēd·ē·ət ¦tōd·əl }

intermittent scanning [ELECTR] Scans of an antenna beam at irregular intervals to increase difficulty of detection by intercept receivers. { ¦in·tər¦mit·ənt 'skan·iŋ }

intermodulation [ELECTR] Modulation of the components of a complex wave by each other, producing new waves whose frequencies are equal to the sums and differences of integral multiples of the component frequencies of the original complex wave. { ,in·tər,mäj·ə'lā·shən }

intermodulation distortion [ELECTR] Nonlinear distortion characterized by the appearance of output frequencies equal to the sums and differences of integral multiples of the input frequency components; harmonic components also present in the output are usually not included as part of the intermodulation distortion. { ,in·tər ,mäj·ə'lā·shən di,stór·shən }

intermodulation interference [ELECTR] Interference that occurs when the signals from two undesired stations differ by exactly the intermediate-frequency value of a superheterodyne receiver, and both signals are able to pass through the preselector due to poor selectivity. { ,in·tər ,mäj·ə'lā·shən ,in·tər'fir·əns }

internal arithmetic [COMPUT SCI] Arithmetic operations carried out in a computer's arithmetic unit within the central processing unit. { in'tərn·əl ə'rith·mə,tik }

internal buffer [COMPUT SCI] A portion of a computer's main storage used to temporarily hold data that is being transferred into and out of main storage. { in'tərn·əl 'bəf·ər }

internal cache *See* primary cache. { in¦tərn·əl 'kash }

internal clocking [COMPUT SCI] Synchronization of the electronic circuitry of a device by a timing clock within the device itself. { in'tərn·əl 'kläk·iŋ }

internal cycle time [COMPUT SCI] The time required to change the information in a single register of a computer, usually a fraction of the cycle time of the main memory. Also known as clock time. { in'tərn·əl 'sī·kəl ˌtīm }

internal data transfer [COMPUT SCI] The movement of data between registers in a computer's central processing unit or between a register and main storage. { in'tərn·əl 'dad·ə ˌtranz·fər }

internal hemorrhage [COMPUT SCI] A condition in which a computer program continues to run following an error but produces dubious results and may adversely affect other programs or the performance of the entire system. { in'tərn·əl 'hem·rij }

internal interrupt [COMPUT SCI] A signal for attention sent to a computer's central processing unit by another component of the computer. { in'tərn·əl 'int·əˌrəpt }

internal label [COMPUT SCI] An identifier providing a name for data that is recorded with the data in a storage medium. { in'tərn·əl 'lā·bəl }

internal loss *See* loss. { in'tərn·əl 'lȯs }

internally stored program [COMPUT SCI] A sequence of instructions stored inside the computer in the same storage facilities as the computer data, as opposed to external storage on tape, disk, or drum. { in'tərn·əl·ē ¦stȯrd 'prō·grəm }

internal memory *See* internal storage. { in 'tərn·əl 'mem·rē }

internal modem *See* integral modem. { in 'tərn·əl 'mōˌdem }

internal schema [COMPUT SCI] The physical configuration of data in a data base. { in'tərn·əl 'skē·mə }

internal sorting [COMPUT SCI] The sorting of a list of items by a computer in which the entire list can be brought into the main computer memory and sorted in memory. { in'tərn·əl 'sȯrd·iŋ }

internal storage [COMPUT SCI] The total memory or storage that is accessible automatically to a computer without human intervention. Also known as internal memory. { in'tərn·əl 'stȯr·ij }

internal storage capacity [COMPUT SCI] The quantity of data that can be retained simultaneously in internal storage. { in'tərn·əl 'stȯr·ij kəˌpas·əd·ē }

internal table [COMPUT SCI] A table or array that is coded directly into a computer program and is compiled along with the rest of the program. { in'tərn·əl 'tā·bəl }

international broadcasting [COMMUN] Radio broadcasting for public entertainment between different countries, on frequency bands between 5950 and 21,750 kilohertz, assigned by international agreement. { ¦in·tər¦nash·ən·əl 'brȯdˌkast·iŋ }

international cable code *See* Morse cable code. { ¦in·tər¦nash·ən·əl 'kā·bəl ˌkōd }

international call sign [COMMUN] Call sign assigned according to the provisions of the International Telecommunication Union to identify a radio station; the nationality of the radio station is identified by the first or the first two characters. { ¦in·tər¦nash·ən·əl 'kȯl ˌsīn }

international code signal [COMMUN] Code adopted by many nations for international communications; it uses combinations of letters in lieu of words, phrases, and sentences; the letters are transmitted by the hoisting of international alphabet flags or by transmitting their dot and dash equivalents in the international Morse code. Also known as international signal code. { ¦in·tər¦nash·ən·əl 'kōd ˌsig·nəl }

international control frequency bands [COMMUN] Radio-frequency bands assigned in the United States to links between stations used for international communication and their associated control centers. { ¦in·tər¦nash·ən·əl kən ¦trōl 'frē·kwən·sē ˌbanz }

international control station [COMMUN] Fixed station in the fixed public control service associated directly with the international fixed public radio communications service. { ¦in tər ¦nash·ən·əl kən'trōl ˌstā·shən }

international fixed public radio communications service [COMMUN] Fixed service, the stations of which are open to public correspondence; this service is intended to provide radio communications between the United States and its territories and foreign or overseas points. { ¦in·tər¦nash·ən·əl ¦fixt ¦pəb·lik ¦rād·ē·ō kəˌmyü·nə'kā·shən ˌsər·vəs }

international Morse code *See* continental code. { ¦in·tər¦nash·ən·əl ¦mȯrs 'kōd }

international radio silence [COMMUN] Three-minute periods of radio silence, on the frequency of 500 kilohertz only, commencing 15 and 45 minutes after each hour, during which all marine radio stations must listen on that frequency for distress signals of ships and aircraft. { ¦in·tər ¦nash·ən·əl ¦rād·ē·ō¦sī·ləns }

international signal code *See* international code signal. { ¦in·tər¦nash·ən·əl 'sig·nəl ˌkōd }

international telegraph alphabet *See* CCIT 2 code. { ¦in·tər¦nash·ən·əl 'tel·əˌgraf ˌal·fəˌbet }

International Telegraphic Consultative Committee code 2 *See* CCIT 2 code. { ¦in·tər ¦nash·ən·əl ¦tel·ə¦graf·ik kən¦səl·tə·div kə¦mid·ē ¦kōd 'tü }

internet [COMMUN] A system of local area networks that are joined together by a common communications protocol. { 'in·tərˌnet }

Internet [COMPUT SCI] A worldwide system of interconnected computer networks, communicating by means of TCP/IP and associated protocols. { 'in·tərˌnet }

Internet Mail Access Protocol [COMPUT SCI] An Internet standard for directly reading and manipulating e-mail messages stored on remote servers. { ¦in·tər,net ,māl 'ak,ses ,prōd·ə,kȯl }

Internet protocol [COMMUN] The set of standards responsible for ensuring that data packets transmitted over the Internet are routed to their intended destinations. Abbreviated IP. { ¦in·tər ,net 'prōd·ə,kȯl }

Internet telephony [COMMUN] Phone calls routed over the Internet by analog-to-digital conversion of speech signals. { ,in·tər,net tə'lef·ə·nē }

internetting [COMPUT SCI] Connections and communications paths between separate data communications networks that allow transfer of messages. { ¦in·tər¦ned·iŋ }

interoffice trunk [COMMUN] A direct trunk between local central offices in the same exchange. { ¦in·tər'ȯf·əs 'trəŋk }

interphone [COMMUN] An intercommunication system using headphones and microphones for communication between adjoining or nearby studios or offices, or between crew locations on an aircraft, vessel, or tank or other vehicle. Also known as talk-back circuit. { 'in·tər,fōn }

interpolation [MATH] A process used to estimate an intermediate value of one (dependent) variable which is a function of a second (independent) variable when values of the dependent variable corresponding to several discrete values of the independent variable are known. { in ,tər·pə'lā·shən }

interposition trunk [COMMUN] Trunk which connects two positions of a large switchboard so that a line on one position can be connected to a line on another position. { ,in·tər·pə'zish·ən ,trəŋk }

interpreter [COMPUT SCI] **1.** A program that translates and executes each source program statement before proceeding to the next one. Also known as interpretive routine. **2.** *See* conversational compiler. { in'tər·prəd·ər }

interpretive code *See* interpretive language. { in'tər·prəd·iv ,kōd }

interpretive language [COMPUT SCI] A computer programming language in which each instruction is immediately translated and acted upon by the computer, as opposed to a compiler which decodes a whole program before a single instruction can be executed. Also known as interpretive code. { in'tər·prəd·iv 'laŋ·gwij }

interpretive programming [COMPUT SCI] The writing of computer programs in an interpretive language, which generally uses mnemonic symbols to represent operations and operands and must be translated into machine language by the computer at the time the instructions are to be executed. { in'tər·prəd·iv 'prō,gram·iŋ }

interpretive trace program [COMPUT SCI] An interpretive routine that provides a record of the machine code into which the source program is translated and of the result of each step, or of selected steps, of the program. { in'tər·prəd·iv 'trās ,prō·grəm }

interprocedure metric [COMPUT SCI] A software metric that estimates the complexity of a module or computer program based on the way that the data are used, organized, and allocated in relationship with some other modules. { ,in·tər·prə ¦sē·jər 'me·trik }

interprocess communication [COMPUT SCI] The communication between computer programs running concurrently under the control of the same operating system. { ¦in·tər,prä·səs kə ,myü·ə'kā·shən }

interrecord gap *See* record gap. { ¦in·tər'rek·ərd ,gap }

interrogation [COMMUN] The transmission of a radio-frequency pulse, or combination of pulses, intended to trigger a transponder or group of transponders, a racon system, or an IFF system, in order to elicit an electromagnetic reply. Also known as challenging signal. { in,ter·ə'gā·shən }

interrogation suppressed time delay [COMMUN] Overall fixed time delay between transmission of an interrogation and reception of the reply to this interrogation at zero distance. { in ,ter·ə'gā·shən sə¦prest 'tīm di,lā }

interrogator [ELECTR] **1.** A radar transmitter which sends out a pulse that triggers a transponder; usually combined in a single unit with a responsor, which receives the reply from a transponder and produces an output suitable for actuating a display of some navigational parameter. Also known as challenger; interrogator-transmitter. **2.** *See* interrogator-responsor. { in'ter·ə,gād·ər }

interrogator-responsor [ELECTR] A transmitter and receiver combined, used for sending out pulses to interrogate a radar beacon and for receiving and displaying the resulting replies. Also known as interrogator. { in'ter·ə,gād·ər ri'spän·sər }

interrogator-transmitter *See* interrogator. { in 'ter·ə,gād·ər tranz'mid·ər }

interrupt [COMPUT SCI] **1.** To stop a running program in such a way that it can be resumed at a later time, and in the meanwhile permit some other action to be performed. **2.** The action of such a stoppage. { 'int·ə,rəpt }

interrupt-driven system [COMPUT SCI] An operating system in which the interrupt system is the mechanism for reporting all changes in the states of hardware and software resources, and such changes are the events that induce new assignments of these resorces to meet work-load demands. { 'int·ə,rəpt ,driv·ən ,sis·təm }

interrupted continuous wave [COMMUN] A continuous wave that is interrupted at a constant audio-frequency rate high enough to give several interruptions for each keyed code dot. Abbreviated ICW. { 'int·ə,rəp·təd kən¦tin·yə·wəs 'wāv }

interrupt handler [COMPUT SCI] A section of a computer program or of the operating system that takes control when an interrupt is received and performs the operations required to service the interrupt. { 'int·ə,rəpt ,hand·lər }

interrupt mask [COMPUT SCI] A technique of suppressing certain interrupts and allowing the control program to handle these masked interrupts at a later time. { 'int·ə,rəpt ,mask }

interrupt mode *See* hold mode. { 'int·ə,rəpt ,mōd }

interrupt priorities [COMPUT SCI] The sequence of importance assigned to attending to the various interrupts that can occur in a computer system. { 'in·tə,rəpt prī,är·ə·dēz }

interrupt routine [COMPUT SCI] A program that responds to an interrupt by carrying out prescribed actions. { 'int·ə,rəpt rü,tēn }

interrupt signal [COMPUT SCI] A control signal which requests the immediate attention of the central processing unit. { 'int·ə,rəpt ,sig·nəl }

interrupt system [COMPUT SCI] The means of interrupting a program and proceeding with it at a later time; this is accomplished by saving the contents of the program counter and other specific registers, storing them in reserved areas, starting the new instruction sequence, and upon completion, reloading the program counter and registers to return to the original program, and reenabling the interrupt. { 'int·ə,rəpt ,sis·təm }

interrupt trap [COMPUT SCI] A program-controlled technique which either recognizes or ignores an interrupt, depending upon a switch setting. { 'int·ə,rəpt ,trap }

interrupt vector [COMPUT SCI] A list comprising the locations of various interrupt handlers. { 'int·ə,rəpt ,vek·tər }

intersection data [COMPUT SCI] Data which are meaningful only when associated with the concatenation of two segments. { ,in·tər'sek·shən ,dad·ə }

interstation noise suppression *See* intercarrier noise suppression. { ¦in·tər'stā·shən 'nȯiz sə ,presh·ən }

intersymbol interference [COMMUN] In a transmission system, extraneous energy from the signal in one or more keying intervals which tends to interfere with the reception of the signal in another keying interval, or the disturbance which results. { ¦in·tər'sim·bəl ,in·tər'fir·əns }

intersystem communications [COMPUT SCI] The ability of two or more computer systems to share input, output, and storage devices, and to send messages to each other by means of shared input and output channels or by channels that directly connect central processors. { ¦in·tər ¦sis·təm kə,myü·nə'kā·shənz }

intertoll trunk [COMMUN] A trunk between toll offices in different telephone exchanges. { 'in·tər,tōl 'trəŋk }

interval arithmetic [COMPUT SCI] A method of numeric computation in which each variable is specified as lying within some closed interval, and each arithmetic operation computes an interval containing all values that can result from operating on any numbers selected from the intervals associated with the operands. Also known as range arithmetic. { 'in·tər·vəl ə'rith·mə·tik }

intra-coded picture [COMMUN] A MPEG-2 picture that is coded using information present only in the picture itself and not depending on information from other pictures; provides a mechanism for random access into the compressed video data; employs transform coding of the pel blocks and provides only moderate compression. Also known as I-frame; I-picture. { ¦in·trə 'kōd·əd 'pik·chər }

intranet [COMPUT SCI] A private network, based on Internet protocols, that is accessible only within an organization. Intranets are set up for many purposes, including e-mail, access to corporate databases and documents, and videoconferencing, as well as buying and selling goods and services. { 'in·trə,net }

intraprocedure metric [COMPUT SCI] A software metric that determines the complexity of a computer program as a function of the relationships of the different modules constituting the program, generally by constructing a flow graph and deriving the complexity from this graph. { ,in·trə·prə¦sē·jər 'me·trik }

intrinsic-barrier diode [ELECTR] A *pin* diode, in which a thin region of intrinsic material separates the *p*-type region and the *n*-type region. { in'trin·sik ¦bar·ē·ər 'dī,ōd }

intrinsic-barrier transistor [ELECTR] A *pnip* or *npin* transistor, in which a thin region of intrinsic material separates the base and collector. { in'trin·sik ¦bar·ē·ər tran'zis·tər }

intrinsic layer [ELECTR] A layer of semiconductor material whose properties are essentially those of the pure undoped material. { in'trin·sik 'lā·ər }

intrinsic procedure *See* built-in function. { in 'trin·sik prə'sē·jər }

inverse feedback *See* negative feedback. { 'in ,vərs 'fēd,bak }

inverse neutral telegraph transmission [COMMUN] Form of transmission in which marking signals are zero current intervals and spacing signals are current pulses of either polarity. { 'in ,vərs ,nü·trəl 'tel·ə,graf tranz,mish·ən }

inverse problem [CONT SYS] The problem of determining, for a given feedback control law, the performance criteria for which it is optimal. { 'in ,vərs 'präb·ləm }

inverse-synthetic-aperture radar [ENG] A radar of such high resolution in both range and Doppler sensing that a moving target's slight rotation around the radar line of sight permits the structure of the target to be discerned and displayed. { 'in,vərs sin'thed·ik ¦ap·ə·chər 'rā ,där }

inverse video *See* reverse video. { 'in,vərs ¦vid·ē·ō }

inversion [COMMUN] The process of scrambling speech for secrecy by beating the voice signal with a fixed, higher audio frequency and using only the difference frequencies. [OPTICS] The formation of an inverted image by an optical system. { in'vər·zhən }

inverted file [COMPUT SCI] **1.** A file, or method of file organization, in which labels indicating the locations of all documents of a given type are placed in a single record. **2.** A file whose usual order has been inverted. { in'vərd·əd 'fīl }

inverted L antenna [ELECTROMAG] An antenna consisting of one or more horizontal wires to which a connection is made by means of a vertical wire at one end. { in'vərd·əd ¦el an,ten·ə }

inverted vee [ELECTROMAG] **1.** A directional antenna consisting of a conductor which has the form of an inverted V, and which is fed at one end and connected to ground through an appropriate termination at the other. **2.** A center-fed horizontal dipole antenna whose arms have ends bent downward 45°. { in'vərd·əd 'vē }

inverter circuit *See* NOT circuit. { in'vərd·ər ,sər·kət }

inverting function [ELECTR] A logic device that inverts the input signal, so that the output is out of phase with the input. { in'vərd·iŋ ,fəŋk·shən }

inverting terminal [ELECTR] The negative input terminal of an operational amplifier; a positive-going voltage at the inverting terminal gives a negative-going output voltage. { in'vərd·iŋ 'tər·mən·əl }

inward-outward dialing system [COMMUN] Dialing system whereby calls within the local exchange area may be dialed directly to and from base private branch exchange telephone stations without the assistance of the base private branch exchange operator; CENTREX, a service offered by some telephone companies, is a form of inward-outward dialing. { ¦in·wərd ¦au̇t,wərd 'dī·liŋ ,sis·təm }

I/O *See* input/output.

IOCS *See* input/output control system.

IOGEN *See* input/output generation. { 'ī¦ō,jen }

ionosphere [GEOPHYS] That part of the earth's upper atmosphere which is sufficiently ionized by solar ultraviolet radiation so that the concentration of free electrons affects the propagation of radio waves; its base is at about 40 or 50 miles (70 or 80 kilometers) and it extends to an indefinite height. { ī'an·ə,sfir }

ionospheric error [COMMUN] Variation in the character of the ionospheric transmission path or paths used by the radio waves of electronic navigation systems which, if not compensated, will produce an error in the information generated by the system. { ,ī,än·ə'sfir·ik 'er·ər }

ionospheric propagation [COMMUN] Propagation of radio waves over long distances by reflection from the ionosphere, useful at frequencies up to about 25 megahertz. { ,ī,än·ə'sfir·ik ,präp·ə'gā·shən }

ionospheric scatter [COMMUN] A form of scatter propagation in which radio waves are scattered by the lower E layer of the ionosphere to permit communication over distances from 600 to 1400 miles (1000 to 2250 kilometers) when using the frequency range of about 25 to 100 megahertz. { ,ī,än·ə'sfir·ik 'skad·ər }

ionospheric wave *See* sky wave. { ,ī,än·ə'sfir·ik 'wāv }

IP address [COMPUT SCI] A computer's numeric address, such as 128.201.86.290, by which it can be located within a network. { ¦ī'pē ə,dres }

I-picture *See* intra-coded picture. { 'ī¦pik·chər }

IPL [COMPUT SCI] **1.** Collective term for a series of list-processing languages developed principally by A. Newell, H. A. Simon, and J. C. Shaw. Derived from Information Processing Language. **2.** *See* initial program load.

IPL button *See* bootstrap button. { ¦ī¦pē'el 'bət·ən }

IRG *See* record gap.

iris [ELECTROMAG] A conducting plate mounted across a waveguide to introduce impedance; when only a single mode can be supported, an iris acts substantially as a shunt admittance and may be used for matching the waveguide impedance to that of a load. Also known as diaphragm; waveguide window. { 'ī·rəs }

ISB modulation *See* independent-sideband modulation. { ¦ī¦es¦bē ,mäj·ə'lā·shən }

I-scan *See* I-display. { 'ī ,skan }

I-scope *See* I-display. { 'ī ,skōp }

ISDN *See* integrated services digital network.

ISDN modem [ELECTR] A device that converts signals used in a computer to signals that can be transmitted over the integrated services digital network, and vice versa. { ,ī,es,dē,en 'mō,dem }

I signal [ELECTR] The in-phase component of the chrominance signal in color television, having a bandwidth of 0 to 1.5 megahertz, and consisting of +0.74(R − Y) and −0.27(B − Y), where Y is the luminance signal, R is the red camera signal, and B is the blue camera signal. { 'ī ,sig·nəl }

isobits [COMPUT SCI] Binary digits having the same value. { 'ī·sə,bits }

isochronous communications [COMMUN] Synchronization of a data communications network from timing signals provided by the network itself. { ī'sä·krə·nəs kə,myü·nə'kā·shənz }

isocirculator [ELECTROMAG] A circulator that has an absorber in one of its terminals and thereby acts as an isolator. { ,ī·sō'sər·kyə ,lād·ər }

isolated location [COMPUT SCI] A location in a computer memory which is protected by some hardware device so that it cannot be addressed by a computer program and its contents cannot be accidentally altered. { 'ī·sə,lād·əd lō'kā·shən }

isolation [COMPUT SCI] The ability of a logic circuit having more than one input to ensure that each input signal is not affected by any of the others. { ,ī·sə'lā·shən }

isolation amplifier [ELECTR] An amplifier used to minimize the effects of a following circuit on the preceding circuit. { ,ī·sə'lā·shən 'am·plə ,fī·ər }

isolation diode [ELECTR] A diode used in a circuit to allow signals to pass in only one direction. { ,ī·sə'lā·shən 'dī,ōd }

isolator [ELECTR] A passive attenuator in which the loss in one direction is much greater than that in the opposite direction; a ferrite isolator for waveguides is an example. { 'ī·sə,lād·ər }

isolith [ELECTR] Integrated circuit of components formed on a single silicon slice, but with the various components interconnected by beam leads and with circuit parts isolated by removal of the silicon between them. { 'ī·sə,lith }

isopulse system [COMMUN] In adaptive communications, a pulse coding system wherein the number of information pulses transmitted is indicated by special inserted pulses. { 'ī·sə,pəls ,sis·təm }

isotropic antenna *See* unipole. { ¦ī·sə¦trä·pik an'ten·ə }

isotropic gain of an antenna *See* absolute gain of an antenna. { ¦ī·sə¦trä·pik 'gān əv ən an'ten·ə }

item [COMPUT SCI] A set of adjacent digits, bits, or characters which is treated as a unit and conveys a single unit of information. { 'īd·əm }

item advance [COMPUT SCI] A technique of efficiently grouping records to optimize the overlap of read, write, and compute times. { 'īd·əm əd ,vans }

item design [COMPUT SCI] The specification of what fields make up an item, the order in which the fields are to be recorded, and the number of characters to be allocated to each field. { 'īd·əm di,zīn }

item size [COMPUT SCI] The length of an item expressed in characters, words, or blocks. { 'īd·əm ,sīz }

iteration process [COMPUT SCI] The process of repeating a sequence of instructions with minor modifications between successive repetitions. { ,īd·ə'rā·shən ,prä·səs }

iterations per second [COMPUT SCI] In computers, the number of approximations per second in iterative division; the number of times an operational cycle can be repeated in 1 second. { ,īd·ə'rā·shənz pər 'sek·ənd }

iterative array [COMPUT SCI] In a computer, an array of a large number of interconnected identical processing modules, used with appropriate driver and control circuits to permit a large number of simultaneous parallel operations. { 'īd·ə,rād·iv ə'rā }

iterative division [COMPUT SCI] In computers, a method of dividing by use of the operations of addition, subtraction, and multiplication; a quotient of specified precision is obtained by a series of successively better approximations. { 'īd·ə,rād·iv di'vizh·ən }

iterative filter [ELECTR] Four-terminal filter that provides iterative impedance. { 'īd·ə,rād·iv 'fil·tər }

iterative impedance [ELECTR] Impedance that, when connected to one pair of terminals of a four-terminal transducer, will cause the same impedance to appear between the other two terminals. { 'īd·ə,rād·iv im'pēd·əns }

iterative routine [COMPUT SCI] A computer program that obtains a result by carrying out a series of operations repetitiously until some specified condition is met. { 'īd·ə,rād·iv rü'tēn }

ITV *See* industrial television.

J

jammer [ELECTR] One who, or equipment which, transmits electromagnetic signals to obscure or deceive radio or radar receivers, preventing them from receiving the intended signals clearly. { 'jam·ər }

jammer finder [ELECTR] Radar which attempts to obtain the range of the target by training a highly directional pencil beam on a jamming source. Also known as burn-through. { 'jam·ər ˌfīn·dər }

jamming [ELECTR] Radiation, reradiation, or reflection of electromagnetic waves so as to impair the usefulness of the radio spectrum for military purposes including communication and radar. Also known as active jamming; electronic jamming. { 'jam·iŋ }

J antenna [ELECTROMAG] Antenna having a configuration resembling a J, consisting of a half-wave antenna end-fed by a parallel-wire quarter-wave section. { 'jāantˌen·ə }

Java [COMPUT SCI] An object-oriented programming language based on C++ that was designed to run in a network such as the Internet; mostly used to write programs, called applets, that can be run on Web pages. { 'jäv·ə }

JavaScript [COMPUT SCI] A scripting language that is added to standard HTML to create interactive documents. { 'jäv·əˌskript }

Java virtual machine [COMPUT SCI] An interpreter that translates Java bytecode into actual machine instructions in real time. Abbreviated JVM. { ¦jäv·ə ˌvər·chə·wəl mə'shēn }

J-display [ELECTR] A radar display format in which range is presented as a reference circle with radial projections from it indicating echo strength; a circular A-display. Also known as J-indicator; J-scan; J-scope. { 'jādiˌsplā }

JFET *See* junction field-effect transistor. { 'jā ˌfet }

J-indicator *See* J-display. { 'jā ˌin·dəˌkād·ər }

jitter [COMMUN] In facsimile, distortion in the received copy caused by momentary errors in synchronism between the scanner and recorder mechanisms; does not include slow errors in synchronism due to instability of the frequency standards used in the facsimile transmitter and recorder. [ELECTR] Small, rapid variations in a waveform due to mechanical vibrations, fluctuations in supply voltages, control-system instability, and other causes. { 'jid·ər }

jittered pulse recurrence frequency [COMMUN] Random variation of the pulse repetition period; provides a discrimination capability against repeater-type jammers. { 'jid·ərd ¦pəls ri¦kə·rəns 'frē·kwən·sē }

J-K flip-flop [ELECTR] A storage stage consisting only of transistors and resistors connected as flip-flops between input and output gates, and working with charge-storage transistors; gives a definite output even when both inputs are 1. { ¦jā¦kā'flipˌfläp }

job [COMPUT SCI] A unit of work to be done by the computer; it is a single entity from the standpoint of computer installation management, but may consist of one or more job steps. { jäb }

job class [COMPUT SCI] The set of jobs on a computer system whose resource requirements (for the central processing unit, memory, and peripheral devices) fall within specified ranges. { 'jäb ˌklas }

job control block [COMPUT SCI] A group of data containing the execution-control data and the job identification when the job is initiated as a unit of work to the operating system. { 'jäb kən ˌtrōl ˌbläk }

job control language *See* command language. { 'jäb kənˌtrōl ˌlaŋ·gwij }

job control statement [COMPUT SCI] Any of the statements used to direct an operating system in its functioning, as contrasted to data, programs, or other information needed to process a job but not intended directly for the operating system itself. Also known as control statement. { 'jäb kənˌtrōl ˌstāt·mənt }

job entry system [COMPUT SCI] A part of the operating system of a large computer system that accepts and schedules jobs for execution and controls the printing of output. { 'jäb ¦en·trē ˌsis·təm }

job family *See* job class. { 'jäb ˌfam·lē }

job flow control [COMPUT SCI] Control over the order in which jobs are handled by a computer in order to use the central processing units and the units under the computer's control as efficiently as possible. { 'jäb 'flō kənˌtrōl }

job grade *See* job class. { 'jäb ˌgrād }

job library [COMPUT SCI] A partitioned data set, or a concatenation of partitioned data sets, used as the primary source of object programs (load modules) for a particular job, and more generally, as a source of runnable programs from which all or most of the programs for a given job will be selected. { 'jäb ,li,brer·ē }

job management program [COMPUT SCI] A control program in a computer's operating system that initials and schedules jobs. { 'jäb ¦man·ij·mənt ,prō·grəm }

job mix [COMPUT SCI] The distribution of the jobs handled by a computer system among the various job classes. { 'jäb, miks }

job-oriented terminal [COMPUT SCI] A terminal, such as a point-of-sale terminal, at which data taken directly from a source can enter a communication network directly. { ¦jäb ¦ȯr·ē,ent·əd 'tər·mən·əl }

job processing control [COMPUT SCI] The section of the control program responsible for initiating operations, assigning facilities, and proceeding from one job to the next. { 'jäb ¦prä·ses·iŋ kən,trōl }

job queue [COMPUT SCI] A set of computer programs that are ready to be executed in a prescribed order. { 'jäb ,kyü }

job stacking [COMPUT SCI] The presentation of jobs to a computer system, each job followed by another. { 'jäb ,stak·iŋ }

job step [COMPUT SCI] A unit of work in a job stream. { 'jäb ,step }

job swapping [COMPUT SCI] Temporary suspension of job processing by a computer so that higher-priority jobs can be handled. { 'jäb ,swäp·iŋ }

Johnson noise *See* thermal noise. { 'jän·sən ,nȯiz }

join [COMPUT SCI] A portion of a robotic control program that directs an activity to resume after it has been interrupted. { jȯin }

Joint Photographic Experts Group [COMPUT SCI] An international group that sets standards for continuous-tone image (still and video) coding. { ¦jȯint ,fōd·ə,graf·iks 'ek,spərts ,grüp }

journaling [COMPUT SCI] Recording processes or transactions for backup or accounting purposes. { 'jər·nəl·iŋ }

JPEG [GRAPHICS] Graphics file format for compressed still images, particularly photographic images found on the World Wide Web; developed by the Joint Photographic Experts Group. { 'jā ,peg }

J-scan *See* J-display. { 'jā ,skan }

J-scope *See* J-display. { 'jā ,skōp }

jump [COMPUT SCI] A transfer of control which terminates one sequence of instructions and begins another sequence at a different location. Also known as branch; transfer. { jəmp }

jumper [ELEC] A short length of conductor used to close a circuit between two electircal terminals. { 'jəm·pər }

jumping trace routine [COMPUT SCI] A trace routine which is primarily concernedwith providing a record of jump instructions in order to show the sequence of program steps that the computer followed. { 'jəm·piŋ ¦trās rü,tēn }

jump resonance [CONT SYS] A jump discontinuity occurring in the frequency response of a nonlinear closed-loop control system with saturation in the loop. { 'jəmp ,rez·ən·əns }

jump vector [COMPUT SCI] A list of entry-point addresses for various sections of a computer program; used by the program to branch to a section that performs a desired function. Also known as vector; vector table. { 'jəmp ,vek·tər }

junction [ELECTR] A region of transition between two different semiconducting regions in a semiconductor device, such as a *pn* junction, or between a metal and a semiconductor. [ELECTROMAG] A fitting used to join a branch waveguide at an angle to a main waveguide, as in a tee junction. Also known as waveguide junction. { 'jəŋk·shən }

junction diode [ELECTR] A semiconductor diode in which the rectifying characteristics occur at an alloy, diffused, electrochemical, or grown junction between *n*-type and *p*-type semiconductor materials. Also known as junction rectifier. { 'jəŋk·shən ¦dī,ōd }

junction field-effect transistor [ELECTR] A field-effect transistor in which there is normally a channel of relatively low-conductivity semiconductor joining the source and drain, and this channel is reduced and eventually cut off by junction depletion regions, reducing the conductivity, when a voltage is applied between the gate electrodes. Abbreviated JFET. { 'jəŋk·shən 'fēld i,fekt tran¦zis·tər }

junction filter [ELECTR] A combination of a high-pass and a low-pass filter that is used to separate frequency bands for transmission over separate paths. { 'jəŋk·shən ,fil·tər }

junction loss [COMMUN] In telephone circuits, that part of the repetition equivalent assignable to interaction effects arising at trunk terminals. { 'jəŋk·shən ,lȯs }

junction point *See* branch point. { 'jəŋk·shən ,pȯint }

junction rectifier *See* junction diode. { 'jəŋk·shən ¦rek·tə,fī·ər }

junction station [ELECTR] Microwave relay station that joins a microwave radio leg or legs to the main or through route. { 'jəŋk·shən ,stā·shən }

junction transistor [ELECTR] A transistor in which emitter and collector barriers are formed between semiconductor regions of opposite conductivity type. { 'jəŋk·shən tran¦zis·tər }

justify [COMPUT SCI] To shift data so that they assume a particular position relative to one or more reference points, lines, or marks in a storage medium. { 'jəs·tə,fī }

JVM *See* Java virtual machine.

K

k *See* kilobit.

K *See* kilobyte.

Ka band [COMMUN] A band of frequencies extending from 33 to 36 gigahertz, corresponding to wavelengths of 9.09 to 8.34 millimeters. { kā 'ā ,band }

Kalman filter [CONT SYS] A linear system in which the mean squared error between the desired output and the actual output is minimized when the input is a random signal generated by white noise. { 'kal·mən ,fil·tər }

Kanji [COMPUT SCI] A set of Chinese characters that are employed by users of the Chinese language to code information in computer programs and on visual displays. { 'kän·jē }

Karnaugh map [ELECTR] A truth table that has been rearranged to show a geometrical pattern of functional relationships for gating configurations; with this map, essential gating requirements can be recognized in their simplest form. { 'kär·nō ,map }

K band [COMMUN] A band of radio frequencies extending from 10,900 to 36,000 megahertz, corresponding to wavelengths of 2.75 to 0.834 centimeters. { 'kā,band }

K-band single-access service [COMMUN] A service provided by the Tracking and Data Relay Satellite System, with return-link data rates up to 300 and 800 megabits per second for the Ku and Ka bands, respectively, and forward-link data at 25 megabits per second in both bands. Abbreviated KSA. { ¦kā,band ¦siŋ·gəl 'ak ,ses ,sər·vəs }

Kbit *See* kilobit. { 'kā,bit }

kbit *See* kilobit. { 'kā,bit }

Kbyte *See* kilobyte. { 'kā,bīt }

kbyte *See* kilobyte. { 'kā,bīt }

KDD *See* knowledge discovery in databases.

K-display [ELECTR] A radar display format in which an echo signal appears on side-by-side A-displays, as from a two-beam tracking radar antenna, with equal amplitudes indicating no pointing error. Also known as K-indicator; K-scan; K-scope. { 'kā di,splā }

Kendall effect [COMMUN] A spurious pattern or other distortion in a facsimile record caused by unwanted modulation products arising from the transmission of a carrier signal; occurs principally when the width of one side band is greater than half the facsimile carrier frequency. { 'kend·əl i,fekt }

Kennelly-Heaviside layer *See* E layer. { 'ken·əl·ē 'hev·ē,sīd ,lā·ər }

kernel [COMPUT SCI] **1.** A computer program that must be modified before it can be used on a particular computer. **2.** The programs that form the most essential part of a computer's operating system. { 'kərn·əl }

Kerr cell [OPTICS] A glass cell containing a dielectric liquid that exhibits the Kerr effect, such as nitrobenzene, in which is inserted the two plates of a capacitor, used to observe the Kerr effect on light passing through the cell. { 'kər ,sel }

Kerr effect *See* electrooptical Kerr effect. { 'kər i,fekt }

Kerr magnetooptical effect *See* magnetooptic Kerr effect. { 'kər mag,ned·ō'äp·tə·kəl i,fekt }

key [COMMUN] A telephone term for an on-off switch in the subscriber loop, either at a manual switchboard or in the telephone set. [COMPUT SCI] A data item that serves to uniquely identify a data record. { kē }

key access [COMPUT SCI] Locating data in a file by using the value of a key. { 'kē ,ak,ses }

key auto-key cipher [COMMUN] A stream cipher in which the cryptographic bit stream generated at a given time is determined by the cryptographic bit stream generated at earlier times. { 'kē 'ȯd·ō ,kē ,sī·fər }

keyboard enhancer [COMPUT SCI] Software that expands the functions of a computer keyboard by allowing the user to implement functions or enter predefined segments of text with a single keystroke. Also known as keyboard processor. { 'kē,bȯrd in,han·sər }

keyboard entry [COMPUT SCI] A piece of information fed manually into a computing system by means of a set of keys, such as a typewriter. { 'kē ,bȯrd 'en·trē }

keyboard inquiry [COMPUT SCI] A question asked a computer concerning the status of a program being run, or concerning the value achieved by a specific variable, by means of a console typewriter. { 'kē,bȯrd ¦in·kwə·rē }

keyboard lockout [COMPUT SCI] An arrangement for preventing transmission from a particular keyboard while other transmissions are taking place on the same circuit. { 'kē,bȯrd 'läk,au̇t }

keyboard lockup [COMPUT SCI] A condition in which entries typed on a keyboard are ignored by a terminal. { 'kē,bȯrd 'läk,əp }

keyboard mapping [COMPUT SCI] The process of assigning the meaning of keys on a computer keyboard. { 'kē,bȯrd ,map·iŋ }

keyboard printer [COMPUT SCI] A computer input device that includes a keyboard and a printer that prints the keyed-in data and often also prints computer output information. { 'kē,bȯrd 'print·ər }

keyboard processor [COMPUT SCI] **1.** The circuitry in a computer keyboard that converts keystrokes into the appropriate character codes. **2.** *See* keyboard enhancer. { 'kē,bȯrd ,prä,ses·ər }

keyboard template [COMPUT SCI] A card that is placed adjacent to the function keys of a computer keyboard and identifies their use for a particular software environment. { 'kē,bȯrd ,tem·plət }

key cabinet [ELECTR] A case, installed on a customer's premises, to permit different lines to the control office to be connected to various telephone stations; it has signals to indicate originating calls and busy lines. { 'kē ,kab·ə·nət }

key change [COMPUT SCI] The occurrence, in a file of records which have been sorted according to their keys and are being read into a computer, of a record whose key differs from that of its immediate predecessor. { 'kē ,chānj }

key compression [COMPUT SCI] A technique used to reduce the number of bits contained in a key. { 'kē kəm,presh·ən }

key-disk machine [COMPUT SCI] A keyboard machine used to record data directly on a magnetic disk. { 'kē ,disk mə,shēn }

keyed sequential access method [COMPUT SCI] A method for locating data in a file either directly, by using the value of a key within a particular record, or sequentially, according to the values of the keys in all the records of the file. Abbreviated KSAM. { 'kēd si'kwen·chəl 'ak,ses ,meth·əd }

key entry [COMPUT SCI] The entering of data into a computer by means of a keyboard. { 'kē ,en·trē }

keyer [ELECTR] Device which changes the output of a transmitter from one condition to another according to the intelligence to be transmitted. { 'kē·ər }

keyer adapter [ELECTR] Device which detects a modulated signal and produces the modulating frequency as a direct-current signal of varying amplitude. { 'kē·ər ə,dap·tər }

key field [COMPUT SCI] A field in a segment or record that holds the value of a key to that record. { 'kē ,fēld }

keying [COMMUN] The forming of signals by modulating a direct currect or other carrier between discrete values of some characteristic. { 'kē·iŋ }

keying error rate [COMMUN] The ratio of the number of characters incorrectly transmitted to the total number of characters in a message. { 'kē·iŋ 'er·ər ,rāt }

keying frequency [COMMUN] In facsimile, the maximum number of times a second that a black-line signal occurs when scanning the subject copy. { 'kē·iŋ ,frē·kwən·sē }

keying interval [COMMUN] In a periodically keyed transmission system, one of the set of intervals starting from a change in state and equal in length to the shortest time between changes of state. { 'kē·iŋ ,int·ər·vəl }

keying sequence [COMMUN] A sequence of letters or numbers that enciphers or deciphers a polyalphabetic substitution cipher character by character. { 'kē·iŋ ,sē·kwəns }

keyless ringing [COMMUN] Form of machine ringing on a manual telephone switchboard which is started automatically by the insertion of the calling plug into the jack of the called line. { 'kē·ləs 'riŋ·iŋ }

keypad [COMPUT SCI] A cluster of special-purpose keys to one side of the regular typing keys on a terminal keyboard. { 'kē,pad }

key pulse [COMMUN] System of signaling where numbered keys are depressed instead of using a dial. { 'kē ,pəls }

key punch [COMPUT SCI] A keyboard-actuated device that punches holes in a card; it may be a hand-feed punch or an automatic feed punch. { 'kē ,pənch }

keystone distortion [COMMUN] Distortion produced by scanning in a rectilinear manner, with constant-amplitude sawtooth waves, a plane target area which is not normal to the average direction of the beam. { 'kē,stōn di'stȯr·shən }

keystoning [ELECTR] Producing a keystone-shaped (wider at the top than at the bottom, or vice versa) scanning pattern resulting from an off-axis condition between an image-projection device and the display surface. { 'kē,stōn·iŋ }

keyswitch [COMPUT SCI] A switch that is operated by depressing a key on the keyboard of a data entry terminal. { 'kē,swich }

key telephone system [COMMUN] A telephone system consisting of phones with several keys, connecting cables, and relay switching apparatus, which does not need a special operator to handle incoming or outgoing calls and which generally permits users to select one of several possible lines and to hold calls. { 'kē 'tel·ə,fōn ,sis·təm }

key telephone unit [COMMUN] A small mounting plate with relays which performs pickup and hold switching functions in a key telephone system. { 'kē 'tel·ə,fōn ,yü·nət }

key-to-disk system [COMPUT SCI] A data-entry system in which information entered on several keyboards is collected on different sections of a magnetic disk, and the data are extracted from the disk when complete, and are copied onto a magnetic tape or another disk for further processing on the main computer. { 'kē tə 'disk ,sis·təm }

key-to-tape system [COMPUT SCI] A data-entry system, predecessor to the modern key-to-disk system, consisting of several keyboards connected to a central controlling unit, typically a minicomputer, which collected information from each keyboard and then directed it to a magnetic tape. { 'kē tə 'tāp ˌsis·təm }

key transformation [COMPUT SCI] A function that assigns integer values to keys. { 'kē ˌtranz·fər'mā·shən }

key value [COMPUT SCI] The actual characters contained in a key. { 'kē 'val·yü }

keyword [COMPUT SCI] A group of letters and numbers in a specific order that has special significance in a computer system. { 'kēˌwərd }

keyword-in-context index [COMPUT SCI] A computer-generated listing of titles of documents, produced on a line printer, with the keywords lined up vertically in a fixed position within the title and arranged in alphabetical order. Abbreviated KWIC index. { 'kēˌwərd in 'känˌtekst ˌinˌdeks }

keyword-out-of-context index [COMPUT SCI] A computer-generated listing of document titles with their keywords listed separately, arranged in the alphabetical order of the keywords. Abbreviated KWOC index. { 'kēˌwərd au̇t əv 'känˌtekst ˌinˌdeks }

keyword parameter [COMPUT SCI] A parameter whose significance is indicated by a keyword, usually with an equal sign linking the two. { 'kē ˌwərd pə'ram·əd·ər }

keyword search [COMPUT SCI] A method of filing and locating information through the use of keywords that describe the content of records. { 'kēˌwərd ˌsərch }

KHN filter *See* state-variable filter. { ¦kā¦āch'en ˌfil·tər }

kidney joint [ELECTROMAG] Flexible joint, or air-gap coupling, used in the waveguide of certain radars and located near the transmitting-receiving position. { 'kid·nē ˌjȯint }

killer stage *See* color killer circuit. { 'kil·ər ˌstāj }

kilobit [COMPUT SCI] A unit of information content equal to 1024 bits. Abbreviated kbit; Kbit. Symbolized k. { 'kil·əˌbit }

kilobyte [COMPUT SCI] A unit of information content equal to 1024 bytes. Abbreviated kbyte; Kbyte. Symbolized K. { 'kil·əˌbīt }

kilomegacycle *See* gigahertz. { ¦kil·ə'meg·ə ˌsī·kəl }

kilomegahertz *See* gigahertz. { ¦kil·ə'meg·ə ˌhərtz }

K-indicator *See* K-display. { 'kāˌin·dəˌkād·ər }

kinescope *See* picture tube. { 'kin·əˌskōp }

Kiviat graph [COMPUT SCI] A circular diagram used in computer performance evaluation, in which variables are plotted on axes of the circle with 0% at the center of the circle and 100% at the circumference, and variables which are "good" and "bad" as they approach 100% are plotted on alternate axes. { 'kiv·ē·ət ˌgraf }

kludge [COMPUT SCI] A poorly designed data-processing system composed of ill-fitting mismatched components. { klüj }

klystron [ELECTR] A type of beam power tube, used often in radar and other microwave applications, in which the beam of electrons passes through radio-frequency resonant cavities, or variations, to effect the interaction between the electrons and the signal being amplified or produced. { 'klīˌsträn }

klystron generator [ELECTR] Klystron tube used as a generator, with its second cavity or catcher directly feeding waves into a waveguide. { 'klī ˌsträn ¦jen·əˌrād·ər }

klystron oscillator *See* velocity-modulated oscillator. { 'klīˌsträn ¦äs·əˌlād·ər }

klystron repeater [ELECTR] Klystron tube operated as an amplifier and inserted directly in a waveguide in such a way that incoming waves velocity-modulate the electron stream emitted from a heated cathode; a second cavity converts the energy of the electron clusters into waves of the original type but of greatly increased amplitude and feeds them into the outgoing guide. { 'klīˌsträn ri¦pēd·ər }

knee frequency *See* break frequency. { 'nē ˌfrē·kwən·sē }

knife-edge refraction [ELECTROMAG] Radio propagation effect in which the atmospheric attenuation of a signal is reduced when the signal passes over and is diffracted by a sharp obstacle such as a mountain ridge. { 'nīf ˌej ri'frak·shən }

Knill-Laflamme bound [COMPUT SCI] In the theory of quantum computation, a necessary condition for an algorithm that encodes N logical qubits into N' carrier qubits (with N' larger than N) to correct any error on any M carrier qubits; namely, that N' be equal to or larger than 4M + N. { kə¦nil lə¦fläm ¦bau̇nd }

knot *See* deadlock. { nät }

knowbot [COMPUT SCI] A program which, when given a request, searches and retrieves information on the Internet. Also known as intelligent agent; knowledge robot. { 'nōˌbät }

knowledge base [COMPUT SCI] A collection of facts, assumptions, beliefs, and heuristics that are used in combination with a database to achieve desired results, such as a diagnosis, an interpretation, or a solution to a problem. { 'näl·ij ˌbās }

knowledge-based system [COMPUT SCI] A computer system whose usefulness derives primarily from a data base containing human knowledge in a computerized format. { 'näl·ij ˌbāst ˌsis·təm }

knowledge discovery in databases [COMPUT SCI] The process of identifying valid, novel, potentially useful, and ultimately understandable structure in data. Abbreviated KDD. { ¦näl·ij di¦skəv·ə·rē in 'dad·əˌbās·əs }

knowledge engineer [COMPUT SCI] An individual who constructs the knowledge base of an expert system. { 'näl·ij ˌen·jəˌnir }

knowledge robot *See* knowbot. { 'näl·ij ¦rō ˌbät }

Korn shell [COMPUT SCI] An enhanced version of the Bourne shell used in Unix systems. { 'kȯrn ˌshel }

KSAM *See* keyed sequential access method. { 'kāˌsam }

KSA service *See* K-band single-access service. { ¦kā¦es'ā ˌsər·vəs }

K-scan *See* K-display. { 'kā ˌskan }

K-scope *See* K-display. { 'kā ˌskōp }

Ku band [COMMUN] A band of frequencies extending from 15.35 to 17.25 gigahertz, corresponding to wavelengths of 1.95 to 1.74 centimeters. { 'kyü ˌband *or* ¦kā¦yü ˌband }

Ku-band fixed satellite service [COMMUN] Satellite communication at and near the Ku band, with the uplink frequency in bands from 12.75 to 13.25 gigahertz and 14.0 to 14.5 gigahertz and the downlink frequency in a band from 10.7 to 11.7 gigahertz. { 'kyü band ˌfikst 'sad·əlˌīt sər·vis }

Kundt effect [OPTICS] **1.** The occurrence of a very large magnetic rotation when polarized light passes through very thin films of pure ferromagnetic materials. **2.** *See* Faraday effect. { 'kůnt iˌfekt }

KWIC index *See* keyword-in-context index. { 'kwik ˌinˌdeks }

KWOC index *See* keyword-out-of-context index. { 'kwäk ˌinˌdeks }

L

LAAS *See* Local-Area Augmentation System.

label [COMPUT SCI] A data item that serves to identify a data record (much in the same way as a key is used), or a symbolic name used in a program to mark the location of a particular instruction or routine. { ˈlā·bəl }

label alignment [COMPUT SCI] The manner in which text is aligned in the cells of a particular spreadsheet. { ˈlā·bəl əˌlīn·mənt }

label constant *See* location constant. { ˈlā·bəl ˌkän·stənt }

label data type [COMPUT SCI] A scalar data type that refers to locations in the computer program. { ˈlā·bəl ˈdad·ə ˌtīp }

label record [COMPUT SCI] A tape record containing information concerning the file on that tape, such as format, record length, and block size. { ˈlā·bəl ˌre·kərd }

labile oscillator [ELECTR] An oscillator whose frequency is controlled from a remote location by wire or radio. { ˈlāˌbīl ˈäs·əˌlād·ər }

labor grade *See* job class. { ˈlā·bər ˌgrād }

ladder attenuator [ELECTR] A type of ladder network designed to introduce a desired, adjustable loss when working between two resistive impedances, one of which has a shunt arm that may be connected to any of various switch points along the ladder. { ˈlad·ər əˈten·yəˌwād·ər }

ladder network [ELECTR] A network composed of a sequence of H, L, T, or pi networks connected in tandem; chiefly used as an electric filter. Also known as series-shunt network. { ˈlad·ər ˈnet ˌwərk }

lag [ELECTR] A persistence of the electric charge image in a camera tube for a small number of frames. { lag }

lagging network *See* integral network. { ˈlag·iŋ ˌnet,wərk }

lag-lead network *See* lead-lag network. { ˈlag ˈlēd ˌnet,wərk }

lag network *See* integral network. { ˈlag ˌnet ˌwərk }

LAN *See* local-area network. { lan }

land [ELECTR] **1.** One of the regions between pits on a track on an optical disk. **2.** *See* terminal area. { land }

land-earth station [COMMUN] A facility that routes calls from mobile stations via satellite to and from terrestrial telephone networks. Abbreviated LES. { ¦land ¦ərth ˌstā·shən }

land effect *See* coastal refraction. { ˈland iˌfekt }

landing aid [NAV] A lamp, searchlight, radio beacon, radar device, communicating device, or any system of such devices for aiding aircraft in an approach and landing. Also known as landing system. { ˈland·iŋ ˌād }

landing system *See* landing aid. { ˈland·iŋ ˌsis·təm }

landing zone [COMPUT SCI] The data-free area on the surface of a hard disk over which the read-write head comes to rest when the computer is shut off and the disk stops rotating. { ˈland·iŋ ˌzōn }

land mobile-satellite service [COMMUN] A mobile-satellite service in which the mobile earth stations are located on land. Abbreviated LMSS. { ¦land ˌmō·bəl ˈsad·əlˌīt ˌsər·vəs }

land mobile service [COMMUN] Mobile service between base stations and mobile stations, or between land mobile stations. { ˈland ¦mō·bəl ¦sər·vəs }

land mobile station [COMMUN] Mobile station in the land mobile service, capable of surface movement within the geographical limits of a country or continent. { ˈland ¦mō·bəl ¦stā·shən }

land return *See* ground clutter. { ˈland riˌtərn }

land station [COMMUN] Station in the mobile service not intended for operation while in motion. { ˈland ˌstā·shən }

land transportation frequency bands [COMMUN] A group of radio-frequency bands between 25 megahertz and 30,000 megahertz allocated for use by taxicabs, railroads, buses, and trucks. { ˈland ˌtranz·pər¦tā·shən ˈfrē·kwən·sē ˌbanz }

land transportation radio services [COMMUN] Any service of radio communications operated by and for the sole use of certain land transportation carriers, the radio transmitting facilities of which are defined as fixed, land, or mobile stations. { ˈland ˌtranz·pər¦tā·shən ˈrād·ē·ōˌsər·vəs·əz }

language [COMPUT SCI] The set of words and rules used to construct sentences with which to express and process information for handling by computers and associated equipment. { ˈlaŋ·gwij }

language converter [COMPUT SCI] A device which translates a form of data (such as that on microfilm) into another form of data (such as that on magnetic tape). { ˈlaŋ·gwij kənˌvərd·ər }

language subset [COMPUT SCI] A portion of a programming language that can be used alone; usually applied to small computers that do not have the capability of handling the complete language. { ˈlaŋ·gwij ˈsəb,set }

language translator [COMPUT SCI] **1.** Any assembler or compiler that accepts human-readable statements and produces equivalent outputs in a form closer to machine language. **2.** A program designed to convert one computer language to equivalent statements in another computer language, perhaps to be executed on a different computer. **3.** A routine that performs or assists in the performance of natural language translations, such as Russian to English, or Chinese to Russian. { ˈlaŋ·gwij ˌtranz,lād·ər }

L antenna [ELECTROMAG] An antenna that consists of an elevated horizontal wire having a vertical down-lead connected at one end. { ˈel an,ten·ə }

lap dissolve [ELECTR] Changeover from one video scene to another so that the new picture appears gradually as the previous picture simultaneously disappears. { ˈlap di,zälv }

laptop computer *See* notebook computer. { ˈlap,täp kəm,pyüd·ər }

large-scale integrated circuit [ELECTR] A very complex integrated circuit, which contains well over 100 interconnected individual devices, such as basic logic gates and transistors, placed on a single semiconductor chip. Abbreviated LSI circuit. Also known as chip circuit; multiple-function chip. { ˈlärj ¦skāl ˌint·ə,grād·əd ˈsər·kət }

large-scale integrated memory *See* semiconductor memory. { ˈlärj ¦skāl ˌint·ə,grād·əd ˈmem·rē }

large-systems control theory [CONT SYS] A branch of the theory of control systems concerned with the special problems that arise in the design of control algorithms (that is, control policies and strategies) for complex systems. { ˈlärj ˌsis·təmz kənˈtrōl ˌthē·ə·rē }

laser [OPTICS] An active electron device that converts input power into a very narrow, intense beam of coherent visible or infrared light; the input power excites the atoms of an optical resonator to a higher energy level, and the resonator forces the excited atoms to radiate in phase. Derived from light amplification by stimulated emission of radiation. { ˈlā·zər }

laser altimeter [NAV] An altimeter in which a laser beam, modulated by radio frequencies, is directed downward from an aircraft, and the laser light reflected from the terrain is picked up by a telescope system, sensed by a photomultiplier, and phase-compared with the transmitted signal to obtain the round-trip propagation time; the height above ground equals one-half the product of the propagation time and the speed of light. { ˈlā·zər al¦tim·əd·ər }

laser communication [COMMUN] Optical communication in which the light source is a laser whose beam is modulated for voice, video, or data communication over wide information bandwidths, typically 1 gigahertz or more. { ˈlā·zər kə,myü·nəˈkā·shən }

laser diode *See* semiconductor laser. { ˈlā·zər ¦dī,ōd }

laser disk storage *See* optical disk storage. { ˈlā·zər ¦disk ˌstȯr·ij }

laser-holography storage [COMPUT SCI] A computer storage technology in which information is stored in microscopic spots burned in a holographic substrate by a laser beam, and is read by sensing a lower-energy laser beam that is transmitted through these spots. { ˈlā·zər hō ¦läg·rə·fē ˌstȯr·ij }

laser memory [COMPUT SCI] A computer memory in which a controlled laser beam acts on individual and extremely small areas of a photosensitive or other type of surface, for storage and subsequent readout of digital data or other types of information. { ˈlā·zər ˌmem·rē }

laser recorder [COMMUN] An image reproducer that resembles a facsimile system, in which a laser beam is initially modulated by the video signal and swept over photographic film or paper to reproduce an image received over wire or radio communication systems. { ˈlā·zər riˈkȯrd·ər }

last-mask read-only memory [COMPUT SCI] A read-only memory in which the final mask used in the fabrication process determines the connections to the internal transistors, and these connections in turn determine the data pattern that will be read out when the cell is accessed. Also known as contact-mask read-only memory. { ˈlast ¦mask ¦rēd ¦ōn·lē ˈmem·rē }

latch [ELECTR] An electronic circuit that reverses and maintains its state each time that power is applied. { lach }

late binding [COMPUT SCI] The assignment of data types (such as integer or string) to variables at the time of execution of a computer program, rather than during the compilation phase. { ¦lāt ˈbīnd·iŋ }

latency [COMPUT SCI] The waiting time between the order to read/write some information from/to a specified place and the beginning of the data-read/write operation. { ˈlat·ən·sē }

lateral parity check [COMPUT SCI] The number of one bits counted across the width of the magnetic tape; this number plus a one or a zero must always be odd (or even), depending upon the manufacturer. { ˈlad·ə·rəl ˈpar·əd·ē ˌchek }

launching [ELECTROMAG] The process of transferring energy from a coaxial cable or transmission line to a waveguide. { ˈlȯn·chiŋ }

lawnmower [ENG] A helix-type recorder mechanism. { ˈlȯn,mō·ər }

Lawrence tube *See* chromatron. { ˈlär·əns ˌtüb }

layer [COMPUT SCI] One of the divisions within which components or functions are isolated in a computer system with layered architecture or a communications system with layered protocols. { ˈlā·ər }

layered architecture [COMPUT SCI] A technique used in designing computer software, hardware, and communications in which system or network components are isolated in layers so that changes can be made in one layer without affecting the others. { 'lā·ərd 'är·kə,tek·chər }

layered-protocols technique [COMMUN] A technique for isolating the functions required in a data communications network so that these functions can be set up in a modular fashion and changes can be made in one area without affecting the others. { 'lā·ərd 'prōd·ə,kȯlz tek'nēk }

layout character [COMPUT SCI] A control character that determines the form in which the output data generated by a printer or display device are arranged. Also known as format effector. { 'lā,au̇t ,kar·ik·tər }

lazy evaluation *See* demand-driven execution. { 'lā·zē i,val·yə'wā·shən }

lazy H antenna [ELECTROMAG] An antenna array in which two or more dipoles are stacked one above the other to obtain greater directivity. { ¦lā·zē 'āch an,ten·ə }

L band [COMMUN] A band of radio frequencies between 1 and 2 gigahertz. { 'el ,band }

LCD *See* liquid crystal display.

L-display [ELECTR] A radar display format in which an echo signal appears as two horizontal deflections, left and right, from a vertical reference line, as associated with a two-beam tracking radar antenna, equal amplitudes indicating no pointing error and position on the vertical axis indicating range. Also known as L-indicator; L-scan; L-scope. { 'el di,splā }

LDM *See* limited-distance modem.

lead compensation [CONT SYS] A type of feedback compensation primarily employed for stabilization or for improving a system's transient response; it is generally characterized by a series compensation transfer function of the type

$$G_c(s) = K\frac{(s-z)}{(s-p)}$$

where $z < p$ and K is a constant. { 'lēd ,käm·pən'sā·shən }

leader [COMPUT SCI] A record which precedes a group of detail records, giving information about the group not present in the detail records; for example, "beginning of batch 17." [ENG] The unrecorded length of magnetic tape that enables the operator to thread the tape through the drive and onto the take-up reel without losing data or recorded music, speech, or such. { 'lēd·ər }

leader label [COMPUT SCI] A record appearing at the beginning of a magnetic tape to uniquely identify the tape as one required by the system. { 'lēd·ər ,lā·bəl }

leading character elimination [COMPUT SCI] A method of data compression used for dictionaries that are stored in alphabetical order, in which the coding for each word has two parts: the number of characters in common with the previous word, and the unique suffix. { ¦lēd·iŋ 'kar·ik·tər i,lim·ə'nā·shən }

leading pad [COMPUT SCI] Characters that fill unused space at the left end of a data field. { 'lēd·iŋ 'pad }

lead-lag network [CONT SYS] Compensating network which combines the characteristics of the lag and lead networks, and in which the phase of a sinusoidal response lags a sinusoidal input at low frequencies and leads it at high frequencies. Also known as lag-lead network. { 'lēd 'lag 'net,wərk }

lead network *See* derivative network. { 'lēd ,net ,wərk }

leaf *See* terminal vertex. { lēf }

leakage radiation [ELECTROMAG] In a radio transmitting system, radiation from anything other than the intended radiating system. { 'lēk·ij ,rād·ē'ā·shən }

leaky-wave antenna [ELECTROMAG] A wide-band microwave antenna that radiates a narrow beam whose direction varies with frequency; it is fundamentally a perforated waveguide, thin enough to permit flush mounting for aircraft and missile radar applications. { 'lēk·ē ¦wāv an'ten·ə }

leapfrog test [COMPUT SCI] A computer test using a special program that performs a series of arithmetical or logical operations on one group of storage locations, transfers itself to another group, checks the correctness of the transfer, then begins the series of operations again; eventually, all storage positions will have been tested. { 'lēp,fräg ,test }

learning control [CONT SYS] A type of automatic control in which the nature of control parameters and algorithms is modified by the actual experience of the system. { 'lərn·iŋ kən,trōl }

learning machine [COMPUT SCI] A machine that is capable of improving its future actions as a result of analysis and appraisal of past actions. { 'lər·niŋ mə,shēn }

leased facility [COMMUN] A collection of communication lines dedicated to a particular service; sometimes the lines have a predetermined path through system switching equipment. { 'lēst fə'sil·əd·ē }

least frequently used [COMPUT SCI] A technique for using main storage efficiently, in which new data replace data in storage locations that have been used least often, as determined by an algorithm. { 'lēst ¦frē·kwənt·lē 'yüzd }

least recently used [COMPUT SCI] A technique for using main storage efficiently, in which new data replace data in storage locations that have not been accessed for the longest period, as determined by an algorithm. { 'lēst ¦rē·sənt·lē ,yüzed }

least significant bit [COMPUT SCI] The bit that carries the lowest value or weight in binary notation for a numeral; for example, when 13 is represented by binary 1101, the 1 at the right is the least significant bit. Abbreviated LSB. { ¦lēst sig¦nif·i·kənt 'bit }

least significant character [COMPUT SCI] The character in the rightmost position in a number or word. { ¦lēst sig¦nif·i·kənt 'kar·ik·tər }

LED *See* light-emitting diode.

left-justify [COMPUT SCI] To shift the contents of a register so that the left, or most significant, digit is at some specified position. { 'left 'jəs·tə·fī }

left value [COMPUT SCI] The memory address of a symbolic variable in a computer program. Abbreviated lvalue. { 'left ˌval·yü }

leg [COMPUT SCI] The sequence of instructions that is followed in a computer routine from one branch point to the next. { leg }

legacy system [COMPUT SCI] A computer system that has been in operation for a long time, and whose functions are too essential to be disrupted by upgrading or integration with another system. { 'leg·ə·sē ˌsis·təm }

length block [COMPUT SCI] The total number of records, words, or characters contained in one block. { 'leŋkth ˌbläk }

lengthened dipole [ELECTROMAG] An antenna element with lumped inductance to compensate an end loss. { 'leŋk·thənd 'dīˌpōl }

lens [COMMUN] A dielectric or metallic structure that is highly transparent to radio waves and can bend them to produce a desired radiation pattern; used with antennas for radar and microwave relay systems. { lenz }

lens antenna [ELECTROMAG] A microwave antenna in which a dielectric lens is placed in front of the dipole or horn radiator to concentrate the radiated energy into a narrow beam or to focus received energy on the receiving dipole or horn. { 'lenz anˌten·ə }

LEO *See* low-altitude earth orbit. { 'lē·ō *or* ¦el ¦ē'ō }

LES *See* land-earth station.

letter [COMMUN] A character used in an alphabet generally representing one or more sounds of a spoken language. { 'led·ər }

letter code [COMPUT SCI] A Baudot code function which cancels errors by causing the receiving terminal to print nothing. { 'led·ər ˌkōd }

letter-perfect printer *See* letter-quality printer. { 'led·ər ¦pər·fekt 'print·ər }

letter-quality printer [COMPUT SCI] A printer that produces high-quality output. Also known as correspondence printer; letter-perfect printer. { 'led·ər ¦kwäl·əd·ē 'print·ər }

letters shift [COMMUN] **1.** A movement of a teletypewriter carriage which permits printing of alphabetic characters in an appropriate, generally linear sequence. **2.** The control which actuates this movement. Abbreviated LTRS. { 'led·ərz 'shift }

level [COMMUN] **1.** A specified position on an amplitude scale (for example, magnitude) applied to a signal waveform, such as reference white level and reference black level in a video signal. **2.** A range of allowed picture parameters and combinations of picture parameters in the digital television system. [COMPUT SCI] **1.** The status of a data item in COBOL language indicating whether this item includes additional items. **2.** *See* channel. [ELECTR] **1.** The difference between a quantity and an arbitrarily specified reference quantity, usually expressed as the logarithm of the ratio of the quantities. **2.** A charge value that can be stored in a given storage element of a charge storage tube and distinguished in the output from other charge values. { 'lev·əl }

level compensator [ELECTR] **1.** Automatic transmission-regulating feature or device used to minimize the effects of variations in amplitude of the received signal. **2.** Automatic gain control device used in the receiving equipment of a telegraph circuit. { 'lev·əl 'käm·pənˌsād·ər }

level converter [ELECTR] An amplifier that converts nonstandard positive or negative logic input voltages to standard DTL or other logic levels. { 'lev·əl kənˌvərd·ər }

level 1 cache *See* primary cache. { ¦lev·əl 'wən ˌkash }

level set [COMPUT SCI] A revision of a software package in which most or all of the executable programs are replaced with improved versions. { 'lev·əl ˌset }

level shifting [ELECTR] Changing the logic level at the interface between two different semiconductor logic systems. { 'lev·əl ˌshif·tiŋ }

level 2 cache *See* secondary cache. { ˌlev·əl ˌtü 'kash }

LF *See* low-frequency.

LF loran *See* low-frequency loran. { ¦el¦ef 'lȯr ˌan }

Liapunov function *See* Lyapunov function. { 'lyä·pu̇·nȯf ˌfəŋk·shən }

librarian [COMPUT SCI] The program which maintains and makes available all programs and routines composing the operating system. { lī'brer·ē·ən }

library [COMPUT SCI] **1.** A computerized facility containing a collection of organized information used for reference. **2.** An organized collection of computer programs together with the associated program listings, documentation, users' directions, decks, and tapes. { 'līˌbrer·ē }

library routine [COMPUT SCI] A computer program that is part of some program library. { 'lī ˌbrer·ē rüˌtēn }

library software [COMPUT SCI] The collection of programs and routines in the library of a computer system. { 'līˌbrer·ē 'sȯftˌwer }

library tape [COMPUT SCI] A magnetic tape that is kept in a stored, indexed collection for ready use and is made generally available. { 'līˌbrer·ē ˌtāp }

light [OPTICS] **1.** Electromagnetic radiation with wavelengths capable of causing the sensation of vision, ranging approximately from 400 (extreme violet) to 770 nanometers (extreme red). Also known as light radiation; visible radiation. **2.** More generally, electromagnetic radiation of any wavelength; thus, the term is sometimes applied to infrared and ultraviolet radiation. { līt }

light amplification by stimulated emission of radiation *See* laser. { 'līt ˌam·plə·fə'kā·shən bī ¦stim·yəˌlād·əd i¦mish·ən əv ˌrād·ē'ā·shən }

light carrier injection [ELECTR] A method of introducing the carrier in a facsimile system by periodic variation of the scanner light beam, the average amplitude of which is varied by the density changes of the subject copy. Also known as light modulation. { 'līt 'kar·ē·ər in,jek·shən }

light-emitting diode [ELECTR] A rectifying semiconductor device which converts electrical energy into electromagnetic radiation. The wavelength of the emitted radiation ranges from the near-ultraviolet to the near-infrared, that is, from about 400 to over 1500 nanometers. Abbreviated LED. { 'līt i,mid·iŋ 'dī,ōd }

light guide *See* optical fiber. { 'līt ,gīd }

light gun [ELECTR] A light pen mounted in a gun-type housing. { 'līt ,gən }

light modulation *See* light carrier injection. { 'līt ,mäj·ə,lā·shən }

light pen [ELECTR] A tiny photocell or photomultiplier, mounted with or without fiber or plastic light pipe in a pen-shaped housing; it is held against a cathode-ray screen to make measurements from the screen or to change the nature of the display. { 'līt ,pen }

light radiation *See* light. { 'līt ,rād·ē,ā·shən }

light-sensitive cell *See* photodetector. { 'līt ¦sen·səd·iv 'sel }

light-sensitive detector *See* photodetector. { 'līt ¦sen·səd·iv di'tek·tər }

light-sensitive tube *See* phototube. { 'līt ¦sen·səd·iv 'tüb }

light sensor photodevice *See* photodetector. { 'līt ,sen·sər 'fōd·ō·di,vīs }

light stability [COMPUT SCI] In optical character recognition, the ability of an image to retain its spectral appearance when exposed to radiant energy. { 'līt stə,bil·əd·ē }

limit check [COMPUT SCI] A check to determine if a value entered into a computer system is within acceptable minimum and maximum values. { 'lim·ət ,chek }

limited-access data [COMPUT SCI] Data to which only authorized users have access. { 'lim·əd·əd ¦ak·ses 'dad·ə }

limited-distance modem [COMMUN] A modem used only for communications within a building in order to improve the signal quality where a long distance exists between the terminal and the computer. [COMPUT SCI] A device designed to transmit and receive signals over relatively short distances, typically less than 5 miles (8 kilometers). Abbreviated LDM. Also known as line driver. { 'lim·əd·əd ¦dis·təns 'mō,dem }

limited-entry decision table [COMPUT SCI] A decision table in which the condition stub specifies exactly the condition or the value of the variable. { 'lim·əd·əd ¦en·trē di'sizh·ən ,tā·bəl }

limited integrator [ELECTR] A device used in analog computers that has two input signals and one output signal whose value is proportional to the integral of one of the input signals with respect to the other as long as this output signal does not exceed specified limits. { 'lim·əd·əd 'int·ə,grād·ər }

limited-sequence robot *See* fixed-stop robot. { 'lim·əd·əd ¦sē·kwəns 'rō,bät }

limited signal [ELECTR] Radar signal that is intentionally limited in amplitude by the dynamic range of the circuits involved; useful in some radio and radar processing. { 'lim·əd·əd 'sig·nəl }

limited space-charge accumulation mode [ELECTR] A mode of operation of a Gunn diode in which the frequency of operation is set by a resonant circuit to be much higher than the transit-time frequency so that domains have insufficient time to form while the field is above threshold and, as a result, the sample is maintained in the negative conductance state during a large fraction of the voltage cycle. Abbreviated LSA mode. { ¦lim·əd·əd ,spās ,chärj ə,kyü·myə'lā·shən ,mōd }

limiting [ELECTR] A desired or undesired amplitude-limiting action performed on a signal by a limiter. Also known as clipping; peak clipping. { 'lim·əd·iŋ }

limit priority [COMPUT SCI] An upper bound to the dispatching priority that a task can assign to itself or any of its subtasks. { 'lim·ət prī'är·əd·ē }

L-indicator *See* L-display. { 'el ,in·də,kād·ər }

line [ELECTR] **1.** The path covered by the electron beam of a picture tube in one sweep from left to right across the screen. **2.** One horizontal scanning element in a facsimile system. **3.** *See* trace. { līn }

line and trunk group [COMMUN] A group consisting of four-wire circuits, incoming private automatic branch exchange trunks, and intertoll trunk groups. { ¦līn ən ¦trəŋk ,grüp }

linear [CONT SYS] Having an output that varies in direct proportion to the input. { 'lin·ē·ər }

linear array [ELECTROMAG] An antenna array in which the dipole or other half-wave elements are arranged end to end on the same straight line. Also known as collinear array. { 'lin·ē·ər ə'rā }

linear bounded automaton [COMPUT SCI] A nondeterministic, one-tape Turing machine whose read/write head is confined to move only on a restricted section of tape initially containing the input. { 'lin·ē·ər ¦baúnd·əd ȯ'täm·ə,tän }

linear conductor antenna [ELECTROMAG] An antenna consisting of one or more wires which all lie along a straight line. {'lin·ē·ər kən'dək·tər an,ten·ə }

linearization [CONT SYS] **1.** The modification of a system so that its outputs are approximately linear functions of its inputs, in order to facilitate analysis of the system. **2.** The mathematical approximation of a nonlinear system, whose departures from linearity are small, by a linear system corresponding to small changes in the variables about their average values. { ,lin·ē·ər·ə'zā·shən }

linear-logarithmic intermediate-frequency amplifier [ELECTR] Amplifier used to avoid overload or saturation as a protection against jamming in a radar receiver. Also known as lin-log amplifier. { 'lin·ē·ər ,läg·ə'rith·mik ,in·tər¦mē·dē·ət ¦frē·kwən·sē 'am·plə,fī·ər }

linear modulation [COMMUN] Modulation in which the amplitude of the modulation envelope (or the deviation from the resting frequency) is directly proportional to the amplitude of the intelligence signal at all modulation frequencies. { 'lin·ē·ər ˌmäj·ə'lā·shən }

linear polarization [OPTICS] Polarization of an electromagnetic wave in which the electric vector at a fixed point in space remains pointing in a fixed direction, although varying in magnitude. Also known as plane polarization. { 'lin·ē·ər ˌpō·lə·rə'zā·shən }

linear programming [MATH] The study of maximizing or minimizing a linear function $f(x_1, \ldots, x_n)$ subject to given constraints which are linear inequalities involving the variables x_i. { 'lin·ē·ər 'prōˌgram·iŋ }

linear-quadratic-Gaussian problem [CONT SYS] An optimal-state regulator problem, containing Gaussian noise in both the state and measurement equations, in which the expected value of the quadratic performance index is to be minimized. Abbreviated LQG problem. { 'lin·ē·ər kwə'drad·ik 'gau̇s·ē·ən ˌpräb·ləm }

linear regulator problem [CONT SYS] A type of optimal control problem in which the system to be controlled is described by linear differential equations and the performance index to be minimized is the integral of a quadratic function of the system state and control functions. Also known as optimal regulator problem; regulator problem. { 'lin·ē·ər 'reg·yəˌlād·ər ˌpräb·ləm }

linear repeater [ELECTR] A repeater used in communication satellites to amplify input signals a fixed amount, generally with traveling-wave tubes or solid-state devices operating in their linear region. { 'lin·ē·ər ri'pēd·ər }

linear system [CONT SYS] A system in which the outputs are components of a vector which is equal to the value of a linear operator applied to a vector whose components are the inputs. { 'lin·ē·ər 'sis·təm }

linear system analysis [CONT SYS] The study of a system by means of a model consisting of a linear mapping between the system inputs (causes or excitations), applied at the input terminals, and the system outputs (effects or responses), measured or observed at the output terminals. { 'lin·ē·ər ¦sis·təm ə'nal·ə·səs }

linear unit [ELECTR] An electronic device used in analog computers in which the change in output, due to any change in one of two or more input signals, is proportional to the change in that input and does not depend upon the values of the other inputs. { 'lin·ē·ər ¦yü·nət }

line circuit [ELEC] **1.** Equipment associated with each station connected to a dial or manual switchboard. **2.** A circuit to interconnect an individual telephone and a channel terminal. { 'līn ˌsər·kət }

line code [COMPUT SCI] The single instruction required to solve a specific type of problem on a special-purpose computer. { 'līn ˌkōd }

line conditioning [COMMUN] The addition of compensating reactances to a data transmission line to reduce amplitude and phase delays over certain frequency bands. { 'līn kənˌdish·ə·niŋ }

line discipline [COMPUT SCI] The rules that govern exactly how data are transferred between locations in a communications network. { 'līn ˌdis·ə·plən }

line dot matrix [COMPUT SCI] A line printer that uses the dot matrix printing technique. Also known as parallel dot character printer. { 'līn ¦dät 'māˌtriks }

line driver [COMPUT SCI] *See* limited-distance modem. [ELECTR] An integrated circuit that acts as the interface between logic circuits and a two-wire transmission line. { 'līn ˌdrīv·ər }

line-drop signal [COMMUN] Signal associated with a subscriber line on a manual switchboard. { 'līn ¦dräp ˌsig·nəl }

line editor [COMPUT SCI] A text-editing system that stores a file of discrete lines of text to be printed out on the console (or displayed) and manipulated on a line-by-line basis, so that editing operations are limited and are specified for lines identified by a specific number. { 'līn ˌed·əd·ər }

line facility [COMMUN] A transmission line in a communication system, together with amplifiers spaced at regular intervals to offset attenuation in the line. { 'līn fəˌsil·əd·ē }

line feed [COMPUT SCI] **1.** Signal that causes a printer to feed the paper up a discrete number of lines. **2.** Rate at which paper is fed through a printer. { 'līn ˌfēd }

line fill [COMMUN] Ratio of the number of connected main telephone stations on a line to the nominal main station capacity of that line. { 'līn ˌfil }

line filter balance [COMMUN] Network designed to maintain phantom group balance when one side of the group is equipped with a carrier system. { 'līn ¦fil·tər ˌbal·əns }

line finder [COMMUN] A switching device that automatically locates an idle telephone or telegraph circuit going to the desired destination. [COMPUT SCI] A device that automatically advances the platen of a line printer or typewriter. { 'līn ˌfīn·dər }

line-finder switch [COMMUN] In telephony, an automatic switch for seizing selector apparatus which provides dial tone to the calling party. { 'līn ˌfīn·dər ˌswich }

line frequency [ELECTR] The number of times per second that the scanning spot sweeps across the screen in a horizontal direction in a video system. Also known as horizontal frequency; horizontal line frequency. { 'līn ˌfrē·kwən·sē }

line-frequency blanking pulse *See* horizontal blanking pulse. { 'līn ˌfrē·kwən·sē 'blaŋk·iŋ ˌpəls }

line interlace *See* interlaced scanning. { 'līn 'in·tərˌlās }

line item [COMPUT SCI] Any data that is considered to be of equal importance to other data in the same file. { 'līn ˌīd·əm }

line level [COMMUN] Signal level in decibels at a particular position on a transmission line. { 'līn ˌlev·əl }

line loop [COMMUN] Portion of a telephone circuit that includes a user's telephone set and the pair of wires that connect it with the distributing frame of a central office. { 'līn ,lüp }

line-loop resistance [ELEC] Metallic resistance of the line wires that extend from an individual telephone set to the dial central office; usually includes the resistance of the telephone set. { 'līn ,lüp ri,zis·təns }

line misregistration [COMPUT SCI] In character recognition, the improper appearance of a line of characters, on site in a character reader, with respect to a real or imaginary horizontal line. { 'līn ,mis,rej·ə'strā·shən }

line noise [COMMUN] Noise originating in a transmission line from such causes as poor joints and inductive interference from power lines. { 'līn ,nȯiz }

line number [COMPUT SCI] A number at the beginning or end of each line of a computer program that specifies its position in a sequence. { 'līn ,nəm·bər }

line of code [COMPUT SCI] A single statement in a programming language. { 'līn əv 'kōd }

line of sight [ELECTROMAG] The straight line for a transmitting radar antenna in the direction of the beam. { 'līn əv 'sīt }

line printer [COMPUT SCI] A device that prints an entire line in a single operation, without necessarily printing one character at a time. { 'līn ,print·ər }

line printing [COMPUT SCI] The printing of an entire line of characters as a unit. { 'līn ,print·iŋ }

line radiation [ELECTROMAG] Electromagnetic radiation from a power line caused mainly by corona pulses; gives rise to radio interference. { 'līn ,rād·ē,ā·shən }

line reflection [COMMUN] Reflection of a signal at the end of a transmission line, at the junction of two or more lines, or at a substation. { 'līn ri ,flek·shən }

line-sequential color television [COMMUN] An analog color television system in which an entire line is one color, with colors changing from line to line in a red, blue, and green sequence. { 'līn si¦kwən·chəl 'kəl·ər 'tel·ə,vizh·ən }

line skew [COMPUT SCI] In character recognition, a form of line misregistration, when the string of characters to be recognized appears in a uniformly slanted condition with respect to a real or imaginary baseline. { 'līn ,skyü }

line speed [COMMUN] Maximum rate at which signals may be transmitted over a given channel, usually in bauds or bits per second. { 'līn ,spēd }

lines per minute [COMPUT SCI] A measure of the speed of the printer. Abbreviated LPM. { 'līnz pər 'min·ət }

line stretcher [ELECTROMAG] Section of waveguide or rigid coaxial line whose physical length is variable to provide impedance matching. { 'līn ,strech·ər }

line switching [COMMUN] A telephone switching system in which a switch attached to a subscriber line connects an originating call to an idle part of the switching apparatus. [ELECTR] Connecting or disconnecting the line voltage from a piece of electronic equipment. { 'līn ,swich·iŋ }

line switching concentrator [COMMUN] Switching center used between a group of users and the switching center to reduce the number of trunks and increase efficiency of switching equipment usage (sometimes referred to as statistical multiplexing). { 'līn ¦swich·iŋ 'kän·sən,trād·ər }

line synchronizing pulse *See* horizontal synchronizing pulse. { 'līn ,siŋ·krə,nīz·iŋ ,pəls }

line turnaround [COMMUN] The time required for a half-duplex circuit to reverse the direction of transmission. { 'līn 'tərn·ə,raund }

line-use ratio [COMMUN] As applied to facsimile broadcasting, the ratio of the available line to the total length of scanning line. { 'līn ,yüs ,rā·shō }

linguistic model [COMPUT SCI] A method of automatic pattern recognition in which a class of patterns is defined as those patterns satisfying a certain set of relations among suitably defined primitive elements. Also known as syntactic model. { liŋ'gwis·tik 'mäd·əl }

link [COMMUN] General term used to indicate the existence of communications facilities between two points. [COMPUT SCI] *See* hyperlink. { liŋk }

linkage [COMPUT SCI] In programming, coding that connects two separately coded routines. { 'liŋ·kij }

linkage editor [COMPUT SCI] A service routine that converts the output of assemblers and compilers into a form that can be loaded and executed. { 'liŋ·kij ,ed·əd·ər }

link control message [COMMUN] **1.** Message sent over a link of a network to condition the link to handle transmissions in a prearranged manner. **2.** Message used only between a pair of terminals for the conditioning of the link for digital system control. { 'liŋk kən'trōl ,mes·ij }

linked list *See* chained list. { 'liŋkt 'list }

link encryption [COMMUN] The application of on-line crypto-operation to the individual links of relay systems so that all messages passing over the link are encrypted in their entirety. { 'liŋk en'krip·shən }

link field [COMPUT SCI] The first word of a message buffer, used to point to the next buffer on the message queue. { 'liŋk ,fēld }

link group [COMMUN] A collection of links that employ the same multiplex terminal equipment. { 'liŋk ,grüp }

linking loader [COMPUT SCI] A loader which combines the functions of a relocating loader with the ability to combine a number of program segments that have been independently compiled into an executable program. { 'liŋk·iŋ 'lō d·ər }

lin-log amplifier *See* linear-logarithmic intermediate-frequency amplifier. { ¦lin ¦läg 'am·plə ,fī·ər }

Linux [COMPUT SCI] A freely available, open-source operating system kernel capable of running on many different types of computer hardware; first released in 1991. { 'lin·əks }

LIOCS [COMPUT SCI] Set of routines handling buffering, blocking, label checking, and overlap of input/output with processing. Derived from logical input/output control system. { 'lī,äks }

lip-sync [COMMUN] Synchronization of sound and motion picture so that facial movements of speech coincide with the sounds. { 'lip ,siŋk }

liquid crystal display [ELECTR] A digital display that consists of two sheets of glass separated by a sealed-in, normally transparent, liquid crystal material; the outer surface of each glass sheet has a transparent conductive coating such as tin oxide or indium oxide, with the viewing-side coating etched into character-forming segments that have leads going to the edges of the display; a voltage applied between front and back electrode coatings disrupts the orderly arrangement of the molecules, darkening the liquid enough to form visible characters even though no light is generated. Abbreviated LCD. { 'lik·wəd 'krist·əl di'splā }

LISP [COMPUT SCI] An interpretive language developed for the manipulation of symbolic strings of recursive data; can also be used to manipulate mathematical and arithmetic logic. Derived from list processing language. { lisp }

list [COMPUT SCI] **1.** A last-in, first-out storage organization, usually implemented by software, but sometimes implemented by hardware. **2.** In FORTRAN, a set of data items to be read or written. { list }

list processing [COMPUT SCI] A programming technique in which list structures are used to organize memory. { 'list ¦prä,ses·iŋ }

list processing language *See* LISP. { 'list ¦prä ,ses·iŋ ,laŋ·gwij }

listserv [COMPUT SCI] The software (server) used to maintain an electronic mailing list. Also known as list server. { 'list,sərv }

list server *See* listserv. { 'list ,sər·vər }

list structure [COMPUT SCI] A set of data items, connected together because each element contains the address of a successor element (and sometimes of a predecessor element). { 'list ,strək·chər }

literal operand [COMPUT SCI] An operand, usually occurring in a source language instruction, whose value is specified by a constant which appears in the instruction rather than by an address where a constant is stored. { 'lid·ə·rəl ¦äp·ə¦rand }

lithography [ELECTR] A technique used for integrated circuit fabrication in which a silicon slice is coated uniformly with a radiation-sensitive film, the resist, and an exposing source (such as light, x-rays, or an electron beam) illuminates selected areas of the surface through an intervening master template for a particular pattern. { ¦ə'thäg·rə·fē }

little LEO system [COMMUN] A system of small satellites in low earth orbit (LEO) that provides messaging, data, and location services but does not have the capability of voice transmission. { ¦lit·əl ¦lē·o ,sis·təm }

live [COMMUN] Being broadcast directly at the time of production, instead of from recorded or filmed program material. { līv }

live data [COMPUT SCI] Actual data that are employed during the final testing of a computer system, as opposed to test data. { 'līv 'dad·ə }

live system [COMPUT SCI] A computer system on which all testing has been completed so that it is fully operational and ready for production work. { 'līv 'sis·təm }

liveware [COMPUT SCI] The people involved in the operation of a computer system, thought of as a component of the system along with hardware and software. { 'līv,wer }

LLL circuit *See* low-level logic circuit. { ¦el¦el'el ,sər·kət }

load [COMPUT SCI] **1.** To place data into an internal register under program control. **2.** To place a program from external storage into central memory under operator (or program) control, particularly when loading the first program into an otherwise empty computer. **3.** An instruction, or operator control button, which causes the computer to initiate the load action. **4.** The amount of work scheduled on a computer system, usually expressed in hours of work. [ELECTR] The device that receives the useful signal output of an amplifier, oscillator, or other signal source. { lōd }

load-and-go [COMPUT SCI] An operating technique with no stops between the loading and execution phases of a program; may include assembling or compiling. { ¦lōd ən 'gō }

load compensation [CONT SYS] Compensation in which the compensator acts on the output signal after it has generated feedback signals. Also known as load stabilization. { 'lōd käm·pən'sā·shən }

loaded Q [ELECTROMAG] The Q factor of a specific mode of resonance of a microwave tube or resonant cavity when there is external coupling to that mode. { 'lōd·əd kyü }

loader [COMPUT SCI] A computer program that takes some other program from an input or storage device and places it in memory at some predetermined address. { 'lōd·ər }

loading coil [ELECTROMAG] **1.** An iron-core coil connected into a telephone line or cable at regular intervals to lessen the effect of line capacitance and reduce distortion. Also known as Pupin coil; telephone loading coil. **2.** A coil inserted in series with a radio antenna to increase its electrical length and thereby lower the resonant frequency. { 'lōd·iŋ ,kȯil }

loading device [COMPUT SCI] Equipment from which programs or other data can be transferred or copied into a computer. { 'lōd·iŋ di,vīs }

loading disk [ELECTROMAG] Circular metal piece mounted at the top of a vertical antenna to increase its natural wavelength. { 'lōd·iŋ ,disk }

loading program [COMPUT SCI] Program used to load other programs into computer memory. Also known as bootstrap program. { 'lōd·iŋ ,prō·grəm }

loading routine *See* input routine. { 'lōd·iŋ rü ,tēn }

load isolator [ELECTROMAG] Waveguide or coaxial device that provides a good energy path from a signal source to a load, but provides a poor energy path for reflections from a mismatched load back to the signal source. { 'lōd ˌī·sə ˌlād·ər }

load module [COMPUT SCI] A program in a form suitable for loading into memory and executing. { 'lōd ˌmä·jül }

load point [COMPUT SCI] Preset point on a magnetic tape from which reading or writing will start. { 'lōd ˌpȯint }

load stabilization *See* load compensation. { 'lōd ˌstā·bə·lə,zā·shən }

lobe [ELECTROMAG] A part of the radiation pattern of a directional antenna representing an area of stronger radio-signal transmission. Also known as radiation lobe. { lōb }

lobe-half-power width [ELECTROMAG] In a plane containing the direction of the maximum energy of a lobe, the angle between the two directions in that plane about the maximum in which the radiation intensity is one-half the maximum value of the lobe. { ¦lōb ¦haf ¦pau̇·ər ˌwidth }

lobe switching *See* beam switching. { 'lōb ˌswich·iŋ }

lobing [ELECTROMAG] Formation of maxima and minima at various angles of the vertical plane antenna pattern by the reflection of energy from the surface surrounding the radar antenna; these reflections reinforce the main beam at some angles and detract from it at other angles, producing fingers of energy. { 'lōb·iŋ }

Local-Area Augmentation System [NAV] A type of local-area DGPS, developed by the Federal Aviation Administration in the United States, which is intended for aviation users in category II and III precision approaches and which broadcasts in the 108–117.95-megahertz band presently used for very high frequency omindirectional range (VOR) systems and instrument landing systems (ILS). Abbreviated LAAS. { ˌlōk·əl ¦er·ē·ə ˌȯg·mən'tā·shən ˌsis·təm }

local-area DGPS [NAV] A version of differential GPS which provides error corrections in the vicinity of a single reference receiver based on the measurements by that receiver. { ˌlōk·əl ¦er·ē·ə ¦dē¦jē¦pē'es }

local-area network [COMPUT SCI] A communications network connecting various hardware devices together within a building by means of a continuous cable, an in-house voice-data telephone system, or a radio -based system. Abbreviated LAN. { 'lō·kəl¦er·ē·ə 'net,wərk }

local-battery telephone set [ELECTR] Telephone set for which the transmitter current is supplied from a battery, or other current supply circuit, individual to the telephone set; the signaling current may be supplied from a local hand generator or from a centralized power source. { 'lō·kəl 'bad·ə·rē 'tel·ə,fōn ˌset }

local central office [COMMUN] A telephone central office, which terminates subscriber lines and makes connections with other central offices, usually equipped to serve 10,000 main telephones of its immediate community. { 'lō·kəl ¦sen·trəl 'ȯf·əs }

local circuit [COMMUN] Circuit to a main or auxiliary circuit which can be made available at any station or patched from point to point through one or more stations. { 'lō·kəl 'sər·kət }

local control [COMMUN] System or method of radio-transmitter control whereby the control functions are performed directly at the transmitter. { 'lō·kəl kən'trōl }

local controller *See* first-level controller. { 'lō·kəl kən'trōl·ər }

local device [COMPUT SCI] Peripheral equipment that is linked directly to a computer or other supporting equipment, without an intervening communications channel. { 'lō·kəl di'vīs }

local exchange *See* exchange. { 'lō·kəl iks 'chānj }

localization [COMPUT SCI] Imposing some physical order upon a set of objects, so that a given object has a greater probability of being in some particular regions of space than in others. { ˌlō·kə·lə'zā·shən }

local line *See* local loop. { 'lō·kəl 'līn }

local loop [COMMUN] A telephone line from the user's location that terminates at the local central office. Also known as local line. { 'lō·kəl 'lüp }

local oscillator [ELECTR] The oscillator in a superheterodyne receiver, whose output is mixed with the incoming modulated radio-frequency carrier signal in the mixer to give the frequency conversions needed to produce the intermediate-frequency signal. { 'lō·kəl 'äs·ə ˌlād·ər }

local-oscillator injection [ELECTR] Adjustment used to vary the magnitude of the local oscillator signal that is coupled into the mixer. { 'lō·kəl 'äs·ə,lād·ər in¦jek·shən }

local-oscillator radiation [ELECTR] Radiation of the fundamental or harmonics of the local oscillator of a superheterodyne receiver. { 'lō·kəl 'äs·ə,lād·ər ˌrād·ē'ā·shən }

local register [COMPUT SCI] One of a relatively small number (usually less than 32) of high-speed storage elements in a computer system which may be directly referred to by the instructions in the programs. Also known as general register. { 'lō·kəl 'rej·ə·stər }

local side [COMMUN] Terminal connections to an internal or in-station source such as data terminal connections to input or output devices. { 'lō·kəl ˌsīd }

local storage [COMPUT SCI] The collection of local registers in a computer system. { 'lō·kəl 'stȯr·ij }

local trunk [COMMUN] Trunk between local and long-distance switchboards, or between local and private branch exchange switchboards. { 'lō·kəl 'trənk }

local variable [COMPUT SCI] A variable which can be accessed (used or changed) only in one block of a computer program. { 'lō·kəl 'ver·ē·ə·bəl }

locate mode [COMPUT SCI] A method of communicating with an input/output control system (IOCS), in which the address of the data involved, but not the data themselves, is transferred between the IOCS routine and the program. { 'lō,kāt ,mōd }

location [COMPUT SCI] Any place in which data may be stored; usually expressed as a number. { lō'kā·shən }

location constant [COMPUT SCI] A number that identifies an instruction in a computer program, written in a higher-level programming language, and used to refer to this instruction at other points in the program. Also known as label constant. { lō'kā·shən ,kän·stənt }

location counter *See* instruction counter. { lō 'kā·shən ,kaůnt·ər }

lock [ELECTR] **1.** To fasten onto and automatically follow a target by means of a radar beam; similarly to acquire and maintain attention to the signal from a single source by anticipation of some feature of it. **2.** To control the frequency of an oscillator by means of an applied signal of constant frequency. { läk }

locked-in line [COMMUN] A telephone line that remains established after the caller has hung up. { 'läkt¦in ,līn }

locked oscillator [ELECTR] A sine-wave oscillator whose frequency can be locked by an external signal to the control frequency divided by an integer. { 'läkt 'äs·ə,lād·ər }

locked-oscillator detector [ELECTR] A frequency-modulation detector in which a local oscillator follows, or is locked to, the input frequency; the phase difference between local oscillator and input signal is proportional to the frequency deviation, and an output voltage is generated proportional to the phase difference. { 'läkt ¦äs·ə,lād·ər di,tek·tər }

lock-on [ELECTR] **1.** The procedure wherein a target-seeking system (such as some types of radars) is continuously and automatically following a target in one or more coordinates (for example, range, bearing, elevation), or wherein a signal intercept system isolates signals from a single source. **2.** The instant at which radar begins to track a target automatically. { 'läk ¦ȯn }

lockout [COMMUN] **1.** In a telephone circuit controlled by two voice-operated devices, the inability of one or both subscribers to get through, because of either excessive local circuit noise or continuous speech from one or both subscribers. Also known as receiver lockout system. **2.** In mobile communications, an arrangement of control circuits whereby only one receiver can feed the system at one time to avoid distortion. Also known as receiver lockout system. [COMPUT SCI] **1.** In computer communications, the inability of a remote terminal to achieve entry to a computer system until project programmer number, processing authority code, and password have been validated against computer-stored lists. **2.** The precautions taken to ensure that two or more programs executing simultaneously in a computer system do not access the same data at the same time, make unauthorized changes in shared data, or otherwise interfere with each other. **3.** Preventing the central processing unit of a computer from accessing storage because input/output operations are taking place. **4.** Preventing input and output operations from taking place simultaneously. { 'läk,aůt }

lockout circuit [ELECTR] A switching circuit which responds to concurrent inputs from a number of external circuits by responding to one, and only one, of these circuits at any time. Also known as finding circuit; hunting circuit. { 'läk ,aůt ,sər·kət }

lodar [NAV] A direction finder used to determine the direction of arrival of loran signals, free of night effect, by observing the separately distinguishable ground and sky-wave loran signals on a cathode-ray oscilloscope and positioning a loop antenna to obtain a null indication of the component selected to be most suitable. Also known as lorad. { 'lō,där }

log [COMMUN] A record of radio and television station operating data. [COMPUT SCI] A record of computer operating runs, including tapes used, control settings, halts, and other pertinent data. { läg }

logarithmic fast time constant [ELECTR] Constant false alarm rate scheme which has a logarithmic intermediate-frequency amplifier followed by a fast time constant circuit. { 'läg·ə ,rith·mik 'fast ¦tīm ,kän·stənt }

logbook [COMPUT SCI] A bound volume in which operating data of a computer is noted. { 'läg ,bůk }

logic [ELECTR] **1.** The basic principles and applications of truth tables, interconnections of on/off circuit elements, and other factors involved in mathematical computation in a computer. **2.** General term for the various types of gates, flip-flops, and other on/off circuits used to perform problem-solving functions in a digital computer. { 'läj·ik }

logical comparison [COMPUT SCI] The operation of comparing two items in a computer and producing a one output if they are equal or alike, and a zero output if not alike. { 'läj·ə·kəl kəm'par·ə·sən }

logical construction [COMPUT SCI] A simple logical property that determines the type of characters which a particular code represents; for example, the first two bits can tell whether a character is numeric or alphabetic. { 'läj·ə·kəl kən'strək·shən }

logical data independence [COMPUT SCI] A data base structured so that changing the logical structure will not affect its accessibility by the program reading it. { 'läj·ə·kəl ¦dad·ə ,in·də'pen·dəns }

logical data type [COMPUT SCI] A scalar data type in which a data item can have only one of two values: true or false. Also known as Boolean data type. { 'läj·ə·kəl 'dad·ə ,tīp }

logical decision [COMPUT SCI] The ability to select one of many paths, depending upon intermediate programming data. { 'läj·ə·kəl di'sizh·ən }

logical device table [COMPUT SCI] A table that is used to keep track of information pertaining to an input/output operation on a logical unit, and that contains such information as the symbolic name of the logical unit, the logical device type and the name of the file currently attached to it, the logical input/output request currently pending on the device, and a pointer to the buffers currently associated with the device. { 'läj·ə·kəl di¦vīs ,tā·bəl }

logical drive [COMPUT SCI] A data storage unit, such as a subpartition of a hard drive or an array of storage units, recognized and handled according to the logic of the operating system like a single physical drive. { 'läj·i·kəl ¦drīv }

logical expression [COMPUT SCI] Two arithmetic expressions connected by a relational operator indicating whether an expression is greater than, equal to, or less than the other, or connected by a logical variable, logical constant (true or false), or logical operator. { 'läj·ə·kəl ik'spresh·ən }

logical field [COMPUT SCI] A data field whose variables can take on only two values, which are designated yes and no, true and false, or 0 and 1. { 'läj·ə·kəl 'fēld }

logical file [COMPUT SCI] A file as seen by the program accessing it. { 'läj·ə·kəl 'fīl }

logical flow chart [COMPUT SCI] A detailed graphic solution in terms of the logical operations required to solve a problem. { 'läj·ə·kəl 'flō,chärt }

logical gate *See* switching gate. { 'läj·ə·kəl 'gāt }

logical instruction [COMPUT SCI] A digital computer instruction which forms a logical combination (on a bit-by-bit basis) of its operands and leaves the result in a known location. { 'läj·ə·kəl in'strək·shən }

logical network [COMPUT SCI] **1.** A collection of computers that is presented as a single network to the user, although it may encompass more than one physical network. **2.** A part of a network of computers that is set up to function as a separate network. { 'läj·i·kəl ¦net,wərk }

logical page [COMPUT SCI] A unit of computer storage consisting of a specified number of bytes. { 'läj·ə·kəl 'pāj }

logical record [COMPUT SCI] A group of adjacent, logically related data items. { 'läj·ə·kəl 'rek·ərd }

logical security [COMPUT SCI] Mechanisms internal to a computing system that are used to protect against internal misuse of computing time and unauthorized access to data. { 'läj·ə·kəl sə'kyůr·əd·ē }

logical shift [COMPUT SCI] A shift operation that treats the operand as a set of bits, not as a signed numeric value or character representation. { 'läj·ə·kəl 'shift }

logical sum [COMPUT SCI] A computer addition in which the result is 1 when either one or both input variables is 1, and the result is 0 when the input variables are both 0. { 'läj·ə·kəl 'səm }

logical symbol [COMPUT SCI] A graphical symbol used to represent a logic element. { 'läj·ə·kəl 'sim·bəl }

logical unit [COMPUT SCI] An abstraction of an input/output device in the form of an additional name given to the device in a computer program. { 'läj·ə·kəl 'yü·nət }

logic-arithmetic unit *See* arithmetical unit. { 'läj·ik·ə'rith·mə·tik ,yü·nət }

logic bomb [COMPUT SCI] A computer program that destroys data, generally immediately after it has been loaded. { 'läj·ik ,bäm }

logic chip [COMPUT SCI] An integrated circuit that performs logic functions. { 'läj·ik ,chip }

logic circuit [COMPUT SCI] A computer circuit that provides the action of a logic function or logic operation. Also known as logic gate. { 'läj·ik ,sər·kət }

logic design [COMPUT SCI] The design of a computer at the level which considers the operation of each functional block and the relationships between the functional blocks. { 'läj·ik di,zīn }

logic diagram [COMPUT SCI] A graphical representation of the logic design or a portion thereof; displays the existence of functional elements and the paths by which they interact with one another. { 'läj·ik ,dī·ə,gram }

logic element [COMPUT SCI] A hardware circuit that performs a simple, predefined transformation on its input and presents the resulting signal as its output. Occasionally known as functor. { 'läj·ik ,el·ə·mənt }

logic error [COMPUT SCI] An error in programming that is caused by faulty reasoning, resulting in the program's functioning incorrectly if the instructions containing the error are encountered. { 'läj·ik ,er·ər }

logic gate *See* logic circuit. { 'läj·ik ,gāt }

logic high [ELECTR] The electronic representation of the binary digit 1 in a digital circuit or device. { 'läj·ik 'hī }

logic level [ELECTR] One of the two voltages whose values have been arbitrarily chosen to represent the binary numbers 1 and 0 in a particular data-processing system. { 'läj·ik ,lev·əl }

logic low [ELECTR] The electronic representation of the binary digit 0 in a digital circuit or device. { 'läj·ik ,lō }

logic operation [COMPUT SCI] A nonarithmetical operation in a computer, such as comparing, selecting, making references, matching, sorting, and merging, where logical yes-or-no quantities are involved. { 'läj·ik ,äp·ə'rā·shən }

logic operator [COMPUT SCI] A rule which assigns, to every combination of the values "true" and "false" among one or more independent variables, the value "true" or "false" to a dependent variable. { 'läj·ik 'äp·ə,rād·ər }

logic section *See* arithmetical unit. { 'läj·ik ,sek·shən }

logic-seeking printer [COMPUT SCI] A line printer that examines each line to be printed so that it can save time by skipping over blank spaces. { 'läj·ik ¦sēk·iŋ 'print·ər }

logic swing [ELECTR] The voltage difference between the logic levels used for 1 and 0; magnitude is chosen arbitrarily for a particular system and is usually well under 10 volts. { 'läj·ik ˌswiŋ }

logic switch [ELECTR] A diode matrix or other switching arrangement that is capable of directing an input signal to one of several outputs. { 'läj·ik ˌswich }

logic unit [COMPUT SCI] A separate unit which exists in some computer systems to carry out logic (as opposed to arithmetic) operations. { 'läj·ik ˌyü·nət }

logic word [COMPUT SCI] A machine word which represents an arbitrary set of digitally encoded symbols. { 'läj·ik ˌwərd }

log-in *See* log-on. { 'lägˌin }

logo [COMPUT SCI] A high-level, interactive programming language that features a triangular shape called a turtle which can be moved about an electronic display through the use of familiar English-word commands. { 'lō·gō }

log-off [COMPUT SCI] The procedure for a user to disconnect from a computer system, including the release of resources that were assigned to the user. { 'läg ˌȯf }

log-on [COMPUT SCI] The procedure for users to identify themselves to a computer system for authorized access to their programs and information. { 'läg ˌȯn }

log-out *See* log-off. { 'läg ˌau̇t }

log-periodic antenna [ELECTROMAG] A broadband antenna which consists of a sheet of metal with two wedge-shaped cutouts, each with teeth cut into its radii along circular arcs; characteristics are repeated at a number of frequencies that are equally spaced on a logarithmic scale. { 'läg ˌpir·ē¦ad·ik an'ten·ə }

long-base-line system [COMMUN] System in which the distance separating ground stations approximates the distance to the target being tracked. { 'lȯŋ 'bās ˌlīn ˌsis·təm }

long card [COMPUT SCI] A full-size printed circuit board that is plugged into an expansion slot in a microcomputer. { 'loŋ ¦kärd }

long-conductor antenna *See* long-wire antenna. { 'lȯŋ kən¦dək·tər anˌten·ə }

long-distance loop [COMMUN] Line from a subscriber's station directly to a long-distance switchboard. { 'lȯŋ ˌdis·təns 'lüp }

long-distance xerography [COMMUN] A facsimile system that uses a cathode-ray scanner at the microwave transmitting terminal; at the receiving terminal, a lens projects the received cathode-ray image onto the selenium-coated drum of a xerographic copying machine. { 'lȯŋ ˌdis·təns zi'räg·rə·fē }

long-haul carrier system [COMMUN] An intercity telephone communication system; it may use a frequency-division multiplexed signal modulating subcarrier or it may use digital technology. { 'lȯŋ ˌhȯl 'kar·ē·ər ˌsis·təm }

long-haul radio [COMMUN] A microwave radio system capable of transmitting telephone, video, data, and telegraph signals over distances on the order of 4000 miles (6500 kilometers) or more on line-of-sight paths between a series of repeaters that demodulate the signal to an intermediate frequency and then remodulate it. { 'lȯŋ ˌhȯl 'rād·ē·ō }

longitudinal parity [COMMUN] Parity associated with bits recorded on one track in a data block, to indicate whether the number of recorded bits in the block is even or odd. { ˌlän·jə'tüd·ən·əl 'par·əd·ē }

longitudinal parity check [COMMUN] The count for even or odd parity of all the bits in a message as a precaution against transmission error. Also known as horizontal parity check. { ˌlän·jə'tüd·ən·əl 'par·əd·ē ˌchek }

longitudinal redundancy check [COMMUN] A method of checking for errors, in which data are arranged in blocks according to some rule, and the correctness of each character in the block is determined according to the rule. Abbreviated LRC. { ˌlän·jə'tüd·ən·əl ri'dən·dən·sē ˌchek }

Longley-Rice [COMMUN] A model used to predict the long-term median transmission loss over irregular terrain that is applied to predicting signal strength at one or more locations. Longley-Rice computations are employed both by the FCC allocations rules for FM stations to predict signal strength contours and by propagation modeling software to predict signal strengths in a two-dimensional grid on a map. The FCC implementation of Longley-Rice computations employs average terrain computations and an assumed 30-ft receive antenna height. { 'lȯŋ·lē 'rīs }

long-lines engineering [COMMUN] Engineering performed to develop, modernize, or expand long-haul, point-to-point communications facilities using radio, microwave, or wire circuits. { 'lȯŋ ˌlīnz 'en·jə'nir·iŋ }

long-term predictor [COMMUN] An electric filter that removes redundancies in a signal associated with long-term correlations so that information can be transmitted more efficiently. { ¦lȯŋ ˌtərm prə'dik·tər }

long wave [COMMUN] An electromagnetic wave having a wavelength longer than the longest broadcast-band wavelength of about 545 meters, corresponding to frequencies below about 550 kilohertz. { 'lȯŋ ¦wāv }

long-wave radio [COMMUN] A radio which can receive frequencies below the lowest broadcast frequency of 550 kilohertz. { 'lȯŋ ˌwāv 'rād·ē·ō }

long-wire antenna [ELECTROMAG] An antenna whose length is a number of times greater than its operating wavelength, so as to give a directional radiation pattern. Also known as long-conductor antenna. { 'lȯŋ ˌwīr an'ten·ə }

lookahead [COMPUT SCI] A procedure in which a processor is preparing one instruction in a computer program while executing its predecessor. { 'lu̇k·əˌhed }

lookahead tree *See* game tree. { 'lu̇k·əˌhed ˌtrē }

look-through [ELECTR] **1.** When jamming, a technique whereby the jamming emission is interrupted irregularly for extremely short periods to allow monitoring of the victim signal during jamming operations. **2.** When being jammed, the technique of observing or monitoring a desired signal during interruptions in the jamming signal. { 'lu̇k ˌthrü }

look time *See* dwell time. { 'lu̇k ˌtīm }

look-up [COMPUT SCI] An operation or process in which a table of stored values is scanned (or searched) until a value equal to (or sometimes, greater than) a specified value is found. { 'lu̇k ˌəp }

look-up table [COMPUT SCI] A stored matrix of data for reference purpose. { 'lu̇k ˌəp ˌtā·bəl }

loop [COMPUT SCI] A sequence of computer instructions which are executed repeatedly, but usually with address modifications changing the operands of each iteration, until a terminating condition is satisfied. [ELECTROMAG] *See* coupling loop; loop antenna. [ENG] A reel of motion picture film or magnetic tape whose ends are spliced together, so that it may be played repeatedly without interruption. { lüp }

loop antenna [ELECTROMAG] A directional-type antenna consisting of one or more complete turns of a conductor, usually tuned to resonance by a variable capacitor connected to the terminals of the loop. Also known as loop. { 'lüp an ˌten·ə }

loopback check *See* echo check. { 'lüpˌbak ˌchek }

loopback switch [ELECTR] A switch at the end of a telephone line that is used to test the line and, when closed, reflects received signals to the sender. { 'lüpˌbak ˌswich }

loop body [COMPUT SCI] The set of statements to be performed iteratively with the range of a loop. { 'lüp ˌbäd·ē }

loop check *See* echo check. { 'lüp ˌchek }

loop checking [COMMUN] Sending signals from the central office to test the integrity of local loops. { 'lüp ˌchek·iŋ }

loop circuit [COMMUN] Common communications circuit shared by more than two parties; when applied to a teletypewriter operation, all machines print all data entered on the loop. { 'lüp ˌsər·kət }

loop coupling [ELECTROMAG] A method of transferring energy between a waveguide and an external circuit, by inserting a conducting loop into the waveguide, oriented so that electric lines of flux pass through it. { 'lüp ˌkəp·liŋ }

loop dialing [COMMUN] Return-path method of dialing in which the dial pulses are sent out over one side of the interconnecting line or trunk and are returned over the other side; limited to short-haul traffic. { 'lüp ˌdī·liŋ }

loop filter [ELECTR] A low-pass filter, which may be a simple RC filter or may include an amplifier, and which passes the original modulating frequencies but removes the carrier-frequency components and harmonics from a frequency-modulated signal in a locked-oscillator detector. { 'lüp ˌfil·tər }

loop gain [CONT SYS] The ratio of the magnitude of the primary feedback signal in a feedback control system to the magnitude of the actuating signal. [ELECTR] Total usable power gain of a carrier terminal or two-wire repeater; maximum usable gain is determined by, and may not exceed, the losses in the closed path. { 'lüp ˌgān }

loop head [COMPUT SCI] The first instruction of a loop, which contains the mode of execution, induction variable, and indexing parameters. { 'lüp ˌhed }

loop network *See* ring network. { 'lüp 'net ˌwərk }

loop pulsing [COMMUN] Regular, momentary interruptions of the direct-current path at the sending end of a transmission line. Also known as dial pulsing. { 'lüp ˌpəls·iŋ }

loop ratio *See* loop transfer function. { 'lüp ˌrā·shō }

loopstick antenna *See* ferrite-rod antenna. { 'lüpˌstik anˌten·ə }

loop stop [COMPUT SCI] A small closed loop that is entered to stop the progress of a computer program, usually when some condition occurs that requires intervention by the operator or that should be brought to the operator's attention. Also known as stop loop. { 'lüp ˌstäp }

loop transfer function [CONT SYS] For a feedback control system, the ratio of the Laplace transform of the primary feedback signal to the Laplace transform of the actuating signal. Also known as loop ratio. { 'lüp 'tranz·fər ˌfəŋk·shən }

loop transmittance [CONT SYS] **1.** The transmittance between the source and sink created by the splitting of a specified node in a signal flow graph. **2.** The transmittance between the source and sink created by the splitting of a node which has been inserted in a specified branch of a signal flow graph in such a way that the transmittance of the branch is unchanged. { 'lüp tranzˌmit·əns }

loop-type radio range [NAV] An A-N radio range employing two loop antennas placed at right angles to one another. { 'lüp ˌtīp 'rād·ē·ō ˌrānj }

loose list [COMPUT SCI] A list, some of whose cells are empty and thus do not contain records of the file. Also known as thin list. { 'lüs 'list }

loosely coupled computer [COMPUT SCI] A computer that can function by itself and can also be connected to other computers to exchange data when necessary. { 'lüs·lē ¦kəp·əld kəm'pyüd·ər }

lorad *See* lodar. { 'lȯrˌad }

loran [NAV] The designation of a family of radio navigation systems by which hyperbolic lines of position are determined by measuring the difference in the times of reception of synchronized pulse signals from two or more fixed transmitters. Derived from long-range navigation. { 'lȯrˌan }

loran A [NAV] A medium-frequency radio navigation system by which hyperbolic lines of position are determined by measuring the difference in the times of reception of synchronized pulse signals from two fixed transmitters. Also known as standard loran. { 'lȯrˌan 'ā }

loran C [NAV] A low-frequency radio navigation system by which hyperbolic lines of position are determined by measuring the difference in the times of reception of synchronized pulse signals from two fixed transmitters; as compared to loran A, time difference measurements are increased in accuracy through utilizing phase comparison techniques in addition to relatively coarse matches of pulse envelopes of received signals within the loran C receiver. { 'lȯr,an 'sē }

loran chain [NAV] **1.** A system or combination of three or more loran A stations, forming two or more pairs of stations for loran A navigation. **2.** A system or combination of a master loran C station and two or more loran C slave stations, forming two or more pairs of stations for loran C navigation. { 'lȯr,an ,chān }

loran chart [NAV] A chart showing loran lines of position along with a limited amount of topographic detail. { 'lȯr,an ,chärt }

loran D [NAV] Tactical loran system that uses the coordinate converter of low-frequency loran C and can operate in conjunction with inertial systems on aircraft, independently of ground facilities and without radiating radio-frequency energy that could reveal the aircraft's location. { 'lȯr,an 'dē }

loran line [NAV] A line of position on a loran chart; each line is the locus of points whose distances from two fixed stations differ by a constant amount. { 'lȯr,an ,līn }

loran rate [NAV] **1.** The frequency channel and pulse repetition rate by which a pair of loran stations is identified. **2.** By extension, the term refers to a pair of transmitting stations, their signals, and resulting lines of position. { 'lȯr ,an,rāt }

loran tables [NAV] A series of tables containing tabular data for constructing loran hyperbolic lines of position. { 'lȯr,an ,tā·bəlz }

loss [COMMUN] *See* transmission loss. [ENG] Power that is dissipated in a device or system without doing useful work. Also known as internal loss. { lȯs }

lossless data compression [COMMUN] Data compression in which the recovered data are assured to be identical to the source. { ¦lȯs,les 'dad·ə kəm,presh·ən }

lossless junction [ELECTROMAG] A waveguide junction in which all the power incident on the junction is reflected from it. { 'lȯs·ləs ,jəŋk·shən }

loss modulation *See* absorption modulation. { 'lȯs ,mäj·ə,lā·shən }

lossy attenuator [ELECTROMAG] In waveguide technique, a length of waveguide deliberately introducing a transmission loss by the use of some dissipative material. { 'lȯs·ē ə'ten·yə ,wād·ər }

lossy data compression [COMMUN] Data compression in which controlled degradation of the data is allowed. { ¦lȯs·ē 'dad·ə kəm,presh·ən }

lost cluster [COMPUT SCI] Disk records that are not associated with a file name in a disk directory. { 'lȯst 'kləs·tər }

low-altitude earth orbit [AERO ENG] An artificial satellite orbit whose altitude is less than about 1000 miles (1600 kilometers) above the earth's surface. Abbreviated LEO. { ,lō ¦al·tə,tüd 'ərth ,ȯr·bət }

low core [COMPUT SCI] The locations with the lower addresses in a computer's main storage, usually used to store control values needed to run the system and other critical information and instructions. { 'lō ,kȯr }

low-definition television [COMMUN] Television that involves less than about 200 scanning lines per complete image. { 'lō ,def·ə,nish·ən 'tel·ə ,vizh·ən }

lower half-power frequency [ELECTR] The frequency on an amplifier response curve which is smaller than the frequency for peak response and at which the output voltage is $1/\sqrt{2}$ of its midband or other reference value. { 'lō·ər 'haf ,pau̇·ər 'frē·kwən·sē }

lower sideband [COMMUN] The sideband containing all frequencies below the carrier-frequency value that are produced by an amplitude-modulation process. { 'lō·ər 'sīd ,band }

lower-sideband upconverter [ELECTR] Parametric amplifier in which the frequency, power, impedance, and gain considerations are the same as for the nondegenerate amplifier; here, however, the output is taken at the difference frequency, or the lower sideband, rather than the signal-input frequency. { 'lō·ər ¦sīd,band 'əp·kən,vərd·ər }

lowest required radiating power [COMMUN] The smallest power output of an antenna which will suffice to maintain a specified grade of broadcast service. Abbreviated LRRP. { 'lō·əst ri¦kwīrd 'rād·ē,ād·iŋ ,pau̇·ər }

lowest useful high frequency [COMMUN] The lowest high frequency that is effective at a specified time for ionospheric propagation of radio waves between two specified points. Abbreviated LUHF. { 'lō·əst ¦yüs·fəl 'hī ,frē·kwən·sē }

low-frequency [COMMUN] A Federal Communications Commission designation for the band from 30 to 300 kilohertz in the radio spectrum. Abbreviated LF. { 'lō ,frē·kwən·sē }

low-frequency antenna [ELECTROMAG] An antenna designed to transmit or receive radiation at frequencies of less than about 300 kilohertz. { 'lō ,frē·kwən·sē an'ten·ə }

low-frequency compensation [ELECTR] Compensation that serves to extend the frequency range of a broad-band amplifier to lower frequencies. { 'lō ,frē·kwən·sē ,käm·pə'sā·shən }

low-frequency loran [NAV] A modification of standard loran, which operates in the low-frequency range of approximately 100 to 200 kilohertz to increase range over land and during daytime, and which matches cycles rather than envelopes of pulses to obtain a more accurate fix. Abbreviated LF loran. Also known as cycle-matching loran. { 'lō ,frē·kwən·sē 'lȯr,an }

low-frequency padder [ELECTR] In a superheterodyne receiver, a small adjustable capacitor connected in series with the oscillator tuning coil and adjusted during alignment to obtain correct calibration of the circuit at the low-frequency end of the tuning range. { 'lō ˌfrē·kwən·sē 'pad·ər }

low-frequency propagation [ELECTROMAG] Propagation of radio waves at frequencies between 30 and 300 kilohertz. { 'lō ˌfrē·kwən·sē ˌpräp·ə'gā·shən }

low level [ELECTR] The less positive of the two logic levels or states in a digital logic system. { 'lō ˌlev·əl }

low-level language [COMPUT SCI] A computer language consisting of mnemonics that directly correspond to machine language instructions; for example, an assembler that converts the interpreted code of a higher-level language to machine language. { 'lō ˌlev·əl 'laŋ·gwij }

low-level logic circuit [ELECTR] A modification of a diode-transistor logic circuit in which a resistor and capacitor in parallel are replaced by a diode, with the result that a relatively small voltage swing is required at the base of the transistor to switch it on or off. Abbreviated LLL circuit. { 'lō ˌlev·əl 'läj·ik ˌsər·kət }

low-level modulation [ELECTR] Modulation produced at a point in a system where the power level is low compared with the power level at the output of the system. { 'lō ˌlev·əl ˌmäj·ə'lā·shən }

low-noise amplifier [ELECTR] An amplifier having very low background noise when the desired signal is weak or absent; field-effect transistors are used in audio preamplifiers for this purpose. { 'lō ˌnȯiz 'am·plə͵fī·ər }

low-noise preamplifier [ELECTR] A low-noise amplifier placed in a system prior to the main amplifier, sometimes close to the source; used to establish a satisfactory noise figure at an early point in the system. { 'lō ˌnȯiz prē'am·plə͵fī·ər }

low-order [COMPUT SCI] Pertaining to the digit which contributes the smallest amount to the value of a numeral, or to its position, or to the rightmost position of a word. { 'lō ˌȯr·dər }

low-pass band-pass transformation *See* frequency transformation. { 'lō ˌpas 'band ˌpas ˌtranz·fər͵mā·shən }

low-power television station [COMMUN] A television broadcasting facility limited in transmitter output so as to provide reception in only a local area, with a typical service area radius of 3–16 miles (5–26 kilometers). Abbreviated LPTV station. { 'lō ¦pau̇·ər 'tel·ə͵vish·ən ˌstā·shən }

low side [COMPUT SCI] The part of a controller or other remote device that communicates with terminals or other remote devices, rather than with the host computer. { 'lō ˌsīd }

low-tier system [COMMUN] A wireless telephone system that provides high quality and low-delay voice and data capabilities but has small cells. { ˌlō 'tēr ˌsis·təm }

low-voltage relay [COMMUN] A relay that responds to the drop in voltage (increase in current) when a telephone line becomes active; used to activate interception and eavesdropping equipment. { 'lō ¦vōl·tij 'rē͵lā }

LPM *See* lines per minute.

LPTV station *See* low-power television station. { ¦el¦pē¦tē'vē ˌstā·shən }

LQG problem *See* linear-quadratic-Gaussian problem. { ¦el¦kyü¦jē ˌpräb·ləm }

LRC *See* longitudinal redundancy check.

LRRP *See* lowest required radiating power.

LSA mode *See* limited space-charge accumulation mode. { ¦el¦es'ā ˌmōd }

LSB *See* least significant bit.

L-scan *See* L-display. { 'el ˌskan }

L-scope *See* L-display. { 'el ˌskōp }

LSI circuit *See* large-scale integrated circuit. { ¦el¦es'ī ˌsər·kət }

Luenberger observer [CONT SYS] A compensator driven by both the inputs and measurable outputs of a control system. { 'lün͵bərg·ər əb'zər·vər }

LUHF *See* lowest useful high frequency.

Lukasiewicz notation *See* Polish notation. { lü ˌkä·shē'ā͵vits nō͵tā·shən }

luminance carrier *See* picture carrier. { 'lü·mə·nəns ˌkar·ē·ər }

luminance channel [COMMUN] A path intended primarily for the luminance signal in an analog color television system. { 'lü·mə·nəns ˌchan·əl }

luminance primary [COMMUN] One of the three transmission primaries whose amount determines the luminance of a color in a color video system. { 'lü·mə·nəns 'prī͵mer·ē }

luminance signal [COMMUN] The color video signal that is intended to have exclusive control of the luminance of the picture. Also known as Y signal. { 'lü·mə·nəns ˌsig·nəl }

lumped discontinuity [ELECTROMAG] An analytical tool in the study of microwave circuits in which the effective values of inductance, capacitance, and resistance representing a discontinuity in a waveguide are shown as discrete components of equivalent value. { 'ləmpt ˌdis ˌkänt·ən'ü·əd·ē }

lumped element [ELECTROMAG] A section of a transmission line designed so that electric or magnetic energy is concentrated in it at specified frequencies, and inductance or capacitance may therefore be regarded as concentrated in it, rather than distributed over the length of the line. { 'ləmpt 'el·ə·mənt }

lumped impedance [ELECTROMAG] An impedance concentrated in a single component rather than distributed throughout the length of a transmission line. { 'ləmpt im'pēd·əns }

Luneberg lens [ELECTROMAG] A type of antenna consisting of a dielectric sphere whose index of refraction varies with distance from the center of the sphere so that a beam of parallel rays falling on the lens is focused at a point on the lens surface diametrically opposite from the direction of incidence, and, conversely, energy emanating from a point on the surface is focused into a plane wave. Accurately spelled Luneburg lens. { 'lü·nə͵bərg ˌlenz }

Luneburg lens *See* Luneberg lens. { 'lü·nə ˌbərg ˌlenz }

Luxemburg effect [COMMUN] Cross modulation between two radio signals during their passage through the ionosphere, due to the nonlinearity of the propagation characteristics of free charges in space. { 'lu̇k·səm,bərg i,fekt }

l value See left value. { 'el ,val·yü }

Lyapunov function [MATH] A function of a vector and of time which is positive-definite and has a negative-definite derivative with respect to time for nonzero vectors, is identically zero for the zero vector, and approaches infinity as the norm of the vector approaches infinity; used in determining the stability of control systems. Also spelled Liapunov function. { lē'ap·ə,nȯf ,fəŋk·shən }

Lyapunov stability criterion [CONT SYS] A method of determining the stability of systems (usually nonlinear) by examining the sign-definitive properties of an associated Lyapunov function. { lē'ap·ə,nȯf stə'bil·əd·ē krī ,tir·ē·ən }

M

M *See* megabyte.

MAC *See* message authentication code.

machinable *See* machine-sensible. { mə'shēn·ə·bəl }

machine [COMPUT SCI] **1.** A mechanical, electric, or electronic device, such as a computer, tabulator, sorter, or collator. **2.** A simplified, abstract model of an internally programmed computer, such as a Turing machine. { mə'shēn }

machine address [COMPUT SCI] The actual and unique internal designation of the location at which an instruction or datum is to be stored or from which it is to be retrieved. { mə'shēn ə'dres }

machine available time [COMPUT SCI] The time during which a computer has its power turned on, is not undergoing maintenance, and is thought to be operating properly. { mə'shēn ə'vāl·ə·bəl ˌtīm }

machine check [COMPUT SCI] A check that tests whether the parts of equipment are functioning properly. Also known as hardware check. { mə'shēn ˌchek }

machine-check indicator [COMPUT SCI] A protective device which turns on when certain conditions arise within the computer; the computer can be programmed to stop or to run a separate correction routine or to ignore the condition. { mə'shēn ˌchek ˌin·dəˌkād·ər }

machine code [COMPUT SCI] **1.** A computer representation of a character, digit, or action command in internal form. **2.** A computer instruction in internal format, or that part of the instruction which identifies the action to be performed. **3.** The set of all instruction types that a particular computer can execute. { mə'shēn ˌkōd }

machine conditions [COMPUT SCI] A component of a task descriptor that specifies the contents of all programmable registers in the processor, such as arithmetic and index registers. { mə'shēn kən'dish·ənz }

machine cycle [COMPUT SCI] **1.** The shortest period of time at the end of which a series of events in the operation of a computer is repeated. **2.** The series of events itself. { mə'shēn ˌsī·kəl }

machine-dependent [COMPUT SCI] Referring to programming languages, programs, systems, and procedures that can be used only on a particular computer or on a line of computers manufactured by a single company. { mə'shēn diˌpen·dənt }

machine error [COMPUT SCI] A deviation from correctness in computer-processed data, caused by equipment failure. { mə'shēn ˌer·ər }

machine-independent [COMPUT SCI] Referring to programs and procedures which function in essentially the same manner regardless of the machine on which they are carried out. { mə'shēn ˌin·də'pen·dənt }

machine instruction [COMPUT SCI] A set of digits, binary bits, or characters that a computer can recognize and act upon, and that, when interpreted or decoded, indicates the action to be performed and which operand is to be involved in the action. { mə'shēn inˌstrək·shən }

machine instruction statement [COMPUT SCI] A statement consisting usually of a tag, an operating code, and one or more addresses. { mə'shēn in¦strək·shən ˌstāt·mənt }

machine interruption [COMPUT SCI] A halt in computer operations followed by the beginning of a diagnosis procedure, as a result of an error detection. { mə'shēn ˌint·ə'rəp·shən }

machine language [COMPUT SCI] The set of instructions available to a particular digital computer, and by extension the format of a computer program in its final form, capable of being executed by a computer. { mə'shēn ˌlaŋ·gwij }

machine language code [COMPUT SCI] A set of instructions appearing as combinations of binary digits. { mə'shēn ¦laŋ·gwij 'kōd }

machine learning [COMPUT SCI] The process or technique by which a device modifies its own behavior as the result of its past experience and performance. { mə'shēn ˌlərn·iŋ }

machine logic [COMPUT SCI] The structure of a computer, the operation it performs, and the type and form of data used internally. { mə'shēn ˌläj·ik }

machine operator [COMPUT SCI] The person who manipulates the computer controls, brings up and closes down the computer, and can override a number of computer decisions. { mə'shēn ˌäp·əˌrād·ər }

machine-oriented language *See* computer-oriented language. { mə'shēn ˌȯr·ē¦en·təd 'laŋ·gwij }

machine-oriented programming system [COMPUT SCI] A system written in assembly language (or macro code) directly oriented toward the computer's internal language. { mə'shēn ¦ȯr·ē ˌent·əd 'prōˌgram·iŋ ˌsis·təm }

machine processible form [COMPUT SCI] Any input medium such as a punch card, paper tape, or magnetic tape. { mə'shēn ¦präsˌes·ə·bəl 'fȯrm }

machine-readable *See* machine-sensible. { mə 'shēn 'rēd·ə·bəl }

machine-recognizable *See* machine-sensible. { mə'shēn ˌrek·ig'niz·ə·bəl }

machine ringing [COMMUN] In a telephone system, ringing which is started either mechanically or by an operator, after which it continues automatically until the call is answered or abandoned. { mə'shēn ˌriŋ·iŋ }

machine run *See* run. { mə'shēn ¦rən }

machine script [COMPUT SCI] Any data written in a form that can immediately be used by a computer. { mə'shēn ˌskript }

machine-sensible [COMPUT SCI] Capable of being read or sensed by a device, usually by one designed and built specifically for this task. Also known as machinable; machine-readable; machine-recognizable; mechanized. { mə'shēn ¦sen·sə·bəl }

machine-sensible information [COMPUT SCI] Information in a form which can be read by a specified machine. { mə'shēn ¦sen·sə·bəl ˌin·fər'mā·shən }

machine-spoiled time [COMPUT SCI] Computer time wasted on production runs that cannot be completed or whose results are made worthless by a computer malfunction, plus extensions of running time on runs that are hampered by a malfunction. { mə'shēn ¦spȯild ˌtīm }

machine switching system *See* automatic exchange. { mə'shēn 'swich·iŋ ˌsis·təm }

machine-tool control [COMPUT SCI] The computer control of a machine tool for a specific job by means of a special programming language. { mə'shēn ˌtül kənˌtrōl }

machine translation *See* mechanical translation. { mə'shēn tranz'lā·shən }

machine vision *See* computer vision. { mə 'shēn ˌvizh·ən }

machine word [COMPUT SCI] The fundamental unit of information in a word-organized digital computer, consisting of a fixed number of binary bits, decimal digits, characters, or bytes. { mə'shēn ˌwərd }

macro *See* macroinstruction. { 'mak·rō }

macroassembler [COMPUT SCI] A program made up of one or more sequences of assembly language statements, each sequence represented by a symbolic name. { ¦mak·rō·ə'sem·blər }

macrocode [COMPUT SCI] A coding and programming language that assembles groups of computer instructions into single instructions. { 'mak·rəˌkōd }

macrodefinition [COMPUT SCI] A statement that defines a macroinstruction and the set of ordinary instructions which it replaces. { ¦mak·rō ˌdef·ə'nish·ən }

macroexpansion [COMPUT SCI] Instructions generated by a macroinstruction and inserted into an assembly language program. { ¦mak·rō·ik'span·chən }

macro flow chart [COMPUT SCI] A graphical representation of the overall logic of a computer program in which entire segments or subroutines of the program are represented by single blocks and no attempt is made to specify the detailed operation of the program. { 'mak·rō 'flō ˌchärt }

macrogeneration [COMPUT SCI] The creation of many machine instructions from one macroword. { ¦mak·rōˌjen·ə¦rā·shən }

macrogenerator *See* macroprocessor. { ¦mak·rō'jen·əˌrād·ər }

macroinstruction [COMPUT SCI] An instruction in a higher-level language which is equivalent to a specific set of one or more ordinary instructions in the same language. Also known as macro. { ¦mak·rō·in'strək·shən }

macrolanguage [COMPUT SCI] A computer language that manipulates stored strings in which particular sites of the string are marked so that other strings can be inserted in these sites when the stored string is brought forth. { 'mak·rō ˌlaŋ·gwij }

macrolibrary [COMPUT SCI] A collection of prewritten specialized but unparticularized routines (or sets of statements) which reside in mass storage. { 'mak·rōˌlīˌbrer·ē }

macroparameter [COMPUT SCI] The character in a macro operand which will complete an open subroutine created by the macroinstruction. { ˌmak·rō·pə'ram·əd·ər }

macroprocessor [COMPUT SCI] A piece of software which replaces each macroinstruction in a computer program by the set of ordinary instructions which it stands for. Also known as macrogenerator. { ¦mak·rə'präsˌes·ər }

macroprogram [COMPUT SCI] A computer program that consists of macroinstructions. { ma'krō'prōˌgrəm }

macroprogramming [COMPUT SCI] The process of writing machine procedure statements in terms of macroinstructions. { ¦mak·rō'prō ˌgram·iŋ }

macroskeleton [COMPUT SCI] A definition of a macroinstruction in a precise but content-free way, which can be particularized by a processor as directed by macroinstruction parameters. Also known as model. { ¦mak·rō'skel·ə·tən }

macrosystem [COMPUT SCI] A language in which words represent a number of machine instructions. { 'mak·rōˌsis·təm }

macro virus [COMPUT SCI] A virus that hides inside document and spreadsheet files used by popular word processing and spreadsheet applications. { ¦mak·rō 'vī·rəs }

magazine [COMPUT SCI] A holder of microfilm or magnetic recording media strips. { ¦mag·ə¦zēn }

magic tee *See* hybrid tee. { 'maj·ik 'tē }

magnetic bubble memory *See* bubble memory. { mag'ned·ik ¦bəb·əl ˌmem·rē }

magnetic card [COMPUT SCI] A card with a magnetic surface on which data can be stored by selective magnetization. { mag'ned·ik 'kärd }

magnetic card file [COMPUT SCI] A direct-access storage device in which units of data are stored on magnetic cards contained in one or more magazines from which they are withdrawn, when addressed, to be carried at high speed past a read/write head. { mag'ned·ik 'kärd ˌfīl }

magnetic character [COMPUT SCI] A character printed with magnetic ink, as on bank checks, for reading by machines as well as by humans. { mag'ned·ik 'kar·ik·tər }

magnetic character reader [COMPUT SCI] A character reader that reads special type fonts printed in magnetic ink, such as those used on bank checks, and feeds the character data directly to a computer for processing. { mag'ned·ik ¦kar·ik·tər ˌrēd·ər }

magnetic character sorter [COMPUT SCI] A device that reads documents printed with magnetic ink; all data read are stored, and records are sorted on any required field. Also known as magnetic document sorter-reader. { mag'ned·ik ¦kar·ik·tər ˌsȯrd·ər }

magnetic core multiplexer [COMPUT SCI] A device which channels many bit inputs into a single output. { mag'ned·ik ¦kȯr 'məl·təˌplek·sər }

magnetic core storage [COMPUT SCI] A computer storage system in which each of thousands of magnetic cores stores one bit of information; current pulses are sent through wires threading through the cores to record or read out data; used extensively in the 1950s and 1960s, and still used in specialized military applications and in space vehicles. Also known as core memory; core storage. { mag'ned·ik ¦kȯr 'stȯr·ij }

magnetic deflection [ELECTR] Deflection of an electron beam by the action of a magnetic field, as in a television picture tube. { mag'ned·ik di'flek·shən }

magnetic dipole antenna [ELECTROMAG] Simple loop antenna capable of radiating an electromagnetic wave in response to a circulation of electric current in the loop. { mag'ned·ik 'dī ˌpōl anˌten·ə }

magnetic disk [COMPUT SCI] A rotating circular plate having a magnetizable surface on which information may be stored as a pattern of polarized spots on concentric recording tracks. { mag'ned·ik 'disk }

magnetic document sorter-reader *See* magnetic character sorter. { mag'ned·ik ¦däk·yə·mənt 'sȯrd·ər 'rēd·ər }

magnetic domain memory *See* domain-tip memory. { mag'ned·ik də'mān ˌmem·rē }

magnetic drum *See* drum. { mag'ned·ik 'drəm }

magnetic drum storage *See* drum. { mag 'ned·ik ¦drəm 'stȯr·ij }

magnetic head [ELECTR] The electromagnet used for reading, recording, or erasing signals on a magnetic disk, drum, or tape. Also known as magnetic read/write head. { mag'ned·ik 'hed }

magnetic-ink character recognition [COMPUT SCI] That branch of character recognition which involves the sensing of magnetic-ink characters for the purpose of determining the character's most probable identity. Abbreviated MICR. { mag'ned·ik ¦iŋk 'kar·ik·tər ˌrek·ig ˌnish·ən }

magnetic memory *See* magnetic storage. { mag'ned·ik 'mem·rē }

magnetic printing [ELECTR] The permanent and usually undesired transfer of a recorded signal from one section of a magnetic recording medium to another when these sections are brought together, as on a reel of tape. Also known as crosstalk; magnetic transfer. { mag'ned·ik 'print·iŋ }

magnetic random access memory [COMPUT SCI] A nonvolatile memory in which submicrometer-sized magnetic structures store digital information in their magnetic orientation. Abbreviated MRAM. { mag¦ned·ik ˌran·dəm 'ak ˌses ˌmem·rē }

magnetic read/write head *See* magnetic head. { mag'ned·ik ¦rēd ¦rīt ˌhed }

magnetic recording [ELECTR] Recording by means of a signal-controlled magnetic field. { mag'ned·ik ri'kȯrd·iŋ }

magnetic rotation [OPTICS] **1.** In a weak magnetic field, the rotation of the plane of polarization of fluorescent light emitted perpendicular to the field and perpendicular to the propagation direction of the incident light. **2.** *See* Faraday effect. { mag'ned·ik rō'tā·shən }

magnetic shift register [COMPUT SCI] A shift register in which the pattern of settings of a row of magnetic cores is shifted one step along the row by each new input pulse; diodes in the coupling loops between cores prevent backward flow of information. { mag'ned·ik 'shift ˌrej·ə·stər }

magnetic spin transistor *See* magnetic switch. { magˌned·ik ˌspin tran'zis·tər }

magnetic storage [COMPUT SCI] A device utilizing magnetic properties of materials to store data; may be roughly divided into two categories, moving (drum, disk, tape) and static (core, thin film). Also known as magnetic memory. { mag'ned·ik 'stȯr·ij }

magnetic stripe [COMPUT SCI] A small length of magnetic tape on a card or badge, containing data that is machine-readable. { mag¦ned·ik 'strīp }

magnetic striped ledger [COMPUT SCI] A ledger sheet used on a special typing device which stores the coded data on a magnetic strip on the sheet while typing out the data on the sheet; the magnetic strip can be read directly by a special reader linked to a computer. { mag'ned·ik ¦strīpt 'lej·ər }

magnetic switch [ELECTR] A switching device consisting of three metallic layers (a paramagnetic layer between two ferromagnetic layers), whose action is based on electron spin and is controlled by a small magnetic field. Also known as bipolar spin device; bipolar spin switch; magnetic spin transistor; spin transistor; spin valve. { mag¦ned·ik 'swich }

magnetic tape [ELECTR] A plastic, paper, or metal tape that is coated or impregnated with magnetizable iron oxide particles; used in magnetic recording and in computer storage chiefly for archiving and backup. { mag'ned·ik 'tāp }

magnetic tape file operation [COMPUT SCI] All the jobs related to creating, sorting, inputting, and maintenance of magnetic tapes in a magnetic tape environment. { mag'ned·ik ¦tāp 'fīl ,äp·ə,rā·shən }

magnetic tape group [COMPUT SCI] A cabinet containing two or more magnetic tape units, each of which can operate independently, but which sometimes share one or more channels with which they communicate with a central processor. Also known as tape cluster; tape group. { mag'ned·ik ¦tāp ,grüp }

magnetic tape librarian [COMPUT SCI] Routine which provides a computer the means to automatically run a sequence of programs. { mag'ned·ik ¦tāp lī,brer·ē·ən }

magnetic tape master file [COMPUT SCI] A magnetic tape consisting of a set of related elements such as is found in a payroll, an inventory, or an accounts receivable; a master file is, as a rule, periodically updated. { mag'ned·ik¦tāp¦mas·tər 'fīl }

magnetic tape parity [COMPUT SCI] A check performed on the data bits on a tape; usually an odd (or even) condition is expected and the occurrence of the wrong parity indicates the presence of an error. { mag'ned·ik ¦tāp 'par·əd·ē }

magnetic tape station [COMPUT SCI] On-line device that provides write, read, and erase data on magnetic tape to permit high-speed storage of data. { mag'ned·ik ¦tāp ,stā·shən }

magnetic tape storage [COMPUT SCI] Storage of binary information on magnetic tape, generally on 5 to 10 tracks, with up to several thousand bits per inch (more than a thousand bits per centimeter) on each track. { mag'ned·ik ¦tāp ,stȯr·ij }

magnetic tape switching unit [COMPUT SCI] A device which permits the computer operator to bring into play any number of tape drives as required by the system. { mag'ned·ik ¦tāp 'swich·iŋ ,yü·nət }

magnetic tape terminal [COMPUT SCI] Device which converts pulses in series to pulses in parallel while checking for bit parity prior to the entry in buffer storage. { mag'ned·ik ¦tāp 'tər·mən·əl }

magnetic tape unit [COMPUT SCI] A computer unit that usually consists of a tape transport, reading and recording heads, and associated electric and electronic equipment. { mag'ned·ik ¦tāp ,yü·nət }

magnetic transfer *See* magnetic printing. { mag'ned·ik 'tranz·fər }

magnetic tunnel junction [ELECTR] A magnetic storage and switching device in which two magnetic layers are separated by an insulating barrier, typically aluminum oxide, that is only 1–2 nanometers thick, allowing an electronic current whose magnitude depends on the orientation of both magnetic layers to tunnel through the barrier when it is subject to a small electric bias. { mag¦ned·ik 'tən·əl ,jəŋk·shən }

magnetooptical switch [COMPUT SCI] A thin-film modulator which acts on a laser beam by polarization, causing the beam to emerge from the output prism at a different angle. { mag ¦nēd·ō¦äp·tə·kəl 'swich }

magnetooptic disk [COMMUN] A data storage device in which information is stored in small magnetic marks along tracks on a rotating disk; the information is read by sensing the change in polarization of reflected focused light and can be altered by using a higher-power focused light spot to locally heat the medium and, with the application of an external magnetic field, switch the magnetic domains of the material. { mag ,ned·ō,äp·tik 'disk }

magnetooptic Kerr effect [OPTICS] Changes produced in the optical properties of a reflecting surface of a ferromagnetic substance when the substance is magnetized; this applies especially to the elliptical polarization of reflected light, when the ordinary rules of metallic reflection would give only plane polarized light. Also known as Kerr magnetooptical effect. { mag¦nēd·ō ¦äp·tik 'kər i,fekt }

magnetooptic material [OPTICS] A material whose optical properties are changed by an applied magnetic field. { mag¦nēd·ō¦äp·tik mə'tir·ē·əl }

magnetooptics [OPTICS] The study of the effect of a magnetic field on light passing through a substance in the field. { mag¦nēd·ō¦äp·tiks }

magnetoresistance [ELECTR] The change in the electrical resistance of a material when it is subjected to an applied magnetic field, this property has widespread application in sensors and magnetic read heads. [ELECTROMAG] The change in electrical resistance produced in a current-carrying conductor or semiconductor on application of a magnetic field. { mag¦nēd·ō·ri'zis·təns }

magnetoresistive memory [ELECTR] A random-access memory that uses the magnetic state of small ferromagnetic regions to store data, plus magnetoresistive devices to read the data, all integrated with silicon integrated-circuit electronics. { mag,ned·ō·ri,zis·tiv 'mem·rē }

magnetron [ELECTR] One of a family of crossed-field microwave tubes, wherein electrons, generated from a heated cathode, move under the combined force of a radial electric field and an axial magnetic field in such a way as to produce a bunching of electrons and hence microwave radiation. Useful in the frequency range 1–40 gigahertz; a pulsed microwave radiation source for radar and continuous source for microwave cooking. { 'mag·nə,trän }

magnetron oscillator [ELECTR] Oscillator circuit employing a magnetron tube. { 'mag·nə,trän 'äs·ə,lād·ər }

magnetron pulling [ELECTR] Frequency shift of a magnetron caused by factors which vary the standing waves or the standing-wave ratio on the radio-frequency lines. { 'mag·nə,trän 'pu̇l·iŋ }

magnetron pushing [ELECTR] Frequency shift of a magnetron caused by faulty operation of the modulator. { 'mag·nə,trän 'pu̇sh·iŋ }

mail box [COMPUT SCI] **1.** A portion of a computer's main storage that can be used to hold information about other devices. **2.** Computer storage facilities designed to hold electronic mail. { 'māl ˌbäks }

mailbox name [COMPUT SCI] The first part of an electronic mail address, which identifies the storage space that has been set aside in a computer to receive a user's electronic mail messages. Also known as username. { 'māl ˌbäks ˌnām }

mailing list [COMMUN] A list of users of the Internet or another computer network who all receive copies of electronic mail messages. { 'māl·iŋ ˌlist }

mail merge [COMPUT SCI] The process of combining a form letter with a list of names and addresses to produce individualized letters. { 'māl ˌmərj }

main bang [ELECTR] In colloquial usage, a transmitted pulse within a radar system. { 'mān 'baŋ }

main clock *See* master clock. { 'mān 'kläk }

main controller [COMPUT SCI] A control unit assigned to direct the other control units in a computer system. { 'mān kən'trō·lər }

main frame [COMPUT SCI] **1.** A large computer. **2.** The part of a computer that contains the central processing unit, main storage, and associated control circuitry. Also known as frame. { 'mān ˌfrām }

main instruction buffer [COMPUT SCI] A section of storage in the instruction unit, 16 bytes in length, used to hold prefetched instructions. { 'mān in'strək·shən ˌbəf·ər }

main level [COMMUN] A range of allowed picture parameters defined by the MPEG-2 video coding specification. { 'mān 'lev·əl }

main lobe *See* major lobe. { 'mān 'lōb }

main loop [COMPUT SCI] A set of instructions that constitute the primary structure of a repetitive computer program. { 'mān ˌlüp }

main memory *See* main storage. { 'mān 'mem·rē }

main path [COMPUT SCI] The principal branch of a routine followed by a computer in the process of carrying out the routine. { 'mān 'path }

main profile [COMMUN] A subset of the syntax of the MPEG-2 video coding specification that is supported over a large range of applications. { 'mān 'prōˌfīl }

main program [COMPUT SCI] **1.** The central part of a computer program, from which control may be transferred to various subroutines and to which control is eventually returned. Also known as main routine. **2.** *See* executive routine. { 'mān 'prō·grəm }

main routine *See* executive routine; main program. { 'mān rü'tēn }

main station [COMMUN] Telephone station with a distinct call number designation, directly connected to a central office. { 'mān 'stā·shən }

main storage [COMPUT SCI] A digital computer's principal working storage, from which instructions can be executed or operands fetched for data manipulation. Also known as main memory. { 'mān 'stȯr·ij }

maintenance pack [COMPUT SCI] A disk drive that is used to store copies of computer programs for the purpose of applying and testing changes made in the course of software maintenance. { 'mānt·ən·əns ˌpak }

maintenance routine [COMPUT SCI] A computer program designed to detect conditions which may give rise to a computer malfunction in order to assist a service engineer in performing routine preventive maintenance. { 'mānt·ən·əns rü ˌtēn }

maintenance time [COMPUT SCI] The time required for both corrective and preventive maintenance of a computer or other components of a computer system. { 'mānt·ən·əns ˌtīm }

main vector [COMMUN] A pair of numbers that represent the vertical and horizontal displacement of a region of a reference picture for MPEG-2 prediction. { 'mān 'vek·tər }

major cycle [COMPUT SCI] The time interval between successive appearances of a given storage position in a serial-access computer storage. { 'mā·jər 'sī·kəl }

majority carrier [ELECTR] The type of charge carrier, that is, electron or hole, that constitutes more than half the carriers in a semiconductor. { mə'jär·əd·ē 'kar·ē·ər }

majority element *See* majority gate. { mə 'jär·əd·ē 'el·ə·mənt }

majority gate [COMPUT SCI] A logic circuit which has one output and several inputs, and whose output is energized only if a majority of its inputs are energized. Also known as majority element; majority logic. { mə'jär·əd·ē 'gāt }

majority logic *See* majority gate. { mə'jär·əd·ē 'läj·ik }

major key [COMPUT SCI] The primary key for identifying a record. { 'mā·jər ˌkē }

major lobe [ELECTROMAG] Antenna lobe indicating the direction of maximum radiation or reception. Also known as main lobe. { 'mā·jər 'lōb }

major wave *See* long wave. { 'mā·jər 'wāv }

make-break operation [COMMUN] A circuit operation in which there is a cessation of current flow as a pulse transmission occurs. { 'māk 'brāk ˌäp·əˌrā·shən }

make-busy [COMMUN] A switch whose activation makes a dial telephone line or group of telephone lines appear to be busy and thereby prevents completion of incoming calls. { 'māk 'biz·ē }

makeup time [COMPUT SCI] The time required to rerun programs on a computer because of operator errors and other problems. { 'mākˌəp ˌtīm }

malfunction routine [COMPUT SCI] A program used in troubleshooting. { mal'fəŋk·shən rü ˌtēn }

malicious code [COMPUT SCI] Programming code that is capable of causing harm to availability, integrity of code or data, or confidentiality in a computing system; encompasses Trojan horses, viruses, worms, and trapdoors. { mə ¦lish·əs 'kōd }

management information system [COMMUN] A communication system in which data are recorded and processed to form the basis for decisions by top management of an organization. Abbreviated MIS. { 'man·ij·mənt ˌin·fər'mā·shən ˌsis·təm }

MANAV *See* Marine Integrated Navigation.

Manchester coding *See* phase encoding. { 'man·chə·stər ˌkōd·iŋ }

manifest constant [COMPUT SCI] A value that is assigned to a symbolic name at the beginning of a computer program and is not subject to change during execution. { 'man·əˌfest 'kän·stənt }

manipulated variable [COMPUT SCI] Variable whose value is being altered to bring a change in some condition. { mə'nip·yəˌlād·əd 'ver·ē·ə·bəl }

manipulator [CONT SYS] An armlike mechanism on a robotic system that consists of a series of segments, usually sliding or jointed which grasp and move objects with a number of degrees of freedom, under automatic control. { mə'nip·yə ˌlād·ərz }

mantissa [COMPUT SCI] A fixed point number composed of the most significant digits of a given floating-point number. Also known as fixed-point part; floating-point coefficient. { man'tis·ə }

manual central office [COMMUN] Central office of a manual telephone system. { 'man·yə·wəl ¦sen·trəl 'ȯf·əs }

manual control unit [CONT SYS] A portable, hand-held device that allows an operator to program and store instructions related to robot motions and positions. Also known as programming unit. { 'man·yə·wəl kən'trōl ˌyü·nət }

manual exchange [COMMUN] Any exchange where calls are completed by an operator. { 'man·yə·wəl iks'chānj }

manual input [COMPUT SCI] The entry of data by hand into a device at the time of processing. { 'man·yə·wəl 'inˌpu̇t }

manual number generator *See* manual word generator. { 'man·yə·wəl 'nəm·bər ˌjen·əˌrād·ər }

manual operation [COMPUT SCI] Any processing operation performed by hand. { 'man·yə·wəl ˌäp·ə'rā·shən }

manual ringing [COMMUN] Ringing which is started by the manual operation of a key and continues only while the key is held in operation. { 'man·yə·wəl 'riŋ·iŋ }

manual telephone system [COMMUN] A telephone system in which connections between customers are ordinarily established manually by telephone operators in accordance with orders given verbally by calling parties. { 'man·yə·wəl 'tel·əˌfōn ˌsis·təm }

manual word generator [COMPUT SCI] A device into which an operator can enter a computer word by hand, either for direct insertion into memory or to be held until it is read during the execution of a program. Also known as manual number generator. { 'man·yə·wəl 'wərd ˌjen·əˌrād·ər }

many-to-many correspondence [COMPUT SCI] A structure that establishes relationships between items in a data base, such that one unit of data can relate to many units, and many units can relate back to one unit and to other units as well. { 'men·ē tə 'men·ē ˌkär·ə'spän·dəns }

map [COMPUT SCI] **1.** An output produced by an assembler, compiler, linkage editor, or relocatable loader which indicates the (absolute or relocatable) locations of such elements as programs, subroutines, variables, or arrays. **2.** By extension, an index of the storage allocation on a magnetic disk or drum. { map }

mapping radar [NAV] Radar carried on an aircraft as an aid to navigation, which displays on a cathode-ray tube imagery of the ground in the vicinity of the aircraft. { 'map·iŋ ˌrāˌdär }

Marconi antenna [ELECTROMAG] Antenna system of which the ground is an essential part, as distinguished from a Hertz antenna. { mär'kō·nē an'ten·ə }

Marine Integrated Navigation [NAV] A navigation system, designed chiefly for large tankers, that routes and integrates information from remote sensors and presents anticollision and ship-handling information through visual display units cited on a central compact console. Abbreviated MANAV. { mə'rēn ¦int·əˌgrād·əd ˌnav·ə'gā·shən }

marine navigation [NAV] The process of directing the movements of watercraft from one point to another; the process, always present in some form when a vessel is under way and not drifting, varies with the type of craft, its mission, and its area of operation. { mə'rēn ˌnav·ə'gā·shən }

marine radio beacon [NAV] A radio-beacon station which produces service primarily for the guidance of ships. { mə'rēn 'rād·ē·ō ˌbē·kən }

maritime DGPS [NAV] A type of local-area DGPS that makes use of existing radio beacons which broadcast error corrections over a limited range to marine and land users. { ¦mar·əˌtīm ¦dē¦jē ¦pē'es }

maritime frequency bands [COMMUN] In the United States, a collection of radio frequencies allocated for communication between coast stations and ships or between ships. { 'mar·əˌtīm 'frē·kwən·sē ˌbanz }

maritime mobile satellite service [COMMUN] A mobile satellite service in which the mobile earth stations are located on board ships. Abbreviated MMSS. { ˌmar·əˌtīm ¦mō·bəl 'sad·əlˌt ˌsər·vəs }

maritime mobile service [COMMUN] A mobile service between coast stations and ship stations, or between ship stations, in which survival craft stations may also participate. { 'mar·əˌtīm 'mō·bəl ¦sər·vəs }

mark [COMMUN] The closed-circuit condition in telegraphic communication, during which the signal actuates the printer; the opposite of space. [COMPUT SCI] A distinguishing feature used to signal some particular location or condition. { märk }

mark detection [COMPUT SCI] That class of character recognition systems which employs coded documents, in the form of boxes or windows, in order to convey intended information by means of pencil or ink marks made in specific boxes. { 'märk di,tek·shən }

mark-hold [COMMUN] The transmission of a steady mark to indicate that there is no traffic over a telegraph channel; the upper marking frequency of a duplex channel (2225 hertz) is used to disable echo suppressors which may interfere with data communications. { ¦märk ¦hōld }

marking and spacing intervals [COMMUN] Intervals of closed and open conditions in transmission circuits. { ¦märk·iŋ ən ¦spās·iŋ 'in·tər·vəlz }

marking bias [COMMUN] Bias distortion that lengthens the marking impulse. { 'märk·iŋ ,bī·əs }

marking-end distortion [COMMUN] End distortion that lengthens the marking impulse. { 'märk·iŋ ¦end di,stȯr·shən }

Markov-based model [COMPUT SCI] A model that represents a computer system by a Markov chain, which represents the set of all possible states of the system, with the possible transitions between these states. { 'mär,kȯf ,bāst ,mäd·əl }

mark reading [COMPUT SCI] In character recognition, that form of mark detection which employs a photoelectric device to locate and convey intended information; the information appears as special marks on sites (windows) within the document coding area. { 'märk ,rēd·iŋ }

mark sensing [COMPUT SCI] In character recognition, that form of mark detection which depends on the conductivity of graphite pencil marks to locate and convey intended information; the information appears as special marks on sites (windows) within the document coding area. { 'märk ,sens·iŋ }

mark-space multiplier [ELECTR] A multiplier used in analog computers in which one input controls the mark-to-space ratio of a square wave while the other input controls the amplitude of the wave, and the output, obtained by a smoothing operation, is proportional to the average value of the signal. Also known as time-division multiplier. { ¦märk ¦spās 'məl·tə,plī·ər }

mark-space ratio See mark-to-space ratio. { ¦märk ¦spās 'rā·shō }

mark-to-space ratio [ELECTR] The ratio of the duration of the positive-amplitude part of a square wave to that of the negative-amplitude part. Also known as mark-space ratio. { ¦märk ¦tə ¦spās 'rā·shō }

mark-to-space transition [COMMUN] The process of switching from a mark to a space. { ¦märk ¦tə ¦spās tran'zish·ən }

markup [COMPUT SCI] The process of adding information (tags) to an electronic document that are not part of the content but describe its structure or elements. { 'märk,əp }

markup language [COMPUT SCI] A set of rules and procedures for markup. { 'märk,əp ,laŋ·gwij }

maser [PHYS] A device for coherent amplification or generation of electromagnetic waves in which an ensemble of atoms or molecules, raised to an unstable energy state, is stimulated by an electromagnetic wave to radiate excess energy at the same frequency and phase as the stimulating wave. Derived from microwave amplification by stimulated emission of radiation. Also known as paramagnetic amplifier. { 'mā·zər }

MA service See multiple-access service. { ¦em'ā ,sər·vəs }

maskable interrupt [COMPUT SCI] An interrupt that can be allowed to occur or prevented from occurring by software. { ¦mas·kə·bəl 'int·ə,rəpt }

masking [COMPUT SCI] **1.** Replacing specific characters in one register by corresponding characters in another register. **2.** Extracting certain characters from a string of characters. [ELECTR] **1.** Using a covering or coating on a semiconductor surface to provide a masked area for selective deposition or etching. **2.** A programmed procedure for eliminating radar coverage in areas where such transmissions may be of use to the enemy for navigation purposes, by weakening the beam in appropriate directions or by use of additional transmitters on the same frequency at suitable sites to interfere with homing; also used to suppress the beam in areas where it would interfere with television reception. { 'mask·iŋ }

mask matching [COMPUT SCI] In character recognition, a method employed in character property detection in which a correlation or match is attempted between a specimen character and each of a set of masks representing the characters to be recognized. { 'mask ,mach·iŋ }

mask register [COMPUT SCI] Filter which determines the parts of a word which are to be tested. { 'mask ,rej·ə·stər }

mask word [COMPUT SCI] A word modifier used in a logical AND operation. { 'mask ,wərd }

Mason's theorem [CONT SYS] A formula for the overall transmittance of a signal flow graph in terms of transmittances of various paths in the graph. { 'mās·ənz ,thir·əm }

massage [COMPUT SCI] To process data, primarily to convert it into a more useful form or into a form that will simplify processing. { mə'säzh }

mass communication [COMMUN] Communication which is directed to or reaches an appreciable fraction of the population. { 'mas kə ,myü·nə'kā·shən }

mass conversion [COMPUT SCI] The transfer of data from one computer system to another, in which all the data is converted in a single operation, rather than in gradual increments. { 'mas kən,vər·zhən }

mass data multiprocessing [COMPUT SCI] The basic concept of time sharing, with many inquiry stations to a central location capable of on-line data retrieval. { 'mas ¦dad·ə ,məl·ti'prä,ses·iŋ }

mass-memory unit [COMPUT SCI] Drum or disk memory that provides rapid access bulk storage for messages that are awaiting availability of outgoing channels. { 'mas 'mem·rē ,yü·nət }

mass storage [COMPUT SCI] A computer storage with large capacity, especially one whose contents are directly accessible to a computer's central processing unit. { 'mas 'stȯr·ij }

mass-storage system [COMPUT SCI] A computer system containing a large number of storage devices, with one of these devices containing the master file of the operating system, routines, and library routines. { 'mas ¦stȯr·ij ,sis·təm }

master antenna television system [COMMUN] A network that distributes television signals from a common antenna to apartments or dwellings under collective ownership. Abbreviated MATV system. { 'mas·tər an¦ten·ə 'tel·ə,vizh·ən ,sis·təm }

master clock [COMPUT SCI] The electronic or electric source of standard timing signals, often called clock pulses, required for sequencing the operation of a computer. Also known as main clock; master synchronizer; master timer. { 'mas·tər 'kläk }

master console See console. { 'mas·tər 'kän ,sōl }

master control [COMMUN] The control console that contains the main program controls for a radio or television transmission system or network. [COMPUT SCI] A computer program, oriented toward applications, which carries out the highest level of control in a hierarchy of programs, routines, and subroutines. { 'mas·tər kən¦trōl }

master control interrupt [COMPUT SCI] A signal which causes the master control program to take over control of a computer system. { 'mas·tər kən¦trōl 'in·tə,rəpt }

master data [COMPUT SCI] A set of data which are rarely changed, or changed in a known and constant manner. { 'mas·tər 'dad·ə }

master file [COMPUT SCI] **1.** A computer file containing relatively permanent information, usually updated periodically, such as subscriber records or payroll data other than time worked. **2.** A computer file that is used as an authoritative source of data in carrying out a particular job on the computer. { 'mas·tər 'fīl }

master group [COMMUN] In carrier telephony, ten supergroups (600 voice channels) multiplexed together and treated as a unit. { 'mas·tər ,grüp }

master instruction tape [COMPUT SCI] A computer magnetic tape on which all programs for a system of runs are recorded. { 'mas·tər in'strək·shən ,tāp }

master mode [COMPUT SCI] The mode of operation of a computer system exercised by the operating system or executive system, in which a privileged class of instructions, which user programs cannot execute, is permitted. Also known as monitor mode; privileged mode. { 'mas·tər ,mōd }

master multivibrator [ELECTR] Master oscillator using a multivibrator unit. { 'mas·tər ,məl·ti'vī ,brād·ər }

master oscillator [ELECTR] An oscillator that establishes the carrier frequency of the output of an amplifier or transmitter. { 'mas·tər 'äs·ə ,lād·ər }

master-oscillator power amplifier [ELECTR] Transmitter using an oscillator followed by one or more stages of radio-frequency amplification. { 'mas·tər ¦äs·ə,lād·ər 'pau̇·ər ,am·plə,fī·ər }

master plan position indicator [ELECTR] In a radar system, a plan position indicator which controls remote indicators or repeaters. { 'mas·tər 'plan pə¦zish·ən 'in·də,kād·ər }

master program file [COMPUT SCI] The tape record of all programs for a system of runs. { 'mas·tər 'prō·grəm ,fīl }

master record [COMPUT SCI] The basic updated record which will be used for the next run. { 'mas·tər 'rek·ərd }

master routine See executive routine. { 'mas·tər rü'tēn }

master scheduler [COMPUT SCI] A program in a job entry system that assigns priorities to jobs submitted for execution. { 'mas·tər 'sked·yə·lər }

master/slave mode [COMPUT SCI] The feature ensuring the protection of each program when more than one program resides in memory. { 'mas·tər 'slāv ,mōd }

master/slave system [COMPUT SCI] A system of interlinked computers under the control of one computer (master computer). { 'mas·tər 'slāv ,sis·təm }

master synchronization pulse [COMMUN] In telemetry, a pulse distinguished from other telemetering pulses by amplitude and duration, used to indicate the end of a sequence of pulses. { 'mas·tər ,siŋ·krə·nə'zā·shən ,pəls }

master synchronizer See master clock. { 'mas·tər 'siŋ·krə,nīz·ər }

master system tape [COMPUT SCI] A monitor program centralizing the control of program operation by loading and executing any program on a system tape. { 'mas·tər 'sis·təm ,tāp }

master tape [COMPUT SCI] A magnetic tape that contains data which must not be overwritten, such as an executive routine or master file; updating a master tape means generating a new master tape onto which supplementary data have been added. { 'mas·tər 'tāp }

master terminal [COMPUT SCI] A computer terminal that is used to monitor and control a computer system. { 'mas·tər 'tər·mən·əl }

master timer See master clock. { 'mas·tər 'tīm·ər }

match [COMPUT SCI] A data-processing operation similar to a merge, except that instead of producing a sequence of items made up from the input sequences, the sequences are matched against each other on the basis of some key. { mach }

matched filter [COMPUT SCI] In character recognition, a method employed in character property detection in which a vertical projection of the input character produces an analog waveform which is then compared to a set of stored waveforms for the purpose of determining the character's identity. [ELECTR] A filter with the property that, when the input consists of noise in addition to a specified desired signal, the signal-to-noise ratio is the maximum which can be obtained in any linear filter. { 'macht 'fil·tər }

matched load [ELECTR] A load having the impedance value that results in maximum absorption of energy from the signal source. { 'macht 'lōd }

match gate *See* equivalence gate. { 'mach ˌgāt }

matching [COMPUT SCI] A computer problem-solving method in which the current situation is represented as a schema to be mapped into the desired situation by putting the two in correspondence. [NAV] The bringing of two or more signals or indications into suitable position or condition preliminary to making a measurement, as on a loran indicator or a sky compass. { 'mach·iŋ }

matching diaphragm [ELECTROMAG] Diaphragm consisting of a slit in a thin sheet of metal, placed transversely across a waveguide for matching purposes; the orientation of the slit with respect to the long dimension of the waveguide determines whether the diaphragm acts as a capacitive or inductive reactance. { 'mach·iŋ 'dī·əˌfram }

matching section [ELECTROMAG] A section of transmission line, a quarter or half wavelength long, inserted between a transmission line and a load to obtain impedance matching. { 'mach·iŋ ¦sek·shən }

matching stub [ELECTROMAG] Device placed on a radio-frequency transmission line which varies the impedance of the line; the impedance of the line can be adjusted in this manner. { 'mach·iŋ ˌstəb }

match processing [COMPUT SCI] The checking of two or more units of data for common characteristics. { 'mach 'präˌses·iŋ }

math coprocessor *See* numeric processor extension. { ¦math 'kōˌpräˌses·ər }

mathematical check [COMPUT SCI] A programmed computer check of a sequence of operations, using the mathematical properties of that sequence. { ¦math·ə¦mad·ə·kəl 'chek }

mathematical function program [COMPUT SCI] A set of routinely used mathematical functions, such as square root, which are efficiently coded and called for by special symbols. { ¦math·ə ¦mad·ə·kəl 'fəŋk·shən ˌprō·grəm }

mathematical software [COMPUT SCI] The set of algorithms used in a computer system to solve general mathematical problems. { ¦math·ə ¦mad·ə·kəl 'sȯftˌwer }

mathematical subroutine [COMPUT SCI] A computer subroutine in which a well-defined mathematical function, such as exponential, logarithm, or sine, relates the output to the input. { ¦math·ə¦mad·ə·kəl 'səb·rüˌtēn }

matrix [COMPUT SCI] A latticework of input and output leads with logic elements connected at some of their intersections. [ELECTR] **1.** The section of an analog video system that transforms the red, green, and blue source signals into color-difference signals and combines them with the chrominance subcarrier. Also known as color coder; color encoder; encoder. **2.** The section of an analog color television receiver that transforms the color-difference signals into the red, green, and blue signals needed to drive the display device. Also known as color decoder; decoder. { 'mā·triks }

matrix algebra tableau [COMPUT SCI] The current matrix at the end of an iteration while running a linear program. { 'mā·triks ¦al·jə·brə ta'blō }

matrix-array camera [ELECTR] A solid-state video camera that has a rectangular array of light-sensitive elements or pixels. { 'mā·triks ə'rā ˌkam·rə }

matrix printing [COMPUT SCI] High-speed printing in which characterlike configurations of dots are printed through the proper selection of wire ends from a matrix of wire ends. Also known as stylus printing; wire printing. { 'mā·triks 'print·iŋ }

matrix storage [COMPUT SCI] A computer storage in which coordinates are used to address the locations of circuit elements. Also known as coordinate storage. { 'mā·triks ˌstȯr·ij }

mattress array *See* billboard array. { 'ma·trəs ə'rā }

MATV system *See* master antenna television system. { ¦em¦ā¦tē¦vē 'sis·təm }

mavar *See* parametric amplifier. { 'māˌvär }

maximum average power output [ELECTR] In television, the maximum radio-frequency output power that can occur under any combination of signals transmitted, averaged over the longest repetitive modulation cycle. { 'mak·sə·məm ¦av·rəj ¦pau̇·ər 'au̇tˌpu̇t }

maximum keying frequency [ELECTR] In facsimile, the frequency in hertz that is numerically equal to the spot speed divided by twice the horizontal dimension of the spot. { 'mak·sə·məm 'kē·iŋ ˌfrē·kwən·sē }

maximum modulating frequency [ELECTR] Highest picture frequency required for a facsimile transmission system; the maximum modulating frequency and the maximum keying frequency are not necessarily equal. { 'mak·sə·məm 'mäj·əˌlād·iŋ ˌfrē·kwən·sē }

maximum operating frequency [COMPUT SCI] The highest rate at which the modules perform iteratively and reliably. { 'mak·sə·məm 'äp·əˌrā d·iŋ ˌfrē·kwən·sē }

maximum signal level [ELECTR] In an amplitude-modulated facsimile system, the level corresponding to copy black or copy white, whichever has the highest amplitude. { 'mak·sə·məm 'sig·nəl ˌlev·əl }

maximum unambiguous range [ELECTROMAG] The range beyond which the echo from a pulsed radar signal returns after generation of the next pulse, and can thus be mistaken as a short-range echo of the next cycle. { 'mak·sə·məm ˌən·am ¦big·yə·wəs 'rānj }

maximum usable frequency [COMMUN] The upper limit of the frequencies that can be used at a specified time for point-to-point radio transmission involving propagation by reflection from the regular ionized layers of the ionosphere. Abbreviated MUF. { 'mak·sə·məm ¦yü·zə·bəl 'frē·kwən·sē }

Maxwell's electromagnetic theory [ELECTROMAG] A mathematical theory of electric and magnetic fields which predicts the propagation of electromagnetic radiation, and is valid for electromagnetic phenomena where effects on an atomic scale can be neglected. { 'mak,swelz i,lek·trō·mag'ned·ik 'thē·ə·rē }

Maxwell equations *See* Maxwell field equations. { 'mak,swel i,kwā·zhənz }

Maxwell field equations [ELECTROMAG] Four differential equations which relate the electric and magnetic fields to electric charges and currents, and form the basis of the theory of electromagnetic waves. Also known as electromagnetic field equations; Maxwell equations. { 'mak,swel 'fēld i,kwā·zhənz }

Mbit *See* megabit. { 'em,bit }

Mbyte *See* megabyte. { 'em,bīt }

McCabe's cyclomatic number [COMPUT SCI] The total number of decision statements in a computer program plus one; a measure of the complexity of the program. { mə,kābz ,sī·klə ,mad·ik 'nəm·bər }

M-derived filter [ELECTR] A filter consisting of a series of T or pi sections whose impedances are matched at all frequencies, even though the sections may have different resonant frequencies. { 'em di,rīvd 'fil·tər }

M-display [ELECTR] A radar display format in which a trace-deflecting pulse can be moved along the range axis of an A-display to assist the operator in determining and reporting the range of a target. Also known as M-indicator; M-scan; M-scope. { 'em di,splā }

MDS *See* minimum discernible signal.

meaconing [ELECTROMAG] A system for receiving electromagnetic signals and rebroadcasting them with the same frequency so as, for instance, to confuse navigation; a confusion reflector, such as chaff, is an example. { 'mē·kə·niŋ }

Mealy machine [COMPUT SCI] A sequential machine in which the output depends on both the current state of the machine and the input. { 'mē·lē mə,shēn }

mean carrier frequency [ELECTR] Average carrier frequency of a transmitter corresponding to the resting frequency in a frequency-modulated system. { 'mēn 'kar·ē·ər ,frē·kwən·sē }

mean power [ELECTR] For a radio transmitter, the power supplied to the antenna transmission line by a transmitter during normal operation, averaged over a time sufficiently long compared with the period of the lowest frequency encountered in the modulation; a time of 1/10 second during which the mean power is greatest will be selected normally. { 'mēn 'pau̇·ər }

means-ends analysis [COMPUT SCI] A method of problem solving in which the difference between the form of the data in the present and desired situations is determined, and an operator is then found to transform from one into the other, or, if this is not possible, objects between the present and desired objects are created, and the same procedure is then repeated on each of the gaps between them. { 'mēnz 'enz ə,nal·ə·səs }

mean-square-error criterion [CONT SYS] Evaluation of the performance of a control system by calculating the square root of the average over time of the square of the difference between the actual output and the output that is desired. { 'mēn 'skwer 'er·ər krī,tir·ē·ən }

mean time between failures [COMPUT SCI] A measure of the reliability of a computer system, equal to average operating time of equipment between failures, as calculated on a statistical basis from the known failure rates of various components of the system. Abbreviated MTBF. { 'mēn 'tīm bi,twēn 'fāl·yərz }

measurand transmitter [COMMUN] A telemetry transmitter that transmits a signal modulated according to the values of the quantity being measured. { 'mezh·ə,rand tranz,mid·ər }

measured service [COMMUN] Telephone service for which charge is made according to the measured amount of usage. { 'mezh·ərd 'sər·vəs }

mechanical bearing cursor *See* bearing cursor. { mi'kan·ə·kəl 'ber·iŋ ,kər·sər }

mechanical filter [ELECTR] Filter, used in intermediate-frequency amplifiers of highly selective superheterodyne receivers, consisting of shaped metal bars, rods, or disks that act as coupled mechanical resonators when used with piezoelectric or magnetostrictive input and output transducers and coupled by small-diameter wires. Also known as mechanical wave filter. { mi'kan·ə·kəl 'fil·tər }

mechanical jamming *See* passive jamming. { mi'kan·ə·kəl 'jam·iŋ }

mechanical replacement [COMPUT SCI] The replacement of one piece of hardware by another piece of hardware at the instigation of the manufacturer. { mi'kan·ə·kəl ri'plās·mənt }

mechanical scanner [COMPUT SCI] In optical character recognition, a device that projects an input character into a rotating disk, on the periphery of which is a series of small, uniformly spaced apertures; as the disk rotates, a photocell collects the light passing through the apertures. { mi'kan·ə·kəl 'skan·ər }

mechanical translation [COMPUT SCI] Automatic translation of one language into another by means of a computer or other machine that contains a dictionary look-up in its memory, along with the programs needed to make logical choices from synonyms, supply missing words, and rearrange word order as required for the new language. Also known as machine translation. { mi'kan·ə·kəl tranz'lā·shən }

mechanical wave filter *See* mechanical filter. { mi'kan·ə·kəl 'wāv ,fil·tər }

mechanized *See* machine-sensible. { 'mek·ə ˌnīzd }

media conversion [COMPUT SCI] The transfer of data from one storage type (such as magnetic tape) to another storage type (such as magnetic or optical disk). { 'mē·dē·ə kənˌvər·zhən }

media conversion buffer [COMPUT SCI] Large storage area, such as a drum, on which data may be stored at low speed during nonexecution time, to be later transferred at high speed into core memory during execution time. { 'mē·dē·ə kən ˌvər·zhən ˌbəf·ər }

medical frequency bands [COMMUN] A collection of radio frequency bands allocated to medical equipment in the United States. { 'med·ə·kəl 'frē·kwən·sē ˌbanz }

medium [COMPUT SCI] The material, or configuration thereof, on which data are recorded; usually applied to storable, removable media, such as disks and magnetic tape. { 'mē·dē·əm }

medium-altitude earth orbit [AERO ENG] An artificial-satellite orbit whose altitude is between about 1000 and 8000 miles (1600 and 12,900 kilometers) above the earth's surface. Abbreviated MEO. { ˌmēd·ē·əm ¦al·təˌtüd 'ərth ˌȯr·bət }

medium frequency [COMMUN] A Federal Communications Commission designation for the band from 300 to 3000 kilohertz in the radio spectrum. Abbreviated MF. { 'mē·dē·əm 'frē·kwən·sē }

medium-frequency propagation [COMMUN] Radio propagation at broadcast frequencies where skip is not an important factor. { 'mē·dē·əm ¦frē·kwən·sē ˌpräp·ə'gā·shən }

megabit [COMPUT SCI] A unit of information content equal to 1,048,576 (1024 × 1024) bits. Abbreviated Mbit. { 'meg·əˌbit }

megabyte [COMPUT SCI] A unit of information content equal to 1,048,576 (1024 × 1024) bytes. Abbreviated Mbyte. Symbolized M. { 'meg·ə ˌbīt }

megaflops [COMPUT SCI] A unit of computer speed, equal to 10^6 flops. { 'meg·əˌfläps }

megapel display [COMPUT SCI] A computer graphics display that handles 10^6 or more pixels (pels). { 'meg·əˌpel diˌsplā }

membrane keyboard [COMPUT SCI] A flat keyboard, used with microcomputers and hand-held calculators, that consists of two closely spaced membranes separated by a flat sheet called a spacer with holes corresponding to the keys. { 'memˌbrān 'kēˌbȯrd }

memex [COMPUT SCI] A hypothetical machine described by Vannevar Bush, which would store written records so that they would be available almost instantly by merely pushing the right button for the information desired. { 'meˌmeks }

memory [COMPUT SCI] Any apparatus in which data may be stored and from which the same data may be retrieved; especially, the internal, high-speed, large-capacity working storage of a computer, as opposed to external devices. Also known as computer memory. { 'mem·rē }

memory address register [COMPUT SCI] A special register containing the address of a word currently required. { 'mem·rē 'adˌres ˌrej·ə·stər }

memory bank [COMPUT SCI] A physical section of a computer memory, which may be designed to handle information transfers independently of other such transfers in other such sections. { 'mem·rē ˌbaŋk }

memory buffer register [COMPUT SCI] A special register in which a word is stored as it is read from memory or just prior to being written into memory. { 'mem·rē 'bəf·ər ˌrej·ə·stər }

memory capacity *See* storage capacity. { 'mem·rē kə'pas·əd·ē }

memory card [COMPUT SCI] A small card, typically with dimensions of about 2 × 3 inches (5 × 8 centimeters), that can store information, usually in integrated circuits or magnetic strips. { 'mem·rē ˌkärd }

memory cell [COMPUT SCI] A single storage element of a memory, together with associated circuits for storing and reading out one bit of information. { 'mem·rē ˌsel }

memory chip *See* semiconductor memory. { 'mem·rē ˌchip }

memory contention [COMPUT SCI] A situation in which two different programs, or two parts of a program, try to read items in the same block of memory at the same time. { 'mem·rē kən'ten·chən }

memory cycle *See* cycle time. { 'mem·rē ˌsī·kəl }

memory dump *See* storage dump. { 'mem·rē ˌdəmp }

memory dump routine [COMPUT SCI] A debugging routine which produces a listing of a consecutive section of memory, either numbers or instructions, at selected points in a program. { 'mem·rē ¦dəmp rüˌtēn }

memory element [COMPUT SCI] Any component part of core memory. { 'mem·rē ˌel·ə·mənt }

memory expansion card [COMPUT SCI] A printed circuit board that contains additional storage and can be plugged into a computer to increase its storage capacity. { 'mem·rē ik'span·chən ˌkärd }

memory fill *See* storage fill. { 'mem·rē ˌfil }

memory gap [COMPUT SCI] A gulf in access time, capacity, and cost of computer storage technologies between fast, expensive, main-storage devices and slow, high-capacity, inexpensive secondary-storage devices. Also known as access gap. { 'mem·rē ˌgap }

memory guard [COMPUT SCI] Built-in safety devices which prevent a program or a programmer from accessing certain memory areas reserved for the central processor. Also known as memory protect. { 'mem·rē ˌgärd }

memory hierarchy [COMPUT SCI] A ranking of computer memory devices, with devices having the fastest access time at the top of the hierarchy, and devices with slower access times but larger capacity and lower cost at lower levels. { 'mem·rē 'hī·ərˌär·kē }

memory lockout register [COMPUT SCI] A special register containing the limiting addresses of an area in memory which may not be accessed by the program. { 'mem·rē 'läk,au̇t ,rej·ə·stər }

memory management [COMPUT SCI] **1.** The allocation of computer storage in a multiprogramming system so as to maximize processing efficiency. **2.** The collection of routines for placing, fetching, and removing pages or segments into or out of the main memory of a computer system. { 'mem·rē ,man·ij·mənt }

memory map [COMPUT SCI] The list of variables, constants, identifiers, and their memory locations when a FORTRAN program is being run. Also known as memory map list. { 'mem·rē ,map }

memory map list *See* memory map. { 'mem·rē ,map ,list }

memory mapping [COMPUT SCI] The method by which a computer translates between its logical address space and its physical address space. { 'mem·rē ,map·iŋ }

memory overlay [COMPUT SCI] The efficient use of memory space by allowing for repeated use of the same areas of internal storage during the different stages of a program; for instance, when a subroutine is no longer required, another routine can replace all or part of it. { 'mem·rē 'ō·vər,lā }

memory port [COMPUT SCI] A logical connection through which data are transferred in or out of main memory under control of the central processing unit. { 'mem·rē ,pȯrt }

memory power [COMPUT SCI] A relative characteristic pertaining to differences in access time speeds in different parts of memory; for instance, access time from the buffer may be a tenth of the access time from core. { 'mem·rē ,pau̇·ər }

memory print *See* storage dump. { 'mem·rē ,print }

memory printout [COMPUT SCI] A listing of the contents of memory. { 'mem·rē 'print,au̇t }

memory protect *See* memory guard. { 'mem·rē prə,tekt }

memory protection *See* storage protection. { 'mem·rē prə'tek·shən }

memory-reference instruction [COMPUT SCI] A type of instruction usually requiring two machine cycles, one to fetch the instruction, the other to fetch the data at an address (part of the instruction itself) and to execute the instruction. { 'mem·rē ¦ref·rəns in,strək·shən }

memory register *See* storage register. { 'mem·rē ,rej·ə·stər }

memory search routine [COMPUT SCI] A debugging routine which has as an essential feature the scanning of memory in order to locate specified instructions. { 'mem·rē 'sərch rü,tēn }

memory-segmentation control [COMPUT SCI] Address-computing logic to address words in memory with dynamic allocation and protection of memory segments assigned to different users. { 'mem·rē ,seg·mən'tā·shən kən,trōl }

memory sniffer [COMPUT SCI] A diagnostic routine that continually tests the computer memory while the machine is in operation. { 'mem·rē ,snif·ər }

memory storage [COMPUT SCI] The sum total of the computer's storage facilities, that is, core, drum, disk, cards, and paper tape. { 'mem·rē ,stȯr·ij }

memory typewriter *See* electronic typewriter. { 'mem·rē ,tīp·rīd·ər }

MEMS *See* micro-electro-mechanical system. { memz *or* ¦em¦ē¦em'es }

menu [COMPUT SCI] A list of computer functions appearing on a video display terminal which indicates the possible operations that a computer can perform next, only one of which can be selected by the operator. { 'men·yü }

menu bar [COMPUT SCI] **1.** In a graphical user interface, a horizontal strip near the top of the screen or a window, containing the titles of available pull-down menus. **2.** A horizontal or vertical strip containing the names of currently available commands. { 'men·yü ,bär }

menu-driven system [COMPUT SCI] An interactive computer system in which the operator requests the processing to be performed by making selections from a series of menus. { 'men·yü ¦driv·ən 'sis·təm }

MEO *See* medium-altitude earth orbit.

merge [COMPUT SCI] To create an ordered set of data by combining properly the contents of two or more sets of data, each originally ordered in the same manner as the output data set. Also known as mesh. { mərj }

merged-transistor logic *See* integrated injection logic. { ¦mərjd tran¦zis·tər 'läj·ik }

merge search [COMPUT SCI] A procedure for searching a table in which both the table and file records must first be ordered in the same sequence on the key involved, and the table is searched sequentially until a table-record key equal to or greater than the file-record key is found, upon which the file record is processed if its key is equal, and the process is repeated with the next file record, starting at the table position where the previous search terminated. { 'mərj ,sərch }

merge sort [COMPUT SCI] To produce a single sequence of items ordered according to some rule, from two or more previously ordered or unordered sequences, without changing the items in size, structure, or total number; although more than one pass may be required for a complete sort, items are selected during each pass on the basis of the entire key. { 'mərj ,sȯrt }

merging routine [COMPUT SCI] A program that creates a single sequence of items, ordered according to some rule, out of two or more sequences of items, each sequence ordered according to the same rule. { 'mərj·iŋ rü,tēn }

MESFET *See* metal semiconductor field-effect transistor. { 'mes,fet }

mesh *See* merge. { mesh }

mesh network [COMMUN] A communications network in which each node has at least two links to other nodes. { 'mesh ,net,wərk }

message [COMMUN] A series of words or symbols, transmitted with the intention of conveying information. [COMPUT SCI] An arbitrary amount of information with beginning and end defined or implied: usually, it originates in one place and is intended to be transmitted to another place. { 'mes·ij }

message accounting [COMMUN] Use of equipment to make records of telephone calls for billing purposes. { 'mes·ij ə,kaůnt·iŋ }

message authentication [COMMUN] Security measure designed to establish the authenticity of a message by means of an authenticator within the transmission derived from certain predetermined elements of the message itself. { 'mes·ij ȯ,then·tə'kā·shən }

message authentication code [COMPUT SCI] The encrypted personal identification code appended to the message transmitted to a computer; the message is accepted only if the decrypted code is recognized as valid by the computer. Abbreviated MAC. { 'mes·ij ȯ,then·tə'kā·shən ,kōd }

message blocking [COMMUN] The division of messages into blocks having a fixed number of bytes in order to provide consistent work units and thereby simplify the design of data communications networks. { 'mes·ij ,bläk·iŋ }

message buffer [COMPUT SCI] One of a number of sections of computer memory, which contains a message that can be transmitted between tasks in the computer system to request service and receive replies from tasks, and which is stored in a system buffer area, outside the address spaces of tasks. { 'mes·ij ,bəf·ər }

message center [COMMUN] A communications facility charged with the responsibility for acceptance, preparation for transmission, transmission, receipt and delivery of messages. { 'mes·ij ,sen·tər }

message display console [COMPUT SCI] A cathode-ray tube on which is displayed information requested by the user. { 'mes·ij di'splā,kän,sōl }

message exchange [COMPUT SCI] A device which acts as a buffer between a communication line and a computer and carries out communication functions. { 'mes·ij iks,chānj }

message indicator [COMMUN] Element placed within a message to serve as a guide to the selection or derivation and application of the correct key to facilitate the prompt decryption of the message. { 'mes·ij ,in·də'kād·ər }

message interpolation [COMMUN] Data message insertion during intersyllable periods or speech pauses on a busy voice channel without breaking down the voice connection or noticeably affecting the voice transmission. { 'mes·ij in ,tər·pə'lā·shən }

message keying element [COMMUN] That part of the key which changes with every message. { 'mes·ij 'kē·iŋ ,el·ə·mənt }

message-oriented applications [COMMUN] Applications of data communications that involve medium-size data transfers in the range of hundreds to a few thousand bytes or characters, and are usually unidirectional information flows from source to destination. { 'mes·ij ¦ȯr·ē,ent·əd ,ap·lə'kā·shənz }

message queuing [COMPUT SCI] The stacking of messages according to some priority rule as the messages await processing. { 'mes·ij ,kyü·iŋ }

message reference block [COMMUN] A set of signals denoting the beginning or end of a message. { 'mes·ij 'ref·rəns ,bläk }

message registration [COMMUN] A method for counting the number of completed charged calls which originate from a particular telephone line, making one scoring for each local call and more than one scoring for calls between zones. { 'mes·ij ,rej·ə,strā·shən }

message routing [COMMUN] Selection of the communication path over which a message is sent. { 'mes·ij ,rüd·iŋ }

message switching [COMMUN] A system in which data transmitted between stations on different circuits within a network are routed through central points. { 'mes·ij ,swich·iŋ }

message trailer [COMMUN] The last part of a data communications message that signals the end of the message and may also contain control information such as a check character. { 'mes·ij ,trā·lər }

messaging [COMMUN] Electronic communication in which a message is sent directly to its destination without being stored en route. { 'mes·ij·iŋ }

meta character [COMPUT SCI] A character in a computer programming language system that has some controlling role with respect to other characters with which it may be associated. { 'med·ə ,kar·ik·tər }

metacompiler [COMPUT SCI] A compiler that is used chiefly to construct compilers for other programming languages. { 'med·ə·kəm,pī·lər }

metadata [COMPUT SCI] A description of the data in a source, distinct from the actual data; for example, the currency by which prices are measured in a data source for purchasing goods. { 'med·ə,dad·ə }

metalanguage [COMPUT SCI] A programming language that uses symbols to represent the syntax of other programming languages, and is used chiefly to write compilers for those languages. { 'med·ə,laŋ·gwij }

metal antenna [ELECTROMAG] An antenna which has a relatively small metal surface, in contrast to a slot antenna. { 'med·əl an'ten·ə }

metallic insulator [ELECTROMAG] Section of transmission line used as a mechanical support device; the section is an odd number of quarter-wavelengths long at the frequency of interest, and the input impedance becomes high enough so that the section effectively acts as an insulator. { mə'tal·ik 'in·sə,lād·ər }

metal oxide semiconductor field-effect transistor [ELECTR] A field-effect transistor having a gate that is insulated from the semiconductor substrate by a thin layer of silicon dioxide. Abbreviated MOSFET; MOST; MOS transistor. Formerly known as insulated-gate field-effect transistor (IGFET). { 'med·əl ¦äk,sīd 'sem·i·kən,dək·tər 'fēld i,fekt tran'zis·tər }

metal oxide semiconductor integrated circuit [ELECTR] An integrated circuit using metal oxide semiconductor transistors; it can have a higher density of equivalent parts than a bipolar integrated circuit. { 'med·əl ¦äk,sīd 'sem·i·kən ,dək·tər 'int·ə,grād·əd 'sər·kət }

metal semiconductor field-effect transistor [ELECTR] A field-effect transistor that uses a thin film of gallium arsenide, with a Schottky barrier gate formed by depositing a layer of metal directly onto the surface of the film. Abbreviated MESFET. { 'med·əl 'sem·i·kən,dək·tər 'fēld i,fekt tran'zis·tər }

metavariable [COMPUT SCI] One of the elements of a formal language, corresponding to the parts of speech of a natural language. Also known as component name; phrase name. { ¦med·ə'ver·ē·ə·bəl }

meteoric scatter [COMMUN] A form of scatter propagation in which meteor trails serve to scatter radio waves back to earth. { ,mēd·ē'ór·ik 'skad·ər }

meteorological frequency bands [COMMUN] A collection of radio and microwave frequency bands allocated for use by radiosondes and ground-based radars used in weather forecasting in the United States. { ,med·ē·ə·rə'läj·ə·kəl 'frē·kwən·sē ,banz }

metric waves [ELECTROMAG] Radio waves having wavelengths between 1 and 10 meters, corresponding to frequencies between 30 and 300 megahertz (the very-high-frequency band). { 'me·trik 'wāvz }

MF *See* medium frequency.

Mflop *See* million floating-point operations per second. { 'em,fläp }

MFSK *See* multiple-frequency-shift keying.

MIC *See* microwave integrated circuit.

mickey-mouse [COMPUT SCI] To play with something new, such as hardware, software, or a system, until a feel is gotten for it and the proper operating procedure is discovered, understood, and mastered. { ¦mik·ē 'mau̇s }

MICR *See* magnetic-ink character recognition.

microbit [COMPUT SCI] A unit of information equal to one-millionth of a bit. { 'mī·krə,bit }

microchip *See* chip. { 'mī·krō,chip }

microcircuitry [ELECTR] Electronic circuit structures that are orders of magnitude smaller and lighter than circuit structures produced by the most compact combinations of discrete components. Also known as microelectronic circuitry; microminiature circuitry. { ¦mī·krō'sər·kə·trē }

microcode [COMPUT SCI] A code that employs microinstructions; not ordinarily used in programming. { 'mī·krō,kōd }

microcomputer [COMPUT SCI] **1.** A digital computer whose central processing unit resides on a single semiconductor integrated circuit chip, a microprocessor. **2.** An electronic device, typically consisting of a microprocessor central processing unit, semiconductor memory (RAM), graphics display, and keyboard. Typical configurations also include a hard disk for persistent memory, a compact disk drive, a disk drive which allows removable disks to be used to move data in and out of the machine, and a pointing device. { ¦mī·krō·kəm'pyüd·ər }

microcomputer development system [COMPUT SCI] A complete microcomputer system that is used to test both the software and hardware of other microcomputer-based systems. { ¦mī·krō·kəm'pyüd·ər di'vel·əp·mənt ,sis·təm }

microcontroller [ELECTR] A microcomputer, microprocessor, or other equipment used for precise process control in data handling, communication, and manufacturing. { ¦mī·krō·kən 'trōl·ər }

microdiagnostic program [COMPUT SCI] A microprogram that tests a specific hardware component, such as a bus or store location, for faults. { ¦mī·krō,dī·əg'näs·tik 'prō·grəm }

microdisk [COMPUT SCI] A small floppy disk with a diameter between 3 and 4 inches (7 and 10 centimeters). Also known as microfloppy disk. { 'mī·krō,disk }

micro-electro-mechanical system [ENG] A system in which micromechanisms are coupled with microelectronics, most commonly fabricated as microsensors or microactuators. Abbreviated MEMS. Also known as microsystem. { ¦mī·krō i,lek·trə·mə'kan·ə·kəl ,sis·təm }

microelectronic circuitry *See* microcircuitry. { ¦mī·krō·i,lek'trän·ik 'sər·kə·trē }

microelectronics [ELECTR] The technology of constructing circuits and devices in extremely small packages by various techniques. Also known as microminiaturization; microsystem electronics. { ¦mī·krō·i,lek'trän·iks }

microfloppy disk *See* microdisk. { ¦mī·krō,fläp·ē ,disk }

microimage [COMPUT SCI] A single image stored on a microform medium. { 'mī·krō ,im·ij }

microinstruction [COMPUT SCI] The portion of a microprogram that specifies the operation of individual computing elements and such related subunits as the main memory and the input/output interfaces; usually includes a next-address field that eliminates the need for a program counter. { ¦mī·krō·in'strək·shən }

micromainframe [COMPUT SCI] A main frame of a computer placed on one or more integrated circuit chips. { ¦mī·krō'mān,frām }

micromechanical display [ENG] A video display based on an array of mirrors on a silicon chip that can be deflected by electrostatic forces. Abbreviated MMD. { ,mī·krō·mə,kan·i·kəl di'splā }

microminiature circuitry *See* microcircuitry. { ¦mī·krō¦min·ə·char 'sər·kə·trē }

microminiaturization *See* microelectronics. { ¦mī·krō,min·ə·chə·rə'zā·shən }

microoperation [COMPUT SCI] Any clock-timed step of an operation. { ¦mī·krō,äp·ə'rā·shən }

micro-opto-electro-mechanical system [ENG] A microsystem that combines the functions of optical, mechanical, and electronic components in a single, very small package or assembly. Abbreviated MOEMS. { ¦mī·kro¦äp·tōi¦lek·trōmə ¦kan·ə·kəl 'sis·təm }

micro-opto-mechanical system [ENG] A microsystem that combines optical and mechanical functions without the use of electronic devices or signals. Abbreviated MOMS. { ¦mī·krō ¦op·to·mə'kan·ə·kəl ˌsis·təm }

microperf [COMPUT SCI] A type of continuous-feed computer paper having extremely small perforations along the separations and edges which give separated pages the appearance of standard typewriter paper. { 'mī·krəˌpərf }

microprocessing unit [ELECTR] A microprocessor with its external memory, input/output interface devices, and buffer, clock, and driver circuits. Abbreviated MPU. { ¦mī·krō'präˌses·iŋ ˌyü·nət }

microprocessor [ELECTR] A single silicon chip on which the arithmetic and logic functions of a computer are placed. { ¦mī·krō'präˌses·ər }

microprocessor intertie and communication system [COMMUN] A data communications system which provides the communication network with its own dedicated processing resources and reduces in-terminal response time, compensating for the capacity used up by communications terminals. Abbreviated MICS. { mī·krō'prä ˌses·ər 'in·tərˌtī ən kəˌmyü·nə'kā·shən ˌsis·təm }

microprogram [COMPUT SCI] A computer program that consists only of basic elemental commands which directly control the operation of each functional element in a microprocessor. { ¦mī·krō'prō·grəm }

microprogrammable instruction [COMPUT SCI] An instruction that does not refer to a core memory address and that can be microprogrammed, thus specifying various commands within one instruction. { ¦mī·krō·prə'gram·ə·bəl in'strək·shən }

microprogramming [COMPUT SCI] Transformation of a computer instruction into a sequence of elementary steps (microinstructions) by which the computer hardware carries out the instruction. { ¦mī·krō'prōˌgram·iŋ }

microspec function [COMPUT SCI] The set of microinstructions which performs a specific operation in one or more machine cycles. { 'mī·krə ˌspek ˌfəŋk·shən }

microstrip [ELECTROMAG] A strip transmission line that consists basically of a thin-film strip in intimate contact with one side of a flat dielectric substrate, with a similar thin-film ground-plane conductor on the other side of the substrate. { 'mī·krəˌstrip }

microsystem *See* micro-electro-mechanical system. { 'mī·krōˌsis·təm }

microsystem electronics *See* microelectronics. { 'mī·krəˌsis·təm iˌlek'trän·iks }

microwave [ELECTROMAG] An electromagnetic wave which has a wavelength between about 0.3 and 30 centimeters, corresponding to frequencies of 1–100 gigahertz; however, there are no sharp boundaries distinguishing microwaves from infrared and radio waves. { 'mī·krəˌwāv }

microwave amplification by stimulated emission of radiation *See* maser. { 'mī·krəˌwāv ˌam·plə·fə'kā·shən bī 'stim·yəˌlād·əd i'mish·ən əv ˌrād·ē'ā·shən }

microwave amplifier [ELECTR] A device which increases the power of microwave radiation. { 'mī·krəˌwāv 'am·pləˌfī·ər }

microwave antenna [ELECTROMAG] A combination of an open-end waveguide and a parabolic reflector or horn, used for receiving and transmitting microwave signal beams at microwave repeater stations. { 'mī·krəˌwāv an'ten·ə }

microwave attenuator [ELECTROMAG] A device that causes the field intensity of microwaves in a waveguide to decrease by absorbing part of the incident power; usually consists of a piece of lossy material in the waveguide along the direction of the electric field vector. { 'mī·krə ˌwāv ə'ten·yəˌwād·ər }

microwave cavity *See* cavity resonator. { 'mī·krəˌwāv ˌkav·ə·dē }

microwave circuit [ELECTROMAG] Any particular grouping of physical elements, including waveguides, attenuators, phase changers, detectors, wavemeters, and various types of junctions, which are arranged or connected together to produce certain desired effects on the behavior of microwaves. { 'mī·krəˌwāv ˌsər·kət }

microwave circulator *See* circulator. { 'mī·krə ˌwāv 'sər·kyəˌlād·ər }

microwave communication [COMMUN] Transmission of messages using highly directional microwave beams, which are generally relayed by a series of microwave repeaters spaced up to 50 miles (80 kilometers) apart. { 'mī·krəˌwāv kə ˌmyü·nə'kā·shən }

microwave device [ELECTR] Any device capable of generating, amplifying, modifying, detecting, or measuring microwaves, or voltages having microwave frequencies. { 'mī·krəˌwāv diˌvīs }

microwave early warning [ENG] High-power, long-range radar with a number of indicators, giving high resolution, and with a large traffic-handling capacity; used for early warning of missiles. { 'mī·krəˌwāv ¦ər·lē 'wȯr·niŋ }

microwave filter [ELECTROMAG] A device which passes microwaves of certain frequencies in a transmission line or waveguide while rejecting or absorbing other frequencies; consists of resonant cavity sections or other elements. { 'mī·krəˌwāv ˌfil·tər }

microwave generator *See* microwave oscillator. { 'mī·krəˌwāv 'jen·əˌrād·ər }

microwave gyrator *See* gyrator. { 'mī·krəˌwāv 'jīˌrād·ər }

microwave hop [COMMUN] A microwave communications channel between two stations with directive antennas that are aimed at each other. { 'mī·krəˌwāv 'häp }

microwave integrated circuit [ELECTR] A microwave circuit that uses integrated-circuit production techniques involving such features as thin or thick films, substrates, dielectrics, conductors, resistors, and microstrip lines, to build passive assemblies on a dielectric. Abbreviated MIC. { ˈmī·krəˌwāv ˈint·əˌgrād·əd ˈsər·kət }

microwave landing system [NAV] A system of ground equipment which generates guidance beams at microwave frequencies for guiding aircraft to landings; it is intended to replace the present lower-frequency instrument landing system. Abbreviated MLS. { ˈmī·krəˌwāv ˈland·iŋ ˌsis·təm }

microwave link *See* microwave repeater. { ˈmī·krəˌwāv ˌliŋk }

microwave network [COMMUN] A series of microwave repeaters, spaced up to 50 miles (80 kilometers) apart, which relay messages over long distances using highly directional microwave beams. { ˈmī·krəˌwāv ˈnetˌwərk }

microwave oscillator [ELECTR] A type of electron tube or semiconductor device used for generating microwave radiation or voltage waveforms with microwave frequencies. Also known as microwave generator. { ˈmī·krəˌwāv ˈäs·əˌlād·ər }

microwave receiver [ELECTR] Complete equipment that is needed to convert modulated microwaves into useful information. { ˈmī·krə ˌwāv riˈsē·vər }

microwave relay *See* microwave repeater. { ˈmī·krəˌwāv ˈrēˌlā }

microwave repeater [COMMUN] A tower equipped with a receiver and transmitter for picking up, amplifying, and passing on in either direction the signals sent over a microwave network by highly directional microwave beams. Also known as microwave link; microwave relay. { ˈmī·krəˌwāv riˈpēd·ər }

microwave resonance cavity *See* cavity resonator. { ˈmī·krəˌwāv ˈrez·ən·əns ˌkav·əd·ē }

microwave solid-state device [ELECTR] A semiconductor device for the generation or amplification of electromagnetic energy at microwave frequencies. { ˈmī·krəˌwāv ¦säl·əd ¦stāt diˈvīs }

microwave transmission line [ELECTROMAG] A material structure forming a continuous path from one place to another and capable of directing the transmission of electromagnetic energy along this path. { ˈmī·krəˌwāv tranzˈmish·ən ˌlīn }

microwave waveguide *See* waveguide. { ˈmī·krəˌwāv ˈwāvˌgīd }

MICS *See* microprocessor intertie and communication system.

middle core [COMPUT SCI] The locations with medium addresses in a computer's main storage; usually assigned to workspace for application programs. { ˈmid·əl ˈkȯr }

MIDI *See* musical instrument digital interface. { ˈmid·ē }

midicomputer [COMPUT SCI] A computer having greater performance and capacity than a minicomputer and less than that of a mainframe. { ˈmid·ē·kəmˌpyüd·ər }

mid-infrared radiation *See* intermediate-infrared radiation. { ˌmid¦in·frəˌred ˌrad·ēˈā·shən }

mid-square generator [COMPUT SCI] A procedure for generating a sequence of random numbers, in which a member of a sequence is squared and the middle digits of the resulting number form the next member of the sequence. { ˈmid¦skwər ˈjen·əˌrād·ər }

migration [COMPUT SCI] Movement of frequently used data items to more accessible storage locations, and of infrequently used data items to less accessible locations. { mīˈgrā·shən }

Miller code [COMPUT SCI] A code used internally in some computers, in which a binary 1 is represented by a transition in the middle of a bit (either up or down), and a binary 0 is represented by no transition following a binary 1; a transition between bits represents successive 0's; in this code, the longest period possible without a transition is two bit times. { ˈmil·ər ˌkōd }

millimeter wave [ELECTROMAG] An electromagnetic wave having a wavelength between 1 millimeter and 1 centimeter, corresponding to frequencies between 30 and 300 gigahertz. Also known as millimetric wave. { ˈmil·əˌmēd·ər ˈwāv }

millimetric wave *See* millimeter wave. { ¦mil·ə ¦me·trik ˈwāv }

million floating-point operations per second [COMPUT SCI] A unit used to measure the processing speed or throughput of supercomputers or array processors. Abbreviated Mflop. { ˈmil·yən ¦flōd·iŋ ¦pȯint ˌäp·əˈrā·shənz pər ˈsek·ənd }

million instructions per second [COMPUT SCI] A unit used to measure the speed at which a computer's central processing unit can process instructions. Abbreviated MIPS. { ˈmil·yən inˈstrək·shənz pər ˈsek·ənd }

Mills cross [ELECTROMAG] An antenna array that consists of two antennas oriented perpendicular to each other and that produces a narrow pencil beam. { ˈmilz ˈkrȯs }

MIMD [COMPUT SCI] A type of multiprocessor architecture in which several instruction cycles may be active at any given time, each independently fetching instructions and operands into multiple processing units and operating on them in a concurrent fashion. Acronym for multiple-instruction-stream, multiple-data-stream.

MIME [COMPUT SCI] The Multimedia Internet Mail Enhancements standard, describing a way of encoding binary files, such as pictures, videos, sounds, and executable files, within a normal text message in an operating-system-independent manner. { mīm }

M-indicator *See* M-display. { ˈem ˌin·dəˌkād·ər }

minicartridge [COMPUT SCI] A self-contained package of reel-to-reel magnetic tape that resembles a cassette or cartridge but is slightly different in design and dimensions. { ˈmin·ēˌkär·trij }

minicomputer [COMPUT SCI] A relatively small general-purpose digital computer, intermediate in size between a microcomputer and a main frame. { 'min·ē·kəm,pyüd·ər }

minidisk [COMPUT SCI] A floppy disk that has a diameter of 5.25 inches (approximately 13 centimeters). Also known as minifloppy disk. { 'min·ē,disk }

minifloppy disk *See* minidisk. { 'min·ē,fläp·ē ,disk }

minimal-latency coding *See* minimum-access coding. { 'min·ə·məl ¦lāt·ən·sē ,kōd·iŋ }

minimal realization [CONT SYS] In linear system theory, a set of differential equations, of the smallest possible dimension, which have an input/output transfer function matrix equal to a given matrix function G(*s*). { 'min·ə·məl ,rē·ə·lə'zā·shən }

mini-maxi regret [CONT SYS] In decision theory, a criterion which selects that strategy which has the smallest maximum difference between its payoff and that of the best hindsight choice. { ¦min·ē ¦mak·sē ri'gret }

minimize [COMMUN] Condition when normal message and telephone traffic is drastically reduced so messages connected with an actual or simulated emergency will not be delayed. [COMPUT SCI] In a graphical user interface environment, to reduce a window to an icon that represents the application running in the window. { 'min·ə,mīz }

minimum-access coding [COMPUT SCI] Coding in such a way that a minimum time is required to transfer words to and from storage, for a computer in which this time depends on the location in storage. Also known as minimal-latency coding; minimum-delay coding; minimum-latency coding. { 'min·ə·məm ¦ak,ses ,kōd·iŋ }

minimum-access programming [COMPUT SCI] The programming of a digital computer in such a way that minimum waiting time is required to obtain information out of the memory. Also known as forced programming; minimum-latency programming. { 'min·ə·məm ¦ak,ses 'prō,gram·iŋ }

minimum-access routine *See* minimum-latency routine. { 'min·ə·məm ¦ak,ses rü,tēn }

minimum configuration [COMPUT SCI] **1.** A computer system that has only essential hardware components. **2.** The smallest assortment of hardware and software components required to carry out a particular data-processing function. { 'min·ə·məm kən,fig·yə'rā·shən }

minimum-delay coding *See* minimum-access coding. { 'min·ə·məm di,lā'kōd·iŋ }

minimum detectable signal *See* threshold signal. { 'min·ə·məm di'tek·tə·bəl 'sig·nəl }

minimum discernible signal [ELECTR] **1.** Receiver input power level that is just sufficient to produce a discernible signal in the receiver output; a receiver sensitivity test. **2.** In radar, the minimum echo power level at the input to the receiver that results in an output that can be confidently seem relative to the noise, usually determined with test equipment in the field. Abbreviated MDS. { 'min·ə·məm di'sər·nə·bəl 'sig·nəl }

minimum-distance code [COMMUN] A binary code in which the signal distance does not fall below a specified minimum value. { 'min·ə·məm ¦dis·təns 'kōd }

minimum-latency coding *See* minimum-access coding. { 'min·ə·məm ¦lat·ən·sē ,kōd·iŋ }

minimum-latency programming *See* minimum-access programming. { 'min·ə·məm ¦lat·ən·sē 'prō,gram·iŋ }

minimum-latency routine [COMPUT SCI] A computer routine that is constructed so that the latency in serial-access storage is less than the random latency that would be expected if storage locations were chosen without regard for latency. Also known as minimum-access routine. { 'min·ə·məm ¦lat·ən·sē rü,tēn }

minimum-loss attenuator [ELECTR] A section linking two unequal resistive impedances which is designed to introduce the smallest attenuation possible. Also known as minimum-loss pad. { 'min·ə·məm ¦lȯs ə'ten·yə,wād·ər }

minimum-loss matching [ELECTR] Design of a network linking two resistive impedances so that it introduces a loss which is as small as possible. { 'min·ə·məm ¦lȯs ,mach·iŋ }

minimum-loss pad *See* minimum-loss attenuator. { 'min·ə·məm ¦lȯs ,pad }

minimum-phase system [CONT SYS] A linear system for which the poles and zeros of the transfer function all have negative or zero real parts. { 'min·ə·məm 'fāz ,sis·təm }

minimum signal level [ELECTR] In facsimile, level corresponding to the copy white or copy black signal, whichever is the lower. { 'min·ə·məm 'sig·nəl ,lev·əl }

mini-supercomputer [COMPUT SCI] A supercomputer that is about a quarter to a half as fast in vector processing as the most powerful supercomputers. { ,min·ē'sü·pər·kəm,püd·ər }

minor bend [ELECTROMAG] Rectangular waveguide bent so that throughout the length of a bend a longitudinal axis of the guide lies in one plane which is parallel to the narrow side of the waveguide. { 'mīn·ər 'bend }

minor control data [COMPUT SCI] Control data which are at the least significant level used, or which are used to sort records into the smallest groups used; for example, if control data are used to specify state, town, and street, then the data specifying street would be minor control data. { 'mīn·ər kən'trōl ,dad·ə }

minor cycle [COMPUT SCI] The time required for the transmission or transfer of one machine word, including the space between words, in a digital computer using serial transmission. Also known as word time. { 'mīn·ər ,sī·kəl }

minority carrier [SOLID STATE] The type of carrier, electron, or hole that constitutes less than half the total number of carriers in a semiconductor. { mə'när·əd·ē 'kar·ē·ər }

minor key [COMPUT SCI] A secondary key for identifying a record. { 'mīn·ər ,kē }

minor lobe [ELECTROMAG] Any lobe except the major lobe of an antenna radiation pattern. Also known as secondary lobe; side lobe. { 'mīn·ər ¦lōb }

minor loop [CONT SYS] A portion of a feedback control system that consists of a continuous network containing both forward elements and feedback elements. { 'mīn·ər ¦lüp }

minus zone [COMPUT SCI] The bit positions in a computer code that represent the algebraic minus sign. { 'mī·nəs ˌzōn }

MIPS See million instructions per second.

mirage effect [COMMUN] Reception of radio waves at distances far beyond the normally expected range due to abnormal refraction caused by meteorological conditions such as abnormal vertical water-vapor and temperature gradients. { mə'räzh iˌfekt }

mirroring [COMPUT SCI] The recording of the same data in two or more locations to ensure continuous operation of a computer system in the event of a failure. { 'mir·ər·iŋ }

MIS See management information system.

mismatch slotted line [ELECTROMAG] A slotted line linking two waveguides which is not properly designed to minimize the power reflected or transmitted by it. { 'misˌmach 'släd·əd 'līn }

misregistration [COMPUT SCI] In character recognition, the improper state of appearance of a character, line, or document, on site in a character reader, with respect to a real or imaginary horizontal baseline. { ˌmisˌrej·ə'strā·shən }

missing error [COMPUT SCI] The result of calling for a subroutine not available in the library. { ˌmis·iŋ 'er·ər }

mistake [COMPUT SCI] A human action producing an unintended result, in contrast to an error in a computer operation. { mə'stāk }

misuse detection [COMPUT SCI] The technology that seeks to identify an attack on a computer system by its attempted effect on sensitive resources. { mis'yüs diˌtek·shən }

mixed congruential generator [COMPUT SCI] A congruential generator in which the constant b in the generating formula is not equal to zero. { 'mikst ˌkän·grü'en·chəl 'jen·əˌrād·ər }

mixed-entry decision table [COMPUT SCI] A decision table in which the action entries may be either sequenced or unsequenced. { 'mikst ¦en·trē di'sizh·ən ˌtā·bəl }

mixed highs [COMMUN] In analog color television, a method of reproducing very fine picture detail by transmitting high-frequency components as part of luminance signals for achromatic reproduction in color pictures. { 'mikst 'hīz }

mixed-mode expression [COMPUT SCI] An expression involving operands of more than one data type. { 'mikst ¦mōd ik'spresh·ən }

mixer [ELECTR] **1.** A device having two or more inputs, usually adjustable, and a common output; used to combine separate audio or video signals linearly in desired proportions to produce an output signal. **2.** The stage in a superheterodyne receiver in which the incoming modulated radio-frequency signal is combined with the signal of a local r-f oscillator to produce a modulated intermediate-frequency signal. Also known as first detector; heterodyne modulator; mixer-first detector. { 'mik·sər }

mixer-first detector See mixer. { 'mik·sər ¦fərst di'tek·tər }

MLS See microwave landing system.

MMD See micromechanical display.

MMSS See maritime mobile satellite service.

mnemonic code [COMPUT SCI] A programming code that is easy to remember because the codes resemble the original words, such as MPY for multiply and ACC for accumulator. { nə'män·ik 'kōd }

mobile code [COMPUT SCI] Code that can be transmitted across, and executed at the other end of, a network, and is capable of running on multiple platforms, for example, Java. { ˌmō·bəl 'kōd }

mobile digital computer [COMPUT SCI] Large, mobile, fixed-point operation, one-address, parallel-mode type digital computer. { 'mō·bəl ¦dij·əd·əl kəm'pyüd·ər }

mobile earth station [COMMUN] An earth station intended to be used while in motion or at halts at unspecified points. { ˌmō·bəl 'ərth ˌstā·shən }

mobile earth terminal [COMMUN] An antenna small enough to fit in a briefcase or suitcase, used for satellite communications, especially by news-service reporters at locations that cannot be accessed by conventional transportable satellite news-gathering terminals. Abbreviated MET. { ˌmō·bəl 'ərth ˌtər·mən·əl }

mobile radio [COMMUN] Radio communication in which the transmitter is installed in a vessel, vehicle, or airplane and can be operated while in motion. { 'mō·bəl 'rād·ē·ō }

mobile-relay station [COMMUN] Base station in which the base receiver automatically tunes on the base station transmitter and which retransmits all signals received by the base station receiver; used to extend the range of mobile units, and requires two frequencies for operation. { 'mō·bəl 'rēˌlā ˌstā·shən }

mobile satellite service [COMMUN] A radiocommunication service between mobile earth stations by means of one or more space stations. Abbreviated MSS. { ¦mō·bəl 'sad·əlˌīt ˌsər·vəs }

mobile service [COMMUN] A radiocommunication service between mobile and land stations or between mobile stations. { ¦mō·bəl 'sər·vəs }

mobile station [COMMUN] **1.** Station in the mobile service intended to be used while in motion or during halts at unspecified points. **2.** One or more transmitters that are capable of transmission while in motion. { 'mō·bəl 'stā·shən }

mobile systems equipment [COMPUT SCI] Computers located on planes, ships, or vans. { 'mō·bəl 'sis·təmz iˌkwip·mənt }

modal distortion See modal noise. { ¦mōd·əl di'stȯr·shən }

modal noise [COMMUN] Interference of a multimode optical communications fiber with a laser light source when a speckle pattern in the light intensity in the fiber alters because of motion of the fiber or changes in the laser spectrum. Also known as modal distortion. { ¦mōd·əl 'nȯiz }

mode [COMMUN] Form of the information in a communication such as literal language, digital data, and video. [COMPUT SCI] One of several alternative conditions or methods of operation of a device. [ELECTROMAG] A form of propagation of guided waves that is characterized by a particular field pattern in a plane transverse to the direction of propagation. Also known as transmission mode. { mōd }

mode filter [ELECTROMAG] A waveguide filter designed to separate waves of the same frequency but of different transmission modes. { 'mōd ˌfil·tər }

model *See* macroskeleton. { 'mäd·əl }

model-based expert system [COMPUT SCI] An expert system that is based on knowledge of the structure and function of the object for which the system is designed. { ¦mäd·əl ˌbāst 'ek·spərt ˌsis·təm }

model-following problem [CONT SYS] The problem of determining a control that causes the response of a given system to be as close as possible to the response of a model system, given the same input. { 'mäd·əl ¦fäl·ə·wiŋ ˌpräb·ləm }

model reduction [CONT SYS] The process of discarding certain modes of motion while retaining others in the model used by an active control system, in order that the control system can compute control commands with sufficient rapidity. { 'mäd·əl ri'dək·shən }

model reference system [CONT SYS] An ideal system whose response is agreed to be optimum; computer simulation in which both the model system and the actual system are subjected to the same stimulus is carried out, and parameters of the actual system are adjusted to minimize the difference in the outputs of the model and the actual system. { 'mäd·əl 'ref·rəns ˌsis·təm }

model symbol [COMPUT SCI] The standard usage of geometrical figures, such as squares, circles, or triangles, to help illustrate the various working parts of a model; each symbol must, nevertheless, be footnoted for complete clarification. { 'mäd·əl ˌsim·bəl }

modem [ELECTR] A combination modulator and demodulator at each end of a telephone line to convert binary digital information to audio tone signals suitable for transmission over the line, and vice versa. Derived from modulator-demodulator. { 'mōˌdem }

modem eliminator [COMPUT SCI] A device that is used to connect two computers in proximity and that mimics the action of two modems and a telephone line. { 'mōˌdem ə'lim·əˌnād·ər }

moder *See* coder. { 'mōd·ər }

Mode S [NAV] An augmentation of the Air Traffic Control Radar Beacon System in which each aircraft is equipped with a transponder that replies when interrogated with a discrete identity code. Also known as ADSEL (in Britain); discrete address beacon system or DABS (in the United States). { 'mōd 'es }

mode switch [COMPUT SCI] A preset control which affects the normal response of various components of a mechanical desk calculator. [ELECTR] A microwave control device, often consisting of a waveguide section of special cross section, which is used to change the mode of microwave power transmission in the waveguide. { 'mōd ˌswich }

MODFET *See* high-electron-mobility transistor. { 'mädˌfet }

modifier [COMPUT SCI] A quantity used to alter the address of an operand in a computer, such as the cycle index. Also known as index word. { 'mäd·əˌfī·ər }

modifier register *See* index register. { 'mäd·ə ˌfī·ər ˌrej·ə·stər }

modify [COMPUT SCI] **1.** To alter a portion of an instruction so its interpretation and execution will be other than normal; the modification may permanently change the instruction or leave it unchanged and affect only the current execution; the most frequent modification is that of the effective address through the use of index registers. **2.** To alter a subroutine according to a defined parameter. { 'mäd·əˌfī }

modify structure [COMPUT SCI] A statement in a database language that allows changes to be made in the structure of the records in a file. { 'mäd·əˌfī ˌstrək·chər }

modular compilation [COMPUT SCI] The separate translation into machine language of the individual parts of a computer program, which are then combined into a single program by a linkage editor. { 'mäj·ə·lər ˌkäm·pə'lā·shən }

modularity [COMPUT SCI] The property of functional flexibility built into a computer system by assembling discrete units which can be easily joined to or arranged with other parts or units. { ˌmäj·ə'lar·əd·ē }

modular programming [COMPUT SCI] The construction of a computer program from a collection of modules, each of workable size, whose interactions are rigidly restricted. { 'mäj·ə·lər 'prōˌgram·iŋ }

modulate [ELECTR] To vary the amplitude, frequency, or phase of a wave, or vary the velocity of the electrons in an electron beam in some characteristic manner. { 'mäj·əˌlāt }

modulated amplifier [ELECTR] Amplifier stage in a transmitter in which the modulating signal is introduced and modulates the carrier. { 'mäj·ə ˌlād·əd 'am·pləˌfī·ər }

modulated carrier [COMMUN] Radio-frequency carrier wave whose amplitude, phase, or frequency has been varied according to the intelligence to be conveyed. { 'mäj·əˌlād·əd 'kar·ē·ər }

modulated continuous wave [COMMUN] Wave in which the carrier is modulated by a constant audio-frequency tone. { 'mäj·əˌlād·əd kən ¦tin·yə·wəs 'wāv }

modulated stage [ELECTR] Radio-frequency stage to which the modulator is coupled and in which the continuous wave (carrier wave) is modulated according to the system of modulation and the characteristics of the modulating wave. { 'mäj·ə,lād·əd 'stāj }

modulating electrode [ELECTR] Electrode to which a potential is applied to control the magnitude of the beam current. { 'mäj·ə,lād·iŋ i'lek,trōd }

modulating signal [COMMUN] Signal which causes a variation of some characteristics of a carrier. { 'mäj·ə,lād·iŋ 'sig·nəl }

modulation [COMMUN] The process or the result of the process by which some parameter of one wave is varied in accordance with some parameter of another wave. { ,mäj·ə'lā·shən }

modulation code [COMMUN] A code used to cause variations in a signal in accordance with a predetermined scheme; normally used to alter or modulate a carrier wave to transmit data. { ,mäj·ə'lā·shən ,kōd }

modulation crest [COMMUN] The peak amplitude of an amplitude-modulated wave. { ,mäj·ə'lā·shən 'krest }

modulation-doped field-effect transistor *See* high-electron-mobility transistor. { ,mäj·ə'lā·shən ¦dōpt 'fēld i¦fekt tran'zis·tər }

modulation envelope [COMMUN] The peaks of the waveform of a modulated signal. { ,mäj·ə'lā·shən 'en·və,lōp }

modulation factor [COMMUN] **1.** In general, the ratio of the peak variation in the modulation actually used in a transmitter to the maximum variation for which the transmitter was designed. **2.** In an amplitude-modulated wave, the ratio (usually expressed in percent) of the peak variation of the envelope from its reference value, to the reference value. Also known as index of modulation. **3.** In a frequency-modulated wave, the ratio of the actual frequency swing to the frequency swing required for 100% modulation. { ,mäj·ə'lā·shən ,fak·tər }

modulation index [COMMUN] The ratio of the frequency deviation to the frequency of the modulating wave in a frequency-modulation system when using a sinusoidal modulating wave. Also known as ratio deviation. { ,mäj·ə'lā·shən ,in ,deks }

modulation rise [ELECTR] Increase of the modulation percentage caused by nonlinearity of any tuned amplifier, usually the last intermediate-frequency stage of a receiver. { ,mäj·ə'lā·shən ,rīz }

modulator [ELECTR] **1.** The transmitter component that supplies the modulating signal to the amplifier stage or that triggers the modulated amplifier stage to produce pulses at desired instants as in radar. **2.** A device that produces modulation by any means, such as by virtue of a nonlinear characteristic or by controlling some circuit quantity in accordance with the waveform of a modulating signal. { 'mäj·ə,lād·ər }

modulator-demodulator *See* modem. { 'mäj·ə ,lād·ər dē'mäj·ə,lād·ər }

modula-2 [COMPUT SCI] A general-purpose programming language that allows a computer program to be written as separate modules which can be compiled separately but can share a common code. { 'mäj·ə·lə'tü }

module [COMPUT SCI] **1.** A distinct and identifiable unit of computer program for such purposes as compiling, loading, and linkage editing. **2.** One memory bank and associated electronics in a computer. [ELECTR] A packaged assembly of wired components, built in a standardized size and having standardized plug-in or solderable terminations. { 'mäj·ül }

modulo N check [COMPUT SCI] A procedure for verification of the accuracy of a computation by repeating the steps in modulo N arithmetic and comparing the result with the original result (modulo N). Also known as residue check. { 'mäj·ə,lō¦en 'chek }

modulo-two adder [COMPUT SCI] A logical circuit for adding one-digit binary numbers. { 'mäj·ə,lō ¦tü 'ad·ər }

MOEMS *See* micro-opto-electro-mechanical system. { 'mō,emz }

moiré [COMMUN] In a video system, the spurious pattern in the reproduced picture resulting from interference beats between two sets of periodic structures in the image. { mȯ'rā }

molecular electronics [ELECTR] The use of biological or organic molecules for fabricating electronic materials with novel electronic, optical, or magnetic properties; applications include polymer light-emitting diodes, conductive-polymer sensors, pyroelectric plastics, and, potentially, molecular computational devices. { mə'lek·yə·lər i,lek'trän·iks }

Molniya orbit [AERO ENG] An earth satellite orbit designed for communications satellite service coverage at high latitudes, with an orbital period of slightly less than 12 hours (semisynchronous orbit), inclination of 63.4°, and high eccentricity (0.722), so that the apogee (where the satellite lingers over the service coverage area) is at 25,000 miles (40,000 kilometers) and the perigee is at only 300 miles (500 kilometers). { 'mōl·nē·ə ,ȯr·bət }

MOMS *See* micro-opto-mechanical system. { mämz *or* ¦em¦ō¦em'es }

monadic operation [COMPUT SCI] An operation on one operand, such as a negation. { mō 'nad·ik ,äp·ə'rā·shən }

monitor [COMPUT SCI] **1.** To supervise a program, and check that it is operating correctly during its execution, usually by means of a diagnostic routine. **2.** *See* video monitor. { 'män·əd·ər }

monitor board [COMMUN] A console at which a supervising telephone operator sits and from which she or he can intercept calls being handled by other operators. { 'män·əd·ər ,bȯrd }

monitor control dump [COMPUT SCI] A memory dump routinely carried out by the system once a program has been run. { 'män·əd·ər kən'trōl ,dəmp }

monitor display [COMPUT SCI] The facility of stopping the central processing unit and displaying information of main storage and internal registers; after manual intervention, normal instruction execution can be initiated. { 'män·əd·ər di,splā }

monitor mode See master mode. { 'män·əd·ər ,mōd }

monitor operating system [COMPUT SCI] The control of the routines which achieves efficient use of all the hardware components. { 'män·əd·ər 'äp·ə,rād·iŋ ,sis·təm }

monitor printer [COMMUN] A teleprinter used in a technical control facility or communications center for checking incoming teletypewriter signals. [COMPUT SCI] Input-output device, capable of receiving coded signals from the computer, which automatically operates the keyboard to print a hard copy and, when desired, to punch paper tape. { 'män·əd·ər ,print·ər }

monitor routine See executive routine. { 'män·əd·ər rü,tēn }

monochromatic radiation [ELECTROMAG] Electromagnetic radiation having wavelengths confined to an extremely narrow range. { män·ə·krə'mad·ik ,rād·ē'ā·shən }

monochrome signal [ELECTR] **1.** A signal waveform used for controlling luminance values in monochrome television. **2.** The portion of a signal wave that has major control of the luminance values in a color television system, regardless of whether the picture is displayed in color or in monochrome. Also known as M signal. { 'män·ə,krōm ,sig·nəl }

monochrome television [COMMUN] Television in which the final reproduced picture is monochrome, having only shades of gray between black and white. Also known as black-and-white television. { 'män·ə,krōm 'tel·ə,vizh·ən }

monofier [ELECTR] Complete master oscillator and power amplifier system in a single evacuated tube envelope; electrically, it is equivalent to a stable low-noise oscillator, an isolator, and a two- or three-cavity klystron amplifier. { 'män·ə ,fī·ər }

monolithic filter [COMMUN] A device used to separate telephone communications sent simultaneously over the transmission line, consisting of a series of electrodes vacuum-deposited on a crystal plate so that the plated sections are resonant with ultrasonic sound waves, and the effect of the device is similar to that of an electric filter. { ,män·ə'lith·ik 'fil·tər }

monolithic integrated circuit [ELECTR] An integrated circuit having elements formed in place on or within a semiconductor substrate, with at least one element being formed within the substrate. { ,män·ə'lith·ik 'int·ə,grād·əd 'sər·kət }

monopole antenna [ELECTROMAG] An antenna, usually in the form of a vertical tube or helical whip, on which the current distribution forms a standing wave, and which acts as one part of a dipole whose other part is formed by its electrical image in the ground or in an effective ground plane. Also known as spike antenna. { 'män·ə ,pōl an'ten·ə }

monopulse [ELECTR] A radar technique for accurate estimation of target position in angle and range on each pulse transmitted, without requiring sequential transmission as in, for example, conical scanning radars; reduces errors due to echo fluctuations. { 'män·ə,pəls }

monopulse radar [ENG] Radar in which directional information is obtained with high precision by using a receiving antenna system having two or more partially overlapping lobes in the radiation patterns. { 'män·ə,pəls 'rā,där }

monostable multivibrator [ELECTR] A multivibrator with one stable state and one unstable state; a trigger signal is required to drive the unit into the unstable state, where it remains for a predetermined time before returning to the stable state. Also known as one-shot multivibrator; single-shot multivibrator; start-stop multivibrator; univibrator. { ¦män·ō¦stā·bəl ,məl·tə'vī ,brād·ər }

monostatic radar [ENG] A radar with transmitter and receiver in the same place, whether or not it uses a duplexed antenna. { ¦män·ə¦stad·ik 'rā ,där }

monotonicity [ELECTR] In an analog-to-digital converter, the condition wherein there is an increasing output for every increasing value of input voltage over the full operating range. { ,män·ə·tə'nis·əd·ē }

Monte Carlo method [STAT] A technique which obtains a probabilistic approximation to the solution of a problem by using statistical sampling techniques. { 'män·tē 'kär·lō ,meth·əd }

Moore's law [COMPUT SCI] The prediction by Gordon Moore (cofounder of the Intel Corporation) that the number of transistors on a microprocessor would double periodically (approximately every 18 months). { 'mürz ,lȯ }

Moore code [COMMUN] A binary teleprinter code with seven binary digits for each letter. { 'mu̇r ,kōd }

Moore machine [COMPUT SCI] A sequential machine in which the output depends uniquely on the current state of the machine, and not on the input. { 'mu̇r mə,shēn }

Moore-Smith sequence See net. { ¦mür 'smith ,sē·kwəns }

Morse cable code [COMMUN] A code used chiefly in submarine cable telegraphy, in which positive and negative current impulses of equal length represent dots and dashes, and a space is represented by the absence of current. Also known as cable code; International cable code. { 'mȯrs 'kā·bəl ,kōd }

Morse code [COMMUN] **1.** A telegraph code for manual operating, consisting of short (dot) and long (dash) signals and various-length spaces. Also known as American Morse code. **2.** Collective term for Morse code (American Morse code) and continental code (International Morse code). { 'mȯrs 'kōd }

mosaic [ELECTR] A light-sensitive surface used in video camera tubes, consisting of a thin mica sheet coated on one side with a large number of tiny photosensitive silver-cesium globules, insulated from each other. { mō'zā·ik }

MOSFET See metal oxide semiconductor field-effect transistor. { 'mȯs,fet }

MOSFET-C filter [ELECTR] An active integrated-circuit filter in which the resistors of an active-RC filter are replaced with metal oxide semiconductor field-effect transistors (MOSFETs). { ¦mȯs ˌfet 'sē ˌfil·tər }

MOST *See* metal oxide semiconductor field-effect transistor.

MOS transistor *See* metal oxide semiconductor field-effect transistor. { ¦em¦ō'es tran'zis·tər }

most significant bit [COMPUT SCI] The left-most bit in a word. Abbreviated msb. { 'mōst sig 'nif·i·gənt 'bit }

most significant character [COMPUT SCI] The character in the leftmost position in a number or word. { 'mōst sig'nif·i·gənt 'kar·ik·tər }

motherboard [COMPUT SCI] A common pathway over which information is transmitted between the hardware devices (the central processing unit, memory, and each of the peripheral control units) in a microcomputer. { 'məth·ərˌbȯrd }

motion-compensated coding scheme [COMMUN] A form of differential pulse-code modulation in which the motions of objects are estimated and comparisons of intensities are carried out between picture elements in successive frames spatially displaced by an amount equal to the motion of an object. { 'mō·shən ¦käm·pən ˌsād·əd 'kōd·iŋ ˌskēm }

motion register [COMPUT SCI] The register which controls the go/stop, forward/reverse motion of a tape drive. { 'mō·shən ˌrej·ə·stər }

motion vector [COMMUN] A pair of numbers which represent the vertical and horizontal displacement of a region of a reference picture for MPEG-2 prediction. { 'mō·shan 'vek·tər }

mountain effect [ELECTROMAG] The effect of rough terrain on radio-wave propagation, causing reflections that produce errors in radio direction-finder indications. { 'maünt·ən iˌfekt }

mousable interface [COMPUT SCI] A user interface that responds to input from a mouse for various functions. { ¦maüs·ə·bəl 'in·tərˌfās }

mouse [COMPUT SCI] A small device with a rubber-coated ball that is moved about by hand over a flat surface and generates signals to control the position of a cursor or pointer on a computer display. { maüs }

movable-head disk drive [COMPUT SCI] A type of disk drive in which read/write heads are moved over the surface of the disk, toward and away from the center, so that they are correctly positioned to read or write the desired information. { 'mü·və·bəl ¦hed 'disk ˌdrīv }

move mode [COMPUT SCI] A method of communicating between an operating program and an input/output control system in which the data records to be read or written are actually moved into and out of program-designated memory areas. { 'müv ˌmōd }

move operation [COMPUT SCI] An operation in which data is moved from one storage location to another. { 'müv ˌäp·əˌrā·shən }

moving-head disk [COMPUT SCI] A disk-storage device in which one or more read-write heads are attached to a movable arm which allows each head to cover many tracks of information. { 'müv·iŋ ¦hed 'disk }

Moving Picture Experts Group *See* MPEG. { 'müv·iŋ 'pik·chər 'ekˌspərts 'grüp }

moving-target indicator [ELECTR] A feature that limits the display of radar information primarily to moving targets; signals due to reflections from stationary objects are canceled by a memory circuit. Abbreviated MTI. { 'müv·iŋ ¦tär·gət 'in·də ˌkād·ər }

MPEG [COMMUN] Standards for compression of digitized audio and video signals developed by the ISO/IEC JTC1/SC29 WG11; may also refer to the group itself (Moving Picture Experts Group). { 'emˌpeg }

MPEG-1 [COMMUN] Standards for compression of digitized audio and video signals that were developed by the Moving Picture Experts Group with relatively low-bit-rate systems in mind, such as video from a CD-ROM, and continue to be used, although most video applications currently use MPEG-2; they comprise ISO/IEC standards 11172-1 (systems), 11172-2 (video), 11172-3 (audio), 11172-4 (compliance testing), and 11172-5 (technical report). { 'emˌpeg 'wən }

MPEG-2 [COMMUN] Standards for compression of digitized audio and video signals that were developed by the Moving Picture Experts Group primarily for professional video applications, such as video production, and distribution and storage applications on media ranging from tape to DVD, and which form the basis of digital television systems; they comprise ISO/IEC standards 13818-1 (systems), 13818-2 (video), 13818-3 (audio), and 13818-4 (compliance). { 'emˌpeg 'tü }

MPEG-2 AAC [COMMUN] An advanced audio coder; a high-quality, low-bit-rate perceptual audio coding system. { 'emˌpeg 'tü ¦ā¦ā¦sē }

MPU *See* microprocessing unit.

MQ register [COMPUT SCI] Temporary-storage register whose contents can be transferred to or from, or swapped with, the accumulator. { ¦em ¦kyü 'rej·ə·stər }

MRAM *See* magnetic random access memory. { 'emˌram }

MSAS *See* MTSAT Satellite-Based Augmentation System. { 'emˌsas *or* ¦em¦es¦ā'es }

msb *See* most significant bit.

M-scan *See* M-display. { 'em ˌskan }

M-scope *See* M-display. { 'em ˌskōp }

M signal *See* monochrome signal. { 'em ˌsig·nəl }

MSS *See* mobile satellite service.

MTBF *See* mean time between failures.

MTI *See* moving-target indicator.

MTSAT Satellite-Based Augmentation System [NAV] A wide-area DGPS for air navigation whose error corrections are broadcast over geostationary satellites, developed by Japan Civil Aviation Bureau. Abbreviated MSAS. { ¦em¦tē¦sat ¦sad·əl ˌīt ˌbāst ˌȯg·mən 'tā·shən ˌsis·təm }

MUF *See* maximum usable frequency.

multiaccess computer [COMPUT SCI] A computer system in which computational and data resources are made available simultaneously to a number of users who access the system through terminal devices, normally on an interactive or conversational basis. { ¦məl·tē'ak,ses kəm ,pyüd·ər }

multiaccess network *See* multiple-access network. { ¦məl·tē'ak,ses 'net,wərk }

multiaddress [COMPUT SCI] Referring to an instruction that has more than one address part. { ¦məl·tē'a,dres }

multiaspect [COMPUT SCI] Pertaining to searches or systems which permit more than one aspect, or facet, of information to be used in combination, one with the other to effect identifying or selecting operations. { ¦məl·tē'as ,pekt }

multicellular horn [ELECTROMAG] A cluster of horn antennas having mouths that lie in a common surface and that are fed from openings spaced one wavelength apart in one face of a common waveguide. { ¦məl·tē'sel·yə·lər 'hȯrn }

multichannel communication [COMMUN] Communication in which there are two or more communication channels over the same path, such as a communication cable, or a radio transmitter which can broadcast on two different frequencies, either individually or simultaneously. { ¦məl·tē'chan·əl kə,myü·nə'kā·shən }

multichannel loading [COMMUN] Behavior of a multichannel communications system with all channels active. { ¦məl·tē'chan·əl 'lōd·iŋ }

multichannel telephone system [COMMUN] A telephone system in which two or more communications channels are carried over a single telephone cable or radio link. { ¦məl·tē'chan·əl 'tel·ə,fōn ,sis·təm }

multichip microcircuit [ELECTR] Microcircuit in which discrete, miniature, active electronic elements (transistor or diode chips) and thin-film or diffused passive components or component clusters are interconnected by thermocompression bonds, alloying, soldering, welding, chemical deposition, or metallization. { 'məl·tē,chip 'mī·krō,sər·kət }

multicomputer system [COMPUT SCI] A system consisting of more than one computer, usually under the supervision of a master computer, in which smaller computers handle input/output and routine jobs while the large computer carries out the more complex computations. { ¦məl·tē·kəm¦pyüd·ər ,sis·təm }

multicoupler [ELECTR] A device for connecting several receivers to one antenna and properly matching the impedances of the receivers to the antenna. { 'məl·tə,kəp·lər }

multidimensional Turing machine [COMPUT SCI] A variation of a Turing machine in which tapes are replaced by multidimensional structures. { ¦məl·tə·di'men·shən·əl 'tu̇r·iŋ mə,shēn }

multidrop line [COMMUN] A telephone pair which terminates at several locations. { 'məl·tē ,dräp ,līn }

multielement array [ELECTROMAG] An antenna array having a large number of antennas. { ¦məl·tē'el·ə·mənt ə'rā }

multielement parasitic array [ELECTROMAG] Antennas consisting of an array of driven dipoles and parasitic elements, arranged to produce a beam of high directivity. { ¦məl·tē'el·ə·mənt ,par·ə'sid·ik ə,rā }

multifunction array radar [ENG] Electronic scanning radar which will perform target detection and identification, tracking, discrimination, and some interceptor missile tracking on a large number of targets simultaneously and as a single unit. { ¦məl·tə'fəŋk·shən ə'rā 'rā,där }

multihead Turing machine [COMPUT SCI] A variation of a Turing machine in which more than one head is allowed per tape. { 'məl·tē,hed 'tu̇r·iŋ mə,shēn }

multijob operation [COMPUT SCI] The concurrent or interleaved execution of job steps from more than one job. { 'məl·tē,jäb ,äp·ə'rā·shən }

multilayer board [ELECTR] A printed wiring board that contains circuitry on internal layers throughout the cross section of the board as well as on the external layers. { ,məl·tē,lā·ər 'bȯrd }

multilayer optical storage [COMPUT SCI] An extension of optical disk storage technology to the third dimension by stacking data layers one above another, with each layer separated by a spacer region. { ,məl·tə¦lā·ər ¦äp·tl·kəl 'stȯr ij }

multilevel address *See* indirect address. { ¦məl·tə'lev·əl a,dres }

multilevel control theory [CONT SYS] An approach to the control of large-scale systems based on decomposition of the complex overall problem into simpler and more easily managed subproblems, and coordination of the subproblems so that overall system objectives and constraints are satisfied. { ¦məl·tə'lev·əl kən'trōl 'thē·ə·rē }

multilevel indirect addressing [COMPUT SCI] A programming device whereby the address retrieved in the memory word may itself be an indirect address that points to another memory location, which in turn may be another indirect address, and so forth. { ¦məl·tə'lev·əl ,in·də,rekt ə'dres·iŋ }

multilevel transmission [COMMUN] Transmission of digital information in which three or more levels of voltage are recognized as meaningful, as 0,1,2 instead of simply 0,1. { ¦məl·tə'lev·əl tranz'mish·ən }

multiline appearances [COMMUN] **1.** The ability of a telephone to receive or originate additional voice or data calls at the terminal while it is still engaged in the primary voice call. **2.** The ability to bring additional parties to a primary telephone call. { ¦məl·tē'līn ə'pir·ən·səz }

multilist organization [COMPUT SCI] A chained file organization in which each segment is indexed. { 'məl·tē,list ,ȯr·gə·nə'zā·shən }

multimedia technology [COMPUT SCI] The synergistic union of digital video, audio, computer, information, and telecommunication technologies. { ¦məl·tə,mēd·ē·ə tek'näl·ə·jē }

multipactor [ELECTR] A high-power, high-speed microwave switching device in which a thin electron cloud is driven back and forth between two parallel plane surfaces in a vacuum by a radio-frequency electric field. { 'məl·tə,pak·tər }

multipass sort [COMPUT SCI] Computer program designed to sort more data than can be contained within the internal storage of a computer; intermediate storage, such as disk, tape, or drum, is required. { 'məl·tē,pas 'sȯrt }

multipath [COMMUN] A radio frequency reception condition in which a radio signal reaching a receiving antenna arrives by multiple paths due to reflections of the signal off of various surfaces in the environment. By traveling different distances to the receiver, the reflections arrive with different time delays and signal strengths. When multiplath conditions are great enough, analog reception of radio broadcasts is affected in a variety of ways, including stop-light fades, picket fencing, and distortion received audio. [ELECTROMAG] **1.** In radar, a propagation situation wherein the direct-path signal from radar to a target is interfered with by the reflected-path signal; usually refers to reflections from earth's surface. Both surveillance sensitivity and tracking accuracy, particularly at low elevation angles, are affected. **2.** *See* multipath transmission. { 'məl·tə,path }

multipath cancellation [COMMUN] Occurrence of essentially complete cancellation of radio signals because of the relative amplitude and phase differences of the components arriving over separate paths. { 'məl·tə,path ,kan·sə'lā·shən }

multipath system [NAV] An electronic navigation system that measures differences in or otherwise compares transmission times of radio signals. { 'məl·tə,path ,sis·təm }

multipath transmission [ELECTROMAG] The propagation phenomenon that results in signals reaching a radio receiving antenna by two or more paths, causing distortion in radio and ghost images in television. Also known as multipath. { 'məl·tə,path tranz'mish·ən }

multiple *See* parallel. { 'məl·tə·pəl }

multiple access [COMMUN] Multiplexing schemes by which multiple users who are geographically dispersed gain access to a shared telecommunications facility or channel. { ,məl·tə·pəl 'ak,ses }

multiple-access computer [COMPUT SCI] A computer system whose facilities can be made available to a number of users at essentially the same time, normally through terminals, which are often physically far removed from the central computer and which typically communicate with it over telephone lines. { 'məl·tə·pəl ¦ak,ses kəm ,pyüd·ər }

multiple-access network [COMPUT SCI] A computer network that permits every computer on it to communicate with the network at any time during operation. Also known as multiaccess network. { 'məl·tə·pəl ¦ak,ses 'net,wərk }

multiple-access service [COMMUN] One of the services of the Tracking and Data Relay Satellite System, which provides simultaneous return-link service from as many as 20 low-earth-orbiting user spacecraft, with data rates up to 3 megabits per second for each user, and a time-shared forward-link service to the user spacecraft with a maximum data rate of 300 kilobits per second, one user at a time. Abbreviated MA service. { ,məl·tə·pəl 'ak,ses ,sər·vəs }

multiple accumulating registers [COMPUT SCI] Special registers capable of handling factors larger than one computer word in length. { 'məl·tə·pəl ə'kyü·myə,lād·iŋ 'rej·ə,stərz }

multiple-address code [COMPUT SCI] A computer instruction code in which more than one address or storage location is specified; the instruction may give the locations of the operands, the destination of the result, and the location of the next instruction. { 'məl·tə·pəl ¦a ,dres ,kōd }

multiple-address computer [COMPUT SCI] A computer whose instruction contains more than one address, for example, an operation code and three addresses A, B, C, such that the content of A is multiplied by the content of B and the product stored in location C. { 'məl·tə·pəl ¦a,dres kəm ,pyüd·ər }

multiple-address instruction [COMPUT SCI] An instruction which has more than one address in a computer; the addresses give locations of other instructions, or of data or instructions that are to be operated upon. { 'məl·tə·pəl ¦a,dres in ,strək·shən }

multiple-beam antenna [ELECTROMAG] An antenna or antenna array which radiates several beams in different directions. { 'məl·tə·pəl ¦bēm an'ten·ə }

multiple computer operation [COMPUT SCI] The utilization of any one computer of a group of computers by means of linkages provided by multiplexor channels, all computers being linked through their channels or files. { 'məl·tə·pəl kəm'pyüd·ər ,äp·ə,rā·shən }

multiple decay *See* branching. { 'məl·tə·pəl di'kā }

multiple disintegration *See* branching. { 'məl·tə·pəl di,sin·tə'grā·shən }

multiple-frequency-shift keying [COMMUN] A modulation scheme in which a number of carrier frequencies (2, 4, 8, and so forth) are transmitted according to a group of consecutive data bits (n bits producing 2^n frequencies). Abbreviated MFSK. { 'məl·tə·pəl ¦frē·kwən·sē ¦shift ,kē·iŋ }

multiple-function chip *See* large-scale integrated circuit. { 'məl·tə·pəl ¦fəŋk·shən ,chip }

multiple-instruction-stream, multiple-data-stream *See* MIMD. { ¦məl·tə·pəl in'strək·shən ,strēm ¦məl·tə·pəl 'dad·ə ,strēm }

multiple-key access [COMPUT SCI] A technique for locating stored data in a computer system by using the values contained in two or more separate key fields. { 'məl·tə·pəl ¦kē 'ak,ses }

multiple-length arithmetic [COMPUT SCI] Arithmetic performed by a computer in which two or more machine words are used to represent each number in the calculations, usually to achieve higher precision in the result. { 'məl·tə·pəl ¦leŋkth ə'rith·mə,tik }

multiple-length number [COMPUT SCI] A number having two or more times as many digits as are ordinarily used in a given computer. { 'məl·tə·pəl ¦leŋkth 'nəm·bər }

multiple-length working [COMPUT SCI] Any processing of data by a computer in which two or more machine words are used to represent each data item. { 'məl·tə·pəl ¦leŋkth 'wərk·iŋ }

multiple modulation [COMMUN] A succession of modulating processes in which the modulated wave from one process becomes the modulating wave for the next. Also known as compound modulation. { 'məl·tə·pəl ,mäj·ə'lā·shən }

multiple module access [COMPUT SCI] Device which establishes priorities in storage access in a multiple computer environment. { 'məl·tə·pəl ¦mäj·yül ,ak,ses }

multiple precision arithmetic [COMPUT SCI] Method of increasing the precision of a result by increasing the length of the number to encompass two or more computer words in length. { 'məl·tə·pəl prə¦sizh·ən ə'rith·mə,tik }

multiple programming [COMPUT SCI] The execution of two or more operations simultaneously. { 'məl·tə·pəl 'prō,gram·iŋ }

multiple target generator [ELECTR] An electronic countermeasures device that produces several false responses in a hostile radar set. { 'məl·tə·pəl 'tär·gət ,jen·ə,rād·ər }

multiple-tuned antenna [ELECTROMAG] Low-frequency antenna having a horizontal section with a multiplicity of tuned vertical sections. { 'məl·tə·pəl ¦tünd an'ten·ə }

multiple-unit steerable antenna *See* musa. { 'məl·tə·pəl ¦yü·nət 'stir·ə·bəl an'ten·ə }

multiplexer [ELECTR] A device for combining two or more signals, as for multiplex, or for creating the composite color video signal from its components in color television. Also spelled multiplexor. { 'məl·tə,plek·sər }

multiplexing [COMMUN] **1.** A set of techniques that enable the sharing of the usable electromagnetic spectrum of a telecommunications channel (the channel pass-band) among multiple users for the transfer of individual information streams. **2.** In particular, the case in which the user information streams join at a common access point to the channel. { 'məl·tə,pleks·iŋ }

multiplex mode [COMPUT SCI] The utilization of differences in operating speeds between a computer and transmission lines; the multiplexor channel scans each line in sequence, and any transmitted pulse on a line is assembled in an area reserved for this line; consequently, a number of users can be handled by the computer simultaneously. Also known as multiplexor channel operation. { 'məl·tə,pleks ,mōd }

multiplex operation [COMMUN] Simultaneous transmission of two or more messages in either or both directions over a carrier channel. { 'məl·tə,pleks ,äp·ə'rā·shən }

multiplexor *See* multiplexer. { 'məl·tə,plek·sər }

multiplexor channel operation *See* multiplex mode. { 'məl·tə,plek·sər ¦chan·əl ,äp·ə,rā·shən }

multiplexor terminal unit [COMPUT SCI] Device which permits a large number of data transmission lines to access a single computer. { 'məl·tə ,plek·sər 'ter·mən·əl ,yü·nət }

multiplex transmission [COMMUN] The simultaneous transmission of two or more programs or signals over a single radio-frequency channel, such as by time division, frequency division, code division, or phase division. { 'məl·tə,pleks tranz'mish·ən }

multiplication table [COMPUT SCI] In certain computers, a part of memory holding a table of numbers in which the computer looks up values in order to perform the multiplication operation. { ,məl·tə·pli'kā·shən ,tā·bəl }

multiplication time [COMPUT SCI] The time required for a computer to perform a multiplication; for a binary number it will be equal to the total of all the addition times and all the shift times involved in the multiplication. { ,məl·tə·pli'kā·shən ,tīm }

multiplicative congruential generator [COMPUT SCI] A congruential generator in which the constant *b* in the generating formula is equal to zero. { ,məl·tə¦plik·əd·iv ,kän,grü'en·chəl 'jen·ə ,rād·ər }

multiplier [ELECTR] **1.** A device that has two or more inputs and an output that is a representation of the product of the quantities represented by the input signals;voltages are the quantities commonly multiplied. **2.** *See* electron multiplier; frequency multiplier. { 'məl·tə,plī·ər }

multiplier field [COMPUT SCI] The area reserved for a multiplication, equal to the length of multiplier plus multiplicand plus one character. { 'məl·tə,plī·ər ,fēld }

multiplier-quotient register [COMPUT SCI] A register equal to two words in length in which the quotient is developed and in which the multiplier is entered for multiplication. { 'məl·tə,plī·ər 'kwō·shənt ,rej·ə·stər }

multipling [COMMUN] Use of multidrop lines to provide for changes in telephone service patterns or requirements; unused terminals afford convenient access to wiretappers. { 'məl·tə·pliŋ }

multiply defined symbol [COMPUT SCI] Common assembler or compiler error printout indicating that a label has been used more than once. { 'məl·tə·plē di¦fīnd 'sim·bəl }

multipoint line [COMMUN] A line which is shared by two or more different tributary stations. { 'məl·tə,pȯint ,līn }

multiport memory [COMPUT SCI] A memory shared by many processors to communicate among themselves. { 'məl·tə,pȯrt 'mem·rē }

multiprecision arithmetic [COMPUT SCI] A form of arithmetic similar to double precision arithmetic except that two or more words may be used to represent each number. { ¦məl·tə·prə'sizh·ən ə'rith·mə,tik }

multiprocessing [COMPUT SCI] Carrying out of two or more sequences of instructions at the same time in a computer. { ,məl·tə'prä,ses·iŋ }

multiprocessing system *See* multiprocessor. { ,məl·tə'prä,ses·iŋ ,sis·təm }

multiprocessor [COMPUT SCI] A data-processing system that can carry out more than one program, or more than one arithmetic operation, at the same time. Also known as multiprocessing system. { ˌməl·tə'präˌses·ər }

multiprocessor interleaving [COMPUT SCI] Technique used to speed up processing time; by splitting banks of memory each with *x* microseconds access time and accessing each one in sequence 1/*n*-th of a cycle later, a reference to memory can be had every *x/n* microseconds; this speed is achieved at the cost of hardware complexity. { ˌməl·tə'präˌses·ər ˌin·tər'lēv·iŋ }

multiprogramming [COMPUT SCI] The interleaved execution of two or more programs by a computer, in which the central processing unit executes a few instructions from each program in succession. { ˌməl·tə'prōˌgram·iŋ }

multiprogramming executive control [COMPUT SCI] Control program structure required to handle multiprogramming with either a fixed or a variable number of tasks. { ˌməl·tə'prōˌgram·iŋ ig¦zek·yəd·iv kən'trōl }

multistatic radar [ENG] Radar in which successive antenna lobes are sequentially engaged to provide a tracking capability without physical movement of the antenna. { 'məl·tēˌstad·ik 'rā ˌdär }

multistation [COMMUN] Pertaining to a network in which each station can communicate with each of the other stations. { 'məl·tēˌstā·shən }

multistrip coupler [ELECTR] A series of parallel metallic strips placed on a surface acoustic wave filter between identical apodized interdigital transducers; it converts the spatially nonuniform surface acoustic wave generated by one transducer into a spatially uniform wave received at the other transducer, and helps to reject spurious bulk acoustic modes. { 'məl·təˌstrip 'kəp·lər }

multisync monitor [COMPUT SCI] A video display monitor that automatically adjusts to the synchronization frequency of the video source from which it is receiving signals. { ¦məl·tiˌsiŋk 'män·ə·tər }

multisystem coupling [COMPUT SCI] The electronic connection of two or more computers in proximity to make them act as a single logical machine. { 'məl·təˌsis·təm 'kəp·liŋ }

multisystem network [COMPUT SCI] A data communications network that has two or more host computers with which the various terminals in the system can communicate. { 'məl·təˌsis·təm 'netˌwərk }

multitape Turing machine [COMPUT SCI] A variation of a Turing machine in which more than one tape is permitted, each tape having its own read-write head. { 'məl·tēˌtāp 'tu̇r·iŋ mə ˌshēn }

multitasking [COMPUT SCI] The simultaneous execution of two or more programs by a single central processing unit. { ¦məl·tē'task·iŋ }

multitask operation [COMPUT SCI] A sophisticated form of multijob operation in a computer which allows a single copy of a program module to be used for more than one task. { ¦məl·tē'task ˌäp·ə'rā·shən }

multithreading [COMPUT SCI] A processing technique that allows two or more of the same type of transaction to be carried out simultaneously. { ¦məl·tə'thred·iŋ }

multitrack operation [COMPUT SCI] The selection of the next read/write head in a cylinder, usually indicated by bit zero of the operation code in the channel command word. { ¦məl·tē'trak ˌäp·ə'rā·shən }

multiuser system [COMPUT SCI] A computer system with multiple terminals, enabling several users, each at their own terminal, to use the computer. { ¦məl·tē¦yü·zər 'sis·təm }

multivibrator [ELECTR] A relaxation oscillator using two tubes, transistors, or other electron devices, with the output of each coupled to the input of the other through resistance-capacitance elements or other elements to obtain in-phase feedback voltage. { ˌməl·tə'vīˌbrād·ər }

multivolume file [COMPUT SCI] A file that consists of more than one physical unit of storage medium. { ¦məl·tē¦väl·yəm 'fīl }

multiway merge [COMPUT SCI] A computer operation in which three or more lists are merged into a single list. { 'məl·tēˌwā'mərj }

Murray code [COMMUN] A binary code with five binary digits per letter which was developed to be used with a typewriterlike device which would punch holes in paper tape, and is now the basis of the widely used CCIT 2 code. { 'mər·ē ˌkōd }

musa [ELECTROMAG] An electrically steerable receiving antenna whose directional pattern can be rotated by varying the phases of the contributions of the individual units. Derived from multiple-unit steerable antenna. { 'myü·sə }

musical instrument digital interface [COMPUT SCI] **1.** The digital standard for connecting computers, musical instruments, and synthesizers. **2.** A compression format for encoding music. Abbreviated MIDI. { ˌmyü·zi·kəl ¦in·strə·mənt ˌdij·ə·dəl 'in·tərˌfās }

muting circuit [ELECTR] **1.** Circuit which cuts off the output of a receiver when no radio-frequency carrier greater than a predetermined intensity is reaching the first detector. **2.** Circuit for making a receiver insensitive during operation of its associated transmitter. { 'myüd·iŋ ˌsər·kət }

mutual deadlock [COMPUT SCI] A condition in which deadlocked tasks are awaiting resource assignments, and each task on a list awaits release of a resource held by the following task, with the last task awaiting release of a resource held by the first task. Also known as circular wait. { 'myü·chə·wəl 'dedˌläk }

mutual interference [COMMUN] Interference from two or more electrical or electronic systems which affects these systems on a reciprocal basis. { 'myü·chə·wəl ˌin·tər'fir·əns }

myriametric waves [ELECTROMAG] Electromagnetic waves having wavelengths between 10 and 100 kilometers, corresponding to the very low frequency band. { ¦mir·ē·ə¦me·trik 'wāvz }

N

NAK *See* negative acknowledgement. { nak *or* ˌenˌāˈkā }

nancy receiver *See* infrared receiver. { ˈnan·sē riˌsēv·ər }

NAND circuit [ELECTR] A logic circuit whose output signal is a logical 1 if any of its inputs is a logical 0, and whose output signal is a logical 0 if all of its inputs are logical 1. { ˈnand ˌsər·kət }

NAPLPS *See* North American presentation-level protocol syntax. { ˈnapˌlips }

narrow-band amplifier [ELECTR] An amplifier which increases the magnitude of signals over a band of frequencies whose bandwidth is small compared to the average frequency of the band. { ˈnar·ō¦band ˈam·pləˌfī·ər }

narrow-band frequency modulation [COMMUN] Frequency-modulated broadcasting system used primarily for two-way voice communication, typically having a maximum deviation of 15 kilohertz or less. { ˈnar·ō¦band ˈfrē·kwən·sē ˌmäj·əˈlā·shən }

narrow-band-pass filter [ELECTR] A band-pass filter in which the band of frequencies transmitted by the filter has a bandwidth which is small compared to the average frequency of the band. { ˈnar·ō¦ban ˌpas ˌfil·tər }

narrow-band path [COMMUN] A communications path having a bandwith typically of less than 20 kilohertz. { ˈnar·ō¦band ˈpath }

narrow-beam antenna [ELECTROMAG] An antenna which radiates most of its power in a cone having a radius of only a few degrees. { ˈnar·ō ¦bēm anˈten·ə }

N-ary code [COMMUN] Code employing N distinguishable types of code elements. { ˈen·ə·rē ˌkōd }

N-ary pulse-code modulation [COMMUN] Pulse-code modulation in which the code for each element consists of any one of N distinguishable types of elements. { ˈen·ə·rē ˈpəls ˌkōd ˌmäj·əˈlā·shən }

National Radio Systems Committee [COMMUN] Abbreviated NRSC. A technical standards setting body of the radio broadcasting industry, cosponsored by the Consumer Electronics Association (CEA) and the National Association of Broadcasters (NAB). { ¦nash·ən·əl ¦rād·ēˌō ¦sis·təmz kəˌmid·ē }

National Television Systems Committee [COMMUN] Abbreviated NTSC. The organization that developed the transmission standard for color television broadcasting in the United States, and the black-and-white system that preceded it. The NTSC color system was adopted by the Federal Communications Commission in 1953 and remains in the FCC Rules today. NTSC was also adopted in a number of other countries around the world for distribution of color video programming. The NTSC standard provides for a screen density of 525 scan lines per picture. For U.S. television service, NTSC has a field repetition rate of just under 60 fields/second, and a frame rate of just under 30 frames/second; one frame is composed of two fields for interlace scanning systems. { ¦nash·ən·əl ˈtel·əˌvizh·ən ¦sis·təmz kəˌmid·ē }

native language [COMPUT SCI] Machine language that is executed by the computer for which it is specifically designed, in contrast to a computer using an emulator. { ˈnād·iv ˈlaŋ·gwij }

native mode [COMPUT SCI] **1.** The mode of operation of a software product that is being used on a computer for which it was specifically designed, without use of an emulator. **2.** The mode of operation of a device that is carrying out the function for which it was designed and is not emulating another device. { ˈnād·iv ˈmōd }

natural antenna frequency [ELECTROMAG] Lowest resonant frequency of an antenna without added inductance or capacitance. { ˈnach·rəl an ˈten·ə ˌfrē·kwən·sē }

natural binary coded decimal system [COMPUT SCI] A particular binary coded decimal system that uses the first ten binary numbers in sequence to represent the digits 0 through 9. { ˈnach·rəl ˈbīˌner·ē ¦kōd·əd ¦des·məl ˌsis·təm }

natural function generator *See* analytical function generator. { ˈnach·rəl ˈfəŋk·shən ˌjen·ə ˌrād·ər }

natural interference [COMMUN] Electromagnetic interference arising from natural terrestrial phenomena (called atmospheric interference), or electromagnetic interference caused by natural disturbances originating outside the atmosphere of the earth (called galactic and solar noise). { ˈnach·rəl ˌin·tərˈfir·əns }

natural language [COMPUT SCI] A computer language whose rules reflect and describe current rather than prescribed usage; it is often loose and ambiguous in interpretation, meaning different things to different hearers. { 'nach·rəl 'laŋ·gwij }

natural language interaction [COMPUT SCI] The interaction of users with computer systems through the medium of natural languages. { 'nach·rəl ¦laŋ·gwij ,in·tər'ak·shən }

natural language processing [COMPUT SCI] Computer analysis and generation of natural language text; encompasses natural language interaction and natural language text processing. { 'nach·rəl ¦laŋ·gwij 'prä,ses·iŋ }

natural language text processing [COMPUT SCI] Computer processing of natural language text into a more useful form, as in automatic text translation or text summarization. { 'nach·rəl ¦laŋ·gwij 'tekst ,prä,ses·iŋ }

natural wavelength [ELECTROMAG] Wavelength corresponding to the natural frequency of an antenna or circuit. { 'nach·rəl 'wāv,leŋkth }

navigation [COMPUT SCI] In a database management system, the techniques provided for locating information within the system. [ENG] The process of directing the movement of a craft so that it will reach its intended destination; subprocesses are position fixing, dead reckoning, pilotage, and homing. { ,nav·ə'gā·shən }

navigational aid [NAV] An instrument, device, chart, method, or such, intended to assist in the navigation of a craft; this expression should not be confused with "aid to navigation," which refers only to devices external to a craft. { ,nav·ə'gā·shən·əl 'ād }

NAVSTAR [NAV] A global system of up to 24 navigation satellites developed to provide instantaneous and highly accurate worldwide three-dimensional location by air, sea, and land vehicles equipped with suitable receivers. Derived from navigation system using time and ranging. { 'nav,stär }

n-channel metal-oxide semiconductor *See* NMOS. { ¦en ,chan·əl ,med·əl ¦äk,sīd 'sem·i·kən ,dək·tər }

N-display [ELECTR] A radar display format in which a trace-deflecting pulse can be moved along the vertical range axis of an L-display to assist an operator in determining and reporting the range of a target. Also known as N-indicator; N-scan; N-scope. { 'en di,splā }

NDRO *See* nondestructive readout.

near-end crosstalk [COMMUN] A type of interference that may occur at carrier telephone repeater stations when output signals of one repeater leak into the same end of the other repeater. { 'nir ,end 'krȯs,tȯk }

near field [ELECTROMAG] The electromagnetic field that exists within one wavelength of a source of electromagnetic radiation, such as a transmitting antenna. { 'nir ,fēld }

near-infrared radiation [ELECTROMAG] Infrared radiation having a relatively short wavelength, between 0.75 and about 2.5 micrometers (some scientists place the upper limit from 1.5 to 3 micrometers), at which radiation can be detected by photoelectric cells, and which corresponds in frequency range to the lower electronic energy levels of molecules and semiconductors. Also known as photoelectric infrared radiation. { 'nir ,in·frə'red ,rād·ē'ā·shən }

necessary bandwidth [COMMUN] For a given class of emission, the minimum value of the occupied bandwidth sufficient to ensure the transmission of information at the rate and with the quality required for the system employed, under specified conditions. { 'nes·ə,ser·ē 'band ,width }

negative acknowledgement [COMPUT SCI] In a data communications network, a control character returned from a receiving machine to a sending machine to indicate the presence of errors in the preceding block of data. Abbreviated NAK. { 'neg·əd·iv ik'näl·ij·mənt }

negative effective mass amplifiers and generators [ELECTR] Class of solid-state devices for broad-band amplification and generation of electrical waves in the microwave region; these devices use the property of the effective masses of charge carriers in semiconductors becoming negative with sufficiently high kinetic energies. { 'neg·əd·iv i¦fek·tiv ¦mas 'am·plə,fī·ərz ən 'jen·ə,rād·ərz }

negative electrode *See* cathode. { 'neg·əd·iv i'lek,trōd }

negative feedback [CONT SYS] Feedback in which a portion of the output of a circuit, device, or machine is fed back 180° out of phase with the input signal, resulting in a decrease of amplification so as to stabilize the amplification with respect to time or frequency, and a reduction in distortion and noise. Also known as inverse feedback; reverse feedback; stabilized feedback. { 'neg·əd·iv 'fēd,bak }

negative impedance [ELECTR] An impedance such that when the current through it increases, the voltage drop across the impedance decreases. { 'neg·əd·iv im'pēd·əns }

negative-impedance repeater [ELECTR] A telephone repeater that provides an effective gain for voice-frequency signals by insertion into the line of a negative impedance that cancels out line impedances responsible for transmission losses. { 'neg·əd·iv im¦pēd·əns ri'pēd·ər }

negative logic [ELECTR] Logic circuitry in which the more positive voltage (or current level) represents the 0 state; the less positive level represents the 1 state. { 'neg·əd·iv ,läj·ik }

negative modulation [ELECTR] **1.** Television modulation system in which an increase in scene brightness corresponds to a decrease in amplitude-modulated transmitter power; used in United States analog television transmitters. **2.** Modulation in which an increase in brightness corresponds to a decrease in the frequency of a frequency-modulated facsimile transmitter. Also known as negative transmission. { 'neg·əd·iv ,mäj·ə'lā·shən }

negative picture phase [ELECTR] The video signal phase in which the signal voltage swings in a negative direction for an increase in brilliance. { 'neg·əd·iv 'pik·chər ,fāz }

negative transmission *See* negative modulation. { 'neg·əd·iv tranz'mish·ən }

negentropy *See* information content. { nə'gen·trə·pē }

nepit *See* nit. { 'nep·ət }

nest [COMPUT SCI] To include data or subroutines in other items of a similar nature with a higher hierarchical level so that it is possible to access or execute various levels of data or routines recursively. { nest }

nesting [COMPUT SCI] **1.** Inclusion of a routine wholly within another routine. **2.** Inclusion of a DO statement within a DO statement in FORTRAN. { 'nest·iŋ }

nesting storage *See* push-down storage. { 'nest·iŋ 'stȯr·ij }

net [COMMUN] A number of communication stations equipped for communicating with each other, often on a definite time schedule and in a definite sequence. { net }

net call sign [COMMUN] A call sign that represents all stations within a net. { 'net 'kȯl ,sīn }

net control station [COMMUN] Communications station having the responsibility of clearing traffic and exercising circuit discipline within a net. { 'net kən'trōl ,stā·shən }

net loss [COMMUN] The ratio of the power at the input of a transmission system to the power at the output; expressed in nepers, it is one-half the natural logarithm of this ratio, and in decibels it is 10 times the common logarithm of the ratio. { 'net 'lȯs }

network [COMMUN] A number of radio or television broadcast stations connected by fiber-optic cable, coaxial cable, radio, or wire lines, so all stations can broadcast the same program simultaneously. [COMPUT SCI] *See* computer network. { 'net,wərk }

network analyzer [COMPUT SCI] An analog computer in which networks are used to simulate power line systems or physical systems and obtain solutions to various problems before the systems are actually built. { 'net,wərk 'an·ə,līz·ər }

network architecture [COMMUN] The high-level design of a communications system, including the choice of hardware, software, and protocols. { 'net,wərk 'är·kə,tek·chər }

network control program [COMPUT SCI] A computer program that controls communications between multiple terminals and a mainframe. { ¦net,wərk kən'trōl ,prō·grəm }

network data structure [COMPUT SCI] The arrangement of data in a computer system into interconnected groupings of information according to relationships between groupings. { 'net,wərk 'dad·ə ,strək·chər }

networking [COMPUT SCI] The use of transmission lines to join geographically separated computers. { 'net,wərk·iŋ }

network operating system [COMPUT SCI] The system software of a local-area network, which manages the network's resources, handling multiple inputs concurrently and providing necessary security. Abbreviated NOS. { ,net,wərk 'äp·ə,rād·iŋ ,sis·təm }

network server *See* file server. { 'net,wərk ,sər·vər }

network system [COMPUT SCI] A type of database management system in which data records can be related in more general structures than in a hierarchical file, permitting a given record to have more than one parent. { 'net,wərk ,sis·təm }

network terminal protocol [COMMUN] A set of standards that allows the user of a computer connected to a network to log in on any other computer on the network. Also known as TELNET. { ¦net,wərk ¦tərm·ən·əl 'prōd·ə,kȯl }

network vulnerability scan [COMPUT SCI] The process of determining the connectivity of the protected subnetwork within a security perimeter of a distributed computing system, and then testing the strength of protection at all access points to the subnetwork. { ¦net,wərk ,vəl·nər·ə'bil·əd·ē ,skan }

neural network [COMPUT SCI] An information-processing device that utilizes a very large number of simple modules, and in which information is stored by components that at the same time effect connections between these modules. { 'nu̇r·əl 'net,wərk }

neutral operation [COMMUN] System whereby marking signals are formed by current impulses of one polarity, either positive or negative, and spacing signals are formed by reducing the current to zero or nearly zero. { 'nü·trəl ,äp·ə'rā·shən }

neutral stability [CONT SYS] Condition in which the natural motion of a system neither grows nor decays, but remains at its initial amplitude. { 'nü·trəl stə'bil·əd·ē }

new-band service [COMMUN] A broadcasting service that is allocated a portion of the radio frequency spectrum that was not previously used. { 'nü ,band ,sər·vis }

newsgroup [COMPUT SCI] A collection of computers on a wide-area network that form a discussion group on a particular topic, such that a message generated by any computer in the group is automatically distributed over the network to all the others. Also known as forum. { 'nüz ,grüp }

next-event file [COMPUT SCI] A portion of a computer simulation program which maintains a list of all events to be processed and updates the simulated time. { 'nekst i'vent ,fīl }

nexus [COMMUN] A connection or interconnection of a communications system, such as a data link or a network of branches and nodes. { 'nek·səs }

nibble [COMPUT SCI] A unit of computer storage or information equal to one-half a byte. { 'nib·əl }

NIF *See* noise improvement factor.

N-indicator *See* N-display. { 'en ,in·də,kād·ər }

nine's complement [COMPUT SCI] The radix-minus-1 complement of a numeral whose radix is 10. { 'nīnz 'käm·plə·mənt }

Nipkow disk [COMPUT SCI] In optical character recognition, a disk having one or more spirals of holes around the outer edge, with successive openings positioned so that rotation of the disk provides mechanical scanning, as of a document. { 'nip·kō,disk }

nit [COMMUN] A unit of information content such that the information content of a symbol or message in nits is the negative of the natural logarithm of the probability of selecting that symbol or message from all the symbols or messages which could have been chosen. Also known as nepit. { nit }

n-key rollover [COMPUT SCI] The ability of a computer-terminal keyboard to remember the order in which keys were operated and pass this information to the computer even when several keys are depressed before other keys have been released. { 'en ,kē 'rōl,ō·vər }

N-level address [COMPUT SCI] A multilevel address specifying N levels of addressing. { 'en ,lev·əl 'ad,res }

N-level logic [ELECTR] An arrangement of gates in a digital computer in which not more than N gates are connected in series. { 'en ,lev·əl 'läj·ik }

N-modular redundancy [COMPUT SCI] A generalization of triple modular redundancy in which there are N identical units, where N is any odd number. { 'en ¦mäj·ə·lər ri'dən·dən·sē }

NMOS [ELECTR] Metal-oxide semiconductors that are made on *p*-type substrates, and whose active carriers are electrons that migrate between *n*-type source and drain contacts. Derived from *n*-channel metal-oxide semiconductor. { 'en ,mȯs }

no-address instruction [COMPUT SCI] An instruction which a computer can carry out without using an operand from storage. { 'nō 'ad,res in ,strək·shən }

node [ELECTR] A junction point within a network. { nōd }

noise [COMMUN] Unwanted electrical signal disturbances. { nȯiz }

noise digit [COMPUT SCI] A digit, usually 0, inserted into the rightmost position of the mantissa of a floating point number during a left-shift operation associated with normalization. Also known as noisy digit. { 'nȯiz ,dij·ət }

noise distortion [COMMUN] Noise on a communications facility which exceeds standards governing acceptable levels and which negatively affects the signal. { 'nȯiz di,stȯr·shən }

noise factor [ELECTR] The ratio of the total noise power per unit bandwidth at the output of a system to the portion of the noise power that is due to the input termination, at the standard noise temperature of 290 K. Also known as noise figure. { 'nȯiz ,fak·tər }

noise figure *See* noise factor. { 'nȯiz ,fig·yər }

noise generator [ELECTR] A device which produces (usually random) electrical noise, for use in tests of the response of electrical systems to noise, and in measurements of noise intensity. Also known as noise source. { 'nȯiz ,jen·ə ,rād·ər }

noise grade [COMMUN] Number which defines the relative noise at a particular location with respect to other locations throughout the world. { 'nȯiz ,grād }

noise improvement factor [COMMUN] In pulse modulation, the receiver output signal-to-noise ratio divided by the receiver input signal-to-noise ratio. Abbreviated NIF. Also known as improvement factor; signal-to-noise improvement factor. { 'nȯiz im'prüv·mənt ,fak·tər }

noise jammer [ELECTR] An electronic jammer that emits a carrier modulated with recordings or synthetic reproductions of natural atmospheric noise; the radio-frequency carrier may be suppressed; used to discourage the enemy by simulating naturally adverse communications conditions. { 'nȯiz ,jam·ər }

noise jamming [ELECTR] The emission of a radio-frequency carrier modulated with a white noise signal, derived from a gas-discharge tube of other broadband noise source, appearing in an enemy radar as background noise, tending to mask the radar echo or, in communications, the radio signal of interest. { 'nȯiz ,jam·iŋ }

noise killer [ELECTR] **1.** Device installed in a circuit to reduce its interference to other circuits. **2.** *See* noise suicide circuit. { 'nȯiz ,kil·ər }

noiseless channel [COMMUN] In information theory, a communications channel in which the effects of random influences are negligible, and there is essentially no random error. { 'nȯiz·ləs 'chan·əl }

noise limiter [ELECTR] A limiter circuit that cuts off all noise peaks that are stronger than the highest peak in the desired signal being received, thereby reducing the effects of atmospheric or human-produced interference. Also known as noise silencer; noise suppressor. { 'nȯiz ,lim·əd·ər }

noise-metallic [ELECTR] In telephone communications, weighted noise current in a metallic circuit at a given point when the circuit is terminated at that point in the nominal characteristic impedance of the circuit. { 'nȯiz mə'tal·ik }

noise-reducing antenna system [ELECTROMAG] Receiving antenna system so designed that only the antenna proper can pick up signals; it is placed high enough to be out of the noise-interference zone, and is connected to the receiver with a shielded cable or twisted transmission line that is incapable of picking up signals. { 'nȯiz ri¦düs·iŋ an'ten·ə ,sis·təm }

noise silencer *See* noise limiter. { 'nȯiz ,sī·lən·sər }

noise source *See* noise generator. { 'nȯiz ,sȯrs }

noise suicide circuit [ELECTR] A circuit which reduces the gain of an amplifier for a short period whenever a sufficiently large noise pulse is received. Also known as noise killer. { 'nȯiz 'sü·ə,sīd ,sər·kət }

noise suppression [ELECTR] Any method of reducing or eliminating the effects of undesirable electrical disturbances, as in frequency modulation whenever the signal carrier level is greater than the noise level. { 'nȯiz sə,presh·ən }

noise suppressor [ELECTR] **1.** A circuit that blocks the audio-frequency amplifier of a radio receiver automatically when no carrier is being received, to eliminate background noise. Also known as squelch circuit. **2.** A circuit that reduces record surface noise when playing phonograph records, generally by means of a filter that blocks out the higher frequencies where such noise predominates. **3.** *See* noise limiter. { 'nȯiz sə,pres·ər }

noise weighting [ELECTR] Use of an electrical network to obtain a weighted average over frequency of the noise power, which is representative of the relative disturbing effects of noise in a communications system at various frequencies. { 'nȯiz ,wād·iŋ }

noisy channel [COMMUN] In information theory, a communications channel in which the effects of random influences cannot be dismissed. { 'nȯiz·ē 'chan·əl }

noisy digit *See* noise digit. { 'nȯiz·ē 'dij·ət }

noisy mode [COMPUT SCI] A floating-point arithmetic procedure associated with normalization in which "1" bits, rather than "0" bits, are introduced in the low-order bit position during the left shift. { 'nȯiz·ē ,mōd }

nominal band [COMMUN] Frequency band of a facsimile-signal wave equal in width to that between zero frequency and maximum modulating frequency; the frequency band occupied in the transmitting medium will, in general, be greater than the nominal band. { 'näm·ə·nəl ,band }

nominal bandwidth [COMMUN] The interval between the assigned frequency limits of a channel. [ENG] The difference between the nominal upper and lower cutoff frequencies of an acoustic or electric filter. { 'näm·ə·nəl 'band,width }

nonacoustic coupler [ELECTR] A type of modem that is built into a microcomputer or terminal and connects it directly to a telephone line. { ¦nän·ə'kü·stik 'kəp·lər }

nonambiguity [COMMUN] The property of a code in which any character can be recognized uniquely without reference to preceding characters or the spatial position of a character. { ¦nän ,am·bə'gyü·əd·ē }

nonanticipatory system *See* causal system. { ¦nän·an'tis·ə·pə,tȯr·ē ,sis·təm }

nonarithmetic shift *See* cyclic shift. { ¦nän,a ,rith'med·ik 'shift }

nonblocking access [COMMUN] Connection of the incoming line or trunk made within the switching center at all times, provided that the required outgoing line or trunk is not busy. { 'nän,bläk·iŋ 'ak,ses }

noncoherent integration [ELECTR] A radar signal processing technique in which the amplitudes of successive pulses from a single scene or target location are added for increased sensitivity; such integration in the excited phosphor of a cathode-ray-tube radar display representing one point in space is an elementary example of this process. { ¦nän·kō'hir·ənt 'in·tə'grā·shən }

noncomposite color picture signal [COMMUN] The signal in analog color television transmission that represents complete color picture information but excludes the line- and field-synchronizing signals. { ¦nän·kəm'päz·ət ¦kəl·ər 'pik·chər ,sig·nəl }

noncontacting piston *See* choke piston. { ¦nän 'kän,tak·tiŋ 'pis·tən }

noncontacting plunger *See* choke piston. { ¦nän'kän,tak·tiŋ 'plən·jər }

nondegenerative basic feasible solution [COMPUT SCI] In linear programming, a basic feasible solution with exactly *m* positive variables x_i, where *m* is the number of constraint equations. { ¦nän·di'jen·rəd·iv 'bā·sik 'fēz·ə·bəl sə'lü·shən }

nondeletable message [COMPUT SCI] A message that appears on a computer display which can be removed only by entering a specific command. { ,nän·di'lēd·ə·bəl 'mes·ij }

nondestructive read [COMPUT SCI] A reading process that does not erase the data in memory; the term sometimes includes a destructive read immediately followed by a restorative write-back. Also known as nondestructive readout (NDRO). { ¦nän di'strək·div 'rēd }

nondestructive readout *See* nondestructive read. { ¦nän·di'strək·div 'rēd,au̇t }

nondirectional *See* omnidirectional. { ¦nän·di 'rek·shən·əl }

nondirectional antenna *See* omnidirectional antenna. { ¦nän·di'rek·shən·əl an'ten·ə }

nonerasable storage *See* read-only memory. { ¦nän·i'rās·ə·bəl 'stȯr·ij }

nonexecutable statement [COMPUT SCI] A statement in a higher-level programming language which cannot be related to the instructions in the machine language program ultimately produced, but which provides the compiler with essential information from which it may determine the allocation of storage and other organizational characteristics of the final program. { ¦nän,ek·sə'kyüd·ə·bəl 'stāt·mənt }

nonfatal error [COMPUT SCI] An error in a computer program which does not result in termination of execution, but which causes the processor to invent an interpretation, issue a warning, and continue processing. { 'nän,fād·əl 'er·ər }

nonfunctional packages software [COMPUT SCI] General-purpose software which permits the user to handle her or his particular applications requirements with little or no additional program or systems design work, or to perform certain specialized computational functions. { nän'fəŋk·shən·əl 'pak·ij·əz 'sȯft ,wer }

nongraphic character [COMPUT SCI] A set of signals that, when sent to a printer, results in a control action, such as carriage return, line feed, or tab, rather than the generation of a printed character. { ¦nän'graf·ik 'kar·ik·tər }

nonintelligible crosstalk [COMMUN] Crosstalk which cannot be understood regardless of its received volume, but which because of its syllabic nature is more annoying subjectively than thermal-type noise. { ˌnän·in'tel·ə·jə·bəl 'krȯs ˌtȯk }

noninverting amplifier [ELECTR] An operational amplifier in which the input signal is applied to the ungrounded positive input terminal to give a gain greater than unity and make the output voltage change in phase with the input voltage. { ¦nän·in'vərd·iŋ 'am·plə,fī·ər }

nonlinear control system [CONT SYS] A control system that does not have the property of superposition, that is, one in which some or all of the outputs are not linear functions of the inputs. { 'nän,lin·ē·ər kən'trōl ˌsis·təmz }

nonlinear crosstalk [COMMUN] Interaction between channels occupying different wavelengths in a wavelength-division-multiplexed system because of optical nonlinearities in the transmission medium. { ˌnän,lin·ē·ər 'krȯs,tȯk }

nonlinear distortion [ELECTR] Distortion in which the output of a system or component does not have the desired linear relation to the input. { 'nän,lin·ē·ər di'stȯr·shən }

nonlinear feedback control system [CONT SYS] Feedback control system in which the relationships between the pertinent measures of the system input and output signals cannot be adequately described by linear means. { 'nän ˌlin·ē·ər 'fēd,bak kən'trōl ˌsis·təm }

nonlinear fiber amplifier [COMMUN] An optical amplifier in which nonlinear interactions (stimulated Raman and Brillouin scattering and four-wave mixing) between pump light and the signal cause transfer of power to the signal, resulting in fiber gain. { ˌnän¦lin·ē·ər ˌfī·bər 'am·plə,fī·ər }

nonlinear optical device [OPTICS] A device based on one of a class of optical effects that result from the interaction of electromagnetic radiation from lasers with nonlinear materials. { 'nän,lin·ē·ər 'äp·tə·kəl di,vīs }

nonlinear optics [OPTICS] The study of the interaction of radiation with matter in which certain variables describing the response of the matter (such as electric polarization or power absorption) are not proportional to variables describing the radiation (such as electric field strength or energy flux). { 'nän,lin·ē·ər 'äp,tiks }

nonlinear oscillator [ELECTR] A radio-frequency oscillator that changes frequency in response to an audio signal; it is the basic circuit used in eavesdropping devices. { 'nän,lin·ē·ər 'äs·ə ˌlād·ər }

nonlinear programming [MATH] A branch of applied mathematics concerned with finding the maximum or minimum of a function of several variables, when the variables are constrained to yield values of other functions lying in a certain range, and either the function to be maximized or minimized, or at least one of the functions whose value is constrained, is nonlinear. { 'nän ˌlin·ē·ər 'prō,gram·iŋ }

nonmaintenance time [COMPUT SCI] The elapsed time during scheduled working hours between the determination of a machine failure and placement of the equipment back into operation. { ¦nän'mānt·ən·əns ˌtīm }

nonnumeric character [COMPUT SCI] Any character except a digit. { ¦nän·nü'mer·ik 'kar·ik·tər }

nonnumeric programming [COMPUT SCI] Computer programming that deals with objects other than numbers. { ¦nän·nü,mer·ik 'prō,gram·iŋ }

nonpreemptive multitasking *See* cooperative multitasking. { ˌnän·prē¦em·tiv 'məl·tē,task·iŋ }

nonprint code [COMPUT SCI] A bit combination which is interpreted as no printing, no spacing. { ¦nän¦print 'kōd }

nonpriority interrupt [COMPUT SCI] Any one of a group of interrupts which may be disregarded by the central processing unit. { ¦nän·pri'är·əd·ē 'int·ə,rəpt }

nonprocedural language [COMPUT SCI] A programming language in which the program does not follow the actual steps a computer follows in executing a program. { ˌnän·prə'sē·jə·rəl 'laŋ ˌgwij }

nonrecoverable error [COMPUT SCI] An error detected during computer processing that cannot be handled by the computer system and therefore causes processing to be interrupted. { ¦nän·ri'kəv·rə·bəl 'er·ər }

nonrecursive filter [ELECTR] A digital filter that lacks feedback; that is, its output depends on present and past input values only and not on previous output values. { ˌnän·ri,kər·siv 'fil·tər }

nonredundant system [COMPUT SCI] A computer system designed in such a way that only the absolute minimum amount of hardware is utilized to implement its function. { ¦nän·ri'dən·dənt 'sis·təm }

nonreproducing code [COMPUT SCI] A code which normally does not appear as such in a generated output but will result in a function such as paging or spacing. { ¦nän,rē·prə'dü·siŋ 'kōd }

nonresident routine [COMPUT SCI] Any computer routine which is not stored permanently in the memory but must be read into memory from a data carrier or external storage device. { ¦nän'rez·ə·dənt rü,tēn }

nonresonant antenna [ELECTROMAG] A long-wire or traveling-wave antenna which does not have natural frequencies of oscillation, and responds equally well to radiation over a broad range of frequencies. { ¦nän'rez·ən·ənt an'ten·ə }

non-return-to-zero [COMPUT SCI] A mode of recording and readout in which it is not necessary for the signal to return to zero after each item of recorded data. Abbreviated NRZ. { 'nän ri,tərn tə 'zir·ō }

nonrotating disk *See* semiconductor disk. { ¦nän'rō,tād·iŋ 'disk }

nonscrollable message [COMPUT SCI] A message on a computer display that does not scroll off the top of the display as new information is written at the bottom. { ¦nän'skrō·lə·bəl 'mes·ij }

nonservo robot *See* fixed-stop robot. { ¦nän 'sər·vō 'rō,bät }

nonshared control unit [COMPUT SCI] A control unit relating to only one device. Also known as unipath. { ¦nän¦sherd kən'trōl ˌyü·nət }

nonstop computer [COMPUT SCI] A computer system that is equipped with duplicate components or excess capacity so that a hardware or software failure will not interrupt processing. { ¦nän'stäp kəm'pyüd·ər }

nonstorage camera tube [ELECTR] Television camera tube in which the picture signal is, at each instant, proportional to the intensity of the illumination on the corresponding area of the scene. { ¦nän'stȯr·ij 'kam·rə ˌtüb }

nonswappable program [COMPUT SCI] A program that is given priority status so that its execution cannot be suspended to allow execution of other programs. { ¦nän'swäp·ə·bəl 'prō·grəm }

nonsynchronous timer [ELECTR] A circuit at the receiving end of a communications link which restores the time relationship between pulses when no timing pulses are transmitted. { ¦nän'siŋ·krə·nəs 'tīm·ər }

nonsynchronous transmission [ELECTR] A data transmission process in which a clock is not used to control the unit intervals within a block or a group of data signals. { ¦nän'siŋ·krə·nəs tranz'mish·ən }

nonuniform memory access machine [COMPUT SCI] A multiprocessor in which the memory is spread out over memory modules, which are attached to the processors, so that each processor has its own memory module. Abbreviated NUMA machine. { ˌnän¦yün·iˌfȯrm ¦mem·rē 'ak·ses mə ˌshēn }

nonvolatile memory *See* nonvolatile storage. { ¦nän'väl·ə·təl 'mem·rē }

nonvolatile random-access memory [COMPUT SCI] A semiconductor storage device which has two memory cells for each bit, one of which is volatile, as in a static RAM (random-access memory), and provides unlimited read and write operations, while the other is nonvolatile, and provides the ability to retain information when power is removed. Abbreviated NV RAM. { ¦nän'väl·ə·təl ¦ran·dəm ¦akˌses 'mem·rē }

nonvolatile storage [COMPUT SCI] A computer storage medium that retains information in the absence of power, such as a magnetic tape, drum, or core. Also known as nonvolatile memory. { ¦nän'väl·ə·təl 'stȯr·ij }

NO OP [COMPUT SCI] An instruction telling the computer to do nothing, except to proceed to the next instruction in sequence. Also known as do-nothing instruction; no-operation instruction. { 'nō ˌäp }

no-operation instruction *See* NO OP. { ¦nō ˌäp·ə'rā·shən in'strək·shən }

NOR circuit [ELECTR] A circuit in which output voltage appears only when signal is absent from all of its input terminals. { 'nȯr ˌsər·kət }

normal direction flow [COMPUT SCI] The direction from left to right or top to bottom in flow charting. { 'nȯr·məl di'rek·shən ˌflō }

normal form [COMPUT SCI] The form of a floating-point number whose mantissa lies between 0.1 and 1.0. { 'nȯr·məl 'fȯrm }

normalization [COMPUT SCI] Breaking down of complex data structures into flat files. { ˌnȯr·mə·lə'zā·shən }

normalize [COMPUT SCI] **1.** To adjust the representation of a quantity so that this representation lies within a prescribed range. **2.** In particular, to adjust the exponent and mantissa of a floating point number so that the mantissa falls within a prescribed range. { 'nȯr·məˌlīz }

normal mode [COMPUT SCI] Operation of a computer in which it executes its own instructions rather than those of a different computer. { 'nȯr·məl ˌmōd }

normal-mode helix [ELECTROMAG] A type of helical antenna whose diameter and electrical length are considerably less than a wavelength, and which has a radiation pattern with greatest intensity normal to the helix axis. { 'nȯr·məl ˌmōd 'hēˌliks }

normal orientation [COMPUT SCI] In optical character recognition, that determinate position which indicates that the line elements of an inputted source document appear parallel with the document's leading edge. { 'nȯr·məl ˌȯr·ē·ən'tā·shən }

normal range [COMPUT SCI] An interval within which results are expected to fall during normal operations. { 'nȯr·məl 'rānj }

North American presentation-level protocol syntax [COMMUN] A format for transmitting text and graphics that allows the transmission of large amounts of information over narrow-bandwidth transmission lines. Abbreviated NAPLPS. { ¦nȯrth ə¦mer·ə·kən ˌprē·zən¦tā·shən ˌlev·əl ˌprōd·əˌkȯl 'sinˌtaks }

NOS *See* network operating system. { ¦en¦ō'es *or* 'näs }

notation *See* positional notation. { nō'tā·shən }

notch antenna [ELECTROMAG] Microwave antenna in which the radiation pattern is determined by the size and shape of a notch or slot in a radiating surface. { 'näch anˌten·ə }

notch filter [ELECTR] A band-rejection filter that produces a sharp notch in the frequency response curve of a system; used in television transmitters to provide attenuation at the low-frequency end of the channel, to prevent possible interference with the sound carrier of the next lower channel. { 'näch ˌfil·tər }

NOT circuit [ELECTR] A logic circuit with one input and one output that inverts the input signal at the output; that is, the output signal is a logical 1 if the input signal is a logical 0, and vice versa. Also known as inverter circuit. { 'nät ˌsər·kət }

notebook computer [COMPUT SCI] A portable computer typically weighing less than 6 pounds (3 kilograms) that has a flat-panel display and miniature hard disk drives, and is powered by rechargeable batteries. Also known as laptop computer. { 'nōtˌbük kəmˌpyüd·ər }

nought state *See* zero condition. { 'nȯt ˌstāt }

NP [COMPUT SCI] The class of decision problems for which solutions can be checked in polynomial time.

NP-complete problem [COMPUT SCI] One of the hardest problems in class NP, such that, if there are any problems in class NP but not in class P, this is one of them. { ¦en¦pē kəm'plēt ˌpräb·ləm }

NP-hard [COMPUT SCI] Referring to problems at least as hard as or harder than any problem in NP. Given a method for solving an NP-hard problem, any problem in NP can be solved with only polynomially more work. { ¦en¦pē 'härd }

N-plus-one address instruction [COMPUT SCI] An instruction with N + 1 address parts, one of which gives the location of the next instruction to be carried out. { 'en pləs 'wən 'adˌres in ˌstrək·shən }

NPX *See* numeric processor extension.

NRSC *See* National Radio Systems Committee.

NRZ *See* non-return-to-zero.

N-scan *See* N-display. { 'en ˌskan }

N-scope *See* N-display. { 'en ˌskōp }

nt *See* nit.

NTSC *See* National Television System Committee.

n-type conduction [ELECTR] The electrical conduction associated with electrons, as opposed to holes, in a semiconductor. { 'en ˌtīp kən ˌdək·shən }

n-type semiconductor [ELECTR] An extrinsic semiconductor in which the conduction electron density exceeds the hole density. { 'en ˌtīp 'sem·i·kənˌdək·tər }

nuclear converter *See* converter. { 'nü·klē·ər kən'vərd·ər }

nucleus [COMPUT SCI] **1.** That portion of the control program that must always be present in main storage. **2.** The main storage area used in the nucleus (first definition) and other transient control program routines. { 'nü·klē·əs }

null character [COMPUT SCI] A control character used as a filler in data processing; may be inserted or removed from a sequence of characters without affecting the meaning of the sequence, but may affect format or control equipment. { 'nəl ˌkar·ik·tər }

null modem cable [COMPUT SCI] A cable that connects two local computers via serial ports without the use of a modem. { ¦nəl 'mōˌdem ˌkā·bəl }

NUMA machine *See* nonuniform memory access machine. { 'nü·mə mə'shēn }

number cruncher [COMPUT SCI] A computer with great power to carry out computations, designed to maximize this ability rather than to process large amounts of data. { 'nəm·bər ˌkrən·chər }

number record printer [COMMUN] A printer in a relay station that provides a complete automatic written record of channel numbers and the fixed routing line associated with each message that is relayed through that particular station. { 'nəm·bər ¦rek·ərd ˌprint·ər }

numeric [COMPUT SCI] In computers, pertaining to data composed wholly or partly of digits, as distinct from alphabetic. { nü'mer·ik }

numerical analysis [MATH] The study of approximation techniques using arithmetic for solutions of mathematical problems. { nü'mer·i·kəl ə'nal·ə·səs }

numerical decrement *See* decrement. { nü 'mer·i·kəl 'dek·rə·mənt }

numerical tape [COMPUT SCI] The tape required by a computer operating a machine tool. { nü'mer·i·kəl 'tāp }

numeric character *See* digit. { nü'mer·ik 'kar·ik·tər }

numeric character set [COMPUT SCI] A character set that includes only digits and certain special characters, such as plus and minus signs and control characters. { nü'mer·ik 'kar·ik·tər ˌset }

numeric coding [COMPUT SCI] Code in which only digits are used, usually binary or octal. { nü'mer·ik 'kōd·iŋ }

numeric control [COMPUT SCI] The action of programs written for specialized computers which operate machine tools. { nü'mer·ik kən'trōl }

numeric coprocessor *See* numeric processor extension. { nü¦mer·ik kō'präˌses·ər }

numeric data [COMPUT SCI] Data consisting of digits and not letters of the alphabet or special characters. { nü'mer·ik 'dad·ə }

numeric format [COMPUT SCI] The manner in which numbers are displayed in the cells of a particular spreadsheet. { nü¦mer·ik 'fȯrˌmat }

numeric keypad [COMPUT SCI] A section of a computer keyboard that contains a group of keys, usually about 12, arranged in compact fashion for entering numeric characters efficiently. Also known as numeric pad. { nü'mer·ik 'kēˌpad }

numeric pad *See* numeric keypad. { nü'mer·ik 'pad }

numeric pager [COMMUN] A receiver in a radio paging system that contains a display device that can show numeric messages, most commonly a telephone number. { nü'mer·ik 'pā·jər }

numeric printer [COMPUT SCI] Old type of printer which positioned its keys to print a field in one operation, rather than one digit at a time. { nü'mer·ik 'print·ər }

numeric processor extension [COMPUT SCI] A specialized integrated circuit that is added to a computer to perform high-speed floating-point mathematical calculations. Abbreviated NPX. Also known as arithmetic processor; math coprocessor; numeric coprocessor. { nü¦mer·ik 'präˌses·ər ikˌsten·chən }

numeric variable [COMPUT SCI] The symbolic name of a data element whose value changes during the carrying out of a computer program. { nü'mer·ik 'ver·ē·ə·bəl }

N unit [OPTICS] A unit of index of refraction; a mathematical simplification designed to replace rather awkward numbers involved in the values of the index of refraction *n* for the atmosphere;

it is defined by the relation $N = (n - 1)10^6$. { 'en ,yü·nət }

nutating antenna [ENG] An antenna system used in conical scan radar, in which a dipole or feed horn moves in a small circular orbit about the axis of a paraboloidal reflector without changing its polarization. { 'nü,tād·iŋ an'ten·ə }

nutator [ENG] A mechanical or electrical device used to move a radar beam in a circular, conical, spiral, or other manner periodically to obtain greater air surveillance than could be obtained with a stationary beam. { 'nü,tād·ər }

NV RAM *See* nonvolatile random-access memory. { ¦en¦vē 'ram }

nybble [COMPUT SCI] A string of bits, smaller than a byte, operated on as a unit. { 'nib·əl }

Nyquist contour [CONT SYS] A directed closed path in the complex frequency plane used in constructing a Nyquist diagram, which runs upward, parallel to the whole length of the imaginary axis at an infinitesimal distance to the right of it, and returns from $+j\infty$ to $-j\infty$ along a semicircle of infinite radius in the right half-plane. { 'nī,kwist ,kän,tu̇r }

Nyquist diagram [CONT SYS] A plot in the complex plane of the open loop transfer function as the complex frequency is varied along the Nyquist contour; used to determine stability of a control system. { 'nī,kwist ,dī·ə,gram }

Nyquist interval [COMMUN] Maximum separation in time which can be given to regularly spaced instantaneous samples of a wave of specified bandwidth for complete determination of the waveform of the signal. { 'nī,kwist ,in·tər·vəl }

Nyquist rate [COMMUN] The maximum rate at which code elements can be unambiguously resolved in a communications channel with a limited range of frequencies; equal to twice the frequency range. { 'nī,kwist ,rāt }

Nyquist sampling [COMMUN] The periodic sampling of audio or video signals, in order to preserve their information content, at a rate equal to twice the highest frequency to be preserved. { 'nī,kwist ,sam·pliŋ }

Nyquist stability criterion *See* Nyquist stability theorem. { 'nī,kwist stə'bil·əd·ē krī,tir·ē·ən }

Nyquist stability theorem [CONT SYS] The theorem that the net number of counterclockwise rotations about the origin of the complex plane carried out by the value of an analytic function of a complex variable, as its argument is varied around the Nyquist contour, is equal to the number of poles of the variable in the right half-plane minus the number of zeros in the right half-plane. Also known as Nyquist stability criterion. { 'nī,kwist stə'bil·əd·ē ,thir·əm }

O

object [COMPUT SCI] **1.** Any collection of related items. **2.** The name of a single element in an object-oriented programming language. { 'äb·jekt }

object code [COMPUT SCI] The statements generated from source code by a compiler, constituting an intermediate step in the translation of source code into executable machine language. { 'äb,jekt ,kōd }

object computer [COMPUT SCI] The computer processing an object program; the same computer compiling the source program could, therefore, be called the source computer, such terminology is seldom used in practice. { 'äb·jekt kəm,pyüd·ər }

object deck [COMPUT SCI] The set of machine-readable computer instructions produced by a compiler, either in absolute format (that is, containing only fixed addresses) or, more frequently, in relocatable format. { 'äb·jekt ,dek }

objective testing [ENG] Using test equipment to directly measure the performance of a system under a test; for example, the power output of a transmitter can be objectively measured using a wattmeter. { äb'jek·tiv ,test·iŋ }

object language [COMPUT SCI] The intended and desired output language in the translation or conversion of information from one language to another. { 'äb·jekt ,laŋ·gwij }

object library *See* object program library. { 'äb·jekt'lī,brer·ē }

Object Management Group object model [COMPUT SCI] A model that defines common object semantics in an object-oriented computer system. Abbreviated OMG object model. { ¦äb·jikt ¦man·ij·mənt ¦grüp 'äb·jikt ,mäd·əl }

object module [COMPUT SCI] The computer language program prepared by an assembler or a compiler after acting on a programmer-written source program. { 'äb·jekt ,mäj·ül }

object-oriented graphics *See* vector graphics. { ¦äb,jekt ,ȯr·ē,en·təd 'graf·iks }

object-oriented interface [COMPUT SCI] A user interface that employs icons and a mouse. { ¦äb ,jekt ,ȯr·ē,en·təd 'in·tər,fās }

object-oriented language [COMPUT SCI] A programming language consisting of a sequence of commands directed at objects. { ¦äb,jekt ,ȯr·ē ,en·təd 'laŋ·gwij }

object-oriented programming [COMPUT SCI] A computer programming methodology that focuses on data rather than processes, with programs composed of self-sufficient modules (objects) containing all the information needed to manipulate a data structure. Abbreviated OOP. { ¦äb,jekt ,ȯr·ē,en·təd 'prō,gram·iŋ }

object program [COMPUT SCI] The computer language program prepared by an assembler or a compiler after acting on a programmer-written source program. Also known as object routine; target program, target routine. { 'äb·jekt ,prō ,gram }

object program library [COMPUT SCI] A collection of computer programs in the form of relocatable instructions, which reside on, and may be read from, a mass storage device. Also known as object library. { 'äb·jekt ,prō,gram ,lī ,brer·ē }

object request broker [COMPUT SCI] The central component of CORBA, which passes requests from clients to the objects on which they are invoked. Abbreviated ORB. { 'äb·jikt ri¦kwest ,brō·kər }

object routine *See* object program. { 'äb·jekt rü ,tēn }

object time [COMPUT SCI] The time during which execution of an object program is carried out. { 'äb·jekt ,tīm }

oblique-incidence transmission [COMMUN] Transmission of a radio wave obliquely up to the ionosphere and down again. { ə'blēk ¦in·sə·dəns tranz'mish·ən }

observability [CONT SYS] Property of a system for which observation of the output variables at all times is sufficient to determine the initial values of all the state variables. { əb ,zər·və'bil·əd·ē }

observation spillover [CONT SYS] The part of the sensor output of an active control system caused by modes that have been omited from the control algorithm in the process of model reduction. { ,äb·zər'vā·shən 'spil,ō·vər }

observer [CONT SYS] A linear system B driven by the inputs and outputs of another linear system A which produces an output that converges to some linear function of the state of system A. Also known as state estimator; state observer. { əb'zər·vər }

occlusion [COMPUT SCI] In computer vision, the obstruction of a view. { ə'klü·zhən }

occupied bandwidth [COMMUN] Frequency bandwidth such that, below its lower and above its upper frequency limits, the mean powers radiated are each equal to 0.5% of the total mean power radiated by a given emission. { 'äk·yə,pīd 'band,width }

OCR *See* optical character recognition.

octal debugger [COMPUT SCI] A simple debugging program which permits only octal (instead of symbolic) address references. { 'äkt·əl dē'bəg·ər }

octonary signaling [COMMUN] A communications mode in which information is passed by the presence and absence of plus and minus variation of eight discrete levels of one parameter of the signaling medium. { 'äk·tə,ner·ē 'sig·nə·liŋ }

odd-even check [COMPUT SCI] A means of detecting certain kinds of errors in which an extra bit, carried along with each word, is set to zero or one so that the total number of zeros or ones in each word is always made even or always made odd. Also known as parity check. { 'äd 'ē·vən ,chek }

odd parity [COMPUT SCI] Property of an expression in binary code which has an odd number of ones. { 'äd 'par·əd·ē }

odd parity check [COMPUT SCI] A parity check in which the number of 0's or 1's in each word is expected to be odd; if the number is even, the check bit is 1, and if the number is odd, the check bit is 0. { 'äd 'par·əd·ē ,chek }

O-display [ELECTR] A radar display format in which an adjustable notch, absenting any trace, is moved in an A-display to assist the operator in determining and reporting the range of a target. Also known as O-indicator; O-scan; O-scope. { 'ō di,splā }

OEM [ELECTR] Abbreviation for original equipment manufacturer. Generally describes original, factory-installed equipment.

off-center plan position indicator [ELECTR] A plan position indicator in which the center of the display that represents the location of the radar can be moved from the center of the screen to any position on the face of the PPI. { 'ȯf ,sent·ər 'plan pə,zish·ən 'in·də,kād·ər }

off-hook [COMMUN] The active state (closed loop) of a subscriber or PBX user loop. { 'ȯf ,hu̇k }

off-hook service [COMMUN] Priority telephone service for key personnel that affords a connection from caller to receiver by the simple expedient of removing the phone from its cradle or hook. { 'ȯf ,hu̇k ,sər·vəs }

off-line [COMPUT SCI] Describing equipment not connected to a computer, or temporarily disconnected from one. { 'ȯf ¦līn }

off-line cipher [COMMUN] Method of encrypting which is not associated with a particular transmission system and in which the resulting encrypted message can be transmitted by any means. { 'ȯf ¦līn 'sī·fər }

off-line equipment [COMPUT SCI] Peripheral equipment or devices not in direct communication with the central processing unit of a computer. Also known as auxiliary equipment. { 'ȯf ¦līn i'kwip·mənt }

off-line mode [COMPUT SCI] Any operation, such as printing, which does not involve the main computer. { 'ȯf ¦līn 'mōd }

off-line operation [COMPUT SCI] Operation of peripheral equipment in conjunction with, but not under the control of, the central processing unit. { 'ȯf ¦līn ,äp·ə'rā·shən }

off-line processing [COMPUT SCI] Any processing which takes place independently of the central processing unit. { 'ȯf ¦līn 'prä,ses·iŋ }

off-line storage [COMPUT SCI] A storage device not under control of the central processing unit. { 'ȯf ¦līn 'stȯr·ij }

off-line unit [COMPUT SCI] Any operation device which is not attached to the main computer. { 'ȯf ¦līn 'yü·nət }

offload [COMPUT SCI] To transfer operations from one computer to another, usually from a large computer to a smaller one. { 'ȯf,lōd }

offset [COMPUT SCI] *See* displacement. [CONT SYS] The steady-state difference between the desired control point and that actually obtained in a process control system. { 'ȯf,set }

offset-center plan position indicator *See* off-center plan position indicator. { 'ȯf,set ¦sen·tər 'plan pə,zish·ən 'in·də,kād·ər }

offset plan position indicator *See* off-center plan position indicator. { 'ȯf,set 'plan pə ,zish·ən 'in·də,kād·ər }

offset voltage [ELECTR] The differential input voltage that must be applied to an operational amplifier to return the zero-frequency output voltage to zero volts, due to device mismatching at the input stage. { 'ȯf,set ,vōl·tij }

O-indicator *See* O-display. { 'ō,in·də,kād·ər }

OL *See* only loadable.

OLRT system *See* on-line real-time system. { ,ō,el,är'tē ,sis·təm }

OMG object model *See* Object Management Group object model. { ¦ō¦em¦jē 'äb·jikt ,mäd·əl }

omission factor [COMPUT SCI] In information retrieval, the ratio obtained in dividing the number of nonretrieved relevant documents by the total number of relevant documents in the file. { ō'mish·ən ,fak·tər }

omnibearing distance navigation [NAV] Navigation based upon polar coordinates relative to a reference point. Also known as r-theta navigation; rho-theta navigation. { 'äm·nə,ber·iŋ ¦dis·təns ,nav·ə'gā·shən }

omnidirectional [ELECTR] Radiating or receiving equally well in all directions. Also known as nondirectional. { ¦äm·nə·di'rek·shən·əl }

omnidirectional antenna [ELECTROMAG] An antenna that has an essentially circular radiation pattern in azimuth and a directional pattern in elevation. Also known as nondirectional antenna. { ¦äm·nə·di'rek·shən·əl an'ten·ə }

OMR *See* optical mark reading.

onboard [COMPUT SCI] Referring to a computer hardware component that is built directly into the computer. { 'ȯn'bȯrd }

on-call circuit [COMMUN] A permanently designated circuit that is activated only upon request of the user; this type of circuit is usually provided when a full-period circuit cannot be justified and the duration of use cannot be anticipated; during unactivated periods, the communications facilities required for the circuit are available for other requirements. { 'ȯn 'kȯl ˌsər·kət }

one-address code [COMPUT SCI] In computers, a code using one-address instructions. { 'wən əˌdres 'kōd }

one-address instruction [COMPUT SCI] A digital computer programming instruction that explicitly describes one operation and one storage location. Also known as single-address instruction. { 'wən əˌdres in'strək·shən }

one condition [COMPUT SCI] The state of a magnetic core or other computer memory element in which it represents the value 1. Also known as one state. { 'wən kənˌdish·ən }

one-digit subtracter *See* half-subtracter. { 'wən ˌdij·ət səb'trak·tər }

one-dimensional array [COMPUT SCI] A group of related data elements arranged in a single row or column. { ¦wən də|men·chən·əl ə'rā }

one-ended tape Turing machine [COMPUT SCI] A variation of a Turing machine in which the tape can be extended to the right, but not to the left. { 'wən ¦end·əd ¦tāp 'tu̇r·iŋ məˌshēn }

one-level address [COMPUT SCI] In digital computers, an address that directly indicates the location of an instruction or some data. { 'wən ˌlev·əl ə'dres }

one-level code [COMPUT SCI] Any code using absolute addresses and absolute operation codes. { 'wən ˌlev·əl 'kōd }

one-level subroutine [COMPUT SCI] A subroutine that does not use other subroutines during its execution. { 'wən ˌlev·əl 'səb·rüˌtēn }

one-line adapter [COMPUT SCI] A unit connecting central processes and permitting high-speed transfer of data under program control. { 'wən ˌlīn ə'dap·tər }

one-part code [COMMUN] Code in which the plain text elements are arranged in alphabetical or numerical order, accompanied by their code groups also arranged in alphabetical, numerical, or other systematic order. { 'wən ˌpärt 'kōd }

one-pass operation [COMPUT SCI] An operating method, now standard, which produces an object program from a source program in one pass. { 'wən ˌpas ˌäp·ə'rā·shən }

one-plus-one address instruction [COMPUT SCI] A digital computer instruction whose format contains two address parts; one address designates the operand to be involved in the operation; the other indicates the location of the next instruction to be executed. { 'wən ˌpləs 'wən ə¦dres inˌstrək·shən }

one-quadrant multiplier [ELECTR] Of an analog computer, a multiplier in which operation is restricted to a single sign of both input variables. { 'wən ˌkwä·drənt 'məl·təˌplī·ər }

one's complement [COMPUT SCI] A numeral in binary notation, derived from another binary number by simply changing the sense of every digit. { 'wənz 'käm·plə·mənt }

ones-complement code [COMPUT SCI] A number coding system used in some computers, where, for any number x, $x = (1 - 2^{n-1})\cdot a_0 + 2^{n-2}a_1 + \cdots + a_{n-1}$, where $a_i = 1$ or 0. { 'wənz 'käm·plə·mənt ˌkōd }

one-shot multivibrator *See* monostable multivibrator. { 'wən ˌshät ˌməl·tə'vīˌbrād·ər }

one-shot operation *See* single-step operation. { 'wən ˌshät ˌäp·ə'rā·shən }

one state *See* one condition. { 'wən ˌstāt }

one-step operation *See* single-step operation. { 'wən ˌstep ˌäp·ə'rā·shən }

one-time pad [COMMUN] A keying sequence based on random numbers that is used to code a single message and is then destroyed. { ¦wən ¦tīm 'pad }

one-to-many correspondence [COMPUT SCI] A structure that establishes relationships between two types of items in a data base such that one item of the first type can relate to several items of the second type, but items of the second type can relate back to only one item of the first type. { ¦wən tə ¦men·ē ˌkär·ə'spän·dəns }

one-to-one assembler [COMPUT SCI] An assembly program which produces a single instruction in machine language for each statement in the source language. Also known as one-to-one translator. { ¦wən tə ¦wən ə'sem·blər }

one-to-one translator *See* one-to-one assembler. { ¦wən tə ¦wən 'tranzˌlād·ər }

on-hook [COMMUN] The idle state (open loop) of a subscriber or PBX user loop. { 'ȯn ˌhu̇k }

on-line [COMPUT SCI] Pertaining to equipment capable of interacting with a computer. [ELECTR] The state in which a piece of equipment or a subsystem is connected and powered to deliver its proper output to the system. { 'ȯn ˌlīn }

on-line central file [COMPUT SCI] An organized collection of data, such as an on-line disk file, in a storage device under direct control of a central processing unit, that serves as a continually available source of data in applications where real-time or direct-access capabilites are required. { 'ȯn ˌlīn 'sen·trəl 'fīl }

on-line cipher [COMMUN] A method of encryption directly associated with a particular transmission system, whereby messages may be encrypted and simultaneously transmitted from one station to one or more stations where reciprocal equipment is automatically operated. { 'ȯn ˌlīn 'sī·fər }

on-line computer system [COMPUT SCI] A computer system which is adapted to on-line operation. { 'ȯn ˌlīn kəm'pyüd·ər ˌsis·təm }

on-line cryptographic operation *See* on-line operation. { 'ȯn ˌlīn ˌkrip·tə'graf·ik ˌäp·ə'rā·shən }

on-line data reduction [COMPUT SCI] The processing of information as rapidly as it is received by the computing system. { 'ȯn ˌlīn ˌdad·ə ri ˌdək·shən }

on-line disk file [COMPUT SCI] A magnetic disk directly connected to the central processing unit, thereby increasing the memory capacity of the computer. { 'ȯn ˌlīn 'disk ˌfīl }

on-line equipment [COMPUT SCI] The equipment or devices in a system whose operation is under control of the central processing unit, and in which information reflecting current activity is introduced into the data-processing system as soon as it occurs. { 'ȯn ˌlīn i'kwip·mənt }

on-line inquiry [COMPUT SCI] A level of computer processing that results from adding to an expanded batch system the capability to immediately access, from any terminal, any record that is stored in the disk files attached to the computer. { 'ȯn ˌlīn 'in·kwə·rē }

on-line mode [COMPUT SCI] Mode of operation in which all devices are responsive to the central processor. { 'ȯn ˌlīn ˌmōd }

on-line operation [COMMUN] A method of operation whereby messages are encrypted and simultaneously transmitted from one station to one or more other stations where reciprocal equipment is automatically operated to permit reception and simultaneous decryptment of the message. Also known as on-line cryptographic operation. [COMPUT SCI] Computer operation in which input data are fed into the computer directly from observing instruments or other input equipment, and computer results are obtained during the progress of the event. { 'ȯn ˌlīn ˌäp·ə'rā·shən }

on-line real-time system [COMPUT SCI] A computer system that communicates interactively with users, and immediately returns to them the results of data processing during an interaction. Abbreviated OLRT system. { 'ȯn ˌlīn 'rēl ˌtīm 'sis·təm }

on-line secured communications system [COMMUN] Any combination of interconnected communications centers partially or wholly equipped for on-line cryptographic operation and capable of relaying or switching message traffic using on-line cryptographic procedures. { 'ȯn ˌlīn si'kyu̇rd kəˌmyü·nə'kā·shənz ˌsis·təm }

on-line storage [COMPUT SCI] Storage controlled by the central processing unit of a computer. { 'ȯn ˌlīn 'stȯr·ij }

on-line tab setting [COMPUT SCI] A feature in some computer printers which allows the computer that controls the printer to issue commands to set and change the tab stops. { 'ȯn ˌlīn 'tab ˌsed·iŋ }

on-line typewriter [COMPUT SCI] A typewriter which transmits information into and out of a computer, and which is controlled by the central processing unit and thus by whatever program the computer is carrying out. { 'ȯn ˌlīn 'tīp ˌrīd·ər }

only loadable [COMPUT SCI] Attribute of a load module which can be brought into main memory only by a LOAD macroinstruction given from another module. Abbreviated OL. { 'ōn·lē 'lōd·ə·bəl }

on-off keying [COMMUN] Binary form of amplitude modulation in which one of the states of the modulated wave is the absence of energy in the keying interval. { 'ȯn 'ȯf ˌkē·iŋ }

OOP *See* object-oriented programming.

open architecture [COMPUT SCI] A computer architecture whose specifications are made widely available to allow third parties to develop add-on peripherals for it. { 'ō·pən 'ar·kəˌtek·chər }

open-bus system [COMPUT SCI] A computer with an expansion bus that is designed to easily accept expansion boards. { ¦ō·pən 'bəs ˌsis·təm }

open-center plan position indicator [ENG] A plan position indicator on which no signal is displayed within a set distance from the center. { 'ō·pən ˌsen·tər 'plan pəˌzish·ən 'in·dəˌkād·ər }

open-circuit signaling [COMMUN] Type of signaling in which no current flows while the circuit is in the idle condition. { 'ō·pən ¦sər·kət 'sig·nə·liŋ }

open-ended [COMPUT SCI] Of techniques, designed to facilitate or permit expansion, extension, or increase in capability; the opposite of closed-in and artificially constrained. { ¦ō·pən ¦en·dəd }

open-ended system [COMPUT SCI] In character recognition, a system in which the input data to be read are derived from sources other than the computer with which the character reader is associated. { ¦ō·pən ¦en·dəd 'sis·təm }

open file [COMPUT SCI] A file that can be accessed for reading, writing, or both. { 'ō·pən 'fīl }

open-loop control system [CONT SYS] A control system in which the system outputs are controlled by system inputs only, and no account is taken of actual system output. { ¦ō·pən ¦lüp kən'trōl ˌsis·təm }

open routine [COMPUT SCI] **1.** A routine which can be inserted directly into a larger routine without a linkage or calling sequence. **2.** A computer program that changes the state of a file from closed to open. { 'ō·pən rüˌtēn }

open shop [COMPUT SCI] A data-processing-center organization in which individuals from outside the data-processing community are permitted to implement their own solutions to problems. { 'ō·pən ¦shäp }

open source software [COMPUT SCI] Software that is written in such a way that others are encouraged to freely redistribute it, and all changes to the code must be made freely available. { ¦ō·pən ˌsȯrs 'sȯfˌtwer }

open standard [COMPUT SCI] Freely distributed. { ¦ō·pən 'stan·dərd }

open subroutine [COMPUT SCI] A set of computer instructions that collectively perform some particular function and are inserted directly into the program each and every time that particular function is required. { 'ō·pən 'səb·rüˌtēn }

open system [COMPUT SCI] A computer system whose key software interfaces are specified, documented, and made publicly available. { 'ō·pən 'sis·təm }

open-system architecture [COMPUT SCI] The structure of a computer network that allows different types of computers and peripheral devices from different manufacturers to be connected together. { 'ō·pən ¦sis·təm 'är·kə ˌtek·chər }

open-wire carrier system [COMMUN] A system for carrier telephony using an open-wire line. { 'ō·pən ¦wir 'kar·ē·ər ˌsis·təm }

operand [COMPUT SCI] Any one of the quantities entering into or arising from an operation. { 'äp·əˌrand }

operate time [COMPUT SCI] The phase of computer operation when an instruction is being carried out. { 'äp·əˌrāt ˌtīm }

operating delay [COMPUT SCI] Computer time lost because of mistakes or inefficiency of operating personnel or users of the system, excluding time lost because of defects in programs or data. { 'äp·əˌrād·iŋ diˌlā }

operating instructions [COMPUT SCI] A detailed description of the actions that must be carried out by a computer operator in running a program or group of interrelated programs, usually included in the documentation of a program supplied by a programmer or systems analyst, along with the source program and flow charts. { 'äp·əˌrād·iŋ inˌstrək·shənz }

operating position [COMMUN] Terminal of a communications channel which is attended by an operator; usually the term refers to a single operator, such as a radio operator's position or a telephone operator's position; however, certain terminals may require more than one operating position. { 'äp·əˌrād·iŋ pəˌzish·ən }

operating power [ELECTROMAG] Power that is actually supplied to a radio transmitter antenna. { 'äp·əˌrād·iŋ ˌpau̇·ər }

operating ratio [COMPUT SCI] The time during which computer hardware operates and gives reliable results divided by the total time scheduled for computer operation. { 'äp·əˌrād·iŋ ˌrā·shō }

operating system [COMPUT SCI] A set of programs and routines which guides a computer or network in the performance of its tasks, assists the programs (and programmers) with certain supporting functions, and increases the usefulness of the computer or network hardware. { 'äp·əˌrād·iŋ ˌsis·təm }

operating system supervisor [COMPUT SCI] The control program of a set of programs which guide a computer in the performance of its tasks and which assist the program with certain supporting functions. { 'äp·əˌrād·iŋ ˌsis·təm 'sü·pərˌvīz·ər }

operation [COMPUT SCI] **1.** A process or procedure that obtains a unique result from any permissible combination of operands. **2.** The sequence of actions resulting from the execution of one digital computer instruction. { ˌäp·ə'rā·shən }

operational amplifier [ELECTR] An amplifier having high direct-current stability and high immunity to oscillation, generally achieved by using a large amount of negative feedback; used to perform analog-computer functions such as summing and integrating. { ˌäp·ə'rā·shən·əl 'am·pləˌfī·ər }

operational label [COMPUT SCI] A combination of letters and digits at the beginning of the tape which uniquely identify the tape required by the system. { ˌäp·ə'rā·shən·əl 'lā·bəl }

operational standby program [COMPUT SCI] The program operating in the standby computer when in the duplex mode of operation. { ˌäp·ə'rā·shən·əl 'standˌbī ˌprōˌgram }

operation code [COMPUT SCI] A field or portion of a digital computer instruction that indicates which action is to be performed by the computer. Also known as command code. { ˌäp·ə'rā·shən ˌkōd }

operation cycle [COMPUT SCI] The portion of a memory cycle required to perform an operation; division and multiplication usually require more than one memory cycle to be completed. { ˌäp·ə'rā·shən ˌsī·kəl }

operation decoder [COMPUT SCI] A device that examines the operation contained in an instruction of a computer program and sends signals to the circuits required to carry out the operation. { ˌäp·ə'rā·shən dē'kōd·ər }

operation number [COMPUT SCI] **1.** Number designating the position of an operation, or its equivalent subroutine, in the sequence of operations composing a routine. **2.** Number identifying each step in a program stated in symbolic code. { ˌäp·ə'rā·shən ˌnəm·bər }

operation part [COMPUT SCI] That portion of a digital computer instruction which is reserved for the operation code. { ˌäp·ə'rā·shən ˌpärt }

operation register [COMPUT SCI] A register used to store and decode the operation code for the next instruction to be carried out by a computer. { ˌäp·ə'rā·shən ˌrej·ə·stər }

operations research [MATH] The mathematical study of systems with input and output from the viewpoint of optimization subject to given constraints. { ˌäp·ə'rā·shənz riˌsərch }

operation time [COMPUT SCI] The time elapsed during the interpretation and execution of an arithmetic or logic operation by a computer. { ˌäp·ə'rā·shən ˌtīm }

operator's console [COMPUT SCI] Equipment which provides for manual intervention and monitoring computer operation. { 'äp·əˌrād·ərz 'känˌsōl }

operator [COMPUT SCI] Anything that designates an action to be performed, especially the operation code of a computer instruction. { 'äp·əˌrād·ər }

operator hierarchy [COMPUT SCI] A sequence of mathematical operators which designates the order in which these operators are to be applied to any mathematical expression in a given programming language. { 'äp·əˌrād·ər 'hī·ərˌär·kē }

operator interrupt [COMPUT SCI] A step whereby control is passed to the monitor, and a message, usually requiring a typed answer, is printed on the console typewriter. { 'äp·əˌrād·ər 'in·təˌrəpt }

operator subgoaling [COMPUT SCI] A computer problem-solving method in which the inability of the computer to take the desired next step at any point in the problem-solving process leads to a subgoal of making that step feasible. { 'äp·ə ˌrād·ər ˌsəb'gōl·iŋ }

optical amplifier [ENG] An optoelectronic amplifier in which the electric input signal is converted to light, amplified as light, then converted back to an electric signal for the output. { 'äp·tə·kəl 'am·plə,fī·ər }

optical bar-code reader [COMPUT SCI] A device which uses any of various photoelectric methods to read information which has been coded by placing marks in prescribed boxes on documents with ink, pencil, or other means. { 'äp·tə·kəl 'bär ˌkōd ˌrēd·ər }

optical character recognition [COMPUT SCI] That branch of character recognition concerned with the automatic identification of handwritten or printed characters by any of various photoelectric methods. Abbreviated OCR. Also known as electrooptical character recognition. { 'äp·tə·kəl 'kar·ik·tər ˌrek·ig,nish·ən }

optical communication [COMMUN] The use of electromagnetic waves in the region of the spectrum near visible light for the transmission of signals representing speech, pictures, data pulses, or other information, usually in the form of a laser beam modulated by the information signal. { 'äp·tə·kəl kə,myü·nə'kā·shən }

optical computer [COMPUT SCI] A computer that uses various combinations of holography, lasers, and mass-storage memories for such applications as ultra-high-speed signal processing, image deblurring, and character recognition. { 'äp·tə·kəl kəm'pyüd·ər }

optical coupler See optoisolator. { 'äp·tə·kəl 'kəp·lər }

optical coupling [ELECTR] Coupling between two circuits by means of a light beam or light pipe having transducers at opposite ends, to isolate the circuits electrically. { 'äp·tə·kəl 'kəp·liŋ }

optical data storage [COMPUT SCI] The technology of placing information in a medium so that, when a light beam scans the medium, the reflected light can be used to recover the information. { ¦äp·tə·kəl 'dad·ə ˌstȯr·ij }

optical disk [COMPUT SCI] A type of video disk storage device consisting of a pressed disk with a spiral groove at the bottom of which are submicrometer-sized depressions that are sensed by a laser beam. { 'äp·tə·kəl 'disk }

optical disk storage [COMPUT SCI] A computer storage technology in which information is stored in submicrometer-sized holes on a rotating disk, and is recorded and read by laser beams focused on the disk. Also known as laser disk storage; video disk storage. { 'äp·tə·kəl ¦disk 'stȯr·ij }

optical fiber [OPTICS] A long, thin thread of fused silica, or other transparent substance, used to transmit light. Also known as light guide. { 'äp·tə·kəl 'fī·bər }

optical-fiber amplifier [COMMUN] A device for amplifying signals transmitted over optical fibers, consisting of a low-loss single-mode fiber made of basic silica glass, along whose length gain is generated by coupling pump light at either or both fiber ends, or at periodic locations in between. { ˌäp·tə·kəl ˌfī·bər 'am·plə,fī·ər }

optical-fiber cable See optical waveguide. { 'äp·tə·kəl ¦fī·bər 'kā·bəl }

optical filter See filter. { 'äp·tə·kəl 'fil·tər }

optical information processor See optical information system. { 'äp·tə·kəl ˌin·fər'mā·shən ˌprä ˌses·ər }

optical information system [COMPUT SCI] A device that uses light to process information; consists of one or several light sources, a one- or two-dimensional plane of data such as a film transparency, lens, or other optical component, and a detector. Also known as optical information processor. { 'äp·tə·kəl ˌin·fər'mā·shən ˌsis·təm }

optical isolator See optoisolator. { 'äp·tə·kəl 'ī·sə,lād·ər }

optically coupled isolator See optoisolator. { 'äp·tə·klē ¦kup·əld 'ī·sə,lād·ər }

optical mark reading [COMPUT SCI] Optically sensing information encoded as a series of marks, such as lines or filled-in boxes on a test answer sheet, or some special pattern, such as the Universal Product Code. Abbreviated OMR. { 'äp·tə·kəl ¦märk ˌrēd·iŋ }

optical memory [COMPUT SCI] A computer memory that uses optical techniques which generally involve an addressable laser beam, a storage medium which responds to the beam for writing and sometimes for erasing, and a detector which reacts to the altered character of the medium when it uses the beam to read out stored data. { 'äp·tə·kəl 'mem·rē }

optical modulator [COMMUN] A device used for impressing information on a light beam. { 'äp·tə·kəl 'mäj·ə,lād·ər }

optical mouse [COMPUT SCI] A mouse that emits a light signal and uses its reflection from a reflective grid to determine position and movement. { 'äp·tə·kəl 'mau̇s }

optical phase conjugation [OPTICS] The use of nonlinear optical effects to precisely reverse the direction of propagation of each plane wave in an arbitrary beam of light, thereby causing the return beam to exactly retrace the path of the incident beam. Also known as time-reversal reflection; wavefront reversal. { 'äp·tə·kəl ¦fāz ˌkän·jə'gā·shən }

optical processing [COMPUT SCI] The use of light, including visible and infrared, to handle data-processing information. { 'äp·tə·kəl 'prä ˌses·iŋ }

optical reader [COMPUT SCI] A computer data-entry machine that converts printed characters, bar or line codes, and pencil-shaded areas into a computer-input code format. { 'äp·tə·kəl 'rēd·ər }

optical scanner See flying-spot scanner. { 'äp·tə·kəl 'skan·ər }

optical storage [COMPUT SCI] Storage of large amounts of data in permanent form on photographic film or its equivalent, for nondestructive readout by means of a light source and photodetector. { 'äp·tə·kəl 'stȯr·ij }

optical tape storage [COMMUN] A data storage technology in which information is stored on a tape that is wound on a spool and has a large number of parallel channels, and information is retrieved by sensing the reflected light when a light beam scans the medium. { ¦äp·tə·kəl 'tāp ˌstȯr·ij }

optical type font [COMPUT SCI] A special type font whose characters are designed to be easily read by both people and optical character recognition machines. { 'äp·tə·kəl ¦tīp ˌfänt }

optical waveguide [ELECTROMAG] A waveguide in which a light-transmitting material such as a glass or plastic fiber is used for transmitting information from point to point at wavelengths somewhere in the ultraviolet, visible-light, or infrared portions of the spectrum. Also known as fiber waveguide; optical-fiber cable. { 'äp·tə·kəl 'wāvˌgīd }

optimal control theory [CONT SYS] An extension of the calculus of variations for dynamic systems with one independent variable, usually time, in which control (input) variables are determined to maximize (or minimize) some measure of the performance (output) of a system while satisfying specified constraints. { 'äp·tə·məl kən'trōl ˌthē·ə·rē }

optimal feedback control [CONT SYS] A subfield of optimal control theory in which the control variables are determined as functions of the current state of the system. { 'äp·tə·məl 'fēd ˌbak kənˌtrōl }

optimal programming [CONT SYS] A subfield of optimal control theory in which the control variables are determined as functions of time for a specified initial state of the system. { 'äp·tə·məl 'prōˌgram·iŋ }

optimal regulator problem *See* linear regulator problem. { 'äp·tə·məl 'reg·yəˌlād·ər ˌpräb·ləm }

optimal smoother [CONT SYS] An optimal filer algorithm which generates the best estimate of a dynamical variable at a certain time based on all available data, both past and future. { 'äp·tə·məl 'smüth·ər }

optimization [SYS ENG] **1.** Broadly, the efforts and processes of making a decision, a design, or a system as perfect, effective, or functional as possible. **2.** Narrowly, the specific methodology, techniques, and procedures used to decide on the one specific solution in a defined set of possible alternatives that will best satisfy a selected criterion. Also known as system optimization. { ˌäp·tə·mə'zā·shən }

optimize [COMPUT SCI] To rearrange the instructions or data in storage so that a minimum number of time-consuming jumps or transfers are required in the running of a program. { 'äp·təˌmīz }

optimized code [COMPUT SCI] A machine-language program that has been revised to remove inefficiencies and unused or unnecessary instructions so that the program is executed more quickly and occupies less storage space. { 'äp·təˌmīzd 'kōd }

optimizer [COMPUT SCI] A utility program that processes machine-language programs and generates optimized code. { 'äp·təˌmīz·ər }

optimum array current [ELECTROMAG] The current distribution in a broadside antenna array which is such that for a specified side-lobe level the beam width is as narrow as possible, and for a specified first null the side-lobe level is as small as possible. { 'äp·tə·məm ə'rā ˌkə·rənt }

optimum code [COMPUT SCI] A computer code which is particularly efficient with regard to a particular aspect; for example, minimum time of execution, minimum or efficient use of storage space, and minimum coding time. { 'äp·tə·məm 'kōd }

optimum programming [COMPUT SCI] Production of computer programs that maximize efficiency with respect to some criteria such as least cost, least use of storage, least time, or least use of time-sharing peripheral equipment. { 'äp·tə·məm 'prōˌgram·iŋ }

optimum traffic frequency *See* optimum working frequency. { 'äp·tə·məm 'traf·ik ˌfrē·kwən·sē }

optimum working frequency [COMMUN] The most effective frequency at a specified time for ionospheric propagation of radio waves between two specified points. Also known as frequency optimum traffic; optimum traffic frequency. { 'äp·tə·məm 'wərk·iŋ ˌfrē·kwən·sē }

optional halt instruction [COMPUT SCI] A halt instruction that can cause a computer program to stop either before or after the instruction is obeyed if certain criteria are met. Also known as optional stop instruction. { 'äp·shən·əl 'hȯlt in ˌstrək·shən }

optional product [COMPUT SCI] Any of various forms of documentation that may be made available with a software product, such as source code, manuals, and instructions. { 'äp·shən·əl 'präd·əkt }

optional stop instruction *See* optional halt instruction. { 'äp·shən·əl 'stäp inˌstrək·shən }

option switch [COMPUT SCI] **1.** A DIP switch or jumper that activates an optional feature. **2.** A software parameter that overrides a default value and thereby activates an optional feature. Also known as option toggle. { 'äp·shən ˌswich }

option toggle *See* option switch. { 'äp·shən ˌtäg·əl }

optoacoustic modulator *See* acoustooptic modulator. { ¦äp·tō·ə¦küs·tik 'mäj·əˌlād·ər }

optocoupler *See* optoisolator. { ¦äp·tō'kəp·lər }

optoelectronic amplifier [ENG] An amplifier in which the input and output signals and the method of amplification may be either electronic or optical. { ¦äp·tō·iˌlek'trän·ik 'am·pləˌfī·ər }

optoelectronic isolator *See* optoisolator. { ¦äp·tō·iˌlek'trän·ik 'ī·səˌlād·ər }

optoelectronics [ELECTR] **1.** The branch of electronics that deals with solid-state and other electronic devices for generating, modulating, transmitting, and sensing electromagnetic radiation in the ultraviolet, visible-light, and infrared portions of the spectrum. **2.** *See* photonics. { ¦äp·tō·i¸lek'trän·iks }

optoisolator [ELECTR] A coupling device in which a light-emitting diode, energized by the input signal, is optically coupled to a photodetector such as a light-sensitive output diode, transistor, or silicon controlled rectifier. Also known as optical coupler; optical isolator; optically coupled isolator; optocoupler; optoelectronic isolator; photocoupler; photoisolator. { ¦äp·tō 'ī·sə¸lād·ər }

or [COMPUT SCI] An instruction which performs the logical operation "or" on a bit-by-bit basis for its two or more operand words, usually storing the result in one of the operand locations. Also known as OR function. { ȯr }

ORB *See* object request broker. { ȯrb *or* ¦ō ¦är'bē }

ORB core [COMPUT SCI] The part of an object request broker that is responsible for the communication of requests. { 'ȯrb ¸kȯr *or* ¦ō¦är'bē ¸kȯr }

OR circuit *See* OR gate. { 'ȯr ¸sər·kət }

ordered array [COMPUT SCI] A set of data elements that has been arranged in rows and columns in a specified order so that each element can be individually accessed. { 'ȯrd·ərd ə'rā }

ordered list [COMPUT SCI] A set of data items that has been arranged in a specified sequence to aid in processing its contents. { 'ȯrd·ərd 'list }

orderly shutdown [COMPUT SCI] The procedures for shutting off a computer system in an organized manner, normally after all work in progress has been completed, permitting restarting of the systems without loss of transactions or data. { 'ȯrd·ər·lē 'shət¸dau̇n }

order tone [COMMUN] Tone sent over a trunk to indicate that the trunk is ready to receive an order or, to the receiving operator, that an order is about to arrive. { 'ȯrd·ər ¸tōn }

ordinal type [COMPUT SCI] A data type whose possible values are sequential in the manner of the integers 1, 2, 3, and so forth; for example, the months January, February, and so forth. { 'ȯrd·nəl 'tīp }

OR function *See* or. { 'ȯr ¸fəŋk·shən }

OR gate [ELECTR] A multiple-input gate circuit whose output is energized when any one or more of the inputs is in a prescribed state; performs the function of the logical inclusive-or; used in digital computers. Also known as OR circuit. { 'ȯr ¸gāt }

orient [COMPUT SCI] To change relative and symbolic addresses to absolute form. { 'ȯr·ē·ənt }

orientation [ELECTROMAG] The physical positioning of a directional antenna or other device having directional characteristics. { ¸ȯr·ē·ən'tā·shən }

orifice [ELECTROMAG] Opening or window in a side or end wall of a waveguide or cavity resonator through which energy is transmitted. { 'ȯr·ə·fəs }

origin [COMPUT SCI] Absolute storage address in relative coding to which addresses in a region are referenced. { 'är·ə·jən }

original document *See* source document. { ə'rij·ən·əl 'däk·yə·mənt }

original equipment manufacturer *See* OEM. { ə·'rij·ə·nəl i'kwip·mənt man·yə·'fak·chər·ər }

orthicon [ELECTR] A camera tube in which a beam of low-velocity electrons scans a photoemissive mosaic that is capable of storing a pattern of electric charges; has higher sensitivity than the iconoscope. { 'ȯr·thə¸kän }

orthogonal [COMPUT SCI] **1.** An area of a computer display in which units of distance are the same horizontally and vertically so that there is no distortion. **2.** A viewing area in which positions are determined by using a cartesian coordinate system with horizontal and vertical axes. { ȯr'thäg·ən·əl }

orthogonal antennas [ELECTROMAG] In radar, a pair of transmitting and receiving antennas, or a single transmitting-receiving antenna, designed for the detection of a difference in polarization between the transmitted energy and the energy returned from the target. { ȯr'thäg·ən·əl an'ten·əz }

orthogonal parity check [COMPUT SCI] A parity checking system involving both a lateral and a longitudinal parity check. { ȯr'thäg·ən·əl 'par·əd·ē ¸chek }

orthotronic error control [COMPUT SCI] An error check carried out to ensure correct transmission, which uses lateral and longitudinal parity checks. { ¦ȯr·thə¦trän·ik 'er·ər kən¸trōl }

O-scan *See* O-display. { 'ō ¸skan }

oscillator [ELECTR] **1.** An electronic circuit that converts energy from a direct-current source to a periodically varying electric output. **2.** The stage of a superheterodyne receiver that generates a radio-frequency signal of the correct frequency to mix with the incoming signal and produce the intermediate-frequency value of the receiver. **3.** The stage of a transmitter that generates the carrier frequency of the station or some fraction of the carrier frequency. { 'äs·ə¸lād·ər }

oscillator harmonic interference [ELECTR] Interference occurring in a superheterodyne receiver due to the interaction of incoming signals with harmonics (usually the second harmonic) of the local oscillator. { 'äs·ə¸lād·ər här'män·ik ¸in·tər'fir·əns }

oscillator-mixer-first detector *See* converter. { 'äs·ə¸lād·ər 'mik·sər ¸fərst di'tek·tər }

O-scope *See* O-display. { 'ō ¸skōp }

OTH radar *See* over-the-horizon radar. { ō¦tē¦āch 'rā¸där }

outline processor [COMPUT SCI] A software system that organizes notes in ordinary English into an outline that serves as the basis for a document. { 'au̇t¸līn ¸prä¸ses·ər }

out-of-line coding [COMPUT SCI] Instructions in a routine that are stored in a different part of computer storage from the rest of the instructions. { 'au̇t əv ¦līn 'kōd·iŋ }

out-plant system [COMPUT SCI] A data-processing system that has one or more remote terminals from which information is transmitted to a central computer. { 'au̇t ˌplant ˌsis·təm }

output [COMPUT SCI] **1.** The data produced by a data-processing operation, or the information that is the objective or goal in data processing. **2.** The data actively transmitted from within the computer to an external device, or onto a permanent recording medium (paper, microfilm). **3.** The activity of transmitting the generated information. **4.** The readable storage medium upon which generated data are written, as in hard-copy output. [ELECTR] **1.** The current, voltage, power, driving force, or information which a circuit or device delivers. **2.** Terminals or other places where a circuit or device can deliver current, voltage, power, driving force, or information. { 'au̇tˌpu̇t }

output area [COMPUT SCI] A part of storage that has been reserved for output data. Also known as output block. { 'au̇tˌpu̇t ˌer·ē·ə }

output block [COMPUT SCI] **1.** A portion of the internal storage of a computer that is reserved for receiving, processing, and transmitting data to be transferred out. **2.** *See* output area. { 'au̇t ˌpu̇t ˌbläk }

output-bound computer [COMPUT SCI] A computer that is slowed down by its output functions. { 'au̇tˌpu̇t ˌbau̇nd kəmˌpyüd·ər }

output bus driver [ELECTR] A device that power-amplifies output signals from a computer to allow them to drive heavy circuit loads. { 'au̇t ˌpu̇t 'bəs ˌdrīv·ər }

output class [COMPUT SCI] An indicator of the priority of output from a computer that determines the order in which it is printed from a spool file. { 'au̇tˌpu̇t ˌklas }

output device *See* output unit. { 'au̇tˌpu̇t di ˌvīs }

output link [COMMUN] The last link in a communications chain. { 'au̇tˌpu̇t ˌliŋk }

output monitor interrupt [COMPUT SCI] A data-processing step in which control is passed to the monitor to determine the precedence order for two requests having the same priority level. { 'au̇tˌpu̇t ˌman·əd·ər 'int·əˌrəpt }

output program *See* output routine. { 'au̇tˌpu̇t ˌprōˌgram }

output rating *See* carrier power output rating. { 'au̇tˌpu̇t ˌrād·iŋ }

output record [COMPUT SCI] **1.** A unit of data that has been transcribed from a computer to an external medium or device. **2.** The unit of data that is currently held in the output area of a computer before being transcribed to an external medium or device. { 'au̇tˌpu̇t ˌrek·ərd }

output routine [COMPUT SCI] A series of computer instructions which organizes and directs all operations associated with the transcription of data from a computer to various media and external devices by various types of output equipment. Also known as output program. { 'au̇tˌpu̇t rüˌtēn }

output unit [COMPUT SCI] In computers, a unit which delivers information from the computer to an external device or from internal storage to external storage. { 'au̇tˌpu̇t ˌyü·nət }

output word [COMPUT SCI] Any running word into which an input word is to be translated. { 'au̇tˌpu̇t ˌwərd }

outside extension [COMMUN] Telephone extension on premises separated from the main station. { 'au̇tˌsīd ik'sten·chən }

overflow [COMPUT SCI] **1.** The condition that arises when the result of an arithmetic operation exeeds the storage capacity of the indicated result-holding storage. **2.** That part of the result which exceeds the storage capacity. { 'ō·vər ˌflō }

overflow bucket [COMPUT SCI] A unit of storage in a direct-access storage device used to hold an overflow record. { 'ō·vərˌflō ˌbək·ət }

overflow check indicator *See* overflow indicator. { 'ō·vərˌflō'chek ˌin·dəˌkād·ər }

overflow error [COMPUT SCI] The condition in which the numerical result of an operation exceeds the capacity of the register. { 'ō·vərˌflō 'er·ər }

overflow indicator [COMPUT SCI] A bistable device which changes state when an overflow occurs in the register associated with it, and which is designed so that its condition can be determined, and its original condition restored. Also known as overflow check indicator. { 'ō·vərˌflōˌin·də ˌkād·ər }

overflow record [COMPUT SCI] A unit of data whose length is too great for it to be stored in an assigned section of a direct-access storage, and which must be stored in another area from which it may be retrieved by means of a reference stored in the original assigned area in place of the record. { 'ō·vərˌflō ˌrek·ərd }

overflow storage [COMMUN] Additional storage provided in a store-and-forward-switching center to prevent the loss of messages (or parts of messages) offered to the switching center when it is fulfilled. [COMPUT SCI] Extra storage capacity in a computer or calculator that allows a small amount of overflow. { 'ō·vərˌflō ˌstȯr·ij }

overhead [COMPUT SCI] The time a computer system spends doing computations that do not contribute directly to the progress of any user tasks in the system, such as allocation of resources, responding to exceptional conditions, providing protection and reliability, and accounting. { 'ō·vərˌhed }

overlap [COMMUN] **1.** In teletypewriter practice, the selecting of another code group while the printing of a previously selected code group is taking place. **2.** Amount by which the effective height of the scanning facsimile spot exceeds the nominal width of the scanning line. [COMPUT SCI] To perform some or all of an operation concurrently with one or more other operations. { 'ō·vərˌlap }

overlapped memories [COMPUT SCI] An arrangement of computer memory banks in which, to cut down access time, successive words are taken from different memory banks, rewriting in one bank being overlapped by logic operations in another bank, with memory access in still another bank. { 'ō·vər,lapt 'mem·rēz }

overlapping [COMPUT SCI] An operation whereby, if the processor determines that the current instruction and the next instruction lie in different storage modules, the two words may be retrieved in parallel. { ¦ō·vər¦lap·iŋ }

overlapping input/output [COMPUT SCI] A procedure in which a computer system works on several programs, suspending work on a program and moving to another when it encounters an instruction for input/output operation, which is then executed when input/output operations from other programs have been carried out. { 'ō·vər¦lap·iŋ 'in,pu̇t 'au̇t,pu̇t }

overlap radar [ENG] Radar located in one sector whose area of useful radar coverage includes a portion of another sector. { 'ō·vər,lap 'rā,där }

overlay [COMPUT SCI] A technique for bringing routines into high-speed storage from some other form of storage during processing, so that several routines will occupy the same storage locations at different times; overlay is used when the total storage requirements for instructions exceed the available main storage. { 'ō·vər,lā }

overloading [COMPUT SCI] The use, in some advanced programming languages, of two or more variables or subroutines with the same name; the compiler determines by inference which entity is referred to each time the name occurs. { ¦ō·vər ¦lōd·iŋ }

overmodulation [COMMUN] Amplitude modulation greater than 100%, causing distortion because the carrier voltage is reduced to zero during portions of each cycle. { ¦ō·vər,mäj·ə'lā·shən }

overrun [COMPUT SCI] The arrival of an amount of data greater than the space allocated to it. { 'ō·və,rən }

overshoot [ELECTROMAG] The reception of microwave signals where they were not intended, due to an unusual atmospheric condition that sets up variations in the index of refraction. { 'ō·vər,shüt }

over-the-horizon propagation *See* scatter propagation. { 'ō·vər t͟hə hə'rīz·ən ,präp·ə'gā·shən }

over-the-horizon radar [ELECTROMAG] Radar operating in such a way that targets otherwise shielded from view by earth's curvature are detected; the use of carrier frequencies at which the ionosphere is particularly reflective, so that radar signals are reflected back to the surface at great ranges, or use of signal characteristics exploiting surface-coupled propagation are example techniques. Abbreviated OTH radar. { 'ō·vər t͟hə hə'rīz·ən 'rā,där }

overthrow distortion [COMMUN] Distortion caused when the maximum amplitude of the signal wavefront exceeds the steady state of amplitude of the signal wave. { 'ō·vər,thrō di ,stȯr·shən }

overwrite [COMPUT SCI] To enter information into a storage location and destroy the information previously held there. { ¦ō·vər¦rīt }

own coding [COMPUT SCI] A series of instructions added to a standard software routine to change or extend the routine so that it can carry out special tasks. { 'ōn 'kōd·iŋ }

owned program *See* proprietary program. { 'ōnd 'prō,gram }

P

PABX *See* private automatic branch exchange.

PAC *See* perceptual audio coding.

pack [COMPUT SCI] To reduce the amount of storage required to hold information by changing the method of encoding the data. { pak }

package [COMPUT SCI] A program that is written for a general and widely used application in such a way that its usefulness is not impaired by the problems of data or organization of a particular user. { pak·ij }

pack assembly *See* pack. { 'pak ə,sem·blē }

packed decimal [COMPUT SCI] A means of representing two digits per character, to reduce space and increase transmission speed. { 'pakt 'des·məl }

packed file [COMPUT SCI] A file that has been encoded so that it takes up less space in storage. Also known as compressed file. { ¦akt 'fīl }

packet [COMMUN] A short section of data of fixed length that is transmitted as a unit; consists of a header followed by a number of contiguous bytes from an elementary data stream. { 'pak·ət }

packetized elementary stream [COMMUN] A generic term for a coded bit stream in a digital transport system. In a digital television system, one coded video, coded audio, or other coded elementary stream is carried in a sequence of PES packets with one stream identification code. { 'pak·ə¦tīzd ,el·ə'men·trē 'strēm }

packet switching *See* packet transmission. { 'pak·ət ,swich·iŋ }

packet transmission [COMMUN] Transmission of standardized packets of data over transmission lines rapidly by networks of high-speed switching computers that have the message packets stored in fast-access core memory. Also known as packet switching. { 'pak·ət tranz,mish·ən }

packing density [COMPUT SCI] The amount of information per unit of storage medium, as characters per inch on tape, bits per inch or drum, or bits per square inch in photographic storage. [ELECTR] The number of devices or gates per unit area of an integrated circuit. { 'pak·iŋ ,den·səd·ē }

packing routine [COMPUT SCI] A subprogram which compresses data so as to eliminate blanks and reduce the storage needed for a file. { 'pak·iŋ rü,tēn }

pad [ELECTR] **1.** An arrangement of fixed resistors used to reduce the strength of a radio-frequency or audio-frequency signal by a desired fixed amount without introducing appreciable distortion. Also known as fixed attenuator. **2.** *See* terminal area. { pad }

padder [ELECTR] A trimmer capacitor inserted in series with the oscillator tuning circuit of a superheterodyne receiver to control calibration at the low-frequency end of a tuning range. { 'pad·ər }

padding [COMPUT SCI] The adding of meaningless data (usually blanks) to a unit of data to bring it up to some fixed size. { 'pad·iŋ }

page [COMPUT SCI] **1.** A standard quantity of main-memory capacity, usually 512 to 4096 bytes or words, used for memory allocation and for partitioning programs into control sections. **2.** A standard quantity of source program coding, usually 8 to 64 lines, used for displaying the coding on a cathode-ray tube. { pāj }

pageable memory [COMPUT SCI] The part of a computer's main storage that is subject to paging in a virtual storage system. { 'pāj·ə·bəl 'mem·rē }

page boundary [COMPUT SCI] The address of the first (lowest) word or byte within a page of memory. { 'pāj ,bau̇n·drē }

page data set [COMPUT SCI] A file for storing images of pages in a virtual storage system, so that they can be returned to main storage for further processing when needed. { 'pāj 'dad·ə ,set }

page description language [COMPUT SCI] A high-level language that specifies the format of a page generated by a printer; it is translated into specific codes by any printer that supports the language. Abbreviated PDL. { 'pāj di,skrip·shən ,laŋ·gwij }

page fault [COMPUT SCI] An interruption that occurs while a page which is referred to by the program is being read into memory. { 'pāj ,fȯlt }

page printer [COMPUT SCI] A computer output device which composes a full page of characters before printing the page. { 'pāj ,print·ər }

pager [COMMUN] A receiver in a radio paging system. { 'pāj·ər }

page reader [COMPUT SCI] In character recognition, a character reader capable of processing cut-form documents of varying sizes; sometimes capable of reading information in reel forms. { 'pāj ,rēd·ər }

page skip [COMPUT SCI] A control character that causes a printer to skip over the remainder of the current page and move to the beginning of the following page. { 'pāj ˌskip }

page table [COMPUT SCI] A key element in the virtual memory technique; a table of addresses where entries are adjusted for easy relocation of pages. { 'pāj ˌtā·bəl }

page turning [COMPUT SCI] **1.** The process of moving entire pages of information between main memory and auxiliary storage, usually to allow several concurrently executing programs to share a main memory of inadequate capacity. **2.** In conversational time-sharing systems, the moving of programs in and out of memory on a round-robin, cyclic schedule so that each program may use its allotted share of computer time. { 'pāj ˌtərn·iŋ }

paging [COMPUT SCI] The scheme used to locate pages, to move them between main storage and auxiliary storage, or to exchange them with pages of the same or other computer programs; used in computers with virtual memories. { 'pāj·iŋ }

paging rate [COMPUT SCI] The number of pages per second moved by virtual storage between main storage and the page data set. { 'pāj·iŋ ˌrāt }

paging system [COMMUN] A system which gives an indication to a particular individual that he or she is wanted at the telephone, such as by sounding a number, calling by name over a loudspeaker, or producing an audible signal in a radio receiver carried in the individual's pocket. { 'pāj·iŋ ˌsis·təm }

paint [COMPUT SCI] To fill an area of a display screen or printed output with a color, shade of gray, or image. [ELECTR] In radar, a colloqual term for an echo signal or its display; sometimes called the "skin paint," as of an aircraft. { pānt }

paint program [COMPUT SCI] A graphics program that maintains images in raster format, allowing the user to simulate painting with the aid of a mouse or a graphics tablet. { 'pānt ˌprō·grəm }

paired synchronous detection [ELECTR] The arrangement of two homodyne channels in a radar receiver such that both the phase and the amplitude of a received signal is preserved in the two video signals produced. { ¦perd 'siŋ·krə·nəs diˌtek·shən }

palette [COMPUT SCI] In computer graphics, the set of colors that can be shown on a display monitor. { 'pal·ət }

Palmer scan [ELECTR] Combination of circular or raster and conical radar scans; the beam is swung around the horizon, and at the same time a conical scan is performed. { 'päm·ər ˌskan }

palmtop *See* hand-held computer. { 'pämˌtäp }

PAL system *See* phase-alternation line system. { 'pal ˌsis·təm }

PAM *See* pulse-amplitude modulation.

pan [COMMUN] To tilt or otherwise move a video or motion picture camera vertically and horizontally to keep it trained on a moving object or to secure a panoramic effect. { pan }

panadapter *See* panoramic adapter. { 'pan·ə ˌdap·tər }

panel [COMPUT SCI] The face of the console, which is normally equipped with lights, switches, and buttons to control the machine, correct errors, determine the status of the various CPU (central processing unit) parts, and determine and revise the contents of various locations. Also known as control panel; patch panel. { 'pan·əl }

panel display [ELECTR] An electronic display in which a large orthogonal array of display devices, such as electroluminescent devices or light-emitting diodes, form a flat screen. Also known as flat-panel display. { 'pan·əl diˌsplā }

panoramic adapter [ELECTR] A device designed to operate with a search receiver to provide a visual presentation on an oscilloscope screen of a band of frequencies extending above and below the center frequency to which the search receiver is tuned. Also known as panadapter. { ¦pan·ə ¦ram·ik ə'dap·tər }

panoramic display [ELECTR] A display that simultaneously shows the relative amplitudes of all signals received at different frequencies. { ¦pan·ə¦ram·ik di'splā }

panoramic radar [ENG] Nonscanning radar which transmits signals over a wide beam in the direction of interest. { ¦pan·ə¦ram·ik 'rāˌdär }

panoramic receiver [ELECTR] Radio receiver that permits continuous observation on a cathode-ray-tube screen of the presence and relative strength of all signals within a wide frequency range. { ¦pan·ə¦ram·ik ri'sē·vər }

paper-tape Turing machine [COMPUT SCI] A variation of a Turing machine in which a blank square can have a nonblank symbol written on it, but this symbol cannot be changed thereafter. { 'pā·pər ¦tāp 'tu̇r·iŋ məˌshēn }

paper throw [COMPUT SCI] The movement of paper through a computer printer for a purpose other than printing, in which the distance traveled, and usually the speed, is greater than that of a single line spacing. { 'pā·pər ˌthrō }

PAR *See* precision approach radar.

paraballoon [ELECTROMAG] Air-inflated radar antenna. { ¦par·ə·bə'lün }

parabolic antenna [ELECTROMAG] Antenna with a radiating element and a parabolic reflector that concentrates the radiated power into a beam. { ¦par·ə¦bäl·ik an'ten·ə }

parabolic reflector [ELECTROMAG] **1.** An antenna having a concave surface which is generated either by translating a parabola perpendicular to the plane in which it lies (in a cylindrical parabolic reflector), or rotating it about its axis of symmetry (in a paraboloidal reflector). Also known as dish. **2.** *See* paraboloidal reflector. { ¦par·ə¦bäl·ik ri'flek·tər }

paraboloidal antenna *See* paraboloidal reflector. { pə¦rab·ə¦lȯid·əl an'ten·ə }

paraboloidal reflector [ELECTROMAG] An antenna having a concave surface which is a paraboloid of revolution; it concentrates radiation from a source at its focal point into a beam. Also known as paraboloidal antenna.

[OPTICS] A concave mirror which is a paraboloid of revolution and produces parallel rays of light from a source located at the focus of the parabola. Also known as parabolic reflector. { pə¦rab·ə¦lȯid·əl ri'flek·tər }

paragraph [COMPUT SCI] A complete, logical sequence of instructions in the COBOL programming language, required to carry out a definable program or task. { 'par·ə,graf }

parallel [COMPUT SCI] Simultaneous transmission of, storage of, or logical operations on the parts of a word, character, or other subdivision of a word in a computer, using separate facilities for the various parts. { 'par·ə,lel }

parallel access [COMPUT SCI] Transferral of information to or from a storage device in which all elements in a unit of information are transferred simultaneously. Also known as simultaneous access. { 'par·ə,lel 'ak,ses }

parallel addition [COMPUT SCI] A method of addition by a computer in which all the corresponding pairs of digits of the addends are processed at the same time during one cycle, and one or more subsequent cycles are used for propagation and adjustment of any carries that may have been generated. { 'par·ə,lel ə'dish·ən }

parallel algorithm [COMPUT SCI] An algorithm in which several computations are carried on simultaneously. { 'par·ə,lel 'al·gə,rith·əm }

parallel buffer [ELECTR] Electronic device (magnetic core or flip-flop) used to temporarily store digital data in parallel, as opposed to series storage. { 'par·ə,lel 'bəf·ər }

parallel by character [COMPUT SCI] The handling of all the characters of a machine word simultaneously in separate lines, channels, or storage cells. { 'par·ə,lel bī 'kar·ik·tər }

parallel communications [COMMUN] The simultaneous transmission of data over two or more communications channels. { 'par·ə,lel kə ,myü·nə'kā·shənz }

parallel compensation *See* feedback compensation. { 'par·ə,lel ,käm·pən'sā·shən }

parallel computation [COMPUT SCI] The simultaneous computation of several parts of a problem. { 'par·ə,lel ,käm·pyü'tā·shən }

parallel computer [COMPUT SCI] **1.** A computer that can carry out more than one logic or arithmetic operation at one time. **2.** *See* parallel digital computer. { 'par·ə,lel kəm'pyüd·ər }

parallel conversion [COMPUT SCI] The process of transferring operations from one computer system to another, during which both systems are run together for a period of time to ensure that they are producing identical results. { 'par·ə,lel kən'vər·zhən }

parallel digital computer [COMPUT SCI] Computer in which the digits are handled in parallel; mixed serial and parallel machines are frequently called serial or parallel, according to the way arithmetic processes are performed; an example of a parallel digital computer is one which handles decimal digits in parallel, although it might handle the bits constituting a digit either serially or in parallel. { 'par·ə,lel 'dij·əd·əl kəm 'pyüd·ər }

parallel dot character printer *See* line dot matrix. { 'par·ə,lel ¦dät 'kar·ik·tər ,print·ər }

parallel element-processing ensemble [COMPUT SCI] A powerful electronic computer used by the U.S. Army to simulate tracking and discrimination of reentry vehicles as part of the ballistic missile defense research program. Abbreviated PEPE. { 'par·ə,lel 'el·ə·mənt ¦prä ,ses·iŋ än,säm·bəl }

parallel input/output [COMPUT SCI] Data that are transmitted into and out of a computer over several conductors simultaneously. { 'par·ə,lel 'in,pu̇t 'au̇t,pu̇t }

parallel interface [ELECTR] A link between two devices in which all the information transferred between them is transmitted simultaneously over separate conductors. Also known as parallel port. { 'par·ə,lel 'in·tər,fās }

parallel operation [COMPUT SCI] Performance of several actions, usually of a similar nature, by a computer system simultaneously through provision of individual similar or identical devices. { 'par·ə,lel ,äp·ə'rā·shən }

parallel-plate waveguide [ELECTROMAG] Pair of parallel conducting planes used for propagating uniform circularly cylindrical waves having their axes normal to the plane. { 'par·ə,lel ¦plāt 'wāv ,gīd }

parallel port *See* parallel interface. { 'par·ə,lel ,pȯrt }

parallel processor *See* multiprocessor. { 'par·ə,lel 'prä,ses·ər }

parallel programming [COMPUT SCI] A method for performing simultaneously the normally sequential steps of a computer program, using two or more processors. { 'par·ə,lel 'prō,gram·iŋ }

parallel radio tap [COMMUN] A telephone tapping procedure in which a battery-powered miniature radio transmitter is bridged across the target pair. { 'par·ə,lel 'rād·ē·ō ,tap }

parallel reliability [SYS ENG] Property of a system composed of functionally parallel elements in such a way that if one of the elements fails, the parallel units will continue to carry out the system function. { 'par·ə,lel ri'lī·ə,bil·əd·ē }

parallel representation [COMPUT SCI] The simultaneous appearance of the different bits of a digital variable on parallel bus lines. { 'par·ə ,lel ,rep·ri,zen'tā·shən }

parallel running [COMPUT SCI] **1.** The running of a newly developed system in a data-processing area in conjunction with the continued operation of the current system. **2.** The final step in the debugging of a system; this step follows a system test. { 'par·ə,lel 'rən·iŋ }

parallel search storage [COMPUT SCI] A device for very rapid search of a volume of stored data to permit finding a specific item. { 'par·ə,lel ¦sərch ,stȯr·ij }

parallel storage [COMPUT SCI] A storage device in which words (or characters or digits) can be read in or out simultaneously. { 'par·ə,lel 'stȯr·ij }

parallel transfer [COMPUT SCI] Simultaneous transfer of all bits in a storage location constituting a character or word. { 'par·ə,lel 'tranz·fər }

parallel transmission [COMPUT SCI] The transmission of characters of a word over different lines, usually simultaneously; opposed to serial transmission. { 'par·ə,lel tranz'mish·ən }

paramagnetic amplifier *See* maser. { ¦par·ə·mag'ned·ik 'am·plə,fī·ər }

parameter-driven system [COMPUT SCI] A software system whose functions and operations are controlled mainly by parameters. { pə'ram·əd·ər ¦driv·ən 'sis·təm }

parameter identification [SYS ENG] The problem of estimating the values of the parameters that govern a dynamical system from data on the observed behavior of the system. { 'pə'ram·əd·ər ī,dent·ə·fə'kā·shən }

parameter tags [COMPUT SCI] Constants that are used by several computer programs. { pə'ram·əd·ər ,tagz }

parameter word [COMPUT SCI] A word in a computer storage containing one or more parameters that specify the action of a routine or subroutine. { pə'ram·əd·ər ,wərd }

parametric amplifier [ELECTR] A highly sensitive ultra-high-frequency or microwave amplifier having as its basic element an electron tube or solid-state device whose reactance can be varied periodically by an alternating-current voltage at a pumping frequency. Also known as mavar; paramp; reactance amplifier. { ¦par·ə¦me·trik 'am·plə,fī·ər }

parametric converter [ELECTR] Inverting or noninverting parametric device used to convert an input signal at one frequency into an output signal at a different frequency. { ¦par·ə¦me·trik kən'vərd·ər }

parametric device [ELECTR] Electronic device whose operation depends essentially upon the time variation of a characteristic parameter usually understood to be a reactance. { ¦par·ə ¦me·trik di'vīs }

parametric down-converter [ELECTR] Parametric converter in which the output signal is at a lower frequency than the input signal. { ¦par·ə ¦me·trik 'daủn kən,vərd·ər }

parametric programming [COMPUT SCI] A programming approach in which data are stored in external tables or files, rather than within the program itself, and accessed by the program when needed, so that the values of these data can be changed with relative ease. { ¦par·ə¦me·trik 'prō,gram·iŋ }

parametric up-converter [ELECTR] Parametric converter in which the output signal is at a higher frequency than the input signal. { ¦par·ə ¦me·trik 'əp kən,vərd·ər }

paramp *See* parametric amplifier. { 'par,amp }

parasitic [ELECTR] An undesired and energy-wasting signal current, capacitance, or other parameter of an electronic circuit. { ¦par·ə¦sid·ik }

parasitic antenna *See* parasitic element. { ¦par·ə¦sid·ik an'ten·ə }

parasitic element [ELECTROMAG] An antenna element that serves as part of a directional antenna array but has no direct connection to the receiver or transmitter and reflects or reradiates the energy that reaches it, in a phase relationship such as to give the desired radiation pattern. Also known as parasitic antenna; parasitic reflector; passive element. { ¦par·ə¦sid·ik 'el·ə·mənt }

parasitic oscillation [ELECTR] An undesired self-sustaining oscillation or a self-generated transient impulse in an oscillator or amplifier circuit, generally at a frequency above or below the correct operating frequency. { ¦par·ə¦sid·ik ,äs·ə'lā·shən }

parasitic reflector *See* parasitic element. { ¦par·ə¦sid·ik ri'flek·tər }

parasitic suppressor [ELECTR] A suppressor, usually in the form of a coil and resistor in parallel, inserted in a circuit to suppress parasitic high-frequency oscillations. { ¦par·ə¦sid·ik sə'pres·ər }

parent [COMPUT SCI] An element that precedes a given element in a data structure. { 'per·ənt }

parenthesis-free notation *See* Polish notation. { pə'ren·thə·səs ¦frē nō'tā·shən }

parity [COMPUT SCI] The use of a self-checking code in a computer employing binary digits in which the total number of 1's or 0's in each permissible code expression is always even or always odd. { 'par·əd·ē }

parity bit [COMMUN] An additional nondata bit that is attached to a set of data bits to check their validity; it is set so that the sum of one-bits in the augmented set is always odd or always even. { 'par·əd·ē ,bit }

parity check *See* odd-even check. { 'par·əd·ē ,chek }

parity error [COMPUT SCI] A machine error in which an odd number of bits are accidentally changed, so that the error can be detected by a parity check. { 'par·əd·ē ,er·ər }

parity transformation [COMMUN] A change in value of a transmitted character denoting the number of one-bits. { 'par·əd·ē ,tranz·fər 'mā·shən }

parser [COMPUT SCI] The portion of a computer program that carries out parsing operations. { 'pär·sər }

parsing [COMPUT SCI] A process whereby phrases in a string of characters in a computer language are associated with the component names of the grammar that generated the string. { 'pärs·iŋ }

partial carry [COMPUT SCI] A word composed of the carries generated at each position when adding many digits in parallel. { 'pär·shəl 'kar·ē }

partial common battery [COMMUN] Type of telephone system in which the talking battery is supplied by each individual telephone, and the signaling and supervisory battery is supplied by the switchboard. { 'pär·shəl 'käm·ən 'bad·ə·rē }

partial function [COMPUT SCI] A partial function from a set A to a set B is a correspondence between some subset of A and B which associates with each element of the subset of A a unique element of B. { 'pär·shəl 'fəŋk·shən }

partially populated board [COMPUT SCI] A printed circuit board on which some but not all of the possible electronic components are mounted, leaving room for additional components. { 'pär·shə·lē ¦päp·yə,lād·əd 'bȯrd }

partial-response maximum-likelihood technique [COMMUN] A method of constructing a digital data stream from an analog signal by using information acquired by sampling the analog waveform at selected instants of time rather than using the entire waveform, and then applying the Viterbi algorithm to find the most likely sequence of bits. Abbreviated PRML technique. { ¦pär·shəl ri,späns 'mak·sə·məm ¦līk·lē,hu̇d tek ,nēk }

partition [COMPUT SCI] **1.** A reserved portion of a computer memory, sometimes used for the execution of a single computer program. **2.** One of a number of fixed portions into which a computer memory is divided in certain multiprogramming systems. { pär'tish·ən }

partitioned data set [COMPUT SCI] A single data set, divided internally into a directory and one or more sequentially organized subsections called members, residing on a direct access for each device, and commonly used for storage or program libraries. { pär'tish·ənd 'dad·ə ,set }

partitioned display [COMPUT SCI] An electronic display that can be divided into two or more viewing areas under user or program control. Also known as split screen. { pär'tish·ənd di'splā }

partitioned file [COMPUT SCI] A file on disk storage that is divided into subdivisions, each of which constitutes a complete file. { pär'tish·ənd 'fīl }

part operation [COMPUT SCI] The part in an instruction that specifies the kind of arithmetical or logical operation to be performed, but not the address of the operands. { 'pärt ,äp·ə,rā·shən }

party line [COMMUN] A subscriber line arranged to serve more than one station, with discriminatory ringing for each station. { 'pärd·ē 'līn }

party-line bus [COMPUT SCI] Parallel input/output bus lines to which are wired all external devices, connected to a processor register by suitable logic. { 'pärd·ē ¦līn 'bəs }

party-line carrier system [COMMUN] A single-frequency carrier telephone system in which the carrier energy is transmitted directly to all other carrier terminals of the same channel. { 'pärd·ē ¦līn 'kar·ē·ər ,sis·təm }

Pascal [COMPUT SCI] A procedure-oriented programming language whose highly structured design facilitates the rapid location and correction of coding errors. { pa'skal }

pass [AERO ENG] The period of time in which a satellite is within telemetry range of a data acquisition station. [COMPUT SCI] A complete cycle of reading, processing, and writing in a computer. { pas }

passband [ELECTR] A frequency band in which the attenuation of a filter is essentially zero. { 'pas,band }

passivation [ELECTR] Growth of an oxide layer on the surface of a semiconductor to provide electrical stability by isolating the transistor surface from electrical and chemical conditions in the environment; this reduces reverse-current leakage, increases breakdown voltage, and raises power dissipation rating. { ,pas·ə'vā·shən }

passive AND gate *See* AND gate. { 'pas·iv 'and ,gāt }

passive antenna [ELECTROMAG] An antenna which influences the directivity of an antenna system but is not directly connected to a transmitter or receiver. { 'pas·iv an'ten·ə }

passive corner reflector [ELECTROMAG] A corner reflector that is energized by a distant transmitting antenna; used chiefly to improve the reflection of radar signals from objects that would not otherwise be good radar targets. { 'pas·iv 'kȯr·nər ri,flek·tər }

passive device [COMPUT SCI] A unit of a computer which cannot itself initiate a request for communication with another device, but which honors such a request from another device. { 'pas·iv di'vīs }

passive double reflector [ELECTROMAG] A combination of two passive reflectors positioned to bend a microwave beam over the top of a mountain or ridge, generally without appreciably changing the general direction of the beam. { 'pas·iv 'dəb·əl ri'flek·tər }

passive electronic countermeasures [ELECTR] Electronic countermeasures that do not radiate energy, including reconnaissance or surveillance equipment that detects and analyzes electromagnetic radiation from radar and communications transmitters, and devices such as chaff which return confusing or obscuring echoes to enemy radar; passive electronic attack. { 'pas·iv i,lek 'trän·ik 'kau̇nt·ər,mezh·ərz }

passive element *See* parasitic element. { 'pas·iv 'el·ə·mənt }

passive jamming [ELECTR] Use of confusion reflectors to return spurious and confusing signals to enemy radars. Also known as mechanical jamming. { 'pas·iv 'jam·iŋ }

passive-matrix liquid-crystal display *See* supertwisted nematic liquid-crystal display. { ¦pas·iv ¦mā·triks ¦lik·wəd ¦krist·əl di'splā }

passive radar [ENG] A technique for detecting objects at a distance by picking up the microwave electromagnetic energy that is both radiated and reflected by all bodies. { 'pas·iv 'rā,där }

passive reflector [ELECTROMAG] A flat reflector used to change the direction of a microwave or radar beam; often used on microwave relay towers to permit placement of the transmitter, repeater, and receiver equipment on the ground, rather than at the tops of towers. Also known as plane reflector. { 'pas·iv ri'flek·tər }

passive sonar [ENG] Sonar that uses only underwater listening equipment, with no transmission of location-revealing pulses. { 'pas·iv 'sō ,när }

passive termination [COMPUT SCI] The simplest means of ending a chain of peripheral devices connected to a small computer system interface (SCSI) port, suitable for chains with no more than four devices. { ˌpas·iv ˌtər·məˈnā·shən }

passthrough [COMPUT SCI] A procedure that allows a user to communicate with a computer through the use of the operating system of a second computer. { ˈpasˌthrü }

password [COMPUT SCI] A unique word or string of characters that must be supplied to meet security requirements before a program, computer operator, or user can gain access to data. { ˈpas ˌwərd }

password guessing [COMPUT SCI] A method of gaining unauthorized access to a computing system by using computers and dictionaries or large word lists to try likely passwords. { ˈpas ˌwərd ˌges·iŋ }

patch [COMPUT SCI] **1.** To modify a program or routine by inserting a machine language correction in an object deck, or by inserting it directly into the computer through the console. **2.** The section of coding inserted in this way. { pach }

patch panel *See* control panel; panel. { ˈpach ˌpan·əl }

path [COMPUT SCI] **1.** The logical sequence of instructions followed by a computer in carrying out a routine. **2.** A series of physical or logical connections between records or segments in a database management system, generally involving the use of pointers. { path }

path attenuation [COMMUN] Power loss between transmitter and receiver, due to any cause. { ˈpath əˌten·yəˈwā·shən }

path length *See* physical path length; software path length. { ˈpath ˌleŋkth }

path plotting [ELECTROMAG] In laying out a microwave system, the plotting of the path followed by the microwave beam on a profile chart which indicates the earth's curvature. { ˈpath ˌpläd·iŋ }

pattern analysis [COMPUT SCI] The phase of pattern recognition that consists of using whatever is known about the problem at hand to guide the gathering of data about the patterns and pattern classes, and then applying techniques of data analysis to help uncover the structure present in the data. { ˈpad·ərn əˌnal·ə·səs }

pattern generator [ELECTR] A signal generator used to produce a test waveform for service work on a display device. { ˈpad·ərn ˌjen·əˌrād·ər }

pattern recognition [COMPUT SCI] The automatic identification of figures, characters, shapes, forms, and patterns without active human participation in the decision process. { ˈpad·ərn ˌrek·igˈnish·ən }

pattern-sensitive fault [COMPUT SCI] A fault that appears only in response to one pattern or sequence of data, or certain patterns or sequences. { ˈpad·ərn ¦sen·səd·iv ˈfȯlt }

PAX *See* private automatic exchange. { paks }

payload [COMMUN] Referring to the bytes which follow the header byte in a packet; the transport stream packet header and adaptation fields are not payload. { ˈpāˌlōd }

pay television *See* subscription television. { ˈpā ˈtel·əˌvizh·ən }

P band [COMMUN] A band of radio frequencies extending from 225 to 390 megahertz, corresponding to wavelengths of 133.3 to 76.9 centimeters. { ˈpē ˌband }

PBX *See* private branch exchange.

p-channel metal-oxide semiconductor *See* PMOS. { ¦pē ˌchan·əl ˌmed·əl ¦äkˌsīd ˈsem·i·kən ˌdək·tər }

PCI *See* peripheral component interconnect.

P class [COMPUT SCI] The class of decision problems that can be solved in polynomial time. { ¦pē klas }

PCM *See* pulse-code modulation.

PCN *See* personal communications network.

PCP *See* primary control program.

PCR *See* program clock reference.

PCS *See* personal communications service.

PCSB *See* pulse-coded scanning beam.

PDF *See* portable document format.

PDL *See* page description language.

PDM *See* pulse-duration modulation.

PDU *See* power distribution unit.

peak attenuation [COMMUN] The diminution of response to a modulated wave experienced on modulation crests. { ˈpēk əˌten·yəˈwā·shən }

peak clipping *See* limiting. { ˈpēk ˌklip·iŋ }

peak detector [ELECTR] A detector whose output voltage approximates the true peak value of an applied signal; the detector tracks the signal in its sample mode and preserves the highest input signal in its hold mode. { ˈpēk diˌtek·tər }

peak distortion [COMMUN] Largest total distortion of telegraph signals noted during a period of observation. { ˈpēk diˈstȯr·shən }

peak envelope power [ELECTR] Of a radio transmitter, the average power supplied to the antenna transmission line by a transmitter during one radio-frequency cycle at the highest crest of the modulation envelope, taken under conditions of normal operation. { ˈpēk ˈen·vəˌlōp ˌpau̇·ər }

peak second algorithm [COMMUN] A set of mathematical procedures for attempting to predict the number of transmissions that will be carried out in a communications system during the busiest 1-second interval during some study period. { ˈpēk ˈsek·ənd ˈal·gəˌrith·əm }

peak-to-valley ratio [COMMUN] The ratio of the largest amplitude of a modulated wave to its smallest value. { ¦pēk tə ˈval·ē ˌrā·shō }

pedestal *See* blanking level. { ˈped·əst·əl }

pedestal level *See* blanking level. { ˈped·əst·əl ˌlev·əl }

peek [COMPUT SCI] An instruction that causes the contents of a specific storage location in a computer to be displayed. { pēk }

peephole masks [COMPUT SCI] In character recognition, a set of characters (each character residing in the character reader in the form of strategically placed points) which theoretically render all input characters as being unique regardless of their style. { ˈpēpˌhōl ˌmasks }

peer [COMMUN] A functional unit in a communications system that is in the same protocol layer as another such unit. { pir }

peer-to-peer network [COMMUN] A local-area network in which there is no central controller and all the nodes have equal access to the resources of the network. { ¦pir tə ¦pir 'net,wərk }

pel *See* pixel. { pel }

pencil beam [ELECTROMAG] A beam of radiant energy concentrated in an approximately conical or cylindrical portion of space of relatively small diameter; this type of beam is used for many revolving navigational lights and radar beams. { 'pen·səl ,bēm }

pencil beam antenna [ELECTROMAG] Unidirectional antenna designed so that cross sections of the major lobe formed by planes perpendicular to the direction of maximum radiation are approximately circular. { 'pen·səl ,bēm an,ten·ə }

pencil follower [COMPUT SCI] A device for converting graphic images to digital form; the information to be analyzed appears on a reading table where a reading pencil is made to follow the trace, and a mechanism beneath the table surface transmits position signals from the pencil to an electronic console for conversion to digital form. { 'pen·səl ,fäl·ə·wər }

pending input/output [COMPUT SCI] An input/output operation that has been initiated but not yet carried out, so that the central processing unit either is temporarily idle or services other programs and tasks until the operation is completed. { 'pend·iŋ 'in,pu̇t 'au̇t,pu̇t }

penetration testing [COMPUT SCI] An activity that is intended to determine if there is a way to cause a computer program to fail to perform in the expected manner; it involves hypothesizing flaws that would prevent the program from enforcing security, and conducting experiments to confirm or refute the hypothesized flaws. { ,pen·ə'trā·shən ,test·iŋ }

PEPE *See* parallel element-processing ensemble. { 'pe,pē }

percentage modulation *See* percent modulation. { pər'sen·tij ,maj·ə'lā·shən }

percent distortion [COMMUN] The ratio of the amplitude of a harmonic component to the fundamental component multiplied by 100. { pər 'sent di'stȯr·shən }

percent modulation [COMMUN] The modulation factor expressed as a percentage. Also known as percentage modulation. { pər'sent ,mäj·ə'lā ·shən }

perceptron [COMPUT SCI] A pattern recognition machine, based on an analogy to the human nervous system, capable of learning by means of a feedback system which reinforces correct answers and discourages wrong ones. { pər'sep ,trän }

perceptual audio coding [COMMUN] The process of representing an audio signal with fewer bits while still preserving audio quality. The coding schemes are based on the perceptual characteristics of the human ear; some examples of these coders are PAC, AAC, MPEG-2, and AC-3. Also known as audio bit rate reduction; audio compression. Abbreviated PAC. { pər ¦sep·chə·wəl 'ȯd·ē·ō ,kōd·iŋ }

percolation [COMPUT SCI] The transfer of needed data back from secondary storage devices to main storage. { pər·kə'lā·shən }

perforator [COMMUN] In telegraph practice, a device for punching code signals in paper tape for application to a tape transmitter. { 'pər·fə ,rād·ər }

perform [COMPUT SCI] A subroutine in the COBOL programming language that allows a portion of a program to be executed on command by other portions of the same program. { pər'fȯrm }

performance failure [COMPUT SCI] Failure of a computer system in which the system operates correctly but fails to deliver the results in a timely fashion. { pər'fȯr·məns ,fāl·yər }

perfory [COMPUT SCI] The removable edges of computer paper containing holes engaged by the pin-feed mechanism. { 'pər·fə·rē }

periodic antenna [ELECTROMAG] An antenna in which the input impedance varies as the frequency is altered. { ¦pir·ē¦äd·ik an'ten·ə }

peripheral *See* peripheral device. { pə'rif·ə·rəl }

peripheral buffer [COMPUT SCI] A device acting as a temporary storage when transmission occurs between two devices operating at different transmission speeds. { pə'rif·ə·rəl 'bəf·ər }

peripheral component interconnect [COMPUT SCI] A bus standard for connecting additional input/output devices (such as graphics or modem cards) to a personal computer. Abbreviated PCI. { pə,rif·ə·rəl kəm,pō·nənt 'in·tər,kə·nek }

peripheral control unit [COMPUT SCI] A device which connects a unit of peripheral equipment with the central processing unit of a computer and which interprets and responds to instructions from the central processing unit. { pə'rif·ə·rəl kən'trōl ,yü·nət }

peripheral device [COMPUT SCI] Any device connected internally or externally to a computer and used to enter or display data, such as the keyboard, mouse, monitor, scanner, and printer. { pə'rif·ərel di,vīs }

peripheral equipment [COMPUT SCI] Equipment that works in conjunction with a computer but is not part of the computer itself. { pə'rif·ə·rəl i'kwip·mənt }

peripheral interface channel [COMPUT SCI] A path along which information can flow between a unit of peripheral equipment and the central processing unit of a computer. { pə'rif·ə·rəl 'in·tər,fās ,chan·əl }

peripheral-limited [COMPUT SCI] Property of a computer system whose processing time is determined by the speed of its peripheral equipment rather than by the speed of its central processing unit. { pə'rif·ə·rəl ¦lim·əd·əd }

peripheral operation [COMPUT SCI] An operation in which an input or output device is used, and which is not directly controlled by a computer while the operation is being carried out. { pə'rif·ə·rəl ,äp·ə'rā·shən }

peripheral processing [COMPUT SCI] Processing that is carried out by peripheral equipment or by an auxiliary computer. { pə'rif·ə·rəl 'prä ˌses·iŋ }

peripheral processor [COMPUT SCI] Auxiliary computer performing specific operations under control of the master computer. { pə'rif·ə·rəl 'präˌses·ər }

peripheral transfer [COMPUT SCI] The transmission of data between two units of peripheral equipment or between a peripheral unit and the central processing unit of a computer. { pə'rif·ə·rəl 'tranz·fər }

peripheral units *See* peripheral equipment. { pə'rif·ə·rəl ˌyü·nəts }

Perl *See* Practical Extraction and Reporting Language. { pərl }

permanent echo [ELECTR] *See* fixed echo. [ELECTROMAG] A signal reflected from an object that is fixed with respect to a radar site. { 'pər·mə·nənt 'ek·ō }

permanent error [COMPUT SCI] An error that occurs when a sector mark on disk pack or floppy disk is incorrectly modified by writing data over it, and that can be corrected only by clearing the entire disk and rewriting the track and sector marks. { 'pər·mə·nənt 'er·ər }

permanent fault [COMPUT SCI] A hardware malfunction that always occurs when a particular set of conditions exists, and that can be made to occur deliberately, in contrast to a sporadic fault. { 'pər·mə·nənt 'fölt }

permanent-magnet focusing [ELECTR] Focusing of the electron beam in a cathode-ray tube by means of the magnetic field produced by one or more permanent magnets mounted around the neck of the device. { 'pər·mə·nənt ¦mag·nət 'fō·kəs·iŋ }

permanent storage [COMPUT SCI] A means of storing data for rapid retrieval by a computer; does not permit changing the stored data. { 'pər·mə·nənt 'stör·ij }

permutation modulation [COMMUN] Proposed method of transmitting digital information by means of band-limited signals in the presence of additive white gaussian noise; pulse-code modulation and pulse-position modulation are considered simple special cases of permutation modulation. { ˌpər·myə'tā·shən ˌmäj·əˌlā·shən }

permutation table [COMMUN] In computers, a table designed for the systematic construction of code groups; it may also be used to correct garbles in groups of code text. { ˌpər·myə'tā ·shən ˌtā·bəl }

perpendicular recording *See* vertical recording. { ˌpər·pən¦dik·yə·lər ri'körd·iŋ }

personal communications network [COMMUN] The series of small low-power antennas that support a personal communications service, and are linked to a master telephone switch that is connected to the main telephone network. Abbreviated PCN. { ˌpərs·ən·əl kəˌmyü·nə¦kā·shənz ˌnetˌwərk }

personal communications service [COMMUN] A mobile telephone service in which pocket-sized telephones carried by the users communicate via small low-power transmitter-receiver antennas that are installed throughout a city or community. Abbreviated PCS. { ˌpərs·ən·əl kəˌmyü·nə'kā·shənz ˌsər·vəs }

personal computer [COMPUT SCI] A computer for home or personal use. { 'pər·sən·əl kəm'pyüd·ər }

personal digital assistant *See* hand-held computer. { ˌpərs·ən·əl ˌdij·əd·əl ə'sis·tənt }

personal identification code [COMPUT SCI] A special number up to six characters in length on a strip of magnetic tape embedded in a plastic card which identifies a user accessing a special-purpose computer. Abbreviated PIC. { 'pər·sən·əl īˌden·tə·fə'kā·shən ˌkōd }

personal information manager [COMPUT SCI] Software that combines the functions of word-processing, database, and desktop accessory programs, making it possible to organize information that is relatively loosely structured. Abbreviated PIM. { ¦pər·sən·əl ˌin·fər'mā·shən ˌman·ij·ər }

pertinency factor [COMPUT SCI] In information retrieval, the ratio obtained in dividing the total number of relevant documents retrieved by the total number of documents retrieved. { 'pər·tə·nən·sē ˌfak·tər }

PES *See* packetized elementary stream. { ¦pē¦ē ¦es }

PES packet [COMMUN] The data structure used to carry elementary stream data; consists of a packet header followed by PES packet payload. { ¦pē¦ē¦es ¦pak·ət }

PES stream [COMMUN] Referring to a stream consisting of PES packets, all of whose payloads consists of data from a single elementary stream, and all of which have the same stream ID number. { ¦pē¦ē¦es ¦strēm }

Petri net [COMMUN] An abstract, formal model of information flow, which is used as a graphical language for modeling systems with interacting concurrent components; in mathematical terms, a structure with four parts or components: a finite set of places, a finite set of transitions, an input function, and an output function. { 'pē·trē ˌnet }

PF key *See* programmed function key. { ˌpē'ef ˌkē }

PFM *See* pulse-frequency modulation.

P-frame *See* predicted picture. { 'pē ˌfrām }

phantom circuit [COMMUN] A communication circuit derived from two other communication circuits or from one other circuit and ground, with no additional wire lines. { 'fan·təm 'sər·kət }

phantom signals [ELECTR] Signals appearing on a radar display, the cause of which cannot readily be determined and which may be caused by circuit fault, interference, propagation anomalies, measurement ambiguities, jamming, and so on. { 'fan·təm 'sig·nəlz }

phase-alternation line system [COMMUN] A color television system used in Europe and other parts of the world, in which the phase of the color subcarrier is changed from scanning line to scanning line, requiring transmission of a line switching signal as well as a color burst. Abbreviated PAL system. { 'fāz ˌöl·tər¦nā·shən ˌlīn ˌsis·təm }

phase-change coefficient *See* phase constant. { ¦fāz ˌchānj ˌkō·iˌfish·ənt }

phase-change recording [COMPUT SCI] An optical recording technique that uses a laser to alter the crystalline structure of a metallic surface to create bits that reflect or absorb light when they are illuminated during the read operation. { ¦fāz ˌchānj ri'kȯrd·iŋ }

phase comparator [COMPUT SCI] A comparator that accepts two radio-frequency input signals of the same frequency and provides two video outputs which are proportional, respectively, to the sine and cosine of the phase difference between the two inputs. { 'fāz kəmˌpar·əd·ər }

phase constant [ELECTROMAG] A rating for a line or medium through which a plane wave of a given frequency is being transmitted; it is the imaginary part of the propagation constant, and is the space rate of decrease of phase of a field component (or of the voltage or current) in the direction of propagation, in radians per unit length. Also known as phase-change coefficient; wavelength constant. { 'fāz ˌkän·stənt }

phase control *See* hue control. { 'fāz kənˌtrōl }

phase-correcting network *See* phase equalizer. { 'fāz kə¦rek·tiŋ 'netˌwərk }

phase correction [COMMUN] Process of keeping synchronous telegraph mechanisms in substantially correct phase relationship. { 'fāz kə ˌrek·shən }

phase crossover [CONT SYS] A point on the plot of the loop ratio at which it has a phase angle of 180°. { 'fāz 'krȯsˌō·vər }

phased array [ELECTROMAG] An array of dipoles on a radar antenna in which the signal feeding each dipole is varied so that antenna beams can be formed in space and scanned very rapidly in azimuth and elevation. { 'fāzd ə'rā }

phased-array radar [ENG] Radar using an antenna of the multiple-element array type in which the relative phasing of the elements, electronically controlled, positions the main beam in angle without need of moving the antenna. { 'fāzd ə'rā 'rāˌdär }

phase delay [COMMUN] Ratio of the total phase shift (radians) of a sinusoidal signal in transmission through a system or transducer, to the frequency (radians/second) of the signal. { 'fāz diˌlā }

phase detector [ELECTR] **1.** A circuit that provides a direct-current output voltage which is related to the phase difference between an oscillator signal and a reference signal, for use in controlling the oscillator to keep it in synchronism with the reference signal. Also known as phase discriminator. **2.** A circuit or device in a radar receiver giving a voltage output dependent upon the phase difference of two inputs; used in Doppler sensing in a coherent radar. { 'fāz di ˌtek·tər }

phase deviation [COMMUN] The peak difference between the instantaneous angle of a modulated wave and the angle of the sine-wave carrier. { 'fāz ˌdē·vē'ā·shən }

phase discriminator *See* phase detector. { 'fāz diˌskrim·əˌnād·ər }

phase distortion [COMMUN] **1.** The distortion which occurs in an instrument when the relative phases of the input signal differ from those of the output signal. **2.** *See* phase-frequency distortion. { 'fāz diˌstȯr·shən }

phase encoding [COMPUT SCI] A method of recording data on magnetic tape in which a logical 1 is defined as the transition from one magnetic polarity to another positioned at the center of the bit cell, and 0 is defined as the transition in the opposite direction, also at the center of the cell. Also known as Manchester coding. { 'fāz in'kō d·iŋ }

phase equalizer [ELECTR] A network designed to compensate for phase-frequency distortion within a specified frequency band. Also known as phase-correcting network. { 'fāz 'ē·kwəˌlīz·ər }

phase excursion [COMMUN] In angle modulation, the difference between the instantaneous angle of the modulated wave and the angle of the carrier. { 'fāz ikˌskər·zhən }

phase-frequency distortion [COMMUN] Distortion occurring because phase shift is not proportional to frequency over the frequency range required for transmission. Also known as phase distortion. { 'fāz ¦frē·kwən·sē diˌstȯr·shən }

phase jitter [ELECTR] Jitter that undesirably shortens or lengthens pulses intermittently during data processing or transmission. { 'fāz ˌjid·ər }

phase lock [ELECTR] Technique of making the phase of an oscillator signal follow exactly the phase of a reference signal by comparing the phases between the two signals and using the resultant difference signal to adjust the frequency of the reference oscillator. { 'fāz ˌläk }

phase-locked communication [COMMUN] Systems in which oscillators at the receiver and transmitter are locked in phase. { 'fāz ¦läkt kə ˌmyü·nə'kā·shən }

phase-locked loop [ELECTR] A circuit that consists essentially of a phase detector which compares the frequency of a voltage-controlled oscillator with that of an incoming carrier signal or reference-frequency generator; the output of the phase detector, after passing through a loop filter, is fed back to the voltage-controlled oscillator to keep it exactly in phase with the incoming or reference frequency. Abbreviated PLL. { 'fāz ¦läkt 'lüp }

phase magnet [COMMUN] Magnetically operated latch used to phase a facsimile transmitter or recorder. Also known as trip magnet. { 'fāz ˌmag·nət }

phase margin [CONT SYS] The difference between 180° and the phase of the loop ratio of a stable system at the gain-crossover frequency. { 'fāz ˌmär·jən }

phase modulation [COMMUN] Modulation in which the linearly increasing angle of a sine wave has added to it a phase angle that is proportional to the instantaneous value of the modulating signal (message to be communicated). Abbreviated PM. { 'fāz ˌmäj·əˌlā·shən }

phase-modulation detector [ELECTR] A device which recovers or detects the modulating signal from a phase-modulated carrier. { 'fāz ˌmäj·ə ˌlā·shən diˌtek·tər }

phase-modulation transmitter [ELECTR] A radio transmitter used to broadcast a phase-modulated signal. { 'fāz ˌmäj·əˌlā·shən tranz ˌmid·ər }

phase modulator [ELECTR] An electronic circuit that causes the phase angle of a modulated wave to vary (with respect to an unmodulated carrier) in accordance with a modulating signal. { 'fāz ˌmäj·əˌlād·ər }

phase plane analysis [CONT SYS] A method of analyzing systems in which one plots the time derivative of the system's position (or some other quantity characterizing the system) as a function of position for various values of initial conditions. { 'fāz ¦plān ə'nal·ə·səs }

phase portrait [CONT SYS] A graph showing the time derivative of a system's position (or some other quantity characterizing the system) as a function of position for various values of initial conditions. { 'fāz ˌpȯr·trət }

phaser [COMMUN] Facsimile device for adjusting equipment so the recorded elemental area bears the same relation to the record sheet as the corresponding transmitted elemental area bears to the subject copy in the direction of the scanning line. [ELECTROMAG] Microwave ferrite phase shifter employing a longitudinal magnetic field along one or more rods of ferrite in a waveguide. { 'fāz·ər }

phase response [ELECTR] A graph of the phase shift of a network as a function of frequency. { 'fāz riˌspäns }

phase reversal modulation [COMMUN] Form of pulse modulation in which reversal of signal phase serves to distinguish between the two binary states used in data transmission. { 'fāz riˌvər·səl ˌmäj·ə'lā·shən }

phase-sensitive detector [ELECTR] An electronic circuit that consists essentially of a multiplier and a low-pass circuit and that produces a direct-current output signal that is proportional to the product of the amplitudes of two alternating-current input signals of the same frequency and to the cosine of the phase between them. { 'fāz ˌsen·səd·iv diˌtek·tər }

phase shift [ELECTR] The phase angle between the input and output signals of a network or system. { 'fāz ˌshift }

phase-shift discriminator [ELECTR] A discriminator that uses two similarly connected diodes, fed by a transformer that is tuned to the center frequency; when the frequency-modulated or phase-modulated input signal swings away from this center frequency, one diode receives a stronger signal than the other; the net output of the diodes is then proportional to the frequency displacement. Also known as Foster-Seely discriminator. { 'fāz ¦shift diˌskrim·əˌnād·ər }

phase-shift keying [COMMUN] A form of phase modulation in which the modulating function shifts the instantaneous phase of the modulated wave between predetermined discrete values. Abbreviated PSK. { 'fāz ¦shift ˌkē·iŋ }

phasing *See* framing. { 'fāz·iŋ }

phasing line [ELECTR] That portion of the length of scanning line set aside for the phasing signal in a video system. { 'fāz·iŋ ˌlīn }

phasing signal [ELECTR] A signal used to adjust the picture position along the scanning line in a facsimile system. { 'fāz·iŋ ˌsig·nəl }

phasitron [ELECTR] An electron tube used to frequency-modulate a radio-frequency carrier; internal electrodes are designed to produce a rotating disk-shaped corrugated sheet of electrons; audio input is applied to a coil surrounding the glass envelope of the tube, to produce a varying axial magnetic field that gives the desired phase or frequency modulation of the RF carrier input to the tube. { 'fāz·əˌträn }

phasor [PHYS] **1.** A rotating line used to represent a sinusoidally varying quantity; the length of the line represents the magnitude of the quantity, and its angle with the *x*-axis at any instant represents the phase. **2.** Any quantity (such as impedance or admittance) which is a complex number. { 'fāz·ər }

phone patch [ELECTR] A device connecting an amateur or citizens'-band transceiver temporarily to a telephone system. { 'fōn ˌpach }

phonetic alphabet [COMMUN] A list of standard words used for positive identification of letters in a voice message transmitted by radio or telephone. { fə'ned·ik 'al·fəˌbet }

phonetic search [COMPUT SCI] A method of locating information in a file in which an algorithm is used to locate combinations of characters that sound similar to a specified combination. { fə'ned·ik 'sərch }

phosphor dot [ELECTR] One of the tiny dots of phosphor material that are used in groups of three, one group for each primary color, on the screen of a color video picture tube. { 'fäs·fər ˌdät }

photocathode [ELECTR] A photosensitive surface that emits electrons when exposed to light or other suitable radiation; used in phototubes, video camera tubes, and other light-sensitive devices. { ¦fōd·ō'kathˌōd }

photocell [ELECTR] A solid-state photosensitive electron device whose current-voltage characteristic is a function of incident radiation. Also known as electric eye; photoelectric cell. { 'fōd·əˌsel }

photocomposition [COMPUT SCI] Composition of type using electrophotographic techniques such as phototypesetters and laser printers. { ˌfōd·ōˌkäm·pə'zish·ən }

photoconductive cell [ELECTR] A device for detecting or measuring electromagnetic radiation by variation of the conductivity of a substance (called a photoconductor) upon absorption of the radiation by this substance. Also known as photoresistive cell; photoresistor. { ¦fōd·ō·kən'dək·tiv 'sel }

photoconductor diode *See* photodiode { fōd·ō·kən'dək·tər 'dīˌōd }

photocoupler *See* optoisolator. { ¦fōd·ō'kəp·lər }

photodetector [ELECTR] A detector that responds to radiant energy; examples include photoconductive cells, photodiodes, photoresistors, photoswitches, phototransistors, phototubes, and photovoltaic cells. Also known as light-sensitive cell; light-sensitive detector; light sensor photodevice; photodevice; photoelectric detector; photosensor. { ¦fōd·ō·di'tek·tər }

photodevice *See* photodetector. { ¦fōd·ō·di ˌvīs }

photodiode [ELECTR] A semiconductor diode in which the reverse current varies with illumination; examples include the alloy-junction photocell and the grown-junction photocell. Also known as photoconductor diode. { ¦fōd·ō'dī ˌōd }

photoelectric cell *See* photocell. { ¦fōd·ō·i'lek·trik 'sel }

photoelectric detector *See* photodetector. { ¦fōd·ō·i'lek·trik di'tek·tər }

photoelectric effect *See* photoelectricity. { ¦fōd·ō·i'lek·trik iˌfekt }

photoelectric infrared radiation *See* near-infrared radiation. { ¦fōd·ō·i'lek·trik ¦in·frə¦red ˌrā·dē'ā·shən }

photoelectricity [ELECTR] The liberation of an electric charge by electromagnetic radiation incident on a substance; includes photoemission, photoionization, photoconduction, the photovoltaic effect, and the Auger effect (an internal photoelectric process). Also known as photoelectric effect; photoelectric process. { ¦fōd·ōˌi ˌlek'tris·əd·ē }

photoelectric process *See* photoelectricity. { ¦fōd·ō·i'lek·trik 'prä·səs }

photoelectric tube *See* phototube. { ¦fōd·ō·i'lek·trik 'tüb }

photographic recording [COMMUN] Facsimile recording in which a photosensitive surface is exposed to a signal-controlled light beam or spot. { ¦fōd·ə¦graf·ik ri'kȯrd·iŋ }

photoisolator *See* optoisolator. { ¦fōd·ō'ī·sə ˌlād·ər }

photonics [ELECTR] The electronic technology involved with the practical generation, manipulation, analysis, transmission, and reception of electromagnetic energy in the visible, infrared, and ultraviolet portions of the light spectrum. It contributes to many fields, including astronomy, biomedicine, data communications and storage, fiber optics, imaging, optical computing, optoelectronics, sensing, and telecommunications. Also known as optoelectronics. { fō'tän·iks }

photoresistive cell *See* photoconductive cell. { ¦fōd·ō·ri'zis·tiv 'sel }

photoresistor *See* photoconductive cell. { ¦fōd·ō·ri'zis·tər }

photosensor *See* photodetector. { ¦fōd·ō'sen·sər }

phototelegraphy *See* facsimile. { ¦fōd·ō·tə'leg·rə·fē }

phototransistor [ELECTR] A junction transistor that may have only collector and emitter leads or also a base lead, with the base exposed to light through a tiny lens in the housing; collector current increases with light intensity, as a result of amplification of base current by the transistor structure. { ¦fōd·ō·tran'zis·tər }

phototronic photocell *See* photovoltaic cell. { ¦fōd·ə¦trän·ik 'fōd·əˌsel }

phototube [ELECTR] An electron tube containing a photocathode from which electrons are emitted when it is exposed to light or other electromagnetic radiation. Also known as electric eye; light-sensitive tube; photoelectric tube. { 'fōd·ōˌtüb }

photovoltaic cell [ELECTR] A device that detects or measures electromagnetic radiation by generating a potential at a junction (barrier layer) between two types of material, upon absorption of radiant energy. Also known as barrier-layer cell; barrier-layer photocell; boundary-layer photocell; photronic photocell. { ¦fōd·ō·vōl'tā·ik ˌsel }

photronic photocell *See* photovoltaic cell. { fō 'trän·ik 'fōd·əˌsel }

phrase name *See* metavariable. { 'frāz ˌnām }

physical data independence [COMPUT SCI] A file structure such that the physical structure of the data can be modified without changing the logical structure of the file. { 'fiz·ə·kəl ¦dad·ə ˌin·di'pen·dəns }

physical data structure [COMPUT SCI] The manner in which data are physically arranged on a storage medium, including various indices and pointers. { 'fiz·ə·kəl 'dad·ə ˌstrək·chər }

physical device table [COMPUT SCI] A table associated with a physical input/output unit containing such information as the device type, an indication of data paths that may be used to transfer information to and from the device, status information on whether the device is busy, the input/output operation currently pending on the device, and the availability of any storage contained in the device. { 'fiz·ə·kəl di¦vīs ˌtā·bəl }

physical drive [COMPUT SCI] An operational hard disk, which may be formatted to include more than one logical drive. { 'fiz·i·kəl ¦drīv }

physical input/output control system *See* PIOCS. { 'fiz·ə·kəl ¦inˌpu̇t ¦au̇tˌpu̇t kən'trōl ˌsis·təm }

physical network [COMPUT SCI] A system of computers that communicate via cabling, modems, or other hardware, and may include more than one logical network or form part of a logical network. { 'fiz·i·kəl ¦netˌwərk }

physical path length [COMPUT SCI] The physical distance that an electronic signal must travel between two points. Also known as path length. { 'fiz·ə·kəl ¦path ˌleŋkth }

physical record [COMPUT SCI] A set of adjacent data characters recorded on some storage medium, physically separated from other physical records that may be on the same medium by means of some indication that can be recognized by a simple hardware test. Also known as record block. { 'fiz·ə·kəl 'rek·ərd }

physical system *See* causal system. { 'fiz·ə·kəl 'sis·təm }

PIC *See* personal identification code. { ¦pē¦ī¦sē *or* pik }

pick device *See* pointing device. { 'pik di,vīs }

picking [COMPUT SCI] Identification of information displayed on a screen for subsequent computer processing, by pointing to it with a lightpen. { 'pik·iŋ }

pickup tube *See* camera tube. { 'pik,əp ,tüb }

picture [COMMUN] **1.** The image on the screen of a video display. **2.** Source, coded, or reconstructed image data; a source or reconstructed picture consists of three rectangular matrices representing the luminance and two chrominance signals. [COMPUT SCI] In COBOL, a symbolic description of each data element or item according to specified rules concerning numerals, alphanumerics, location of decimal points, and length. { 'pik·chər }

picture black *See* black signal. { 'pik·chər ¦blak }

picture carrier [COMMUN] A carrier frequency located 1.25 megahertz above the lower frequency limit of a standard National Television Systems Committee television signal; in color television, it is used for transmitting color information. Also known as luminance carrier. { 'pik·chər ,kar·ē·ər }

picture compression [COMPUT SCI] The elimination of redundant information from a digital picture through the use of efficient encoding techniques in which frequently occurring gray levels or blocks of gray levels are represented by short codes and infrequently occurring ones by longer codes. { 'pik·chər kəm,presh·ən }

picture element [ELECTR] **1.** That portion, in facsimile, of the subject copy which is seen by the scanner at any instant; it can be considered a square area having dimensions equal to the width of the scanning line. **2.** In video, any segment of a scanning line, the dimension of which along the line is exactly equal to the nominal line width; the area which is being explored at any instant in the scanning process. Also known as critical area; elemental area; pixel; recording spot; scanning spot. { 'pik·chər ,el·ə·mənt }

picture frequency [COMMUN] A frequency that results solely from scanning of subject copy in a facsimile system. [ELECTR] *See* frame frequency. { 'pik·chər ,frē·kwən·sē }

picture grammar [COMPUT SCI] A formalism for carrying out computations on pictures and describing picture structure. { 'pik·chər ,gram·ər }

picture processing *See* image processing. { 'pik·chər ,prä,ses·iŋ }

picture segmentation [COMPUT SCI] The division of a complex picture into parts corresponding to regions or objects, so that the picture can then be described in terms of the parts, their properties, and their spatial relationships. Also known as scene analysis; segmentation. { 'pik·chər ,seg·mən'tā·shən }

picture signal [COMMUN] The signal resulting from the scanning process in a video system. { 'pik·chər ,sig·nəl }

picture synchronizing pulse *See* vertical synchronizing pulse. { 'pik·chər 'siŋ·krə,nīz·iŋ ,pəls }

picture transmission [COMMUN] Electric transmission of a picture having a gradation of shade values. { 'pik·chər tranz'mish·ən }

picture transmitter *See* visual transmitter. { 'pik·chər tranz,mid·ər }

picture tube [ELECTR] A cathode-ray tube used in video displays to produce an image by varying the electron-beam intensity as the beam is deflected from side to side and up and down to scan a raster on the fluorescent screen at the large end of the tube. Also known as kinescope; television picture tube. { 'pik·chər ,tüb }

picture white *See* white signal. { 'pik·chər ¦wīt }

piercing point *See* trace. { 'pirs·iŋ ,pȯint }

piezoelectric oscillator *See* crystal oscillator. { pē¦ā·zō·ə'lek·trik 'äs·ə,lād·ər }

piggyback board [ELECTR] A small printed circuit board that is mounted on a larger board to provide additional circuitry. { 'pig·ē,bak ,bȯrd }

pill [ELECTROMAG] A microwave stripline termination. { pil }

pillbox antenna [ELECTROMAG] Cylindrical parabolic reflector enclosed by two plates perpendicular to the cylinder, spaced to permit the propagation of only one mode in the desired direction of polarization. { 'pil,bäks an'ten·ə }

pilot [COMMUN] **1.** In a transmission system, a signal wave, usually single frequency, transmitted over the system to indicate or control its characteristics. **2.** Instructions, in tape relay, appearing in routing line, relative to the transmission or handling of that message. [COMPUT SCI] A model of a computer system designed to test its design, logic, and data flow under operating conditions. { 'pī·lət }

PILOT [COMPUT SCI] A programming language designed for applications to computer-aided instruction and the question-and-answer type of interaction that occurs in that environment. { 'pī·lət }

pilot system [COMPUT SCI] A system for evaluating new procedures for handling data in which a sample that is representative of the data to be handled is processed. { 'pī·lət ,sis·təm }

pilot test [COMPUT SCI] A test of a computer system under operating conditions and in the environment for which the system was designed. { 'pī·lət ,test }

pilot tone [COMMUN] Single frequency transmitted over a channel to operate an alarm or automatic control. { 'pī·lət ,tōn }

PIM *See* personal information manager. { ¦pē ¦ī'em *or* pim }

pincushion distortion [ELECTR] Distortion in which all four sides of a video image are concave (curving inward). { 'pin,ku̇sh·ən di,stȯr·shən }

pine-tree array [ELECTROMAG] Array of dipole antennas aligned in a vertical plane known as the radiating curtain, behind which is a parallel array of dipole antennas forming a reflecting curtain. { 'pīn ˌtrē əˌrā }

pin-feed printer [COMPUT SCI] A computer printer in which the paper is aligned and advanced by protrusions on two wheels which engage evenly spaced holes along the edges of the paper. Also known as tractor-feed printer. { 'pin ¦fēd 'print·ər }

ping-pong [COMMUN] To switch a transmission so that it travels in the opposite direction. [COMPUT SCI] The programming technique of using two magnetic tape units for multiple reel files and switching automatically between the two units until the complete file is processed. { 'piŋˌpäŋ }

PIOCS [COMPUT SCI] An extension of the hardware, constituting an interface between programs and data channels; opposed to LIOCS, logical input/output control system. Derived from physical input/output control system. { 'pīˌäks }

pip *See* blip. { pip }

pipe [COMPUT SCI] Any software-controlled technique for transfering data from one program or task to another during processing. { pīp }

pipelining [COMPUT SCI] A procedure for processing instructions in a computer program more rapidly, in which each instruction is divided into numerous small stages, and a population of instructions are in various stages at any given time. { 'pīpˌlīn·iŋ }

piston [ELECTROMAG] A sliding metal cylinder used in waveguides and cavities for tuning purposes or for reflecting essentially all of the incident energy. Also known as plunger; waveguide plunger. { 'pis·tən }

piston attenuator [ELECTROMAG] A microwave attenuator inserted in a waveguide to introduce an amount of attenuation that can be varied by moving an output coupling device along its longitudinal axis. { 'pis·tən ə'ten·yəˌwād·ər }

pitch [COMPUT SCI] The distance between the centerlines of adjacent rows of hole positions in punched paper tape. { pich }

pitch-row [COMPUT SCI] The distance between two adjacent holes in a paper tape. { 'pich ˌrō }

pixel [COMPUT SCI] The smallest part of an electronically coded picture image. [ELECTR] The smallest addressable element in an electronic display; a short form for picture element. Also known as pel. { pik'sel }

PLA *See* programmed logic array.

placeholder [COMPUT SCI] A section of computer storage reserved for information that will be provided later. { 'plāsˌhōl·dər }

plaintext [COMMUN] The form of a message in which it can be generally understood, before it has been transformed by a code or cipher into a form in which it can be read only by those privy to the secrets of the cipher. [COMPUT SCI] Data that are to be encrypted. { 'plānˌtekst }

plain vanilla *See* vanilla. { 'plān və'nil·ə }

planar area [COMPUT SCI] In computer graphics, an object with boundaries, such as a circle or polygon. { 'plān·ər ˌer·ē·ə }

planar-array antenna [ELECTROMAG] An array antenna in which the centers of the radiating elements are all in the same plane. { 'plā·nər ə¦rāan'ten·ə }

plane earth [ELECTROMAG] Earth that is considered to be a plane surface as used in ground-wave calculations. { 'plān ˌərth }

plane-earth attenuation [ELECTROMAG] Attenuation of an electromagnetic wave over an imperfectly conducting plane earth in excess of that over a perfectly conducting plane. { 'plān ˌərth əˌten·yə'wā·shən }

plane of polarization [ELECTROMAG] Plane containing the electric vector and the direction of propagation of electromagnetic wave. { 'plān əv ˌpō·lə·rə'zā·shən }

plane polarization *See* linear polarization. { 'plān ˌpō·lə·rə'zā·shən }

plane-polarized wave [ELECTROMAG] An electromagnetic wave whose electric field vector at all times lies in a fixed plane that contains the direction of propagation through a homogeneous isotropic medium. { 'plān ¦pō·ləˌrīzd ˌwāv }

plane reflector *See* passive reflector. { 'plān ri ¦flek·tər }

planetary wave *See* long wave. { 'plan·əˌter·ē 'wāv }

plane wave [PHYS] Wave in which the wavefront is a plane surface; a wave whose equiphase surfaces form a family of parallel planes. { 'plān ˌwāv }

planning by abstraction [COMPUT SCI] A computer problem-solving method in which the task to be accomplished is simplified; the simplified task is solved; and the solution is used as a guide. { 'plan·iŋ bī ab'strak·shən }

plan position indicator [ELECTR] A radar display in which echoes from various targets appear as bright spots at the same locations as they would on a circular map of the area being scanned, the radar antenna being at the center of the map. Variations of the plan position indicator format include limited-sector display with the radar location offset from the center appropriately, the orientation to true or magnetic north or the radar-vehicle heading at the top, and so on. Abbreviated PPI. { 'plan pə'zish·ən 'in·də ˌkād·ər }

plan position indicator repeater [ELECTR] Unit which repeats a plan position indicator (PPI) at a location remote from the radar console. Also known as remote plan position indicator. { 'plan pə'zish·ən 'in·dəˌkād·ər riˌpēd·ər }

plant [COMPUT SCI] To place a number or instruction that has been generated in the course of a computer program in a storage location where it will be used or obeyed at a later stage of the program. { plant }

plant decomposition [CONT SYS] The partitioning of a large-scale control system into subsystems along lines of weak interaction. { 'plant dēˌkäm·pə'zish·ən }

plasma display [ELECTR] A display in which sets of parallel conductors at right angles to each other are deposited on glass plates, with the very small space between the plates filled with a gas; each intersection of two conductors defines a single cell that can be energized to produce a gas discharge forming one element of a dot-matrix display. { 'plaz·mə di'splā }

plate *See* anode. { plāt }

plated wire memory [COMPUT SCI] A nonvolatile magnetic memory utilizing small zones of thin films plated on wires; such memories are characterized by very fast access and nondestructive readout. { 'plād·əd ¦wīr 'mem·rē }

platform [COMPUT SCI] The hardware system and the system software used by a computer program. { 'plat,fórm }

platter [COMPUT SCI] One of the disks in a hard-disk drive or disk pack. { 'plad·ər }

PLL *See* phase-locked loop.

PL/1 [COMPUT SCI] A multipurpose programming language, developed by IBM for the Model 360 systems, which can be used for both commercial and scientific applications. { ¦pē¦el'wən }

plot *See* contact. { plät }

plotter [ENG] A visual display or board on which a dependent variable is graphed by an automatically controlled pen or pencil as a function of one or more variables. { 'pläd·ər }

plugboard *See* control panel. { 'pləg,bórd }

plugboard chart *See* plugging chart. { 'pləg ,bórd ,chärt }

plug-compatible hardware [COMPUT SCI] A piece of equipment which can be immediately connected to a computer manufactured by another company. { 'pləg kəm,pad·ə·bəl 'härd·wer }

plugging chart [COMPUT SCI] A printed chart of the sockets in a plugboard on which may be shown the jacks or wires connecting these sockets. Also known as plugboard chart. { 'pləg·iŋ ,chärt }

plug-in [COMPUT SCI] A small software application that extends the capabilities (such as multimedia, audio, or video) of a browser. { 'pləg ,in }

plug program patching [COMPUT SCI] A relatively small auxiliary plugboard patched with a specific variation of a portion of a program and designed to be plugged into a relatively larger plugboard patched with the main program. { 'pləg 'prō,gram ,pach·iŋ }

plug-to-plug compatibility [COMPUT SCI] Property of a peripheral device that can be made to operate with a computer merely by attachment of a plug or a relatively small number of cables. { 'pləg tə 'pləg kəm,pad·ə'bil·əd·ē }

plunger *See* piston. { 'plən·jər }

plus-90 orientation [COMPUT SCI] In optical character recognition, that determinate position which indicates that the line elements of an inputted source document appear perpendicular with the leading edge of the optical reader. { ¦pləs 'nīn·tē ,ór·ē·ən,tā·shən }

plus zone [COMPUT SCI] The bit positions in a computer code which represent the algebraic plus sign. { 'pləs ,zōn }

PM *See* phase modulation.

PMLCD *See* supertwisted nematic liquid-crystal display.

PMOS [ELECTR] Metal-oxide semiconductors that are made on *n*-type substrates, and whose active carriers are holes that migrate between *p*-type source and drain contacts. Derived from *p*-channel metal-oxide semiconductor. { 'pē ,mós }

PMS notation [COMPUT SCI] A notation that provides a clear, concise description of the physical structure of computer systems, and that contains only a few primitive components, namely symbols for memory, link, switch, data operation, control unit, and transducer. Acronym for processor-memory-switch notation. { ¦pē ¦em¦es nō'tā·shən }

PN code *See* pseudorandom noise code. { ,pē'en 'kōd }

pocket [COMPUT SCI] One of the several receptacles into which punched cards are fed by a card sorter. { 'päk·ət }

pointer [COMPUT SCI] The part of an instruction which contains the address of the next record to be accessed. { 'póint·ər }

pointing device [COMPUT SCI] A handheld device, such as a mouse, puck, or stylus, that controls a position indicator on a display screen. Also known as pick device. { 'póint·iŋ di,vīs }

pointing stick [COMPUT SCI] A small rubberized device located in the center of a computer keyboard, which is moved with a finger tip to position a pointer. { 'póint·iŋ ,stik }

point-mode display [COMPUT SCI] A method of representing information in the form of dots on the face of a cathode-ray tube. { 'póint ,mōd di ,splā }

point-of-origin system [COMPUT SCI] A computer system in which data collection occurs at the point where the data are actually created, as in a point-of-sale terminal. { ¦póint əv 'är·ə·jən ,sis·təm }

point-of-sale terminal [COMPUT SCI] A computer-connected terminal used in place of a cash register in a store, for customer checkout and such added functions as recording inventory data, transferring funds from the customer's bank account to the merchant's bank account, and checking credit on charged or charge-card purchases; the terminals can be modified for many nonmerchandising applications, such as checkout of books in libraries. Abbreviated POS terminal. { 'póint əv ¦sāl 'term·ən·əl }

point target [ELECTROMAG] In radar, an object which returns a target signal by reflection from a relatively simple discrete surface; such targets are ships, aircraft, projectiles, missiles, and buildings. { 'póint ,tär·gət }

point-to-point communication [COMMUN] Radio communication between two fixed stations. { 'póint tə 'póint kə,myü·nə'kā·shən }

Point-to-Point Protocol [COMMUN] A standard governing dial-up connections of computers to the Internet via a telephone modem. Abbreviated PPP. { ,póin· tü ,póint 'prōd·ə,kól }

poke [COMPUT SCI] An instruction that causes a value in a storage location in a microcomputer's main storage to be replaced. { pōk }

polar-coordinate navigation system [NAV] A system in which one or more signals are emitted from a facility (or co-located facilities) to produce simultaneous indication of bearing and distance. { 'pō·lər kō¦ȯrd·ən·ət ˌnav·ə'gā·shən ˌsis·təm }

polarity [COMMUN] **1.** The direction in which a direct current flows, in a teletypewriter system. **2.** The sense of the potential of a portion of a video signal representing a dark area of a scene relative to the potential of a portion of the signal representing a light area. { pə'lar·əd·ē }

polarization diversity [COMMUN] A method of transmission and reception used to minimize the effects of selective fading of the horizontal and vertical components of a radio signal; it is usually accomplished through the use of separate vertically and horizontally polarized receiving antennas. { ˌpō·lə·rə'zā·shən də'ver·səd·ē }

polarization division multiple access [COMMUN] A technique for allowing multiple users at geographically dispersed locations to gain access to a shared communications channel by assigning them electric fields of different polarization. { ˌpō·lə·rə¦zā·shən dəˌvizh·ən ¦məl·tə·pəl 'akˌses }

polarization division multiplexing [COMMUN] The sharing of a communications channel among multiple users by assigning them electric fields of different polarization. { ˌpō·lə·rə¦zā·shən di ˌvizh·ən 'məl·təˌpleks·iŋ }

polarization fading [COMMUN] Fading as the result of changes in the direction of polarization in one or more of the propagation paths of waves arriving at a receiving point. { ˌpō·lə·rə'zā·shən ˌfād·iŋ }

polarized electromagnetic radiation [ELECTROMAG] Electromagnetic radiation in which the direction of the electric field vector is not random. { 'pō·ləˌrīzd i¦lek·trō·mag¦ned·ik ˌrād·ē'ā·shən }

polar keying [COMMUN] Telegraph signal in which circuit current flows in one direction for spacing. { 'pō·lər 'kē·iŋ }

polar modulation [COMMUN] Amplitude modulation in which the positive excursions of the carrier are modulated by one signal and the negative excursions by another. { 'pō·lər ˌmäj·ə'lā·shən }

polar radiation pattern [ELECTROMAG] Diagram showing the relative strength of the radiation from an antenna in all directions in a given plane. { 'pō·lər ˌrād·ē'ā·shən ˌpad·ərn }

polar resolution [COMPUT SCI] Given the *x* and *y* components of a vector, the process of finding the magnitude of the vector and the angle it makes with the *x* axis. { 'pō·lər ˌrez·ə'lü·shən }

polar transmission [COMMUN] **1.** A method of signaling in teletypewriter transmission in which direct currents flowing in opposite directions represent a mark and a space respectively, and absence of current indicates a no-signal condition. **2.** By extension, any system of signaling that uses three conditions, representing a mark, a space, or a no-signal condition. { 'pō·lər tranz'mish·ən }

pole-positioning [CONT SYS] A design technique used in linear control theory in which many or all of a system's closed-loop poles are positioned as required, by proper choice of a linear state feedback law; if the system is controllable, all of the closed-loop poles can be arbitrarily positioned by this technique. { 'pōl pəˌzish·ən·iŋ }

pole-zero configuration [CONT SYS] A plot of the poles and zeros of a transfer function in the complex plane; used to study the stability of a system, its natural motion, its frequency response, and its transient response. { 'pōl ¦zir·ō kənˌfig·yə'rā·shən }

Polish notation [COMPUT SCI] **1.** A notation system for digital-computer or calculator logic in which there are no parenthetical expressions and each operator is a binary or unary operator in the sense that it operates on not more than two operands. Also known as Lukasiewicz notation; parenthesis-free notation. **2.** The version of this notation in which operators precede the operands with which they are associated. Also known as prefix notation. { 'pō·lish nō'tā·shən }

polling [COMMUN] A process that involves interrogating in succession every terminal on a shared communications line to determine which of the terminals require service. { 'pōl·iŋ }

polling list [COMMUN] A roster of transmitting devices sequentially scanned in a time-sharing system. { 'pōl·iŋ ˌlist }

polyalphabetic substitution cipher [COMMUN] A cipher that uses several substitution alphabets in turn. { ˌpäl·ēˌal·fə'bed·ik ˌsəb·stə'tü·shən ˌsī·fər }

polychromatic radiation [ELECTROMAG] Electromagnetic radiation that is spread over a range of frequencies. { ¦päl·i·krō'mad·ik ˌrād·ē 'ā·shən }

polyline [COMPUT SCI] In computer graphics, a series of connected line segments and arcs that are treated as a single entity. { 'päl·ēˌlīn }

polymorphic system [COMPUT SCI] A computer system that is organized around a central pool of shared software modules which are selected as they are needed for processing. { ¦päl·i¦mȯr·fik 'sis·təm }

polymorphism [COMPUT SCI] A property of object-oriented programming that allows many different types of objects to be treated in a uniform manner by invoking the same operation on each object. { ˌpäl·i'mȯrˌfiz·əm }

polynomial time [COMPUT SCI] The property of the time required to solve a problem on a computer for which there exist constants c and k such that, if the input to the problem can be specified in N bits, the problem can be solved in $c \times N^k$ elementary operations. { ¦päl·ə¦nō·mē·əl 'tīm }

polyrod antenna [ELECTROMAG] End-fire directional dielectric antenna consisting of a polystyrene rod energized by a section of waveguide. { 'päl·iˌräd an'ten·ə }

polyvalent number [COMPUT SCI] A number, consisting of several figures, used for description, wherein each figure represents one of the characteristics being described. { ¦päl·i'vā·lənt 'nəm·bər }

pop [COMPUT SCI] To obtain information from the top of a stack and then reset a pointer to the next item in the stack. { päp }

POP *See* Post Office Protocol. { pöp *or* ¦pē¦ō'pē }

pop hole *See* pop. { 'päp ˌhōl }

Popov's stability criterion [CONT SYS] A frequency domain stability test for systems consisting of a linear component described by a transfer function preceded by a nonlinear component characterized by an input-output function, with a unity gain feedback loop surrounding the series connection. { pä'pȯfs stə'bil·əd·ē krīˌtir·ē·ən }

popping [COMPUT SCI] The deletion of the top element of a stack. { 'päp·iŋ }

pop shot *See* pop. { 'päp ˌshät }

populate [COMPUT SCI] To add electronic components, such as memory chips, to a circuit board. { 'päp·yəˌlāt }

population [COMPUT SCI] A collection of records in a data base that share one or more characteristics in common. [ELECTR] The set of electronic components on a printed circuit board. { ˌpäp·yə'lā·shən }

port [COMPUT SCI] **1.** An interface between a communications channel and a unit of computer hardware. **2.** To modify an application program, developed to run with a particular operating system, so that it can run with another operating system. **3.** A designation which a program on a client computer uses to specify a server program on a computer in a network. [ELECTROMAG] An opening in a waveguide component, through which energy may be fed or withdrawn, or measurements made. { pȯrt }

portability [COMPUT SCI] Property of a computer program that is sufficiently flexible to be easily transferred to run on a computer of a type different from the one for which it was designed. { ˌpȯrd·ə'bil·əd·ē }

portable audio terminal [COMPUT SCI] A lightweight, self-contained computer terminal with a typewriter keyboard, which can be attached to a telephone line by placing the telephone handset in a receptacle in the terminal. { 'pȯrd·ə·bəl 'ȯd·ē·ō ˌtərm·ən·əl }

portable data terminal [COMPUT SCI] A computer terminal that can be carried about by hand to collect data from remote locations and to transfer this data to a computer system. { 'pȯrd·ə·bəl 'dad·ə ˌtər·mən·əl }

portable document format [COMPUT SCI] A computer file format for publishing and distributing electronic documents (text, image, or multimedia) with the same layout, formatting, and font attributes as in the original. The files can be opened and viewed on any computer or operating system; however, special software is required. Abbreviated PDF. { ¦pȯrd·ə·bəl ˌdäk·yə·mənt 'fȯrˌmat }

port expander [COMPUT SCI] Equipment that connects links to several other devices to one port in a computer. { 'pȯrt ikˌspan·dər }

porting [COMPUT SCI] The process of converting software to run on a computer other than the one for which it was originally written. { 'pȯrd·iŋ }

port operations service [COMMUN] Maritime mobile communications service in or near a port, between coast stations and ship stations, or between ship stations, in which messages are restricted to those relating to the movement and safety of ships and, in an emergency, to the safety of persons. { 'pȯrt ˌäp·ə'rā·shənz ˌsər·vəs }

positional notation [MATH] Any of several numeration systems in which a number is represented by a sequence of digits in such a way that the significance of each digit depends on its position in the sequence as well as its numeric value. Also known as notation. { pə'zish·ən·əl nō'tā·shən }

positional parameter [COMPUT SCI] One of a number of parameters in a group, whose significance is determined by its position within the group. { pə'zish·ən·əl pə'ram·əd·ər }

positional servomechanism [CONT SYS] A feedback control system in which the mechanical position (as opposed to velocity) of some object is automatically maintained. { pə'zish·ən·əl ¦sər·vō'mek·əˌniz·əm }

position control [CONT SYS] A type of automatic control in which the input commands are the desired position of a body. { pə'zish·ən kən ˌtrōl }

positioning action [CONT SYS] Automatic control action in which there is a predetermined relation between the value of a controlled variable and the position of a final control element. { pə'zish·ən·iŋ ˌak·shən }

positioning time [COMPUT SCI] The time required for a storage medium such as a disk to be positioned and for read/write heads to be properly located so that the desired data can be read or written. { pə'zish·ən·iŋ ˌtīm }

position pulse *See* commutator pulse. { pə 'zish·ən ˌpəls }

positive electrode *See* anode. { 'päz·əd·iv i'lek ˌtrōd }

positive feedback [CONT SYS] Feedback in which a portion of the output of a circuit or device is fed back in phase with the input so as to increase the total amplification. Also known as reaction (British usage); regeneration; regenerative feedback; retroaction (British usage). { 'päz·əd·iv 'fēdˌbak }

positive logic [ELECTR] Logic circuitry in which the more positive voltage (or current level) represents the 1 state; the less positive level represents the 0 state. { 'päz·əd·iv 'läj·ik }

positive modulation [ELECTR] In an amplitude-modulated analog television system, that form of television modulation in which an increase in brightness corresponds to an increase in transmitted power. { 'päz·əd·iv ˌmäj·ə'lā·shən }

positive transmission [COMMUN] Transmission of analog television signals in such a way that an increase in initial light intensity causes an increase in the transmitted power. { 'päz·əd·iv tranz'mish·ən }

positive zero [COMPUT SCI] The zero value reached by counting down from a positive number in the binary system. { 'päz·əd·iv 'zir·ō }

post [COMPUT SCI] To add or update records in a file. { pōst }

postdecrementing *See* autodecrement addressing. { ¦pōst'dek·rə,ment·iŋ }

postedit [COMPUT SCI] To edit the output data of a computer. { 'pōst,ed·ət }

POS terminal *See* point-of-sale terminal. { ¦pē ¦ō'es ,term·ən·əl }

postfix notation *See* reverse Polish notation. { 'pōst,fiks nō'tā·shən }

postincrementing *See* autoincrement addressing. { ¦pōst'in·krə,ment·iŋ }

postindexing [COMPUT SCI] Operation in which the contents of a register indicated by the index bits of an indirect address are added to the indirect address to form the effective address. { pōst'in,dek·siŋ }

posting *See* update. { 'pōst·iŋ }

posting interpreter *See* transfer interpreter. { 'pōst iŋ in'tər prəd·ər }

postmortem [COMPUT SCI] Any action taken after an operation is completed to help analyze that operation. { pōst'mȯrd·əm }

postmortem dump [COMPUT SCI] **1.** The printout showing the state of all registers and the contents of main memory, taken after a computer run terminates normally or terminates owing to fault. **2.** The program which generates this printout. { pōst'mȯrd·əm 'dəmp }

postmortem program *See* postmortem routine. { pōst'mȯrd·əm 'prō·grəm }

postmortem routine [COMPUT SCI] A computer routine designed to provide information about the operation of a program after the program is completed. Also known as postmortem program. { pōst'mȯrd·əm rü,tēn }

post office [COMPUT SCI] The software and files in an electronic mail system that receive messages and deliver them to recipients. { 'pōst ,ȯf·əs }

Post Office Protocol [COMPUT SCI] An Internet standard for delivering e-mail from a server to an e-mail client on a personal computer. Abbreviated POP. { ¦pōst ,ȯf·əs 'prōd·ə,kül }

postprocessor [COMPUT SCI] A program that converts graphical output data to a form that can be used by computing equipment. { ¦pōst'prä ,ses·ər }

power bandwidth [COMMUN] The frequency range for which half the rated power of an audio amplifier is available at rated distortion. { 'paü·ər ¦band,width }

power check [COMPUT SCI] An automatic suspension of computer operations resulting from a significant fluctuation in internal electric power. { 'paü·ər ,chek }

power-density spectrum *See* frequency spectrum. { 'paü·ər ¦den·səd·ē ,spek·trəm }

power detection [ELECTR] Form of detection in which the power output of the detecting device is used to supply a substantial amount of power directly to a device such as a loudspeaker or recorder. { 'paü·ər di,tek·shən }

power detector [ELECTR] Detector capable of handling strong input signals without appreciable distortion. { 'paü·ər di,tek·tər }

power distribution unit [COMPUT SCI] Equipment located in or near a computer room which breaks down electric power from a high-voltage source to appropriate levels for distribution to the central processing unit and peripheral devices. Abbreviated PDU. { 'paü·ər ,di·strə'byü·shən ,yü·nət }

power down [COMPUT SCI] To exit from any running programs and remove floppy- and hard-disk cartridges before switching the computer off. { ¦paü·ər ¦daün }

power-line interference [COMMUN] Interference caused by radiation from high-voltage power lines. { 'paü·ər ,līn ,in·tər,fir·əns }

power spectrum *See* frequency spectrum. { 'paü·ər ,spek·trəm }

power typing [COMPUT SCI] A word-processing technique that allows the automatic typing of repetitious text, such as appears in a form letter. { 'paü·ər ,tīp·iŋ }

power up [COMPUT SCI] To check that the computer memory, peripherals, and input/output channels are working properly before the operating system is loaded. { ¦paü·ər ¦əp }

PPI *See* plan position indicator.

P-picture *See* predicted picture. { 'pē ,pik·chər }

PPM *See* pulse-position modulation.

PPP *See* Point-to-Point Protocol.

P pulse *See* commutator pulse. { 'pē ,pəls }

Practical Extraction and Reporting Language [COMPUT SCI] A scripting language often used for creating CGI programs. Abbreviated Perl. { ¦prak·ti·kəl ik,strak·shən and ri'pȯrt·iŋ ,laŋ·gwij }

pragma [COMPUT SCI] A directive inserted into a computer program to prevent the automatic execution of certain error checking and reporting routines which are no longer necessary when the program has been perfected. { 'prag·mə }

pragmatics [COMMUN] The branch of semiotics that treats the relation of symbols to behavior and the meaning received by the listener or reader of a statement. [COMPUT SCI] The fourth and final phase of natural language processing, following contextual analysis, that takes into account the speaker's goal in uttering a particular thought in a particular way in determining what constitutes an appropriate response. { prag'mad·iks }

preamble [COMMUN] The portion of a commercial radiod ata message that is sent first, containing the message number, office of origin, date, and other numerical data not part of the following message text. { 'prē,am·bəl }

preamplifier [ELECTR] An amplifier whose primary function is to boost the output of a low-level audio-frequency, radio-frequency, or microwave source to an intermediate level so that the signal may be further processed without appreciable degradation of the signal-to-noise ratio of the system. Also known as preliminary amplifier. { prē'am·plə,fī·ər }

precedence [COMPUT SCI] The order in which operators are processed in a programming language. { 'pres·əd·əns }

precedence relation [COMPUT SCI] A rule stating that, in a given programming language, one of two operators is to be applied before the other in any mathematical expression. { 'pres·əd·əns ri,lā·shən }

precipitation attenuation [ELECTROMAG] Loss of radio energy due to the passage through a volume of the atmosphere containing precipitation; part of the energy is lost by scattering, and part by absorption. { prə,sip·ə'tā·shən ə,ten·yə'wā·shən }

precipitation static [COMMUN] Static interference due to the discharge of large charges built up on an aircraft or other object by rain, sleet, snow, or electrically charged clouds. { prə ,sip·ə'tā·shən ,stad·ik }

precision approach radar [NAV] A radar system located on an airfield for observation of the position of an aircraft with respect to an approach path, and specifically intended to provide guidance to the aircraft during its approach to the field; the system consists of a ground radar equipment which is alternately connected to two antenna systems; one antenna system sweeps a narrow beam over a 20° sector in the horizontal plane; the second sweeps a narrow beam over a 7° sector in the vertical plane; course correction is transmitted to the aircraft from the ground. Abbreviated PAR. { prə'sizh·ən ə¦prōch 'rā,där }

precision attribute [COMPUT SCI] A set of one or more integers that denotes the number of symbols used to represent a given number and positional information for determining the base point of the number. { prə'sizh·ən 'a·trə,byüt }

precision sweep [ELECTR] Delayed and expanded sweep as in an analog radar display, or similar selection and timing of a digital display, permitting closer examination of received signals of high resolution. { prə'sizh·ən ,swēp }

precompiled module [COMPUT SCI] A standardized subroutine that is separately developed and compiled for use in many different computer programs. { ¦prē·kəm'pīld 'mäj·yül }

precompiler [COMPUT SCI] A computer program that indentifies syntax errors and other problems in a program before it is converted to machine language by a compiler. { ¦prē·kəm'pīl·ər }

predecessor job [COMPUT SCI] A job whose output is used as input to another job, and which must therefore be completed before the second job is started. { 'pred·ə,ses·ər ,jäb }

predefined function [COMPUT SCI] A sequence of instructions that is identified by name in a computer program but is built into the high-level programming language from which the program is complied or is retrieved from somewhere outside the program, such as a subroutine library. { ¦prē·di'fīnd 'fəŋk·shən }

predetection combining [ELECTR] Method used to produce an optimum signal from multiple receivers involved in diversity reception of signals. { ¦prē·di'tek·shən kəm'bīn·iŋ }

predicate [COMPUT SCI] A statement in a computer program that evaluates an expression in order to arrive at a true or false answer. { 'pred·ə ,kāt }

predicted picture [COMMUN] A MPEG-2 picture that is coded with respect to the nearest previous intra-coded picture. This technique is termed forward prediction. Predicted pictures provide more compression than intra-coded pictures and serve as a reference for future predicted pictures or bidirectional pictures. Predicted pictures can propagate coding errors when they (or bidirectional pictures) are predicted from prior predicted pictures where the prediction is flawed. Also known as P-frame; P-picture. { pri'dikt·əd 'pik·chər }

predicted-wave signaling [COMMUN] Communications system in which detection is optimized in the presence of severe noise by using mechanical resonator filters and other circuits in the detector to take advantage of known information on the arrival and completion times of each pulse, as well as on pulse shape, pulse frequency and spectrum, and possible data content. { prə'dik·təd ,wāv 'sig·nəl·iŋ }

predictive coder [COMMUN] Any technique for compressing audio or video signals in which a synthesizer at the receiver is controlled by signal parameters extracted at the transmitter to remake the signal. Also known as predictive encoder. { prə,dik·tiv 'kō·dər }

predictive coding [COMMUN] In data compression, a method of coding information in which a sample value is presented as the error term formed by the difference between the sample and its prediction. { prə¦dik·tiv 'kōd·iŋ }

predictive encoder *See* predictive coder. { prə ,dik·tiv in'kō·dər }

preedit [COMPUT SCI] To edit data before feeding it to a computer. { prē'ed·ət }

preemphasis [ELECTR] A process which increases the magnitude of some frequency components with respect to the magnitude of others to reduce the effects of noise introduced in subsequent parts of the system. { prē'em·fə·səs }

preemphasis network [ELECTR] An RC (resistance-capacitance) filter inserted in a system to emphasize one range of frequencies with respect to another. Also known as emphasizer. { prē'em·fə·səs ,net,wərk }

preemptive multitasking [COMPUT SCI] A method of running more than one program on a computer at a time, in which control of the processor is decided by the operating system, which allocates each program a recurring time segment. { prē ¦emp·tiv 'məl·tē,task·iŋ }

prefix notation *See* Polish notation. { 'prē,fiks nō,tā·shən }

preindexing [COMPUT SCI] Operation in which the address bits of a word are added to the contents of a specified register to determine the pointer address. { prē'in,deks·iŋ }

preliminary amplifier *See* preamplifier. { pri 'lim·ə,ner·ē 'am·plə,fī·ər }

preprocessor [COMPUT SCI] A program that converts data into a format suitable for computer processing. { ¦prē'prä,ses·ər }

preprogramming [COMPUT SCI] The prerecording of instructions or commands for a machine, such as an automated tool in a factory. { prē'prō ,gram·iŋ }

preread head [COMPUT SCI] A read head that is placed near another read head in such a way that it can read data stored on a moving medium such as a tape or disk before these data reach the second head. { 'prē,rēd ,hed }

preselection [COMPUT SCI] A technique for saving computation time in buffered computers in which a block of data is read into computer storage from the next input tape to be called upon before the data are required in the computer; the selection of the next input tape is determined by instructions to the computer. { ¦prē·si'lek·shən }

preselector [ELECTR] A tuned radio-frequency amplifier stage used ahead of the frequency converter in a superheterodyne receiver to increase the selectivity and sensitivity of the receiver. { ¦prē·si'lek·tər }

presentation graphics program [COMPUT SCI] An application program for creating and enhancing the visual appeal and understandability of charts and graphs, with the aid of a library or predrawn images that can be combined with other artwork. { ,prez·ən¦tā·shən 'graf·iks ,prō·grəm }

preset [COMPUT SCI] **1.** Of a variable, having a value established before the first time it is used. **2.** To initialize a value of a variable before the value of the variable is used or tested. { 'prē ,set }

preset parameter [COMPUT SCI] In computers, a parameter which is fixed for each problem at a value set by the programmer. { 'prē,set pə'ram·əd·ər }

presort [COMPUT SCI] **1.** The first part of a sort program in which data items are arranged into strings that are equal to or greater than some prescribed length. **2.** The sorting of data on off-line equipment before it is processed by a computer. { prē'sȯrt }

press teletype network [COMMUN] A large teletypewriter network employed by a press association or other news distributing organization, usually employing modern carrier telegraph circuits operating over both wire and radio facilities, and transmitting to as many as 2000 stations simultaneously. { 'pres 'tel·ə,tīp ,net,wərk }

prestore [COMPUT SCI] To store a quantity in an available computer location before it is required in a routine. { ¦prē'stȯr }

presumptive address *See* address constant. { pri'zəm·tiv ə'dres }

presumptive instruction *See* basic instruction. { pri'zəm·tiv in'strək·shən }

pretrigger [ELECTR] Trigger used to initiate sweep ahead of transmitted pulse. { prē'trig·ər }

previewing [COMPUT SCI] In character recognition, a process of attempting to gain prior information about the characters that appear on an incoming source document; this information, which may include the range of ink density, relative positions, and so forth, is used as an aid in the normalization phase of character recognition. { 'prē,vyü·iŋ }

previous element coding [COMMUN] System of signal coding, used for digital television transmission, whereby each transmitted picture element is dependent upon the similarity of the preceding picture element. { 'prē·vē·əs 'el·ə·mənt }

PRF *See* pulse repetition rate.

primary cache [COMPUT SCI] A cache memory located within a microprocessor chip itself. Also known as internal cache; level 1 cache. { ¦prī ,mer·ē 'kash }

primary center [COMMUN] A telephone office having lower rank than a sectional center and higher rank than a toll center; connects toll centers and may also serve as a toll center for nearby end offices. { 'prī,mer·ē 'sen·tər }

primary control program [COMPUT SCI] The program which provides the sequential scheduling of jobs and basic operating systems functions. Abbreviated PCP. { 'prī,mer·ē kən'trōl ,prō·grəm }

primary frequency [COMMUN] Frequency assigned for normal use on a particular circuit or communications channel. { 'prī,mer·ē 'frē·kwən·sē }

primary-frequency standard [COMMUN] One of the standards of frequency maintained by various governments; the operating frequency of a radio station is determined by comparison with multiples of this standard frequency. { 'prī,mer·ē 'frē·kwən·sē ,stan·dərd }

primary index [COMPUT SCI] An index that holds the values of primary keys, in sequence. { 'prī ,mer·ē 'in,deks }

primary key [COMPUT SCI] A key that identifies a record or portion of a record and determines the sequence of records in a file or other data structure. { 'prī,mer·ē 'kē }

primary radar [ENG] A radar that receives and interprets the reflected signal from scattering objects (targets and clutter) in its view. { 'prī ,mer·ē 'rā,där }

primary register [COMPUT SCI] A general-purpose register in a central processing unit that is available for direct utilization by computer programs. { 'prī,mer·ē 'rej·ə,stər }

primary service area [COMMUN] The area in which the ground wave of a broadcast station is not subject to objectionable interference or fading. { 'prī,mer·ē 'sər·vəs ,er·ē·ə }

primary skip zone [ELECTROMAG] Area around a transmitter beyond the ground wave but within the skip distance. { 'prī,mer·ē 'skip ,zōn }

primary storage [COMPUT SCI] Main internal storage of a computer. { ˈprī,mer·ē ˈstȯr·ij }

primary surveillance radar *See* primary radar. { ˈprī,mer·ē sərˈvā·ləns ˌrā,där }

primary wave [COMMUN] A radio wave traveling by a direct path, as contrasted with skips. { ˈprī ˌmer·ē ˈwāv }

prime register [COMPUT SCI] One of the registers that is inactive at any given time in a central processing unit with duplicate general-purpose registers. { ˈprīm ˈrej·ə·stər }

primitive [COMPUT SCI] A sketchy specification, omitting details, of some action in a computer program. [CONT SYS] A basic operation of a robot, initialized by a single command statement in the program that controls the robot. { ˈprim·əd·iv }

primitive abstract data type [COMPUT SCI] A simple abstract data type that is typically implemented directly in a high-level programming language; examples include integers and real numbers (with appropriate arithmetic operators), booleans (with appropriate logical operators), text strings, and pointers. { ˈprim·əd·iv ˈab,strakt ˈdad·ə ˌtīp }

principal E plane [ELECTROMAG] Plane containing the direction of radiation of electromagnetic waves and arranged so that the electric vector everywhere lies in the plane. { ˈprin·sə·pəl ˈē ˌplān }

principal H plane [ELECTROMAG] Plane that contains the direction of radiation and the magnetic vector, and is everywhere perpendicular to the E plane. { ˈprin·sə·pəl ˈāch ˌplān }

principal mode *See* fundamental mode. { ˈprin·sə·pəl ˈmōd }

principle of optimality [CONT SYS] A principle which states that for optimal systems, any portion of the optimal state trajectory is optimal between the states it joins. { ˈprin·sə·pəl əv ˌäp·təˈmal·əd·ē }

principle of reciprocity *See* reciprocity theorem. { ˈprin·sə·pəl əv ˌres·əˈpräs·əd·ē }

print driver [COMPUT SCI] The portion of a computer program that directs output to a printer and usually also controls printer functions such as pagination and the setting of the margins and page headers. { ˈprint ˌdrī·vər }

printed circuit [ELECTR] A conductive pattern that may or may not include printed components, formed in a predetermined design on the surface of an insulating base in an accurately repeatable manner. { ˈprint·əd ˈsər·kət }

printed circuit board [ELECTR] A flat board whose front contains slots for integrated circuit chips and connections for a variety of electronic components, and whose back is printed with electrically conductive pathways between the components. Also known as circuit board. { ˈprint·əd ˈsər·kət ˌbȯrd }

printer [COMPUT SCI] A computer output mechanism that prints characters one at a time or one line at a time. { ˈprint·ər }

printer file [COMPUT SCI] **1.** A file that contains the information that the printer driver needs in order to generate the codes required by the printer. **2.** A document in print image format. { ˈprin·tər ˌfīl }

print head [COMPUT SCI] The mechanism that generates the characters to be reproduced by a character printer. { ˈprint ˌhed }

print image format [COMPUT SCI] The format of a document that has been prepared for output on the printer. { ˈprint ˌim·ij ˌfȯr,mat }

printing calculator [COMPUT SCI] A desk-model electronic calculator that provides a printed record on paper tape with or without a digital display. { ˈprint·iŋ ˈkal·kyə,lād·ər }

printing element [COMPUT SCI] The part of the print head mechanism that comes into contact with the paper to print characters or other images. { ˈprint·iŋ ˌel·ə·mənt }

printing-telegraph code [COMMUN] A five- or seven-unit code used for operation of a teleprinter, teletypewriter, and similar telegraph printing devices. { ˈprint·iŋ ¦tel·ə,graf ˌkōd }

printing telegraphy [COMMUN] Method of telegraph operation in which the received signals are automatically recorded in printed characters. { ˈprint·iŋ təˈleg·rə·fē }

print member [COMPUT SCI] The part of a computer printer that determines the form of a printed character, such as a print wheel or type bar. { ˈprint ˌmem·bər }

printout [COMPUT SCI] A printed output of a data-processing machine or system. { ˈprint ˌau̇t }

print position [COMPUT SCI] One of the positions on a printer at which a character can be printed. { ˈprint pə,zish·ən }

print queue [COMPUT SCI] A prioritized list, maintained by the operating system, of the output from a computer system waiting on a spool file to be printed. { ˈprint ˌkyü }

print server [COMPUT SCI] A computer controlling a series of printers. { ˈprint ˌsər·vər }

print train [COMPUT SCI] **1.** The chain in a chain printer or the drum in a drum printer that holds the type slugs used to make impressions on paper. **2.** The electronic character set that serves a similar function in a laser printer. { ˈprint ˌtrān }

print wheel [COMPUT SCI] A disk which has around its rim the letters, numerals, and other characters that are used in printing in a wheel printer. { ˈprint ˌwēl }

priority-arbitration circuit [COMPUT SCI] A logic circuit which combines all interrupts but allows only the highest-priority request to enable its active flipflop. { prīˈär·əd·ē ˌär·bəˈtrā·shən ˌsər·kət }

priority indicator [COMMUN] Data attached to a message to indicate its relative priority and hence the order in which it will be transmitted. [COMPUT SCI] Data attached to a computer program or job which are used to determine the order in which it will be processed by the computer. { prīˈär·əd·ē ˈin·də,kād·ər }

priority interrupt [COMPUT SCI] An interrupt procedure in which control is passed to the monitor, the required operation is initiated, and then control returns to the running program, which never knows that it has been interrupted. { prī'är·əd·ē 'int·ə,rəpt }

priority phase [COMPUT SCI] Phase consisting of execution of operations in response to instruments or process interrupts other than clock interrupts. { prī'är·əd·ē 'fāz }

priority polling [COMMUN] In a data communications network, a system in which nodes with high activity are interrogated more frequently than those with only occasional traffic. { prī'är·əd·ē 'pōl·iŋ }

priority processing [COMPUT SCI] A method of computer time-sharing in which the order in which programs are processed is determined by a system of priorities, involving such factors as the length, nature, and source of the programs. { prī'är·əd·ē 'prä,ses·iŋ }

priority queueing [COMPUT SCI] The arrangement of jobs to be carried out in a list according to their relative importance, with the most important first. { prī'är·əd·ē 'kyü·iŋ }

privacy system [COMMUN] A device or method for scrambling overseas telephone conversations handled by radio links in order to make them unintelligible to outside listeners. Also known as privacy transformation; secrecy system. { 'prī·və·sē ,sis·təm }

privacy transformation *See* privacy system. { 'prī·və·sē ,tranz·fər'mā·shən }

private automatic branch exchange [COMMUN] A private branch exchange in which connections are made by remote-controlled switches. Abbreviated PABX. { 'prī·vət ¦ȯd·ə,mad·ik 'branch iks ,chānj }

private automatic exchange [COMMUN] A private telephone exchange in which connections are made by remote-controlled switches. Abbreviated PAX. { 'prī·vət ¦ȯd·ə,mad·ik iks,chānj }

private branch exchange [COMMUN] A telephone exchange serving a single organization, having a switchboard and associated equipment, usually located on the customer's premises; provides for switching calls between any two extensions served by the exchange or between any extension and the national telephone system via a trunk to a central office. Abbreviated PBX. { 'prī·vət 'branch iks,chānj }

private data [COMPUT SCI] Data that are open to a single user only. { 'prī·vət 'dad·ə }

private exchange [COMMUN] Telephone exchange serving a single organization and having no means for connecting to a public telephone system. { 'prī·vət iks'chānj }

private library [COMPUT SCI] An organized collection of programs and other software that is the property of a single user of a computer system and is not generally available to other users. { 'prī·vət 'lī,brer·ē }

private line [COMMUN] A line, channel, or service reserved solely for one user. { 'prī·vət 'līn }

private line arrangement [COMPUT SCI] The structure of a computer system in which each input/output device has a set of lines leading to the central processing unit for the device's own private use. Also known as radial selector. { 'prī·vət ¦līn ə,rānj·mənt }

private line service [COMMUN] Service provided by United States common carriers engaged in domestic or international wire, radio, and cable communications for the intercity communications purposes of a customer; this service is provided over integrated communications pathways, including facilities or local channels, which are integrated components of intercity private line services, and station equipment between specified locations for a continuous period or for regularly recurring periods at stated hours. { 'prī·vət ¦līn ,sər·vəs }

private pack [COMPUT SCI] A disk pack assigned exclusively to one application or one user so that the operating system does not try to allocate space on the device to others. { 'prī·vət 'pak }

privileged instruction [COMPUT SCI] A class of instructions, usually including storage protection setting, interrupt handling, timer control, input/output, and special processor status-setting instructions, that can be executed only when the computer is in a special privileged mode that is generally available to an operating or executive system, but not to user programs. { 'priv·ə·lijd in'strək·shən }

privileged mode *See* master mode. { 'priv·ə·lijd ,mōd }

PRML technique *See* partial-response maximum-likelihood technique. { ¦pē¦är¦em'el tek,nēk }

probabilistic automaton [COMPUT SCI] A device, with a finite number of internal states, which is capable of scanning input words over a finite alphabet and responding by successively changing its internal state in a probabilistic way. Also known as stochastic automaton. { ,präb·ə·bə'lis·tik ȯ'täm·ə,tän }

probabilistic sequential machine [COMPUT SCI] A probabilistic automaton that has the capability of printing output words probabilistically, over a finite output alphabet. Also known as stochastic sequential machine. { ,präb·ə·bə'lis·tik si'kwen·chəl mə'shēn }

probability of detection [ENG] In radar, the probability that a target will be detected, that is, will cause a contact to be generated by whatever detection mechanism is employed, when in fact it is present. It is the complement of the probability of miss (the failure to declare a target present when in fact it is). { ,präb·ə'bil·əd·ē əv di ,tek·shən }

probability of false alarm [ENG] In radar, the probability of declaring a target to be present when in fact none is. { ,präb·ə'bil·əd·ē əv ¦fȯls ə'lärm }

probe [COMMUN] To determine a radio interference by obtaining the relative interference level in the immediate area of a source by the use of a small, insensitive antenna in conjunction with a receiving device. [ELECTROMAG] A metal rod that projects into but is insulated from a waveguide or resonant cavity; used to provide coupling to an external circuit for injection or extraction of energy or to measure the standing-wave ratio. Also known as waveguide probe. { prōb }

problem check [COMPUT SCI] One or more tests used to assist in obtaining the correct machine solution to a problem. { 'präb·ləm ,chek }

problem-defining language [COMPUT SCI] A programming language that literally defines a problem and may specifically define the input and output, but does not define the method of transforming one to the other. Also known as problem-specification language. { 'präb·ləm di ¦fīn·iŋ ,laŋ·gwij }

problem definition [COMPUT SCI] The art of compiling logic in the form of general flow charts and logic diagrams which clearly explain and present the problem to the programmer in such a way that all requirements involved in the run are presented. { 'präb·ləm ,def·ə,nish·ən }

problem-describing language [COMPUT SCI] A programming language that describes, in the most general way, the problem to be solved, but gives no indication of the problem's detailed characteristics or its solution. { 'präb·ləm di ¦skrīb·iŋ ,laŋ·gwij }

problem file *See* run book. { 'präb·ləm ,fīl }

problem folder *See* run book. { 'präb·ləm ,fō ld·ər }

problem mode [COMPUT SCI] A condition of computer operation in which, in contrast to supervisor mode, the privileged instructions cannot be executed, preventing the program from upsetting the supervisor program or any other program. { 'präb·ləm ,mōd }

problem-oriented language [COMPUT SCI] A language designed to facilitate the accurate expression of problems belonging to specific sets of problem types. { 'präb·ləm ,ȯr·ē¦ent·əd ,laŋ·gwij }

problem-solving language [COMPUT SCI] A programming language that can be used to specify a complete solution to a problem. { 'präb·ləm ¦sälv·iŋ ,laŋ·gwij }

problem-specification language *See* problem-defining language. { 'präb·ləm ,spes·ə·fə¦kā·shən ,laŋ·gwij }

procedural programming [COMPUT SCI] A list of instructions telling a computer, step-by-step, what to do, usually having a linear order of execution from the first statement to the second and so forth with occasional loops and branches. Procedural programming languages include C, C++, Fortran, Pascal, and Basic. { prə,sē·jə·rəl 'prō,gram·iŋ }

procedural representation [COMPUT SCI] The representation of certain concepts in a computer by procedures or programs in some appropriate language, rather than by static data items such as numbers or lists. { prə'sē·jə·rəl ,rep·rə·zen'tā·shən }

procedure [COMPUT SCI] **1.** A sequence of actions (or computer instructions) which collectively accomplish some desired task. **2.** In particular, a subroutine that causes an effect external to itself. { prə'sē·jər }

procedure declaration [COMPUT SCI] A statement that causes a procedure to be given a name and written as a segment of a computer program. { prə'sē·jər ,dek·lə,rā·shən }

procedure division [COMPUT SCI] The section of a program (written in the COBOL language) in which a programmer specifies the operations to be performed with the data names appearing in the program. { prə'sē·jər di,vizh·ən }

procedure library [COMPUT SCI] A collection of job control language routines that are stored on a disk file and can be executed by entering a command naming the routine. Abbreviated PROCLIB. { prə'sē·jər ,lī,brer·ē }

procedure-oriented language [COMPUT SCI] A language designed to facilitate the accurate description of procedures, algorithms, or routines belonging to a certain set of procedures. { prə'sē·jər ,ȯr·ē¦ent·əd ,laŋ·gwij }

proceed-to-select signal [COMMUN] Signal returned from distant automatic equipment over the backward signaling path, in response to a calling signal, to indicate that selecting information can be transmitted; in certain signaling systems, both signals can be the same. { prə'sēd tə si'lekt ,sig·nəl }

proceed-to-transmit signal [COMMUN] Signal returned from a distant manual switchboard over the backward signaling path, in response to a calling signal, to indicate that the teleprinter of the distant operator is connected to the circuit. { prə'sēd tə tranz'mit ,sig·nəl }

process [COMPUT SCI] **1.** To assemble, compile, generate, interpret, compute, and otherwise act on information in a computer. **2.** A program that is running on a computer. { 'prä,ses }

process-bound program *See* CPU-bound program. { 'prä,ses ¦bau̇nd 'prō·grəm }

process control system [CONT SYS] The automatic control of a continuous operation. { 'prä ,səs kən,trōl ,sis·təm }

processing [COMMUN] Further handling, manipulation, consolidation, compositing, and so on, of information to convert it from one format to another or to reduce it to manageable or intelligible information. { 'prä,ses·iŋ }

processing interrupt [COMPUT SCI] The interruption of the batch processing mode in a real-time system when live data are entered in the system. { 'prä,ses·iŋ 'int·ə,rəpt }

processing program [COMPUT SCI] Any computer program that is not a control program, such as an application program, or a noncontrolling part of the operating system, such as a sort-merge program or language translator. { 'prä ,ses·iŋ ,prō,gram }

processing section [COMPUT SCI] The computer unit that does the actual changing of input into output; includes the arithmetic unit and intermediate storage. { 'prä,ses·iŋ ,sek·shən }

process-limited *See* processor-limited. { 'prä ,səs ¦lim·əd·əd }

processor [COMPUT SCI] **1.** A device that performs one or many functions, usually a central processing unit. **2.** A program that transforms some input into some output, such as an assembler, compiler, or linkage editor. { 'prä ˌses·ər }

processor complex [COMPUT SCI] The central portion of a very large computer consisting of several central processing units working in concert. { 'präˌses·ər ˌkämˌpleks }

processor error interrupt [COMPUT SCI] The interruption of a computer program because a parity check indicates an error in a word that has been transferred to or within the central processing unit. { 'präˌses·ər ¦er·ər ˌint·əˌrəpt }

processor-limited [COMPUT SCI] Property of a computer system whose processing time is determined by the speed of its central processing unit rather than by the speed of its peripheral equipment. Also known as process-limited. { 'präˌses·ər ˌlim·əd·əd }

processor-memory-switch notation *See* PMS notation. { 'präˌses·ər 'mem·rē ˌswich nōˌtā·shən }

processor stack pointer [COMPUT SCI] A programmable register used to access all temporary-storage words related to an interrupt-service routine which was halted when a new service routine was called in. { 'präˌses·ər 'stak ˌpȯint·ər }

processor status word [COMPUT SCI] A word comprising a set of flag bits and the interrupt-mask status. { 'präˌses·ər 'stad·əs ˌwərd }

process simulation [COMPUT SCI] The use of computer programming, computer vision, and feedback to simulate manufacturing techniques. { 'präˌses ˌsim·yəˌlā·shən }

PROCLIB *See* procedure library. { 'präkˌlīb }

product demodulator [ELECTR] A receiver demodulator whose output is the product of the input signal voltage and a local oscillator signal voltage at the input frequency. Also known as product detector. { 'präd·əkt diˌmäj·əˌlād·ər }

product detector *See* product demodulator. { 'präd·əkt diˌtek·tər }

production [COMPUT SCI] **1.** The processing of useful work by a computer system, excluding the development and testing of new programs. **2.** A rule in a grammar of a formal language that describes how parts of a string (or word, phrase, or construct) can be replaced by other strings. Also known as rule of inference. { prə'dək·shən }

production program [COMPUT SCI] A proprietary program used primarily for internal processing in a business and not generally made available to third parties for profit. { prə'dək·shən ˌprōˌgram }

production test [COMPUT SCI] A test of a computer system with actual data in the environment where it will be used. { prə'dək·shən ˌtest }

production time [COMPUT SCI] Good computing time, including occasional duplication of one case for a check or rerunning of the test run; also including duplication requested by the sponsor, any reruns caused by misinformation or bad data supplied by sponsor, and error studies using different intervals, covergence criteria, and so on. { prə'dək·shən ˌtīm }

product modulator [ELECTR] Modulator whose modulated output is substantially equal to the carrier and the modulating wave; the term implies a device in which intermodulation between components of the modulating wave does not occur. { 'prä·dəkt ˌmäj·əˌlād·ər }

profile [COMMUN] A defined subset of the syntax specified in the MPEG-2 video coding specification. { 'prōˌfīl }

program [COMMUN] **1.** A sequence of audio signals alone, or audio and video signals, transmitted for entertainment or information. **2.** A collection of program elements. Program elements may be elementary streams, and need not have any defined time base. Those that do have a common time base are intended for synchronized presentation. [COMPUT SCI] A detailed and explicit set of directions for accomplishing some purpose, the set being expressed in some language suitable for input to a computer, or in machine language. { 'prō·grəm *or* 'prōˌgram }

program analysis [COMPUT SCI] The process of determining the functions to be carried out by a computer program. { 'prō·grəm əˌnal·ə·səs }

program block [COMPUT SCI] A division or section of a computer program that functions to a large extent as if it were a separate program. { 'prō·grəm ˌbläk }

program check [COMPUT SCI] A built-in check system in a program to determine that the program is running correctly. { 'prō·grəm ˌchek }

program clock reference [COMMUN] A time stamp in the transport stream from which decoder timing is derived. Abbreviated PCR. { 'prō·grəm ¦kläk 'ref·rəns }

program compatibility [COMPUT SCI] The type of compatibility shared by two computers that can process the identical program or programs written in the same source language or machine language. { 'prō·grəm kəmˌpad·ə'bil·əd·ē }

program control [CONT SYS] A control system whose set point is automatically varied during definite time intervals in order to make the process variable vary in some prescribed manner. { 'prō·grəm kənˌtrōl }

program conversion [COMPUT SCI] The changing of the source language of a computer program from one dialect to another, or the modification of the program to operate with a different operating system or data-base management system. { 'prō·grəm kənˌvər·zhən }

program counter *See* instruction counter. { 'prō·grəm ˌkau̇nt·ər }

program design [COMPUT SCI] The phase of computer program development in which the hardware and software resources needed by the program are identified and the logic to be used by the program is determined. { 'prō·grəm di ˌzīn }

program development time [COMPUT SCI] The total time taken on a computer to produce operating programs, including the time taken to compile, test, and debug programs, plus the time taken to develop and test new procedures and techniques. { 'prō·grəm di'vel·əp·mənt ,tīm }

program editor [COMPUT SCI] A computer routine used in time-sharing systems for on-line modification of computer programs. { 'prō·grəm ,ed·ə·tər }

program element [COMMUN] A generic term for one of the elementary streams or other data streams that may be included in the program of a digital video system. [COMPUT SCI] Part of a central computer system that carries out the instruction sequence scheduled by the programmer. { 'prō·grəm ,el·ə·mənt }

program failure alarm [COMMUN] Signal-operated radio or television relay that gives a visual and/or aural alarm when the program fails on the line being monitored; a time delay is provided to prevent the relay from operating and giving a false alarm during station identification periods or other short periods of silence in program continuity. { 'prō·grəm 'fāl·yər ə,lärm }

program generator [COMPUT SCI] A program that permits a computer to write other programs automatically. { 'prō·grəm ,jen·ə,rād·ər }

program library [COMPUT SCI] An organized set of computer routines and programs. { 'prō·grəm ,lī·brer·ē }

program listing [COMPUT SCI] A list of the statements in a computer program, usually produced as a by-product of the compilation of the program. { 'prō·grəm ,list·iŋ }

program logic [COMPUT SCI] A particular sequence of instructions in a computer program. { 'prō·grəm ,läj·ik }

programmable calculator [COMPUT SCI] An electronic calculator that has some provision for changing its internal program, usually by inserting a new magnetic card on which the desired calculating program has been stored. { prō'gram·ə·bəl 'kal·kyə,lād·ər }

programmable device [COMPUT SCI] Any device whose operation is controlled by a stored program that can be changed or replaced. { prō 'gram·ə·bəl di'vīs }

programmable logic array *See* field-programmable logic array. { prō'gram·ə·bəl ¦läj·ik ə,rā }

programmable read-only memory [COMPUT SCI] An integrated-circuit memory chip which can be programmed only once by the user after which the information stored in the chip cannot be altered. Abbreviated PROM. { prō'gram·ə·bəl ¦rēd ¦ōn·lē 'mem·rē }

program maintenance [COMPUT SCI] The updating of computer programs both by error correction and by alteration of programs to meet changing needs. { 'prō·grəm 'mānt·ən·əns }

programmatic interface *See* application program interface. { ,prō·grə¦mad·ik 'in·tər,fās }

programmed check [COMPUT SCI] **1.** An error-detecting operation programmed by instructions rather than built into the hardware. **2.** A computer check in which a sample problem with known answer, selected for having a program similar to that of the next problem to be run, is put through the computer. { 'prō,gramd 'chek }

programmed dump [COMPUT SCI] A storage dump which results from an instruction in a computer program at a particular point in the program. { 'prō,gramd 'dəmp }

programmed function key [COMPUT SCI] A key on the keyboard of a computer terminal that lacks a predefined function but can be assigned a function by a computer program. Abbreviated PF key. { 'prō,gramd 'fəŋk·shən ,kē }

programmed halt [COMPUT SCI] A halt that occurs deliberately as the result of an instruction in the program. Also known as programmed stop. { 'prō,gramd 'hölt }

programmed logic array [ELECTR] An array of AND/OR logic gates that provides logic functions for a given set of inputs programmed during manufacture and serves as a read-only memory. Abbreviated PLA. { 'prō,gramd ¦läj·ik ə,rā }

programmed marginal check [COMPUT SCI] Computer program that varies its own voltage to check some piece of electronic computer equipment during a preventive maintenance check. { 'prō,gramd 'mär·jən·əl 'chek }

programmed operators [COMPUT SCI] Computer instructions which enable subroutines to be accessed with a single programmed instruction. { 'prō,gramd 'äp·ə,rād·ərz }

programmed stop *See* programmed halt. { 'prō ,gramd 'stäp }

programmer's tool kit [COMPUT SCI] A collection of programs designed to help programmers in developing software, usually oriented toward a particular programming language. { 'prō ,gram·ərz 'tül ,kit }

programmer [COMPUT SCI] A person who prepares sequences of instructions for a computer, without necessarily converting them into the detailed codes. { 'prō,gram·ər }

programmer analyst [COMPUT SCI] A person who both writes computer programs and analyzes and designs information systems. { 'prō ,gram·ər 'an·əl,ist }

programmer-defined macroinstruction [COMPUT SCI] A macroinstruction which is equivalent to a set of ordinary instructions as specified by the programmer for use in a particular computer program. { 'prō,gram·ər di¦fīnd ¦ma·krō·in'strək·shən }

programming [COMPUT SCI] Preparing a detailed sequence of operating instructions for a particular problem to be run on a digital computer. Also known as computer programming. { 'prō,gram·iŋ }

programming language [COMPUT SCI] The language used by a programmer to write a program for a computer. { 'prō,gram·iŋ ,laŋ·gwij }

programming unit *See* manual control unit. { 'prō,gram·iŋ ,yü·nət }

program module [COMPUT SCI] A logically self-contained and discrete part of a larger computer program, for example, a subroutine or a coroutine. { 'prō·grəm ,mäj·yül }

program monitor [COMMUN] A monitor used to observe the quality of a radio or television broadcast. { 'prō·grəm 'män·əd·ər }

program parameter [COMPUT SCI] In computers, an adjustable parameter in a subroutine which can be given a different value each time the subroutine is used. { 'prō·grəm pə'ram·əd·ər }

program register [COMPUT SCI] The register in the control unit of a digital computer that stores the current instruction of the program and controls the operation of the computer during the execution of that instruction. Also known as computer control register. { 'prō·grəm ˌrej·ə·stər }

program-sensitive fault [COMPUT SCI] A hardware malfunction that appears only in response to a particular sequence (or kind of sequence) of program instructions. { 'prō·grəm ¦sen·səd·iv 'fȯlt }

program specification [COMPUT SCI] A statement of the precise functions which are to be carried out by a computer program, including descriptions of the input to be processed by the program, the processing needed, and the output from the program. { 'prō·grəm ˌspes·ə·fə'kā·shən }

program specific information [COMMUN] Normative data that is necessary for the demultiplexing of transport streams and the successful regeneration of programs. Abbreviated PSI. { ¦prō·grəm spə¦sif·ik 'in·fər'mā·shən }

program state [COMPUT SCI] The mode of operation of a computer during the execution of instructions in an application program. { 'prō·grəm ˌstāt }

program status word [COMPUT SCI] An internal register to the central processing unit denoting the state of the computer at a moment in time. { 'prō·grəm 'stad·əs ˌwərd }

program step [COMPUT SCI] In computers, some part of a program, usually one instruction. { 'prō·grəm ˌstep }

program stop [COMPUT SCI] An instruction built into a computer program that will automatically stop the machine under certain conditions, or upon reaching the end of processing or completing the solution of a program. Also known as halt instruction; stop instruction. { 'prō·grəm ˌstäp }

program storage [COMPUT SCI] Portion of the internal storage reserved for the storage of programs, routines, and subroutines; in many systems, protection devices are used to prevent inadvertent alteration of the contents of the program storage; contrasted with temporary storage. { 'prō·grəm ˌstȯr·ij }

program tape [COMPUT SCI] Tape containing the sequence of computer instructions for a given problem. { 'prō·grəm ˌtāp }

program test [COMPUT SCI] A system of checking before running any problem in which a sample problem of the same type with a known answer is run. { 'prō·grəm ˌtest }

program testing time [COMPUT SCI] The machine time expended for program testing, debugging, and volume and compatibility testing. { 'prō·grəm 'test·iŋ ˌtīm }

program time [COMPUT SCI] The phase of computer operation when an instruction is being interpreted so that it can be carried out. { 'prō·grəm ˌtīm }

progressive overflow [COMPUT SCI] Retrieval of a randomly stored overflow record by a forward serial search from the home address. { prə 'gres·iv 'ō·vər,flō }

progressive scanning [COMMUN] Scanning all lines in sequence, without interlace, so all picture elements are included during one vertical sweep of the scanning beam. Also known as sequential scanning. { prə'gres·iv 'skan·iŋ }

progressive-wave antenna *See* traveling-wave antenna. { prə'gres·iv ¦wāv an'ten·ə }

project development methodology [COMPUT SCI] A structured set of procedures designed to control the development of computer programs in a large organization. { 'prä,jekt di¦vel·əp·mənt ˌmeth·ə'däl·ə·jē }

projection cathode-ray tube [ELECTR] A cathode-ray tube designed to produce an intensely bright but relatively small image that can be projected onto a large viewing screen by an optical system. { prə'jek·shən ¦kath,ōd 'rā ˌtüb }

projection net *See* net. { prə'jek·shən ˌnet }

PROLOG [COMPUT SCI] A programming language that is for artificial intelligence applications, and uses problem descriptions to reach solutions, based on precise rules. { 'prō,läg }

PROM *See* programmable read-only memory. { präm }

PROM burner [COMPUT SCI] A special device used to write on a programmable read-only memory (PROM). { 'präm ˌbər·nər }

PROM programmer [ELECTR] A device that holds several programmable read-only memory (PROM) chips and writes instructions and data into them by melting connections in their circuitry. { 'präm 'prō,gram·ər }

prompt [COMPUT SCI] A message or format displayed on the screen of a computer terminal that requires the user to respond in some way before processing can continue. { prämpt }

proof total [COMPUT SCI] One of a group of totals which are compared with each other to check their consistency. { 'prüf ˌtōd·əl }

propagated error [COMPUT SCI] An error which takes place in one operation and spreads through succeeding operations. { 'präp·ə,gād·əd 'er·ər }

propagation constant [ELECTROMAG] A rating for a line or medium along or through which a wave of a given frequency is being transmitted; it is a complex quantity; the real part is the attenuation constant in nepers per unit length, and the imaginary part is the phase constant in radians per unit length. { ˌpräp·ə'gā·shən ˌkän·stənt }

propagation delay [ELECTR] The time required for a signal to pass through a given complete operating circuit; it is generally of the order of nanoseconds, and is of extreme importance in computer circuits. { ˌpräp·ə'gā·shən di,lā }

propagation loss [COMMUN] The attenuation of signals passing between two points of a transmission path. { ˌpräp·əˈgā·shən ˌlȯs }

propagation mode [ELECTROMAG] A form of propagation of electromagnetic radiation in a periodic beamguide in which the field distributions over cross sections of the beam are identical at positions separated by one period of the guide. { ˈpräp·əˈgā·shən ˌmōd }

propagation notice [COMMUN] A forecast of propagation conditions for long-distance radio communications, broadcast at regular intervals over radio stations operated by the National Institute of Standards and Technology. { ˌpräp·əˈgā·shən ˌnōd·əs }

propagation path [COMMUN] A path between receiver and transmitter including direct tropospheric scatter, ionospheric scatter, E-layer skip, and F_1-layer and F_2-layer skip and echo. { ˌpräp·əˈgā·shən ˌpath }

propagation time delay [COMMUN] The time required for a wave to travel between two points of a transmission path. { ˌpräp·əˈgā·shən ˈtīm di ˌlā }

propagation velocity [ELECTROMAG] Velocity of electromagnetic wave propagation in the medium under consideration. { ˌpräp·əˈgā·shən vəˌläs·əd·ē }

property detector [COMPUT SCI] In character recognition, that electronic component of a character reader which processes the normalized signal for the purpose of extracting from it a set of characteristic properties on the basis of which the character can be subsequently identified. { ˈpräp·ərd·ē diˌtek·tər }

property list [COMPUT SCI] A list for describing some object or concept, in which odd-numbered items name a property or attribute of a relevant class of objects, and the item following the property name is the property's value for the described objects. { ˈpräp·ərd·ē ˌlist }

proportional control [CONT SYS] Control in which the amount of corrective action is proportional to the amount of error; used, for example, in chemical engineering to control pressure, flow rate, or temperature in a process system. { prəˈpȯr·shən·əl kənˈtrōl }

proportional controller [CONT SYS] A controller whose output is proportional to the error signal. { prəˈpȯr·shən·əl kənˈtrōl·ər }

proportional-plus-derivative control [CONT SYS] Control in which the control signal is a linear combination of the error signal and its derivative. { prəˈpȯr·shən·əl ˌpləs dəˈriv·əd·iv kənˌtrōl }

proportional-plus-integral control [CONT SYS] Control in which the control signal is a linear combination of the error signal and its integral. { prəˈpȯr·shən·əl ˌpləs ˈint·ə·grəl kənˌtrōl }

proportional-plus-integral-plus-derivative control [CONT SYS] Control in which the control signal is a linear combination of the error signal, its integral, and its derivative. { prəˈpȯr·shən·əl ˌpləs ˈint·ə·grəl ˌpləs dəˈriv·əd·iv kənˌtrōl }

proprietary program [COMPUT SCI] **1.** A computer program that is owned by someone, and whose use may thus be restricted in some manner or entail payment of a fee. Also known as owned program. **2.** More narrowly, a program that is exploited commercially as a separate product. { prəˈprī·əˌter·ē ˈprō·grəm }

protected contour [COMMUN] A representation of the theoretical signal strength of a radio station that appears on a map as a closed polygon surrounding the station's transmitter site. The FCC defines a particular signal strength contour such as 60 dBuV/m, for certain classes of station, as the protected contour. In allocating the facilities of other radio stations, the protected contour of an existing station may not be overlapped by certain interferring contours of other stations. The protected contour coarsely represents the primary coverage area of a station, within which there is little likelihood that the signals of another station will cause interference with its reception. { prəˈtek·təd ˈkän·tu̇r }

protected format [COMPUT SCI] Parts of a computer display that cannot be altered by typing from the keyboard. { prəˈtek·təd ˈfȯrˌmat }

protected location [COMPUT SCI] A storage cell arranged so that access to its contents is denied under certain circumstances, in order to prevent programming accidents from destroying essential programs and data. { prəˈtek·təd lōˈkā·shən }

protected-logic module [COMPUT SCI] A module that stores selected computer programs that must remain unaltered. { prəˈtek·təd ¦läj·ik ˈmäj·yül }

protected subnetwork *See* domain. { prə¦tek·təd səbˈnetˌwərk }

protection code [COMPUT SCI] A component of a task descriptor that specifies the protection domain of the task, that is, the authorizations it has to perform certain actions. { prəˈtek·shən ˌkōd }

protection key [COMPUT SCI] An indicator, usually 1 to 6 bits in length, associated with a program and intended to grant the program access to those sections of memory which the program can use but to deny the program access to all other parts of memory. { prəˈtek·shən ˌkē }

protection profile [COMPUT SCI] A structure for defining the security and functionality requirements of a computing system. { prəˈtek·shən ˌprō·fīl }

protector gap [ELEC] A device designed to limit voltage, usually from lightning strkes, in order to protect telephone and telegraph equipment; consists of two carbon blocks with an air gap between them. { prəˈtek·tər ˌgap }

Proteus *See* advanced signal-processing system. { ˈprōd·ē·əs }

protocol [COMPUT SCI] **1.** A set of hardware and software interfaces in a terminal or computer which allows it to transmit over a communications network, and which collectively forms a communications language. **2.** *See* communication protocol. { ˈprōd·əˌkȯl }

protocol-level timer [COMMUN] A time-measuring unit within a communicating device that issues high-priority interrupts which synchronize and set deadlines for protocol-related activities. { 'prōd·ə,kȯl ¦lev·əl 'tīm·ər }

proving [COMPUT SCI] Testing whether a computer is free of faults and capable of functioning normally, usually by having it carry out a check routine or diagnostic routine. { 'prüv·iŋ }

proximity warning indicator [NAV] An airborne instrument which produces a warning signal indicating the approach of an aircraft on a possible collision course. Abbreviated PWI. { präk'sim·əd·ē ¦wȯrn·iŋ 'in·də,kād·ər }

proxy server [COMPUT SCI] Software for caching and filtering Web content to reduce network traffic on intranets, and for increasing security by filtering content and restricting access. { 'präk·sē ,sər·vər }

PRR *See* pulse repetition rate.

pseudocode [COMPUT SCI] In software engineering, an outline of a program written in English or the user's natural language; it is used to plan the program, and also serves as a source for test engineers doing software maintenance; it cannot be compiled. { 'süd·ō,kōd }

pseudocoloring [COMPUT SCI] A method of assigning arbitrary colors to the gray levels of a black-and-white image. It is popular in thermography (the imaging of heat), where hotter objects (with high pixel values) are assigned one color (for example, red), and cool objects (with low pixel values) are assigned another color (for example, blue), with other colors assigned to intermediate values. { ,süd·ō'kəl·ər·iŋ }

pseudoinstruction [COMPUT SCI] **1.** A symbolic representation in a compiler or interpreter. **2.** *See* quasi-instruction. { 'sü·dō·in,strək·shən }

pseudolite [NAV] A ground-based reference station in a differential global positioning system, situated at a known location, which broadcasts a signal that has the same structure as a satellite signal and also contains differential corrections for the signals of satellites in view. { 'süd·ə,līt }

pseudonoise code *See* pseudorandom noise code. { ¦süd·ō'nȯiz 'kōd }

pseudo-operation [COMPUT SCI] An operation which is not part of the computer's operation repertoire as realized by hardware; hence, an entension of the set of machine operations. { ¦sü·dō ,äp·ə'rā·shən }

pseudorandom noise code [COMMUN] A method of transmitting messages in the presence of interference or noise, in which each binary digit in the original message is encoded by a long series of binary digits with desirable autocorrelation properties. Also known as pseudonoise code. Abbreviated PN code. { ¦süd·ō¦ran·dəm 'nȯiz ,kōd }

pseudorandom numbers [COMPUT SCI] Numbers produced by a definite arithmetic process, but satisfying one or more of the standard tests for randomness. { ,sü·dō'ran·dəm 'nəm·bərz }

PSI *See* program specific information.

PSK *See* phase-shift keying.

psophometric electromotive force [ELECTR] The true noise voltage that exists in a circuit. { ¦säf·ə¦me·trik i¦lek·trə¦mōd·iv 'fȯrs }

psophometric voltage [ELECTR] The noise voltage as actually measured in a circuit under specified conditions. { ¦säf·ə¦me·trik 'vōl·tij }

PSR *See* primary radar.

PSTN *See* public switched telephone network.

PTM *See* pulse-time modulation.

p-type conductivity [ELECTR] The conductivity associated with holes in a semiconductor, which are equivalent to positive charges. { 'pē ¦tīp ,kän,dək'tiv·əd·ē }

p-type semiconductor [ELECTR] An extrinsic semiconductor in which the hole density exeeds the conduction electron density. { 'pē ¦tīp 'sem·i·kən,dək·tər }

public communications service [COMMUN] Telephone or telegraph service provided for the transmission of unofficial communications for the public. { 'pəb·lik kə,myü·nə'kā·shənz ,sər·vəs }

public correspondence [COMMUN] Any telecommunications which offices and stations at the disposal of the public must accept for transmission. { 'pəb·lik ,kär·ə'spän·dəns }

public data [COMPUT SCI] Data that are open to all users, with no security measures necessary as far as reading is concerned. { 'pəb·lik 'dad·ə }

public-key algorithm [COMMUN] A cryptographic algorithm in which one key (usually the enciphering key) is made public and a different key (usually the deciphering key) is kept secret; it must not be possible to deduce the private key from the public key. { 'pəb·lik ¦kē 'al·gə ,rith·əm }

public network [COMMUN] A communications network that can be used by anyone, usually on a fee basis. { 'pəb·lik 'net,wərk }

public pack [COMPUT SCI] A disk pack that can be used by any program and any application in a computer system. { 'pəb·lik 'pak }

public radio communications services [COMMUN] Land, mobile, and fixed services, the stations of which are open to public correspondence. { 'pəb·lik 'rād·ē·ōkə,myü·nə 'kā·shənz ,sər·və·səz }

public-safety frequency bands [COMMUN] Radio-frequency bands allocated in the United States for communication on land between base stations and mobile stations or between mobile stations by police, fire, highway, forestry, and emergency services. { 'pəb·lik ¦sāf·tē 'frē·kwən·sē ,banz }

public-safety radio service [COMMUN] Any service of radio communication essential to either the discharge of non-Federal governmental functions relating to public safety responsibilities or the alleviation of an emergency endangering life or property, the radio transmitting facilities of which are defined as fixed, land, or mobile stations. { 'pəb·lik ¦sāf·tē 'rād·ē·ō ,sər·vəs }

public switched telephone network [COMMUN] The worldwide voice telephone network. Abbreviated PSTN. { ¦pəb·lik ,swicht 'tel·ə,fōn ,net ,wərk }

pull-down menu [COMPUT SCI] A list of options for action that appears near the top of a display screen, usually overlaying the current contents of the screen without disrupting them, and usually in response to an indicator being pointed at an icon. { 'pu̇l ¦dau̇n 'men·yü }

pulse-amplitude modulation [COMMUN] Amplitude modulation of a pulse carrier. Abbreviated PAM. { 'pəls ¦am·plə,tüd ,mäj·ə,lā·shən }

pulse-amplitude modulation-frequency modulation [COMMUN] System in which pulse-amplitude-modulated subcarriers are used to frequency-modulate a second carrier; binary digits are formed by the absence or presence of a pulse in an assigned position. { 'pəls ¦am·plə ,tüd ,mäj·ə,lā·shən 'frē·kwən·sē ,mäj·ə,lā·shən }

pulse bandwidth [COMMUN] The bandwidth outside of which the amplitude of a pulse-frequency spectrum is below a prescribed fraction of the peak amplitude. { 'pəls 'band,width }

pulse cable [COMMUN] A communications cable, capable of transmitting pulses without unacceptable distortion. { 'pəls ,kā·bəl }

pulse carrier [COMMUN] A pulse train used as a carrier. { 'pəls ,kar·ē·ər }

pulse code [COMMUN] A code consisting of various combinations of pulses, such as the Morse code, Baudot code, and the binary code used in computers. { 'pəls ,kōd }

pulse-coded scanning beam [NAV] **1.** A radio or radar beam which is swept over a sector of space and is accompanied by a repeated pattern of pulses that is varied to indicate the position of the beam in space. **2.** A system of ground equipment that generates such beams at microwave frequencies to furnish guidance to aircraft making microwave landings. Abbreviated PCSB. { 'pəls ¦kōd·əd 'skan·iŋ ,bēm }

pulse-code modulation [COMMUN] Modulation in which the peak-to-peak amplitude range of the signal to be transmitted is divided into a number of standard values, each having its own code; each sample of the signal is then transmitted as the code for the nearest standard amplitude. Abbreviated PCM. { 'pəls ¦kōd ,mäj·ə'lā·shən }

pulse coder *See* coder. { 'pəls ,kōd·ər }

pulse coding and correlation [COMMUN] A general technique concerning a variety of methods used to change the transmitted waveform and then decode upon its reception; pulse compression is a special form of pulse coding and correlation. { 'pəls ¦kōd·iŋ ən ,kär·ə'lā·shən }

pulse communication [COMMUN] Radio communication using pulse modulation. { 'pəls kə ,myü·nə,kā·shən }

pulse compression [ELECTR] **1.** A matched filter technique used to discriminate against signals which do not correspond to the transmitted signal. **2.** In radar, a process in which a relatively long pulse is frequency- or phase-modulated so that a properly designed receiver produces an output with a very narrow peak response much as though a very narrow pulse had been transmitted; valuable in achieving high range resolution in long transmitted pulses. { 'pəls kəm,presh·ən }

pulse-compression radar [ENG] A radar system in which the transmitted signal is linearly frequency-modulated or otherwise spread out in time to reduce the peak power that must be handled by the transmitter; signal amplitude is kept constant; the receiver uses a linear filter to compress the signal and thereby reconstitute a short pulse for the radar display. { 'pəls kəm ,presh·ən 'rā,där }

pulse decay time [COMMUN] The interval of time required for the trailing edge of a pulse to decay from 90% to 10% of the peak pulse amplitude. { 'pəls di'kā,tīm }

pulse demodulator [ELECTR] A device that recovers the modulating signal from a pulse-modulated wave. { ¦pəls dē¦mäj·ə,lād·ər }

pulse-density modulation *See* pulse-frequency modulation. { ¦pəls ,den·sət·ē ,mäj·əlā·shən }

pulse discriminator [ELECTR] A discriminator circuit that responds only to a pulse having a particular duration or amplitude. { 'pəls di ,skrim·ə,nād·ər }

pulse Doppler radar [ENG] A coherent radar capable of estimating the radial velocity of targets by sensing the Doppler shift of the echoes. { 'pəls ¦dap·lər 'rā,där }

pulsed transfer function [CONT SYS] The ratio of the *z*-transform of the output of a system to the *z*-transform of the input, when both input and output are trains of pulses. Also known as discrete transfer function; *z*-transfer function. { 'pəlst 'tranz·fər ,faŋk·shən }

pulse duration [COMMUN] The time interval between the first and last instants at which the instantaneous amplitude reaches a stated fraction of the peak pulse amplitude. Also known as pulse length; pulse width (both deprecated usages). { 'pəls du̇'rā·shən }

pulse-duration coder *See* coder. { 'pəls du̇¦rā·shən 'kōd·ər }

pulse-duration modulation [COMMUN] Modulation of a pulse carrier wherein the value of each instantaneous sample of a modulating wave produces a pulse of proportional duration by varying the leading, trailing, or both edges of a pulse. Abbreviated PDM. Also known as pulse-length modulation; pulse-width modulation. { 'pəls du̇¦rā·shən ,mäj·ə,lā·shən }

pulse-duration modulation-frequency modulation [COMMUN] System in which pulse-duration-modulated subcarriers are used to frequency-modulate a second carrier. Also known as pulse-width modulation-frequency modulation. { 'pəls du̇¦rā·shən ,mäj·ə,lā·shən 'frē·kwən·sē ,mäj·ə,lā·shən }

pulse-frequency modulation [COMMUN] A form of pulse-time modulation in which the pulse repetition rate is the characteristic that is varied. Abbreviated PFM. { 'pəls ¦frē·kwən·sē ,mäj·ə ,lā·shən }

pulse generator [ELECTR] A generator that produces repetitive pulses or signal-initiated pulses. { 'pəls ,jen·ə,rād·ər }

pulse improvement threshold [COMMUN] In a constant-amplitude pulse-modulation system, the condition in which the peak pulse voltage is greater than twice the peak noise voltage, after selection and before nonlinear processes such as amplitude clipping and limiting. { 'pəls im ¦prüv·mənt 'thresh,hōld }

pulse interference suppression [ELECTR] Means employed in radar, such as noting asynchronous returns or pulses clearly of unlikely widths or pulses at frequencies other than the operating frequency, to reduce confusion from pulses of other radars or pulsed deceptive countermeasures. { 'pəls ,in·tər ¦fir·əns sə'presh·ən }

pulse interleaving [COMMUN] A process in which pulses from two or more sources are combined in time-division multiplex for transmission over a common path. { 'pəls ,in·tər'lēv·iŋ }

pulse-interval modulation *See* pulse-spacing modulation. { 'pəls ¦in·tər·vəl ,mäj·ə,lā·shən }

pulse jitter [COMMUN] A relatively small variation of the pulse spacing in a pulse train; the jitter may be random or systematic, depending on its origin, and is generally not coherent with any pulse modulation imposed. { 'pəls ,jid·ər }

pulse length *See* pulse duration. { 'pəls ,leŋkth }

pulse-length modulation *See* pulse-duration modulation. { 'pəls ¦leŋkth ,mäj·ə,lā·shən }

pulse-mode multiplexing [COMMUN] A type of time-division multiplexing employing pulse-amplitude modulation in which a sequence of pulses is repeatedly transmitted, and the amplitude of each pulse in the sequence is modulated by a different communication channel. { 'pəls ¦mōd 'məl·tə,pleks·iŋ }

pulse-modulated jamming [COMMUN] Use of jamming pulses of various widths and repetition rates. { 'pəls ,mäj·ə¦lād·əd 'jam·iŋ }

pulse-modulated radar [ENG] Form of radar in which the radiation consists of a series of discrete pulses. { 'pəls ,mäj·ə¦lād·əd 'rā,där }

pulse modulation [COMMUN] A system of modulation in which the amplitude, duration, position, or mere presence of discrete pulses may be so controlled as to represent the message to be communicated. { 'pəls ,mäj·ə,lā·shən }

pulse modulator [ELECTR] A device for carrying out the pulse modulation of a radio-frequency carrier signal. { 'pəls ,mäj·ə,lād·ər }

pulse-numbers modulation [COMMUN] Modulation in which a pulse carrier's pulse density per unit time varies in accordance with a modulating wave, by making systematic omissions without changing the phase or amplitude of the transmitted pulses; as an example, the omission of every other pulse could correspond to zero modulation; the reinsertion of some or all pulses then corresponds to positive modulation, and the omission of more than every other pulse corresponds to negative modulation. { 'pəls ¦nəm·bərz ,mäj·ə,lā·shən }

pulse period [COMMUN] In telephony, time required for one opening and closing of the loop of a calling telephone; for example, the time required to open and close the dial pulse springs once. Also known as impulse period. { 'pəls ,pir·ē·əd }

pulse-phase modulation *See* pulse-position modulation. { 'pəls ¦fāz ,mäj·ə,lā·shən }

pulse-position modulation [COMMUN] Modulation of a pulse carrier wherein the value of each instantaneous sample of a modulating wave varies the position in time of a pulse relative to its unmodulated time of occurrence. Abbreviated PPM. Also known as pulse-phase modulation. { 'pəls pə¦zish·ən ,mäj·ə,lā·shən }

pulse power [ELECTR] In radar, the average power transmitted during a pulse. While often called the radar's peak power, it is not to be confused with the instantaneous peak power in each cycle of the carrier frequency. { 'pəls ,pau̇·ər }

pulser [ELECTR] A modulator of the energy-storage type, using a pulse-forming network, to produce the pulsed voltage and current required by a microwave oscillator, such as a magnetron, in radar transmitters. { 'pəl·sər }

pulse radar [ENG] Radar in which the transmitter sends out high-power pulses that are spaced far apart in comparison with the duration of each pulse; the receiver is active for reception of echoes in the interval following each pulse. { 'pəls 'rā,där }

pulse-rate telemetering [ELECTR] Telemetering in which the number of pulses per unit time is proportional to the magnitude of the measured quantity. { 'pəls ¦rāt ,tel·ə,mēd·ə·riŋ }

pulse recurrence rate *See* pulse repetition rate. { 'pəls ri'kə·rəns ,rāt }

pulse recurrence time [COMMUN] Time elapsing between the start of one transmitted pulse and the next pulse; the reciprocal of the pulse repetition rate. { 'pəls ri'kə·rəns ,tīm }

pulse regeneration [ELECTR] The process of restoring pulses to their original relative timings, forms, and magnitudes. { 'pəls ri,jen·ə,rā·shən }

pulse repeater [ELECTR] Device used for receiving pulses from one circuit and transmitting corresponding pulses into another circuit; it may also change the frequencies and waveforms of the pulses and perform other functions. { 'pəls ri,pēd·ər }

pulse repetition frequency *See* pulse repetition rate. { 'pəls ,rep·ə¦tish·ən ,frē·kwən·sē }

pulse repetition rate [ELECTR] The number of times per second that a pulse is transmitted. Abbreviated PRR. Also known as pulse recurrence rate; pulse repetition frequency (PRF). { 'pəls ,rep·ə¦tish·ən ,rāt }

pulse rise time [COMMUN] The interval of time required for the leading edge of a pulse to rise from 10% to 90% of the peak pulse amplitude. { 'pəls 'rīz ,tīm }

pulse shaper [ELECTR] A transducer used for changing one or more characteristics of a pulse, such as a pulse regenerator or pulse stretcher. { 'pəls ,shāp·ər }

pulse-spacing modulation [COMMUN] A form of pulse-time modulation in which the pulse spacing is varied. Also known as pulse-interval modulation. { 'pəls ¦spās·iŋ ˌmäj·əˌlā·shən }

pulse stretcher [ELECTR] A pulse shaper that produces an output pulse whose duration is greater than that of the input pulse and whose amplitude is proportional to the peak amplitude of the input pulse. { 'pəls ˌstrech·ər }

pulse subcarrier [COMMUN] One of a number of frequency-modulation carriers modulating a radio-frequency carrier, each of which is in turn pulse-modulated. { 'pəls 'səbˌkar·ē·ər }

pulse synthesizer [ELECTR] A circuit used to supply pulses that are missing from a sequence due to interference or other causes. { 'pəls ˌsin·thəˌsīz·ər }

pulse-time modulation [COMMUN] Modulation in which the time of occurrence of some characteristic of a pulse carrier is varied from the unmodulated value; examples include pulse-duration, pulse-interval, and pulse-position modulation. Abbreviated PTM. { 'pəls ¦tīm ˌmäjəˌlā·shən }

pulse-train analysis [COMMUN] A Fourier analysis of a pulse train. { 'pəls ¦trān əˌnal·ə·səs }

pulse transmitter [ELECTR] A pulse-modulated transmitter whose peak-power-output capabilities are usually large with respect to the average-power-output rating. { 'pəls tranzˌmid·ər }

pulse-type altimeter *See* radar altimeter. { 'pəls ¦tīp al'tim·əd·ər }

pulse-type telemetering [COMMUN] Signal transmission system with pulses as a function of time, but independent of electrical magnitude; in a pulse-counting system the number of pulses per unit time corresponds to the measured variable; in pulse-width or pulse-duration types, the length of the pulse is controlled by the measured variable. { 'pəls ¦tīp ˌtel·əˌmēd·ə·riŋ }

pulse width *See* pulse duration. { 'pəls ˌwidth }

pulse-width modulation *See* pulse-duration modulation. { 'pəls ¦width ˌmäj·əˌlā·shən }

pulse-width modulation-frequency modulation *See* pulse-duration modulation - frequency modulation. { 'pəls ¦width ˌmäj·əˌlā·shən 'frē·kwən·sē ˌmäj·əˌlā·shən }

pulsing key [COMMUN] **1.** Method of passing voice frequency pulses over the line under control of a key at the original office; used with E and M supervision on intertoll dialing. **2.** System of signaling where numbered keys are depressed instead of using a dial. { 'pəls·iŋ ˌkē }

punch [COMPUT SCI] **1.** A device for making holes representing information in a medium such as cards or paper tape, in response to signals sent to it. **2.** A hole in a medium such as a card or paper tape, generally made in an array with other holes (or lack of holes) to represent information. { pənch }

punch card [COMPUT SCI] A medium by means of which data are fed into a computer in the form of rectangular holes punched in the card; once the primary data-output medium, it is now largely obsolete. Also known as card; punched card. { 'pənch ˌkärd }

punched card *See* punch card. { 'pəncht ˌkärd }

punctuation bit [COMPUT SCI] A binary digit used to indicate the beginning or end of a variable-length record. { ˌpəŋk·chə'wā·shən ˌbit }

Pupin coil *See* loading coil. { pyü'pēn ˌkȯil }

pure procedure [COMPUT SCI] A procedure that never modifies any part of itself during execution. { 'pyu̇r prə¦sē·jər }

pure vanilla *See* vanilla. { 'pyu̇r və'nil·ə }

purge [COMPUT SCI] To remove data from computer storage so that space occupied by the data can be reused. { pərj }

purge date [COMPUT SCI] The date after which data are released and the storage area can be used for storing other data. { 'pərj ˌdāt }

purify [COMPUT SCI] To remove errors from data. { 'pyu̇r·əˌfī }

purity coil [ELECTR] A coil mounted on the neck of a color picture tube, used to produce the magnetic field needed for adjusting color purity; the direct current through the coil is adjusted to a value that makes the magnetic field orient the three individual electron beams so each strikes only its assigned color of phosphor dots. { 'pyu̇r·əd·ē ˌkȯil }

purity control [ELECTR] A potentiometer or rheostat used to adjust the direct current through the purity coil. { 'pyu̇r·əd·ē kənˌtrōl }

purity magnet [ELECTR] An adjustable arrangement of one or more permanent magnets used in place of a purity coil in a color cathode ray. { 'pyu̇r·əd·ē ˌmag·nət }

push [COMPUT SCI] To add an item to a stack. { pu̇sh }

push button [COMPUT SCI] A small area delineated on a graphical user interface whose selection by the user instructs the computer to perform a specific task. { 'pu̇sh ˌbət·ən }

push-button dialing [ELECTR] Dialing a number by pushing buttons on the telephone rather than turning a circular wheel; each depressed button causes an oscillator to oscillate simultaneously at two different frequencies, generating a pair of audio tones which are recognized by central-office (or PBX) switching equipment as digits of a telephone number. Also known as dual-tone multifrequency dialing; tone dialing; touch call. { 'pu̇sh ¦bət·ən 'dī·liŋ }

push-down automaton [COMPUT SCI] A nondeterministic, finite automaton with an auxiliary tape having the form of a push-down storage. { 'pu̇shˌdau̇n ȯ'täm·əˌtän }

push-down list [COMPUT SCI] An ordered set of data items so constructed that the next item to be retrieved is the item most recently stored; in other words, last-in, first-out (LIFO). { 'pu̇sh ˌdau̇n ˌlist }

push-down storage [COMPUT SCI] A computer storage in which each new item is placed in the first location in the storage and all the other items are moved back one location; it thus follows the principle of a push-down list. Also known as cellar; nesting storage; running accumulator. { 'pu̇shˌdau̇n ˌstȯr·ij }

pushing [COMPUT SCI] The placing of a data element at the top of a stack. { 'pu̇sh·iŋ }

push-pull amplifier [ELECTR] A balanced amplifier employing two similar electron tubes or equivalent amplifying devices working in phase opposition. { 'pu̇sh ¦pu̇l 'am·plə,fī·ər }

push-up list [COMPUT SCI] An ordered set of data items so constructed that the next item to be retrieved will be the item that was inserted earliest in the list, resulting in a first-in, first-out (FIFO) structure. { 'pu̇sh,əp ,list }

put [COMPUT SCI] A programming instruction that causes data to be written from computer storage into a file. { pu̇t }

PWI *See* proximity warning indicator.

pyrotechnic code [COMMUN] Significant arrangement of the various colors and patterns of fireworks, signal lights, or signal smokes used for communication between units or between ground and air. { ¦pī·rə¦tek·nik 'kōd }

pyrotechnic signal [COMMUN] Signal designed for military use to produce a colored light or smoke, for the purpose of transmitting information. { ¦pī·rə¦tek·nik 'sig·nəl }

Q

Q [PHYS] A measure of the ability of a system with periodic behavior to store energy equal to 2π times the average energy stored in the system divided by the energy dissipated per cycle. Also known as Q factor; quality factor; storage factor.

QAM *See* quadrature amplitude modulation.

QBE *See* query by example.

Q factor *See* Q. { 'kyü ˌfak·tər }

QPSK *See* quadrature phase-shift keying.

Q signal [COMMUN] A three-letter abbreviation starting with Q, used in the International List of Abbreviations for radiotelegraphy to represent complete sentences. [ELECTR] The quadrature component of the chrominance signal in analog color television, having a bandwidth of 0 to 0.5 megahertz; it consists of +0.48(R-Y) and +0.41(B-Y), where Y is the luminance signal, R is the red camera signal, and B is the blue camera signal. { 'kyü ˌsig·nəl }

quadded redundancy [COMPUT SCI] A form of redundancy in which each logic gate is quadruplicated, and the outputs of one stage are interconnected to the inputs of the succeeding stage by a connection pattern so that errors made in earlier stages are overridden in later stages, where the original correct signals are restored. { 'kwäd·əd ri'dən·dən·sē }

quad density [COMPUT SCI] A format for floppy-disk storage that holds four times as much data as would normally be contained. { 'kwäd 'den·səd·ē }

quadratic performance index [CONT SYS] A measure of system performance which is, in general, the sum of a quadratic function of the system state at fixed times, and the integral of a quadratic function of the system state and control inputs. { kwä'drad·ik pər'for·məns ˌin ˌdeks }

quadrature amplifier [ELECTR] An amplifier that shifts the phase of a signal 90°; used in an analog color television receiver to amplify the 3.58-megahertz chrominance subcarrier and shift its phase 90° for use in the Q demodulator. { 'kwä·drə·chər ˌam·pləˌfī·ər }

quadrature amplitude modulation [COMMUN] **1.** Quadrature modulation in which the two carrier components are amplitude-modulated. **2.** A digital modulation technique in which digital information is encoded in bit sequences of specified length and these bit sequences are represented by discrete amplitude levels of an analog carrier, by a phase shift of the analog carrier from the phase that represented the previous bit sequence by a multiple of 90°, or by both. Abbreviated QAM. { ¦kwäd·rə·chər ˌam·pləˌtüd ˌmäj·ə'lā·shən }

quadrature modulation [COMMUN] Modulation of two carrier components 90° apart in phase by separate modulating functions. { 'kwä·drə·chər ˌmäj·ə'lā·shən }

quadrature partial-response keying [COMMUN] A modulation technique in which two orthogonally phased carriers are combined; each carrier is modulated by one of the digital bit streams to one of three levels. Abbreviated QPRK. { 'kwä·drə·chər ¦pär·shəl ri'späns ˌkē·iŋ }

quadrature phase-shift keying [COMMUN] Phase-shift keying in which four different phase angles are used, usually spaced 90° apart. Abbreviated QPSK. Also known as quadriphase; quaternary phase-shift keying. { ¦kwäd·rə·chər 'fāzˌshift ˌkē·iŋ }

quadriphase *See* quadrature phase-shift keying. { 'kwäd·rəˌfāz }

quad word [COMPUT SCI] A word 16 bytes long. { 'kwäd ˌwərd }

qualified name [COMPUT SCI] A name that is further identified by associating it with additional names, usually the names of things that contain the thing being named. { 'kwäl·əˌfīd ¦nām }

qualifier [COMPUT SCI] A name that is associated with another name to give additional information about the latter and distinguish it from other things having the same name. { 'kwäl·ə ˌfī·ər }

quality factor *See* Q. { 'kwäl·əd·ē ˌfak·tər }

quality program [COMPUT SCI] A computer program that is correct, reliable, efficient, maintainable, flexible, testable, portable, and reusable. { ¦kwäl·əd·ē 'prō·grəm }

quantity [COMPUT SCI] In computers, a positive or negative real number in the mathematical sense; the term quantity is preferred to the term number in referring to numerical data; the term number is used in the sense of natural number and reserved for "the number of digits," the "number of operations," and so forth. { 'kwän·əd·ē }

quantization [COMMUN] Division of the range of values of a wave into a finite number of subranges, each of which is represented by an assigned or quantized value within the subrange. { ˌkwän·tə'zā·shən }

quantization distortion [COMMUN] Inherent distortion introduced in the process of quantization of a waveform. Also known as quantization noise; quantumization distortion; quantumization noise. { ˌkwän·tə'zā·shən di ˌstȯr·shən }

quantization level [COMMUN] Discrete value of the output designating a particular subrange of the input. { ˌkwän·tə'zā·shən ˌlev·əl }

quantization noise See quantization distortion. { ˌkwän·tə'zā·shən ˌnȯiz }

quantized frequency modulation [COMMUN] Frequency modulation that involves quantization; it uses time and frequency redundancy within a voice frequency channel during each transmitted symbol; used to combat distortion due to multipath, selection fading, and noise spikes. { 'kwänˌtīzd 'frē·kwən·sē ˌmäj·əˌlā·shən }

quantized pulse modulation [COMMUN] Pulse modulation that involves quantization, such as pulse-numbers modulation and pulse-code modulation. { 'kwänˌtīzd 'pəls ˌmäj·əˌlā·shən }

quantizer [COMMUN] A processing step that intentionally reduces the precision of discrete cosine transform coefficients. [ELECTR] A device that measures the magnitude of a time-varying quantity in multiples of some fixed unit, at a specified instant or specified repetition rate, and delivers a proportional response that is usually in pulse code or digital form. { kwän'tīz·ər }

quantum [COMMUN] One of the subranges of possible values of a wave which is specified by quantization and represented by a particular value within the subrange. { 'kwän·təm }

quantum computer [COMPUT SCI] A computer in which the time evolution of the state of the individual switching elements of the computer is governed by the laws of quantum mechanics. { 'kwän·təm kəm¦pyüd·ər }

quantumization distortion See quantization distortion. { ˌkwän·tə·mə'zā·shən diˌstȯr·shən }

quantumization noise See quantization distortion. { ˌkwän·tə·mə'zā·shən ˌnȯiz }

quantum well injection transit-time diode [ELECTR] An active microwave diode that employs resonant tunneling through a gallium arsenide quantum well located between two aluminum gallium arsenide barriers to inject electrons into an undoped gallium arsenide drift region. Abbreviated QWITT diode. { ˌkwän·təm ¦wel inˌjek·shən ¦tranz·it ˌtīm 'dīˌōd }

quarternary phase-shift keying [ELECTR] Modulation of a microwave carrier with two parallel streams of nonreturn-to-zero data in such a way that the data is transmitted as 90° phase shifts of the carrier; this gives twice the message channel capacity of binary phase-shift keying in the same bandwidth. Abbreviated QPSK. { 'kwät·əˌner·ē 'fāz ˌshift ˌkē·iŋ }

quarter-square multiplier [COMPUT SCI] A device used to carry out function multiplication in an analog computer by implementing the algebraic identity $xy = \frac{1}{4}[(x + y)^2 - (x - y)^2]$. { 'kwȯrd·ər ˌskwer 'məl·təˌplī·ər }

quarter-wave [ELECTROMAG] Having an electrical length of one quarter-wavelength. { 'kwȯrd·ər ˌwāv }

quarter-wave antenna [ELECTROMAG] An antenna whose electrical length is equal to one quarter-wavelength of the signal to be transmitted or received. { 'kwȯrd·ər ˌwāv an'ten·ə }

quarter-wave attenuator [ELECTROMAG] Arrangement of two wire gratings, spaced an odd number of quarter-wavelengths apart in a waveguide, used to attenuate waves traveling through in one direction. { 'kwȯrd·ər ˌwāv ə'ten·yəˌwād·ər }

quarter-wave line See quarter-wave stub. { 'kwȯrd·ər ˌwāv ˌlīn }

quarter-wave matching section See quarter-wave transformer. { 'kwȯrd·ər ˌwāv 'mach·iŋ ˌsek·shən }

quarter-wave stub [ELECTROMAG] A section of transmission line that is one quarter-wavelength long at the fundamental frequency being transmitted; when shorted at the far end, it has a high impedance at the fundamental frequency and all odd harmonics, and a low impedance for all even harmonics. Also known as quarter-wave line; quarter-wave transmission line. { 'kwȯrd·ər ˌwāv ¦stəb }

quarter-wave termination [ELECTROMAG] Metal plate and a wire grating spaced about one-fourth of a wavelength apart in a waveguide, with the plate serving as the termination of the guide; waves reflected from the metal plate are canceled by waves reflected from the grating so that all energy is absorbed (none is reflected) by the quarter-wave termination. { 'kwȯrd·ər ˌwāv tər·mə'nā·shən }

quarter-wave transformer [ELECTROMAG] A section of transmission line approximately one quarter-wavelength long, used for matching a transmission line to an antenna or load. Also known as quarter-wave matching section. { 'kwȯrd·ər ˌwāv tranz'fȯr·mər }

quarter-wave transmission line See quarter-wave stub. { 'kwȯrd·ər ˌwāv tranz'mish·ən ˌlīn }

quasi-instruction [COMPUT SCI] An expression in a source program which resembles an instruction in form, but which does not have a corresponding machine instruction in the object program, and is directed to the assembler or compiler. Also known as pseudoinstruction. { ¦kwä·zē in'strək·shən }

quasi-linear system [CONT SYS] A control system in which the relationships between the input and output signals are substantially linear despite the existence of nonlinear elements. { ¦kwä·zē 'lin·ē·ər 'sis·təm }

quasi-parallel execution [COMPUT SCI] The execution of a collection of coroutines by a single processor that can work on only one coroutine at a time; the order of execution is arbitrary and each coroutine is executed independently of the rest. { ¦kwä·zē 'par·əˌlel ˌek·sə'kyü·shən }

quasi-random code generator [COMMUN] High-speed coded information source used in the design and evaluation of wide-band communications links by providing a means of closed-loop testing. { ¦kwä·zē 'ran·dəm 'kōd ˌjen·əˌrād·ər }

quaternary phase-shift keying *See* quadrature phase-shift keying. { ¦kwät·ərˌner·ē 'fāz ˌshift ˌkē·iŋ }

quaternary signaling [COMMUN] An electrical communications mode in which information is passed by the presence and absence, or plus and minus variations, of four discrete levels of one parameter of the signaling medium. { 'kwät·ən ˌer·ē 'sig·nə·liŋ }

qubit [COMPUT SCI] In quantum computation, a superposition of the ground state and the excited state of an elementary two-level quantum system (such as a two-level atom or a nuclear spin), corresponding to a classical bit that is either 0 (corresponding to the ground state) or 1 (corresponding to the excited state). { 'kyü·bit }

quenching frequency [ELECTR] The frequency of an alternating voltage that is applied to a superregenerative detector stage to prevent sustained oscillation. { 'kwench·iŋ ˌfrē·kwən·sē }

quench oscillator [ELECTR] Circuit in a superregenerative receiver which produces the frequency signal. { 'kwench ˌäs·əˌlād·ər }

query [COMPUT SCI] A computer instruction to interrogate a database. { 'kwir·ē }

query by example [COMPUT SCI] A software product used to search a database for information having formats or ranges of values specified by English-like statements that indicate the desired results. Abbreviated QBE. { 'kwir·ē bī ig'zam·pəl }

query language [COMPUT SCI] A generalized computer language that is used to interrogate a database. { 'kwir·ē ˌlaŋ·gwij }

query layer [COMPUT SCI] A program that mediates between data sources on the World Wide Web and a user's query by breaking the query into subqueries against each information source and then gathering together the results for presentation to the user. { 'kwir·ē ˌlā·ər }

query program [COMPUT SCI] A computer program that allows a user to retrieve information from a database and have it displayed on a terminal or printed out. { 'kwir·ē ˌprō·grəm }

question-answering system [COMPUT SCI] An information retrieval system in which a direct answer is expected in response to a submitted query, rather than a set of references that may contain the answers. { 'kwes·chən 'an·sə·riŋ ˌsis·təm }

queue [COMPUT SCI] **1.** A list of items waiting for attention in a computer system, generally ordered according to some criteria. **2.** A linear list whose elements are inserted and deleted in a first-in-first-out order. { kyü }

queued access method [COMPUT SCI] A set of precedures controlled by queues for efficient transfer of data between a computer and input-output devices. { 'kyüd 'akˌses ˌmeth·əd }

queue-driven system [COMPUT SCI] A software system that uses many queues for tasks in various phases of processing. { 'kyü ¦driv·ən ˌsis·təm }

queueing [ENG] The movement of discrete units through channels, such as programs or data arriving at a computer, or movement on a highway of heavy traffic. { 'kyü·iŋ }

queueing theory [MATH] The area of stochastic processes emphasizing those processes modeled on the situation of individuals lining up for service. { 'kyü·iŋ ˌthē·ə·rē }

queuing network model [COMPUT SCI] A model that represents a computer system by a network of devices through which customers (such as transactions, processes, or server requests) flow, and queues may form at each device due to its finite service rate. { 'kyü·iŋ ˌnetˌwərk ˌmäd·əl }

quibinary [COMPUT SCI] A numeration system, used in data processing, in which each decimal digit is represented by seven binary digits, a group of five which are coefficients of 8, 6, 4, 2, and 0, and a group of two which are coefficients of 1 and 0. { 'kwib·əˌner·ē }

quiesce [COMPUT SCI] To prevent a computer system from starting new jobs so that the system gradually winds down as current jobs are completed, usually in preparation for a planned outage. { kwē'es }

quiescent-carrier telephony [COMMUN] A radiotelephony system in which the carrier is suppressed whenever there are no voice signals to be transmitted. { kwē'es·ənt ¦kar·ē·ər tə'lef·ə·nē }

quiescent period [COMMUN] Resting period, or the period between pulse transmissions. { kwē'es·ənt ¦pir·ē·əd }

quiescent push-pull [ELECTR] Push-pull output stage so arranged in a radio receiver that practically no current flows when an input signal is not present.. { kwē'es·ənt ¦pu̇sh ¦pu̇l }

quiet automatic volume control *See* delayed automatic gain control. { 'kwī·ət ¦ȯd·ə¦mad·ik 'väl·yəm kənˌtrōl }

quiet battery [ELECTR] Source of energy of special design or with added filters which is sufficiently quiet and free from interference that it may be used for speech transmission. Also known as talking battery. { 'kwī·ət 'bad·ə·rē }

quieting sensitivity [ELECTR] Minimum signal input to a frequency-modulated receiver which is required to give a specified output signal-to-noise ratio under specified conditions. { 'kwī·əd·iŋ ˌsen·səˌtiv·əd·ē }

quiet tuning [ELECTR] Circuit arrangement for silencing the output of a radio receiver, except when it is accurately tuned to an incoming carrier wave. { 'kwī·ət 'tün·iŋ }

quinary code [COMPUT SCI] A code based on five possible combinations for representing digits. { 'kwī·nə·rē ˌkōd }

QWITT diode *See* quantum well injection transit-time diode. { ¦kyü¦dəb·əlˌyü¦ī¦tē¦tē 'dīˌōd }

R

racon *See* radar beacon. { 'rā,kän }

radar [ENG] A system using beamed and reflected radio-frequency energy for detecting, locating, and examining objects, measuring distance or altitude, assisting in navigation, military operations, air traffic management, and weather appraisal, and many other military and civil purposes. Timing of the return of reflected energy and examination of its nature are fundamental to all radar applications. Derived from radio detection and ranging. { 'rā,där }

radar absorbent material [MATER] Material that can be applied to surfaces of buildings or aircraft that, because of its specific dimensions, laminar nature, and resistive compounds, reflects very little of incident radar signals, thereby greatly reducing the radar cross section of the structure to which it is applied. { 'rā,där ad'sȯr·bənt mə'tir·ē·əl }

radar advisory [COMMUN] The term used to indicate that the provision of advice and information is based on radar observation. { 'rā,där id'vīz·ə·rē }

radar altimeter [NAV] A radio altimeter, useful at altitudes much greater than the 5000-foot (1500-meter) limit of frequency-modulated radio altimeters, in which simple pulse-type radar equipment is used to send a pulse straight down from an aircraft and to measure its total time of travel to the surface and back to the aircraft. Also known as high-altitude radio altimeter; pulse-type altimeter. { 'rā,där al'tim·əd·ər }

radar altitude [NAV] The altitude of an aircraft or spacecraft as determined by a radio altimeter; thus, the actual vertical distance from the terrain. Also known as radio altitude. { 'rā,där 'al·tə ,tüd }

radar antenna [ELECTROMAG] A device which radiates radio-frequency energy in a radar system, concentrating the transmitted power in the direction of the target, and which provides a large area to collect the echo power of the returning wave. { 'rā,där an'ten·ə }

radar antijamming [ELECTR] Measures taken to counteract radar jamming (electronic attack). { 'rā,där ¦ant·i'jam·iŋ }

radar attenuation [ELECTROMAG] Ratio of the power delivered by the transmitter to the transmission line connecting it with the transmitting antenna, to the power reflected from the target which is delivered to the receiver by the transmission line connecting it with the receiving antenna. { 'rā,där ə,ten·yə'wā·shən }

radar beacon [NAV] A radar receiver-transmitter that transmits a strong coded radar signal whenever its radar receiver is triggered by an interrogating radar on an aircraft or ship; the coded beacon reply can be used by the navigator to determine his own position in terms of bearing and range from the beacon. Also known as racon; radar transponder. { 'rā,där ,bē·kən }

radar beam [ELECTROMAG] The movable beam of radio-frequency energy produced by a radar transmitting antenna; its shape is commonly defined as the loci of all points at which the power has decreased to one-half of that at the center of the beam. { 'rā,där ,bēm }

radar cell [ELECTROMAG] Volume whose dimensions are one radar pulse length by one radar beam width. { 'rā,där ,sel }

radar clutter *See* clutter. { 'rā,där ,kləd·ər }

radar constant [ELECTR] The product of the factors of radar performance equation that describe characteristics of the particular radar to which the equations are applied; these include peak power, antenna gain or aperture, beam width, pulse length, pulse repetition frequency, wavelength, polarization, and noise level of the receiver. { 'rā,där ,kän·stənt }

radar contact [ENG] Recognition and identification of an echo on a radar screen; an aircraft is said to be on radar contact when its radar echo can be seen and identified on a PPI (plan-position indicator) display. { 'rā,där ,kän,takt }

radar control and interface apparatus [ELECTR] That subsystem of a radar that acts on the output of the receiver to provide significant reports to the system using that radar and also to control the radar in ways appropriate to the situation; constituted of a human operator and visual display in elementary radar, and of computer operations and data displays for human management in more modern radar. { 'rā,där kən,trōl ənd in·tər ,fās ,ap·ə,rad·əs }

radar countermeasure [ELECTR] Electronic and electromagnetic actions used against enemy radar, such as jamming and confusion reflectors. Abbreviated RCM. { 'rā,där 'kau̇nt·ər,mezh·ər }

radar cross section [ELECTROMAG] In representing a radar target, a convenient expression of the incident-signal intercept area that, if the intercepted signal were reradiated isotropically, would return to the radar the same signal strength as the target actually does. { 'rā,där 'krȯs ,sek·shən }

radar data filtering [ELECTR] Quality analysis processes, as in a computer, that reject certain radar data and perhaps alert personnel operating the radar to the rejection. { 'rā,där 'dad·ə ,fil·triŋ }

radar display [ELECTR] Visual presentation of the output of a radar receiver produced either on the screen of a cathode-ray tube or in computer-generated displays of symbols and notations based on that output in more automated systems. Also known as radar presentation. { 'rā ,där di,splā }

radar display formats [ELECTR] Any of a variety of visual representations of radar receiver output to assist the operator in interpreting the data, managing the radar, and making reasonable reports to the user system. Many of the formats have been given letter names, such as the A-display (or A-scope), and so on; the PPI (plan position indicator), RHI (range-height indicator), A-scope, and B-scope are among the most frequently used. Also known as display formats. { 'rā,där di,splā ,fȯr,matz }

radar dome [ENG] Weatherproof cover for a primary radiating element of a radar or radio device which is transparent to radio-frequency energy, and which permits active operation of the radiating element, including mechanical rotation or other movement as applicable. { 'rā,där ,dōm }

radar echo *See* echo. { 'rā,där ,ek·ō }

radar equation [ELECTROMAG] An equation that relates the transmitted and received powers and antenna gains of a primary radar system to the echo area and distance of the radar target. { 'rā ,där i,kwā·zhən }

radar frequency band [ELECTROMAG] A frequency band of microwave radiation in which radar operates. { 'rā,där 'frē·kwən·sē ,band }

radar horizon [NAV] The distance to which a radar's operation is limited by the quasi-optical characteristics of the radio waves employed. { 'rā,där hə,rīz·ən }

radar image [ELECTR] The image of an object, a vehicle or an entire scene, which is produced on a radar display or in an appropriate medium. { 'rā ,där ,im·ij }

radar indicator [ELECTR] A cathode-ray tube and associated equipment used to provide a visual indication of the echo signals picked up by a radar set. { 'rā,där ,in·də,kād·ər }

radar intelligence [COMMUN] **1.** Intelligence concerning radar or intelligence derived from the use of radar equipment. **2.** Organization or activity that deals with such intelligence. { 'rā ,där in,tel·ə·jəns }

radar marker [ENG] A fixed facility which continuously emits a radar signal so that a bearing indication appears on a radar display. { 'rā,där ,mär·kər }

radar mile [ELECTROMAG] The time for a radar pulse to travel from the radar to a target 1 mile (1.61 kilometers) distant and return, equal to 10.75 microseconds. { 'rā,där 'mīl }

radar nautical mile [ELECTROMAG] The time interval of approximately 12.355 microseconds that is required for the radio-frequency energy of a radar pulse to travel 1 nautical mile (1852 meters) and return. { 'rā,där 'nȯd·ə·kəl ,mīl }

radar netting [ENG] The linking of several radars to a single center to provide integrated target information. { 'rā,där ,ned·iŋ }

radar netting station [ENG] A center which can receive data from radar tracking stations and exchange these data among other radar tracking stations, thus forming a radar netting system. { 'rā,där ¦ned·iŋ ,stā·shən }

radar pulse [ELECTROMAG] Radio-frequency radiation emitted with high power by a pulse radar installation for a period of time which is brief compared to the interval between such pulses. { 'rā,där ,pəls }

radar range [ELECTROMAG] The maximum distance at which a radar set is ordinarily effective in detecting objects. { 'rā,där ,rānj }

radar range equation [ELECTROMAG] An equation which expresses radar range in terms of transmitted power, minimum detectable signal, antenna gain, and the target's radar cross section. { 'rā,där ,rānj i,kwā·zhən }

radar range marker *See* distance marker. { 'rā ,där 'rānj ,mär·kər }

radar receiver [ELECTR] That subsystem of a radar that is designed to amplify, enhance as appropriate with signal processing, and demodulate radar echo signals and feed them to a radar display or similar data processer. { 'rā,där ri ,sēv·ər }

radar receiver-transmitter [ELECTR] A single component having the dual functions of generating electromagnetic energy for transmission, and of receiving, demodulating, and sometimes presenting intelligence from the reflected electromagnetic energy. { 'rā,där ri ,sēv·ər tranz'mid·ər }

radar reflection [ELECTROMAG] The return of electromagnetic waves, generated by a radar installation, from an object on which the waves are incident. { 'rā,där ri,flek·shən }

radar reflection interval [ELECTROMAG] The time required for a radar pulse to travel from the source to the target and return to the source, taking the velocity of radio propagation to be equal to the velocity of light. { 'rā,där ri ,flek·shən ,in·tər·vəl }

radar reflectivity [ELECTROMAG] The fraction of electromagnetic energy generated by a radar installation which is reflected by an object. { 'rā ,där ,rē,flek'tiv·əd·ē }

radar reflector [ELECTROMAG] A device that reflects or deflects radar waves. { 'rā,där rē ,flek·tər }

radar relay [ENG] **1.** Equipment for relaying the radar video and appropriate synchronizing signal to a remote location. **2.** Process or system by which radar echoes and synchronization data are transmitted from a search radar installation to a receiver at a remote point. { 'rā,där 'rē,lā }

radar repeater [ELECTR] A radar indicator used to reproduce the radar's own display at a remote position; with proper selection, the display of any one of several radar systems can be reproduced. { 'rā,där ri,pēd·ər }

radar return [NAV] The signal indication of an object which has reflected energy that was transmitted by a primary radar. Also known as radio echo. { 'rā,där ri,tərn }

radar scanning [ENG] The process or action of directing a radar beam through a space search pattern for the purpose of locating a target. { 'rā ,där ,skan·iŋ }

radarscope [ELECTR] An older coined term for a radar display, connoting usually the use of a cathode-ray tube serving as an oscilloscope, the face of which is the radar viewing screen. Also known as scope. { 'rā,där,skōp }

radarscope overlay [ENG] A transparent overlay placed on a radarscope for comparison and identification of radar returns. { 'rā,där,skōp 'ō·vər,lā }

radar set [ENG] A complete assembly of radar equipment, consisting of a transmitter, antenna, receiver and signal processor, and appropriate control and interface apparatus. The term radar alone is often used. { 'rā,där ,set }

radar shadow [ELECTROMAG] A region shielded from radar illumination by an intervening reflecting or absorbing medium such as a hill. { 'rā ,där ,shad·ō }

radar signal spectrograph [ELECTR] An electronic device in the form of a scanning filter which provides a frequency analysis of the amplitude-modulated back-scattered signal. { 'rā,där ¦sig·nəl 'spek·trə,graf }

radar station [ENG] The place, position, or location from which, or at which, a radar set transmits or receives signals. { 'rā,där ,stā·shən }

radar station pointer [NAV] A transparent plotting chart inscribed with radial lines from 0° to 360°; used to plot radar echoes to the scale of the chart in use; it is also used to assist in identifying radar responses with charted features. { 'rā,där ,stā·shən ,pȯint·ər }

radar target [ELECTROMAG] An object belonging to a desired class which reflects back a signal sufficient to produce a fluorescent mark on the radar screen. { 'rā,där ,tär·gət }

radar transmitter [ELECTR] That subsystem of a radar that converts electrical power to the radio-frequency electromagnetic signals desired, then sends them to the antenna. { 'rā,där tranz ,mid·ər }

radar transponder *See* radar beacon. { 'rā,där tranz'pän·dər }

radar triangulation [ENG] A radar system of locating targets, usually aircraft, in which two or more separate radars are employed to measure range only; the target is located by automatic trigonometric solution of the triangle composed of a pair of radars and the target in which all three sides are known. { 'rā,där trī,aŋ·gyə'lā·shən }

radar volume [ELECTROMAG] The volume in space that is irradiated by a given radar; for a continuous-wave radar it is equivalent to the antenna radiation pattern; for a pulse radar it is a function of the cross-section area of the beam of the antenna and the pulse length of the transmitted pulse. { 'rā,där ,väl·yəm }

radar warning receiver [ELECTR] An electronic countermeasure system, carried on a tactical or transport aircraft, which is programmed to alert a pilot when his or her aircraft is being illuminated by a specific radar signal above predetermined power thresholds. { 'rā,där 'wȯrn·iŋ ri,sē·vər }

radial grating [ELECTROMAG] Conformal wire grating consisting of wires arranged radially in a circular frame, like the spokes of a wagon wheel, and placed inside a circular waveguide to obstruct E waves of zero order while passing the corresponding H waves. { 'rād·ē·əl 'grād·iŋ }

radial selector *See* private line arrangement. { 'rād·ē·əl si'lek·tər }

radiant reflectance [ELECTROMAG] Ratio of reflected radiant power to incident radiant power. { 'rād·ē·ənt ri'flek·təns }

radiant transmittance [ELECTROMAG] Ratio of transmitted radiant power to incident radiant power. { 'rād·ē·ənt tranz'mit·əns }

radiated interference [COMMUN] Interference which is transmitted through the atmosphere according to the laws of electromagnetic wave propagation; the term is generally considered to include the transfer of interfering energy in inductive or capacitive coupling. { 'rād·ē,ād·əd ,in·tər'fir·əns }

radiated power [ELECTROMAG] The total power emitted by a transmitting antenna. { 'rād·ē ,ād·əd 'paú·ər }

radiating curtain [ELECTROMAG] Array of dipoles in a vertical plane, positioned to reinforce each other; it is usually placed one-fourth wavelength ahead of a reflecting curtain of corresponding half-wave reflecting antennas. { 'rād·ē,ād·iŋ 'kərt·ən }

radiating element [ELECTROMAG] Basic subdivision of an antenna which in itself is capable of radiating or receiving radio-frequency energy. { 'rād·ē,ād·iŋ 'el·ə·mənt }

radiating guide [ELECTROMAG] Waveguide designed to radiate energy into free space; the waves may emerge through slots or gaps in the guide, or through horns inserted in the wall of the guide. { 'rād·ē,ād·iŋ 'gīd }

radiation angle [ELECTROMAG] The vertical angle between the line of radiation emitted by a directional antenna and the horizon. { ,rād·ē 'ā·shən ,aŋ·gəl }

radiation characteristic [COMMUN] One of the identifying features of a radiating signal, such as frequency and pulse width. { ,rād·ē'ā·shən ,kar·ik·tə'ris·tik }

radiation efficiency [ELECTROMAG] Of an antenna, the ratio of the power radiated to the total power supplied to the antenna at a given frequency. { ,rād·ē'ā·shən i,fish·ən·sē }

radiation field [ELECTROMAG] The electromagnetic field that breaks away from a transmitting antenna and radiates outward into space as electromagnetic waves; the other type of electromagnetic field associated with an energized antenna is the induction field. { ˌrād·ēˈā·shən ˌfēld }

radiation hardening [ENG] Improving the ability of a device or piece of equipment to withstand nuclear or other radiation; applies chiefly to dielectric and semiconductor materials. { ˌrād·ēˈā·shən ˈhärd·ən·iŋ }

radiation intensity [ELECTROMAG] The power radiated from an antenna per unit solid angle in a given direction. { ˌrād·ēˈā·shən inˌten·səd·ē }

radiation lobe *See* lobe. { ˌrād·ēˈā·shən ˌlōb }

radiation pattern [ELECTROMAG] Directional dependence of the radiation of an antenna. Also known as antenna pattern; directional pattern; field pattern. { ˌrād·ēˈā·shən ˌpad·ərn }

radiation zone *See* Fraunhofer region. { ˌrād·ēˈā·shən ˌzōn }

radiator [ELECTROMAG] **1.** The part of an antenna or transmission line that radiates electromagnetic waves either directly into space or against a reflector for focusing or directing. **2.** A body that emits radiant energy. { ˈrād·ēˌād·ər }

radio- [ELECTROMAG] A prefix denoting the use of radiant energy, particularly radio waves. { ˈrād·ē·ō }

radio [COMMUN] The transmission of signals through space by means of electromagnetic waves. *See* radio receiver. { ˈrād·ē·ō }

radioacoustics [COMMUN] Study of the production, transmission, and reproduction of sounds carried from one place to another by radiotelephony. { ¦rād·ē·ō·əˈküs·tiks }

radio aid to navigation [ELECTR] An aid to navigation which utilizes the propagation characteristics of radio waves to furnish navigation information. { ˈrād·ē·ō ˈäd tə ˌnav·əˈgā·shən }

radio altitude *See* radar altitude. { ˈrād·ē·ō ˈal·təˌtüd }

radio and wire integration [COMMUN] The combining of wire circuits with radio facilities. { ˈrād·ē·ō ən ˈwīr ˌint·əˈgrā·shən }

radio antenna *See* antenna. { ˈrād·ē·ō anˈten·ə }

radio attenuation [ELECTROMAG] For one-way propagation, the ratio of the power delivered by the transmitter to the transmission line connecting it with the transmitting antenna to the power delivered to the receiver by the transmission line connecting it with the receiving antenna. { ˈrād·ē·ō əˌten·yəˈwā·shən }

radio aurora *See* artificial radio aurora. { ˈrād·ē·ō əˈrȯr·ə }

radio beacon [NAV] A nondirectional radio transmitting station in a fixed geographic location, emitting a characteristic signal from which bearing information can be obtained by a radio direction finder on a ship or aircraft. Also known as aerophare; radiophare. { ˈrād·ē·ō ˈbē·kən }

radio-beacon monitor station [COMMUN] A station which monitors the signal from one or more remotely located marine radio beacons. { ˈrād·ē·ō ¦bē·kən ˈman·ə·tər ˌstā·shən }

radio bearing [NAV] The bearing of a radio transmitter from a receiver as determined by a radio direction finder. { ˈrād·ē·ō ˌber·iŋ }

radio blackout [COMMUN] A fadeout that may last several hours or more at a particular frequency. Also known as blackout. { ˈrād·ē·ō ˈblak ˌau̇t }

radio broadcasting [COMMUN] Radio transmission intended for general reception. { ˈrād·ē·ō ˈbrȯdˌkast·iŋ }

radio button [COMPUT SCI] In a graphical user interface, one of a group of small circles that represent a set of choices (indicated by text next to the circles) from which only one can be selected; the selected choice is indicated by a partly filled circle. { ˈrād·ē·ō ˌbət·ən }

radio communication [COMMUN] Communication by means of radio waves. { ˈrād·ē·ō kə ˌmyü·nəˈkā·shən }

radiocommunication service [COMMUN] A service involving the emission, transmission, or reception of radio waves for specific telecommunications purposes. { ˌrād·ē·ō·kəˌmyü·nə ˈkā·shən ˌsər·vəs }

radio communications guard *See* radio guard. { ˈrād·ē·ō kəˌmyü·nəˈkā·shənz ˌgärd }

radio data system [COMMUN] The radio data system (RDS) signal is a low-bit-rate data stream transmitted on the 57-kHz subcarrier of an FM radio signal. Radio listeners know that radio data system through its ability to permit RDS radios to display call letters and search for stations based on their programming format. Special traffic announcements can be transmitted to RDS radios, as well as emergency alerts. { ˈrād·ē·ō ˈdad·ə ˌsis·təm }

radio deception [COMMUN] The use of radio signals to deceive an enemy. { ˈrād·ē·ō di ˈsep·shən }

radio detection and ranging *See* radar. { ˈrād·ē·ō diˈtek·shən ən ˈrānj·iŋ }

radiodetermination satellite service [COMMUN] A system that employs at least two geosynchronous satellites, a central ground station, and hand-held or vehicle-mounted transceivers to enable users to determine and transmit their precise position. Abbreviated RDSS. { ˈrād·ē·ō diˌtər·məˈnā·shən ˈsad·əlˌīt ˌsər·vəs }

radio direction finder [NAV] A radio aid to navigation that uses a rotatable loop or other highly directional antenna arrangement to determine the direction of arrival of a radio signal. Abbreviated RDF. Also known as direction finder. { ˈrād·ē·ō diˈrek·shən ˌfīn·dər }

radio direction-finder station [COMMUN] A land-based radio station equipped with special apparatus for determining the direction of radio signals transmitted by ships and other stations. { ˈrād·ē·ō diˈrek·shən ˌfīn·dər ˌstā·shən }

radio duct [GEOPHYS] An atmospheric layer, typically shallow and almost horizontal, in which radio waves propagate in an anomalous fashion; ducts occur when, due to sharp inversions of temperature or humidity, the vertical gradient of the radio index of refraction exceeds a critical value. { 'rād·ē·ō ,dəkt }

radio echo *See* radar return. { 'rād·ē·ō ,ek·ō }

radio engineering [ENG] The field of engineering that deals with the generation, transmission, and reception of radio waves and with the design, manufacture, and testing of associated equipment. { 'rād·ē·ō ,en·jə'nir·iŋ }

radio facsimile system [COMMUN] A facsimile system in which signals are transmitted by radio rather than by wire. { 'rād·ē·ō fak'sim·ə·lē ,sis·təm }

radio fadeout [COMMUN] Increased absorption of radio waves passing through the lower layers of the ionosphere due to a sudden and abnormal increase in ionization in these regions; signals at receivers then fade out or disappear. { 'rād·ē·ō 'fād,au̇t }

radio fan-marker beacon *See* fan-marker beacon. { 'rād·ē·ō 'fan ,mär·kər ,bē·kən }

radio fix [COMMUN] Determination of the position of the source of radio signals by obtaining cross bearings on the transmitter with two or more radio direction finders in different locations, then computing the position by triangulation. [NAV] **1.** Determination of the position of a vessel or aircraft equipped with direction-finding equipment by ascertaining the direction of radio signals received from two or more transmitting stations of known location and then computing the position by triangulation. **2.** Determination of position of an aircraft in flight by identification of a radio beacon or by locating the intersection of two radio beams. { 'rād·ē·ō ,fiks }

radio-frequency amplifier [ELECTR] An amplifier that amplifies the high-frequency signals commonly used in radio communications. { 'rād·ē·ō ¦frē·kwən·sē 'am·plə,fī·ər }

radio-frequency bandwidth [COMMUN] Band of frequencies comprising 99% of the total radiated power of the signal transmission extended to include any discrete frequency on which the power is at least 0.25% of the total radiated power. { 'rād·ē·ō ¦frē·kwən·sē 'band,width }

radio-frequency cable [ELECTROMAG] A cable having electric conductors separated from each other by a continuous homogeneous dielectric or by touching or interlocking spacer beads; designed primarily to conduct radio-frequency energy with low losses. Also known as RG line. { 'rād·ē·ō ¦frē·kwən·sē ,kā·bəl }

radio-frequency component [COMMUN] Portion of a signal or wave which consists only of the radio-frequency alternations, and not including its audio rate of change in amplitude frequency. { 'rād·ē·ō ¦frē·kwən·sē kəm,pō·nənt }

radio-frequency filter [ELECTR] An electric filter which enhances signals at certain radio frequencies or attenuates signals at undesired radio frequencies. { 'rād·ē·ō ¦frē·kwən·sē ,fil·tər }

radio-frequency head [ENG] Unit consisting of a radar transmitter and part of a radar receiver, the two contained in a package for ready removal and installation. { 'rād·ē·ō ¦frē·kwən·sē 'hed }

radio-frequency interference [COMMUN] Interference from sources of energy outside a system or systems, as contrasted to electromagnetic interference generated inside systems. Abbreviated RFI. { 'rād·ē·ō ¦frē·kwən·sē ,in·tər'fir·əns }

radio-frequency oscillator [ELECTR] An oscillator that generates alternating current at radio frequencies. { 'rād·ē·ō ¦frē·kwən·sē 'äs·ə,lād·ər }

radio-frequency pulse [COMMUN] A radio-frequency carrier that is amplitude-modulated by a pulse; the amplitude of the modulated carrier is zero before and after the pulse. Also known as radio pulse. { 'rād·ē·ō ¦frē·kwən·sē ,pəls }

radio-frequency shift *See* frequency shift. { 'rād·ē·ō ¦frē·kwən·sē ,shift }

radio-frequency signal generator [ELECTR] A test instrument that generates the various radio frequencies required for alignment and servicing of electronics equipment. Also known as service oscillator. { 'rād·ē·ō ¦frē·kwən·sē 'sig·nəl ,jen·ə ,rād·ər }

radio-frequency spectrum *See* radio spectrum. { 'rād·ē·ō ¦frē·kwən·sē 'spek·trəm }

radio guard [COMMUN] Ship, aircraft, or radio station designed to listen for and record transmission, and to handle traffic on a designated frequency for a certain unit or units. Also known as radio communications guard. { 'rād·ē·ō ,gärd }

radio homing beacon *See* homing beacon. { 'rād·ē·ō 'hōm·iŋ ,bē·kən }

radio horizon [COMMUN] The locus of points at which direct rays from a transmitter become tangential to the surface of the earth; the distance to the radio horizon is affected by atmospheric refraction. { 'rād·ē·ō hə'rīz·ən }

radio intelligence [COMMUN] Military information regarding an enemy obtained by interception and interpretation of enemy radio transmissions. { 'rād·ē·ō in'tel·ə·jəns }

radio interception [COMMUN] Tuning in on a radio message not intended for the listener. { 'rād·ē·ō ,in·tər'sep·shən }

radio interference *See* interference. { 'rād·ē·ō ,in·tər'fir·əns }

radio landing beam [NAV] A radio beam used for vertical guidance of aircraft during descent to a landing surface. { 'rād·ē·ō 'land·iŋ ,bēm }

radio link [COMMUN] A radio system used to provide a communication or control channel between two specific points. { 'rād·ē·ō ,liŋk }

radiolocation [ENG] Determination of relative position of an object by means of equipment operating on the principle that propagation of radio waves is at a constant velocity and rectilinear. { ¦rād·ē·ō·lō'kā·shən }

radio log [COMMUN] A log of radio messages sent and received, together with other pertinent information, maintained by radio operators. { 'rād·ē·ō ,läg }

radio mast [ENG] A tower, pole, or other structure for elevating an antenna. { 'rād·ē·ō 'mast }

radio-meteorograph [COMMUN] A device for the automatic radio transmission of the indications of a set of meteorological instruments. { 'rā d·ē·ō ˌmēd·ē'ȯr·əˌgraf }

radio mirage [ELECTROMAG] The detection of radar targets at phenomenally long range due to radio ducting. { ¦rād·ē·ō mə'räzh }

radio net [COMMUN] System of radio stations operating with each other; a military net usually consists of a radio station of a superior unit and stations of all subordinate or supporting units. { 'rād·ē·ō ˌnet }

radio-paging system [COMMUN] A system consisting of personal paging receivers, radio transmitters, and an encoding device, designed to alert an individual, or group of individuals, and deliver a short message. { 'rād·ē·ō ¦pāj·iŋ ˌsis·təm }

radiophare *See* radio beacon. { 'rād·ē·ōˌfer }

radiophone *See* radiotelephone. { 'rād·ē·ō ˌfōn }

radiophoto *See* facsimile. { ¦rād·ē·ō'fōd·ō }

radio position finding [ENG] Process of locating a radio transmitter by plotting the intersection of its azimuth as determined by two or more radio direction finders. { 'rād·ē·ō pə'zish·ən ˌfīnd·iŋ }

radio-positioning land station [COMMUN] Station in the radiolocation service, other than a radio-navigation station, not intended for operation while in motion. { 'rād·ē·ō pə¦zish·ən·iŋ 'land ˌstā·shən }

radio-positioning mobile station [COMMUN] Station in the radiolocation service, other than a radio-navigation station, intended to be used while in motion or during halts at unspecified points. { 'rād·ē·ō pə¦zish·ən·iŋ 'mō·bəl ˌstā·shən }

radio pulse *See* radio-frequency pulse. { 'rād·ē·ō ˌpəls }

radio range [NAV] A radio facility emitting signals that, when received by appropriate companion equipment, provide a direct indication of the bearing of the facility from the vehicle. Also known as range. { 'rad·ē·ō ˌrānj }

radio receiver [ELECTR] A device that converts radio waves into intelligible sounds or other perceptible signals. Also known as radio; radio set; receiving set. { 'rād·ē·ō riˌsēv·ər }

radio recognition [COMMUN] Determination by radio means of the friendly or enemy character, or the individuality, of another radio station. { 'rād·ē·ō ˌrek·ig'nish·ən }

radio relay satellite *See* communications satellite. { 'rād·ē·ō 'rēˌlā ˌsad·əlˌīt }

radio relay system [COMMUN] A radio transmission system in which intermediate radio stations or radio repeaters receive and retransmit radio signals. Also known as relay system. { 'rād·ē·ō 'rēˌlā ˌsis·təm }

radio repeater [COMMUN] A repeater that acts as an intermediate station in transmitting radio communications signals or radio programs from one fixed station to another; serves to extend the reliable range of the originating station; a microwave repeater is an example. { 'rād·ē·ō ri ˌpēd·ər }

radio scanner *See* scanning radio. { 'rād·ē·ō 'skan·ər }

radio scattering *See* scattering. { 'rād·ē·ō 'skad·ə·riŋ }

radio set *See* radio transmitter. { 'rād·ē·ō ˌset }

radio signal [COMMUN] A signal transmitted by radio. { 'rād·ē·ō ˌsig·nəl }

radio-signal reporting code [COMMUN] A code for reporting the quality of radiotelephone or radiotelegraph transmission, consisting of a code word followed by a group of numbers rating various characteristics. Also known as signal reporting code. { 'rād·ē·ō ˌsig·nəl ri'pȯrd·iŋ ˌkōd }

radio silence [COMMUN] Period during which all or certain radio equipment capable of radiation is kept inoperative. { 'rād·ē·ō 'sī·ləns }

radio sonobuoy *See* sonobuoy. { 'rād·ē·ō 'sän·əˌbȯi }

radio spectrum [COMMUN] The entire range of frequencies in which useful radio waves can be produced, extending from the audio range to about 300,000 megahertz. Also known as radio-frequency spectrum. { 'rād·ē·ō 'spek·trəm }

radio spectrum allocation [COMMUN] The specification of the frequencies of the radio spectrum which are available for use by the various radio services. { 'rād·ē·ō ¦spek·trəm ˌal·ə'kā·shən }

radio station [COMMUN] A station equipped to engage in radio communication or radio broadcasting. { 'rād·ē·ō ˌstā·shən }

radiotelemetry [COMMUN] The reception of data at a location remote from the source of the data, using radio-frequency electromagnetic radiation as the means of transmission. { 'rād·ē·ō·tə'lem·ə·trē }

radiotelephone [COMMUN] **1.** Pertaining to telephony over radio channels. **2.** A radio transmitter and radio receiver used together for two-way telephone communication by radio. Also known as radiophone. { ¦rād·ē·ō'tel·əˌfōn }

radiotelephony [COMMUN] Two-way transmission of sounds by means of modulated radio waves, without interconnecting wires. { ¦rād·ē·ō·tə'lef·ə·nē }

radio time signal [COMMUN] A time signal sent by radio broadcast. { 'rād·ē·ō 'tīm ˌsig·nəl }

radio tower [COMMUN] A tower, usually several hundred meters tall, either guyed or freestanding, on which a transmitting antenna is mounted to increase the range of radio transmission; in some cases, the tower itself may be the antenna. { 'rād·ē·ō ˌtau̇·ər }

radio tracking [ENG] The process of keeping a radio or radar beam set on a target and determining the range of the target continuously. { 'rād·ē·ō 'trak·iŋ }

radio transmission [COMMUN] The transmission of signals through space at radio frequencies by means of radiated electromagnetic waves. { 'rād·ē·ō tranz'mish·ən }

radio transmitter [ELECTR] The equipment used for generating and amplifying a radio-frequency carrier signal, modulating the carrier signal with intelligence, and feeding the modulated carrier to an antenna for radiation into space as electromagnetic waves. Also known as radio set; transmitter. { 'rād·ē·ō 'tranz,mid·ər }

radio transponder [ELECTR] A transponder which receives and transmits radio waves. { 'rād·ē·ō tran'spän·dər }

radio tube *See* electron tube. { 'rād·ē·ō ,tüb }

radio watch *See* watch. { 'rād·ē·ō ,wäch }

radio wave [ELECTROMAG] An electromagnetic wave produced by reversal of current in a conductor at a frequency in the range from about 10 kilohertz to about 300,000 megahertz. { 'rād·ē·ō ,wāv }

radix *See* root. { 'rād·iks }

radix transformation [COMPUT SCI] A method of transformation that involves changing the radix or base of the original key and either discarding excess high-order digits (that is, digits in excess of the number desired in the key) or extracting some part of the transformed number. { 'rād·iks ,tranz·fər'mā·shən }

radome [ELECTROMAG] A strong, thin shell, made from a dielectric material that is transparent to radio-frequency radiation, and used to house a radar antenna, or a space communications antenna of similar structure. { 'rā,dōm }

RAID [COMPUT SCI] A group of hard disks that operate together to improve performance or provide fault tolerance and error recovery through data striping, mirroring, and other techniques. Derived from redundant array of inexpensive disks. { rād }

rail-fence jammer *See* continuous-wave jammer. { 'rāl ¦fens ,jam·ər }

railing [ELECTR] Simple radar pulse jamming at high recurrence rates (50 to 150 kilohertz); it results in an image on a radar indicator resembling fence railing. { 'rāl·iŋ }

rain attenuation [COMMUN] Attenuation of radio waves when passing through moisture-bearing cloud formations or areas in which rain is falling; increases with the density of the moisture in the transmission path. { 'rān ə,ten·yə,wā·shən }

rainbow [ELECTR] Technique which applies pulse-to-pulse frequency changing to identifying and discriminating against decoys and chaff. { 'rān,bō }

RAM *See* radar absorbent material; random-access memory. { ram }

Rambus dynamic random-access memory [COMPUT SCI] High-performance memory that can transfer data at rates of 800 megahertz and higher. Abbreviated RDRAM. { ¦ram,bəs dī ¦nam·ik ,ran·dəm 'ak,ses ,mem·rē }

RAM disk *See* RAM drive. { 'ram ,disk }

RAM drive [COMPUT SCI] A portion of a computer's random-access memory (RAM) that is made to simulate a disk drive. Also known as RAM disk. { 'ram ,drīv }

rampage through core [COMPUT SCI] Action of a computer program that writes data in incorrect locations or otherwise alters storage locations improperly, because of a program error. { 'ram ,pāj thrü 'kȯr }

RAM resident [COMPUT SCI] A program that remains stored in a computer's random-access memory (RAM) at all times. Also known as terminate and stay resident (TSR). { ¦ram 'rez·ə·dənt }

random access [COMPUT SCI] **1.** The ability to read or write information anywhere within a storage device in an amount of time that is constant regardless of the location of the information accessed and of the location of the information previously accessed. Also known as direct access. **2.** A process in which data are accessed in nonsequential order and possibly at irregular intervals of time. Also known as single reference. [COMMUN] The process of beginning to read and decode the coded bit stream at an arbitrary point. { 'ran·dəm 'ak,ses }

random-access discrete address [COMMUN] Communications technique in which radio users share one wide band instead of each user getting an individual narrow band. { 'ran·dəm ¦ak,ses di'skrēt ə'dres }

random-access disk file [COMPUT SCI] A file which is contained on a disk having one head per track and in which consecutive records are not necessarily in consecutive locations. { 'ran·dəm ¦ak,ses 'disk ,fīl }

random-access input/output [COMPUT SCI] A technique which minimizes seek time and overlaps with processing. { 'ran·dəm ¦ak,ses 'in,pu̇t 'au̇t,pu̇t }

random-access memory [COMPUT SCI] A data storage device having the property that the time required to access a randomly selected datum does not depend on the time of the last access or the location of the most recently accessed datum. Abbreviated RAM. Also known as direct-access memory; direct-access storage; random-access storage; random storage; uniformly accessible storage. { 'ran·dəm ¦ak,ses 'mem·rē }

random-access programming [COMPUT SCI] Programming without regard for the time required for access to the storage positions called for in the program, in contrast to minimum-access programming. { 'ran·dəm ¦ak,ses 'prō ,gram·iŋ }

random-access storage *See* random-access memory. { 'ran·dəm ¦ak,ses 'stȯr·ij }

randomizing scheme [COMPUT SCI] A technique of distributing records among storage modules to ensure even distribution and seek time. { 'ran·də,mīz·iŋ ,skēm }

random number generator [COMPUT SCI] **1.** A mathematical program which generates a set of numbers which pass a randomness test. **2.** An analog device that generates a randomly fluctuating variable, and usually operates from an electrical noise source. { 'ran·dəm 'nəm·bər ,jen·ə,rād·ər }

random pulsing [COMMUN] Continuous, varying, pulse-repetition rate, accomplished by noise modulation or continuous frequency change. { 'ran·dəm 'pəls·iŋ }

random storage *See* random-access memory. { 'ran·dəm 'stȯr·ij }

random superimposed coding [COMPUT SCI] A system of coding in which a set of random numbers is assigned to each concept to be encoded; with punched cards, each number corresponds to some one hole to be punched in a given field. { 'ran·dəm ¦sü·pər·im'pōzd 'kōd·iŋ }

range [COMMUN] **1.** In printing telegraphy, that fraction of a perfect signal element through which the time of selection may be varied to occur earlier or later than the normal time of selection without causing errors while signals are being received; the range of a printing telegraph receiving device is commonly measured in percent of a perfect signal element by adjusting the indicator. **2.** Upper and lower limits through which the index arm of the range-finder mechanism of a teletypewriter may be moved and still receive correct copy. [ENG] **1.** The distance capability of a radio or radar system. **2.** A line defined by two fixed landmarks, used for missile or vehicle testing and other test purposes. **3.** In radar measurement, the distance to a target measured usually by the time elapsed between the transmission of a pulse and the receipt of the target's echo. [NAV] *See* radio range. { rānj }

range arithmetic *See* interval arithmetic. { 'rānj ə¦rith·mə·tik }

range attenuation [ELECTROMAG] In radar terminology, the decrease in power density (flux density) caused by the divergence of the flux lines with distance, this decrease being in accordance with the inverse-square law. { 'rānj ə,ten·yə¦wā·shən }

range-bearing display *See* B display. { 'rānj 'ber·iŋ di,splā }

range calibrator [ELECTR] **1.** A device with which the operator of a transmitter calculates the distance over which the signal will extend intelligibly. **2.** A device for adjusting radar range indications by use of known range targets or delayed signals; particularly useful in radars using analog echo timing. { 'rānj ¦kal·ə,brād·ər }

range check [COMPUT SCI] A method of checking the validity of input data by determining whether the values fall within an expected range. { 'rānj ,chek }

range delay [ELECTROMAG] A control used in radars which permits the operator to present on the radarscope only those echoes from targets which lie beyond a certain distance from the radar; by using range delay, undesired echoes from nearby targets may be eliminated while the indicator range is increased. { 'rānj di,lā }

range discrimination *See* distance resolution. { 'rānj di,skrim·ə¦nā·shən }

rangefinder [COMMUN] A movable, calibrated unit of the receiving mechanism of a teletypewriter by means of which the selecting interval may be moved with respect to the start signal. [ELECTR] A device which determines the distance to an object by measuring the time it takes for a radio wave to travel to the object and return. { 'rānj,fīnd·ər }

range gate [ELECTR] A gate voltage that is used to select radar echoes from a very narrow interval of ranges. { 'rānj ¦gāt }

range gating [ELECTR] The process of selecting, for further use, only those radar echoes that lie within a small interval of ranges. { 'rānj ,gād·iŋ }

range-height indicator display [ELECTR] A radar display showing the distance between a reference point, usually the radar, and a target, along with the vertical distance between a horizontal reference plane, usually containing the radar, and the target. Abbreviated RHI. { 'rānj 'hīt 'in·də,kād·ər di,splā }

range-imaging sensor [ENG] A robotic device that makes precise measurements, by using the principles of algebra, trigonometry, and geometry, of the distance from a robot's end effector to various parts of an object, in order to form an image of the object. { 'rānj ¦im·ij·iŋ ,sen·sər }

range marker *See* distance marker. { 'rānj ,mär·kər }

range of a loop [COMPUT SCI] The set of instructions contained between the opening and closing statements of a do loop. { 'rānj əv ə 'lüp }

range rate [ELECTR] The rate at which the distance from the measuring equipment to the target or signal source that is being tracked is changing with respect to time. { 'rānj ,rāt }

range resolution *See* distance resolution. { 'rānj ,rez·ə,lü·shən }

range ring [ELECTR] Accurate, adjustable ranging mark on a plan position indicator; such marks at set range intervals are displayed as concentric rings as the display is generated. { 'rānj ,riŋ }

range strobe [ELECTROMAG] An index mark which may be displayed on various types of radar indicators to assist in the determination of the exact range of a target. { 'rānj ,strōb }

range-tracking element [ELECTR] An element in a radar set that measures range and its time derivative, by means of which a range gate is actuated slightly before the predicted instant of signal reception. { 'rānj ¦trak·iŋ ,el·ə·mənt }

range unit [ELECTR] Radar system component used for control and indication (usually counters) of range measurements. { 'rānj ,yü·nət }

range zero [ELECTR] Alignment of start sweep trace with zero range. { 'rānj ,zir·ō }

rapid access loop [COMPUT SCI] A small section of storage, particularly in drum, tape, or disk storage units, which has much faster access than the remainder of the storage. { 'rap·əd ¦ak,ses ,lüp }

rapid memory *See* rapid storage. { 'rap·əd 'mem·rē }

rapid selector [COMPUT SCI] A device which scans codes recorded on microfilm; microimages of the documents associated with the codes may also be recorded on the film. { 'rap·əd si'lek·tər }

rapid storage [COMPUT SCI] In computers, storage with a very short access time; rapid access is generally gained by limiting storage capacity. Also known as high-speed storage; rapid memory. { 'rap·əd 'stór·ij }

rare-earth-doped fiber amplifier [COMMUN] An optical fiber amplifier whose fiber core is lightly doped with trivalent rare-earth ions, which absorb light at certain pump wavelengths and emit it at some signal wavelength through stimulated emission. { ¦rār ˌərth ˌdōpt ˌfī·bər 'am·pləˌfī·ər }

raster [ELECTR] A predetermined pattern of scanning lines that provides substantially uniform coverage of an area; in video the raster is seen as closely spaced parallel lines, most evident when there is no picture. { 'ras·tər }

raster graphics [COMPUT SCI] A computer graphics coding technique which codes each picture element of the picture area in digital form. Also known as bit-mapped graphics. { 'ras·tər ¦graf·iks }

rasterization [COMPUT SCI] The conversion of graphics objects composed of vectors or line segments into dots for transmission to raster graphics displays and to dot matrix and laser printers. { ˌras·tə·rə'zā·shən }

rate action *See* derivative action. { 'rāt ˌak·shən }

rate control [CONT SYS] A form of control in which the position of a controller determines the rate or velocity of motion of a controlled object. Also known as velocity control. { 'rāt kənˌtrōl }

rated speed [COMPUT SCI] The maximum operating speed that can be sustained by a data-processing device or communications line, not allowing for periodic pauses for various reasons such as carriage return on a printer. { 'rād·əd 'spēd }

rate multiplier [COMPUT SCI] An integrator in which the quantity to be integrated is held in a register and is added to the number standing in an accumulator in response to pulses which arrive at a constant rate. { 'rāt ˌməl·təˌplī·ər }

rate test [COMPUT SCI] A test that verifies that the time constants of the integrators are correct; used in analog computers. { 'rāt ˌtest }

ratio detector [ELECTR] A frequency-modulation detector circuit that uses two diodes and requires no limiter at its input; the audio output is determined by the ratio of two developed intermediate-frequency voltages whose relative amplitudes are a function of frequency. { 'rā·shō diˌtek·tər }

ratio deviation *See* modulation index. { 'rā·shō ˌdē·vē'ā·shən }

rat race [ELECTR] A hybrid network in the form of a ring in microwave circuitry. { 'rat ˌrās }

Rayleigh video [ELECTR] Referring to the video and its particular probability density produced by an amplitude detector (demodulato) when a Gaussian radio noise is incident to it. { 'rā·lē ¦vid·ē·ō }

ray path [COMMUN] Geometric path between signal transmitting and receiving locations. { 'rā ˌpath }

ray tracing [COMPUT SCI] The creation of reflections, refractions, and shadows in a graphics image by following a series of rays from a light source and determining the effect of light on each pixel in the image. { 'rā ˌtrās·iŋ }

R-DAT system *See* rotary digital audio tape system. { 'är ˌdat ˌsis·təm *or* ¦är ¦dē¦ā'tē ˌsis·təm }

RDF *See* radio direction finder.

R-display [ELECTR] A radar display format in which only the display around a target of interest is expanded in range in an A-display format, to improve the accuracy of range estimation and to permit closer examination of the target signal. Also known as R-indicator; R-scan; R-scope. { 'är diˌsplā }

RDRAM *See* Rambus dynamic random-access memory. { ¦är¦dē'ram }

RDS *See* radio data system.

RDSS *See* radiodetermination satellite service.

reactance amplifier *See* parametric amplifier. { rē'ak·təns 'am·pləˌfī·ər }

reaction *See* positive feedback. { rē'ak·shən }

read [COMPUT SCI] **1.** To acquire information, usually from some form of storage in a computer. **2.** To convert magnetic spots, characters, or punched holes into electrical impulses. { rēd }

read-around number *See* read-around ratio. { 'rēd əˌraund ˌnəm·bər }

read-around ratio [COMPUT SCI] The number of times that a particular bit in electrostatic storage may be read without seriously affecting nearby bits. Also known as read-around number. { 'rēd əˌraund ˌrā·shō }

read-back check *See* echo check. { 'rēd ˌbak ˌchek }

reader [COMPUT SCI] A device that converts information from one form to another, as from punched paper tape to magnetic tape. { 'rēd·ər }

reader-interpreter [COMPUT SCI] A service routine that reads an input string, stores programs and data on random-access storage for later processing, identifies the control information contained in the input string, and stores this control information separately in the appropriate control lists. { 'rēd·ər in'tər·prəd·ər }

read error [COMPUT SCI] A condition in which the content of a storage device cannot be electronically identified. { 'rēd ˌer·ər }

read head [COMPUT SCI] A device that converts digital information stored on a magnetic tape, drum, or disk into electrical signals usable by the computer arithmetic unit. { 'rēd ˌhed }

read-in [COMPUT SCI] To sense information contained in some source and transmit this information to an internal storage. { 'rēd ¦in }

readiness review [COMPUT SCI] An on-site examination of the adequacy of preparations for effective utilization upon installation of a computer, and to identify any necessary corrective actions. { 'red·i·nəs riˌvyü }

reading rate [COMPUT SCI] Number of characters, words, or fields sensed by an input sensing device per unit of time. { 'rēd·iŋ ˌrāt }

read-in program [COMPUT SCI] Computer program that can be put into a computer in a simple binary form and allows other programs to be read into the computer in more complex forms. { 'rēd ˌin ˌprō·grəm }

read-only memory [COMPUT SCI] A device for storing data in permanent, or nonerasable, form; usually an optical, static electronic, or magnetic device allowing extremely rapid access to data. Abbreviated ROM. Also known as nonerasable storage; read-only storage. { 'rēd ¦ōn·lē 'mem·rē }

read-only storage See read-only memory. { 'rēd ¦ōn·lē 'stȯr·ij }

read-only terminal [COMPUT SCI] A peripheral device, such as a printer, that can only receive signals. { 'rēd ¦ōn·lē 'tər·mən·əl }

readout [COMPUT SCI] **1.** The presentation of output information by means of lights a display, printout, or other methods. **2.** To sense information contained in some computer internal storage and transmit this information to a storage external to the computer. { 'rēd,au̇t }

readout station [COMMUN] A recording or receiving radio station at which data are received. { 'rēd,au̇t ,stā·shən }

read screen [COMPUT SCI] In optical character recognition (OCR), the transparent component part of most character readers through which appears the input document to be recognized. { 'rēd ,skrēn }

read time [COMPUT SCI] The time interval between the instant at which information is called for from storage and the instant at which delivery is completed in a computer. { 'rēd ,tīm }

read-while-writing [COMPUT SCI] The reading of a record or group of records into storage from tape at the same time another record or group of records is written from storage to tape. { 'rēd ,wīl 'rīd·iŋ }

read/write channel [COMPUT SCI] A path along which information is transmitted between the central processing unit of a computer and an input, output, or storage unit under the control of the computer. { 'rēd 'rīt ,chan·əl }

read/write check indicator [COMPUT SCI] A device incorporated in certain computers to indicate upon interrogation whether or not an error was made in reading or writing; the machine can be made to stop, retry the operation, or follow a special subroutine, depending upon the result of the interrogation. { 'rēd 'rīt 'chek ,in·də,kād·ər }

read/write comb [COMPUT SCI] The set of arms mounted with magnetic heads that reach between the disks of a disk storage device to read and record information. { 'rēd 'rīt ,kōm }

read/write head [COMPUT SCI] A magnetic head that both senses and records data. Also known as combined head. { 'rēd 'rīt ,hed }

read/write memory [COMPUT SCI] A computer storage in which data may be stored or retrieved at comparable intervals. { 'rēd 'rīt ,mem·rē }

read/write random-access memory [COMPUT SCI] A random access memory in which data can be written into memory as well as read out of memory. { 'rēd 'rīt 'ran·dəm 'ak,ses ,mem·rē }

ready-to-receive signal [COMMUN] Signal sent back to a facsimile transmitter to indicate that a facsimile receiver is ready to accept the transmission. { 'red·ē tə ri'sēv ,sig·nəl }

real data type [COMPUT SCI] A scalar data type which contains a normalized fraction (mantissa) and an exponent (characteristic) and is used to represent floating-point data, usually decimal. { 'rēl 'dad·ə ,tīp }

realizability [CONT SYS] Property of a transfer function that can be realized by a network that has only resistances, capacitances, inductances, and ideal transformers. { ,rē·ə,līz·ə'bil·əd·ē }

real storage [COMPUT SCI] Actual physical storage of data and instructions. { 'rēl 'stȯr·ij }

real-time [COMPUT SCI] Pertaining to a data-processing system that controls an ongoing process and delivers its outputs (or controls its inputs) not later than the time when these are needed for effective control; for instance, airline reservations booking and chemical processes control. { 'rēl ,tīm }

real-time clock [COMPUT SCI] A pulse generator which operates at precise time intervals to determine time intervals between events and initiate specific elements of processing. { 'rēl ,tīm 'kläk }

real-time control system [COMPUT SCI] A computer system which controls an operation in real time, such as a rocket flight. { 'rēl ,tīm kən'trōl ,sis·təm }

real-time operation [COMPUT SCI] **1.** Of a computer or system, an operation or other response in which programmed responses to an event are essentially simultaneous with the event itself. **2.** An operation in which information obtained from a physical process is processed to influence or control the physical process. { 'rēl ,tīm ,äp·ə'rā·shən }

real-time processing [COMPUT SCI] The handling of input data at a rate sufficient to ensure that the instructions generated by the computer will influence the operation under control at the required time. { 'rēl ,tīm 'prä,ses·iŋ }

real-time programming [COMPUT SCI] Programming for a situation in which results of computations will be used immediately to influence the course of ongoing physical events. { 'rēl ,tīm 'prō,gram·iŋ }

real-time system [COMPUT SCI] A system in which the computer is required to perform its tasks within the time restraints of some process or simultaneously with the system it is assisting. { 'rēl ¦tīm 'sis·təm }

rear-projection [ELECTR] Pertaining to video system in which the picture is projected on a ground-glass screen for viewing from the opposite side of the screen. { 'rir prə'jek·shən }

reasonableness [COMPUT SCI] A measure of the extent to which data processed by a computer falls within an acceptable allowance for errors, as determined by quantitative tests. { 'rēz·nə·bəl·nəs }

reboot [COMPUT SCI] To reload systems software into a computer so that it makes a new start. { rē'büt }

rebroadcast [COMMUN] Repetition of a radio or television program at a later time. { rē'brōd ,kast }

recall factor [COMPUT SCI] A measure of the efficiency of an information retrieval system, equal to the number of retrieved relevant documents divided by the total number of relevant documents in the file. { 'rē,kȯl ,fak·tər }

received power [ELECTROMAG] **1.** The total power received at an antenna from a signal, such as a radar target signal. **2.** In a mobile communications system, the root-mean-square value of power delivered to a load which properly terminates an isotropic reference antenna. { ri'sēvd 'pau̇·ər }

receive-only [COMMUN] A teleprinter which has no keyboard, and thus can receive but not transmit. Abbreviated RO. { ri'sēv 'ōn·lē }

receiver [ELECTR] The complete equipment required for receiving modulated radio waves and converting them into the original intelligence, such as into sounds or pictures, or converting to desired useful information as in a radar receiver. { ri'sē·vər }

receiver bandwidth [ELECTR] Spread, in frequency, between the halfpower points on the receiver response curve. { ri'sē·vər 'band,width }

receiver gating [ELECTR] Application of operating voltages to one or more stages of a receiver only during that part of a cycle of operation when reception is desired. { ri'sē·vər ,gād·iŋ }

receiver incremental tuning [ELECTR] Control feature to permit receiver tuning (of a transceiver) up to 3 kilohertz to either side of the transmitter frequency. { ri'sē·vər ,in·krə,ment·əl 'tün·iŋ }

receiver lockout system *See* lockout. { ri'sē·vər 'läk,au̇t ,sis·təm }

receiver noise threshold [ELECTR] External noise appearing at the front end of a receiver, plus the noise added by the receiver itself, whichdetermines a noise threshold that has to be exceeded by the minimum discernible signal. { ri'sē·vər 'nȯiz ,thresh,hōld }

receiver radiation [ELECTROMAG] Radiation of interfering electromagnetic fields by the oscillator of a receiver. { ri'sē·vər ,rād·ē'ā·shən }

receiving antenna [ELECTROMAG] An antenna used to convert electromagnetic waves to modulated radio-frequency currents. { ri'sēv·iŋ an ,ten·ə }

receiving area [ELECTROMAG] The factor by which the power density must be multiplied to obtain the received power of an antenna, equal to the gain of the antenna times the square of the wavelength divided by 4π. { ri'sēv·iŋ ,er·ē·ə }

receiving loop loss [COMMUN] In telephones, that part of the repetition equivalent assignable to the station set, subscriber line, and battery supply circuit that are on the receiving end. { ri'sēv·iŋ ,lüp ,lȯs }

receiving set *See* radio receiver. { ri'sēv·iŋ ,set }

reception [COMMUN] The conversion of modulated electromagnetic waves or electric signals, transmitted through the air or over wires or cables, into the original intelligence, or into desired useful information (as in radar), by means of antennas and electronic equipment. { ri'sep·shən }

reciprocal ferrite switch [ELECTROMAG] A ferrite switch that can be inserted in a waveguide to switch an input signal to either of two output waveguides; switching is done by a Faraday rotator when acted on by an external magnetic field. { ri'sip·rə·kəl 'fe,rīt ,swich }

reciprocity theorem [ELECTROMAG] Given two loop antennas, a and b, then $I_{ab}/V_a = I_{ba}/V_b$, where I_{ab} denotes the current received in b when a is used as transmitter, and V_a denotes the voltage applied in a; I_{ba} and V_b are the corresponding quantities when b is the transmitter, a the receiver; it is assumed that the frequency and impedances remain unchanged. Also known as principle of reciprocity. { ,res·ə'präs·əd·ē ,thir·əm }

reclaimer [COMPUT SCI] A device that performs dynamic storage allocation, periodically searching memory to locate cells whose contents are no longer useful for computation, and making them available for other uses. { rē'klām·ər }

recognition [COMPUT SCI] The act or process of identifying (or associating) an input with one of a set of possible known alternatives, as in character recognition and pattern recognition. { ,rek·ig'nish·ən }

recognition gate [COMPUT SCI] A logic circuit used to select devices identified by a binary address code. Also known as decoding gate. { ,rek·ig'nish·ən ,gāt }

reconditioned carrier reception [ELECTR] Method of reception in which the carrier is separated from the sidebands to eliminate amplitude variations and noise, and is then added at an increased level to the sideband, to obtain a relatively undistorted output. { ,rē·kən'dish·ənd 'kar·ē·ər ri,sep·shən }

reconstitution [COMPUT SCI] The conversion of tokens back to the keywords they represent in a programming language, before generation of the output of an interpreted program. { rē ,kän·stə'tü·shən }

record [COMPUT SCI] A group of adjacent data items in a computer system, manipulated as a unit. Also known as entity. { 'rek·ərd }

record block *See* physical record. { 'rek·ərd ,bläk }

record density *See* bit density; character density. { 'rek·ərd ,den·səd·ē }

record gap [COMPUT SCI] An area in a storage medium, such as magnetic tape or disk, which is devoid of information; it delimits records, and, on tape, allows the tape to stop and start between records without loss of data. Also known as interrecord gap (IRG). { 'rek·ərd ,gap }

record head *See* recording head. { ri'kȯrd ,hed }

recording density [COMPUT SCI] The amount of data that can be stored in a unit length of magnetic tape, usually expressed in bits per inch or characters per inch. { ri'kȯrd·iŋ ,den·səd·ē }

recording head [ELECTR] A magnetic head used only for recording. Also known as record head. { ri'kȯrd·iŋ ,hed }

recording spot *See* picture element. { ri'kȯrd·iŋ ,spät }

record layout [COMPUT SCI] A form showing how fields are positioned within a record, usually with information about each field. { 'rek·ərd ,lā ,au̇t }

record length [COMPUT SCI] The number of characters required for all the information in a record. { 'rek·ərd ,leŋkth }

record locking [COMPUT SCI] Action of a computer system that makes a record that is being processed by one user unavailable to other users, to prevent more than one user from attempting to update the same information simultaneously. { 'rek·ərd ,läk·iŋ }

record mark [COMPUT SCI] A symbol that signals a record's beginning or end. { 'rek·ərd ,märk }

record variable [COMPUT SCI] A group of related but dissimilar data items that can be worked on as a single unit. Also known as structured variable. { 'rek·ərd ,ver·ē·ə·bəl }

recovery interrupt [COMPUT SCI] A type of interruption of program execution which provides the computer with access to subroutines to handle an error and, if successful, to continue with the program execution. { ri'kəv·ə·rē 'int·ə,rəpt }

recovery routine [COMPUT SCI] A computer routine that attempts to resolve automatically conditions created by errors, without causing the computer system to shut down or otherwise do serious damage. { ri'kəv·ə·rē rü,tēn }

recovery system [COMPUT SCI] A system for recognizing a malfunction in a database management system, reporting it, reconstructing the damaged part of the database, and resuming processing. { ri'kəv·ə·rē ,sis·təm }

recovery time [ELECTR] **1.** The time required for the control electrode of a gas tube to regain control after anode-current interruption. **2.** The time required for a fired TR (transmit-receive) or pre-TR tube to deionize to such a level that the attenuation of a low-level radio-frequency signal transmitted through the tube is decreased to a specified value. **3.** The time required for a fired ATR (anti-transmit-receive) tube to deionize to such a level that the normalized conductance and susceptance of the tube in its mount are within specified ranges. **4.** The interval required, after a sudden decrease in input signal amplitude to a system or component, to attain a specified percentage (usually 63%) of the ultimate change in amplification or attenuation due to this decrease. **5.** The time required for a radar receiver to recover to half sensitivity after the end of the transmitted pulse, so it can effectively receive a return echo; a consequence of duplexed operation. { ri'kəv·ə·rē ,tīm }

rectangular scanning [ELECTR] Two-dimensional sector scanning in which a slow sector scanning in one direction is superimposed on a rapid sector scanning in a perpendicular direction. { rek'taŋ·gyə·lər 'skan·iŋ }

rectangular wave [ELECTR] A periodic wave that alternately and suddenly changes from one to the other of two fixed values. Also known as rectangular wave train. { rek'taŋ·gyə·lər 'wāv }

rectangular waveguide [ELECTROMAG] A waveguide having a rectangular cross section. { rek'taŋ·gyə·lər 'wāv,gīd }

rectangular wave train *See* rectangular wave. { rek'taŋ·gyə·lər 'wāv ,trān }

recuperability [COMMUN] Ability to continue to operate after a partial or complete loss of the primary communications facility resulting from sabotage, enemy attack, or other disaster. { rē ,küp·rə'bil·əd·ē }

recurrence rate *See* repetition rate. { ri'kər·əns ,rāt }

recursion [COMPUT SCI] A technique in which an apparently circular process is used to perform an iterative process. { ri'kər·zhən }

recursive filter [ELECTR] A digital filter that has feedback; that is, its output depends not only on present and past input values but on past output values as well. { ri,kər·siv 'fil·tər }

recursive macro call [COMPUT SCI] A call to a macroinstruction already called when used in conjunction with conditional assembly. { ri'kər·siv ¦mak·rō ,kȯl }

recursive procedure [COMPUT SCI] A method of calculating a function by deriving values of it which become more accurate at each step; recursive procedures are explicitly outlawed in most systems with the exception of a few which use languages such as ALGOL and LISP. { ri'kər·siv prə'sē·jər }

recursive subroutine [COMPUT SCI] A reentrant subroutine whose partial results are stacked, with a processor stack pointer advancing and retracting as the subroutine is called and completed. { ri'kər·siv 'səb·rü,tēn }

redefine [COMPUT SCI] A procedure used in certain programming languages to specify different utilizations of the same storage area at different times. { ¦rē·di'fīn }

red-tape operation *See* bookkeeping operation. { 'red ¦tāp ,äp·ə,rā·shən }

reduced instruction set computer [COMPUT SCI] A computer in which the compiler and hardware are interlocked, and the compiler takes over some of the hardware functions of conventional computers and translates high-level-language programs directly into low-level machine code. Abbreviated RISC. { ri¦düst in'strək·shən ,set kəm'pyüd·ər }

reduced-order controller [CONT SYS] A control algorithm in which certain modes of the structure to be controlled are ignored, to enable control commands to be computed with sufficient rapidity. { ri'düst ¦ȯr·dər kən'trōl·ər }

reduced telemetry [COMMUN] Raw telemetry data transformed into a usable form. { ri 'düst tə'lem·ə·trē }

reduction [COMPUT SCI] Any process by which data are condensed, such as changing the encoding to eliminate redundancy, extracting significant details from the data and eliminating the rest, or choosing every second or third out of the totality of available points. { ri'dək·shən }

reduction rule [COMPUT SCI] The principal computation rule in the lambda calculus; it states that an operator-operand combination of the form (λ*x*MA) may be transformed into the expression $S^x{}_A M$, obtained by substituting the lambda expression A for all instances of *x* in M, provided there are no conflicts of variable names. Also known as beta rule. { ri'dək·shən ˌrül }

reductive grammar [COMPUT SCI] A set of syntactic rules for the analysis of strings to determine whether the strings exist in a language. { ri'dək·tiv 'gram·ər }

redundancy [COMMUN] In the transmission of information, the fraction of the gross information content of a message which can be eliminated without loss of essential information. [COMPUT SCI] Any deliberate duplication or partial duplication of circuitry or information to decrease the probability of a system or communication failure. { ri'dən·dən·sē }

redundancy bit [COMPUT SCI] A bit which carries no information but which is added to the information-carrying bits of a character or stream of characters to determine their accuracy. { ri'dən·dən·sē ˌbit }

redundancy check [COMPUT SCI] A forbidden-combination check that uses redundant digits called check digits to detect errors made by a computer. { ri'dən·dən·sē ˌchek }

redundant array of inexpensive disks *See* RAID. { riˌdən·dənt ə¦rā əv ˌin·ikˌspen·siv 'disks }

redundant character [COMPUT SCI] A character specifically added to a group of characters to ensure conformity with certain rules which can be used to detect computer malfunction. { ri'dən·dənt 'kar·ik·tər }

redundant code [COMMUN] A code which uses more signal elements than are needed to represent the information it transmits. { ri'dən·dənt 'kōd }

redundant digit [COMPUT SCI] Digit that is not necessary for an actual computation but serves to reveal a malfunction in a digital computer. { ri'dən·dənt 'dij·it }

Reed-Solomon code [COMMUN] A linear, block-based error-correcting code with wide-ranging applications, which is based on the mathematics of finite fields. { ¦rēd 'säl·ə·mən ˌkōd }

reel number [COMPUT SCI] A number identifying a reel of magnetic tape in a file containing more than one reel and indicating the order in which the reel is to be used. Also known as reel sequence number. { 'rēl ˌnəm·bər }

reel sequence number *See* reel number. { 'rēl 'sē·kwəns ˌnəm·bər }

reenterable [COMPUT SCI] The attribute that describes a program or routine which can be shared by several tasks concurrently. { rē'en·trə·bəl }

reentrant code *See* reentrant program. { rē 'en·trənt ˌkōd }

reentrant program [COMPUT SCI] A subprogram in a time-sharing or multiprogramming system that can be shared by a number of users, and can therefore be applied to a given user program, interrupted and applied to some other user program, and then reentered at the point of interruption of the original user program. Also known as reentrant code. { rē'en·trənt ˌprō ˌgram }

reentry point [COMPUT SCI] The instruction in a computer program at which execution is resumed after the program has jumped to another place. { rē'en·trē ˌpȯint }

reentry system *See* turnaround system. { rē 'en·trē ˌsis·təm }

reference address *See* address constant. { 'ref·rəns 'adˌres }

reference block [COMPUT SCI] A block within a computer program governing a numerically controlled machine which has enough data to allow resumption of the program following an interruption. { 'ref·rəns ˌbläk }

reference burst *See* color burst. { 'ref·rəns ˌbərst }

reference frequency [COMMUN] Frequency having a fixed and specified position with respect to the assigned frequency. { 'ref·rəns ˌfrē·kwən·sē }

reference listing [COMPUT SCI] A list printed by a compiler showing the instructions in the machine language program which it generates. { 'ref·rəns ˌlist·iŋ }

reference monitor [COMPUT SCI] A means of checking that a particular user is allowed access to a specified object in a computing system. Also known as access-control mechanism; reference validation mechanism. { 'ref·rəns ˌmän·əd·ər }

reference record [COMPUT SCI] Output of a compiler that lists the operations and their positions in the final specific routine and contains information describing the segmentation and storage allocation of the routine. { 'ref·rəns ˌrek·ərd }

reference validation mechanism *See* reference monitor. { ¦ref·rəns ˌval·ə'dā·shən ˌmek·ə ˌniz·əm }

reference white [COMMUN] **1.** In a scene viewed by video camera, the color of light from a nonselective diffuse reflector that is lighted by the normal illumination of the scene. **2.** The color by which this color is simulated on a video screen or other display device. { 'ref·rəns ˌwīt }

reference white level [ELECTR] In television, the level at the point of observation corresponding to the specified maximum excursion of the picture signal in the white direction. { 'ref·rəns 'wīt ˌlev·əl }

reflectance [COMPUT SCI] In optical character recognition, the relative brightness of the inked area that forms the printed or handwritten character; distinguished from background reflectance and brightness. { ri'flek·təns }

reflected binary [COMPUT SCI] A particular form ofGray code which is constructed according to the following rule: Let the first 2^N code patterns be given, for any N greater than 1; the next 2^N code patterns are derived by changing the (N + 1)-th bit from the right from 0 to 1 and repeating the original 2^N patterns in reverse order in the N rightmost positions. Also known as reflected code. { ri'flek·təd 'bī,ner·ē }

reflected code *See* reflected binary. { ri'flek·təd 'kōd }

reflecting antenna [ELECTROMAG] An antenna used to achieve greater directivity or desired radiation patterns, in which a dipole, slot, or horn radiates toward a larger reflector which shapes the radiated wave to produce the desired pattern; the reflector may consist of one or two plane sheets, a parabolic or paraboloidal sheet, or a paraboloidal horn. { ri'flek·tiŋ an'ten·ə }

reflecting curtain [ELECTROMAG] A vertical array of half-wave reflecting antennas, generally used one quarter-wavelength behind a radiating curtain of dipoles to form a high-gain antenna. { ri'flek·tiŋ 'kərt·ən }

reflecting grating [ELECTROMAG] Arrangement of wires placed in a waveguide to reflect one desired wave while allowing one or more other waves to pass freely. { ri'flek·tiŋ 'grād·iŋ }

reflection lobes [ELECTROMAG] Three-dimensional sections of the radiation pattern of a directional antenna, such as a radar antenna, which results from reflection of radiation from the earth's surface. { ri'flek·shən ,lōbz }

reflective binary code *See* reflected binary. { ri'flek·tiv 'bī,ner·ē 'kōd }

reflective code *See* Gray code. { ri'flek·tiv 'kōd }

reflective spot [COMPUT SCI] A piece of metallic foil that is embedded in a magnetic tape to indicate the end of a reel. { ri'flek·tiv ,spät }

reflector [ELECTROMAG] **1.** A single rod, system of rods, metal screen, or metal sheet used behind an antenna to increase its directivity. **2.** A metal sheet or screen used as a mirror to change the direction of a microwave radio beam. { ri'flek·tər }

reflexive processing [COMPUT SCI] Information processing in which two or more computers connected by communications channels run identical programs and take the same actions at the same time, so that users in different locations can work on the same programs at the same time. { ri'flek·siv 'prä,ses·iŋ }

reformat [COMPUT SCI] To change the arrangement of data in a storage device. { rē'fór·mat }

refraction [COMMUN] That property of earth's atmosphere that, due to its density profile, causes radio waves to propagate generally with a downward curve, sometimes rivaling the curvature of the earth; in radar height estimation, corrections for estimated refraction must be made. [ELECTROMAG] The change in direction of lines of force of an electric or magnetic field at a boundary between media with different permittivities or permeabilities. { ri'frak·shən }

refraction loss [ELECTROMAG] Portion of the transmission loss that is due to refraction resulting from nonuniformity of the medium. { ri'frak·shən ,lós }

refractive constant *See* index of refraction. { ri 'frak·tiv 'kän·stənt }

refractive index *See* index of refraction. { ri 'frak·tiv ,in,deks }

refresh [COMPUT SCI] A process of periodically replacing data to prevent the data from decaying, as on a cathode-ray-tube display or in a dynamic random-access memory. { ri'fresh }

regeneration *See* positive feedback. { rē,jen·ə'rā·shən }

regenerative feedback *See* positive feedback. { rē'jen·rəd·iv 'fēd,bak }

regenerative read [COMPUT SCI] A read operation in which the data are automatically written back into the locations from which they are taken. { rē'jen·rəd·iv 'rēd }

regenerative repeater [COMMUN] A repeater that performs pulse regeneration to restore the original shape of a pulse signal used in teletypewriter and other code circuits. { rē'jen·rəd·iv ri'pēd·ər }

regenerator [ELECTR] **1.** A circuit that repeatedly supplies current to a display or memory device to prevent data from decaying. **2.** *See* repeater. { rē'jen·ə,rād·ər }

region [COMPUT SCI] A group of machine addresses which refer to a base address. { 'rē·jən }

regional address [COMPUT SCI] An address of a machine instruction within a series of consecutive addresses; for example, R18 and R19 are specific addresses in an R region of N consecutive addresses, where all addresses must be named. { 'rēj·ən·əl ə'dres }

regional center [COMMUN] A long-distance telephone office which has the highest rank in routing of telephone calls. { 'rēj·ən·əl 'sen·tər }

register [COMMUN] **1.** The accurate matching or superimposition of two or more images, such as the three color images on the screen of a color display. **2.** The alignment of positions relative to a specified reference or coordinate, such as hole alignments in punched cards, or positioning of images in an optical character recognition device. **3.** Part of an automatic switching telephone system that receives and stores the dialing pulses that control the further operations necessary in establishing a telephone connection. [COMPUT SCI] The computer hardware for storing one machine word. Also known as registration. { 'rej·ə·stər }

register capacity [COMPUT SCI] The upper and lower limits of the numbers which may be processed in a register. { 'rej·ə·stər kə'pas·əd·ē }

register circuit [ELECTR] A switching circuit with memory elements that can store from a few to millions of bits of coded information; when needed, the information can be taken from the circuit in the same code as the input, or in a different code. { 'rej·ə·stər ,sər·kət }

register length [COMPUT SCI] The number of digits, characters, or bits, which a register can store. { 'rej·ə·stər ¦leŋkth }

register-level compatibility [COMPUT SCI] Property of hardware components that are totally compatible, having registers with the same type, size, and names. { ¦rej·ə·stər ¦lev·əl kəm ˌpad·ə'bil·əd·ē }

register-sender [COMMUN] A unit that generates and recognizes the supervisory signals to make connection to a circuit switching unit. { 'rej·ə·stər 'sen·dər }

register variable [COMPUT SCI] A variable in a computer program that is assigned to a register in the central processing unit instead of to a location in main storage. { 'rej·ə·stər ˌver·ē·ə·bəl }

registration See register. { ˌrej·ə'strā·shən }

registration mark [COMPUT SCI] In character recognition, a preprinted indication of the relative position and direction of various elements of the source document to be recognized. { ˌrej·ə'strā·shən ˌmärk }

regular [ELECTROMAG] In a definite direction; not diffused or scattered, when applied to reflection, refraction, or transmission. { 'reg·yə·lər }

regular expression [COMPUT SCI] A formal description of a language acceptable by a finite automaton or for the behavior of a sequential switching circuit. { 'reg·yə·lər ik'spresh·ən }

regulator problem See linear regulator problem. { 'reg·yəˌlād·ər ˌpräb·ləm }

regulatory control function [CONT SYS] That level in the functional decomposition of a large-scale control system which interfaces with the plant to implement the decisions of the optimizing controller inputted in the form of set points, desired trajectories, or targets. Also known as direct control function. { 'reg·yə·ləˌtór·ē kən 'trōl ˌfəŋk·shən }

reimbursed time [COMPUT SCI] The machine time which is loaned or rented to another office, agency, or organization, either on a reimbursable or reciprocal basis. { 'rē·əmˌbərst 'tīm }

reinitialize [COMPUT SCI] To return a computer program to the condition it was in at the start of processing, so that nothing remains from previous executions of the program. { ¦rē·i'nish·əlˌīz }

reinsertion of carrier [ELECTR] Combining a locally generated carrier signal in a receiver with an incoming signal of the suppressed carrier type. { ¦rē·ən'sər·shən əv 'kar·ē·ər }

rejection band [ELECTROMAG] The band of frequencies below the cutoff frequency in a uniconductor waveguide. Also known as stop band. { ri'jek·shən ˌband }

rejector See trap. { ri'jek·tər }

rejector circuit See band-stop filter. { ri'jek·tər ˌsər·kət }

relation [COMPUT SCI] A two-dimensional table in which data are arranged in a relational data structure. { ri'lā·shən }

relational algebraic language [COMPUT SCI] A low-level procedural language for carrying out fundamental algebraic operations on a database of relations. { ri'lā·shən·əl 'al·jə ˌbrā·ik ˌlaŋ·gwij }

relational calculus language [COMPUT SCI] A higher-level nonprocedural language for operating on a database of relations, containing statements that can be mapped to the fundamental algebraic operations on the database. { ri'lā·shən·əl 'kal·kyə·ləs ˌlaŋ·gwij }

relational capability [COMPUT SCI] Property of two or more data files that can be joined together for viewing, editing, or creation of reports. { ri ¦lā·shən·əl ˌkāp·ə'bil·əd·ē }

relational database See relational system. { ri 'lā·shən·əl 'dad·əˌbās }

relational data structure [COMPUT SCI] A type of data structure in which data are represented as tables in which no entry contains more than one value. { ri'lā·shən·əl 'dad·ə ˌstrək·chər }

relationally complete [COMPUT SCI] Property of a programming language that provides for the construction of all relations derivable from some set of base relations by the application of the primitive algebraic operations. { ri'lā·shən·əl·ē kəm'plēt }

relational operator [COMPUT SCI] An operator that indicates whether one quantity is equal to, greater than, or less than another. { ri'lā·shən·əl 'äp·əˌrād·ər }

relational spreadsheet [COMPUT SCI] A spreadsheet whose data are stored in a central database and are copied from the database into the spreadsheet when the spreadsheet is called up. { ri¦lā·shən·əl 'spredˌshēt }

relational system [COMPUT SCI] A database management system in which a relational data structure is used. Also known as relational database. { ri'lā·shən·əl ˌsis·təm }

relative address [COMPUT SCI] The numerical difference between a desired address and a known reference address. { 'rel·əd·iv ə'dres }

relative byte address [COMPUT SCI] A relative address expressed as the number of bytes from a point of reference to the desired address. { 'rel·ə·tiv 'bīt ˌadˌres }

relative coding [COMPUT SCI] A form of computer programming in which the address part of an instruction indicates not the desired address but the difference between the location of the instruction and the desired address. { 'rel·əd·iv 'kōd·iŋ }

relative gain [ELECTROMAG] The gain of an antenna in a given direction when the reference antenna is a half-wave, loss-free dipole isolated in space whose equatorial plane contains the given direction. { 'rel·əd·iv ¦gān }

relative gain array [CONT SYS] An analytical device used in process control multivariable applications, based on the comparison of single-loop control to multivariable control; expressed as an array (for all possible input-output pairs) of the ratios of a measure of the single-loop behavior between an input-output variable pair, to a related measure of the behavior of the same input-output pair under some idealization of multivariable control. { 'rel·əd·iv ¦gān əˌrā }

relative power gain [ELECTROMAG] Of one transmitting or receiving antenna over another, the measured ratio of the signal power one produces at the receiver input terminals to that produced by the other, the transmitting power level remaining fixed. { 'rel·əd·iv 'paů·ər ˌgān }

relative triple precision [COMPUT SCI] The retention of three times as many digits of a quantity as the computer normally handles; for example, a computer whose basic word consists of 10 decimal digits is called upon to handle 30 decimal digit quantities. { 'rel·əd·iv 'trip·əl prə'sizh·ən }

relative vector [COMPUT SCI] In computer graphics, a vector whose end points are given in relative coordinates. { 'rel·əd·iv 'vek·tər }

relaxation oscillator [ELECTR] An oscillator whose fundamental frequency is determined by the time of charging or discharging a capacitor or coil through a resistor, producing waveforms that may be rectangular or sawtooth. { ˌrēˌlak'sā·shən ˌäs·əˌlād·ər }

relay [COMMUN] A microwave or other radio system used for passing a signal from one radio communication link to another. { 'rēˌlā }

relay center [COMMUN] A switching center in which messages are automatically routed according to data contained in the messages or message headers. { 'rēˌlā ˌsen·tər }

relay satellite *See* communications satellite. { 'rēˌlā ˌsad·əlˌīt }

relay station *See* repeater station. { 'rēˌlā ˌstā·shən }

relay system *See* radio relay system. { 'rēˌlā ˌsis·təm }

reliability [ENG] The probability that a component part, equipment, or system, including computer hardware and software, will satisfactorily perform its intended function under given circumstances, such as environmental conditions, limitations as to operating time, and frequency and thoroughness of maintenance for a specified period of time. { riˌlī·ə'bil·əd·ē }

relocatable code [COMPUT SCI] A code generated by an assembler or compiler, and in which all memory references needing relocation are either specially marked or relative to the current program-counter reading. { ¦rē·lō¦kād·ə·bəl 'kōd }

relocatable emulator [COMPUT SCI] An emulator which does not require a stand-alone machine but executes in a multiprogramming environment. { ¦rē·lō¦kād·ə·bəl 'em·yəˌlād·ər }

relocatable program [COMPUT SCI] A program coded in such a way that it may be located and executed in any part of memory. { ¦rē·lō ¦kād·ə·bəl 'prōˌgram }

relocate [COMPUT SCI] To establish or change the location of a program routine while adjusting or modifying the address references within the instructions to correctly indicate the new locations. { rē'lōˌkāt }

relocating loader [COMPUT SCI] A loader in which some of the addresses in the program to be loaded are expressed relative to the start of the program rather than in absolute form. { ¦rē·lō ¦kād·iŋ 'lōd·ər }

relocation hardware [COMPUT SCI] Equipment in a multiprogramming system which allows a computer program to be run in any available space in memory. { ˌrē·lō'kā·shən ˌhärdˌwer }

relocation register [COMPUT SCI] A hardware element that holds a constant to be added to the address of each memory location in a computer program running in a multiprogramming system, as determined by the location of the area in memory assigned to the program. { ˌrē·lō'kā·shən ˌrej·ə·stər }

remedial maintenance *See* corrective maintenance. { ri'mēd·ē·əl 'mānt·ən·əns }

remember condition [ELECTR] Condition of a flip-flop circuit in which no change takes place between a given internal state and the next state. { ri'mem·bər kənˌdish·ən }

remodulator [ELECTR] A circuit that converts amplitude modulation to audio frequency-shift modulation for transmission of data signals over a radio channel. Also known as converter. { rē'mäj·əˌlād·ər }

remote access [COMPUT SCI] Ability to gain entry to a computer system from a location some distance away. { ri'mōt 'akˌses }

remote batch computing [COMPUT SCI] The running of programs, usually during nonprime hours, or whenever the demands of real-time or time-sharing computing slacken sufficiently to allow less pressing programs to be run. { ri 'mōt 'bach kəmˌpyüd·iŋ }

remote batch processing [COMPUT SCI] Batch processing in which an input device is located at a distance from the main installation and has access to a computer through a communication link. { ri'mōt 'bach ˌpräˌses·iŋ }

remote calculator [COMPUT SCI] A keyboard device that can be connected to the central processing unit of a distant computer over an ordinary telephone channel, enabling the user to present programs to the computer. { ri'mōt 'kal·kyə ˌlād·ər }

remote communications software [COMPUT SCI] Software that allows a microcomputer to control or duplicate the operation of another microcomputer at a distant location, using the standard telephone system. { ri¦mōt kəˌmyü·nə'kā·shənz 'söfˌwer }

remote computing system [COMPUT SCI] A data-processing system that has terminals distant from the central processing unit, from which users can communicate with the central processing unit and compile, debug, test, and execute programs. { ri'mōt kəm'pyüd·iŋ ˌsis·təm }

remote computing system exchange [COMPUT SCI] A device that handles communications between the central processing unit and remote consoles of a remote computing system, and enables several remote consoles to operate at the same time without interfering with each other. { ri'mōt kəm'pyüd·iŋ ˌsis·təm iksˌchānj }

remote computing system language [COMPUT SCI] A computer language used for communications between the central processing unit and remote consoles of a remote computer system, generally incorporating a procedure-oriented language such as FORTRAN, but also containing operating statements, such as instructions to debug or execute programs. { ri'mōt kəm'pyüd·iŋ ˌsis·təm ˌlaŋ·gwij }

remote computing system log [COMPUT SCI] A record of the volumes of data transmitted and of the frequency of various types of events during the operation of remote consoles in a remote computing system. { ri'mōt kəm'pyüd·iŋ ˌsis·təm ˌläg }

remote console [COMPUT SCI] A terminal in a remote computing system that has facilities for communicating with, and exerting control over, the central processing unit, and which may have any of various types of display units, printers, and data entry devices for direct communication with the central processing unit. { ri'mōt 'känˌsōl }

remote control [CONT SYS] Control of a quantity which is separated by an appreciable distance from the controlling quantity; examples include telemetering, telephone, and television. { ri 'mōt kən'trōl }

remote debugging [COMPUT SCI] **1.** The testing and correction of computer programs at a remote console of a remote computing system. **2.** *See* remote testing. { ri'mōt dē'bəg·iŋ }

remote inquiry [COMPUT SCI] Interrogation of the content of an automatic data processing equipment storage unit from a device remotely displaced from the storage unit site. { ri'mōt 'iŋˌkwə·rē }

remote metering *See* telemetering. { ri'mōt 'mēd·ə·riŋ }

remote pickup [COMMUN] Picking up a radio or television program at a remote location and relaying it to the studio or transmitter over wire lines or a radio link. { ri'mōt 'pikˌəp }

remote plan position indicator *See* plan position indicator repeater. { ri'mōt ¦plan pə¦zish·ən 'in·dəˌkād·ər }

remote sensing [ENG] The gathering and recording of information without actual contact with the object or area being investigated. { ri'mōt 'sens·iŋ }

remote subscriber [COMMUN] Subscriber to a network that does not have direct access to the switching center, but has access to the circuit through a facility such as a base message center. { ri'mōt səb'skrīb·ər }

remote terminal [COMPUT SCI] A computer terminal which is located away from the central processing unit of a data-processing system, at a location convenient to a user of the system. { ri'mōt 'tər·mən·əl }

remote testing [COMPUT SCI] A method of testing and correcting computer programs; programmers do not go to the computer center but provide detailed instructions to be carried out by computer operators along with the programs and associated test data. Also known as remote debugging. { ri'mōt 'test·iŋ }

removable medium [COMPUT SCI] A data storage medium, such as magnetic tape or floppy disk, that can be physically removed from the unit that reads and writes on it. { ri'müv·ə·bəl 'mē·dē·əm }

removable plugboard *See* detachable plugboard. { ri'müv·ə·bəl 'pləgˌbȯrd }

REM statement [COMPUT SCI] A statement in a computer program that consists of remarks or comments that document the program, and contains no executable code. { 'rem ˌstāt·mənt }

repeater [ELECTR] An amplifier or other device that receives weak signals and delivers corresponding stronger signals with or without reshaping of waveforms; may be either a one-way or two-way repeater. Also known as regenerator. { ri'pēd·ər }

repeater jammer [ELECTR] A jammer that intercepts an enemy radar signal and reradiates the signal after modifying it to incorporate erroneous data on azimuth, range, or number of targets. { ri'pēd·ər ˌjam·ər }

repeater station [COMMUN] A station containing one or more repeaters. Also known as relay station. { ri'pēd·ər ˌstā·shən }

repeat key [COMPUT SCI] A key on a typewriter or computer keyboard that, when depressed at the same time as a character key, causes repeated printing or generation of the character until one of the keys is released. { ri'pēt ˌkē }

repeat operator [COMPUT SCI] A pseudo instruction using two arguments, a count p and an increment n: the word immediately following the instruction is repeated p times, with the values 0, n, $2n$, . . . , $(p - 1)n$ added to the successive words. { ri'pēt ˌäp·əˌrād·ər }

repetition equivalent [COMMUN] In a complete telephone connection, a measure of the grade of transmission experienced by the subscribers using the connection; it includes the combined effects of volume, distortion, noise, and all other subscriber reactions and usages. { ˌrep·ə'tish·ən i'kwiv·ə·lənt }

repetition frequency *See* repetition rate. { ˌrep·ə'tish·ən ˌfrē·kwən·sē }

repetition instruction [COMPUT SCI] An instruction that causes one or more other instructions to be repeated a specified number of times, usually with systematic address modification occurring between repetitions. { ˌrep·ə'tish·ən inˌstrək·shən }

repetition rate [COMMUN] The rate at which recurrent signals are produced or transmitted. Also known as recurrence rate; repetition frequency. { ˌrep·ə'tish·ən ˌrāt }

repetitive addressing [COMPUT SCI] A system used on some computers in which, under certain conditions, an instruction is written without giving the address of the operand, and the operand address is automatically that of the location addressed by the last previous instruction. { rə'ped·əd·iv ə'dres·iŋ }

repetitive analog computer [COMPUT SCI] An analog computer which repeatedly carries out the solution of a problem at a rapid rate (10 to 60 times a second) while an operator may vary parameters in the problem. { rə'ped·əd·iv 'an·ə ˌläg kəm'pyüd·ər }

repetitive statement [COMPUT SCI] A statement in a computer program that is repeatedly executed for a specified number of times or for as long as a specified condition holds true. { ri'ped·əd·iv 'stāt·mənt }

repetitive unit [COMPUT SCI] A type of circuit which appears more than once in a computer. { rə'ped·əd·iv yü·nət }

reply [COMMUN] A radio-frequency signal or combination of signals transmitted by a transponder in response to an interrogation. Also known as response. { ri'plī }

report [COMPUT SCI] An output document prepared by a data-processing system. { ri'pȯrt }

report generator [COMPUT SCI] A routine which produces a complete data-processing report, given only a description of the desired content and format, plus certain information concerning the input file. Also known as report writer. { ri'pȯrt ˌjen·əˌrād·ər }

reporting time interval [COMMUN] The time for transmission of data or a report from the originating terminal to the end receiver. { ri'pȯrd·iŋ 'tīm ˌin·tər·vəl }

report program [COMPUT SCI] A program that prints out an analysis of a file of records, usually arranged by keys, each analysis or total being produced when a key change takes place. { ri'pȯrt ˌprōˌgram }

report program generator [COMPUT SCI] A nonprocedural programming language that provides a convenient method of producing a wide variety of reports. Abbreviated RPG. { ri'pȯrt ¦prōˌgram ˌjen·əˌrād·ər }

report writer *See* report generator. { ri'pȯrt ˌrīd·ər }

representation condition [COMPUT SCI] The condition that, if one software entity is less than another entity in terms of a selected attribute, then any software metric for that attribute must associate a smaller number to the first entity than it does to the second entity. { ˌrep·rə·zen'tā·shən kənˌdish·ən }

representative calculating time [COMPUT SCI] The time required to perform a specified operation or series of operations. { ¦rep·ri¦zen·təd·iv 'kal·kyəˌlād·iŋ ˌtīm }

reproduction speed [COMMUN] Area of copy recorded per unit time in facsimile transmission. { ¦rē·prə¦dək·shən ˌspēd }

request/grant logic [COMPUT SCI] Logic circuitry which, in effect, selects the interrupt line with highest priority. { ri'kwest 'grant ˌläj·ik }

request repeat system [COMMUN] System using an error-detecting code, and so arranged that a signal detected as being in error automatically initiates a request for retransmission. { ri'kwest ri¦pēt ˌsis·təm }

reradiation [COMMUN] Undesirable radiation of signals generated locally in a radio receiver, causing interference or revealing the location of the receiver. { rēˌrā·dē'ā·shən }

rerun [COMPUT SCI] To run a program or a portion of it again on a computer. Also known as rollback. { 'rēˌrən }

rerun point [COMPUT SCI] A location in a program from which the program may be started anew after an interruption of the computer run. { 'rēˌrən ˌpȯint }

rerun routine [COMPUT SCI] A routine designed to be used in the wake of a computer malfunction or a coding or operating mistake to reconstitute a routine from the last previous rerun point. { 'rē ˌrən ˌrüˌtēn }

rescue dump [COMPUT SCI] The copying of the entire contents of a computer memory into auxiliary storage devices, carried out periodically during the course of a computer program so that in case of a machine failure the program can be reconstituted at the last point at which this operation was executed. { 'res·kyü ˌdəmp }

reserve [COMPUT SCI] To assign portions of a computer memory and of input/output and storage devices to a specific computer program in a multiprogramming system. { ri'zərv }

reserved word [COMPUT SCI] A word which cannot be used in a programming language to represent an item of data because it has some particular significance to the compiler, or which can be used only in a particular context. { ri'zərvd 'wərd }

reset *See* clear. { 'rēˌset }

reset condition [ELECTR] Condition of a flip-flop circuit in which the internal state of the flip-flop is reset to zero. { 'rēˌset kənˌdish·ən }

reset cycle [COMPUT SCI] The return of a cycle index counter to its initial value. { 'rēˌset ˌsī·kəl }

reset input [COMPUT SCI] The act of resetting the original conditions of a problem after a program is run on an analog computer. { 'rēˌset 'inˌpu̇t }

reset mode [COMPUT SCI] The phase of operation of an analog computer during which the required initial conditions are entered into the system and the computing units are inoperative. Also known as initial condition mode. { 'rēˌset ˌmōd }

resident executive [COMPUT SCI] The portion of the executive routine that is permanently stored in a computer's main memory. Also known as resident monitor. { 'rez·ə·dənt ig'zek·yəd·iv }

resident module *See* resident routine. { 'rez·ə·dənt 'mä·jəl }

resident monitor *See* resident executive. { 'rez·ə·dənt 'män·əd·ər }

resident routine [COMPUT SCI] Any computer routine which is stored permanently in the memory, such as the resident executive. Also known as resident module. { 'rez·ə·dənt rü'tēn }

residual error rate *See* undetected error rate. { rə'zij·ə·wəl 'er·ər ˌrāt }

residual modulation *See* carrier noise. { rə'zij·ə·wəl ˌmäj·ə'lā·shən }

residue check *See* modulo N check. { 'rez·əˌdü ˌchek }

residue system [COMPUT SCI] A number system in which each digit position corresponds to a different radix, all pairs of radices are relatively prime, and the value of a digit with radix *r* for an integer A is equal to the remainder when A is divided by *r*. { 'rez·əˌdü ˌsis·təm }

resilience [COMPUT SCI] The ability of computer software to be used for long periods of time. { rə'zil·yəns }

resistance noise *See* thermal noise. { ri'zis·təns ˌnȯiz }

resistor-capacitor-transistor logic [ELECTR] A resistor-transistor logic with the addition of capacitors that are used to enhance switching speed. { ri'zis·tər kə'pas·əd·ər tran'zis·tər ˌläj·ik }

resistor-transistor logic [ELECTR] One of the simplest logic circuits, having several resistors, a transistor, and a diode. Abbreviated RTL. { ri'zis·tər tran'zis·tər ˌläj·ik }

resolution [CONT SYS] The smallest increment in distance that can be distinguished and acted upon by an automatic control system. [ELECTR] In television, the maximum number of lines that can be discerned on the screen at a distance equal to screen height. [ELECTROMAG] In radar, the minimum separation between two targets or features thereof, in angle, range, cross range, or range rate, at which they can be distinguished on a radar display or in the data processing. Also known as resolving power. { ˌrez·ə'lü·shən }

resolution chart *See* test pattern. { ˌrez·ə'lü·shən ˌchärt }

resolution error [COMPUT SCI] An error of an analog computing unit that results from its inability to respond to changes of less than a given magnitude. { ˌrez·ə'lü·shən ˌer·ər }

resolution factor [COMPUT SCI] In information retrieval, the ratio obtained in dividing the total number of documents retrieved (whether relevant or not to the user's needs) by the total number of documents available in the file. { ˌrez·ə'lü·shən ˌfak·tər }

resolution in azimuth [ENG] The angle by which two targets must be separated in azimuth in order to be distinguished by a radar set when the targets are at the same range. { ˌrez·ə'lü·shən in 'az·ə·məth }

resolution in range [ENG] Distance by which two targets must be separated in range in order to be distinguished by a radar set when the targets are on the same azimuth line. { ˌrez·ə'lü·shən in 'rānj }

resolution wedge [COMMUN] On a video test pattern, a group of gradually converging lines used to measure resolution. { ˌrez·ə'lü·shən ˌwej }

resolving cell [ELECTROMAG] In radar, volume in space whose diameter is the product of slant range and beam width, and whose length is the pulse length. { ri'zälv·iŋ ˌsel }

resolving power *See* resolution. { ri'zälv·iŋ ˌpau̇·ər }

resolving time [COMPUT SCI] In computers, the shortest permissible period between trigger pulses for reliable operation of a binary cell. [ENG] Minimum time interval, between events, that can be detected; resolving time may refer to an electronic circuit, to a mechanical recording device, or to a counter tube. { ri'zälv·iŋ ˌtīm }

resonant antenna [ELECTROMAG] An antenna for which there is a sharp peak in the power radiated or intercepted by the antenna at a certain frequency, at which electric currents in the antenna form a standing-wave pattern. { 'res·ən·ənt an 'ten·ə }

resonant cavity *See* cavity resonator. { 'res·ən·ənt 'kav·əd·ē }

resonant chamber *See* cavity resonator. { 'res·ən·ənt 'chām·bər }

resonant diaphragm [ELECTROMAG] A diaphragm, in waveguide technique, so proportioned as to introduce no reactive impedance at the design frequency. { 'res·ən·ənt 'dī·əˌfram }

resonant element *See* cavity resonator. { 'res·ən·ənt 'el·ə·mənt }

resonant helix [ELECTROMAG] An inner helical conductor in certain types of transmission lines and resonant cavities, which carries currents with the same frequency as the rest of the line or cavity. { 'res·ən·ənt 'hē·liks }

resonant iris [ELECTROMAG] A resonant window in a circular waveguide; it resembles an optical iris. { 'res·ən·ənt 'ī·rəs }

resonant line [ELECTROMAG] A transmission line having values of distributed inductance and distributed capacitance so as to make the line resonant at the frequency it is handling. { 'res·ən·ənt 'līn }

resonant-line tuner [ELECTR] A device in which resonant lines are used to tune the antenna, radio-frequency amplifier, or radio-frequency oscillator circuits; tuning is achieved by moving shorting contacts that change the electrical lengths of the lines. { 'res·ən·ənt ¦līn 'tün·ər }

resonant wavelength [ELECTROMAG] The wavelength in free space of electromagnetic radiation having a frequency equal to a natural resonance frequency of a cavity resonator. { 'res·ən·ənt 'wāvˌleŋkth }

resonant window [ELECTROMAG] A parallel combination of inductive and capacitive diaphragms, used in a waveguide structure to provide transmission at the resonant frequency and reflection at other frequencies. { 'res·ən·ənt 'win·dō }

resonating cavity [ELECTROMAG] Short piece of waveguide of adjustable length, terminated at either or both ends by a metal piston, an iris diaphragm, or some other wave-reflecting device; it is used as a filter, as a means of coupling between guides of different diameters, and as impedance networks corresponding to those used in radio circuits. { 'rez·ənˌād·iŋ 'kav·əd·ē }

responder [ELECTR] The transmitter section, including the appropriate encoder, of a radar transponder. { ri'spän·dər }

response [COMMUN] *See* reply. [CONT SYS] A quantitative expression of the output of a device or system as a function of the input. Also known as system response. { ri'späns }

response characteristic [CONT SYS] The response as a function of an independent variable, such as direction or frequency, often presented in graphical form. { ri'späns ˌkar·ik·təˌris·tik }

response time [COMPUT SCI] The delay experienced in time sharing between request and answer, a delay which increases when the number of users on the system increases. [CONT SYS] The time required for the output of a control system or element to reach a specified fraction of its new value after application of a step input or disturbance. { ri'späns ˌtīm }

responsor [ELECTR] The receiving section of an interrogator-responsor. { ri'spän·sər }

restart [COMPUT SCI] To go back to a specific planned point in a routine, usually in the case of machine malfunction, for the purpose of rerunning the portion of the routine in which the error occurred; the length of time between restart points in a given routine should be a function of the mean free error time of the machine itself. { 'rēˌstärt }

resting frequency *See* carrier frequency. { 'rest·iŋ ˌfrē· kwən·sē }

restore [COMPUT SCI] In computers, to regenerate, to return a cycle index or variable address to its initial value, or to store again. [ELECTR] Periodic charge regeneration of volatile computer storage systems. { ri'stȯr }

restorer pulses [ELECTR] In computers, pairs of complement pulses, applied to restore the coupling-capacitor charge in an alternating-current flip-flop. { ri'stȯr·ər ˌpəls·əz }

restoring logic [ELECTR] Circuitry designed so that even with an imperfect input pulse a standard output occurs at the exit of each successive logic gate. { ri'stȯr·iŋ ˌläj·ik }

restricted function [COMPUT SCI] A function of the operating system that cannot be used by application programs. { ri'strik·təd ˌfəŋk·shən }

retard transmitter [ELECTR] Transmitter in which a delay period is introduced between the time of actuation and the time of transmission. { ˌri'tärd tranzˌmid·ər }

retention period [COMPUT SCI] The length of time that data must be kept on a reel of magnetic tape before it can be destroyed. { ri'ten·chən ˌpir·ē·əd }

retina [COMPUT SCI] In optical character recognition, a scanning device. { 'ret·ən·ə }

retina character reader [COMPUT SCI] A character reader that operates in the manner of the human retina in recognizing identical letters in different type fonts. { 'ret·ən·ə 'kar·ik·tər ˌrēd·ər }

retrace *See* flyback. { 'rēˌtrās }

retrace blanking [ELECTR] Blanking a video display during vertical retrace intervals to prevent retrace lines from showing on the screen. { 'rē ˌtrās ˌblaŋk·iŋ }

retrace line [ELECTR] The line traced by the electron beam in a cathode-ray tube in going from the end of one line or field to the start of the next line or field. Also known as return line. { 'rē ˌtrās ˌlīn }

retransmission unit [ELECTR] Control unit used at an intermediate station for feeding one radio receiver-transmitter unit for two-way communication. { ¦rē·tranz'mish·ən ˌyü·nət }

retrieve [COMPUT SCI] To find and select specific information. { ri'trēv }

retroaction *See* positive feedback. { ¦re·trō 'ak·shən }

retry [COMPUT SCI] When a central processing unit error is detected during execution of an instruction, the computer will execute this instruction unless a register was altered by the operation. { 'rēˌtrī }

return [COMPUT SCI] **1.** To return control from a subroutine to the calling program. **2.** To go back to a planned point in a computer program and rerun a portion of the program, usually when an error is detected; rerun points are usually not more than 5 minutes apart. [ELECTR] *See* echo. { ri'tərn }

return address [COMPUT SCI] The address in storage to which a computer program is directed upon completion of a subroutine. { ri'tərn 'ad ˌres }

return busy tone [COMMUN] A signal returned to the register-sender that, in turn, returns a busy indication to the calling station. { ri'tərn 'biz·ē ˌtōn }

return code [COMPUT SCI] An indicator that is issued by a computer upon completion of a subroutine or function, or of the entire program, that indicates the result of the processing and, in particular, whether the processing was successful or ended abnormally because of an error. { ri'tərn ˌkōd }

return difference [CONT SYS] The difference between 1 and the loop transmittance. { ri'tərn ˌdif·rəns }

return jump [COMPUT SCI] A jump instruction in a subroutine which passes control to the first statement in the program which follows the instruction called the subroutine. { ri'tərn ˌjəmp }

return key [COMPUT SCI] A key on a typewriter or a computer keyboard that, when depressed, causes a print mechanism or cursor to move to the beginning of the next line. { ri'tərn ˌkē }

return line *See* retrace line. { ri'tərn ˌlīn }

return loss [COMMUN] **1.** The difference between the power incident upon a discontinuity in a transmission system and the power reflected from the discontinuity. **2.** The ratio in decibels of the power incident upon a discontinuity to the power reflected from the discontinuity. { ri'tərn ˌlȯs }

return to zero mode [COMPUT SCI] Computer readout mode in which the signal returns to zero between each bit indication. { ri'tərn tə 'zir·ō ˌmōd }

return trace *See* flyback. { ri'tərn ˌtrās }

reusable [COMPUT SCI] Of a program, capable of being used by several tasks without having to be reloaded; it is a generic term, including reenterable and serially reusable. { rē'yü·zə·bəl }

reverse code dictionary [COMPUT SCI] Alphabetic or alphanumeric arrangement of codes associated with their corresponding English words or terms. { ri'vərs ¦kōd 'dik·shəˌner·ē }

reverse-direction flow [COMPUT SCI] A logical path that runs upward or to the left on a flowchart. { ri'vərs di¦rek·shən ˌflō }

reverse feedback *See* negative feedback. { ri 'vərs 'fēdˌbak }

reverse Polish notation [COMPUT SCI] The version of Polish notation, used in some calculators, in which operators follow the operators with which they are associated. Abbreviated RPN. Also known as postfix notation; suffix notation. { ri'vərs 'pō·lish nō'tā·shən }

reverse video [COMPUT SCI] An electronic display mode in which the normal properties of the display are reversed; for example, normally white characters on a black background will appear as black characters on a white background. Also known as inverse video. { ri'vərs 'vid·ē·ō }

reversible counter [COMPUT SCI] A counter which stores a number whose value can be decreased or increased in response to the appropriate control signal. { ri'vər·sə·bəl 'kau̇nt·ər }

reversible transducer [ELECTR] Transducer whose loss is independent of transmission direction. { ri'vər·sə·bəl tranz'düs·ər }

rewrite [COMPUT SCI] The process of restoring a storage device to its state prior to reading; used when the information-storing state may be destroyed by reading. { 'rēˌrīt }

RFI *See* radio-frequency interference.

RGB monitor [COMPUT SCI] A video display screen that requires separate red, green, and blue signals from a computer or other source. { ¦är ¦jē¦bē 'män·əd·ər }

RG line *See* radio-frequency cable. { ¦är'jē ˌlīn }

RHI display *See* range-height indicator display. { ¦är¦āch'ī diˌsplā }

rhombic antenna [ELECTROMAG] A horizontal antenna having four conductors forming a diamond or rhombus; usually fed at one apex and terminated with a resistance or impedance at the opposite apex. Also known as diamond antenna. { 'räm·bik an'ten·ə }

rho-theta navigation *See* omnibearing distance navigation. { 'rō'thād·ə ˌnav·ə'gā·shən }

rhumbatron *See* cavity resonator. { 'rəm·bə ˌträn }

Rice video [ELECTR] Referring to the video and its particular probability density produced by an amplitude detector (demodulator) when the Gaussian radio noise and a signal of a known and constant amplitude are together incident to it. { 'rīs ¦vid·ē·ō }

ridge waveguide [ELECTROMAG] A circular or rectangular waveguide having one or more longitudinal internal ridges that serve primarily to increase transmission bandwidth by lowering the cutoff frequency. { 'rij 'wāvˌgīd }

right-justify [COMPUT SCI] To shift the contents of a register so that the right or least significant digit is at some specified position. { 'rīt 'jəs·tə ˌfī }

right value [COMPUT SCI] The actual data content of a symbolic variable in a computer program; it is one of two components of the symbolic variable, the other being the memory address. Abbreviated rvalue. { 'rīt 'val·yü }

rigid copper coaxial line [ELECTROMAG] A coaxial cable in which the central conductor and outer conductor are formed by joining rigid pieces of copper. { 'rij·id 'käp·ər kō'ak·sē·əl ˌlīn }

R-indicator *See* R-display. { 'är ˌin·dəˌkād·ər }

ring [COMPUT SCI] A cyclic arrangement of data elements, usually including a specified entry pointer. { riŋ }

ring-around [COMMUN] **1.** Improper routing of a call back through a switching center already trying to complete the same call, thus tying up the trunks by repeating the cycle. **2.** Oscillation of a repeater caused by leakage of the transmitter signal into the receiver. { 'riŋ əˌrau̇nd }

ring circuit [ELECTROMAG] In waveguide practice, a hybrid T junction having the physical configuration of a ring with radial branches. { 'riŋ ˌsər·kət }

ring data structure [COMPUT SCI] Stored data that is organized by a chain of pointers so that the last pointer is directed back to the beginning of the chain. { 'riŋ 'dad·ə ˌstrək·chər }

ring head [ELECTR] A recording and playback head in a magnetic recording system which has the form of a ring with a gap at one point, and on which the coils are wound. { 'riŋ ˌhed }

ringing [COMMUN] The production of an audible or visible signal at a station or switchboard by means of an alternating or pulsating current. [CONT SYS] An oscillatory transient occurring in the output of a system as a result of a sudden change in input. { 'riŋ·iŋ }

ring modulator [ELECTR] A modulator in which four diode elements are connected in series to form a ring around which current flows readily in one direction; input and output connections are made to the four nodal points of the ring; used as a balanced modulator, demodulator, or phase detector. { 'riŋ 'mäj·əˌlād·ər }

ring network [COMMUN] A communications network in which the nodes can be considered to be on a circle, about which messages must be routed. Also known as loop network. { 'riŋ 'net ˌwərk }

ring shift *See* cyclic shift. { 'riŋ ¦shift }

ring structure [COMPUT SCI] A chained file organization such that the end of the chain points to its beginning. { 'riŋ ˌstrək·chər }

ring time [ELECTR] The length of time in microseconds required for a pulse of energy transmitted into an echo box to die out; a measurement of the performance of radar. { 'riŋ ˌtīm }

ripple-carry adder [COMPUT SCI] A device for addition of two *n*-bit binary numbers, formed by connecting *n* full adders in cascade, with the carry output of each full adder feeding the carry input of the following full adder. { 'rip·əl ¦kar·ē ˌad·ər }

RISC *See* reduced instruction set computer. { risk }

rise time [CONT SYS] The time it takes for the output of a system to change from a specified small percentage (usually 5 or 10) of its steady-state increment to a specified large percentage (usually 90 or 95). { 'rīz ˌtīm }

Rivest-Shamir-Adleman algorithm [COMMUN] A public-key algorithm whose strength is based on the fact that factoring large composite prime numbers into their prime factors involves an overwhelming amount of computation. Abbreviated RSA algorithm. { ri'vest shə'mir 'ad·əl·mən ˌal·gəˌrith·əm }

RLL code See run-length-limited code. { ¦är ¦el'el ˌkōd }

rms value See root-mean-square value. { ¦är ¦em'es ˌval·ü }

RNAV See area navigation. { 'ärˌnav }

RO See receive-only.

robot [CONT SYS] A mechanical device that can be programmed to perform a variety of tasks of manipulation and locomotion under automatic control. { 'rōˌbät }

robust program [COMPUT SCI] **1.** A computer program using an iterative process that converges rapidly to the solution being sought. **2.** A computer program that performs well even under unusual conditions. { ¦rō·bəst 'prō·grəm }

rocket antenna [ELECTROMAG] An antenna carried on a rocket, to receive signals controlling the rocket or to transmit measurements made by instruments aboard the rocket. { 'räk·ət an ˌten·ə }

role indicator [COMPUT SCI] In information retrieval, a code assigned to a key word to indicate its part of speech, nature, or function. { 'rōl ˌin·dəˌkād·ər }

rollback See rerun. { 'rōlˌbak }

roll in [COMPUT SCI] To restore to main memory a section of program or data that had previously been rolled out. { 'rōl ˌin }

roll-off [ELECTR] Gradually increasing loss or attenuation with increase or decrease of frequency beyond the substantially flat portion of the amplitude-frequency response characteristic of a system or transducer. { 'rōl ˌȯf }

roll out [COMPUT SCI] **1.** To make available additional main memory for one task by copying another task onto auxiliary storage. **2.** To read a computer register or counter by adding a one to each digit column simultaneously until all have returned to zero, with a signal being generated at the instant a column returns to zero. { 'rōl ˌau̇t }

rollover [COMPUT SCI] A keyboard feature that allows more than one key to be depressed simultaneously, enabling the keys to be depressed more rapidly in sequence. { 'rōlˌō·vər }

roll your own See user program. { 'rōl yər 'ōn }

ROM See read-only memory. { räm }

ROMable code [COMPUT SCI] A computer program developed to be stored permanently in a read-only memory (ROM). { 'räm·ə·bəl 'kōd }

roof filter [ELECTR] Low-pass filter used in carrier telephone systems to limit the frequency response of the equipment to frequencies needed for normal transmission, thereby blocking unwanted higher frequencies induced in the circuit by external sources; improves runaround crosstalk suppression and minimizes high-frequency singing. { 'rüf ˌfil·tər }

room noise [COMMUN] Ambient noise in a telephone station. { 'rüm ˌnȯiz }

room power [ELECTR] The electric power that is fed to the machinery in a computer room after passing through a power distribution unit, motor-generator set, or other conditioning and isolating device. { 'rüm ˌpau̇·ər }

root [COMPUT SCI] The origin or most fundamental point of a tree diagram. Also known as base. { rüt }

root component See root symbol. { 'rüt kəm 'pō·nənt }

root directory [COMPUT SCI] The starting point in a hierarchical file system, where the system operates when it is first started. { 'rüt diˌrek·trē }

root locus plot [CONT SYS] A plot in the complex plane of values at which the loop transfer function of a feedback control system is a negative number. { 'rüt ¦lō·kəs ˌplät }

root-mean-square value [PHYS] The square root of the time average of the square of a quantity; for a periodic quantity, such as a sine wave used for audio measurements, the average is taken over one complete cycle. Abbreviated rms value. Also known as effective value. { 'rüt ˌmēn 'skwer 'val·yü }

root segment [COMPUT SCI] The master or controlling segment of an overlay structure which always resides in the main memory of a computer. { 'rüt ˌseg·mənt }

root sum square [COMMUN] A method of combining the power of multiple signals by taking the square root of the sum of the squares of all the signals. Abbreviated RSS. { 'rüt ˌsəm 'skwər }

root-sum-square value [PHYS] The square root of the sum of the squares of a series of related values; commonly used to express total harmonic distortion. { 'rüt ˌsəm 'skwər 'val·yü }

root symbol [COMPUT SCI] An element of a formal language, generally unique, that is not derivable from other language elements. Also known as root component. { 'rüt ˌsim·bəl }

root task [COMPUT SCI] The initial program on a parallel machine from which one or more child processes branch out in the fork-join model. { 'rüt ˌtask }

rotary beam [ELECTROMAG] Short-wave antenna system highly directional in azimuth and altitude, mounted in such a manner that it can be rotated to any desired position, either manually or by an electric motor drive. { 'rōd·ə·rē 'bēm }

rotary coupler See rotating joint. { 'rōd·ə·rē 'kəp·lər }

rotary digital audio tape system [ELECTR] A digital audio tape system that uses the helical-scan technology developed for video systems, with a rotating drum containing two metal-in-gap heads. Abbreviated R-DAT system. { ¦rōd·ə·rē ˌdij·əd·əl ˌȯd·ē·ō 'tāp ˌsis·təm }

rotary joint See rotating joint. { 'rōd·ə·rē 'jȯint }

rotary system [COMMUN] A telephone switching system that uses unidirectional, rotary switches that carry ten sets of brushes (wipers), only one of which is tripped as part of the control and selection process. { 'rōd·ə·rē 'sis·təm }

rotary-vane attenuator [ELECTROMAG] Device designed to introduce attenuation into a waveguide circuit by varying the angular position of a resistive material in the guide. { 'rōd·ə·rē ¦vān ə'ten·yə,wād·ər }

rotating joint [ELECTROMAG] A joint that permits one section of a transmission line or waveguide to rotate continuously with respect to another while passing radio-frequency energy. Also known as rotary coupler; rotary joint. { 'rō ,tād·iŋ 'jȯint }

rotation [COMPUT SCI] An operation performed on data in a register of the central processing unit, in which all the bits in the register are shifted one position to the right or left, and the endmost bit, which is shifted out of the register, is carried around to the position at the opposite end of the register. { rō'tā·shən }

rotational delay *See* rotational latency. { rō'tā·shən·əl di'lā }

rotational latency [COMPUT SCI] The time required, following an order to read or write information in disk storage, for the location of the information to revolve beneath the appropriate read/write head. Also known as rotational delay. { rō'tā·shən·əl 'lāt·ən·sē }

rotational position sensing [COMPUT SCI] A fast disk search method whereby the control unit looks for a specified sector, and then receives the sector number required to access the record. { rō'tā·shən·əl pə'zish·ən ,sens·iŋ }

rotator [ELECTROMAG] A device that rotates the plane of polarization of a plane-polarized electromagnetic wave, such as a twist in a waveguide. { 'rō,tād·ər }

rotoflector [ELECTROMAG] In radar, elliptically shaped, rotating reflector used to reflect a vertically directed radar beam at right angles so that it radiates in a horizontal direction. { 'rōd·ə,flek·tər }

rotor [COMMUN] **1.** Disk with a set of input contacts and a set of output contacts, connected by any prearranged scheme designed to rotate within an electrical cipher machine. **2.** Disk whose rotation produces a variation of some cryptographic element in a cipher machine usually by means of lugs (or pins) in or on its periphery. { 'rōd·ər }

round-robin scheduling [COMPUT SCI] A scheduling algorithm which repeatedly runs through a list of users, giving each user the opportunity to use the central processing unit in succession. { 'raủnd ¦räb·ən 'skej·ə·liŋ }

round-the-world echo [COMMUN] A signal occurring every 1/7 second when a radio wave repeatedly encircles the earth at its speed of 186,000 miles (300,000 kilometers) per second. { 'raủnd thə 'wərld 'ek·ō }

round-trip echoes [ELECTROMAG] Multiple reflection echoes produced when a radar pulse is reflected from a target strongly enough so that the echo is reflected back to the target where it produces a second echo. { 'raủnd ¦trip 'ek·ōz }

router [COMMUN] A device that selects an appropriate pathway for a message and routes the message accordingly. { 'raủd·ər }

routine [COMPUT SCI] A set of digital computer instructions designed and constructed so as to accomplish a specified function. { rü'tēn }

routine library [COMPUT SCI] Ordered set of standard and proven computer routines by which problems or parts of problems may be solved. { rü'tēn ,lī,brer·ē }

routing [COMMUN] The assignment of a path by which a message will travel to its destination. { 'rüd·iŋ }

routing indicator [COMMUN] **1.** A group of letters, engineered and assigned, to identify a station within a digital communications network. **2.** A group of letters assigned to indicate the geographic location of a station; a fixed headquarters of a command, activity, or unit at a geographic location; or the general location of a tape relay or tributary station to facilitate the routing of traffic over tape relay networks. { 'rüd·iŋ ,in də ,kād·ər }

routing message [COMMUN] The function performed at a central message processor of selecting the route, or alternate route required, by which a message will proceed to the next point in reaching its destination. { 'rüd·iŋ ,mes·ij }

row [COMPUT SCI] **1.** The characters, or corresponding bits of binary-coded characters, in a computer word. **2.** Equipment which simultaneously processes the bits of a character, the characters of a word, or corresponding bits of binary-coded characters in a word. **3.** Corresponding positions in a group of columns. { rō }

row address [COMPUT SCI] An index array entry field which contains the main storage address of a data block. { 'rō 'ad,res }

row order [COMPUT SCI] The storage of a matrix $a(m,n)$ as $a(1,1), a(1,2), \ldots, a(1,n), a(2,1), a(2,2), \ldots$. { 'rō ,ȯr·dər }

RPG *See* report program generator.

RPN *See* reverse Polish notation.

RSA algorithm *See* Rivest-Shamir-Adleman algorithm. { ¦är¦es¦ā 'al·gə,rith·əm }

R-scan *See* R-display. { 'är ,skan }

R-scope *See* R-display. { 'är ,skōp }

RSS *See* root sum square.

RS-232 [COMMUN] A standard developed by the Electronic Industries Association that governs the interface between data processing and data communications equipment, and is widely used to connect microcomputers to peripheral devices.

r-theta navigation *See* omnibearing distance navigation. { ¦är 'thād·ə ,nav·ə'gā·shən }

RTL *See* resistor-transistor logic.

rubber banding [COMPUT SCI] In computer graphics, the moving of a line or object, with one end held fixed in position. { ¦rəb·ər 'band·iŋ }

ruggedized computer [COMPUT SCI] A computer built so as to reduce vibrations, resist moisture, and remain unaffected by electromagnetic interferences such as are found in factory, military, or mobile environments. { ˈrəg·ə‚dīzd kəmˈpyüd·ər }

rule-based control system *See* direct expert control system. { ¦rül ‚bāst kənˈtrōl ‚sis·təm }

rule-based expert system [COMPUT SCI] An expert system based on a collection of rules that a human expert would follow in dealing with a problem. { ¦rül ¦bāst ˈek‚spərt ‚sis·təm }

rule of inference *See* production. { ˈrül əv ˈin·frəns }

run [COMPUT SCI] A single, complete execution of a computer program, or one continuous segment of computer processing, used to complete one or more tasks for a single customer or application. Also known as machine run. { rən }

runaround crosstalk [COMMUN] Crosstalk resulting from coupling between the high-level end of one repeater and the low-level end of another repeater, as at a carrier telephone repeater station. { ˈrən·ə‚raủnd ˈkrȯs‚tȯk }

runaway tape [COMPUT SCI] A tape reel that spins rapidly and out of control as the result of a hardware malfunction. { ¦rən·ə¦wā ˈtāp }

run book [COMPUT SCI] The collection of materials necessary to document a program run on a computer. Also known as problem file; problem folder. { ˈrən ‚bủk }

run chart [COMPUT SCI] A flow chart for one or more computer runs which shows input, output, and the use of peripheral units, but no details of the execution of the run. Also known as run diagram. { ˈrən ‚chärt }

run diagram *See* run chart. { ˈrən ‚dī·ə‚gram }

run documentation [COMPUT SCI] Detailed instructions to the operator on how to run a particular computer program. { ˈrən ‚däk·yə·menˈtā·shən }

run-length encoding [COMPUT SCI] A method of data compression that encodes strings of the same character as a single number. { ˈrən ¦leŋkth inˈkōd·iŋ }

run-length-limited code [COMMUN] A binary code in which a 1 is inserted after a certain number of 0's, in order to avoid long strings of 0's, which would require very accurate clocking in order to ensure that a bit was not lost. Abbreviated RLL code. { ¦rən ‚leŋkth ‚lim·əd·əd ˈkōd }

running accumulator *See* push-down storage. { ˈrən·iŋ əˈkyü·mə‚lād·ər }

run-time error [COMPUT SCI] An error in a computer program that is not detected until the program is executed, and then causes a processing error to occur. { ˈrən ¦tīm ˈer·ər }

run-time error handler [COMPUT SCI] A system control program that detects and diagnoses run-time errors and issues messages concerning them. { ˈrən ¦tīm ˈer·ər ‚hand·lər }

run-time library [COMPUT SCI] A collection of general-purpose routines that form part of a language translator and allow computer programs to be run with a particular operating system. { ˈrən ¦tīm ˈlī‚brer·ē }

rural radio service [COMMUN] A radio service used to provide public message communication service between a central office and subscribers located in rural areas to which it is impracticable or uneconomic to run wire lines. { ˈrủr·əl ˈrād·ē·ō ‚sər·vəs }

rvalue *See* right value. { ˈär‚val·yü }

RWR *See* radar warning receiver.

S

safety service [COMMUN] Radio communications service used permanently or temporarily for safeguarding human life and property. { 'sāf·tē ˌsər·vəs }

Salisbury dark box [ELECTR] Isolating chamber used for test work in connection with radar equipment; the walls of the chamber are specially constructed to absorb all impinging microwave energy at a certain frequency. { 'sȯlzˌber·ē 'därk 'bäks }

Sallen-Key filter [ELECTR] An electric filter that uses a single amplifier of positive low gain, realized by an operational amplifier and two feedback resistors. { ¦sal·ən 'kē ˌfil·tər }

sample-data tracking [ELECTR] Radar operation in which a target is detected and tracked with subsequent observations made at a rate appropriate to the track and independent of regular search scanning; possible with phased array radars. { 'sam·pəl 'dad·ə 'trak·iŋ }

sampled-data control system [CONT SYS] A form of control system in which the signal appears at one or more points in the system as a sequence of pulses or numbers usually equally spaced in time. { 'sam·pəld ¦dad·ə kən'trōl ˌsis·təm }

sampler [CONT SYS] A device, used in sampled-data control systems, whose output is a series of impulses at regular intervals in time; the height of each impulse equals the value of the continuous input signal at the instant of the impulse. { 'sam·plər }

sampling gate [ELECTR] A gate circuit that extracts information from the input waveform only when activated by a selector pulse. { 'sam·pliŋ ˌgāt }

sampling theorem [COMMUN] The theorem that a signal that varies continuously with time is completely determined by its values at an infinite sequence of equally spaced times if the frequency of these sampling times is greater than twice the highest frequency component of the signal. Also known as Shannon's sampling theorem. { 'sam·pliŋ ˌthir·əm }

sand load [ELECTROMAG] An attenuator used as a power-dissipating terminating section for a coaxial line or waveguide; the dielectric space in the line is filled with a mixture of sand and graphite that acts as a matched-impedance load, preventing standing waves. { 'san ˌlōd }

SAR See synthetic-aperture radar.

SASAR See segmented aperture-synthetic aperture radar. { 'sāˌsär }

satellite-based augmentation system [NAV] A type of wide-area DGPS, intended for aviation users in enroute, terminal, nonprecision approach, and category I (or near category I) precision approach phases of flight, in which the error corrections are broadcast from geostationary satellites and are useful throughout the geographic areas in which these satellites are visible. Abbreviated SBAS. { ¦sad·əlˌīt ˌbāst ˌȯg·mən'tā·shən ˌsis·təm }

satellite communication [COMMUN] Communication that involves the use of an active or passive satellite to extend the range of a communications, radio, television, or other transmitter by returning signals to earth from an orbiting satellite. { 'sad·əlˌīt kəˌmyü·nəˌkā·shən }

satellite computer [COMPUT SCI] A computer which, under control of the main computer, handles the input and output routines, thereby allowing the main computer to be fully dedicated to computations. { 'sad·əlˌīt kəmˌpyüd·ər }

Satellite Digital Audio Radio Service [COMMUN] Referring to satellite-delivered digital audio systems. The digital audio data rate in these systems is specified as being 64 kbits/s. Abbreviated SDARS. { 'sad·əlˌīt 'dij·əd·əl 'ȯd·ē·ō 'rād·ē·ō ˌsər·vəs }

satellite master antenna television system [COMMUN] A master antenna television system equipped with a television receive-only antenna and associated electronics to receive broadcasts relayed by geostationary satellites. Abbreviated SMATV system. { 'sad·əlˌīt ¦mas·tər an¦ten·ə 'tel·əˌvizh·ən ˌsis·təm }

satellite processor [COMPUT SCI] One of the outlying processors in a hierarchical distributed processing system, typically placed at or near point-of-transaction locations, and designed to serve the users at those locations. { 'sad·əlˌīt ˌpräˌses·ər }

saturation signal [ELECTROMAG] A radio signal (or radar echo) which exceeds a certain power level fixed by the design of the receiver equipment; when a receiver or indicator is "saturated," the limit of its power output has been reached. { ˌsach·ə'rā·shən ¦sig·nəl }

sawtooth waveform [ELECTR] A waveform characterized by a slow rise time and a sharp fall, resembling a tooth of a saw. { 'sȯ,tüth 'wāv ,fȯrm }

saxophone [ELECTROMAG] Vertex-fed linear array antenna giving a cosecant-squared radiation pattern. { 'sak·sə,fōn }

S band [COMMUN] A band of radio frequencies extending from 1550 to 5200 megahertz, corresponding to wavelengths of 19.37 to 5.77 centimeters. { 'es ,band }

S-band single-access service [COMMUN] One of the services provided by the Tracking and Data Relay Satellite System, which provides return-link data rates up to 6 megabits per second for each user spacecraft and forward-link data at 300 kilobits per second. Abbreviated SSA { ¦es ,band ,siŋ·gəl 'ak,ses ,sər·vəs }

SBAS *See* satellite-based augmentation system. { 'es,bās or ¦es,bē,ā'es }

SC *See* sectional center.

scalar [COMPUT SCI] A single value or item. { 'skā·lər }

scalar data type [COMPUT SCI] The manner in which a sequence of bits represents a single data item in a computer program. Also known as aggregate data type. { 'skā·lər 'dad·ə ,tīp }

scalar processor [COMPUT SCI] A computer that carries out computations on one number at a time. { 'skā·lər 'prä,ses·ər }

scalar quantization [COMPUT SCI] A data compression technique in which a value is presented (in approximation) by the closest, in some mathematical sense, of a predefined set of allowable values. { ¦skā·lər ,kwän·tə'zā·shən }

scale-of-ten circuit *See* decade scaler. { ¦skāl əv ¦ten 'sər·kət }

scale-of-two circuit *See* binary scaler. { ¦skāl əv ¦tü 'sər·kət }

scaler [ELECTR] A circuit that produces an output pulse when a prescribed number of input pulses is received. Also known as counter; scaling circuit. { skāl·ər }

scaling circuit *See* scaler. { 'skāl·iŋ ,sər·kət }

scan [COMPUT SCI] To examine information, following a systematic, predetermined sequence, for some particular purpose. [ELECTR] The motion, usually periodic, given to the major lobe of an antenna; the process of directing the radio-frequency beam successively over all points in a given region of space. [ENG] **1.** To examine an area, a region in space, or a portion of the radio spectrum point by point in an ordered sequence; for example, conversion of a scene or image to an electric signal or use of radar to monitor an airspace for detection, navigation, or traffic control purposes. **2.** One complete circular, up-and-down, or left-to-right sweep of the radar, light, or other beam or device used in making a scan. { skan }

scan converter [ELECTR] **1.** Equipment that converts radar date images to data at a sampling rate suitable for transmission over telephone lines or narrow-band radio circuits for use at remote locations. Scan converters may work digitally with quantized data; analog ones often use a "memory" scope, a cathode-ray tube of long persistence, permitting nondestructive readout of radar, television, and data displays. **2.** A cathode-ray tube that is capable of storing radar, television, and data displays for nondestructive readout over prolonged periods of time. { 'skan kən,vərd·ər }

scan line [ELECTR] A horizontal row of pixels on a video screen that are examined or refreshed in succession in one sweep across the screen during the scanning process. { 'skan ,līn }

scanner [COMMUN] That part of a facsimile transmitter which systematically translates the densities of the elemental areas of the subject copy into corresponding electric signals. [COMPUT SCI] A device that converts an image of something outside a computer, such as text, a drawing, or a photograph, into a digital image that it sends into the computer for display or further processing. { 'skan·ər }

scanner selector [COMPUT SCI] An electronic device interfacing computer and multiplexers when more than one multiplexer is used. { 'skan·ər si,lek·tər }

scanning circuit *See* sweep circuit. { 'skan·iŋ ,sər·kət }

scanning frequency *See* stroke speed. { 'skan·iŋ ,frē·kwən·sē }

scanning line [COMMUN] **1.** In a video system, a single, continuous, narrow strip which is determined by the process of scanning. **2.** Path traced by the scanning or recording spot in one sweep across the subject copy or record sheet. { 'skan·iŋ ,līn }

scanning linearity [ELECTR] In a video system, the uniformity of scanning speed during the trace interval. { 'skan·iŋ ,lin·ē'ar·əd·ē }

scanning line frequency *See* stroke speed. { 'skan·iŋ ¦līn ,frē·kwən·sē }

scanning loss [ELECTROMAG] In a radar system employing a scanning antenna, the reduction in sensitivity (usually expressed in decibels) due to scanning across the target, compared with that obtained when the beam is directed constantly at the target. { 'skan·iŋ ,lȯs }

scanning radio [ELECTR] A radio receiver that automatically scans across public service, emergency service, or other radio bands and stops at the first preselected station which is on the air. Also known as radio scanner. { 'skan·iŋ 'rād·ē·ō }

scanning speed *See* spot speed. { 'skan·iŋ ,spēd }

scanning spot *See* picture element. { 'skan·iŋ ,spät }

scanning yoke *See* deflection yoke. { 'skan·iŋ ,yōk }

scatter band [COMMUN] In pulse interrogation systems, the total bandwidth occupied by the frequency spread by numerous interrogations operating on the same nominal radio frequency. { 'skad·ər ,band }

scatterer [ELECTROMAG] Object in an otherwise relatively homogeneous propagation medium that intercepts electromagnetic waves such as radar signals and reflects them in directions associated with the shape and composition of the object. Examples include individual raindrops, earth surface features, sea-wave crests, buildings, and vehicles. { 'skad·ər·ər }

scattering [ELECTROMAG] Diffusion of electromagnetic waves in a random manner by air masses in the upper atmosphere, permitting long-range reception, as in scatter propagation. Also known as radio scattering. { 'skad·ə·riŋ }

scattering coefficient [ELECTROMAG] One of the elements of the scattering matrix of a waveguide junction; that is, a transmission or reflection coefficient of the junction. { 'skad·ə·riŋ ˌkō·iˌfish·ənt }

scattering cross section [ELECTROMAG] The power of electromagnetic radiation scattered by an antenna divided by the incident power. { 'skad·ə·riŋ 'kròs ˌsek·shən }

scattering matrix [ELECTROMAG] A square array of complex numbers consisting of the transmission and reflection coefficients of a waveguide junction. { 'skad·ə·riŋ ˌmā·triks }

scatter loading [COMPUT SCI] The process of loading a program into main memory such that each section or segment of the program occupies a single, connected memory area but the several sections of the program need not be adjacent to each other. { 'skad·ər ˌlōd·iŋ }

scatter propagation [ELECTROMAG] Transmission of radio waves far beyond line-of-sight distances by using high power and a large transmitting antenna to beam the signal upward into the atmosphere and by using a similar large receiving antenna to pick up the small portion of the signal that is scattered by the atmosphere. Also known as beyond-the-horizon communication; forward-scatter propagation; over-the-horizon propagation. { 'skad·ər ˌpräp·əˌgā·shən }

scatter read [COMPUT SCI] An input operation that places various segments of an input record into noncontiguous areas in central memory. { 'skad·ər ˌrēd }

scatter reflections [ELECTROMAG] Reflections from portions of the ionosphere having different virtual heights, which mutually interfere and cause rapid fading. { 'skad·ər riˌflek·shənz }

scene analysis *See* picture segmentation. { 'sēn əˌnal·ə·səs }

scheduled down time [COMPUT SCI] A period of time designated for closing down a computer system for preventive maintenance. { 'skej·əld 'daůnˌtīm }

scheduler [COMPUT SCI] A system control program that determines the sequence in which programs will be processed by a computer and automatically submits them for execution at predetermined times. { 'skej·ə·lər }

scheduling algorithm [COMPUT SCI] A systematic method of determining the order in which tasks will be performed by a computer system, generally incorporated into the operating system. { 'skej·ə·liŋ ˌal·gəˌrith·əm }

schema [COMPUT SCI] A logical description of the data in a data base, including definitions and relationships of data. { 'skē·mə }

Schmitt circuit [ELECTR] A bistable pulse generator in which an output pulse of constant amplitude exists only as long as the input voltage exceeds a certain value. Also known as Schmitt limiter; Schmitt trigger. { 'shmit ˌsər·kət }

Schmitt limiter *See* Schmitt circuit. { 'shmit 'lim·əd·ər }

Schmitt trigger *See* Schmitt circuit. { 'shmit 'trig·ər }

Schottky barrier diode [ELECTR] A semiconductor diode formed by contact between a semiconductor layer and a metal coating; it has a nonlinear rectifying characteristic; hot carriers (electrons for *n*-type material or holes for *p*-type material) are emitted from the Schottky barrier of the semiconductor and move to the metal coating that is the diode base; since majority carriers predominate, there is essentially no injection or storage of minority carriers to limit switching speeds. Also known as hot-carrier diode; Schottky diode. { 'shät·kē ¦bar·ē·ər 'dīˌōd }

Schottky diode *See* Schottky barrier diode. { 'shät·kē 'dīˌōd }

Schottky-diode FET logic [ELECTR] A logic gate configuration used with gallium-arsenide field-effect transistors operating in the depletion mode, in which very small Schottky diodes at the gate input provide the logical OR function and the level shifting required to make the input and output voltage levels compatible. Abbreviated SDFL. { 'shät·kē ¦dīˌōd ¦ef¦ē¦tē 'läj·ik }

Schottky noise *See* shot noise. { 'shät·kē ˌnòiz }

Schottky transistor-transistor logic [ELECTR] A transistor-transistor logic circuit in which a Schottky diode with forward diode voltage is placed across the base-collector junction of the output transistor in order to improve the speed of the circuit. { 'shät·kē tran¦zis·tər tran¦zis·tər 'läj·ik }

scientific calculator [COMPUT SCI] An electronic calculator that has provisions for handling exponential, trigonometric, and sometimes other special functions in addition to performing arithmetic operations. { ˌsī·ən'tif·ik 'kal·kyə ˌlād·ər }

scientific computer [COMPUT SCI] A computer which has a very large memory and is capable of handling extremely high-speed arithmetic and a very large variety of floating-point arithmetic commands. { ˌsī·ən'tif·ik kəm'pyüd·ər }

scientific notation [COMPUT SCI] The display of numbers in which a base number, representing the significant digits, is followed by a number representing the power of 10 to which the base number is raised. { ˌsī·ən¦tif·ik nō'tā·shən }

scientific system [COMPUT SCI] A system devoted principally to computations as opposed to business and data-processing systems, the main emphasis of which is on the updating of data records and files rather than the performance of calculations. { ˌsī·ən'tif·ik 'sis·təm }

scintillation [ELECTROMAG] **1.** Fluctuation in radar echo amplitude, usually that associated with atmospheric irregularities in the propagation path. **2.** Random fluctuation, in radio propagation, of the received field about its mean value, the deviations usually being relatively small. { ˌsint·əlˈā·shən }

scissoring [COMPUT SCI] In computer graphics, the deletion of those parts of an image that fall outside a window that has been placed over the original image. Also known as clipping. { ˈsiz·ər·iŋ }

scope [COMPUT SCI] For a variable in a computer program, the portion of the computer program within which the variable can be accessed (used or changed). [ELECTR] *See* radarscope. { skōp }

SCR *See* system clock reference.

scramble [COMMUN] To mix, in cryptography, in random or quasi-random fashion. { ˈskram·bəl }

scrambler [ELECTR] A circuit that divides speech frequencies into several ranges by means of filters, then inverts and displaces the frequencies in each range so that the resulting reproduced sounds are unintelligible; the process is reversed at the receiving apparatus to restore intelligible speech. Also known as speech inverter; speech scrambler. { ˈskram·blər }

scrambling [COMMUN] The alteration of the characteristics of a video, audio, or coded data stream in order to prevent unauthorized reception of the information in a clear form. { ˈskram·bliŋ }

scratch [COMPUT SCI] To remove data or to set up its identifying labels so that new data can be written over it. { skrach }

scratch file [COMPUT SCI] A temporary file for future use, created by copying all or part of a data set to an auxiliary memory device. { ˈskrach ˌfīl }

scratch-pad memory [COMPUT SCI] A very fast intermediate storage (in the form of flip-flop register or semiconductor memory) which often supplements main core memory. { ˈskrach ˌpad ˌmem·rē }

scratch tape [COMPUT SCI] A reel of magnetic tape containing data that may now be destroyed. { ˈskrach ˌtāp }

screen [COMPUT SCI] To make a preliminary selection from a set of entities, selection criteria being based on a given set of rules or conditions. [ELECTR] **1.** The surface on which an image is made visible for viewing; it may be a fluorescent screen with a phosphor layer that converts the energy of an electron beam to visible light, or a translucent or opaque screen on which the optical image is projected, or a display surface of the types commonly used in computers. **2.** *See* screen grid. [ELECTROMAG] Metal partition or shield which isolates a device from external magnetic or electric fields. { skrēn }

screen angle [ELECTROMAG] Vertical angle bounded by a straight line from the radar antenna to the horizon and the horizontal at the antenna assuming a $\frac{4}{3}$ earth's radius. { ˈskrēn ˌaŋ·gəl }

screen capture *See* screen shot. { ˈskrēn ˌkap·chər }

screen dump [COMPUT SCI] **1.** The printing of everything that appears on a computer screen. **2.** The printed copy that results from this action. { ˈskrēn ˌdəmp }

screen format [COMPUT SCI] The manner in which information is arranged and presented on a cathode-ray tube or other electronic display. { ˈskrēn ˌfȯrˌmat }

screen formatter [COMPUT SCI] A computer program that enables the user to design and set up screen formats. Also known as screen generator; screen painter. { ˈskrēn ˌfȯrˌmad·ər }

screen generator *See* screen formatter. { ˈskrēn ˌjen·əˌrād·ər }

screen grid [ELECTR] A grid placed between a control grid and an anode of an electron tube, and usually maintained at a fixed positive potential, to reduce the electrostatic influence of the anode in the space between the screen grid and the cathode. Also known as screen. { ˈskrēn ˌgrid }

screen image buffer [COMPUT SCI] A section of computer storage that contains a representation of the information that appears on an electronic display. Abbreviated SIB. { ˈskrēn ¦im·ij ˌbəf·ər }

screening *See* electric shielding. { ˈskrēn·iŋ }

screen memory [COMPUT SCI] The portion of a microcomputer storage that is reserved for setting up screen formats. { ˈskrēn ¦mem·rē }

screen overlay [COMPUT SCI] **1.** An array of cells on a video display screen that allow a user to command a computer by touching buttons displayed on the screen at the locations of the cells. **2.** A window of data that is temporarily displayed on a screen, leaving the original display intact when the window is removed. { ¦skrēn ˈō·vərˌlā }

screen painter *See* screen formatter. { ˈskrēn ˌpān·tər }

screen saver [COMPUT SCI] A program that launches when a computer is not in use for a predetermined period, displaying various transient or moving images on a computer screen. Originally used to prevent computer screen damage from prolonged display of a static image, screen savers are now more of an amusement or security feature as modern monitors are less susceptible to screen burning. { ˈskrēn ˌsāv·ər }

screen shot [COMPUT SCI] A digital image or file containing all or part of what is seen on a computer display. Also known as screen capture. { ˈskrēn ˌshät }

script [COMPUT SCI] An executable list of commands written in a programming language. { skript }

scripting language [COMPUT SCI] An interpreted language (for example, JavaScript and Perl) used to write simple programs, called scripts. { ˈskrip·tiŋ ˌlaŋ·gwij }

scroll [COMPUT SCI] To move information in an electronic display up, down, left, or right, so that new information appears and some of the existing information is moved away. { skrōl }

scroll arrow [COMPUT SCI] An arrow on a video display screen that is clicked in order to scroll the screen in the corresponding direction. { 'skrōl ˌa·rō }

scroll bar [COMPUT SCI] A horizontal or vertical bar that contains a box that is clicked and dragged up, down, left, or right in order to scroll the screen. { 'skrōl ˌbär }

scrolling [COMPUT SCI] The continuous movement of information either vertically or horizontally on a video screen. { 'skrōl·iŋ }

scrub [COMPUT SCI] To examine a large amount of data and eliminate duplicate or unneeded items. { skrəb }

SCSI *See* small computer system interface. { 'skəz·ē }

scuzzy *See* small computer system interface. { 'skəz·ē }

SDARS *See* Satellite Digital Audio Radio Service.

SDFL *See* Schottky-diode FET logic.

SDHT *See* high-electron-mobility transistor.

SDMA *See* space-division multiple access.

SDRAM *See* synchronous dynamic random access memory. { ¦es¦dē'ram }

SDTV *See* standard definition television.

sea clutter [ELECTROMAG] A clutter on an airborne radar due to reflection of signals from the sea. Also known as sea return; wave clutter. { 'sē ˌkləd·ər }

seamless integration [COMPUT SCI] The addition of a routine or program that works smoothly with an existing system and can be activated and used as if it had been built into the system when the system was put together. { ¦sēm·ləs ˌint·ə'grā·shən }

search [COMPUT SCI] To seek a desired item or condition in a set of related or similar items or conditions, especially a sequentially organized or nonorganized set, rather than a multidimensional set. [ENG] To explore a region in space with radar. { sərch }

search-and-rescue coordination center [COMMUN] A primary search and rescue facility suitably staffed by supervisory personnel and equipped for coordinating and controlling search and rescue operations. { ¦sərch ən ¦res ˌkyü kō'ȯrd·ənˌā·shən ˌsen·tər }

search antenna [ELECTROMAG] A radar antenna or antenna system designed for search. { 'sərch anˌten·ə }

search argument [COMPUT SCI] The item or condition that is desired in a search procedure. { 'sərch ˌar·gyə·mənt }

search engine [COMPUT SCI] **1.** Any software that locates and retrieves information in a database. **2.** A server with a stored index of Web pages that is capable of returning lists of pages that match keyword queries. { 'sərch ˌen·jən }

search field [COMPUT SCI] A field in a record or segment whose value is examined in a search. { 'sərch ˌfēld }

search key [COMPUT SCI] A data item, or the value of a data item, that is used in carrying out a search. { 'sərch ˌkē }

search radar [ENG] A radar the purpose of which is to detect targets of interest in its surroundings not previously detected, report the presence and location of these to a user system, and continue to do so while the targets remain in the radar's field of view. { 'sərch ˌrāˌdär }

search time [COMPUT SCI] Time required to locate a particular field of data in a computer storage device; requires a comparison of each field with a predetermined standard until an identity is obtained. { 'sərch ˌtīm }

sea return *See* sea clutter. { 'sē riˌtərn }

seasonal factors [COMMUN] Factors that are used to adjust skywave absorption data for seasonal variations; these variations are due primarily to seasonal fluctuations in the heights of the ionospheric layers. { 'sēz·ən·əl 'fak·tərz }

secondary allocation [COMPUT SCI] An area of disk storage that is assigned to a file which has become too large for the area originally assigned to it. { 'sek·ənˌder·ē ˌal·ə'kā·shən }

secondary cache [COMPUT SCI] High-speed memory between the primary cache and main memory that supplies the processor with the most frequently requested data and instructions. Also known as level 2 cache. { ˌsek·ənˌder·ē 'kash }

secondary cell *See* storage cell. { 'sek·ənˌder·ē 'sel }

secondary index [COMPUT SCI] An index that provides an alternate method of accessing records or portions of records in a data base or file. Also known as alternate index. { 'sek·ən ˌder·ē 'inˌdeks }

secondary key [COMPUT SCI] A key that holds the physical location of a record or a portion of a record in a file or database, and provides an alternative means of accessing data. Also known as alternate key. { 'sek·ənˌder·ē 'kē }

secondary lobe *See* minor lobe. { 'sek·ənˌder·ē 'lōb }

secondary radar [ELECTR] A radar system in which the transmitted signal from its interrogator causes a transponder borne by a cooperative aircraft to transmit a response on a separate frequency that is received and interpreted by the interrogating radar. { 'sek·ənˌder·ē 'rāˌdär }

secondary station [COMMUN] Any station in a radio network other than the net control station. { 'sek·ənˌder·ē 'stā·shən }

secondary storage [COMPUT SCI] Any means of storing and retrieving data external to the main computer itself but accessible to the program. { 'sek·ənˌder·ē 'stȯr·ij }

secondary surveillance radar [NAV] The secondary radar that operates in conjunction with the airborne transponder of the air-traffic control radar beacon system (ATCRBS). { 'sek·ənˌder·ē sər'vā·ləns ˌrāˌdär }

second-channel interference *See* alternate-channel interference. { 'sek·ənd ¦chan·əl ˌin·tər'fir·əns }

second detector [ELECTR] The detector that separates the intelligence signal from the intermediate-frequency signal in a superheterodyne receiver. { 'sek·ənd di'tek·tər }

second-generation computer [COMPUT SCI] A computer characterized by the use of transistors rather than vacuum tubes, the execution of input/output operations simultaneously with calculations, and the use of operating systems. { 'sek·ənd ˌjen·ə¦rā·shən kəm'pyüd·ər }

second-level controller [CONT SYS] A controller which influences the actions of first-level controllers, in a large-scale control system partitioned by plant decomposition, to compensate for subsystem interactions so that overall objectives and constraints of the system are satisfied. Also known as coordinator. { 'sek·ənd ¦lev·əl kən'trōl·ər }

second-order subroutine [COMPUT SCI] A subroutine that is entered from another subroutine, in contrast to a first-order subroutine; it constitutes the second level of a two-level or higher-level routine. Also known as second-remove subroutine. { 'sek·ənd ¦ȯr·dər 'səb·rü,tēn }

second-remove subroutine *See* second-order subroutine. { 'sek·ənd ri¦müv 'səb·rü,tēn }

second-time-around echo [ELECTR] A radar echo received from one pulse after the transmission of a subsequent pulse and liable to be associated with the latter, giving an erroneous indication of range. { 'sek·ənd ¦tīm ə'raùnd ˌek·ō }

second-trip echo *See* second-time-around echo. { 'sek·ənd ¦trip ˌek·ō }

secrecy system *See* privacy system. { 'sē·krə·sē ˌsis·təm }

secret-key algorithm [COMPUT SCI] A cryptographic algorithm which uses the same cryptographic key for encryption and decryption, requiring that the key first be transmitted from the sender to the recipient via a secure channel. { ˌsē·krət ˌkē 'al·gəˌrith·əm }

section [COMMUN] Each individual transmission span in a radio relay system; a system has one more section than it has repeaters. { 'sek·shən }

sectional center [COMMUN] A long-distance telephone office which connects several primary centers and which is in class number 2; only a regional center has greater importance in routing telephone calls. Abbreviated SC. { 'sek·shən·əl 'sen·tər }

sectionalized vertical antenna [ELECTROMAG] Vertical antenna that is insulated at one or more points along its length; the insertion of suitable reactances or applications of a driving voltage across the insulated points results in a modified current distribution giving a more desired radiation pattern in the vertical plane. { 'sek·shən·əl,īzd 'vərd·ə·kəl anˌten·ə }

sector [COMPUT SCI] **1.** A portion of a track on a magnetic disk or a band on a magnetic drum. **2.** A unit of data stored in such a portion. [ELECTROMAG] Coverage of a radar as measured in azimuth. { 'sek·tər }

sectoral horn [ELECTROMAG] Horn with two opposite sides parallel and the two remaining sides which diverge. { 'sek·tə·rəl 'hȯrn }

sector display [ELECTR] A display in which only a sector of the total service area of a radar system is shown; usually the sector is selectable. { 'sek·tər diˌsplā }

sector interleave [COMPUT SCI] A sequence indicating the order in which sectors are arranged on a hard disk, generally so as to minimize access times. Also known as sector map. { 'sek·tər 'in·tərˌlēv }

sector map *See* sector interleave. { 'sek·tər ˌmap }

sector mark [COMPUT SCI] A location on each sector of each track of a disk pack or floppy disk that gives the sector's address, tells whether the sector is in use, and gives other control information. { 'sek·tər ˌmärk }

sector scan [ELECTR] A radar scan through a limited angle, as distinguished from complete rotation. { 'sek·tər ˌskan }

secure visual communications [COMMUN] The transmission of an encrypted digital signal consisting of animated visual and audio information; the distance may vary from a few hundred feet to thousands of miles. { si'kyůr 'vizh·ə·wəl kə ˌmyü·nə'kā·shənz }

secure voice [COMMUN] Voice message that is scrambled or coded, therefore not transmitted in the clear. { si'kyůr 'vȯis }

security [COMPUT SCI] The existence and enforcement of techniques which restrict access to data, and the conditions under which data may be obtained. { si'kyůr·əd·ē }

security kernel [COMPUT SCI] A portion of an operating system into which all security-related functions have been concentrated, forming a small, certifiably secure nucleus which is separate from the rest of the system. { si'kyůr·əd·ē ˌkər·nəl }

security perimeter [COMPUT SCI] A logical boundary of a distributed computer system, surrounding all the resources that are controlled and protected by the system. { sə'kyůr·əd·ē pə ˌrim·əd·ər }

security reporting/alerting system [COMMUN] A rapid communications procedure that integrates all U.S. Air Force bases and commands, so that a significant happening at one location, or a pattern of seemingly unrelated happenings at several locations, can serve as a basis for swift security alerting or warning throughout the system. { si'kyůr·əd·ē ri¦pȯrd·iŋ ə¦lərd·iŋ ˌsis·təm }

security target [COMPUT SCI] A description of a product meeting the security and functionality requirements of a computing system. { sə'kyůr·əd·ē ˌtär·gət }

seed [COMPUT SCI] An initial number used by an algorithm such as a random number generator. { sēd }

seek [COMPUT SCI] **1.** To position the access mechanism of a random-access storage device at a designated location or position. **2.** The command that directs the positioning to take place. { sēk }

seek area [COMPUT SCI] An area of a direct-access storage device, such as a magnetic disk file, assigned to hold records to which rapid access is needed, and located so that the physical characteristics of the device permit such access. Also known as cylinder. { 'sēk ,er·ē·ə }

seek time [COMPUT SCI] The time required for the access mechanism of a random-access storage device to be properly positioned. { 'sēk ,tīm }

segment [COMPUT SCI] **1.** A single section of an overlay program structure, which can be loaded into the main memory when and as needed. **2.** In some direct-access storage devices, a hardware-defined portion of a track having fixed data capacity. { 'seg·mənt }

segmentation [COMMUN] The division of a long communications message into smaller messages that can be transmitted intermittently. [COMPUT SCI] **1.** The division of virtual storage into identifiable functional regions, each having enough addresses so that programs or data stored in them will not assign the same addresses more than once. **2.** The division of a large computer program into smaller units, called segments. **3.** *See* picture segmentation. { ,seg·mən'tā·shən }

segmented aperture-synthetic aperture radar [ENG] An enhancement of synthetic aperture radar that overcomes restrictions on the effective length of the receiving antenna by using a receiving antenna array composed of a set of contiguous subarrays and employing signal processing to provide the proper phase corrections for each subarray. Abbreviated SASAR. { 'seg ,ment·əd ¦ap·ə·chər sin'thed·ik ¦ap·ə·chər 'rā,där }

segment mark [COMPUT SCI] A special character written on tape to separate one section of a tape file from another. { 'seg·mənt ,märk }

select [COMPUT SCI] **1.** To choose a needed subroutine from a file of subroutines. **2.** To take one alternative if the report on a condition is of one state, and another alternative if the report on the condition is of another state. **3.** To pull from a mass of data certain items that require special attention. { si'lekt }

select bit [COMPUT SCI] The bit (or bits) in an input/output instruction word which selects the function of a specified device. Also known as subdevice bit. { si'lekt ,bit }

selection [COMMUN] The process of addressing a call to a specific station in a selective calling system. { si'lek·shən }

selection check [COMPUT SCI] Electronic computer check, usually automatic, to verify that the correct register, or other device, is selected in the performance of an instruction. { si'lek·shən ,chek }

selection sort [COMPUT SCI] A sorting routine that scans a list of items repeatedly and, on each pass, selects the item with the lowest value and places it in its final position. { si'lek·shən ,sòrt }

selective absorption [ELECTROMAG] A greater absorption of electromagnetic radiation at some wavelengths (or frequencies) than at others. { si'lek·tiv ab'sòrp·shən }

selective calling system [COMMUN] A radio communications system in which the central station transmits a coded call that activates only the receiver to which that code is assigned. { si'lek·tiv ,kòl·iŋ ,sis·təm }

selective dump [COMPUT SCI] An edited or nonedited listing of the contents of selected areas of memory or auxiliary storage. { si'lek·tiv 'dəmp }

selective fading [COMMUN] Fading that is different at different frequencies in a frequency band occupied by a modulated wave, causing distortion that varies in nature from instant to instant. { si'lek·tiv 'fād·iŋ }

selective identification feature [ELECTR] Airborne pulse-type transponder which provides automatic selective identification of aircraft in which it is installed to ground, shipboard, or airborne recognition installations. { si'lek·tiv ī,den·tə·fə'kā·shən ,fē·chər }

selective interference [COMMUN] Interference whose energy is concentrated in a narrow band of frequencies. { si'lek·tiv ,in·tər'fir·əns }

selective jamming [ELECTR] Jamming in which only a single radio channel is jammed. { si'lek·tiv 'jam·iŋ }

selectively doped heterojunction transistor *See* high-electron-mobility transistor. { si'lek·tiv·lē ¦dōpt ¦hed·ə·rō¦jəŋk·shən tran'zis·tər }

selective reflection [ELECTROMAG] Reflection of electromagnetic radiation more strongly at some wavelengths (or frequencies) than at others. { si'lek·tiv ri'flek·shən }

selective ringing [COMMUN] Telephone arrangement on party lines, in which only the bell of the called subscriber rings, with other bells on the party line remaining silent. { si'lek·tiv 'riŋ·iŋ }

selective scattering [ELECTROMAG] Scattering of electromagnetic radiation more strongly at some wavelengths than at others. { si'lek·tiv 'skad·ə·riŋ }

selective trace [COMPUT SCI] A tracing routine wherein only instructions satisfying certain specified criteria are subject to tracing. { si'lek·tiv 'trās }

selectivity [ELECTR] **1.** The ability of a radio receiver to separate a desired signal frequency from other signal frequencies, some of which may differ only slightly from the desired value. **2.** The inverse of the shape factor of a bandpass filter. { sə,lek'tiv·əd·ē }

selector [COMPUT SCI] Computer device which interrogates a condition and initiates a particular operation dependent upon the report. { si'lek·tər }

selector channel [COMPUT SCI] A unit which connects high-speed input/output devices, such as magnetic tapes, disks, and drums, to a computer memory. { si'lek·tər ,chan·əl }

self-adjusting communications *See* adaptive communications. { ¦self ə¦jəst·iŋ kə,myü·nə'kā·shənz }

self-checking code [COMPUT SCI] An encoding of data so designed and constructed that an invalid code can be rapidly detected; this permits the detection, but not the correction, of almost all errors. Also known as error-checking code; error-detecting code. { 'self ¦chek·iŋ 'kōd }

self-checking number [COMPUT SCI] A number with a suffix figure related to the figure of the number, used to check the number after it has been transferred from one medium or device to another. { 'self ¦chek·iŋ 'nəm·bər }

self-complementing code [COMPUT SCI] A binary-coded-decimal code in which the combination for the complement of a digit is the complement of the combination for that digit. { ¦self ¦käm·plə,ment·iŋ 'kōd }

self-contained database management system [COMPUT SCI] A database management system that is in no way an extension of any programming language, and is usually quite independent of any language. { ¦self kən¦tānd ¦dad·ə¦bās 'man·ij·mənt ,sis·təm }

self-diagnostic routine [COMPUT SCI] A test of an electronic device that is performed automatically, usually when the device is turned on. Also known as self-test. { ¦self ,dī·əg¦näs·tik rü'tēn }

self-documenting code [COMPUT SCI] A sequence of programming statements that are simple and straightforward and can be readily implemented by another programmer. { ¦self ,däk·yə,ment·iŋ 'kōd }

self-extracting file [COMPUT SCI] A compressed (zipped) file that unzips itself when it is executed. { ,self ik,strak·tiŋ 'fīl }

self-optimizing communications See adaptive communications. { ¦self ¦äp·tə,mīz·iŋ kə,myü·nə'kā·shənz }

self-repair [COMPUT SCI] Any type of hardware redundancy in which faults are selectively masked and are detected, located, and subsequently corrected by the replacement of the failed unit by an unfailed replica. { ¦self ri¦per }

self-resetting loop [COMPUT SCI] A loop whose termination causes the numbers stored in all locations affected by the loop to be returned to the original values which they had upon entry into the loop. { ¦self ri¦sed·iŋ 'lüp }

self-steering microwave array [ELECTROMAG] An antenna array used with electronic circuitry that senses the phase of incoming pilot signals and positions the antenna beam in their direction of arrival. { 'self ¦stir·iŋ 'mī·krō,wāv ə'rā }

self-test See self-diagnostic routine. { 'self ¦test }

self-triggering program [COMPUT SCI] A computer program which automatically commences execution as soon as it is fed into the central processing unit. { ¦self ¦trig·ə·riŋ 'prō·grəm }

self-tuning regulator [CONT SYS] A type of adaptive control system composed of two loops, an inner loop which consists of the process and an ordinary linear feedback regulator, and an outer loop which is composed of a recursive parameter estimator and a design calculation, and which adjusts the parameters of the regulator. Abbreviated STR. { ¦self ¦tün·iŋ 'reg·yə,lād·ər }

semantic analysis [COMPUT SCI] A phase of natural language processing, following parsing, that involves extraction of context-independent aspects of a sentence's meaning, including the semantic roles of entities mentioned in the sentence, and quantification information, such as cardinality, iteration, and dependency. { si'man·tik ə'nal·ə·səs }

semantic error [COMPUT SCI] The use of an incorrect symbolic name in a computer program. { si'man·tik 'er·ər }

semantic extension [COMPUT SCI] An extension mechanism which introduces new kinds of objects into an extensible language, such as additional data types or operations. { si'man·tik ik'sten·shən }

semantic gap [COMPUT SCI] The difference between a data or language structure and the objects that it models. { si'man·tik 'gap }

semaphore [COMPUT SCI] A memory cell that is shared by two parallel processes which rely on each other for their continued operation, and that provides an elementary form of communication between them by indicating when significant events have taken place. { 'sem·ə,fȯr }

semialgorithm [COMPUT SCI] A procedure for solving a problem that will continue endlessly if the problem has no solution. { ,sem·ē'al·gə ,rith·əm }

semiautomatic telephone system [COMMUN] Telephone system that limits automatic dialing to only those subscribers who are served by the same exchange as the calling subscriber. { ¦sem·ē,ȯd·ə'mad·ik 'tel·ə,fōn ,sis·təm }

semiconductor [ELECTR] A solid crystalline material whose conductivity is intermediate between that of a metal and an insulator and may depend on temperature or voltage; by making suitable contacts to the material or by making the material suitably inhomogenous, elctrical rectification and amplification may be obtained. { ¦sem·i·kən¦dək·tər }

semiconductor device [ELECTR] Electronic device in which the characteristic distinguishing electronic conduction takes place within a semiconductor. { ¦sem·i·kən¦dək·tər di,vīs }

semiconductor diode [ELECTR] **1.** A two-electrode semiconductor device that utilizes the rectifying properties of a *pn* junction or a point contact. **2.** More generally, any two-terminal electronic device that utilizes the properties of the semiconductor from which it is constructed. Also known as crystal diode; crystal rectifier; diode. { ¦sem·i·kən¦dək·tər 'dī,ōd }

semiconductor disk [COMPUT SCI] A large semiconductor memory that imitates a disk drive in that the operating system can read and write to it as though it were an ordinary disk, but at a much faster rate. Also known as nonrotating disk. { 'sem·i·kən,dək·tər ,disk }

semiconductor doping See doping. { ¦sem·i·kən¦dək·tər 'dōp·iŋ }

semiconductor laser [OPTICS] A laser in which stimulated emission of coherent light occurs at a *pn* junction when electrons and holes are driven into the junction by carrier injection, electron-beam excitation, impact ionization, optical excitation, or other means; used as light transmitters and modulators in optical communications and integrated optics. Also known as diode laser; laser diode. { ¦sem·i·kən¦dək·tər ′lā·zər }

semiconductor memory [COMPUT SCI] A device for storing digital information that is fabricated by using integrated circuit technology. Also known as integrated-circuit memory; large-scale integrated memory; memory chip; semiconductor storage; transistor memory. { ¦sem·i·kən ¦dək·tər ˌmem·rē }

semiconductor storage See semiconductor memory. { ′sem·i·kənˌdək·tər ˌstȯr·ij }

semidense list [COMPUT SCI] A list that can be divided into two contiguous portions, with all the cells in the larger portion filled and all the other cells empty. { ¦sem·i′dens ′list }

seminumerical algebraic manipulation language [COMPUT SCI] The most elementary type of algebraic manipulation language, constructed to manipulate data from rigid classes of mathematical objects possessing strictly canonical forms. { ¦sem·i·nü′mer·ə·kəl ˌal·jə′brā·ik məˌnip·yə′lā·shən ˌlaŋ·gwij }

semiselective ringing [COMMUN] In telephone service, party line ringing wherein the bells of two stations are rung simultaneously; the differentiation is made by the number of rings. { ¦sem·i·si′lek·tiv ′riŋ·iŋ }

semisynchronous satellite [AERO ENG] An artificial earth satellite that makes one revolution in exactly one-half of a sidereal day (11 hours 58 minutes 2 seconds). { ˌsem·ēˌsiŋ·krə·nəs ′sad·əlˌīt }

sender [COMMUN] Part of an automatic-switching telephone system that receives pulses from a dial or other source and, in accordance with them, controls the further operations necessary in establishing a telephone connection. { ′sen·dər }

sense [NAV] The general direction from which a radio signal arrives; if a radio bearing is received by a simple loop antenna, there are two possible readings approximately 180° apart; the resolving of this ambiguity is called sensing of the bearing. { sens }

sense amplifier [ELECTR] Circuit used to determine either a phase or voltage change in communications-electronics equipment and to provide automatic control function. { ′sens ˌam·pləˌfī·ər }

sense antenna [ELECTROMAG] An auxiliary antenna used with a directional receiving antenna to resolve a 180° ambiguity in the directional indication. Also known as sensing antenna. { ′sens anˌten·ə }

sense light [COMPUT SCI] A light which can be turned on or off, its status being the determinant as to which path a program will select. { ′sens ˌlīt }

sensing antenna See sense antenna. { ′sens·iŋ anˌten·ə }

sensing signal [COMMUN] A special signal that is transmitted to alert the receiving station at the beginning of a message. { ′sens·iŋ ˌsig·nəl }

sensitive data [COMPUT SCI] Data that can be read or processed in specified transactions by a specified program, device, or user. { ′sen·səd·iv ′dad·ə }

sensitivity [ELECTR] **1.** The minimum input signal required to produce a specified output signal, for a radio receiver or similar device. **2.** Of a camera tube, the signal current developed per unit incident radiation, that is, per watt per unit area. { ˌsen·sə′tiv·əd·ē }

sensitivity function [CONT SYS] The ratio of the fractional change in the system response of a feedback-compensated feedback control system to the fractional change in an open-loop parameter, for some specified parameter variation. { ˌsen·sə′tiv·əd·ē ˌfəŋk·shən }

sensitivity time control [ELECTR] A controlled reduction in sensitivity of a radar receiver immediately after the transmission of a pulse, with a programmed restoration of full sensitivity as returns come from greater ranges; done to prevent the reception of a multitude of tiny targets close to the radar, such as birds and insects, and to prevent receiver saturation by large targets at very short range. { ˌsen·sə′tiv·əd·ē ′tīm kən ˌtrōl }

sentence [COMPUT SCI] An entire instruction in the COBOL programming language. { ′sent·əns }

sentinel [COMPUT SCI] Symbol marking the beginning or end of an element of computer information such as an item or a tape. { ′sent·ən·əl }

separation theorem [CONT SYS] A theorem in optimal control theory which states that the solution to the linear quadratic Gaussian problem separates into the optimal deterministic controller (that is, the optimal controller for the corresponding problem without noise) in which the state used is obtained as the output of an optimal state estimator. { ˌsep·ə′rā·shən ˌthir·əm }

separator [COMPUT SCI] A datum or character that denotes the beginning or ending of a unit of data. [ELECTR] A circuit that separates one type of signal from another by clipping, differentiating, or integrating action. { ′sep·əˌrād·ər }

separator page [COMPUT SCI] A page preceding or following a report in a computer printout giving all information needed to identify the report. { ′sep·əˌrād·ər ˌpāj }

sepatrix [CONT SYS] A curve in the phase plane of a control system representing the solution to the equations of motion of the system which would cause the system to move to an unstable point. { ′sep·əˌtriks }

septate coaxial cavity [ELECTROMAG] Coaxial cavity having a vane or septum, added between the inner and outer conductors, so that it acts as a cavity of a rectangular cross section bent transversely. { 'sep,tāt kō'ak·sē·əl 'kav·əd·ē }

septate waveguide [ELECTROMAG] Waveguide with one or more septa placed across it to control microwave power transmission. { 'sep,tāt 'wāv ,gīd }

septum [ELECTROMAG] A metal plate placed across a waveguide and attached to the walls by highly conducting joints; the plate usually has one or more windows, or irises, designed to give inductive, capacitive, or resistive characteristics. { 'sep·təm }

sequence [COMPUT SCI] To put a set of symbols into an arbitrarily defined order; that is, to select A if A is greater than or equal to B, or to select B if A is less than B. { 'sē·kwəns }

sequence calling [COMPUT SCI] The instructions used for linking a closed subroutine with a main routine; that is, standard linkage and a list of the parameters. { 'sē·kwəns ,kȯl·iŋ }

sequence check [COMPUT SCI] To verify that correct precedence relationships are obeyed, usually by checking for ascending sequence numbers. { 'sē·kwəns ,chek }

sequence checking routine [COMPUT SCI] In computers, a checking routine which records specified data regarding the operations resulting from each instruction. { 'sē·kwəns ¦chek·iŋ rü ,tēn }

sequence counter *See* instruction counter. { 'sē·kwəns ,kau̇nt·ər }

sequence error [COMPUT SCI] An error that arises when the arrangement of items in a set does not follow some specified order. { 'sē·kwəns ,er·ər }

sequence monitor [COMPUT SCI] The automatic step-by-step check by a computer of the manual actions required for the starting and shutdown of a computer. { 'sē·kwəns ,män·əd·ər }

sequence number [COMPUT SCI] A number assigned to an item to indicate its relative position in a series of related items. { 'sē·kwəns ,nəm·bər }

sequence pointer [COMPUT SCI] For a list that is stored in computer memory, the portion of a list item that gives the storage location of the subsequent item on the list (or the locations of the subsequent and previous items of a symmetric list). Also known as sequencing pointer. { 'sē·kwəns 'pȯint·ər }

sequencer [COMPUT SCI] A machine which puts items of information into a particular order, for example, it will determine whether A is greater than, equal to, or less than B, and sort or order accordingly. Also known as sorter. { 'sē·kwən·sər }

sequence register [COMPUT SCI] A counter which contains the address of the next instruction to be carried out. { 'sē·kwəns ,rej·ə·stər }

sequencing equipment [COMMUN] Special selecting device that permits messages received from several teletypewriter circuits to be subsequently selected and retransmitted over a reduced number of trunks or circuits. { 'sē·kwəns·iŋ i,kwip·mənt }

sequencing pointer *See* sequence pointer. { 'sē·kwəns·iŋ 'pȯint·ər }

sequential access [COMPUT SCI] A process that involves reading or writing data serially and, by extension, a data-recording medium that must be read serially, as a magnetic tape. { si'kwen·chəl 'ak,ses }

sequential batch operating system [COMPUT SCI] Software equipment that automatically begins running a new job on a computer system as soon as the current job is completed. { si'kwen·chəl 'bach 'äp·ə,rād·iŋ ,sis·təm }

sequential color television [COMMUN] A color television system in which the primary color components of a picture are transmitted one after the other; the three basic types are the line-sequential, dot-sequential, and field-sequential color television systems. Also known as sequential system. { si'kwen·chəl ¦kəl·ər 'tel·ə,vizh·ən }

sequential control [COMPUT SCI] Manner of operating a computer by feeding orders into the computer in a given order during the solution of a problem. { si'kwen·chəl kən'trōl }

sequential logic element [ELECTR] A circuit element having at least one input channel, at least one output channel, and at least one internal state variable, so designed and constructed that the output signals depend on the past and present states of the inputs. { si'kwen·chəl ¦läj·ik ,el·ə·mənt }

sequential machine [COMPUT SCI] A mathematical model of a certain type of sequential circuit, which has inputs and outputs that can each take on any value from a finite set and are of interest only at certain instants of time, and in which the output depends on previous inputs as well as the concurrent input. { si'kwen·chəl mə'shēn }

sequential network [COMPUT SCI] An idealized model of a sequential circuit that reflects its logical but not its electronic properties. { si'kwen·chəl 'net,wərk }

sequential operation [COMPUT SCI] The consecutive or serial execution of operations, without any simultaneity or overlap. { si'kwen·chəl ,äp·ə'rā·shən }

sequential organization [COMPUT SCI] The write and read of records in a physical rather than a logical sequence. { si'kwen·chəl ,ȯr·gə·nə'zā·shən }

sequential processing [COMPUT SCI] Processing items in a collection of data according to some specified sequence of keys, in contrast to serial processing. { si'kwen·chəl 'prä,ses·iŋ }

sequential scanning *See* progressive scanning. { si'kwen·chəl 'skan·iŋ }

sequential scheduling system [COMPUT SCI] A first-come, first-served method of selecting jobs to be run. { si'kwen·chəl 'skej·ə·liŋ ,sis·təm }

sequential search [COMPUT SCI] A procedure for searching a table that consists of starting at some table position (usually the beginning) and comparing the file-record key in hand with each table-record key, one at a time, until either a match is found or all sequential positions have been searched. { si'kwen·chəl 'sərch }

sequential selection [COMMUN] The selection of the elements of a message (such as letters) from a set of possible elements (such as the alphabet), one after another. { si'kwen·chəl si'lek·shən }

sequential system *See* sequential color television. { si'kwen·chəl 'sis·təm }

serial [COMPUT SCI] Pertaining to the internal handling of data in sequential fashion. { 'sir·ē·əl }

serial-access [COMPUT SCI] **1.** Pertaining to memory devices having structures such that data storage sites become accessible for read/write in time-sequential order; circulating memories and magnetic tapes are examples of serial-access memories. **2.** Pertaining to a particular process or program that accesses data items sequentially, without regard to the capability of the memory hardware. **3.** Pertaining to character-by-character transmission from an on-line real-time keyboard. { 'sir·ē·əl 'ak,ses }

serial addition [COMPUT SCI] An arithmetic operation in which two numbers are added one digit at a time. { 'sir·ē·əl ə'dish·ən }

serial bit [COMPUT SCI] Digital computer storage in which the individual bits that make up a computer word appear in time sequence. { 'sir·ē·əl ,bit }

serial communications [COMMUN] The transmission of digital data over a single channel. { 'sir·ē·əl kə,myü·nə'kā·shənz }

serial digital computer [COMPUT SCI] A digital computer in which the digits are handled serially, although the bits that make up a digit may be handled either serially or in parallel. { 'sir·ē·əl 'dij·əd·əl kəm'pyüd·ər }

serial dot character printer [COMPUT SCI] A computer printer in which the dot matrix technique is used to print characters, one at a time, with a movable print head that is driven back and forth across the page. { 'sir·ē·əl ¦dät 'kar·ik·tər ,print·ər }

serial file [COMPUT SCI] The simplest type of file organization, in which no subsets are defined, no directories are provided, no particular file order is specified, and a search is performed by sequential comparison of the query with identifiers of all stored items. { 'sir·ē·əl 'fīl }

serial input/output [COMPUT SCI] Data that are transmitted into and out of a computer over a single conductor, one bit at a time. { 'sir·ē·əl 'in,pu̇t 'au̇t,pu̇t }

serial interface [COMPUT SCI] A link between a microcomputer and a peripheral device in which data is transmitted over a single conductor, one bit at a time. Also known as serial port. { 'sir·ē·əl 'in·tər,fās }

serialize [COMPUT SCI] To convert a signal suitable for parallel transmission into a signal suitable for serial transmission, consisting of a sequence of bits. { 'sir·ē·ə,līz }

serially reusable [COMPUT SCI] An attribute possessed by a program that can be used for several tasks in sequence without having to be reloaded into main memory for each additional use. { 'sir·ē·ə·lē rē'yü·zə·bəl }

serial memory [COMPUT SCI] A computer memory in which data are available only in the same sequence as originally stored. { 'sir·ē·əl 'mem·rē }

serial operation [COMPUT SCI] The flow of information through a computer in time sequence, using only one digit, word, line, or channel at a time. { 'sir·ē·əl ,äp·ə'rā·shən }

serial-parallel [COMPUT SCI] **1.** A combination of serial and parallel; for example, serial by character, parallel by bits comprising the character. **2.** Descriptive of a device which converts a serial input into a parallel output. { 'sir·ē·əl 'par·ə,lel }

serial-parallel conversion [COMPUT SCI] The transformation of a serial data representation as found on a disk or drum into the parallel data representation as exists in core. { 'sir·ē·əl ¦par·ə,lel kən'vər·zhən }

serial port *See* serial interface. { 'sir·ē·əl ,pȯrt }

serial processing [COMPUT SCI] Processing items in a collection of data in the order that they appear in a storage device, in contrast to sequential processing. { 'sir·ē·əl 'prä,ses·iŋ }

serial processor [COMPUT SCI] A computer in which data are handled sequentially by separate units of the system. { 'sir·ē·əl 'prä,ses·ər }

serial programming [COMPUT SCI] In computers, programming in which only one operation is executed at one time. { 'sir·ē·əl 'prō,gram·iŋ }

serial storage [COMPUT SCI] Computer storage in which time is one of the coordinates used to locate any given bit, character, or word; access time, therefore, includes a variable waiting time, ranging from zero to many word times. { 'sir·ē·əl 'stȯr·ij }

serial transfer [COMPUT SCI] Transfer of the characters of an element of information in sequence over a single path in a digital computer. { 'sir·ē·əl 'tranz·fər }

serial transmission [COMMUN] Transmission of groups of elements of a signal in time intervals that follow each other without overlapping. { 'sir·ē·əl tranz'mish·ən }

series compensation *See* cascade compensation. { 'sir·ēz ,käm·pən'sā·shən }

series-fed vertical antenna [ELECTROMAG] Vertical antenna which is insulated from the ground and energized at the base. { 'sir·ēz ¦fed 'vərd·i·kəl an'ten·ə }

series radio tap [COMMUN] A telephone tapping procedure in which a miniature radio transmitter is inserted in series with one wire of the target pair so that the transmitter derives its power from the telephone central battery. { 'sir·ēz 'rād·ē·ō ,tap }

series reliability [SYS ENG] Property of a system composed of elements in such a way that failure of any one element causes a failure of the system. { 'sir·ēz ri,lī·ə'bil·əd·ē }

series-shunt network *See* ladder network. { 'sir·ēz ¦shənt 'net,wərk }

series T junction See E-plane T junction. { 'sir·ēz 'tē ,jəŋk·shən }

serrated pulse [ELECTR] Vertical and horizontal synchronizing pulse divided into a number of small pulses, each of which acts for the duration of half a line in an analog television system. { 'se,rād·əd 'pəls }

server [COMPUT SCI] A computer or software package that sends requested information to a client or clients in a network. { 'sər·vər }

service area [COMMUN] The area that is effectively served by a given radio or television transmitter, navigation aid, or other type of transmitter. Also known as coverage. { 'sər·vəs ,er·ē·ə }

service band [COMMUN] Band of frequencies allocated to a given class of radio service. { 'sər·vəs ,band }

service bit [COMMUN] A bit used in data transmission to monitor the transmission rather than to convey information, such as a request that part of a message be repeated. { 'sər·vəs ,bit }

service bureau [COMPUT SCI] An organization that offers time sharing and software services to its users who communicate with a computer in the bureau from terminals on their premises. { 'sər·vəs ,byu̇r·ō }

service program [COMPUT SCI] A computer program that is used in a computer system to support the functioning of the system, such as a librarian or a utility program. { 'sər·vəs ,prō ,gram }

service provider [COMPUT SCI] An organization that provides access to a wide-area network, such as the Internet. { 'sər·vəs prə,vīd·ər }

service routine [COMPUT SCI] A section of a computer code that is used in so many different jobs that it cannot belong to any one job. { 'sər·vəs rü,tēn }

servicing time [COMPUT SCI] Machine downtime necessary for routine testing, for machine servicing due to breakdown, or for preventive servicing measures; includes all test time (good or bad) following breakdown and subsequent repair or preventive servicing. { 'sər·vəs·iŋ ,tīm }

servo See servomotor. { 'sər·vō }

servo loop See single-loop servomechanism. { 'sər·vō ,lüp }

servomechanism [CONT SYS] An automatic feedback control system for mechanical motion; it applies only to those systems in which the controlled quantity or output is mechanical position or one of its derivatives (velocity, acceleration, and so on). Also known as servo system. { ¦sər·vō'mek·ə,niz·əm }

servomotor [CONT SYS] The electric, hydraulic, or other type of motor that serves as the final control element in a servomechanism; it receives power from the amplifier element and drives the load with a linear or rotary motion. Also known as servo. { 'sər·vō,mōd·ər }

servo system See servomechanism. { 'sər·vō ,sis·təm }

sesquisideband transmission [COMMUN] Transmission of a carrier modulated by one full sideband and half of the other sideband. { ¦ses·kwē'sīd,band tranz'mish·ən }

set [COMPUT SCI] A collection of record types. [ELECTR] The placement of a storage device in a prescribed state, for example, a binary storage cell in the high or 1 state. [ENG] **1.** A combination of units, assemblies, and parts connected or otherwise used together to perform an operational function, such as a radar set. **2.** Saw teeth bent out of the plane of the saw body, resulting in a wide cut in the workpiece. { set }

set analyzer See analyzer. { 'set ,an·ə,līz·ər }

set-associative cache [COMPUT SCI] A cache memory in which incoming data are distributed in sequence to each of two to eight areas or sets, and is generally read out in the same manner, allowing each set to prepare for the next input/output operation. { 'set ə,sōs·ē,ād·iv ,kash }

set class [COMPUT SCI] The collection of set occurrences that have been or may be created in accordance with a particular set description. { 'set ,klas }

set condition [ELECTR] Condition of a flip-flop circuit in which the internal state of the flip-flop is set to 1. { 'set kən,dish·ən }

set description [COMPUT SCI] For a specified data set, a definition of the set class name, set-owner selection criteria, set-member eligibility rules, and set-member ordering rules. { 'set di ,skrip·shən }

set occurrence [COMPUT SCI] An instance of a set created in accordance with a set description. { 'set ə,kə·rəns }

set point [CONT SYS] The value selected to be maintained by an automatic controller. { 'set ,pȯint }

set pulse [ELECTR] An electronic pulse designed to place a memory cell in a specified state. { 'set ,pəls }

settling time See correction time. { 'set·liŋ ,tīm }

setup [ELECTR] The ratio between the reference black level and the reference white level in analog television, both measured from the blanking level; usually expressed as a percentage. { 'sed ,əp }

SGML See Standard Generalized Markup Language.

shading [ELECTR] Television process of compensating for the spurious signal generated in a camera tube during trace intervals. { 'shād·iŋ }

shading signal [ELECTR] Television camera signal that serves to increase the gain of the amplifier in the camera during those intervals of time when the electron beam is on an area corresponding to a dark portion of the scene being televised. { 'shād·iŋ ,sig·nəl }

shadow attenuation [ELECTROMAG] Attenuation of radio waves over a sphere in excess of that over a plane when the distance over the surface and other factors are the same. { 'shad·ō ə,ten·yə'wā·shən }

shadow batch system [COMPUT SCI] An online data collection system that initially only stores transactions in the computer system for reference, and updates the master files only at the end of the day or processing period. { 'shad·ō ˌbach ˌsis·təm }

shadow effect [COMMUN] Reduction in the strength of an ultra-high-frequency signal caused by some object (such as a mountain or a tall building) between the points of transmission and reception. { 'shad·ō iˌfekt }

shadow factor [ELECTROMAG] The ratio of the electric-field strength that would result from propagation of waves over a sphere to that which would result from propagation over a plane under comparable conditions. { 'shad·ō ˌfak·tər }

shadow mask [ELECTR] A thin, perforated metal mask mounted just back of the phosphor-dot faceplate in a three-gun color picture tube; the holes in the mask are positioned to ensure that each of the three electron beams strikes only its intended color phosphor dot. Also known as aperture mask. { 'shad·ō ˌmask }

shadow region [ELECTROMAG] Region in which, under normal propagation conditions, the field strength from a given transmitter is reduced by some obstruction which renders effective radio reception of signals or radar detection of objects in this region improbable. { 'shad·ō ˌrē·jən }

Shannon's sampling theorem *See* sampling theorem. { 'shan·ənz 'sam·pliŋ ˌthir·əm }

shannon [COMMUN] A unit of information content, equal to the designation of one of two possible and equally likely values or states of anything used to store or convey information. { 'shan·ən }

Shannon formula [COMMUN] A theorem in information theory which states that the highest number of binary digits per second which can be transmitted with arbitrarily small frequency of error is equal to the product of the bandwidth and $\log_2 (1 + R)$, where R is the signal-to-noise ratio. { 'shan·ən ˌfór·myə·lə }

Shannon limit [COMMUN] Maximum signal-to-noise ratio improvement which can be achieved by the best modulation technique as implied by Shannon's theorem relating channel capacity to signal-to-noise ratio. { 'shan·ən ˌlim·ət }

shaped-beam antenna [ELECTROMAG] Antenna with a directional pattern which, over a certain angular range, is of special shape for some particular use. { 'shāpt ¦bēm an'ten·ə }

shape factor [ELECTR] The ratio of the 60-decibel bandwidth of a bandpass filter to the 3-decibel bandwidth. { 'shāp ˌfak·tər }

shape-fill [COMPUT SCI] The filled-in areas on a graphic electronic display. { 'shāp ˌfil }

shared control unit [COMPUT SCI] A control unit which controls several devices with similar characteristics, such as tape devices. { 'sherd kən ¦trōl ˌyü·nət }

shared file [COMPUT SCI] A direct-access storage device that is used by more than one computer or data-processing system. { 'sherd 'fīl }

shared load [COMPUT SCI] A workload that can be shared by more than one computer, particularly during peak periods. { 'sherd 'lōd }

shared logic [COMPUT SCI] **1.** The simultaneous use of a single computer by multiple users. **2.** An arrangement of computers or computerized equipment in which the processing capabilities of one computer, including the ability to use peripheral devices, can be distributed to the other computers. { 'sherd 'läj·ik }

shared-logic cluster word processor [COMPUT SCI] A system of terminals lacking word-processing capability and printers joined to a single computer designed to carry out word-processing functions. { 'sherd ¦läj·ik ¦kləs·tər 'wərd ˌpräˌses·ər }

shared resource [COMPUT SCI] Peripheral equipment that is simultaneously shared by several users. { 'sherd 'rēˌsórs }

shareware [COMPUT SCI] Copyrighted software that can be tried before buying. { 'sherˌwer }

sharing device [COMPUT SCI] A small, inexpensive multiplexer that combines two independent data signals, which are then transmitted over the same communications line. { 'sher·iŋ diˌvīs }

sheath [ELECTR] A space charge formed by ions near an electrode in a gas tube. [ELECTROMAG] The metal wall of a waveguide. { shēth }

sheath-reshaping converter [ELECTROMAG] In a waveguide, a mode converter in which the change of wave pattern is achieved by gradual reshaping of the sheath of the waveguide and of conducting metal sheets mounted longitudinally in the guide. { 'shēth rē¦shāp·iŋ kən'vərd·ər }

sheet feeder [COMPUT SCI] A device that feeds noncontinuous forms or sheets of paper into a printer. { 'shēt ˌfēd·ər }

sheet grating [ELECTROMAG] Three-dimensional grating consisting of thin, longitudinal, metal sheets extending along the inside of a waveguide for a distance of about a wavelength, and used to stop all waves except one predetermined wave that passes unimpeded. { 'shēt ˌgrād·iŋ }

shell [COMPUT SCI] A program that provides an interface between a user and the computer's operating system by reading commands and sending them to the operating system for execution. { shel }

shell account [COMPUT SCI] A type of limited access to the Internet in which the user is connected to the Internet indirectly through a second computer on which the user has established an account. { 'shel əˌkaủnt }

SHF *See* superhigh frequency.

shielded line [ELECTROMAG] Transmission line, the elements of which confine the propagated waves to an essentially finite space; the external conducting surface is called the sheath. { 'shēl·dəd 'līn }

shield factor [COMMUN] Ratio of noise (or induced current or voltage) in a telephone circuit when a source of shielding is present to the corresponding quantity when the shielding is absent. { 'shēld ˌfak·tər }

shielding *See* electric shielding. { 'shēld·iŋ }

shielding ratio [ELECTROMAG] The ratio of a field in a specified region when electrical shielding is in place to the field in that region when the shielding is removed. { 'shēld·iŋ ,rā·shō }

shift [COMPUT SCI] A movement of data to the right or left, in a digital-computer location, usually with the loss of characters shifted beyond a boundary. { shift }

shift register [COMPUT SCI] A computer hardware element constructed to perform shifting of its contained data. { 'shift ,rej·ə·stər }

shift-register generator [COMPUT SCI] A random-number generator which consists of a sequence of shift operations and other operations, such as no-carry addition. { 'shift ¦rej·ə·stər 'jen·ə,rād·ər }

Shor's algorithm [COMPUT SCI] An algorithm for factoring a large number within a reasonable amount of time, using a quantum computer. { ¦shȯrz 'al·gə,rith·əm }

shore effect [ELECTROMAG] Bending of radio waves toward the shoreline when traveling over water near a shoreline, due to the slightly greater velocity of radio waves over water than over land; this effect causes errors in radio-direction-finder indications. { 'shȯr i,fekt }

short antenna [ELECTROMAG] An antenna shorter than about one-tenth of a wavelength, so that the current may be assumed to have constant magnitude along its length, and the antenna may be treated as an elementary dipole. { 'shȯrt an,ten·ə }

short card [COMPUT SCI] A printed circuit board that is plugged into an expansion slot in a microcomputer and is only half the length of a full-size card. { ¦shȯrt ¦kärd }

short-haul [COMMUN] Pertaining to devices capable of transmitting and receiving signals over distances up to about 1 mile (1.6 kilometers). { 'shȯrt ,hȯl }

short-line seeking [COMPUT SCI] A method of accelerating the operation of a computer printer, in which the printer is sent directly to the beginning of the next line to be printed without going to the left margin of the paper. { 'shȯrt ¦līn 'sēk·iŋ }

short-precision number *See* single-precision number. { 'shȯrt pri¦sizh·ən 'nəm·bər }

short-term predictor [COMMUN] An electric filter that removes redundancies in a signal associated with short-term correlations so that information can be transmitted more efficiently. { ,shȯrt ,tərm prə'dik·tər }

shortwave broadcasting [COMMUN] Radio broadcasting at frequencies in the range from about 1600 to 30,000 kilohertz, above the standard broadcast band. { 'shȯrt'wāv 'brȯd ,kast·iŋ }

shortwave converter [ELECTR] Electronic unit designed to be connected between a receiver and its antenna system to permit reception of frequencies higher than those the receiver ordinarily handles. { 'shȯrt'wāv kən'vərd·ər }

short waveguide isolator [ELECTR] A device that functions as an isocirculator in a miniature microwave circuit and consists of a waveguide T junction with a magnetized cylinder of ferrite at the center and an absorber on the side arm of the T. Also known as flange isolator. { 'shȯrt ¦wāv ,gīd 'ī·sə,lād·ər }

shortwave propagation [COMMUN] Propagation of radio waves at frequencies in the range from about 1600 to 30,000 kilohertz. { 'shȯrt 'wāv ,präp·ə'gā·shən }

short word [COMPUT SCI] The fixed word of lesser length in computers capable of handling words of two different lengths; in many computers this is referred to as a half-word because the length is exactly the half-length of the full word. { 'shȯrt 'wərd }

shot effect *See* shot noise. { 'shät i,fekt }

shot noise [ELECTR] Noise voltage developed in a thermionic tube because of the random variations in the number and the velocity of electrons emitted by the heated cathode; the effect causes sputtering or popping sounds in radio receivers and snow effects in analog television pictures. Also known as Schottky noise; shot effect. { 'shät ,nȯiz }

shunt-excited antenna [ELECTROMAG] A tower antenna, not insulated from the ground at the base, whose feeder is connected at a point about one-fifth of the way up the antenna and usually slopes up to this point from a point some distance from the antenna's base. { 'shənt ik ¦sīd·əd an'ten·ə }

shunt-fed vertical antenna [ELECTROMAG] Vertical antenna connected to the ground at the base and energized at a point suitably positioned above the grounding point. { 'shənt ¦fed ¦vərd·ə·kəl an'ten·ə }

shunt T junction *See* H-plane T junction. { 'shənt 'tē ,jəŋk·shən }

SIB *See* screen image buffer.

sideband [ELECTROMAG] **1.** The frequency band located either above or below the carrier frequency, within which fall the frequency components of the wave produced by the process of modulation. **2.** The wave components lying within such bands. { 'sīd,band }

side circuit [COMMUN] One of the circuits arranged to derive a phantom circuit. { 'sīd ,sər·kət }

side echo [ELECTROMAG] Echo due to a side lobe of an antenna. { 'sīd ,ek·ō }

side effect [COMPUT SCI] A consistent result of a procedure that is in addition to or peripheral to the basic result. { 'sīd i,fekt }

side lobe *See* minor lobe. { 'sīd ,lōb }

side-lobe blanking [ELECTR] Radar technique that compares the signal strength in the main antenna with the echo received in an auxiliary antenna of gain between the side-lobe level and the main-beam gain of the main antenna; done to determine if the echo is coming from the main-beam direction, and blanking the echoes whenever they are stronger in the auxiliary channel. { 'sīd ¦lōb 'blaŋk·iŋ }

side-lobe suppression [ELECTR] Design or techniques in radar intended to reduce the effect of side lobes in the antenna's pattern. { 'sīd ¦lōbe sə,presh·ən }

side-looking radar [ENG] A high-resolution airborne radar having antennas aimed to the right and left of the flight path; used to provide high-resolution strip maps with photographlike detail, to map unfriendly territory while flying along its perimeter, and to detect submarine snorkels against a background of sea clutter. { 'sīd ¦lu̇k·iŋ 'rā·där }

sidetone [COMMUN] The sound of the speaker's own voice as heard in his or her telephone receiver; the effect is undesirable if excessive and is usually reduced by special circuits. { 'sīd ˌtōn }

sidetone level [COMMUN] The ratio of the volume of the sidetone to the volume of the speaker's voice, usually expressed in decibels. { 'sīd,tōn ˌlev·əl }

sidetone ranging [COMMUN] A method of measuring time delay, and thereby range, by sending a radio signal to a satellite, in which several audio tones of different frequencies are broadcast, and the phases of the tones transmitted from the satellite are compared with the sent tone phases. { 'sīd,tōn 'rānj·iŋ }

sift [COMPUT SCI] To extract certain desired information items from a large quantity of data. { sift }

sigma-delta analog-to-digital converter [ELECTR] A converter that uses an analog circuit to generate a single-valued pulse stream in which the frequency of pulses is determined by the analog source, and then uses a digital circuit to repeatedly sum the number of these pulses over a fixed time interval, converting the pulses to numeric values. { ¦sig·mə ¦del·tə ˌan·əˌläg tü ˌdij·əd·əl kənˌvərd·ər }

sigma-delta converter [ELECTR] A class of electronic systems containing both analog and digital subsystems whose most common application is the conversion of analog signals to digital form, and vice versa, using pulse density modulation to create a high-rate stream of single-amplitude pulses in either case. Also known as delta-sigma converter. { ˌsig·mə ˌdel·tə kən'vərd·ər }

sigma-delta digital-to-analog converter [ELECTR] A converter that uses a digital circuit to convert numeric values from a digital processor to a pulse stream and then uses an analog low-pass filter to produce an analog waveform. { ¦sig·mə ¦del·tə ˌdij·əd·əl tü ˌan·əˌläg kən'vərd·ər }

sigma-delta modulator [ELECTR] The circuit used to generate a pulse stream in a sigma-delta converter. Also known as delta-sigma modulator. { ˌsig·mə ˌdel·tə 'mäj·əˌlād·ər }

signal [COMMUN] **1.** A visual, aural, or other indication used to convey information. **2.** The intelligence, message, or effect to be conveyed over a communication system. **3.** See signal wave. { 'sig·nəl }

signal bias [COMMUN] Form of teletypewriter signal distortion brought about by the lengthening or shortening of pulses during transmission; when marking pulses are all lengthened, a marking signal bias results; when marking pulses are all shortened, a spacing signal bias results. { 'sig·nəl ˌbī·əs }

signal carrier See carrier. { 'sig·nəl ˌkar·ē·ər }

signal channel [COMMUN] A signal path for transmitting electric signals; such paths may be separated by frequency division or time division. { 'sig·nəl ˌchan·əl }

signal conditioning [COMMUN] Processing the form or mode of a signal so as to make it intelligible to or compatible with a given device, such as a data transmission line, including such manipulation as pulse shaping, pulse clipping, digitizing, and linearizing. { 'sig·nəl kən ˌdish·ən·iŋ }

signal distance [COMPUT SCI] The number of bits that are not the same in two binary words of equal length. Also known as hamming distance. { 'sig·nəl ˌdis·təns }

signal distortion generator [ELECTR] Instrument designed to apply known amounts of distortion on a signal for the purpose of testing and adjusting communications equipment such as teletypewriters. { 'sig·nəl di'stȯr·shən ˌjen·ə ˌrād·ər }

signal-flow graph [SYS ENG] An abbreviated block diagram in which small circles, called nodes, represent variables of the system, and the nodes are connected by lines, called branches, which represent one-way signal multipliers; an arrow on the line indicates direction of signal flow, and a letter near the arrow indicates the multiplication factor. Also known as flow graph. { 'sig·nəl ¦flō 'graf }

signal in band [COMMUN] To send control signals at frequencies within the frequency range of the data signal. { ¦sig·nəl in ¦band }

signaling rate [COMMUN] The rate at which signals are transmitted. { 'sig·nə·liŋ ˌrāt }

signal intensity [COMMUN] The electric-field strength of the electromagnetic wave transmitting a signal. { 'sig·nəl inˌten·səd·ē }

signal level [COMMUN] The difference between the level of a signal at a point in a transmission system and the level of an arbitrarily specified reference signal. { 'sig·nəl ˌlev·əl }

signal light [COMMUN] A light specifically designed for the transmission of code messages by means of visible light rays that are interrupted or deflected by electric or mechanical means. { 'sig·nəl ˌlīt }

signal normalization See signal standardization. { 'sig·nəl ˌnȯr·mə·lə'zā·shən }

signal out of band [COMMUN] To send control signals at frequencies outside the frequency range of the data signal. { ¦sig·nəl au̇t əv ¦band }

signal processing [COMMUN] The extraction of information from complex signals in the presence of noise, generally by conversion of the signals into digital form followed by analysis using various algorithms. Also known as digital signal processing (DSP). { 'sig·nəl ˌpräˌses·iŋ }

signal regeneration [COMMUN] The restoration of a waveform representing a signal to approximate its original amplitude and shape. Also known as signal reshaping. { 'sig·nəl rē,jen·ə 'rā·shən }

signal reporting code *See* radio-signal reporting code. { 'sig·nəl ri'pȯrd·iŋ ,kōd }

signal reshaping *See* signal regeneration. { 'sig·nəl rē,shāp·iŋ }

signal speed [COMMUN] The rate at which code elements are transmitted by a communications system. { 'sig·nəl ,spēd }

signal standardization [COMMUN] The use of one signal to generate another which meets specified requirements for shape, amplitude, and timing. Also known as signal normalization. { 'sig·nəl ,stan·dər·də'zā·shən }

signal strength [ELECTROMAG] The strength of the signal produced by a radio transmitter at a particular location, usually expressed as microvolts or millivolts per meter of effective receiving antenna height. { 'sig·nəl ,strəŋkth }

signal-strength meter [ELECTR] A meter that is connected to the automatic volume-control circuit of a communication receiver and calibrated in decibels or arbitrary S units to read the strength of a received signal. Also known as S meter; S-unit meter. { 'sig·nəl ¦strəŋkth ,mēd·ər }

signal-to-interference ratio [ELECTR] The relative magnitude of signal waves and waves which interfere with signal-wave reception. { 'sig·nəl tü ,in·tər'fir·əns ,rā·shō }

signal-to-noise improvement factor *See* noise improvement factor. { 'sig·nəl tə 'nȯiz im'prüv·mənt ,fak·tər }

signal-to-noise ratio [ELECTR] The ratio of the amplitude of a desired signal at any point to the amplitude of noise signals at that same point; often expressed in decibels; the peak value is usually used for pulse noise, while the root-mean-square (rms) value is used for random noise. Abbreviated S/N; SNR. { 'sig·nəl tə 'nȯiz ,rā·shō }

signal tracer [ELECTR] An instrument used for tracing the progress of a signal through a radio receiver or an audio amplifier to locate a faulty stage. { 'sig·nəl ,trā·sər }

signal wave [COMMUN] A wave whose characteristics permit some intelligence, message, or effect to be conveyed. Also known as signal. { 'sig·nəl ,wāv }

signal-wave envelope [COMMUN] Contour of a signal wave which is composed of a series of wave cycles. { 'sig·nəl ¦wāv 'en·və,lōp }

sign-and-magnitude code [COMPUT SCI] The representation of an integer X by $(-1)^{a_0}(2^{n-2}a_1 + 2^{n-3}a_2 + \cdots + a_{n-1})$, where a_0 is 0 for X positive, and a_0 is 1 for X negative, and any a_j is either 0 or 1. { 'sīn ən 'mag·nə,tüd ,kōd }

signature [ELECTR] The characteristic pattern of a target as displayed by detection and classification equipment. { 'sig·nə·chər }

sign bit [COMPUT SCI] A sign digit consisting of one bit. { 'sīn ,bit }

sign check indicator [COMPUT SCI] An error checking device, indicating no sign or improper signing of a field used for arithmetic processes; the machine can, upon interrogation, be made to stop or enter into a correction routine. { 'sīn ¦chek 'in·də,kād·ər }

sign digit [COMPUT SCI] A digit containing one to four binary bits, associated with a data item and used to denote an algebraic sign. { 'sīn ,dij·ət }

signed decimal [COMPUT SCI] A form of packed decimal representation in which the low-order nibble of the last byte has a sign bit that specifies whether the number is positive or negative. { 'sīnd 'des·məl }

signed field [COMPUT SCI] A field of data that contains a number which includes a sign digit indicating the number's sign. { 'sīnd 'fēld }

signed integer [COMPUT SCI] A whole number whose value lies anywhere in a domain that extends from a negative to a positive integer, and which therefore carries a sign. { 'sīnd 'int·ə·jər }

sign flag [COMPUT SCI] A bit in a status byte in a computer's central processing unit that indicates whether the result of an arithmetic operation is positive or negative. { 'sīn ,flag }

significance arithmetic [COMPUT SCI] A rough technique for estimating the numbers and positions of the significant digits of the radix approximation that results when an arithmetic operation is applied to operands in radix approximation form. { sig'nif·i·kəns ə,rith·mə·tik }

sign position [COMPUT SCI] That position, always at or near the left or right end of a numeral, in which the algebraic sign of the number is represented. { 'sīn pə,zish·ən }

silent period [COMMUN] Period during each hour in which ship and shore radio stations must remain silent and listen for distress calls. { 'sī·lənt 'pir·ē·əd }

silicon detector *See* silicon diode. { 'sil·ə·kən di'tek·tər }

silicon diode [ELECTR] A crystal diode that uses silicon as a semiconductor; used as a detector in ultra-high- and super-high-frequency circuits. Also known as silicon detector. { 'sil·ə·kən 'dī ,ōd }

silicon homojunction *See* bipolar junction transistor. { ¦sil·ə·kən 'hä·mə,jəŋk·shən }

silicon image sensor [ELECTR] A video camera in which the image is focused on an array of individual light-sensitive elements formed from a charge-coupled-device semiconductor chip. Also known as silicon imaging device. { 'sil·ə·kən 'im·ij ,sen·sər }

silicon imaging device *See* silicon image sensor. { 'sil·ə·kən 'im·ij·iŋ di,vīs }

silicon-on-insulator [ELECTR] A semiconductor manufacturing technology in which thin films of single-crystalline silicon are grown over an electrically insulating substrate. { 'sil·ə·kən ȯn 'in·sə,lād·ər }

silicon-on-sapphire [ELECTR] A semiconductor manufacturing technology in which metal oxide semiconductor devices are constructed in a thin single-crystal silicon film grown on an electrically insulating synthetic sapphire substrate. Abbreviated SOS. { 'sil·ə·kən ȯn 'sa,fīr }

silicon retina [ELECTR] An analog very large scale integrated circuit chip that performs operations which resemble some of the functions performed by the retina of the human eye. { ˌsil·əˌkän 'ret·ən·ə }

SIMD [COMPUT SCI] A type of multiprocessor architecture in which there is a single instruction cycle, but multiple sets of operands may be fetched to multiple processing units and may be operated upon simultaneously within a single instruction cycle. Acronym for single-instruction-stream, multiple-data-stream. { ¦es¦ī¦em¦dē }

SIMM [COMPUT SCI] A printed circuit board that holds several semiconductor memory chips and is used to add memory to a computer. Acronym for single in-line memory module. { sim }

simple buffering [COMPUT SCI] A technique for obtaining simultaneous performance of input/output operations and computing; it involves associating a buffer with only one input or output file (or data set) for the entire duration of the activity on that file (or data set). { 'sim·pəl 'bəf·ə·riŋ }

simple data structure [COMPUT SCI] An arrangement of data in a database or file in which each grouping of data, such as a record, is of equal importance or significance. { 'sim·pəl 'dad·ə ˌstrək·chər }

Simple Mail Transfer Protocol [COMPUT SCI] An Internet standard for sending e-mail messages. Abbreviated SMTP. { ¦sim·pəl 'māl ˌtranz·fər ˌprōd·əˌkȯl }

simple path *See* path. { 'sim·pəl 'path }

simplex channel [COMMUN] A channel which permits transmission in one direction only. { 'simˌpleks ¦chan·əl }

simplex structure [COMPUT SCI] The structure of an information processing system designed in such a way that only the minimum amount of hardware is utilized to implement its function. { 'simˌpleks ¦strək·chər }

simplex transmission [COMMUN] A mode of radio transmission in which communication takes place between two stations in only one direction at a time. { 'simˌpleks tranz¦mish·ən }

SIMSCRIPT [COMPUT SCI] A high-level programming language used in simulation, in which systems are described in terms of sets, entities, which are groups of sets, and attributes, which are properties associated with entities. { 'sim ˌskript }

simulate [ENG] To mimic some or all of the behavior of one system with a different, dissimilar system, particularly with computers, models, or other equipment. { 'sim·yəˌlāt }

simulation [COMPUT SCI] The development and use of computer models for the study of actual or postulated dynamic systems. { ˌsim·yə'lā·shən }

simulation language [COMPUT SCI] A computer language used to write programs for the simulation of the behavior through time of such things as transportation and manufacturing systems; SIMSCRIPT is an example. { ˌsim·yə'lā·shən ˌlaŋ·gwij }

simulator [COMPUT SCI] A routine which is executed by one computer but which imitates the operations of another computer. [ENG] A computer or other piece of equipment that simulates a desired system or condition and shows the effects of various applied changes, such as a flight simulator. { 'sim·yəˌlād·ər }

simultaneous access *See* parallel access. { ˌsī·məl'tā·nē·əs 'akˌses }

simultaneous computer [COMPUT SCI] **1.** A computer, usually of the analog or hybrid type, in which separate units of hardware are used to carry out the various parts of a computation, the execution of different parts usually overlap in time, and the various hardware units are interconnected in a manner determined by the computation. **2.** A computer that serves to back up another computer and can replace it when it is not operating effectively. { ˌsī·məl'tā·nē·əs kəm'pyüd·ər }

simultaneous lobing [ELECTR] A radar direction-finding technique in which the signals received by two partly overlapping antenna beams are compared in phase or power to obtain a measure of the angular displacement of a target from the equisignal direction; arrangement of (usually) four such beams to effect measurement in both angle directions. { ˌsī·məl'tā·nē·əs 'lōb·iŋ }

simultaneous peripheral operations on line *See* spooling. { ˌsī·məl'tā·nē·əs pə'rif·ə·rəl ˌäp·ə'rā·shənz ȯn 'līn }

sine-wave oscillator *See* sinusoidal oscillator. { 'sīn ¦wāv 'as·əˌlād·ər }

sine-wave response *See* frequency response. { 'sīn ¦wāv ri'späns }

singing [CONT SYS] An undesired, self-sustained oscillation in a system or component, at a frequency in or above the passband of the system or component; generally due to excessive positive feedback. { 'siŋ·iŋ }

singing margin [CONT SYS] The difference in level, usually expressed in decibels, between the singing point and the operating gain of a system or component. { 'siŋ·iŋ ˌmär·jən }

singing point [CONT SYS] The minimum value of gain of a system or component that will result in singing. { 'siŋ·iŋ ˌpȯint }

single-address instruction *See* one-address instruction. { 'siŋ·gəl ¦adˌres in'strək·shən }

single-board computer [COMPUT SCI] A computer consisting of a processor and memory on a single printed circuit board. { 'siŋ·gəl ¦bȯrd kəm'pyüd·ər }

single-channel simplex [COMMUN] Simplex operation that provides nonsimultaneous radio communications between stations using the same frequency channel. { 'siŋ·gəl ¦chan·əl 'sim ˌpleks }

single-chip computer [COMPUT SCI] A computer whose processor consists of a single integrated circuit. { 'siŋ·gəl ¦chip kəm'pyüd·ər }

single-current transmission [COMMUN] Telegraph transmission in which a current flows, in only one direction, during marking intervals, and no current flows during spacing intervals. { 'siŋ·gəl ¦kə·rənt tranz'mish·ən }

single density [COMPUT SCI] Property of computer storage which holds the standard amount of data per unit of storage space. { 'siŋ·gəl 'den·səd·ē }

single-ended signal [ELECTR] A circuit signal that is the voltage difference between two nodes, one of which can be defined as being at ground or reference voltage. { ¦siŋ·gəl ¦en·dəd 'sig·nəl }

single-event upset [ELECTR] A change in the state of a logic device from 0 to 1 or vice versa, as the result of the passage of a single cosmic ray. { ¦siŋ·gəl i¦vent 'əp,set }

single-frequency duplex [COMMUN] Duplex carrier communications that provide communications in opposite directions, but not simultaneously, over a single-frequency carrier channel, the transfer between transmitting and receiving conditions being automatically controlled by the voices or other signals of the communicating parties. { 'siŋ·gəl 'frē·kwən·sē 'dü,pleks }

single-frequency simplex [COMMUN] Single-frequency carrier communications in which manual rather than automatic switching is used to change over from transmission to reception. { 'siŋ·gəl 'frē·kwən·sē 'sim,pleks }

single-gun color tube [ELECTR] A color picture tube having only one electron gun and one electron beam; the beam is sequentially deflected across phosphors for the three primary colors to form each color picture element, as in the chromatron. { 'siŋ·gəl ¦gən 'kəl·ər ,tüb }

single-hop transmission [COMMUN] Radio transmission in which radio waves are reflected from the ionosphere only once along their path from the transmitter to the receiver. { 'siŋ·gəl ¦häp trans'mish·ən }

single in-line memory module *See* SIMM. { ¦siŋ·gəl ¦in ,līn 'mem·rē ,mä·jəl }

single-instruction-stream, multiple-data-stream *See* SIMD. { ¦siŋ·gəl in¦strək·shən ,strēm ¦məl·tə·pəl 'dad·ə ,strēm }

single-instruction-stream, single-data-stream *See* SISD. { ¦siŋ·gəl in¦strək·shən ,strēm ¦siŋ·gəl 'dad·ə ,strēm }

single-keyboard point-of-sale system [COMPUT SCI] A point-of-sale system based upon electronic cash registers as stand-alone units, each equipped with a few internal registers and some programming capability. { 'siŋ·gəl ¦kē ,bȯrd ¦pȯint əv 'sāl ,sis·təm }

single-length [COMPUT SCI] Pertaining to the expression of numbers in binary form in such a way that they can be included in a single computer word. { 'siŋ·gəl 'leŋkth }

single-loop feedback [CONT SYS] A system in which feedback may occur through only one electrical path. { 'siŋ·gəl ¦lüp 'fēd,bak }

single-loop servomechanism [CONT SYS] A servomechanism which has only one feedback loop. Also known as servo loop. { 'siŋ·gəl ¦lüp 'sər·vō,mek·ə,niz·əm }

single-polarity pulse-amplitude modulation *See* unidirectional pulse-amplitude modulation. { 'siŋ·gəl pə¦lar·əd·ē 'pəls 'am·plə,tüd ,mäj·ə'lā·shən }

single-precision number [COMPUT SCI] A number having as many digits as are ordinarily used in a given computer, in contrast to a double-precision number. Also known as short-precision number. { 'siŋ·gəl prə¦sizh·ən 'nəm·bər }

single-program, multiple-data *See* SPMD. { ¦siŋ·gəl ¦prō·grəm ¦məl·tə·pəl 'dad·ə }

single reference *See* random access. { 'siŋ·gəl 'ref·rəns }

singlesheet feed [COMPUT SCI] Equipment for feeding one sheet of paper to a computer printer at a time. { 'siŋ·gəl,shēt 'fēd }

single-shot multivibrator *See* monostable multivibrator. { 'siŋ·gəl ¦shät ¦məl·ti'vī,brād·ər }

single-shot operation *See* single-step operation. { 'siŋ·gəl ¦shät ,äp·ə'rā·shən }

single-sideband [COMMUN] Pertaining to single-sideband communication. Abbreviated SSB. { 'siŋ·gəl 'sīd,band }

single-sideband communication [COMMUN] A communication system in which one of the two sidebands used in amplitude-modulation is suppressed; the carrier wave may be either transmitted, suppressed, or partially suppressed. { 'siŋ·gəl ¦sīd,band kə,myü·nə'kā·shən }

single-sideband modulation [COMMUN] Modulation resulting from elimination of all components of one sideband from an amplitude-modulated wave. { 'siŋ·gəl ¦sīd,band ,mäj·ə'lā·shən }

single-sideband transmission [COMMUN] Transmission of a carrier and substantially only one sideband of modulation frequencies, as in television where only the upper sideband is transmitted completely for the picture signal; the carrier wave may be either transmitted or suppressed, partially or totally. { 'siŋ·gəl ¦sīd ,band tranz'mish·ən }

single-sided [COMPUT SCI] Pertaining to storage media that use only one of two sides for recording data. { 'siŋ·gəl 'sīd·əd }

single-signal receiver [ELECTR] A highly selective superheterodyne receiver for code reception, having a crystal filter in the intermediate-frequency amplifier. { 'siŋ·gəl ¦sig·nəl ri'sē·vər }

single-step operation [COMPUT SCI] A method of computer operation, used in debugging or detecting computer malfunctions, in which a program is carried out one instruction at a time, each instruction being performed in response to a manual control device such as a switch or button. Also known as one-shot operation; one-step operation; single-shot operation; step-by-step operation. { 'siŋ·gəl ¦step ,äp·ə'rā·shən }

single-stub transformer [ELECTROMAG] Shorted section of a coaxial line that is connected to a main coaxial line near a discontinuity to provide impedance matching at the discontinuity. { 'siŋ·gəl ¦stəb tranz'fȯr·mər }

single-stub tuner [ELECTROMAG] Section of transmission line terminated by a movable short-circuiting plunger or bar, attached to a main transmission line for impedance-matching purposes. { 'siŋ·gəl ¦stəb 'tün·ər }

single threading [COMPUT SCI] Transaction processing in which one transaction is completed before another is begun. { 'siŋ·gəl 'thred·iŋ }

single-tone keying [COMMUN] Form of keying in which the modulating function causes the carrier to be modulated with a single tone for one condition, which may be either marking or spacing, and the carrier is unmodulated for the other condition. { 'siŋ·gəl ¦tōn 'kē·iŋ }

singly linked ring [COMPUT SCI] A cyclic arrangement of data elements in which searches may be performed in either a clockwise or a counterclockwise direction, but not both. { 'siŋ·glē ¦liŋkt 'riŋ }

singular arc [CONT SYS] In an optimal control problem, that portion of the optimal trajectory in which the Hamiltonian is not an explicit function of the control inputs, requiring higher-order necessary conditions to be applied in the process of solution. { 'siŋ·gyə·lər 'ärk }

sink [COMMUN] Equipment at the end of a communications channel that receives signals and may perform other functions such as error detection. [ELECTROMAG] The region of a Rieke diagram where the rate of change of frequency with respect to phase of the reflection coefficient is maximum for an oscillator; operation in this region may lead to unsatisfactory performance by reason of cessation or instability of oscillations. { siŋk }

sinusoidal angular modulation *See* angle modulation. { ˌsī·nə'sȯid·əl 'aŋ·gyə·lər ˌmäj·ə'lā·shən }

sinusoidal oscillator [ELECTR] An oscillator circuit whose output voltage is a sine-wave function of time. Also known as harmonic oscillator; sine-wave oscillator. { ˌsī·nə'sȯid·əl 'as·əˌlād·ər }

SISD [COMPUT SCI] A type of computer architecture in which there is a single instruction cycle, and operands are fetched in serial fashion into a single processing unit before execution. Acronym for single-instruction-stream, single-data-stream. { ¦es¦ī¦es'dē }

site [COMPUT SCI] A position available for the symbols of an inscription, for example, a digital place. { sīt }

16 VSB [COMMUN] Vestigial sideband modulation with 16 discrete amplitude levels.

size control [ELECTR] A control provided on a video display device for changing the size of a picture either horizontally or vertically. { 'sīz kənˌtrōl }

skeletal coding [COMPUT SCI] A set of incomplete instructions in symbolic form, intended to be completed and specialized by a processing program written for that purpose. { 'skel·əd·əl 'kōd·iŋ }

skew [COMPUT SCI] In character recognition, a condition arising at the read station whereby a character or a line of characters appears in a "twisted" manner in relation to a real or imaginary horizontal baseline. [ELECTR] **1.** The deviation of a received facsimile frame from rectangularity due to lack of synchronism between scanner and recorder; expressed numerically as the tangent of the angle of this deviation. **2.** The degree of nonsynchronism of supposedly parallel bits when bit-coded characters are read from magnetic tape. { skyü }

skew failure [COMPUT SCI] In character recognition, the condition that exists during document alignment whereby the document reference edge is not parallel to that of the read station. { 'skyü ˌfāl·yər }

skin antenna [ELECTROMAG] Flush-mounted aircraft antenna made by using insulating material to isolate a portion of the metal skin of the aircraft. { 'skin anˌten·ə }

skin depth [ELECTROMAG] The depth beneath the surface of a conductor, which is carrying current at a given frequency due to electromagnetic waves incident on its surface, at which the current density drops to one neper below the current density at the surface. { 'skin ˌdepth }

skin tracking [ELECTROMAG] Tracking of an object by means of radar without using a beacon or other signal device on board the object being tracked. { 'skin ˌtrak·iŋ }

skip [COMPUT SCI] **1.** In fixed-instruction-length digital computers, to bypass or ignore one or more instructions in an otherwise sequential process. **2.** Action of a computer printer that moves rapidly over a line so that a blank line appears in the printout. { skip }

skip chain [COMPUT SCI] A programming technique which matches a word against a set of test words; if there is a match, control is transferred (skipped) to a routine, otherwise the word is matched with the next test word in sequence. { 'skip ˌchān }

skip distance [ELECTROMAG] The minimum distance that radio waves can be transmitted between two points on the earth by reflection from the ionosphere, at a specified time and frequency. { 'skip ˌdis·təns }

skip effect [COMMUN] The existence of a circular-shaped area around a radio transmitter within which no radio signals are received, because ground signals are received only inside the oval and sky-wave signals are received only outside the oval. { 'skip iˌfekt }

skip fading [ELECTROMAG] Fading due to fluctuations of ionization density at the place in the ionosphere where the wave is reflected which causes the skip distance to increase or decrease. { 'skip ˌfād·iŋ }

skip flag [COMPUT SCI] The thirty-fifth bit of a channel command word which suppresses the transfer of data to main storage. { 'skip ˌflag }

skip-searched chain [COMPUT SCI] A chain which has pointers and can therefore be searched without examining each link. { 'skip ˌsərcht ˌchān }

skip zone [COMMUN] The area between the outer limit of reception of radio high-frequency ground waves and the inner limit of reception of sky waves, where no signal is received. { 'skip ,zōn }

sky wave [ELECTROMAG] A radio wave that travels upward into space and may or may not be returned to earth by reflection from the ionosphere. Also known as ionospheric wave. { 'skī ,wāv }

sky-wave correction [ELECTR] The correction to be applied to the time difference readings of received sky waves to convert them to an equivalent ground-wave reading. { 'skī ¦wāv kə'rek·shən }

sky-wave-synchronized loran [NAV] A type of loran in which the transmitting stations are synchronized by signals reflected from the ionosphere; used to obtain greater range and more accurate nighttime navigation. Abbreviated ss loran. { 'skī ¦wāv ,siŋ·krə¦nīzd 'lȯr,an }

sky-wave transmission delay [ELECTROMAG] Amount by which the time of transit from transmitter to receiver of a pulse carried by sky waves reflected once from the E layer exceeds the time of transit of the same pulse carried by ground waves. { 'skī ¦wāv tranz'mish·ən di,lā }

slave [COMPUT SCI] A terminal or computer that is controlled by another computer. [CONT SYS] A device whose motions are governed by instructions from another machine. { slāv }

slave antenna [ELECTROMAG] A directional antenna positioned in azimuth and elevation by a servo system; the information controlling the servo system is supplied by a tracking or positioning system. { 'slāv an,ten·ə }

slave mode *See* user mode. { 'slāv ,mōd }

slave tube [ELECTR] A display monitor that is connected to another monitor and provides an identical display. { 'slāv ,tüb }

sleep [COMPUT SCI] State of a computer system that halts, or a program that appears to be doing nothing because the program is caught in an endless loop. { 'slēp }

sleeve antenna [ELECTROMAG] A single vertical half-wave radiator, the lower half of which is a metallic sleeve through which the concentric feed line runs; the upper radiating portion, one quarter-wavelength long, connects to the center of the line. { 'slēv an,ten·ə }

sleeve dipole antenna [ELECTROMAG] Dipole antenna surrounded in its central portion by a coaxial cable. { 'slēv 'dī,pōl an'ten·ə }

slew rate [COMPUT SCI] The speed at which a logic-seeking print head advances to the succeeding line and finds the position where it is to start printing. [CONT SYS] The maximum rate at which a system can follow a command. [ELECTR] The maximum rate at which the output voltage of an operational amplifier changes for a square-wave or step-signal input; usually specified in volts per microsecond. { 'slü ,rāt }

slip [ELECTR] Distortion produced in the recorded facsimile image which is similar to that produced by skew but is caused by slippage in the mechanical drive system. { slip }

slit scan [COMPUT SCI] In character recognition, a magnetic or photoelectric device that obtains the horizontal structure of an inputted character by vertically projecting its component elements at given intervals. { 'slit ,skan }

slot [COMPUT SCI] A connection to a computer bus into which printed ciruit boards or integrated circuit boards can be inserted. { slät }

slot antenna [ELECTROMAG] An antenna formed by cutting one or more narrow slots in a large metal surface fed by a coaxial line or waveguide. { 'slät an,ten·ə }

slot-bound [COMPUT SCI] Condition of a computer when all the slots in the machine's bus are filled with printed circuit boards, so that it is not possible to expand the machine's capacity by plugging in additional boards. { 'slät ,bau̇nd }

slot coupling [ELECTROMAG] Coupling between a coaxial cable and a waveguide by means of two coincident narrow slots, one in a waveguide wall and the other in the sheath of the coaxial cable. { 'slät ,kəp·liŋ }

slot-mask picture tube [ELECTR] An in-line gun-type color picture tube in which the shadow mask is perforated by short, vertical slots, and the screen is painted with vertical phosphor stripes. { 'slät ,mask 'pik·chər ,tüb }

slot radiator [ELECTROMAG] Primary radiating element in the form of a slot cut in the walls of a metal waveguide or cavity resonator or in a metal plate. { 'slät ,rād·ē,ād·ər }

slotted line *See* slotted section. { 'släd·əd 'līn }

slotted section [ELECTROMAG] A section of waveguide or shielded transmission line in which the shield is slotted to permit the use of a movable probe for examination of standing waves. Also known as slotted line; slotted waveguide. { 'släd·əd ,sek·shən }

slotted waveguide *See* slotted section. { 'släd·əd 'wāv,gīd }

slow memory *See* slow storage. { 'slō 'mem·rē }

slow-scan television [COMMUN] Television system that uses a slow rate of horizontal scanning, requiring typically 8 seconds for each complete scan of the scene; suitable for transmitting printed matter, photographs, and illustrations. Abbreviated SSTV. { 'slō ¦skan 'tel·ə,vizh·ən }

slow storage [COMPUT SCI] In computers, storage with a relatively long access time. Also known as slow memory. { 'slō 'stȯr·ij }

slow time scale [COMPUT SCI] In simulation by an analog computer, a time scale in which the time duration of a simulated event is greater than the actual time duration of the event in the physical system under study. Also known as extended time scale. { 'slō 'tīm ,skāl }

slow wave [ELECTROMAG] A wave having a phase velocity less than the velocity of light, as in a ridge wave guide. { 'slō 'wāv }

SLSI circuit *See* super-large-scale integrated circuit. { ,es,el,es'ī ,sər·kət }

slug tuner [ELECTROMAG] Waveguide tuner containing one or more longitudinally adjustable pieces of metal or dielectric. { 'sləg ¦tün·ər }

slug tuning [ELECTROMAG] Means of varying the frequency of a resonant circuit by introducing a slug of material into either the electric field or magnetic field, or both. { 'sləg ¦tün·iŋ }

small computer system interface [COMPUT SCI] An interface standard or format for personal computers that allows the connection of up to seven peripheral devices. Abbreviated SCSI (scuzzy). { ¦smȯl kəm¦pyüd·ər ˌsis·təm ˈin·tər ˌfās }

small talk [COMPUT SCI] A high-level, user-friendly programming language that incorporates the functions of an operating system. { ˈsmȯl ˌtȯk }

smart card [COMPUT SCI] A plastic card in which is embedded a microprocessor that is usually programmed to hold information about the card holder or user. Also known as chip card. { ˈsmärt ˌkärd }

smart sensor [ENG] A microsensor integrated with signal-conditioning electronics such as analog-to-digital converters on a single silicon chip to form an integrated microelectromechanical component that can process information itself or communicate with an embedded microprocessor. Also known as intelligent sensor. { ˌsmärt ˈsen·sər }

smart terminal *See* intelligent terminal. { ˈsmärt ˈtər·mən·əl }

smart tool [CONT SYS] A robot end effector or fixed tool that uses sensors to measure the tool's position relative to reference markers or a workpiece or jig, and an actuator to adjust the tool's position with respect to the workpiece. { ˈsmärt ˌtül }

SMATV system *See* satellite master antenna television system. { ¦esˌemˌāˌtēˈvē ˌsis·təm }

smear [ELECTR] A video picture defect in which objects appear to be extended horizontally beyond their normal boundaries in a blurred or smeared manner; one cause is excessive attenuation of high video frequencies in an analog television receiver. { smir }

S meter *See* signal-strength meter. { ˈes ˌmēd·ər }

smiley *See* emoticon. { ˈsmī·lē }

SMPT *See* Simple Mail Transfer Protocol.

S/N *See* signal-to-noise ratio.

snapshot [COMPUT SCI] The storing of the entire contents of the memory, including status indicators and hardware registers. { ˈsnapˌshät }

snapshot dump [COMPUT SCI] An edited printout of selected parts of the contents of main memory, performed at one or more times during the execution of a program without materially affecting the operation of the program. { ˈsnap ˌshät ˌdəmp }

snapshot program [COMPUT SCI] A program that provides dumps of certain portions of memory when certain instructions are executed or when certain conditions are fulfilled. { ˈsnap ˌshät ˌprō·grəm }

sneak path [COMPUT SCI] In computers, an undesired circuit through a series-parallel configuration. { ˈsnēk ˌpath }

SNOBOL [COMPUT SCI] A computer programming language that has significant applications in program compilation and generation of symbolic equations. Derived from String-Oriented-Symbolic Language. { ˈsnōˌbȯl }

snow [ELECTR] Small, random, white spots produced on an analog television or radar screen by inherent noise signals originating in the receiver. { snō }

snow static [ELECTROMAG] Precipitation static caused by falling snow. { ˈsnō ˌstad·ik }

SNR *See* signal-to-noise ratio.

soft computing [COMPUT SCI] A family of methods that imitate human intelligence with the goal of creating tools provided with some human-like capabilities (such as learning, reasoning, and decision making), and are based on fuzzy logic, neural networks, and probabilistic reasoning techniques such as genetic algorithms. { ˌsȯft kəmˈpyüd·iŋ }

soft copy [COMPUT SCI] Information that is displayed on a screen, given by voice, or stored in a form that cannot be read directly by a person, as on magnetic tape, disk, or microfilm. { ˈsȯft ˈkäp·ē }

soft-copy terminal [COMPUT SCI] A computer terminal that presents its output through an electronic display, rather than printing it on paper. { ˈsȯft ¦käp·ē ˈtər·mən·əl }

soft crash [COMPUT SCI] A halt in computer operations in which the computer operator has enough warning time to take action to minimize the effects of the stoppage. { ˈsȯft ˈkrash }

soft edit [COMPUT SCI] A checking and correction process that allows data in which problems have been identified to be accepted by a computer system. { ˈsȯft ˈed·it }

soft error [COMPUT SCI] An error that occurs in automatic operations but does not recur when the operation is attempted a second time. { ˈsȯft ˈer·ər }

soft failure [COMPUT SCI] A failure that can be overcome without the assistance of a person with specialized knowledge to repair the device. { ˈsȯft ˈfāl·yər }

soft font [COMPUT SCI] A typeface or set of typefaces that is contained in the software of a computer system and is transmitted to the printer before printing. Also known as downloadable font. { ˈsȯft ˈfänt }

soft page break [COMPUT SCI] A page break that is inserted in a document by a word-processing program, and can move if text is added, deleted, or reformatted above it. { ¦sȯft ˈpāj ˌbrāk }

soft patch [COMPUT SCI] A temporary change in a computer program's machine language that is carried out while the program is in memory, and thus prevails only for the duration of a single run of the program. { ˈsȯft ˈpach }

soft return [COMPUT SCI] A control code that is automatically entered into a text document by the word-processing program to mark the end of a line, based on the current right margin. { ˈsȯft riˈtərn }

soft sector [COMPUT SCI] A disk or drum format in which the locations of sectors are determined by control information written on the storage medium rather than by some physical means. { ˈsȯft ˈsek·tər }

software [COMPUT SCI] The totality of programs usable on a particular kind of computer, together with the documentation associated with a computer or program, such as manuals, diagrams, and operating instructions. { 'sȯf,wer }

software compatibility [COMPUT SCI] Property of two computers, with respect to a particular programming language, in which a source program from one machine in that language will compile and execute to produce acceptably similar results in the other. { 'sōf,wer kəm,pad·ə'bil·əd·ē }

software driver [COMPUT SCI] Software that is designed to handle the interaction between a computer and its peripheral equipment, changing the format of data as necessary. { 'sȯf,wer 'drīv·ər }

software engineering [COMPUT SCI] The systematic application of scientific and technological knowledge, through the medium of sound engineering principles, to the production of computer programs, and to the requirements definition, functional specification, design description, program implementation, and test methods that lead up to this code. { 'sȯf,wer ,en·jə'nir·iŋ }

software flexibility [COMPUT SCI] The ability of software to change easily in response to different user and system requirements. { 'sȯf,wer ,flek·sə'bil·əd·ē }

software floating point [COMPUT SCI] Special routines that allow high-level programming languages to perform floating-point arithmetic on computer hardware designed for integer arithmetic. { 'sȯf,wer 'flōd·iŋ 'pȯint }

software interface [COMPUT SCI] A computer language whereby computer programs can communicate with each other, and one language can call upon another for assistance. { 'sȯf,wer 'in·tər·fās }

software maintenance [COMPUT SCI] The correction of errors in software systems and the remedying of inadequacies in running the software. { 'sōf,wer ,mānt·ən,əns }

software metric [COMPUT SCI] **1.** A rule for quantifying some characteristic or attribute of a computer software entity. **2.** One of a set of techniques whose aim is to measure the quality of a computer program. { ¦sȯf,wer 'me·trik }

software monitor [COMPUT SCI] A system, used to evaluate the performance of computer software, that is similar to accounting packages, but can collect more data concerning usage of various components of a computer system and is usually part of the control program. { 'sōf,wer ,män·əd·ər }

software multiplexing [COMPUT SCI] A procedure used in a time-sharing or multiprogrammed system in which the central processing unit, acting under control of a software algorithm, interleaves its attention between a family of programs waiting for service, in such a way that the programs appear to be processed in parallel. { 'sȯf,wer 'məl·ti,pleks·iŋ }

software package [COMPUT SCI] A program for performing some specific function or calculation which is useful to more than one computer user and is sufficiently well documented to be used without modification on a defined configuration of some computer system. { 'sȯf,wer ,pak·ij }

software path length [COMPUT SCI] The number of machine-language instructions required to carry out some specified task. Also known as path length. { 'sȯf,wer 'path ,leŋkth }

software piracy [COMPUT SCI] The process of copying commercial software without the permission of the originator. { 'sȯf,wer 'pir·ə·sē }

software protection [COMPUT SCI] The use of various techniques to prevent the unauthorized duplication of software. Also known as copy protection. { 'sȯf,wer prə,tek·shən }

soft-wired numerical control *See* computer numerical control. { 'sȯf ,wīrd nü'mer·ə·kəl kən 'trōl }

solar noise *See* solar radio noise. { 'sō·lər 'nȯiz }

solar radio noise [ELECTROMAG] Radio noise originating at the sun, and increasing greatly in intensity during sunspots and flares; it is heard as a hissing noise on shortwave radio receivers. Also known as solar noise. { 'sō·lər 'rād·ē·ō ,nȯiz }

solder-ball flip chip *See* flip chip. { ¦säd·ər ,bȯl 'flip ,chip }

solenoid [ELECTROMAG] Also known as electric solenoid. **1.** An electrically energized coil of insulated wire which produces a magnetic field within the coil. **2.** In particular, a coil that surrounds a movable iron core which is pulled to a central position with respect to the coil when the coil is energized by sending current through it. { 'säl·ə,nȯid }

solid logic technology [ELECTR] A method of computer construction that makes use of miniaturized modules, resulting in faster circuitry because of the reduced distances that current must travel. { 'säl·əd ¦läj·ik tek'näl·ə·jē }

solid state [ENG] Pertaining to a circuit, device, or system that depends on some combination of electrical, magnetic, and optical phenomena within a solid that is usually a crystalline semiconductor material. { 'säl·əd 'stāt }

solid-state laser [OPTICS] A laser in which a semiconductor material produces the coherent output beam. { 'säl·əd ¦stāt 'lā·zər }

solid-state memory [COMPUT SCI] A computer memory whose elements consist of integrated-circuit bistable multivibrators in which bits of information are stored as one of two states. { 'säl·əd ¦stāt 'mem·rē }

Sommerfeld equation *See* Sommerfeld formula. { 'zȯm·ər,felt i,kwā·zhən }

Sommerfeld formula [ELECTROMAG] An approximate formula for the field strength of electromagnetic radiation generated by an antenna at distances small enough so that the curvature of the earth may be neglected, in terms of radiated power, distance from the antenna, and various constants and parameters. Also known as Sommerfeld equation. { 'zȯm·ər,felt ,fȯr·myə·lə }

sonar [ENG] **1.** A system that uses underwater sound, at sonic or ultrasonic frequencies, to detect and locate objects in the sea, or for communication; the commonest type is echo-ranging sonar; other versions are passive sonar, scanning sonar, and searchlight sonar. Derived from sound navigation and ranging. **2.** *See* sonar set. { 'sō,när }

sonar array [ELECTR] An arrangement of several sonar transducers or sonar projectors, appropriately spaced and energized to give proper directional characteristics. { 'sō,när ə,rā }

sonar detector *See* sonar receiver. { 'sō,när di ,tek·tər }

sonar dome [ENG] A streamlined, watertight enclosure that provides protection for a sonar transducer, sonar projector, or hydrophone and associated equipment, while offering minimum interference to sound transmission and reception. { 'sō,när ,dōm }

sonar receiver [ELECTR] A receiver designed to intercept and amplify the sound signals reflected by an underwater target and display the accompanying intelligence in useful form; it may also pick up other underwater sounds. Also known as sonar detector. { 'sō,när ri'sē·vər }

sonar set [ENG] A complete assembly of sonar equipment for detecting and ranging or for communication. Also known as sonar. { 'sō,när ,set }

sonar transmitter [ELECTR] A transmitter that generates electrical signals of the proper frequency and form for application to a sonar transducer or sonar projector, to produce sound waves of the same frequency in water; the sound waves may carry intelligence. { 'sō,när tranz ,mid·ər }

S-100 bus [ELECTR] A bus assembly with 100 conductors; widely used in microcomputer-based systems. { ,es ¦wən'hən·drəd 'bəs }

son file [COMPUT SCI] The master file that is currently being updated. { 'sən ,fīl }

sonobuoy [ENG] An acoustic receiver and radio transmitter mounted in a buoy that can be dropped from an aircraft by parachute to pick up underwater sounds of a submarine and transmit them to the aircraft; several buoys are dropped and a computer determines the location of the submarine by comparison of the received signals and triangulation of the resulting time-delay data. Also known as radio sonobuoy. { 'sän·ə ,bȯi }

sophisticated vocabulary [COMPUT SCI] An advanced and elaborate set of instructions; a computer with a sophisticated vocabulary can go beyond the more common mathematical calculations such as addition, multiplication, and subtraction, and perform operations such as linearize, extract square root, and select highest number. { sə'fis·tə,kād·əd və'kab·yə,ler·ē }

sort [COMPUT SCI] **1.** To rearrange a set of data items into a new sequence, governed by specific rules of precedence. **2.** The program designed to perform this activity. { sȯrt }

sort algorithm [COMPUT SCI] The methods followed in arranging a set of data items into a sequence according to precise rules. { 'sȯrt ¦al·gə,rit͟h·əm }

sorter *See* sequencer. { 'sȯrd·ər }

sort field [COMPUT SCI] A field in a record that is used in determining the final sorted sequence of the records. { 'sȯrt ,fēld }

sort generator [COMPUT SCI] A computer program that produces other programs which arrange collections of items into sequences as specified by parameters in the original program. { 'sȯrt ,jen·ə,rād·ər }

sort key [COMPUT SCI] A key used as a basis for determining the sequence of items in a set. { 'sȯrt ,kē }

sort/merge [COMPUT SCI] To combine two or more similar files, with the records arranged in the appropriate order, according to precise rules. { 'sȯrt 'mərj }

sort/merge package [COMPUT SCI] A set of programs capable of sorting and merging data files. { 'sȯrt 'mərj ,pak·ij }

sort order [COMPUT SCI] The sequence into which a collection of records are arranged after they have been sorted. { 'sȯrt ,ȯr·dər }

sort pass [COMPUT SCI] Any one of a collection of similar procedures carried out during a sort operation in which a part of the sort is completed. { 'sȯrt ,pas }

sortworker [COMPUT SCI] A file created temporarily by a computer program to hold intermediate results when the amount of data to be sorted exceeds the available storage space. { 'sȯrt,wər·kər }

SOS [COMMUN] The distress signal in radiotelegraphy, consisting of the letters S, O, and S of the international Morse code.

sound board [COMPUT SCI] An adapter which provides a computer with the capability of reproducing and recording digitally encoded sound. Also known as audio adapter; sound card. { 'sau̇n ,bȯrd }

sound card *See* sound board. { 'sau̇n kärd }

sound carrier [COMMUN] The analog television carrier that is frequency-modulated by the sound portion of a television program; the unmodulated center frequency of the sound carrier is 4.5 megahertz higher than the video carrier frequency for the same television channel. { 'sau̇nd ,kar·ē·ər }

sound channel [ELECTR] The series of stages that handles only the sound signal in a television receiver. { 'sau̇nd ,chan·əl }

sound navigation and ranging *See* sonar. { 'sau̇nd ,nav·ə'gā·shən ən 'rānj·iŋ }

sound trap [ELECTR] A wave trap in an analog television receiver circuit that prevents sound signals from entering the picture channels. { 'sau̇nd ,trap }

source [ELECTR] The terminal in a field-effect transistor from which majority carriers flow into the conducting channel in the semiconductor material. { sȯrs }

source address [COMPUT SCI] The first address of a two-address instruction (the sound address is known as the destination address). { 'sȯrs 'ad ,res }

source code [COMPUT SCI] The statements in which a computer program is initially written before translation into machine language. { 'sȯrs ˌkōd }

source data automation equipment [COMPUT SCI] Equipment (except paper tape and magnetic tape cartridge typewriters acquired separately and not operated in support of a computer) which, as a by-product of its operation, produces a record in a medium which is acceptable by automatic data-processing equipment. { 'sȯrs ¦dad·ə ˌȯd·ə'mā·shən iˌkwip·mənt }

source data capture [COMPUT SCI] The procedures for entering source data into a computer system. { 'sȯrs 'dad·ə ˌkap·chər }

source data entry [COMPUT SCI] Entry of data into a computer system directly from its source, without transcription. { 'sȯrs 'dad·ə ˌen·trē }

source document [COMPUT SCI] The original medium containing the basic data to be used by a data-processing system, from which the data are converted into a form which can be read into a computer. Also known as original document. { 'sȯrs ˌdäk·yə·mənt }

source-follower amplifier *See* common-drain amplifier. { 'sȯrs 'fäl·ə·wər 'am·pləˌfī·ər }

source language [COMPUT SCI] The language in which a program (or other text) is originally expressed. { 'sȯrs ˌlaŋ·gwij }

source library [COMPUT SCI] A collection of computer programs in compiler language or assembler language. { 'sȯrs ¦līˌbrer·ē }

source listing [COMPUT SCI] A printout of a source program. { 'sȯrs ˌlist·iŋ }

source module [COMPUT SCI] An organized set of statements in any source language recorded in machine-readable form and suitable for input to an assembler or compiler. { 'sȯrs ˌmäj·ül }

source program [COMPUT SCI] The form of a program just as the programmer has written it, often on coding forms or machine-readable media; a program expressed in a source-language form. { 'sȯrs ˌprōˌgram }

source program optimizer [COMPUT SCI] A routine for examining the source code of a program under development and providing information about use of the various portions of the code, enabling the programmer to modify those sections of the target program that are most heavily used in order to improve performance of the final, operational program. { 'sȯrs ˌprōˌgram ˌäp·tə ˌmīz·ər }

source stream [COMMUN] A single, nonmultiplexed stream of samples before compression coding. { 'sȯrs ˌstrēm }

source time [COMPUT SCI] The time involved in fetching the contents of the register specified by the first address of a two-address instruction. { 'sȯrs ˌtīm }

space [COMMUN] The open-circuit condition or the signal causing the open-circuit condition in telegraphic communication; the closed-circuit condition is called the mark. { spās }

space character *See* blank character. { 'spās ˌkar·ik·tər }

space-charge layer *See* depletion layer. { 'spās ¦chärj ˌlā·ər }

space communication [COMMUN] Communication between a vehicle in outer space and the earth, using high-frequency electromagnetic radiation. { 'spās kəˌmyü·nə'kā·shən }

spacecraft ground instrumentation [ENG] Instrumentation located on the earth for monitoring, tracking, and communicating with manned spacecraft, satellites, and space probes. Also known as ground instrumentation. { 'spāsˌkraft 'grau̇nd ˌin·strəˌmən'tā·shən }

spaced antenna [ELECTROMAG] Antenna system consisting of a number of separate antennas spaced a considerable distance apart, used to minimize local effects of fading at short-wave receiving stations. { 'spāst an'ten·ə }

space diversity reception [ELECTROMAG] Radio reception involving the use of two or more antennas located several wavelengths apart, feeding individual receivers whose outputs are combined; the system gives an essentially constant output signal despite fading due to variable propagation characteristics, because fading affects the spaced-out antennas at different instants of time. { 'spās di'vər·səd·ē ri'sep·shən }

space-division multiple access [COMMUN] The use of the same portion of the electromagnetic spectrum over two or more transmission paths; in most applications, the paths are formed by multibeam antennas, and each beam is directed toward a different geographic area. Abbreviated SDMA. { ¦spās dəˌvizh·ən ˌməl·tə·pəl 'akˌses }

space-hold [COMMUN] The transmission of a steady space signal over a transmission line which is carrying no traffic. { 'spās ¦hōld }

space reflection symmetry *See* parity. { 'spās ri¦flek·shən 'sim·ə·trē }

space request [COMPUT SCI] A parameter that specifies the amount of storage space required by a new file at the time the file is created. { 'spās riˌkwest }

space suppression [COMPUT SCI] Prevention of the normal movement of paper in a computer printer after the printing of a line of characters. { 'spās səˌpresh·ən }

space-time adaptive processing [ELECTR] Radar techniques in which the antenna is subject to automatic pattern shaping to counter angularly displace noise sources (such as jammers), and the coherent signal processing is subject to automatic processes in which Doppler filters are optimally shaped to counter nonuniform distribution of background signals (such as surface clutter in airborne radar) in Doppler. { 'spās 'tīm ə'dap·tiv 'präsˌes·iŋ }

space-to-mark transition [COMMUN] The transition from the space condition to the mark condition in telegraphic communication. { ¦spās tə ¦märk tran'zish·ən }

space wave [ELECTROMAG] The component of a ground wave that travels more or less directly through space from the transmitting antenna to the receiving antenna; one part of the space wave goes directly from one antenna to the other;

another part is reflected off the earth between the antennas. { 'spās ˌwāv }

spacing pulse [COMMUN] In teletypewriter operation, the signal interval during which the selector unit is not operated. { 'spās·iŋ ˌpəls }

spaghetti code [COMPUT SCI] Computer program code that lacks a coherent structure, and in which the sequence of program execution frequently jumps to a distant instruction in the program listing, making the program very difficult to follow. { spə'ged·ē ˌkōd }

spam [COMPUT SCI] Unsolicited commercial e-mail. { spam }

spanned record [COMPUT SCI] A logical record which covers more than one block, used when the size of a data buffer is fixed or limited. { 'spand 'rek·ərd }

spark transmitter [ELECTR] A radio transmitter that utilizes the oscillatory discharge of a capacitor through an inductor and a spark gap as the source of radio-frequency power. { 'spärk tranz'mid·ər }

spatial data management [COMPUT SCI] A technique whereby users retrieve information in databases, document files, or other sources by making contact with picture symbols displayed on the screen of a video terminal through the use of such devices as light pens, joy sticks, and heat-sensitive screens for finger-touch activation. { 'spā·shəl 'dad·ə ˌman·ij·mənt }

SPC *See* stored-program control.

special character [COMPUT SCI] A computer-representable character that is not alphabetic, numeric, or blank. { 'spesh·əl 'kar·ik·tər }

special-purpose computer [COMPUT SCI] A digital or analog computer designed to be especially efficient in a certain class of applications. { 'spesh·əl ¦pər·pəs kəm'pyüd·ər }

special-purpose language [COMPUT SCI] A programming language designed to solve a particular type of problem. { 'spesh·əl ¦pər·pəs 'laŋ·gwij }

specific cryptosystem [COMMUN] A general cryptosystem and a cryptographic key or set of keys for controlling the cryptographic process. { spə'sif·ik 'krip·tōˌsis·təm }

specific repetition rate [ELECTR] The pulse repetition rate of a pair of transmitting stations of an electronic navigation system using various rates differing slightly from each other, as in loran. { spə'sif·ik ˌrep·ə'tish·ən ˌrāt }

specific routine [COMPUT SCI] Computer routine to solve a particular data-handling problem in which each address refers to explicitly stated registers and locations. { spə'sif·ik rü'tēn }

spectral density *See* frequency spectrum. { 'spek·trəl 'den·səd·ē }

spectrum level [COMMUN] The level of the part of a specified signal at a specified frequency that is contained within a specified frequency bandwidth, centered at the particular frequency. { 'spek·trəm ˌlev·əl }

spectrum-selectivity characteristic [ELECTR] Measure of the increase in the minimum input signal power over the minimum detectable signal required to produce an indication on a radar indicator, if the received signal has a spectrum different from that of the normally received signal. { 'spek·trəm ˌsi ˌlek'tiv·əd·ē ˌkar·ik·təˌris·tik }

spectrum signature [ELECTR] The spectral characteristics of the transmitter, receiver, and antenna of an electronic system, including emission spectra, antenna patterns, and other characteristics. { 'spek·trəm ˌsig·nə·chər }

spectrum signature analysis [ELECTR] The evaluation of electromagnetic interference from transmitting and receiving equipment to determine operational and environment compatibility. { 'spek·trəm ˌsig·nə·chər əˌnal·ə·səs }

speech bandwidth [COMMUN] The range of speech frequencies that can be transmitted by a carrier telephone system. { 'spēch 'band ˌwidth }

speech coder [COMMUN] A device that uses data-compression techniques to convert a high-bit-rate signal resulting from digital pulse-code modulation of speech to a low-rate digital signal that can be transmitted or stored. { 'spēch ˌkōd·ər }

speech compression [COMMUN] Modulation technique that takes advantage of certain properties of the speech signal to permit adequate information quality, characteristics, and the sequential pattern of a speaker's voice to be transmitted over a narrower frequency band than would otherwise be necessary. { 'spēch kəmˌpresh·ən }

speech frequency *See* voice frequency. { 'spēch ˌfrē·kwən·sē }

speech intelligibility *See* intelligibility. { 'spēch inˌtel·ə·jə'bil·əd·ē }

speech interpolation [COMMUN] Method of obtaining more than one voice channel per voice circuit by giving each subscriber a speech path in the proper direction only at times when the subscriber's speech requires it. { 'spēch ˌin·tər·pəl¦ā·shən }

speech inverter *See* scrambler. { 'spēch in ˌvərd·ər }

speech recognition [ENG ACOUS] The process of analyzing an acoustic speech signal to identify the linguistic message that was intended, so that a machine can correctly respond to spoken commands. { 'spēch ˌrek·ig'nish·ən }

speech scrambler *See* scrambler. { 'spēch ˌskram·blər }

speech synthesis *See* voice response. { 'spēch 'sin·thə·səs }

speed-matching buffer [COMPUT SCI] A small computer storage unit that connects two devices operating at different data transfer rates; each device writes into and reads from the buffer at its own rate. { 'spēd ¦mach·iŋ 'bəf·ər }

speed of light [ELECTROMAG] The speed of propagation of electromagnetic waves in a vacuum, which is a physical constant equal to exactly 299,792.458 kilometers per second. Also known as electromagnetic constant; velocity of light. { 'spēd əv 'līt }

spelling checker [COMPUT SCI] A program, used in conjunction with word-processing software, which automatically checks words in a text against a dictionary of commonly used words and identifies words that appear to be misspelled. { 'spel·iŋ ,chek·ər }

spherical-earth attenuation [ELECTROMAG] Attenuation over an imperfectly conducting spherical earth in excess of that over a perfectly conducting plane. { 'sfir·ə·kəl ¦ərth ə,ten·yə ,wā·shən }

spherical-earth factor [ELECTROMAG] The ratio of the electric field strength that would result from propagation over an imperfectly conducting spherical earth to that which would result from propagation over a perfectly conducting plane. { 'sfir·ə·kəl ¦ərth ,fak·tər }

spider [COMPUT SCI] A program that searches the Internet for new, publicly accessible resources and transmits its findings to a database that is accessible to search engines. { 'spīd·ər }

spiderweb antenna [ELECTROMAG] All-wave receiving antenna having several different lengths of doublets connected somewhat like the web of a spider to give favorable pickup characteristics over a wide range of frequencies. { 'spīd·ər,web an,ten·ə }

spike antenna *See* monopole antenna. { 'spīk an,ten·ə }

spillover [COMMUN] The receiving of a radio signal of a different frequency from that to which the receiver is tuned, due to broad tuning characteristics. { 'spil,ō·vər }

spillover positions [COMMUN] When a transmitting channel is unusually busy or inoperative, the resulting backlogged traffic can be switched to spillover (storage) positions where it is held for immediate transmission when a channel becomes available. { 'spil,ō·vər pə,zish·ənz }

spin transistor *See* magnetic switch. { 'spin tran,zis·tər }

spin valve *See* magnetic switch. { 'spin ,valv }

spiral delay line [ELECTROMAG] A transmission line which has a helical inner conductor. { 'spī·rəl di'lā ,līn }

splatter [COMMUN] Distortion due to overmodulation of a transmitter by peak signals of short duration, particularly sounds containing high-frequency harmonics; it is a form of adjacent-channel interference. { 'splad·ər }

splicing [COMMUN] The concatenation, performed on the system level, of two different elementary streams. { 'splīs·iŋ }

split [COMPUT SCI] To divide a database, file, or other data set into two or more separate parts. { split }

split screen *See* partitioned display. { 'split 'skrēn }

split-word operation [COMPUT SCI] A computer operation performed with portions of computer words rather than whole words as is normally done. { 'split ¦wərd ,äp·ə'rā·shən }

SPMD [COMPUT SCI] A type of programming on a multiprocessor in which parallel programs all run the same subroutine but operate on different data. Acronym for single-program, multiple-data.

spoiler [ELECTROMAG] Rod grating mounted on a parabolic reflector to change the pencil-beam pattern of the reflector to a cosecant-squared pattern; rotating the reflector and grating 90° with respect to the feed antenna changes one pattern to the other. { 'spȯi·lər }

spoofing [COMPUT SCI] A method of gaining unauthorized access to computers or networkds by sending messages with someone else's IP address, so that the message appears, to the targeted system, to be coming from a trusted host. [ELECTR] Deceiving or misleading an enemy in electronic operations, as by continuing transmission on a frequency after it has been effectively jammed by the enemy, using decoy radar transmitters to lead the enemy into a useless jamming effort, or transmitting radio messages containing false information for intentional interception by the enemy. { 'spüf·iŋ }

spooling [COMPUT SCI] The temporary storage of input and output on high-speed input-output devices, typically magnetic disks and drums, in order to increase throughput. Acronym for simultaneous peripheral operations on line. { 'spül·iŋ }

sporadic E layer [GEOPHYS] A layer of intense ionization that occurs sporadically within the E layer; it is variable in time of occurrence, height, geographical distribution, penetration frequency, and ionization density. { spə'rad·ik 'ē ,lā·ər }

sporadic fault [COMPUT SCI] A hardware malfunction that occurs intermittently and at unpredictable times. { spə'rad·ik 'fȯlt }

sporadic reflections [ELECTROMAG] Sharply defined reflections of substantial intensity from the sporadic E layer at frequencies greater than the critical frequency of the layer; they are variable with respect to time of occurrence, geographic location, and range of frequencies at which they are observed. { spə'rad·ik ri'flek·shənz }

spot beam [COMMUN] A beam generated by a communications satellite antenna of sufficient size that the angular spread of energy in the beam is small, always smaller than the earth's angular beam width as seen from the satellite. { 'spät ,bēm }

spot jammer [ELECTR] A jammer that interferes with reception of a specific channel or frequency. { 'spät ,jam·ər }

spot jamming [ELECTR] An electronic attack technique in which a continuous narrow-band signal is transmitted, giving a stronger jamming signal to a particular victim radar than had a wide-band transmission been used. { 'spät ,jam·iŋ }

spot speed [COMMUN] **1.** In a video system, the product of the length (in units of elemental area, that is, in spots) of scanning line by the number of scanning lines per second. **2.** In facsimile transmission, the speed of the scanning or recording spot within the available line. Also known as scanning speed. { 'spät ,spēd }

spreadsheet program [COMPUT SCI] A computer program that simulates an accountant's worksheet on screen as an array of rows (usually numbered) and columns (usually assigned alphabetical letters) whose intersections are called cells; the program allows the user to enter data in the cells and to embed formulas which relate the values in different cells. { 'spred ˌshēt 'prō·grəm }

spread spectrum transmission [ELECTR] Communications technique in which many different signal waveforms are transmitted in a wide band; power is spread thinly over the band so narrow-band radios can operate within the wide-band without interference; used to achieve security and privacy, prevent jamming, and utilize signals buried in noise. { 'spred ¦spek·trəm tranzˌmish·ən }

sprocket pulse [COMPUT SCI] **1.** A pulse generated by a magnetized spot which accompanies every character recorded on magnetic tape; this pulse is used during read operations to regulate the timing of the read circuits, and also to provide a count on the number of characters read from the tape. **2.** A pulse generated by the sprocket or driving hole in paper tape which serves as the timing pulse for reading or punching the paper tape. { 'spräk·ət ˌpəls }

spurious emission *See* spurious radiation. { 'spyu̇r·ē·əs i'mish·ən }

spurious modulation [ELECTR] Undesired modulation occurring in an oscillator, such as frequency modulation caused by mechanical vibration. { 'spyu̇r·ē·əs ˌmäj·ə'lā·shən }

spurious radiation [ELECTROMAG] Any emission from a radio transmitter at frequencies outside its frequency band. Also known as spurious emission. { 'spyu̇r·ē·əs ˌrād·ē'ā·shən }

spurious response [ELECTR] Response of a radio receiver to a frequency different from that to which the receiver is tuned. { 'spyu̇r·ē·əs ri'späns }

spurt tone [COMMUN] Short audio-frequency tone used for signaling or dialing selection. { 'spərt ˌtōn }

SQL *See* Structured Query Language.

square-law demodulator *See* square-law detector. { 'skwer ¦lȯ dē'mäj·əˌlād·ər }

square-law detector [ELECTR] A demodulator whose output voltage is proportional to the square of the amplitude-modulated input voltage. Also known as square-law demodulator. { 'skwer ¦lȯ diˌtek·tər }

squealing [ELECTR] A condition in which a radio receiver produces a high-pitched note or squeal along with the desired radio program, due to interference between stations or to oscillation in some receiver circuit. { 'skwēl·iŋ }

squeezable waveguide [ELECTROMAG] A waveguide whose dimensions can be altered periodically; used in rapid scanning. { 'skwēz·ə·bəl 'wāvˌgīd }

squeeze section [ELECTROMAG] Length of waveguide constructed so that alteration of the critical dimension is possible with a corresponding alteration in the electrical length. { 'skwēz ˌsek·shən }

squegger *See* blocking oscillator. { 'skweg·ər }

squegging oscillator *See* blocking oscillator. { 'skweg·iŋ ˌäs·əˌlād·ər }

squelch [ELECTR] To automatically quiet a receiver by reducing its gain in response to a specified characteristic of the input. { skwelch }

squelch circuit *See* noise suppressor. { 'skwelch ˌsər·kət }

squint [ELECTROMAG] **1.** The angle between the two major lobe axes in a radar lobe-switching antenna. **2.** The angular difference between the axis of radar antenna radiation and a selected geometric axis, such as the axis of the reflector. **3.** The angle between the full-right and full-left positions of the beam of a conical-scan radar antenna. { skwint }

squishing *See* compaction. { 'skwish·iŋ }

squitter [ELECTR] Random firing, intentional or otherwise, of the transponder transmitter in the absence of interrogation. { 'skwid·ər }

SRAM *See* static random-access memory. { 'es ˌram }

SRC *See* stored response chain.

SSA Service *See* S-band single-access service. { ¦es¦es'ā ˌsər·vəs }

SSB *See* single-sideband.

ss loran *See* sky-wave-synchronized loran. { ¦es ¦es 'lȯrˌan }

SSTV *See* slow-scan television.

stability [CONT SYS] The property of a system for which any bounded input signal results in a bounded output signal. { stə'bil·əd·ē }

stability criterion [CONT SYS] A condition which is necessary and sufficient for a system to be stable, such as the Nyquist criterion, or the condition that poles of the system's overall transmittance lie in the left half of the complex-frequency plane. { stə'bil·əd·ē krīˌtir·ē·ən }

stabilization [CONT SYS] *See* compensation. [ELECTR] Feedback introduced into transistor amplifier stages to reduce distortion by making the amplification substantially independent of electrode voltages. { ˌstā·bə·lə'zā·shən }

stabilized feedback *See* negative feedback. { 'stā·bəˌlīzd 'fēdˌbak }

stable local oscillator *See* stalo. { 'stā·bəl 'lō·kəl 'äs·əˌlād·ər }

stack [COMPUT SCI] A portion of a computer memory used to temporarily hold information, organized as a linear list for which all insertions and deletions, and usually all accesses, are made at one end of the list. { stak }

stack automaton [COMPUT SCI] A variation of a pushdown automaton in which the read-only head of the input tape is allowed to move both ways, and the read-write head on the pushdown storage is allowed to scan the entire pushdown list in a read-only mode. { 'stak ȯ'täm·əˌtän }

stacked array [ELECTROMAG] An array in which the antenna elements are stacked one above the other and connected in phase to increase the gain. { 'stakt ə'rā }

stacked-beam radar [ENG] Three-dimensional radar system that derives elevation by emitting narrow beams stacked vertically to cover a vertical segment, azimuth information from horizontal scanning of the beam, and range information from echo-return time. { 'stakt ¦bēm 'rā,där }

stacked-dipole antenna [ELECTROMAG] Antenna in which directivity is increased by providing a number of identical dipole elements, excited either directly or parasitically; the resultant radiation pattern depends on the number of dipole elements used, the spacing and phase difference between the elements, and the relative magnitudes of the currents. { 'stakt 'dī,pōl an,ten·ə }

stacked-job processing [COMPUT SCI] A technique of automatic job-to-job transition, with little or no operator intervention. { 'stakt 'jäb ,prä,ses·iŋ }

stacked loops [ELECTROMAG] Two or more loop antennas arranged above each other on a vertical supporting structure and connected in phase to increase the gain. Also known as vertically stacked loops. { 'stakt 'lüps }

stacking [ELECTROMAG] The placing of antennas one above the other, connecting them in phase to increase the gain. { 'stak·iŋ }

stack model [COMPUT SCI] A model for describing the run-time execution of programs written in block-structured languages, consisting of a program component, which remains unchanged throughout the execution of the program; a control component, consisting of an instruction pointer and an environment pointer; and a stack of records containing all the data the program operates on. { 'stak ,mäd·əl }

stack operation [COMPUT SCI] A computer system in which flags, return address, and all temporary addresses are saved in the core in sequential order for any interrupted routine so that a new routine (including the interrupted routine) may be called in. { 'stak ,äp·ə,rā·shən }

stack pointer [COMPUT SCI] A register which contains the last address of a stack of addresses. { 'stak ,pȯint·ər }

stadiometry [COMPUT SCI] In computer vision, the determination of the distance to an object based on the size of its image. { ,stād·ē'äm·ə·trē }

stagger [COMMUN] Periodic error in the position of the recorded spot along a recorded facsimile line. { 'stag·ər }

staggered tuning [ELECTR] Alignment of successive tuned circuits to slightly different frequencies in order to widen the overall amplitude-frequency response curve. { 'stag·ərd 'tün·iŋ }

staggering [COMMUN] Offsetting of two channels of different carrier systems from exact sideband frequency coincidence to avoid mutual interference. { 'stag·ə·riŋ }

staggering advantage [COMMUN] Effective reduction of interference between carrier channels, due to staggering. { 'stag·ə·riŋ ad,van·tij }

stagger-tuned amplifier [ELECTR] An amplifier that uses staggered tuning to give a wide bandwidth. { 'stag·ər ¦tünd 'am·plə,fī·ər }

stagger-tuned filter [ELECTR] A filter consisting of a cascade of amplifier stages with tuned coupling networks whose resonant frequencies and bandwidths may be easily adjusted to achieve an overall transmission function of desired shape (maximally flat or equal ripple). { 'stag·ər ¦tünd 'fil·tər }

staging [COMPUT SCI] Moving blocks of data from one storage device to another. { 'stāj·iŋ }

staircase signal [COMMUN] In analog television transmissions, a waveform that consists of a series of discrete steps resembling a staircase. { 'ster,kās ,sig·nəl }

stale link [COMPUT SCI] A hyperlink to a document that has been erased or removed from the World Wide Web. Also known as black hole. { ¦stāl 'liŋk }

stalo [ELECTR] A highly stable local radio-frequency oscillator used in coherent radar both for up-converting the transmit signal to the carrier frequency and down-converting the received signals to the intermediate frequency. { 'stā ,lō }

stand-alone machine [COMPUT SCI] A machine capable of functioning independently of a master computer, either part of the time or all of the time. { 'stand ə¦lōn mə'shēn }

standard antenna [ELECTROMAG] An open single-wire antenna (including the lead-in wire) having an effective height of 4 meters. { 'stan·dərd an'ten·ə }

standard blocked F-format data set *See* FBS data set. { 'stan·dərd 'bläkt ¦ef'fȯr,mat 'dad·ə ,set }

standard broadcast band *See* broadcast band. { 'stan·dərd 'brȯd,kast ,band }

standard broadcast channel [COMMUN] Band of frequencies occupied by the carrier and two side bands of a radio broadcast signal, with the carrier frequency at the center. { 'stan·dərd 'brȯd,kast ,chan·əl }

standard broadcasting [COMMUN] Radio broadcasting using amplitude modulation in the band of frequencies from 535 to 1605 kilohertz; carrier frequencies are placed 10 kilohertz apart. { 'stan·dərd 'brȯd,kast·iŋ }

standard definition television [COMMUN] Term used to signify a digital television system in which the quality is approximately equivalent to that of NTSC. Also called standard digital television. Abbreviated SDTV. { 'stan·dərd def·ə¦nish·ən 'tel·ə,vizh·ən }

standard digital television *See* standard definition television. { 'stan·dərd 'dij·əd·əl 'tel·ə ,vizh·ən }

standard form [COMPUT SCI] The form of a floating point number whose mantissa lies within a standard specified range of values. { 'stan·dərd 'fȯrm }

standard-frequency signal [COMMUN] One of the highly accurate signals broadcast by government radio stations and used for testing and calibrating radio equipment all over the world; in the United States signals are broadcast by the National Bureau of Standards' radio stations WWV, WWVH, WWVB, and WWVL. { 'stan·dərd ¦frē·kwən·sē ˌsig·nəl }

standard function *See* built-in function. { 'stan·dərd 'fəŋk·shən }

Standard Generalized Markup Language [COMPUT SCI] A system that encodes the logical structure and content of a document rather than its display formatting, or even the medium in which the document will be displayed; widely used in the publishing business and for producing technical documentation. Abbreviated SGML. { ¦stan·dərd ˌjen·rəˌlīzd 'märkˌəp ˌlaŋ·gwij }

standard interface [COMPUT SCI] **1.** A joining place of two systems or subsystems that has a previously agreed-upon form, so that two systems may be readily connected together. **2.** In particular, a system of uniform circuits and input/output channels connecting the central processing unit of a computer with various units of peripheral equipment. { 'stan·dərd 'in·tər ˌfās }

standardize [COMPUT SCI] To replace any given floating point representation of a number with its representation in standard form; that is, to adjust the exponent and fixed-point part so that the new fixed-point part lies within a prescribed standard range. { 'stan·dərˌdīz }

standard loran *See* loran A. { 'stan·dərd 'lȯ ˌran }

standard parallel port [COMPUT SCI] A parallel port that can transfer data in only one direction. { ¦stan·dərd ˌpar·əˌlel 'pȯrt }

standard preemphasis [COMMUN] Preemphasis in frequency-modulation and analog television aural broadcasting whose level lies between upper and lower limits specified by the Federal Communications Commission. { 'stan·dərd prē'em·fə·səs }

standard propagation [ELECTROMAG] Propagation of radio waves over a smooth spherical earth of specified dielectric constant and conductivity, under conditions of standard refraction in the atmosphere. { 'stan·dərd ˌpräp·ə'gā·shən }

standard refraction [ELECTROMAG] Refraction which would occur in an idealized atmosphere in which the index of refraction decreases uniformly with height at a rate of 39×10^{-6} per kilometer; standard refraction may be included in ground wave calculations by use of an effective earth radius of 8.5×10^6 meters, or $\frac{4}{3}$ the geometrical radius of the earth. { 'stan·dərd ri'frak·shən }

standard subroutine [COMPUT SCI] In computers, a subroutine which is applicable to a class of problems. { 'stan·dərd 'səb·rüˌtēn }

standard target [ELECTROMAG] A radar target which will produce an echo of known power under various conditions; smooth metal spheres or corner reflectors of known dimensions are such targets, and they may be used to calibrate a radar or check its performance. { 'stan·dərd 'tär·gət }

standby computer [COMPUT SCI] A computer in a duplex system that takes over when the need arises. { 'stand¦bī kəmˌpyüd·ər }

standby register [COMPUT SCI] In computers, a register into which information can be copied to be available in case the original information is lost or mutilated in processing. { 'stand ¦bī ˌrej·ə·stər }

standby replacement redundancy [COMPUT SCI] A form of redundancy in which there is a single active unit and a reserve of spare units, one of which replaces the active unit if it fails. { 'stand¦bī ri'plās·mənt riˌdən·dən·sē }

standby time [COMPUT SCI] **1.** The time during which two or more computers are tied together and available to answer inquiries or process intermittent actions on stored data. **2.** The elapsed time between inquiries when the equipment is operating on an inquiry application. { 'stand¦bī ˌtīm }

standing-on-nines carry [COMPUT SCI] In high-speed parallel addition of decimal numbers, an arrangement that causes carry digits to pass through one or more nine digits, while signaling that the skipped nines are to be reset to zero. { ¦stand·iŋ ȯn ¦nīnz 'kar·ē }

standing wave [PHYS] A wave in which the ratio of an instantaneous value at one point to that at any other point does not vary with time. Also known as stationary wave. { 'stand·iŋ 'wāv }

standing-wave loss factor [ELECTROMAG] The ratio of the transmission loss in an unmatched waveguide to that in the same waveguide when matched. { 'stand·iŋ ¦wāv 'lȯs ˌfak·tər }

standstill feature [CONT SYS] A device which insures that false signals such as fluctuations in the power supply do not cause a controller to be altered. { 'stanˌstil ˌfē·chər }

star connection *See* star network. { 'stär kə ¦nek·shən }

star-free expression [COMPUT SCI] An expression containing only Boolean operations and concatenation, used to define the language corresponding to a counter-free machine. { 'stär ¦frē ik'spresh·ən }

star network [COMMUN] A communications network in which all communications between any two points must pass through a central node. Also known as centralized configuration. { 'stär ¦netˌwərk }

start bit [COMPUT SCI] The first bit transmitted in asynchronous data transmission to unequivocally indicate the start of the word. { 'stärt ˌbit }

start codes [COMMUN] 32-bit codes embedded in the coded bit stream that are unique; used for several purposes including identifying some of the layers in the coding syntax. { 'stärt ˌkōdz }

start dialing signal [COMMUN] Signal transmitted from the incoming end of a circuit, following the receipt of a seizing signal, to indicate that the necessary circuit conditions have been established for receiving the numerical routine information. { 'stärt 'dīl·iŋ ˌsig·nəl }

started task [COMPUT SCI] A computer program that is kept permanently in main storage and, though not a part of the operating system, is treated as though it were. { 'stärd·əd 'task }

start element [COMMUN] The first element of a character in certain serial transmissions, used to permit synchronization. { 'stärt ˌel·ə·mənt }

startover [COMPUT SCI] Program function that causes a computer that is not active to become active. { 'stärˌdō·vər }

startover data transfer and processing program [COMPUT SCI] Program which controls the transfer of startover data from the active to the standby machine and their subsequent processing by the standby machine. { 'stärˌdō·vər 'dad·ə ˌtranz·fər ən 'präˌses·iŋ ˌprōˌgram }

start-stop multivibrator *See* monostable multivibrator. { 'stärt 'stäp ˌməl·ti'vīˌbrād·ər }

start-stop printing telegraph [COMMUN] Form of printing telegraph in which the signal-receiving mechanisms, normally at rest, are started in operation at the beginning and stopped at the end of each character transmitted over the channel. { 'stärt 'stäp 'print·iŋ 'tel·ə·graf }

start-stop system [COMMUN] A telegraph system in which each group of code elements corresponding to a character is preceded by a start signal that prepares the receiving mechanism to receive and register a character, and is followed by a stop signal that brings the receiving mechanism to rest in preparation for the reception of the next character. { 'stärt 'stäp ˌsis·təm }

state [CONT SYS] A minimum set of numbers which contain enough information about a system's history to enable its future behavior to be computed. { stāt }

state equations [CONT SYS] Equations which express the state of a system and the output of a system at any time as a single valued function of the system's input at the same time and the state of the system at some fixed initial time. { 'stāt iˌkwā·zhənz }

state estimator *See* observer. { 'stāt ˌes·təˌmād·ər }

state feedback [CONT SYS] A class of feedback control laws in which the control inputs are explicit memoryless functions of the dynamical system state, that is, the control inputs at a given time t_a are determined by the values of the state variables at t_a and do not depend on the values of these variables at earlier times $t \geq t_a$. { 'stāt 'fēdˌbak }

state graph [COMPUT SCI] A directed graph whose nodes correspond to internal states of a sequential machine and whose edges correspond to transitions among these states. { 'stāt ˌgraf }

statement [COMPUT SCI] An elementary specification of a computer action or process, complete and not divisible into smaller meaningful units; it is analogous to the simple sentence of a natural language. { 'stāt·mənt }

statement editor [COMPUT SCI] A text editor in which the text is divided into superlines, that is, units greater than ordinary lines, resulting in easier editing and freedom from truncation problems. { 'stāt·mənt ˌed·əd·ər }

state observer *See* observer. { 'stāt əbˌzər·vər }

state space [CONT SYS] The set of all possible values of the state vector of a system. { 'stāt ˌspās }

state table [COMPUT SCI] A table that represents a sequential machine, in which the rows correspond to the internal states, the columns to the input combinations, and the entries to the next state. { 'stāt ˌtā·bəl }

state transition equation [CONT SYS] The equation satisfied by the $n \times n$ state transition matrix $\Phi(t,t_0)$: $\partial\Phi(t,t_0)/\partial t = A(t)\,\Phi(t,t_0)$, $\Phi(t_0,t_0) = I$; here I is the unit $n \times n$ matrix, and $A(t)$ is the $n \times n$ matrix which appears in the vector differential equation $dx(t)/dt = A(t)x(t)$ for the n-component state vector $x(t)$. { 'stāt tran'zish·ən iˌkwā·zhən }

state transition matrix [CONT SYS] A matrix $\Phi(t,t_0)$ whose product with the state vector x at an initial time t_0 gives the state vector at a later time t; that is, $x(t) = \Phi(t,t_0)x(t_0)$. { 'stāt tran'zish·ən ˌmā·triks }

state variable [CONT SYS] One of a minimum set of numbers which contain enough information about a system's history to enable computation of its future behavior. { 'stāt ˌver·ē·ə·bəl }

state-variable filter [ELECTR] A multiple-amplifier active filter that has three outputs for high-pass, band-pass, and low-pass transfer functions respectively. Also known as KHN filter. { 'stāt ˌver·ē·ə·bəl ˌfil·tər }

state vector [COMPUT SCI] *See* task descriptor. [CONT SYS] A column vector whose components are the state variables of a system. { 'stāt ˌvek·tər }

static [COMMUN] A hissing, crackling, or other sudden sharp sound that tends to interfere with the reception, utilization, or enjoyment of desired signals or sounds. { 'stad·ik }

static algorithm [COMPUT SCI] An algorithm whose operation is known in advance. Also known as deterministic algorithm. { 'stad·ik 'al·gəˌri<u>th</u>·əm }

static check [COMPUT SCI] Of a computer, one or more tests of computing elements, their interconnections, or both, performed under static conditions. { 'stad·ik 'chek }

static debugging routine [COMPUT SCI] A debugging routine which is used after the program being checked has been run and has stopped. { 'stad·ik dē'bəg·iŋ rüˌtēn }

static dump [COMPUT SCI] An edited printout of the contents of main memory or of the auxiliary storage, performed in a fixed way; it is usually taken at the end of a program run either automatically or by operator intervention. { 'stad·ik 'dəmp }

static eliminator [ELECTR] Device intended to reduce the effect of atmospheric static interference in a radio receiver. { 'stad·ik iˌlim·ə ˌnād·ər }

staticize [COMPUT SCI] **1.** To capture transient data in stable form, thus converting fleeting events into examinable information. **2.** To extract an instruction from the main computer memory and store the various component parts of it in the appropriate registers, preparatory to interpreting and executing it. { 'stad·ə,sīz }

static random-access memory [COMPUT SCI] A read-write random-access memory that uses either four transistors and two resistors to form a passive-load flip-flop, or six transistors to form a flip-flop with dynamic loads, for each cell in an array. Once data are loaded into the flip-flop storage elements, the flip-flop will indefinitely remain in that state until the information is intentionally changed or the power to the memory circuit is shut off. Abbreviated SRAM. { 'stad·ik 'rand·əm ¦ak,ses 'mem·rē }

static regulator [ELECTR] Transmission regulator in which the adjusting mechanism is in self-equilibrium at any setting and requires control power to change the setting. { 'stad·ik ,reg·yə ,lād·ər }

static storage [COMPUT SCI] Computer storage such that information is fixed in space and available at any time, as in flip-flop circuits, electrostatic memories, and coincident-current magnetic-core storage. { 'stad·ik 'stȯr·ij }

static subroutine [COMPUT SCI] In computers, a subroutine which involves no parameters other than the addresses of the operands. { 'stad·ik 'səb·rü,tēn }

static variable [COMPUT SCI] A local variable that does not cease to exist upon termination of the block in which it can be accessed, but instead retains its most recent value until the next execution of this block. { 'stad·ik 'ver·ē·ə·bəl }

station [COMMUN] *See* broadcast station. [COMPUT SCI] One of a series of essentially similar positions or facilities occurring in a data-processing system. [ELECTR] A location at which radio, television, radar, or other electric equipment is installed. { stā·shən }

stationary noise [ELECTR] A random noise for which the probability that the noise voltage lies within any given interval does not change with time. { 'stā·shə,ner·ē 'nȯiz }

stationary orbit [AERO ENG] A circular, equatorial orbit in which the satellite revolves about the primary body at the angular rate at which the primary body rotates on its axis; from the primary body, the satellite thus appears to be stationary over a point on the primary body; a stationary orbit must be synchronous, but the reverse need not be true. { 'stā·shə,ner·ē 'ȯr·bət }

stationary satellite [AERO ENG] A satellite in a stationary orbit. { 'stā·shə,ner·ē 'sad·əl,īt }

stationary wave *See* standing wave. { 'stā·shə ,ner·ē 'wāv }

station authentication [COMMUN] Security measure designed to establish the authenticity of a transmitting or receiving station. { 'stā·shən ȯ ,then·tə'kā·shən }

statistical monitor [COMPUT SCI] A software monitor that collects information by periodically sampling activity in the system. { stə'tis·tə·kəl 'män·əd·ər }

statistical multiplexer [ELECTR] A device which combines several low-speed communications channels into a single high-speed channel, and which can manage more communications traffic than a standard multiplexer by analyzing traffic and choosing different transmission patterns. { stə'tis·tə·kəl 'məl·tə,plek·sər }

statistical multiplexing [COMMUN] Time-division multiplexing in which time on a communications channel is assigned to multiple users on a demand basis, rather than periodically to each user. { stə,tis·ti·kəl 'məl·tə,pleks·iŋ }

status byte [COMPUT SCI] A byte of storage whose contents indicate the activities currently taking place in some part of the computer or various conditions governing the execution of a computer program; often, each bit is assigned a particular meaning. { 'stad·əs ,bīt }

status check [COMPUT SCI] The detection of software failures and verification of programs through the use of redundant computers. { 'stad·əs ,chek }

status line [COMPUT SCI] A conductor on the bus of a computer over which an addressed storage location or component transmits its status to the central processing unit. { 'stad·əs ,līn }

status register [COMPUT SCI] A register maintained by the central processing unit that contains a status byte with information about activities currently taking place there. { 'stad·əs ,rej·ə·stər }

status word [COMPUT SCI] A word indicating the state of the system or the diagnosis of a state into which the system has entered. { 'stad·əs ,wərd }

STD *See* system target decoder.

STD input buffer [COMMUN] A first-in, first-out buffer at the input of a system target decoder for storage of compressed data from elementary streams before decoding. { ¦es¦tē¦dē 'in,pu̇t ,bəf·ər }

STDM *See* synchronous time-division multiplexing.

steadiness [CONT SYS] Freedom of a robot arm or end effector from high-frequency vibrations and jerks. { 'sted·ē·nəs }

steady-state error [CONT SYS] The error that remains after transient conditions have disappeared in a control system. { 'sted·ē ¦stāt 'er·ər }

stealth [ENG] A term in radar describing special constructional techniques, usually involving shaping and special materials, to reduce greatly the radar cross section of targets, making it very difficult for radars to detect them at useful ranges. { stelth }

steerable antenna [ELECTROMAG] A directional antenna whose major lobe can be readily shifted in direction. { 'stir·ə·bəl an'ten·ə }

steganography [COMPUT SCI] The art and science of hiding a message in a medium, such as a digital picture or audio file, so as to defy detection. { ,steg·ə'näg·rə·fē }

stenode circuit [ELECTR] Superheterodyne receiving circuit in which a piezoelectric unit is used in the intermediate-frequency amplifier to balance out all frequencies except signals at the crystal frequency, thereby giving very high selectivity. { 'ste,nōd ,sər·kət }

step [COMPUT SCI] A single computer instruction or operation. { step }

step-by-step operation *See* single-step operation. { ¦step bī ¦step ,äp·ə'rā·shən }

step-by-step system [COMMUN] *See* Strowger system. [CONT SYS] A control system in which the drive motor moves in discrete steps when the input element is moved continuously. { ¦step bī ¦step 'sis·təm }

step counter [COMPUT SCI] In computers, a counter in the arithmetic unit used to count the steps in multiplication, division, and shift operations. { 'step ,kau̇nt·ər }

stepping *See* zoning. { 'step·iŋ }

step response [CONT SYS] The behavior of a system when its input signal is zero before a certain time and is equal to a constant nonzero value after this time. { 'step ri,späns }

step strobe marker [ELECTR] Form of strobe marker in which the discontinuity is in the form of a step in the time base. { 'step 'strōb ,mär·kər }

step tablet *See* density step tablet. { 'step ,tab·lət }

sterba curtain [ELECTROMAG] Type of stacked dipole antenna array consisting of one or more phased half-wave sections with a quarter-wave section at each end; the array can be oriented for either vertical or horizontal radiation, and can be either center or end fed. { 'stər·bə ,kərt·ən }

stereo broadcasting [COMMUN] Broadcasting two sound channels for reproduction by a stereo sound system having a stereo tuner at its input, to afford a listener a sense of the spatial distribution of the sound sources. { 'ster·ē·ō 'bród,kast·iŋ }

stereographic net *See* net. { ¦ster·ē·ə¦graf·ik 'net }

stereo multiplex [COMMUN] Stereo broadcasting by a frequency-modulation station, in which the outputs of two channels are transmitted on the same carrier by frequency-division multiplexing. { 'ster·ē·ō 'məl·tə,pleks }

stereo subcarrier [COMMUN] A subcarrier whose frequency is the second harmonic of the pilot subcarrier frequency used in frequency-modulation stereo broadcasting. { 'ster·ē·ō ¦səb'kar·ē·ər }

sterile field *See* field. { 'ster·əl 'fēld }

sticking [COMPUT SCI] In computers, the tendency of a flip-flop to remain in, or to spontaneously switch to, one of its two stable states. { 'stik·iŋ }

stimulus [CONT SYS] A signal that affects the controlled variable in a control system. { 'stim·yə·ləs }

STN LCD *See* supertwisted nematic liquid-crystal display.

stochastic automaton *See* probabilistic automaton. { stō'kas·tik ȯ'täm·ə,tän }

stochastic control theory [CONT SYS] A branch of control theory that aims at predicting and minimizing the magnitudes and limits of the random deviations of a control system through optimizing the design of the controller. { stō 'kas·tik kən'trōl ,thē·ə·rē }

stochastic sequential machine *See* probabilistic sequential machine. { stō'kas·tik si'kwen·chəl mə'shēn }

stop band *See* rejection band. { 'stäp ,band }

stop bits [COMPUT SCI] The last two bits transmitted in asynchronous data transmission to unequivocally indicate the end of a word. { 'stäp ,bits }

stop code [COMPUT SCI] A character that is placed in a storage medium and, when encountered, causes the computer system to cease processing until it is directed to continue. { 'stäp ,kōd }

stop element [COMMUN] The last element of a character in certain serial transmissions, used to ensure the recognition of the next start element. { 'stäp ,el·ə·mənt }

stop instruction [COMPUT SCI] An instruction in a computer program that causes execution of the program to stop. { 'stäp in,strək·shən }

stop loop *See* loop stop. { 'stäp ,lüp }

stop signal [COMMUN] Signal that initiates the transfer of facsimile equipment from active to standby conditions. { 'stäp ,sig·nəl }

storage [COMPUT SCI] Any device that can accept, retain, and read back one or more times; the means of storing data may be chemical, electrical, magnetic, mechanical, or sonic. { 'stȯr·ij }

storage address register [COMPUT SCI] A register used to hold the address of a location in storage containing data that is being processed. { 'stȯr·ij ¦ad,res ,rej·ə·stər }

storage allocation [COMPUT SCI] The process of assigning storage locations to data or instructions in a digital computer. { 'stȯr·ij ,al·ə'kā·shən }

storage and retrieval system [COMPUT SCI] An organized method of putting items away in a manner which permits their recall or retrieval from storage. Also known as storetrieval system. { 'stȯr·ij ən ri'trē·vəl ,sis·təm }

storage area [COMPUT SCI] A specified set of locations in a storage unit. Also known as zone. { 'stȯr·ij ,er·ē·ə }

storage block [COMPUT SCI] A contiguous area of storage whose contents can be handled in a single operation. { 'stȯr·ij ,bläk }

storage buffer register [COMPUT SCI] A register used in some microcomputers during input or output operations to temporarily hold a copy of the contents of a storage location. { 'stȯr·ij ¦bəf·ər ,rej·ə·stər }

storage capacity [COMPUT SCI] The quantity of data that can be retained simultaneously in a storage device; usually measured in bits, digits, characters, bytes, or words. Also known as capacity; memory capacity. { 'stȯr·ij kə,pas·əd·ē }

storage cell [COMPUT SCI] An elementary (logically indivisible) unit of storage; the storage cell can contain one bit, character, byte, digit (or sometimes word) of data. { 'stȯr·ij ˌsel }

storage compacting [COMPUT SCI] The practice, followed on multiprogramming computers which use dynamic allocation, of assigning and reassigning programs so that the largest possible area of adjacent locations remains available for new programs. { 'stȯr·ij kəmˌpakt·iŋ }

storage cycle [COMPUT SCI] **1.** Periodic sequence of events occurring when information is transferred to or from the storage device of a computer. **2.** Storing, sensing, and regeneration from parts of the storage sequence. { 'stȯr·ij ˌsī·kəl }

storage cycle time [COMPUT SCI] The time required to read and restore one word from a computer storage, or to write one word in computer storage. { 'stȯr·ij ¦sī·kəl ˌtīm }

storage density [COMPUT SCI] The number of characters stored per unit-length of area of storage medium (for example, number of characters per inch of magnetic tape). { 'stȯr·ij ˌden·səd·ē }

storage device [COMPUT SCI] A mechanism for performing the function of data storage: accepting, retaining, and emitting (unchanged) data items. Also known as computer storage device. { 'stȯr·ij diˌvīs }

storage dump [COMPUT SCI] A printout of the contents of all or part of a computer storage. Also known as memory dump; memory print. { 'stȯr·ij ˌdəmp }

storage element [COMPUT SCI] Smallest part of a digital computer storage used for storing a single bit. { 'stȯr·ij ˌel·ə·mənt }

storage factor *See* Q. { 'stȯr·ij ˌfak·tər }

storage fill [COMPUT SCI] Storing a pattern of characters in areas of a computer storage that are not intended for use in a particular machine run; these characters cause the machine to stop if one of these areas is erroneously referred to. Also known as memory fill. { 'stȯr·ij ˌfil }

storage hierachy [COMPUT SCI] The sequence of storage devices, characterized by speed, type of access, and size for the various functions of a computer; for example, core storage for programs and data, disks or drums for temporary storage of massive amounts of data, magnetic tapes and disks for backup storage. { 'stȯr·ij 'hī·ərˌär·kē }

storage integrator [COMPUT SCI] In an analog computer, an integrator used to store a voltage in the hold condition for future use while the rest of the computer assumes another computer control state. { 'stȯr·ij ˌint·əˌgrād·ər }

storage key [COMPUT SCI] A special set of bits associated with every word or character in some block of storage, which allows tasks having a matching set of protection key bits to use that block of storage. { 'stȯr·ij ˌkē }

storage location [COMPUT SCI] A digital-computer storage position holding one machine word and usually having a specific address. { 'stȯr·ij lōˌkā·shən }

storage mark [COMPUT SCI] The name given to a point location which defines the character space immediately to the left of the most significant character in accumulator storage. { 'stȯr·ij ˌmärk }

storage medium [COMPUT SCI] Any device or recording medium into which data can be copied and held until some later time, and from which the entire original data can be obtained. { 'stȯr·ij ˌmēd·ē·əm }

storage pool [COMPUT SCI] A collection of similar data storage devices. { 'stȯr·ij ˌpül }

storage print [COMPUT SCI] In computers, a utility program that records the requested core image, core memory, or drum locations in absolute or symbolic form either on the line-printer or on the delayed-printer tape. { 'stȯr·ij ˌprint }

storage protection [COMPUT SCI] Any restriction on access to storage blocks, with respect to reading, writing, or both. Also known as memory protection. { 'stȯr·ij prəˌtek·shən }

storage register [COMPUT SCI] A register in the main internal memory of a digital computer storing one computer word. Also known as memory register. { 'stȯr·ij rej·ə·stər }

storage ripple [COMPUT SCI] A hardware function, used during maintenance periods, which reads or writes zeros or ones through available storage locations to detect a malfunctioning storage unit. { 'stȯr·ij ˌrip·əl }

storage surface [COMPUT SCI] In computers, the surface (screen), in an electrostatic storage tube, on which information is stored. { 'stȯr·ij ˌsər·fəs }

storage-to-register instruction [COMPUT SCI] A machine-language instruction to move a word of data from a location in main storage to a register. { 'stȯr·ij tə 'rej·ə·stər inˌstrək·shən }

storage-to-storage instruction [COMPUT SCI] A machine-language instruction to move a word of data from one location in main storage to another. { 'stȯr·ij tə 'stȯr·ij inˌstrək·shən }

store [COMPUT SCI] **1.** To record data into a (static) data storage device. **2.** To preserve data in a storage device. { stȯr }

store and forward [COMMUN] A procedure in data communications in which data are stored at some point between the sender and the receiver and are later forwarded to the receiver. { 'stȯr ən 'fȯr·wərd }

stored program [COMPUT SCI] A computer program that is held in a computer's main storage and carried out by a central processing unit that reads and acts on its instructions. { 'stȯrd 'prō ˌgram }

stored-program computer [COMPUT SCI] A digital computer which executes instructions that are stored in main memory as patterns of data. { 'stȯrd ¦prōˌgram kəm'pyüd·ər }

stored-program control [COMMUN] Electronic control of a telecommunications switching system by means of a program of instructions stored in bulk electronic memory. Abbreviated SPC. { 'stȯrd 'prōˌgram kən'trōl }

stored-program logic [COMPUT SCI] Program that is stored in a memory unit containing logical commands in order to perform the same processes on all problems. { 'stȯrd 'prō,gram ,läj·ik }

stored-program numerical control *See* computer numerical control. { 'stȯrd ¦prō,gram nü'mer·ə·kəl kən,trōl }

stored response chain [COMPUT SCI] A fixed sequence of instructions that are stored in a file and acted on by an interactive computer program at a point where it would normally request instructions from the user, in order to save the user the trouble of repeatedly keying the same commands for a frequently used function. Abbreviated SRC. { 'stȯrd ri'späns ,chān }

stored routine [COMPUT SCI] In computers, a series of instructions in storage to direct the step-by-step operation of the machine. { 'stȯrd rü'tēn }

stored word [COMPUT SCI] The actual linear combination of letters (or their machine equivalents) to be placed in the machine memory; this may be physically quite different from a dictionary word. { 'stȯrd 'wərd }

storethrough [COMPUT SCI] The process of updating data in main memory each time the central processing unit writes into a cache. { 'stȯr,thrü }

storetrieval system *See* storage and retrieval system. { 'stȯ·ri,trē·vəl ,sis·təm }

STR *See* self-tuning regulator.

straightforward circuit [COMMUN] Circuit in which signaling is automatic and in one direction. { 'strāt¦fȯr·wərd 'sər·kət }

straight-line coding [COMPUT SCI] A digital computer program or routine (section of program) in which instructions are executed sequentially, without branching, looping, or testing. { 'strāt ¦līn 'kōd·iŋ }

strapping option [COMPUT SCI] The rearrangement of jumpers on a printed circuit board to render a hardware feature operative or inoperative. { 'strap·iŋ ,äp·shən }

stream [COMPUT SCI] A collection of binary digits that are transmitted in a continuous sequence, and from which extraneous data such as control information or parity bits are excluded. { strēm }

stream cipher [COMMUN] A cipher that makes use of an algorithmic procedure to produce an unending sequence of binary digits which is then combined either with plaintext to produce ciphertext or with ciphertext to recover plaintext. { 'strēm ,sī·fər }

stream editor [COMPUT SCI] A modification of a statement editor to allow superlines that expand and contract as necessary; the most powerful type of text editor. Also known as string editor. { 'strēm ,ed·əd·ər }

streaming [COMPUT SCI] A malfunction in which a communicating device constantly transmits worthless data and thereby locks out all other devices on the line. { 'strēm·iŋ }

streaming media [COMPUT SCI] Audio or video files that can begin playing as they are being downloaded to a computer. { ¦strēm·iŋ 'mēd·ē·ə }

streaming tape [COMPUT SCI] A type of high-speed magnetic tape that is used as a backup storage for disks, particularly hard disks in microcomputer systems. { 'strēm·iŋ 'tāp }

STRESS [COMPUT SCI] A problem-oriented programming language used to solve structural engineering problems. Derived from structural engineering system solver. { stres }

stress test [COMPUT SCI] A test of new software or hardware under unusually heavy work loads. { 'stres ,test }

string [COMPUT SCI] A set of consecutive, adjacent items of similar type; normally a bit string or a character string. { striŋ }

string break [COMPUT SCI] In the sorting of records, the situation that arises when there are no records having keys with values greater than the highest key already written in the sequence of records currently being processed. { 'striŋ ,brāk }

string constant [COMPUT SCI] An arbitrary combination of letters, digits, and other symbols that is treated in a manner completely analogous to numeric constants. { 'striŋ ,kän·stənt }

string editor *See* stream editor. { 'striŋ ,ed·əd·ər }

string manipulation [COMPUT SCI] The handling of strings of characters in a computer storage as though they were single units of data. { 'striŋ mə,nip·yə,lā·shən }

string manipulation language *See* string processing language. { 'striŋ mə,nip·yə,lā·shən ,laŋ·gwij }

String-Oriented-Symbolic Language *See* SNOBOL. { 'striŋ ¦ȯr·ē,ent·əd sim'bäl·ik 'laŋ·gwij }

string processing language [COMPUT SCI] A higher-level programming language equipped with facilities to synthesize and decompose character strings, search them in response to arbitrarily complex criteria, and perform a variety of other manipulations. Also known as string manipulation language. { 'striŋ 'prä,ses·iŋ ,laŋ·gwij }

stringy floppy [COMPUT SCI] A peripheral storage device for microcomputers that uses a removable magnetic tape cartridge with a $\frac{1}{16}$-inch-wide (1.5875-millimeter) loop of magnetic tape. { 'striŋ·ē 'fläp·ē }

strip-line circuit [ELECTROMAG] A circuit in which one or more strip transmission lines serve as filters or other circuit components. { 'strip ¦līn ,sər·kət }

strip transmission line [ELECTROMAG] A microwave transmission line consisting of a thin, narrow, rectangular metal strip that is supported above a ground-plane conductor or between two wide ground-plane conductors and is usually separated from them by a dielectric material. { 'strip tranz'mish·ən ,līn }

strobe [ELECTR] **1.** Intensified spot in the sweep of a deflection-type indicator, used as a reference mark for ranging or expanding the presentation. **2.** Intensified sweep on a radar's plan-position indicator or B-scope; such a strobe may result from certain types of interference, or it may be purposely applied as a bearing or heading marker, or to show the estimated azimuth of a jamming source, as a "jam strobe." **3.** A signaling pulse of very short duration. { strōb }

strobe circuit [ELECTR] A circuit that produces an output pulse only at certain times or under certain conditions, such as a gating circuit or a coincidence circuit. { 'strōb ˌsər·kət }

strobing [COMPUT SCI] The technique required to time-synchronize data appearing as pulses at the output of a computer memory. { 'strōb·iŋ }

stroke [COMPUT SCI] **1.** In optical character recognition, straight or curved portion of a letter, such as is commonly made with one smooth motion of a pen. Also known as character stroke. **2.** That segment of a printed or handwritten character which has been temporarily isolated from other segments for the purpose of analyzing it, particularly with regard to its dimensions and relative reflectance. Also known as character stroke. [ELECTR] The penlike motion of a focused electron beam in cathode-ray-tube diplays. { strōk }

stroke analysis [COMPUT SCI] In character recognition, a method employed in character property detection in which an input specimen is dissected into certain prescribed elements; the sequence, relative positions, and number of detected elements are then used to identify the characters. { 'strōk əˌnal·ə·səs }

stroke center line [COMPUT SCI] In character recognition, a line midway between the two average-edge lines; the center line describes the stroke's direction of travel. Also known as center line. { 'strōk 'sen·tər ˌlīn }

stroke edge [COMPUT SCI] In character recognition, a continuous line, straight or otherwise, which traces the outermost part of intersection of the stroke along the two sides of its greatest dimension. { 'strōk ˌej }

stroke speed [COMMUN] Number of times per minute that a fixed line, perpendicular to the direction of scanning, is crossed in one direction by a scanning or recording spot in a facsimile system. Also known as scanning frequency; scanning line frequency. { 'strōk ˌspēd }

stroke width [COMPUT SCI] In character recognition, the distance that obtains, at a given location, between the points of intersection of the stroke edges and a line drawn perpendicular to the stroke center line. { 'strōk ˌwidth }

strong algorithm [COMMUN] A cryptographic algorithm for which the cost or time required to obtain the message or key is prohibitively great in practice even though the message may be obtainable in theory. { 'strȯŋ 'al·gəˌrith·əm }

strongly typed language [CONT SYS] A high-level programming language in which the type of each variable must be declared at the beginning of the program, and the language itself then enforces rules concerning the manipulation of variables according to their types. { 'strȯŋ·lē ¦tīpt 'laŋ·gwij }

Strowger system [COMMUN] An automatic telephone switching system that uses successive step-by-step selector switches actuated by current pulses produced by rotation of a telephone dial. Also known as step-by-step system. { 'strō·gər ˌsis·təm }

structural engineering system solver *See* STRESS. { 'strək·chə·rəl ˌen·jə'nir·iŋ 'sis·təm ˌsäl·vər }

structural information [COMPUT SCI] Information specifying the number of independently variable features or degrees of freedom of a pattern. { 'strək·chə·rəl ˌin·fər'mā·shən }

structure [COMPUT SCI] For a data-processing system, the nature of the chain of command, the origin and type of data collected, the form and destination of results, and the procedures used to control operations. { 'strək·chər }

structured data type [COMPUT SCI] The manner in which a collection of data items, which may have the same or different scalar data types, is represented in a computer program. { 'strək·chərd 'dad·ə ˌtīp }

structured programming [COMPUT SCI] The use of program design and documentation techniques that impose a uniform structure on all computer programs. { 'strək·chərd 'prō ˌgram·iŋ }

Structured Query Language [COMPUT SCI] The standard language for accessing relational databases. Abbreviated SQL. { ˌstrək·chərd 'kwir·ē ˌlaŋ·gwij }

structured variable *See* record variable. { 'strək·chərd 'ver·ē·ə·bəl }

structured walkthrough [COMPUT SCI] A formal method of debugging a computer system or program, involving a systematic review to search for errors and inefficiencies. { 'strək·chərd 'wȯk ˌthrü }

stub [COMPUT SCI] **1.** The left-hand portion of a decision table, consisting of a single column, and comprising the condition stub and the action stub. **2.** A program module that is only partly completed, to the extent needed to fulfill the requirements of other modules in the computer system. [ELECTROMAG] **1.** A short section of transmission line, open or shorted at the far end, connected in parallel with a transmission line to match the impedance of the line to that of an antenna or transmitter. **2.** A solid projection one-quarter-wavelength long, used as an insulating support in a waveguide or cavity. { stəb }

stub angle [ELECTROMAG] Right-angle elbow for a coaxial radio-frequency transmission line which has the inner conductor supported by a quarter-wave stub. { 'stəb ˌaŋ·gəl }

stub matching [ELECTROMAG] Use of a stub to match a transmission line to an antenna or load; matching depends on the spacing between the two wires of the stub, the position of the shorting bar, and the point at which the transmission line is connected to the stub. { 'stəb ˌmach·iŋ }

stub-supported line [ELECTROMAG] A transmission line that is supported by short-circuited quarter-wave sections of coaxial line; a stub exactly a quarter-wavelength long acts as an insulator because it has infinite reactance. { 'stəb sə¦pȯrd·əd 'līn }

stub tuner [ELECTROMAG] Stub which is terminated by movable short-circuiting means and used for matching impedance in the line to which it is joined as a branch. { 'stəb ˌtün·ər }

studio [COMMUN] A facility in which video or audio programs are produced. { 'stüd·ē·ō }

stutter [COMMUN] Series of undesired black and white lines sometimes produced when a facsimile signal undergoes a sharp amplitude change. { 'stəd·ər }

stylus [COMPUT SCI] The pointed device used to draw images on a graphics tablet. { 'stī·ləs }

stylus printing *See* matrix printing. { 'stī·ləs ˌprint·iŋ }

subalphabet [COMPUT SCI] A subset of an alphabet. { səb'al·fəˌbet }

subcarrier **1.** A carrier that is applied as a modulating wave to modulate another carrier. **2.** *See* chrominance subcarrier. { ¦səb'kar·ē·ər }

subcarrier oscillator [ELECTR] **1.** The crystal oscillator that operates at the chrominance subcarrier or burst frequency of 3.579545 megahertz in an analog color television receiver; this oscillator, synchronized in frequency and phase with the transmitter master oscillator, furnishes the continuous subcarrier frequency required for demodulators in the receiver. **2.** An oscillator used in a telemetering system to translate variations in an electrical quantity into variations of a frequency-modulated signal at a subcarrier frequency. { ¦səb'kar·ē·ər 'äs·əˌlād·ər }

subchannel [COMPUT SCI] The portion of an input/output channel associated with a specific input/output operation. { ¦səb'chan·əl }

subclutter visibility [ELECTR] A measure of the effectiveness of moving-target indicator radar, equal to the ratio of the signal from a fixed target that can be canceled to the signal from a just visible moving target; often calculated for a target moving at an optimum velocity (unlike improvement factor). { 'səb¦kləd·ər ˌviz·ə'bil·əd·ē }

subcommutation [COMMUN] In telemetry, commutation of additional channels with output applied to individual channels of the primary commutator. { ¦səbˌkäm·yə'tā·shən }

subcycle generator [ELECTR] Frequency-reducing device used in telephone equipment which furnishes ringing power at a submultiple of the power supply frequency. { 'səbˌsī·kəl 'jen·əˌrād·ər }

subdevice bit *See* select bit. { 'səb·diˌvīs ˌbit }

subframe [COMMUN] In telemetry, a complete sequence of frames during which all subchannels of a specific channel are sampled once. { 'səb ˌfrām }

subharmonic triggering [ELECTR] A method of frequency division which makes use of a triggered multivibrator having a period of one cycle which allows triggering only by a pulse that is an exact integral number of input pulses from the last effective trigger. { ¦səb·här'män·ik 'trig·ə·riŋ }

subjective testing [ENG] Using human subjects to judge the performance of a system; especially useful when testing systems that include components such as perceptual audio coders. Traditional audio measurement techniques, such as signal-to-noise distortion measurements, are often not compatible with the way perceptual audio coders work and therefore cannot characterized their performance in a manner that can be compared with other coders or with traditional analog systems. { səb'jek·tiv 'test·iŋ }

submillimeter wave [ELECTROMAG] An electromagnetic wave whose wavelength is less than 1 millimeter, corresponding to frequencies above 300 gigahertz. { ˌsəb¦mil·əˌmēd·ər ˌwāv }

sub-Nyquist sampling [COMMUN] **1.** Any technique of sampling an analog signal at a rate lower than the Nyquist rate in such a way as to preserve signal content without aliasing distortion. **2.** In particular, the sampling of video signals at a rate lower than the Nyquist rate and at an odd multiple of the frame rate, so that the aliasing components are placed into periodically spaced voids in the video spectrum where they can be removed by a comb filter at the receiver. { ¦səb 'nīˌkwist ˌsam·pliŋ }

suboptimization [SYS ENG] The process of fulfilling or optimizing some chosen objective which is an integral part of a broader objective; usually the broad objective and lower-level objective are different. { ¦səbˌäp·tə·mə'zā·shən }

subprogram [COMPUT SCI] A part of a larger program which can be converted independently into machine language. { ¦səb'prōˌgram }

subrefraction [ELECTROMAG] Atmospheric refraction which is less than standard refraction. { ¦səb·ri'frak·shən }

subroutine [COMPUT SCI] **1.** A body of computer instruction (and the associated constants and working-storage areas, if any) designed to be used by other routines to accomplish some particular purpose. **2.** A statement in FORTRAN used to define the beginning of a closed subroutine (first definition). { 'səb·rüˌtēn }

subroutine library [COMPUT SCI] A collection of subroutines that is stored on a disk or other direct-access storage device and can be used by a programmer through facilities of the computer's operating system. { 'səb·rüˌtēn līˌbrer·ē }

subschema [COMPUT SCI] An individual user's partial view of a database. { 'səbˌskē·mə }

subscriber set *See* subset. { səb'skrīb·ər ˌset }

subscriber station [COMMUN] The connection between a central office and an outside location, including the circuit, some circuit termination equipment, and possibly some associated input/output equipment. { səb'skrīb·ər ˌstā·shən }

subscription database *See* information network. { səb'skrip·shən 'dad·əˌbās }

subscription television [COMMUN] A television service in which programs are broadcast in coded or scrambled form, for reception only by subscribers who make payments for use of the decoding or unscrambling devices required to obtain a clear program. Also known as pay television. { səb'skrip·shən 'tel·ə,vizh·ən }

subset [COMMUN] A telephone or other subscriber equipment connected to a communication system, such as a modem. Derived from subscriber set. { 'səb,set }

substandard propagation [ELECTROMAG] The propagation of radio energy under conditions of substandard refraction in the atmosphere; that is, refraction by an atmosphere or section of the atmosphere in which the index of refraction decreases with height at a rate of less than 12 N units (unit of index of refraction) per 1000 feet (304.8 meters). { ¦səb'stan·dərd ,präp·ə'gā·shən }

substitute mode [COMPUT SCI] One method of exchange buffering, in which segments of storage function alternately as buffer and as program work area. { 'səb·stə,tüt ,mōd }

substitution alphabet [COMMUN] An alphabet used in a coded message in which each letter in the original message is replaced by another letter in the coded message, according to a set of rules. { ,səb·stə¦tü·shən 'al·fə,bet }

substitution cipher [COMMUN] A cipher in which the characters of the original message are replaced by other characters according to a key. { ,səb·stə'tü·shən ,sī·fər }

substring [COMPUT SCI] A sequence of successive characters within a string. { 'səb,striŋ }

subsurface radar *See* ground-probing radar. { ,səb,sər·fəs 'rā·dar }

subsurface wave [ELECTROMAG] Electromagnetic wave propagated through water or land; operating frequencies for communications may be limited to approximately 35 kilohertz due to attenuation of high frequencies. { ¦səb'sər·fəs 'wāv }

subtracter [COMPUT SCI] A computer device that can form the difference of two numbers or quantities. { səb'trak·tər }

subvoice-grade channel [COMMUN] A channel whose bandwidth is smaller than the bandwidth of a voice-grade channel; it is usually a subchannel of a voice-grade line. { ¦səb'vȯis ,grād ,chan·əl }

successive approximation converter [COMPUT SCI] An analog-to-digital converter which operates by successively considering each bit position in the digital output and setting that bit equal to 0 or 1 on the basis of the output of a comparator. { sək'ses·iv ə,präk·sə'mā·shən kən ,vərd·ər }

successor job [COMPUT SCI] A job that uses the output of another job (predecessor) as its input, so that it cannot start until the other job has been successfully completed. { sək'ses·ər ,jäb }

suffix notation *See* reverse Polish notation. { 'səf,iks nō,tā·shən }

suite [COMPUT SCI] A collection of related computer programs run one after another. { swēt }

summary recorder [COMPUT SCI] In computers, output equipment which records a summary of the information handled. { 'səm·ə·rē ri'kȯrd·ər }

summation check [COMPUT SCI] An error-detecting procedure involving adding together all the digits of some number and comparing this sum to a previously computed value of the same sum. { sə'mā·shən ,chek }

summing amplifier [ELECTR] An amplifier that delivers an output voltage which is proportional to the sum of two or more input voltages or currents. { 'səm·iŋ 'am·plə,fī·ər }

S-unit meter *See* signal-strength meter. { 'es ,yü·nət ,mēd·ər }

superchip *See* super-large-scale integrated circuit. { 'sü·pər,chip }

supercomputer [COMPUT SCI] A computer which is among those with the highest speed, largest functional size, biggest physical dimensions, or greatest monetary cost in any given period of time. { 'sü·pər·kəm,pyüd·ər }

superconducting computer [COMPUT SCI] A high-performance computer whose circuits employ superconductivity and the Josephson effect to reduce computer cycle time. { ¦sü·pər·kən'dəkt·iŋ kəm'pyüd·ər }

superemitron camera *See* image iconoscope. { ¦sü·pər'em·ə,trän ,kam·rə }

supergroup [COMMUN] In carrier telephony, five groups (60 voice channels) multiplexed together and treated as a unit; a basic supergroup occupies the band between 312 and 552 kilohertz. { 'sü·pər,grüp }

superhet *See* superheterodyne receiver. { 'sü·pər,het }

superheterodyne receiver [ELECTR] A receiver in which all incoming modulated radio-frequency carrier signals are converted to a common intermediate-frequency carrier value for additional amplification and selectivity prior to demodulation, using heterodyne action; the output of the intermediate-frequency amplifier is then demodulated in the second detector to give the desired audio-frequency signal. Also known as superhet. { ¦sü·pər'he·trə,dīn ri'sē·vər }

superhigh frequency [COMMUN] A frequency band from 3000 to 30,000 megahertz, corresponding to wavelengths from 1 to 10 centimeters. Abbreviated SHF. { ¦sü·pər'hī ¦frē·kwən·sē }

super-large-scale integrated circuit [ELECTR] A very complex integrated circuit that has a high density of transistors and other components, for a total of 10^6 or more components. Also known as superchip. Abbreviated SLSI circuit. { ¦sü·pər ¦lärj ¦skāl ,in·tə'grād·əd 'sər·kət }

superline [COMPUT SCI] A unit of text longer than an ordinary line, used in some of the more powerful text editors. { 'sü·pər,līn }

supermicro [COMPUT SCI] A computer resembling a supermini in design but scaled down to the size of a microcomputer, usually capable of working with a small number of users at once. { ¦sü·pər'mī·krō }

superposed circuit [COMMUN] Additional channel obtained from one or more circuits, normally provided for other channels, in a way that all channels can be used simultaneously without mutual interference. { ¦sü·pər'pōzd 'sər·kət }

superposition integral [CONT SYS] An integral which expresses the response of a linear system to some input in terms of the impulse response or step response of the system; it may be thought of as the summation of the responses to impulses or step functions occurring at various times. { ˌsü·pər·pə'zish·ən 'int·ə·grəl }

superscalar architecture [COMPUT SCI] A design that enables a central processing unit to send several instructions to different execution units simultaneously, allowing it to execute several instructions in each clock cycle. { ¦sü·pər ˌskā·lər 'är·kəˌtek·chər }

superset [COMPUT SCI] A programming language that contains all the features of a given language and has been expanded or enhanced to include other features as well. { 'sü·pərˌset }

superstandard propagation [ELECTROMAG] The propagation of radio waves under conditions of superstandard refraction in the atmosphere, that is, refraction by an atmosphere or section of the atmosphere in which the index of refraction decreases with height at a rate of greater than 12 N units (unit of index of refraction) per 1000 feet (304.8 meters). { ¦sü·pər'stan·dərd ˌpräp·ə'gā·shən }

supertwisted nematic liquid-crystal display [ELECTR] A display in which nematic liquid-crystal molecules are twisted more than 90°, and the picture elements respond to the average (root-mean-square) voltage applied by transistors connected to each row and column to switch the liquid. Abbreviated STN LCD. Also known as passive-matrix liquid-crystal display (PM LCD). { ¦sü·pərˌtwis·təd nə¦mad·ik ˌlik·wəd 'krist·əl diˌsplā }

supervisor [COMPUT SCI] A collection of programs, forming part of the operating system, that provides services for and controls the running of user programs. { 'sü·pərˌvī·zər }

supervisor call [COMPUT SCI] A mechanism whereby a computer program can interrupt the normal flow of processing and ask the supervisor to perform a function for the program that the program cannot or is not permitted to perform for itself. Also known as system call. { 'sü·pər ˌvī·zər ˌkȯl }

supervisor interrupt [COMPUT SCI] An interruption caused by the program being executed which issues an instruction to the master control program. { 'sü·pərˌvī·zər 'int·əˌrəpt }

supervisor mode [COMPUT SCI] A method of computer operation in which the computer can execute all its own instructions, including the privileged instruction not normally allowed to the programmer, in contrast to problem mode. { 'sü·pərˌvī·zər ˌmōd }

supervisory computer [COMPUT SCI] A computer which accepts test results from satellite computers, transmits new programs to the satellite computers, and may further communicate with a larger computer. { ¦sü·pər ¦vīz·ə·rē kəm'pyüd·ər }

supervisory expert control system [CONT SYS] A control system in which an expert system is used to supervise a set of control, identification, and monitoring algorithms. { ˌsü·pər ¦vīz·ə·rē ˌekˌspərt kən'trōl ˌsis·təm }

supervisory program [COMPUT SCI] A program that organizes and regulates the flow of work in a computer system, for example, it may automatically change over from one run to another and record the time of the run. { ¦sü·pər ¦vīz·ə·rē 'prōˌgram }

supervisory routine [COMPUT SCI] A program or routine that initiates and guides the execution of several (or all) other routines and programs; it usually forms part of (or is) the operating system. { ¦sü·pər¦vīz·ə·rē rü'tēn }

suppressed carrier [COMMUN] A carrier in a modulated signal that is suppressed at the transmitter; the chrominance subcarrier in an analog color television transmitter is an example. { sə'prest 'kar·ē·ər }

suppressed-carrier modulation [COMMUN] Modulation resulting from elimination or partial suppression of the carrier component from an amplitude modulated wave. { sə'prest 'kar·ē·ər ˌmäj·ə'lā·shən }

suppressed-carrier transmission [COMMUN] Transmission in which the carrier component of the modulated wave is eliminated or partially suppressed, leaving only the side bands to be transmitted. { sə'prest 'kar·ē·ər tranz'mish·ən }

suppression [COMPUT SCI] **1.** Removal or deletion usually of insignificant digits in a number, especially zero suppression. **2.** Optional function in either on-line or off-line printing devices that permits them to ignore certain characters or groups of characters which may be transmitted through them. [ELECTR] Elimination of any component of an emission, as a particular frequency or group of frequencies in a radio-frequency signal. { sə'presh·ən }

suppressor [ELEC] In general, a device used to reduce or eliminate noise or other signals that interfere with the operation of a communication system, usually at the noise source. { sə'pres·ər }

surface-acoustic-wave device [ELECTR] Any device, such as a filter, resonator, or oscillator, which employs surface acoustic waves with frequencies in the range 10^7-10^9 hertz, traveling on the optically polished surface of a piezoelectric substrate, to process electronic signals. { 'sər·fəs ə'kü·stik 'wāv diˌvīs }

surface-acoustic-wave filter [ELECTR] An electric filter consisting of a piezoelectric bar with a polished surface along which surface acoustic waves can propagate, and on which are deposited metallic transducers, one of which is connected, via thermocompression-bonded leads, to the electric source, while the other drives the load. { 'sər·fəs ə'kü·stik 'wāv ˌfiltər }

surface analysis [COMPUT SCI] A procedure in which a computer program writes a series of test characters onto a magnetic data storage medium and then reads them back to determine the location of any flaws in the medium. { 'sər·fəs əˌnal·ə·səs }

surface movement radar *See* airport surface detection equipment. { 'sər·fəs ,müv·mənt 'rā,där }

surface-penetrating radar *See* ground-probing radar. { ,sər·fəs ,pen·ə,trād·iŋ 'rā,där }

surface wave [COMMUN] *See* ground wave. [ELECTROMAG] A wave that can travel along an interface between two different mediums without radiation; the interface must be essentially straight in the direction of propagation; the commonest interface used is that between air and the surface of a circular wire. { 'sər·fəs ,wāv }

surface-wave transmission line [ELECTROMAG] A single conductor transmission line energized in such a way that a surface wave is propagated along the line with satisfactorily low attenuation. { 'sər·fəs ¦wāv tranz'mish·ən ,līn }

surge impedance *See* characteristic impedance. { 'sərj im,pēd·əns }

surveillance radar [ENG] A search radar that includes significant means of associating detections of targets of interest (contacts) into tracks with additional sorting and labeling of data as the user system may require, normally more highly automated and equipped with data-processing computers than the simpler search radar. { sər'vā·ləns ,rā,där }

survivable route [COMMUN] A communication cable system begun in 1960 in which the cable, main stations, amplifiers, and power feed stations are placed underground; it incorporates the latest techniques of protection against natural disasters and nuclear blasts, and avoids possible target areas. { sər'vī·və·bəl 'rüt }

sustained oscillation [CONT SYS] Continued oscillation due to insufficient attenuation in the feedback path. { sə'stānd ,äs·ə'lā·shən }

swap out [COMPUT SCI] The action of an operating system on a process wherein it blocks the process and writes the contents of its memory onto a disk in order to make available more memory for other current processes. { 'swäp ,au̇t }

swapping [COMPUT SCI] A procedure in which a running program is temporarily suspended and moved onto secondary storage, and primary storage is reassigned to a more pressing job, in order to maximize the efficient use of primary storage. { 'swäp·iŋ }

sweep [ELECTR] **1.** The steady movement of the electron beam across the screen of a cathode-ray tube, producing a steady bright line when no signal is present; the line is straight for a linear sweep and circular for a circular sweep. **2.** The steady change in the output frequency of a signal generator from one limit of its range to the other. { swēp }

sweep amplifier [ELECTR] An amplifier used with a cathode-ray tube, such as in a television receiver or cathode-ray oscilloscope, to amplify the sawtooth output voltage of the sweep oscillator, to shape the waveform for the deflection circuits of a television picture tube, or to provide balanced signals to the deflection plates. { 'swēp ,am·plə,fī·ər }

sweep circuit [ELECTR] The sweep oscillator, sweep amplifier, and any other stage used to produce the deflection voltage or current for a cathode-ray tube. Also known as scanning circuit. { 'swēp ,sər·kət }

sweep generator [ELECTR] **1.** An electronic circuit that generates a voltage or current, usually recurrent, as a prescribed function of time; the resulting waveform is used as a time base to be applied to the deflection system of an electron-beam device, such as a cathode-ray tube. Also known as time-base generator; timing-axis oscillator. **2.** A test instrument that generates a radio-frequency voltage whose frequency varies back and forth through a given frequency range at a rapid constant rate; used to produce an input signal for circuits or devices whose frequency response is to be observed on an oscilloscope. Also known as sweep oscillator. { 'swēp ,jen·ə,rād·ər }

sweeping receivers [ELECTR] Automatically and continuously tuned receivers designed to stop and lock on when a signal is found, or to continually plot band occupancy. { 'swēp·iŋ ri ,sē·vərz }

sweep jamming [ELECTR] Jamming with a relatively narrow-band continuous signal being varied in frequency (swept) so that pulselike signals are produced in a radar as the jamming passes through its passband. { 'swēp ,jam·iŋ }

sweep oscillator *See* sweep generator. { 'swēp ,äs·ə,lad·ər }

sweep rate [ELECTR] The number of times a radar radiation pattern rotates during 1 minute; sometimes expressed as the duration of one complete rotation in seconds. { 'swēp ,rāt }

sweep-through jammer [ELECTR] A jamming transmitter which is swept through a radio-frequency band in short steps to jam each frequency briefly. { 'swēp¦thrü 'jam·ər }

sweep voltage [ELECTR] Periodically varying voltage applied to the deflection plates of a cathode-ray tube to give a beam displacement that is a function of time, frequency, or other data base. { 'swēp ,vōl·tij }

switch [COMPUT SCI] **1.** A hardware or programmed device for indicating that one of several alternative states or conditions have been chosen, or to interchange or exchange two data items. **2.** A symbol used to indicate a branch point, or a set of instructions to condition a branch. { swich }

switchboard [COMMUN] A manually or automatically operated apparatus at a telephone exchange, on which the various circuits from subscribers and other exchanges are terminated to enable communication either between two subscribers on the same exchange, or between subscribers on different exchanges. Also known as telephone switchboard. [ELEC] A single large panel or assembly of panels on which are mounted switches, circuit breakers, meters, fuses, and terminals essential to the operation of electric equipment. Also known as electric switchboard. { 'swich,bȯrd }

switched-capacitor filter [ELECTR] An integrated-circuit filter in which a resistor is simulated by a combination of a capacitor and metal oxide semiconductor switches that are turned on and off periodically at a high frequency. Also known as switched-C filter. { ˌswicht kə'pas·əd·ər ˌfil·tər }

switched-C filter *See* switched-capacitor filter. { ˌswicht 'sē ˌfil·tər }

switched circuit [COMMUN] A communications circuit or channel that can be turned on and off and made to serve various users. { 'swicht 'sər·kət }

switched line [COMMUN] A communications line, such as a dial telephone line, whose path can vary each time the line is used. { 'swicht 'līn }

switched-message network [COMPUT SCI] A data transmission system in which a user can communicate with any other user of the network. { 'swicht ¦mes·ij 'net,wərk }

switched network [COMMUN] A communications network, such as the dial telephone network, in which any station may be connected with any other through the use of switching and control devices. { 'swicht 'net,wərk }

switch function [ELECTR] A circuit having a fixed number of inputs and outputs designed such that the output information is a function of the input information, each expressed in a certain code or signal configuration or pattern. { 'swich ˌfəŋk·shən }

switch hook [ELECTR] A switch on a telephone set that closes the circuit when the receiver is removed from the hook or cradle. { 'swich ˌhu̇k }

switching center [COMMUN] The equipment in a relay station for automatically or semi-automatically relaying communications traffic. { 'swich·iŋ ˌsen·tər }

switching gate [ELECTR] An electronic circuit in which an output having constant amplitude is registered if a particular combination of input signals exists; examples are the OR, AND, NOT, and INHIBIT circuits. Also known as logical gate. { 'swich·iŋ ˌgāt }

switching node [COMMUN] A location in a communications network where messages or lines are routed. { 'swich·iŋ ˌnōd }

switching surface [CONT SYS] In feedback control systems employing bang-bang control laws, the surface in state space which separates a region of maximum control effort from one of minimum control effort. { 'swich·iŋ ˌsər·fəs }

switching system [COMMUN] An assembly of switching and control devices provided so that any station in a communications system may be connected as desired with any other station. { 'swich·iŋ ˌsis·təm }

switching theory [ELECTR] The theory of circuits made up of ideal digital devices; included are the theory of circuits and networks for telephone switching, digital computing, digital control, and data processing. { 'swich·iŋ ˌthē·ə·rē }

switching tube [ELECTR] A gas tube used for switching high-power radio-frequency energy in the antenna circuits of radar and other pulsed radio-frequency systems; examples are those used in some radar modulators (pulsers) and those used for receiver protection in radar duplexers. { 'swich·iŋ ˌtüb }

switch register [COMPUT SCI] A manual switch on the control panel by means of which a bit may be entered in a processor register. { 'swich ˌrej·ə·stər }

switch room [COMMUN] Part of a central office building that houses switching mechanisms and associated apparatus. { 'swich ˌrüm }

switch selectable addressing [COMPUT SCI] The setting of DIP switches in a peripheral or terminal device to determine the address that identifies the device to the computer system. { 'swich si¦lek·tə·bəl 'ad,res·iŋ }

symbolic address [COMPUT SCI] In coding, a programmer-defined symbol that represents the location of a particular datum item, instruction, or routine. Also known as symbolic number. { sim'bäl·ik 'ad,res }

symbolic algebraic manipulation language [COMPUT SCI] An algebraic manipulation language which admits the most general species of mathematical expressions, usually representing them as general tree structures, but which lacks certain special algorithms. { sim'bäl·ik ˌal·jə'brā·ik məˌnip·yə'lā·shən ˌlaŋ·gwij }

symbolic assembly language listing [COMPUT SCI] A list that may be produced by a computer during the compilation of a program showing the source language statements together with the corresponding machine language instructions generated by them. { sim'bäl·ik ə'sem·blē ˌlaŋ·gwij ˌlist·iŋ }

symbolic assembly system [COMPUT SCI] A system for forming programs that can be run on a computer, consisting of an assembly language and an assembler. { sim'bäl·ik ə'sem·blē ˌsis·təm }

symbolic coding [COMPUT SCI] Instruction written in an assembly language, using symbols for operations and addresses. Also known as symbolic programming. { sim'bäl·ik 'kōd·iŋ }

symbolic computation system *See* symbolic system. { sim¦bäl·ik kəm'pyü'tā·shən ˌsis·təm }

symbolic computing [COMPUT SCI] The development and use of symbolic systems. { sim ¦bäl·ik kəm'pyüd·iŋ }

symbolic debugging [COMPUT SCI] A method of correcting known errors in a computer program written in a source language, in which certain statements are compiled together with the program. { sim'bäl·ik dē'bəg·iŋ }

symbolic language [COMPUT SCI] A language which expresses addresses and operation codes of instructions in symbols convenient to humans rather than in machine language. { sim'bäl·ik 'laŋ·gwij }

symbolic mathematical computation [COMPUT SCI] The manipulation of symbols, representing variables, functions, and other mathematical objects, and combinations of these symbols, representing formulas, equations, and expressions, according to mathematical rules, for example, the rules of algebra or calculus. { sim'bäl·ik ¦math·ə¦mad·ə·kəl ˌkäm·pyə'tā·shən }

symbolic name [COMPUT SCI] A name given to some entity that is actually something else; for example, the name of a table in a computer program actually represents the physical storage locations used to hold the data stored in that table, as well as the values stored in those locations. { sim'bäl·ik 'nām }

symbolic number *See* symbolic address. { sim'bäl·ik 'nəm·bər }

symbolic programming *See* symbolic coding. { sim'bäl·ik 'prōˌgram·iŋ }

symbolic system [COMPUT SCI] A computer program that performs computations with constants and variables according to the rules of algebra, calculus, and other branches of mathematics. Also known as algebraic computation system; computer algebra system; symbolic computation system. { sim¦bäl·ik 'sis·təm }

symbol input [COMPUT SCI] Includes all contextual symbols that may appear in a source text. { 'sim·bəl 'inˌpu̇t }

symbol sequence [COMPUT SCI] A sequence of contextual symbols not interrupted by space. { 'sim·bəl 'sē·kwəns }

symbol table [COMPUT SCI] A mapping for a set of symbols to another set of symbols or numbers. { 'sim·bəl 'tā·bəl }

symmetrical architecture [COMPUT SCI] A type of computer design that allows any type of data to be used with any type of instruction. { si'me·trə·kəl 'ärk·əˌtek·chər }

symmetrical band-pass filter [ELECTR] A band-pass filter whose attenuation as a function of frequency is symmetrical about a frequency at the center of the pass band. { sə'me·trə·kəl 'band ˌpas ˌfil·tər }

symmetrical band-reject filter [ELECTR] A band-rejection filter whose attenuation as a function of frequency is symmetrical about a frequency at the center of the rejection band. { sə'me·trə·kəl 'band riˌjekt ˌfil·tər }

symmetrical inductive diaphragm [ELECTROMAG] A waveguide diaphragm which consists of two plates that leave a space at the center of the waveguide, and which introduces an inductance in the waveguide. { sə'me·trə·kəl in'dək·tiv 'dī·əˌfram }

symmetric list [COMPUT SCI] A list with sequencing pointers to previous as well as subsequent items. { sə'me·trik 'list }

sync *See* synchronization. { siŋk }

sync generator *See* synchronizing generator. { 'siŋk ˌjen·əˌrād·ər }

synchronization [ENG] The maintenance of one operation in step with another, as in keeping the electron beam of a television picture tube in step with the electron beam of the television camera tube at the transmitter. Also known as sync. { ˌsiŋ·krə·nə'zā·shən }

synchronized blocking oscillator [ELECTR] A blocking oscillator which is synchronized with pulses occurring at a rate slightly faster than its own natural frequency. { 'siŋ·krəˌnīzd 'bläk·iŋ 'äs·əˌlād·ər }

synchronizer [COMPUT SCI] A computer storage device used to compensate for a difference in rate of flow of information or time of occurrence of events when transmitting information from one device to another. [ELECTR] The component of a radar set which generates the timing voltage for the complete set. { 'siŋ·krəˌnīz·ər }

synchronizing generator [ELECTR] An electronic generator that supplies synchronizing pulses to television studio and transmitter equipment. Also known as sync generator; sync-signal generator. { 'sin·krəˌnīz·iŋ 'jen·əˌrād·ər }

synchronizing pulse [COMMUN] In pulse modulation, a pulse which is transmitted to synchronize the transmitter and the receiver; it is usually distinguished from signal-carrying pulses by some special characteristic. { 'sin·krəˌnīz·iŋ ˌpəls }

synchronizing signal *See* sync signal. { 'sin·krəˌnīz·iŋ ˌsig·nəl }

synchronous communications [COMPUT SCI] The high-speed transmission and reception of long groups of characters at a time, requiring synchronization of the sending and receiving devices. { 'siŋ·krə·nəs kəˌmyü·nə'kā·shənz }

synchronous computer [COMPUT SCI] A digital computer designed to operate in sequential elementary steps, each step requiring a constant amount of time to complete, and being initiated by a timing pulse from a uniformly running clock. { 'siŋ·krə·nəs kəm'pyüd·ər }

synchronous data-link control [COMMUN] A bit-oriented protocol for managing the flow of information in a data-communications system, in full, half-duplex, or multipoint modes, that uses an error-check algorithm. { 'siŋ·krə·nəs 'dad·ə ˌliŋk kənˌtrōl }

synchronous data transmission [COMMUN] Data transmission in which a clock defines transmission times for data; since start and stop bits for each character are not needed, more of the transmission bandwidth is available for message bits. { 'siŋ·krə·nəs 'dad·ə tranz ˌmish·ən }

synchronous demodulator *See* synchronous detector. { 'siŋ·krə·nəs dē'mäj·əˌlād·ər }

synchronous detection [ELECTR] The act of mixing two nearly identical frequencies, such as the oscillator reference signal and the signal received in a coherent radar, producing a voltage output sinusoidally related to the phase difference of the two. { 'siŋ·krə·nəs diˌtek·shən }

synchronous detector [ELECTR] **1.** A detector that inserts a missing carrier signal in exact synchronism with the original carrier at the transmitter; when the input to the detector consists of two suppressed-carrier signals in phase quadrature, as in the chrominance signal of an analog color television receiver, the phase of the reinserted carrier can be adjusted to recover either one of the signals. Also known as synchronous demodulator. **2.** *See* cross-correlator. { 'siŋ·krə·nəs di'tek·tər }

synchronous dynamic random access memory [COMPUT SCI] High-speed memory that is controlled by the system clock and can run at bus speeds up to 100 megahertz. Abbreviated SDRAM. { ¦siŋ·krə·nəs dī,nam·ik ,ran·dəm 'ak ,ses ,mem·rē }

synchronous gate [ELECTR] A time gate in which the output intervals are synchronized with an incoming signal. { 'siŋ·krə·nəs 'gāt }

synchronous operation [ELECTR] **1.** An operation that takes place regularly or predictably with respect to the occurrence of a particular event in another process. **2.** In particular, an operation whose timing is controlled by pulses generated by an electronic clock. { 'siŋ·krə·nəs ,äp·ə'rā·shən }

synchronous satellite *See* geosynchronous satellite. { 'siŋ·krə·nəs 'sad·əl,īt }

synchronous system [COMMUN] A telecommunication system in which transmitting and receiving apparatus operate continuously at substantially the same rate, and correction devices are used, if necessary, to maintain them in a fixed time relationship. { 'siŋ·krə·nəs 'sis·təm }

synchronous time-division multiplexing [COMMUN] A data transmission technique in which several users make use of a single channel by means of a system in which time slots are allotted on a fixed basis, usually in round-robin fashion. Abbreviated STDM. { 'siŋ·krə·nəs 'tīm də,vizh·ən 'məl·tə,pleks·iŋ }

synchronous working [COMPUT SCI] The mode of operation of a synchronous computer, in which the starting of each operation is clock-controlled. { 'siŋ·krə·nəs 'wərk·iŋ }

synchroscope [ELECTR] A cathode-ray oscilloscope designed to show a short-duration pulse by using a fast sweep that is synchronized with the pulse signal to be observed. { 'siŋ·krə ,skōp }

sync separator [ELECTR] A circuit that separates synchronizing pulses from the video signal in an analog television receiver. { 'siŋk ,sep·ə,rād·ər }

sync signal [COMMUN] A signal transmitted after each line and field to synchronize the scanning process in a video system. Also known as synchronizing signal. { 'siŋk ,sig·nəl }

sync-signal generator *See* synchronizing generator. { 'siŋk ¦sig·nəl 'jen·ə,rād·ər }

syntactic analysis [COMPUT SCI] The problem of associating a given string of symbols through a grammar to a programming language, so that the question of whether the string belongs to the language may be answered. { sin'tak·tik ə'nal·ə·səs }

syntactic error *See* syntax error. { sin'tak·tik 'er·ər }

syntactic extension [COMPUT SCI] An extension mechanism which creates new notations for existing or user-defined mechanisms in an extensible language. { sin'tak·tik ik'sten·shən }

syntactic model *See* linguistic model. { sin'tak·tik 'mäd·əl }

syntax [COMPUT SCI] The set of rules needed to construct valid expressions or sentences in a language. { 'sin,taks }

syntax checker *See* syntax scanner. { 'sin,taks ,chek·ər }

syntax diagram [COMPUT SCI] A pictorial diagram showing the rules for forming an instruction in a computer programming language, and how the components of the statement are related. { 'sin,taks 'dī·ə,gram }

syntax-directed compiler [COMPUT SCI] A general-purpose compiler that can service a family of languages by providing the syntactic rules for language analysis in the form of data, typically in tabular form, rather than using a specific parsing algorithm for a particular language. Also known as syntax-oriented compiler. { 'sin,taks di¦rek·təd kəm'pīl·ər }

syntax error [COMPUT SCI] An error in the format of a statement in a computer program that violates the rules of the programming language employed. Also known as syntactic error. { 'sin ,taks ,er·ər }

syntax-oriented compiler *See* syntax-directed compiler. { 'sin,taks ¦ȯr·ē,ent·əd kəm'pīl·ər }

syntax scanner [COMPUT SCI] A subprogram of a compiler or interpreter that checks the source program for syntax errors, and reports any such errors by printing the erroneous statement together with a diagnostic message. Also known as syntax checker. { 'sin,taks ,skan·ər }

synthesis *See* system design. { 'sin·thə·səs }

synthesizer [ELECTR] **1.** An electronic instrument which combines simple elements to generate more complex entities; examples are frequency synthesizer and sound synthesizer. **2.** Circuitry generating multiple frequencies at very low power that are used in radar transmissions, particularly in frequency-agile radars. { 'sin·thə,sīz·ər }

synthetic address *See* generated address. { sin'thed·ik 'ad,res }

synthetic aperture [ENG] A method of increasing the ability of an imaging system, such as radar or acoustical holography, to resolve small details of an object, in which a receiver of large size (or aperture) is in effect synthesized by the motion of a smaller receiver and the proper correlation of the detected signals. { sin'thed·ik 'ap·ə·chər }

synthetic aperture radar [ENG] A radar of such high resolution in range and Doppler sensing that the apparent rotation of a stationary target, often the earth's surface, as the radar moves past it is sufficient to reveal the detailed structure of the target or scene, thereby providing a visual image of it. Abbreviated SAR. { sin'thed·ik ¦ap·ə·chər 'rā,där }

synthetic language [COMPUT SCI] A pseudocode or symbolic language; fabricated language. { sin'thed·ik 'laŋ·gwij }

sysgen *See* system generation. { 'sis,jen }

SYSIN [COMPUT SCI] The principal input stream of an operating system. Derived from system input. { 'sis,in }

system [ENG] An assemblage of interrelated components designed to perform prescribed functions. [PHYS] A region in space or a portion of matter that has a certain amount of one or more substances, ordered in one or more phases. { 'sis·təm }

system analysis [CONT SYS] The use of mathematics to determine how a set of interconnected components whose individual characteristics are known will behave in response to a given input or set of inputs. { 'sis·təm ə,nal·ə·səs }

systematic error-checking code [COMPUT SCI] A type of self-checking code in which a valid character consists of the minimum number of digits needed to identify the character and distinguish it from any other valid character, and a set of check digits which maintain a minimum specified signal distance between any two valid characters. Also known as group code. { ,sis·tə'mad·ik 'er·ər ¦chek·iŋ ,kōd }

system bandwidth [CONT SYS] The difference between the frequencies at which the gain of a system is $\sqrt{2}/2$ (that is, 0.707) times its peak value. { 'sis·təm 'band,width }

system calendar [COMPUT SCI] A register in a computer system that holds the date and year and provides them in response to supervisor calls to the operating system. { 'sis·təm 'kal·ən·dər }

system call *See* supervisor call. { 'sis·təm ,kȯl }

system catalog [COMPUT SCI] An index of all files controlled by the operating system of a large computer. { 'sis·təm 'kad·əl,äg }

system chart [COMPUT SCI] A flowchart that emphasizes the component operations which make up a system. { 'sis·təm ,chärt }

system check [COMPUT SCI] A check on the overall performance of the system, usually not made by built-in computer check circuits; for example, control total, hash totals, and record counts. { 'sis·təm ,chek }

system clock [COMPUT SCI] A circuit that emits regularly timed pulses that are used to synchronize the operations of all the circuits of a computer. { 'sis·təm 'kläk }

system clock reference [COMMUN] A time stamp in the program stream from which decoder timing is derived. Abbreviated SCR. { 'sis·təm ¦kläk 'ref·rəns }

system command [COMPUT SCI] A special instruction to a computer system to carry out a particular processing function, such as allowing a user to gain access to the system, running a program, activating a translator, or issuing a status report. { 'sis·təm kə,mand }

system design [COMPUT SCI] Determination in detail of the exact operational requirements of a system, resolution of these into file structures and input/output formats, and relation of each to management tasks and information requirements. [CONT SYS] A technique of constructing a system that performs in a specified manner, making use of available components. Also known as synthesis. { 'sis·təm di,zīn }

system designer [COMPUT SCI] A person who prepares final system documentation, analyzes findings, and synthesizes new system design. { 'sis·təm di,zīn·ər }

system documentation [COMPUT SCI] Detailed information, in either written or computerized form, about a computer system, including its architecture, design, data flow, and programming logic. { 'sis·təm ,däk·yə·mən'tā·shən }

system engineering *See* systems engineering. { 'sis·təm ,en·jə'nir·iŋ }

system evaluation [COMPUT SCI] A periodic evaluation of the system to assess its status in terms of original or current expectations and to chart its future direction. { 'sis·təm i,val·yə'wā·shən }

system flowchart *See* data flow diagram. { 'sis·təm 'flō,chärt }

system generation [COMPUT SCI] A process that creates a particular and uniquely specified operating system; it combines user-specified options and parameters with manufacturer-supplied general-purpose or nonspecialized program subsections to produce an operating system (or other complex software) of the desired form and capacity. Abbreviated sysgen. { 'sis·təm ,jen·ə'rā·shən }

system header [COMMUN] A data structure that carries information summarizing the system characteristics of the digital television multiplexed bit stream. { 'sis·təm ,hed·ər }

system improvement time [COMPUT SCI] The machine downtime needed for the installation and testing of new components, large or small, and machine downtime necessary for modification of existing components; this includes all programming tests following the above actions to prove the machine is operating properly. { 'sis·təm im'prüv·mənt ,tīm }

system input *See* SYSIN. { 'sis·təm 'in,pu̇t }

system integration [COMPUT SCI] The procedures involved in combining separately developed modules of components so that they work together as a complete computer system. { 'sis·təm ,in·tə'grā·shən }

system-level timer [COMPUT SCI] A hardware device that is set by the operating system to interrupt it after a specified time interval, either to set deadlines for events or to remind the operating system to take some action. { 'sis·təm ¦lev·əl 'tīm·ər }

system library [COMPUT SCI] An organized collection of computer programs that is maintained on-line with a computer system by being held on a secondary storage device and is managed by the operating system. { 'sis·təm 'lī,brer·ē }

system loader [COMPUT SCI] A computer program that loads all the other programs, including the operating system, into a computer's main storage. { 'sis·təm ,lōd·ər }

system master tapes [COMPUT SCI] Magnetic tapes containing programmed instructions necessary for preparing a computer prior to running programs. { 'sis·təm 'mas·tər 'tāps }

system operation [COMPUT SCI] The administration and operation of an automatic data-processing equipment-oriented system, including staffing, scheduling, equipment and service contract administration, equipment utilization practices, and time-sharing. { ˈsis·təm ˌäp·əˈrā·shən }

system optimization *See* optimization. { ˈsis·təm ˌäp·tə·məˈzā·shən }

system response *See* response. { ˈsis·təm ri ˈspäns }

systems analysis [ENG] The analysis of an activity, procedure, method, technique, or business to determine what must be accomplished and how the necessary operations may best be accomplished. { ˈsis·təmz əˌnal·ə·səs }

systems architecting [SYS ENG] The discipline that combines elements which, working together, create unique structural and behavioral capabilities in a system that none could produce alone. Also known as systems architecture. { ¦sis·təmz ˈär·kəˌtek·tiŋ }

systems architecture *See* systems architecting. { ˈsis·təmz ˌär·kəˌtek·chər }

systems definition [COMPUT SCI] A document describing a computer-based system for processing data or solving a problem, including a general description of the aims and benefits of the system and clerical procedures employed, and detailed program specification. Also known as systems specification. { ˈsis·təmz ˌdef·ə ˌnish·ən }

systems engineering [ENG] The design of a complex interrelation of many elements (a system) to maximize an agreed-upon measure of system performance, taking into consideration all of the elements related in any way to the system, including utilization of worker power as well as the characteristics of each of the system's components. Also known as system engineering. { ˈsis·təmz ˌen·jəˌnir·iŋ }

systems integration [SYS ENG] A discipline that combines processes and procedures from systems engineering, systems management, and product development for the purpose of developing large-scale complex systems that involve hardware and software and may be based on existing or legacy systems coupled with totally new requirements to add significant functionality. { ¦sis·təmz ˌin·təˈgrā·shən }

systems management [SYS ENG] The management of information technology systems in an organization or commercial enterprise, including all activities involved in configuring, installing, maintaining, and updating these systems. { ˈsis·təmz ˈman·ij·mənt }

system software [COMPUT SCI] Computer software involved with data and program management, including operating systems, control programs, and database management systems. { ˈsis·təm ˈsȯftˌwer }

systems programming [COMPUT SCI] The development and production of programs that have to do with translation, loading, supervision, maintenance, control, and running of computers and computer programs. { ˈsis·təmz ˌprō ˌgram·iŋ }

systems specification *See* systems definition. { ˈsis·təmz ˌspes·ə·fəˌkā·shən }

systems test [COMPUT SCI] The running of whole computer system against test data; a complete simulation of the actual running system for purposes of testing the adequacy of the system. { ˈsis·təmz ˌtest }

system study [COMPUT SCI] A detailed study to determine whether, to what extent, and how automatic data-processing equipment should be used; it usually includes an analysis of the existing system and the design of the new system, including the development of system specifications which provide a basis for the selection of equipment. { ˈsis·təm ˌstəd·ē }

system supervisor [COMPUT SCI] A control program which ensures an efficient transition in running program after program and accomplishing setups and control functions. { ˈsis·təm ˈsü·pərˌvīz·ər }

system target decoder [COMMUN] A hypothetical reference model of a decoding process used to describe the semantics of the digital television multiplexed bit stream. Abbreviated STD. { ˈsis·təm ¦tär·gət dēˈkōd·ər }

system unit [COMPUT SCI] **1.** An individual card, section of tape, or the like, which is manipulated during operation of the system; class 1 systems have one unit per document; class 2 systems have one unit per vocabulary term or concept. **2.** *See* case. { ˈsis·təm ˌyü·nət }

systolic array [COMPUT SCI] An array of processing elements of cells connected to a memory which pulses data through the array in such a way that each data item can be used effectively at each cell it passes while being pumped from cell to cell along the array. { siˈstäl·ik əˈrā }

T

table [COMPUT SCI] A set of contiguous, related items, each uniquely identified either by its relative position in the set or by some label. { 'tā·bəl }

table-driven compiler [COMPUT SCI] A compiler in which the source language is described by a set of syntax rules. { 'tā·bəl ¦driv·ən kəm'pī·lər }

table-driven program [COMPUT SCI] A computer program that relies on tables stored outside of the program in the computer's memory to furnish data. { 'tā·bəl ¦driv·ən 'prō,gram }

table look-up [COMPUT SCI] A procedure for calculating the location of an item in a table by means of an algorithm, rather than by conducting a search for the item. { 'tā·bəl 'lu̇k ,əp }

table look-up device [ELECTR] A logic circuit in which the input signals are grouped as address digits to a memory device, and, in response to any particular combination of inputs, the memory device location that is addressed becomes the output. { 'tā·bəl ¦lu̇k ,əp di,vīs }

table management program [COMPUT SCI] A computer program that handles the creation and maintenance of tables and access to data stored in them. { 'tā·bəl ¦man·ij·mənt ,prō,gram }

tabular language [COMPUT SCI] A part of a program which represents the composition of a decision table required by the problem considered. { 'tab·yə·lər 'laŋ·gwij }

tabulate [COMPUT SCI] To order a set of data into a table form, or to print a set of data as a table, usually indicating differences and totals, or just totals. { 'tab·yə,lāt }

tabulation character [COMPUT SCI] A character that controls the action of a computer printer and is not itself printed, although it forms part of the data to be printed. { ,tab·yə'lā·shən ,kar·ik·tər }

Tacan *See* tactical air navigation system. { 'tak ,an }

tactical air navigation system [NAV] Short-range ultra-high-frequency air navigation system that provides accurate slant-range distance and bearing information; this information is presented to the pilot in two dimensions, that is, distance and bearing from a selected ground station. Also known as Tacan. { 'tak·tə·kəl 'er ,nav·ə'gā·shən ,sis·təm }

tactical communications system [COMMUN] A system which provides internal communications within tactical air elements, composed of transportable and mobile equipment assigned as unit equipment to the supporting tactical unit. { 'tak·tə·kəl kə,myü·nə'kā·shənz ,sis·təm }

tactical frequency [COMMUN] Radio frequency assigned to a military unit to be used in the accomplishment of a tactical mission. { 'tak·tə·kəl 'frē·kwən·sē }

tactile feedback [COMPUT SCI] In haptics, devices that provide a user with the sensations of heat, pressure, and texture. { ,tak·təl 'fēd,bak }

tag [COMPUT SCI] **1.** A unit of information used as a label or marker. **2.** The symbol written in the location field of an assembly-language coding form, and used to define the symbolic address of the data or instruction written on that line. { 'tag }

tag converting unit [COMPUT SCI] A device capable of reading the perforations of a price tag as input data. { 'tag kən'vərd·iŋ ,yü·nət }

tag field [COMPUT SCI] A data item within a variant record that identifies the format to be used in the record. { 'tag ,fēld }

tag format [COMPUT SCI] The arrangement of data in a short record inserted in a direct-access storage to indicate the location of an overflow record. { 'tag ,fȯr,mat }

tag image file format [COMPUT SCI] File format used for storing bitmap images at any resolution. Abbreviated TIFF. { ¦tag ,im·ij 'fīl ,fȯr,mat }

tag sort [COMPUT SCI] A method of sorting data in which the addresses of records rather than the records themselves are used to determine the sequence. { 'tag ,sȯrt }

takedown [COMPUT SCI] The actions performed at the end of an equipment operating cycle to prepare the equipment for the next setup; for example, to remove the tapes from the tape handlers at the end of a computer run is a takedown operation. { 'tāk,dau̇n }

takedown time [COMPUT SCI] The time required to take down a piece of equipment. { 'tāk,dau̇n ,tīm }

talk-back circuit *See* interphone. { 'tȯk ,bak ,sər·kət }

talking battery *See* quiet battery. { 'tȯk·iŋ ,bad·ə·rē }

tandem central office [COMMUN] A telephone office that makes connections between local offices in an area where there is such a high density of local offices that it would be uneconomical to make direct connections between each of them. Also known as tandem office. { 'tan·dəm 'sen·trəl 'ȯf·əs }

tandem compensation *See* cascade compensation. { 'tan·dəm ˌkäm·pən'sā·shən }

tandem office *See* tandem central office. { 'tan·dəm 'ȯf·əs }

tandem switching [COMMUN] System of routing telephone calls in which calls do not travel directly between local offices, but rather through a tandem central office. { 'tan·dəm 'swich·iŋ }

tandem system [COMPUT SCI] A computing system in which there are two central processing units, usually with one controlling the other, and with data proceeding from one processing unit into the other. { 'tan·dəm ˌsis·təm }

tank [ELECTR] **1.** A unit of acoustic delay-line storage containing a set of channels, each forming a separate recirculation path. **2.** The heavy metal envelope of a large mercury-arc rectifier or other gas tube having a mercury-pool cathode. *See* tank circuit. [ENG] A large container for holding, storing, or transporting a liquid. { taŋk }

tank circuit [ELECTR] A circuit which exhibits resonance at one or more frequencies, and which is capable of storing electric energy over a band of frequencies continuously distributed about the resonant frequency, such as a coil and capacitor in parallel. Also known as electrical resonator; tank. { 'taŋk ˌsər·kət }

T antenna [ELECTROMAG] An antenna consisting of one or more horizontal wires, with a lead-in connection being made at the approximate center of each wire. { 'tē anˌten·ə }

tape [COMPUT SCI] A ribbonlike material used to store data in lengthwise sequential position. { tāp }

tape alternation [COMPUT SCI] The switching of a computer program back and forth between two tape units in order to avoid interruption of the program during mounting and removal of tape reels. { 'tāp ˌȯl·tərˌnā·shən }

tape bootstrap routine [COMPUT SCI] A computer routine stored in the first block of a magnetic tape that instructs the computer to read certain programs from the tape. { 'tāp 'büt ˌstrap rüˌtēn }

tape cluster *See* magnetic tape group. { 'tāp ˌkləs·tər }

tape control unit [COMPUT SCI] A device which senses which tape unit is to be accessed for read or write purpose and opens up the necessary electronic paths. Formerly known as hypertape control unit. { 'tāp kənˌtrōl ˌyü·nət }

tape crease [COMPUT SCI] A fold or wrinkle in a magnetic tape that results in an error in the reading or writing of data at that point. { 'tāp ˌkrēs }

tape drive [COMPUT SCI] A tape reading or writing device consisting of a tape transport, electronics, and controls; it usually refers to magnetic tape exclusively. { 'tāp ˌdrīv }

tape editor [COMPUT SCI] A routine designed to help edit, revise, and correct a routine contained on a tape. { 'tāp ˌed·əd·ər }

tape group *See* magnetic tape group. { 'tāp ˌgrüp }

tape label [COMPUT SCI] A record appearing at the beginning or at the end of a magnetic tape to uniquely identify the tape as the one required by the system. { 'tāp ˌlā·bəl }

tape library [COMPUT SCI] A special area, most often a room within a computer installation, used to store magnetic tapes. { 'tāp ˌlīˌbrer·ē }

tape-limited [COMPUT SCI] Pertaining to a computer operation in which the time required to read and write tapes exceeds the time required for computation. { 'tāp ¦lim·əd·əd }

tape mark [COMPUT SCI] **1.** A special character or coding, an attached piece of reflective material, or other device that indicates the physical end of recording on a magnetic tape. Also known as destination warning mark; end-of-tape mark. **2.** A special character that divides a file of magnetic tape into sections, usually followed by a record with data describing the particular section of the file. Also known as control mark. { 'tāp ˌmärk }

tape operating system [COMPUT SCI] A computer operating system in which source programs and sometimes incoming data are stored on magnetic tape, rather than in the computer memory. Abbreviated TOS. { 'tāp 'äp·əˌrād·iŋ ˌsis·təm }

tape plotting system [COMPUT SCI] A digital incremental plotter in which the digital data are supplied from a magnetic or paper tape. { 'tāp 'pläd·iŋ ˌsis·təm }

tape pool [COMPUT SCI] A collection of tape drives. { 'tāp ˌpül }

tape-processing simultaneity [COMPUT SCI] A feature of some computer systems whereby reading or writing of data can be carried out on all the tape units at the same time, while the central processing unit continues to process data. { 'tāp ¦präˌses·iŋ ˌsī·məl·tə'nē·əd·ē }

tapered transmission line *See* tapered waveguide. { 'tā·pərd tranz'mish·ən ˌlīn }

tapered waveguide [ELECTROMAG] A waveguide in which a physical or electrical characteristic changes continuously with distance along the axis of the waveguide. Also known as tapered transmission line. { 'tā·pərd 'wāvˌgīd }

tape search unit [COMPUT SCI] Small, fully transistorized, special-purpose, digital data-processing system using a stored program to perform logical functions necessary to search a magnetic tape in off-line mode, in response to a specific request. { 'tā·pər 'sərch ˌyü·nət }

tape serial number [COMPUT SCI] A number identifying a magnetic tape which remains unchanged throughout the time the tape is used, even though all other information about the tape may change. { 'tā·pər 'sir·ē·əl ˌnəm·bər }

tape skip [COMPUT SCI] A machine instruction to space forward and erase a portion of tape when a defect on the tape surface causes a write error to persist. { 'tāp ,skip }

tape station [COMPUT SCI] A tape reading or writing device consisting of a tape transport, electronics, and controls; it may use either magnetic tape or paper tape. { 'tāp ,stā·shən }

tape-to-tape conversion [COMPUT SCI] A routine which directs a computer to copy information from one tape to another tape of a different kind; for example, from a seven-track onto a nine-track tape. { ¦tāp tə ¦tāp kən'vər·zhən }

tape transport [COMPUT SCI] The mechanism that physically moves a tape past a stationary head. Also known as transport. { 'tāp ,tranz ,pȯrt }

tape unit [COMPUT SCI] A tape reading or writing device consisting of a tape transport, electronics, controls, and possibly a cabinet; the cabinet may contain one or more magnetic tape stations. { 'tāp ,yü·nət }

target [ELECTR] **1.** In a television camera tube, the storage surface that is scanned by an electron beam to generate an output signal current corresponding to the charge-density pattern stored there. **2.** In radar and sonar, any object capable of reflecting the transmitted beam; depending on context, often connotes an object of interest as opposed to clutter. [ENG] In radar and sonar, any object capable of reflecting the transmitted beam. { 'tär·gət }

target acquisition [ELECTR] **1.** The first appearance of a recognizable and useful echo signal from a new target in radar and sonar. **2.** *See* acquisition. { 'tär·gət ,ak·wə'zish·ən }

target central processing unit [COMPUT SCI] The type of central processing unit for which a language processor (assembler, compiler, or interpreter) generates machine language output. { 'tar·gət ¦sen·trəl 'prä,ses·iŋ ,yü·nət }

target configuration [COMPUT SCI] The combination of input, output, and storage units and the amount of computer memory required to carry out an object program. { 'tär·gət kən,fig·yə ,rā·shən }

target cross section *See* echo area. { 'tär·gət 'krȯs ,sek·shən }

target discrimination [ELECTR] The ability of a detection or guidance system to distinguish a target from its background or to discriminate between two or more targets that are close acquisition. { 'tär·gət di,skrim·ə,nā·shən }

target language [COMPUT SCI] The language into which a program (or text) is to be converted. { 'tär·gət ¦laŋ·gwij }

target pack [COMPUT SCI] A disk pack that is used to maintain systems software and, in particular, to hold a copy of a system control program on which modifications are made and tested. { 'tär·gət ,pak }

target phase [COMPUT SCI] The stage of handling a computer program at which the object program is first carried out after it has been compiled. { 'tär·gət ,fāz }

target program *See* object program. { 'tär·gət ¦prō,gram }

target routine *See* object program. { 'tär·gət rü ,tēn }

target signal [ELECTROMAG] The radio energy returned to a radar by a target. Also known as echo signal; video signal. { 'tär·gət ,sig·nəl }

task [COMPUT SCI] A set of instructions, data, and control information capable of being executed by the central processing unit of a digital computer in order to accomplish some purpose; in a multiprogramming environment, tasks compete with one another for control of the central processing unit, but in a nonmultiprogramming environment a task is simply the current work to be done. { task }

task descriptor [COMPUT SCI] The vital information about a task in a multitask system which must be saved when the task is interrupted. Also known as state vector. { 'task di,skrip·tər }

task management [COMPUT SCI] The functions, assumed by the operating system, of switching the processor among tasks, scheduling, sending messages or timing signals between tasks, and creating or removing tasks. { 'task ,man·ij·mənt }

task programmer [COMPUT SCI] A person who writes applications programs for controlling a robotic system. { 'task ,prō,gram·ər }

task switching [COMPUT SCI] Switching back and forth between two or more active programs without having to close or open any of them. Also known as context switching. { 'task ,swich·iŋ }

T circulator [ELECTROMAG] A circulator in which three identical rectangular waveguides are joined asymmetrically to form a T-shaped structure, with a ferrite post or wedge at its center; power entering any waveguide emerges from only one adjacent waveguide. { 'tē ,sər·kyə,lād·ər }

TCP *See* Transmission Control Protocol.

TCP/IP *See* Transmission Control Protocol/ Internet Protocol.

TDD *See* display device.

TDM *See* time-division multiplexing.

TDMA *See* time-division multiple access.

TDR *See* time-domain reflectometer.

TDRSS *See* Tracking and Data Relay Satellite System.

TEA *See* transferred-electron amplifier.

teach [CONT SYS] To program a robot by guiding it through its motions, which are then recorded and stored in its computer. { tēch }

tears [COMMUN] In an analog television picture, a horizontal disturbance caused by noise, in which the picture appears to be torn apart. { tirz }

TEGFET *See* high-electron-mobility transistor. { 'teg,fet }

telautograph [COMMUN] A writing telegraph instrument, the forerunner of the facsimile machine, in which manual movement of a pen at the transmitting position varies the current in two circuits in such a way as to cause corresponding movements of a pen at the remote receiving instrument; ordinary handwriting can thus be transmitted over wires. { te'lȯd·ə,graf }

telecast [COMMUN] A television broadcast intended for reception by the general public, involving the transmission of the picture and sound portions of the program. { 'tel·ə,kast }

telecine camera [ELECTR] A video camera used in conjunction with film or slide projectors to televise motion pictures and still images. { ¦tel·ə¦sin·ē 'kam·rə }

telecommunicating device for the deaf *See* telecommunications display device. { ¦tel·ə·kə 'myü·nə·kād·iŋ di'vīs ,fȯr thə 'def }

telecommunications [COMMUN] Communication over long distances. { ¦tel·ə·kə,myü·nə 'kā·shənz }

Telecommunications Coordinating Committee [COMMUN] Committee organized by the U.S. State Department and composed of major government departments, agencies, and industrial organizations; makes recommendations on telecommunications matters affecting international telecommunications. { ¦tel·ə·kə,myü·nə'kā·shənz kō'ȯrd·ən,ād·iŋ kə,mid·ē }

telecommunications display device [COMMUN] A telephone equipped with a keyboard and display for users who have hearing or speech impairments. Also known as telecommunications device for the deaf; text telephone. Abbreviated TDD. { ¦tel·ə·kə,myü·nə'kā·shənz di'splā di,vīs }

teleconference [COMMUN] **1.** A two-way interactive meeting between relatively small groups of people remote from one another but linked by telecommunication facilities involving audio communication, and possibly also video, graphics, or facsimile. **2.** More broadly, any of various facilities allowing people to communicate among each other over some distance, encompassing teleseminars and telemeetings. { ¦tel·ə'kän·frəns }

telegram [COMMUN] A message sent by telegraphy. { 'te·lə,gram }

telegraph alphabet *See* telegraph code. { 'tel·ə ,graf 'al·fə,bet }

telegraph bandwidth [COMMUN] The difference between the limiting frequencies of a channel used to transmit telegraph signals. { 'tel·ə,graf 'band,width }

telegraph carrier [COMMUN] The single-frequency wave which is modulated by transmitting apparatus in carrier telegraphy. { 'tel·ə,graf ,kar·ē·ər }

telegraph circuit [COMMUN] The complete wire or radio circuit over which signal currents flow between transmitting and receiving apparatus in a telegraph system. { 'tel·ə,graf ,sər·kət }

telegraph code [COMMUN] A system of symbols for transmitting telegraph messages in which each letter or other character is represented by a set of long and short electrical pulses, or by pulses of opposite polarity, or by time intervals of equal length in which a signal is present or absent. Also known as telegraph alphabet. { 'tel·ə,graf ,kōd }

telegraph emission [COMMUN] The signal transmitted by a telegraph system, classified by type of transmission, type of modulation, bandwidth, and supplementary characteristics. { 'tel·ə,graf i,mish·ən }

telegraph grade [COMMUN] The class of communication circuits that can transmit only telegraphic signals, comprising the lowest types of circuits in regard to speed, accuracy, and cost. { 'tel·ə,graf ,grād }

telegraph interference [COMMUN] Any undesired electrical energy that tends to interfere with the reception of telegraph signals. { 'tel·ə,graf ,in·tər'fir·əns }

telegraph signal distortion [COMMUN] Time displacement of transitions between conditions, such as marking and spacing, with respect to their proper relative positions in perfectly timed signals; the total distortion is the algebraic sum of the bias and the characteristic and fortuitous distortions. { 'tel·ə,graf ¦sig·nəl di,stȯr·shən }

telegraphy [COMMUN] Communication at a distance by means of code signals consisting of current pulses sent over wires or by radio; it is the oldest form of electrical digital communication. { tə'leg·rə·fē }

telemeeting [COMMUN] A meeting between people remote from one another, but linked by audio and video telecommunications facilities that provide primarily one-way communication from a few people at one location to large numbers of people at other locations, and use temporary equipment or circuits. { 'tel·ə,mēd·iŋ }

telemeter [ENG] **1.** The complete measuring, transmitting, and receiving apparatus for indicating or recording the value of a quantity at a distance. Also known as telemetering system. **2.** To transmit the value of a measured quantity to a remote point. { 'tel·ə,mēd·ər }

telemetering [ENG] Transmitting the readings of instruments to a remote location by means of wires, radio waves, or other means. Also known as remote metering; telemetry. { ,tel·ə'mēd·ə·riŋ }

telemetering antenna [ELECTROMAG] A highly directional antenna, generally mounted on a servo-controlled mount for tracking purposes, used at ground stations to receive telemetering signals from a guided missile or spacecraft. { ,tel·ə'mēd·ə·riŋ an'ten·ə }

telemetering receiver [ELECTR] A device in a telemetering system which converts electrical signals into an indication or recording of the value of the quantity being measured at a distance. { ,tel·ə'mēd·ə·riŋ ri'sē·vər }

telemetering system *See* telemeter. { ,tel·ə'mēd·ə·riŋ ,sis·təm }

telemetering transmitter [ELECTR] A device which converts the readings of instruments into electrical signals for transmission to a remote location by means of wires, radio waves, or other means. { ,tel·ə'mēd·ə·riŋ tranz'mid·ər }

telemetry *See* telemetering. { tə'lem·ə·trē }

telephone [COMMUN] A system of converting sound waves into variations in electric current or other electrical quantities that can be transmitted and reconverted into sound waves at a distant point, used primarily for voice communication; it consists essentially of a telephone transmitter and receiver at each station, interconnecting wires, cables, optical fibers, or terrestrial or satellite radio transmission systems, signaling devices, a central power supply, and switching facilities. Also known as telephone system. { 'tel·ə‚fōn }

telephone-answering system [COMMUN] A special type of private branch exchange system used by a telephone-answering service bureau to provide secretarial service for its customers. { 'tel·ə‚fōn ¦an·sə·riŋ ‚sis·təm }

telephone central office *See* central office. { 'tel·ə‚fōn 'sen·trəl 'ȯf·əs }

telephone channel [COMMUN] A one-way or two-way path suitable for the transmission of audio signals between two stations. { 'tel·ə‚fōn ‚chan·əl }

telephone circuit [ELEC] The complete circuit over which audio and signaling currents travel in a telephone system between the two telephone subscribers in communication with each other; the circuit usually consists of insulated conductors, a radio link, or a fiber-optic cable. { 'tel·ə ‚fōn ‚sər·kət }

telephone data set [COMPUT SCI] Equipment interfacing a data terminal with a telephone circuit. { 'tel·ə‚fōn 'dad·ə ‚set }

telephone dial [ENG] **1.** A switch operated by a finger wheel, used to make and break a pair of contacts the required number of times for setting up a telephone circuit to the party being called. **2.** By extension, the push-button apparatus used to generate dual-tone multifrequency (DTMF) signals. { 'tel·ə‚fōn ‚dīl }

telephone emission *See* telephone signal. { 'tel·ə‚fōn i‚mish·ən }

telephone influence factor [COMMUN] A measure of the interference of power-line harmonics with telephone lines, which is derived by weighting the terms in the mathematical expression for the total harmonic distortion of the power-line voltage. { 'tel·ə‚fōn 'in·flü·əns ‚fak·tər }

telephone loading coil *See* loading coil. { 'tel·ə ‚fōn 'lōd·iŋ ‚kȯil }

telephone modem [ELECTR] A piece of equipment that modulates and demodulates one or more separate telephone circuits, each containing one or more telephone channels; it may include multiplexing and demultiplexing circuits, individual amplifiers, and carrier-frequency sources. { 'tel·ə‚fōn 'mō‚dem }

telephone repeater [ELECTR] A repeater inserted at one or more intermediate points in a long telephone line to amplify telephone signals so as to maintain the required current strength. { 'tel·ə‚fōn ri‚pēd·ər }

telephone ringer [ELECTROMAG] **1.** An electromagnetic device that actuates a clapper which strikes one or more gongs to produce a ringing sound; used with a telephone set to signal a called party. **2.** By extension, the electronic device that performs the same function. { 'tel·ə ‚fōn ‚riŋ·ər }

telephone signal [COMMUN] The electrical signal transmitted by a telephone system, classified by type of transmission, type of modulation, bandwidth, and supplementary characteristics. Also known as telephone emission. { 'tel·ə ‚fōn ‚sig·nəl }

telephone switchboard *See* switchboard. { 'tel·ə‚fōn 'swich‚bȯrd }

telephone system *See* telephone. { 'tel·ə‚fōn ‚sis·təm }

telephony [COMMUN] The transmission of speech to a distant point by means of electric signals. { tə'lef·ə·nē }

telephoto *See* facsimile. { ¦tel·ə'fōd·ō }

telephotography *See* facsimile. { ¦tel·ə·fə'täg·rə·fē }

teleport [COMMUN] A planned business development area that features direct and economic access to a large number of domestic and international satellites for users in the surrounding region, with the aid of a regional distribution network. { 'tel·ə‚pȯrt }

teleprinter [COMPUT SCI] Any typewriter-type device capable of being connected to a computer and of printing out a set of messages under computer control. { 'tel·ə‚print·ər }

teleprinting [COMMUN] Telegraphy in which the transmitter and receiver are teletypewriters. { 'tel·ə‚print·iŋ }

teleprocessing [COMPUT SCI] **1.** The use of telecommunications equipment and systems by a computer. **2.** A computer service involving input/output at locations remote from the computer itself. { 'tel·ə‚prä‚ses·iŋ }

teleprocessing monitor [COMPUT SCI] A computer program that manages the transfer of information between local and remote terminals. Abbreviated TP monitor. { 'tel·ə‚prä‚ses·iŋ 'män·əd·ər }

Teleran *See* television and radar navigation system. { 'tel·ə‚ran }

telering [ELECTR] In telephony, a frequency-selector device for the production of ringing power. { 'tel·ə‚riŋ }

teleseminar [COMMUN] A form of long-distance, electronic communication, primarily one-way, to many destinations from one source, for educational purposes, involving audio communication, and possibly also video and some form of graphics. { ¦tel·ə'sem·ə‚när }

teleterminal [COMPUT SCI] An instrument that integrates the functions of a telephone set and a computer terminal with keyboard and video screen. { ¦tel·ə'tər·mən·əl }

teletypewriter [COMMUN] A special electric typewriter that produces coded electric signals corresponding to manually typed characters, and automatically types messages when fed with similarly coded signals produced by another machine; it allows access to telephone services for people who are deaf, or who have a hearing, speech, or communication impairment. Also known as TWX machine. Abbreviated TTY. { ¦tel·ə'tīp‚rīd·ər }

teletypewriter code [COMMUN] Special code in which each code group is made up of five units, or elements, of equal length which are known as marking or spacing impulses; the five-unit start-stop code consists of five signal impulses preceded by a start impulse and followed by a stop impulse. { ¦tel·ə'tīp,rīd·ər ,kōd }

teletypewriter exchange service [COMMUN] A service furnished by telephone companies to subscribers in the United States, whereby any of the subscribers can communicate directly with any other subscriber via teletypewriter. Also known as TWX service. { ¦tel·ə'tīp,rīd·ər iks ¦chānj ,sər·vəs }

teletypewriter signal distortion [COMMUN] Of a start-stop teletypewriter signal, the shifting of the transition points of the signal pulses from their proper positions relative to the beginning of the start pulse; the magnitude of the distortion is expressed in percent of a perfect unit pulse length. { ¦tel·ə'tīp,rīd·ər 'sig·nəl di,stȯr·shən }

televise [COMMUN] To pick up a scene with a video camera and convert it into corresponding electric signals for transmission by a television station. { 'tel·ə,vīz }

television [COMMUN] A system for converting a succession of visual images into corresponding electric signals and transmitting these signals by radio or over wires to distant receivers at which the signals can be used to reproduce the original images. Abbreviated TV. { 'tel·ə,vizh·ən }

television and radar navigation system [NAV] Navigational system which employs ground-based search radar equipment along an airway to locate aircraft flying near that airway; transmits, by television, information pertaining to these aircraft and other information to the pilots of properly equipped aircraft, and provides information to the pilots appropriate for use in the landing approach. Also known as Teleran; television-radar air navigation. { 'tel·ə,vizh·ən ən 'rā,där ,nav·ə'gā·shən ,sis·təm }

television antenna [ELECTROMAG] An antenna suitable for transmitting or receiving television broadcasts; since television transmissions in the United States are horizontally polarized, the most basic type of receiving antenna is a horizontally mounted half-wave dipole. { 'tel·ə,vizh·ən an ,ten·ə }

television bandwidth [COMMUN] The difference between the limiting frequencies of a television channel; in the United States, this is 6 megahertz. { 'tel·ə,vizh·ən 'band,width }

television broadcast band [COMMUN] Several groups of channels, each containing a number of 6-megahertz channels, that are available for assignment to television broadcast stations. { 'tel·ə,vizh·ən 'brȯd,kast ,band }

television broadcasting [COMMUN] Transmission of television programs by means of radio waves for reception by the public. { 'tel·ə,vizh·ən 'brȯd,kast·iŋ }

television camera [ELECTR] The pickup unit used to convert a scene into corresponding electric signals; optical lenses focus the scene to be televised on the photosensitive surface of a camera tube, and the tube breaks down the visual image into small picture elements and converts the light intensity of each element in turn into a corresponding electric signal. Also known as camera. { 'tel·ə,vizh·ən ,kam·rə }

television camera tube *See* camera tube. { 'tel·ə,vizh·ən 'kam·rə ,tüb }

television channel [COMMUN] A band of frequencies 6 megahertz wide in the television broadcast band, available for assignment to a television broadcast station. { 'tel·ə,vizh·ən ,chan·əl }

television emission *See* television signal. { 'tel·ə,vizh·ən i,mish·ən }

television interference [COMMUN] Interference produced in television receivers by other transmitting devices. Abbreviated TVI. { 'tel·ə ,vizh·ən ,in·tər'fir·əns }

television monitor [ELECTR] Adisplay device used to continuously check the image picked up by a television camera and the sound picked up by video camera or other source to provide continuous observation of image content and/or quality. { 'tel·ə,vizh·ən ,man·əd·ər }

television network [COMMUN] An arrangement of communication channels, suitable for transmission of video and accompanying audio signals, which link together groups of television broadcasting stations or closed-circuit television users in different cities so that programs originating at one point can be fed simultaneously to all others. { 'tel·ə,vizh·ən ,net,wərk }

television pickup station [COMMUN] A land mobile station used for the transmission of television program material and related communications from the scene of an event occurring at a point remote to a television broadcast station. { 'tel·ə,vizh·ən 'pik·əp ,stā·shən }

television picture tube *See* picture tube. { 'tel·ə,vizh·ən 'pik·chər ,tüb }

television-radar air navigation *See* television and radar navigation system. { 'tel·ə,vizh·ən 'rā ,där 'er ,nav·ə,gā·shən }

television receive only antenna [COMMUN] A parabolic reflector or dish with sufficient gain to receive signals from geostationary satellites, together with a feed horn that collects the signals reflected by the dish, a low-noise amplifier for preamplification, and a tunable satellite receiver. Abbreviated TVRO. { 'tel·ə,vizh·ən ri'sēv ¦ōn·lē an'ten·ə }

television receiver [ELECTR] A receiver that converts incoming television signals into the original scenes along with the associated sounds. Also known as television set. { 'tel·ə,vizh·ən ri ,sē·vər }

television relay system *See* television repeater. { 'tel·ə,vizh·ən 'rē,lā ,sis·təm }

television repeater [ELECTR] A repeater that transmits television signals from point to point by using radio waves in free space as a medium, such transmission not being intended for direct reception by the public. Also known as television relay system. { 'tel·ə,vizh·ən ri,pēd·ər }

television screen [ELECTR] The fluorescent screen of the picture tube in a television receiver. { 'tel·ə,vizh·ən ,skrēn }

television set *See* television receiver. { 'tel·ə ˌvizh·ən ˌset }

television signal [COMMUN] A general term for the aural and visual signals that are broadcast together to provide the sound and picture portions of an analog television program. Also known as television emission. { 'tel·əˌvizh·ən ˌsig·nəl }

television station [COMMUN] The installation, assemblage of equipment, and location where radio transmissions are sent or received. { 'tel·ə ˌvizh·ən ˌstā·shən }

television studio [COMMUN] A complex of rooms specifically designed for the origination of live or taped television programs. { 'tel·ə ˌvizh·ən ˌstüd·ē·ō }

television tower [ENG] A tall metal structure used as a television transmitting antenna, or used with another such structure to support a television transmitting antenna wire. { 'tel·ə ˌvizh·ən ˌtau̇·ər }

television transmitter [ELECTR] An electronic device that converts the audio and video signals of a television program into modulated radio-frequency energy that can be radiated from an antenna and received on a television receiver. { 'tel·əˌvizh·ən tranzˌmid·ər }

television tuner [ELECTR] A component in a television receiver that selects the desired channel and converts the frequencies received to lower frequencies within the passband of the intermediate-frequency chain. { 'tel·əˌvizh·ən ˌtü·nər }

telewriter [COMMUN] System in which writing movement at the transmitting end causes corresponding movement of a writing instrument at the receiving end. { 'tel·əˌrīd·ər }

Telex [COMMUN] A worldwide teleprinter exchange service providing direct send and receive teleprinter connections between subscribers. Abbreviated TEX. { 'teˌleks }

TELNET *See* network terminal protocol. { 'tel ˌnet }

TEM mode *See* transverse electromagnetic mode. { ˌtēˌē'em ˌmōd }

TE mode *See* transverse electric mode. { ¦tē'ē ˌmōd }

template [COMPUT SCI] **1.** A prototype pattern against which observed patterns are matched in a pattern recognition system. **2.** A computer program that is used in conjunction with an electronic spreadsheet to solve a particular type of problem. { 'tem·plət }

template matching [COMPUT SCI] The comparison of a picture or other data with a stored program or template, for purposes of identification or inspection. { 'tem·plət ˌmach·iŋ }

temporary file [COMPUT SCI] A file that is created during the execution of a computer program to hold interim results and is erased before the program is completed. { 'tem·pəˌrer·ē 'fīl }

temporary storage [COMPUT SCI] The storage capacity reserved or used for retention of temporary or transient data. { 'tem·pəˌrer·ē 'stȯr·ij }

TEM wave *See* transverse electromagnetic wave. { ˌtēˌē'em ˌwāv }

terahertz technology [ENG] The generation, detection, and application (such as in communications and imaging) of electromagnetic radiation roughly in the frequency range from 0.05 to 20 terahertz, corresponding to wavelengths from 6 millimeters down to 15 micrometers. { ˌter·ə ˌhərts tek'näl·ə·jē }

TERCOM *See* terrain contour matching. { 'tər ˌkäm }

terminal [COMPUT SCI] A site or location at which data can leave or enter a system. { 'ter·mən·əl }

terminal area [ELECTR] The enlarged portion of conductor material surrounding a hole for a lead on a printed circuit. Also known as land; pad. { 'tər·mən·əl ¦er·ē·ə }

terminal block [COMMUN] **1.** A cluster of five captive screw terminals at which a telephone pair terminates; the center terminal is for the ground wire, and two other terminals are used for the tip and ring wires. **2.** By extension, a similar cluster of any number of screw terminals. { 'tər·mən·əl ˌbläk }

terminal Doppler weather radar [ENG] Radars specifically designed to detect which conditions around airports, where sudden disturbances such as downdrafts (microbursts) and resulting windshears are particularly hazardous; radars detect wind conditions by the echo signals from particulate matter borne by the wind. { ¦tər·mən·əl 'däp·lər ¦weth·ər ˌrāˌdär }

terminal equipment [COMMUN] **1.** Assemblage of communications-type equipment required to transmit or receive a signal on a channel or circuit, whether it be for delivery or relay. **2.** In radio relay systems, equipment used at points where intelligence is inserted or derived, as distinct from equipment used to relay a reconstituted signal. **3.** Telephone and teletypewriter switchboards and other centrally located equipment at which wire circuits are terminated. { 'tər·mən·əl iˌkwip·mənt }

terminal network [COMPUT SCI] A system that links intelligent terminals through a communications channel. { 'tər·mən·əl 'netˌwərk }

terminal repeater [COMMUN] **1.** Assemblage of equipment designed specifically for use at the end of a communications circuit, as contrasted with the repeater designed for an intermediate point. **2.** Two microwave terminals arranged to provide for the interconnection of separate systems, or separate sections of a system. { 'tər·mən·əl ri'pēd·ər }

terminal room [COMMUN] In telephone practice, a room associated with a central office, private branch exchange, or private exchange, which contains distributing frames, relays, and similar apparatus, except that mounted in the switchboard section. { 'tər·mən·əl ˌrüm }

terminal station [COMMUN] Receiving equipment and associated multiplex equipment used at the ends of a radio-relay system. { 'tər·mən·əl ˌstā·shən }

terminal vertex [MATH] A vertex in a rooted tree that has no successor. Also known as leaf. { 'tər·mən·əl 'vər,teks }

terminate and stay resident *See* RAM resident. { ¦tər·mə,nāt ən ,stā'rez·ə·dənt }

ternary code [COMMUN] Code in which each code element may be any one of three distinct kinds or values. { 'tər·nə·rē 'kōd }

ternary incremental representation [COMPUT SCI] A type of incremental representation in which the value of the change in a variable is defined as +1, −1, or 0. { 'tər·nə·rē ,iŋ·krə'ment·əl ,rep·ri·zən'tā·shən }

ternary pulse code modulation [COMMUN] Pulse code modulation in which each code element may be any one of three distinct kinds or values. { 'tər·nə·rē 'pəls ¦kōd ,mäj·ə'lā·shən }

terrain-avoidance radar [NAV] Airborne radar which provides a display of terrain ahead of a low-flying aircraft to permit horizontal avoidance of obstacles. { tə'rān ə¦vȯid·əns 'rā,där }

terrain contour matching [NAV] A method of correcting errors in the inertial navigation system of a cruise missile, in which predicted terrain profile data stored in the missile's computer are compared with the terrain profile below the missile; calculated by subtracting the radar-altimeter-derived height of the missile above the terrain from the missile's altitude as derived from inertial navigator and barometric measurements. Abbreviated TERCOM. { tə'rān 'kan,tu̇r ,mach·iŋ }

terrain echoes *See* ground clutter. { tə'rān 'ek·ōz }

terrain-following radar [NAV] Airborne radar which provides a display of terrain ahead of a low-flying aircraft to permit manual control, or signals for automatic control, to maintain constant altitude above the ground. { tə'rān ¦fäl·ə·wiŋ 'rā,där }

terrain profile recorder *See* airborne profile recorder. { tə'rān ¦prō,fīl ri,kȯrd·ər }

tertiary storage [COMPUT SCI] Any of several types of computer storage devices, usually consisting of magnetic tape transports and mass storage tape systems, which have slower access times, larger capacity, and lower cost than main storage or secondary storage. { 'tər·shē,er·ē 'stȯr·ij }

test data [COMPUT SCI] A set of data developed specifically to test the adequacy of a computer run or system; the data may be actual data that has been taken from previous operations, or artificial data created for this purpose. { 'test ,dad·ə }

test file [COMPUT SCI] A file consisting of test data. { 'test ,fīl }

test pattern [COMMUN] A chart having various combinations of lines, squares, circles, and graduated shading used to check definition, linearity, and contrast of a video system. Also known as resolution chart. { 'test ,päd·ərn }

test program *See* check routine. { 'test ,prō ,gram }

test record [COMPUT SCI] A record within a test file. { 'test ,rek·ərd }

test routine *See* check routine. { 'test rü,tēn }

test run [COMPUT SCI] The performance of a computer program to check that it is operating correctly, by using test data to generate results that can be compared with expected answers. { 'test ¦rən }

test system [COMPUT SCI] **1.** A computer system that is being tested before being used for production work. **2.** A version of a computer system that is retained, even after a live system is in use, chiefly to diagnose problems without interfering with the work of the live system. { 'test ,sis·təm }

test under mask [COMPUT SCI] A procedure for checking the status of selected bits in a byte by comparing the byte with another byte in which these selected bits are set to one and the other bits are set to zero. { 'test ,ən·dər 'mask }

TE wave *See* transverse electric wave. { ¦tē'ē ,wāv }

text [COMMUN] The part of a message that conveys information, excluding bits or characters needed to facilitate transmission of the message. { tekst }

text-editing system [COMPUT SCI] A computer program, together with associated hardware, for the on-line creation and modification of computer programs and ordinary text. { 'tekst ¦ed·əd·iŋ ,sis·təm }

theory of games *See* game theory. { 'thē·ə·rē əv 'gāmz }

thermal charge *See* entropy. { 'thər·məl ¦chärj }

thermal noise [COMMUN] *See* Gaussian noise. [ELECTR] Electric noise produced by thermal agitation of electrons in conductors and semiconductors. Also known as Johnson noise; resistance noise. { 'thər·məl 'nȯiz }

theta-theta [NAV] The generic term for electronic navigation systems in which position is derived by calculations based on the bearing from two or more emitters located at different but accurately known positions. { 'thād·ə 'thād·ə }

thimble [COMPUT SCI] A cone-shaped, rotating printing element on an impact printer having character slugs around the perimeter and a hammer that drives the appropriate slug forward to print the impression on paper. { 'thim·bəl }

thin-film memory *See* thin-film storage. { 'thin ¦film 'mem·rē }

thin-film storage [COMPUT SCI] A high-speed storage device that is fabricated by depositing layers, one molecule thick, of various materials which, after etching, provide microscopic circuits which can move and store data in small amounts of time. Also known as thin-film memory. { 'thin ¦film 'stȯr·ij }

think time [COMPUT SCI] Idle time between time intervals in which transmission takes place in a real-time system. { 'thiŋk ,tīm }

thin list *See* loose list. { 'thin 'list }

third-generation computer [COMPUT SCI] One of the general purpose digital computers introduced in the late 1960s; it is characterized by integrated circuits and has logical organization and software which permit the computer to handle many programs at the same time, allow one to add or remove units from the computer, permit some or all input/output operations to occur at sites remote from the main processor, and allow conversational programming techniques. { 'thərd ,jen·ə¦rā·shən kəm'pyüd·ər }

thrashing [COMPUT SCI] An undesirable condition in a multiprogramming system, due to overcommitment of main memory, in which the various tasks compete for pages and none can operate efficiently. { 'thrash·iŋ }

thread [COMPUT SCI] A sequence of beads that are strung together. { thred }

threat [COMPUT SCI] An event that can cause harm to computers, to their data or programs, or to computations. { thret }

threat collision avoidance system [NAV] A system, based on air-traffic control transponders installed on aircraft, that issues an evasive maneuver command when it senses a collision threat. { 'thret kə¦lizh·ən ə'vȯid·əns ,sis·təm }

three-address code [COMPUT SCI] In computers, a multiple-address code which includes three addresses, usually two addresses from which data are taken and one address where the result is entered; location of the next instruction is not specified, and instructions are taken from storage in preassigned order. { 'thrē 'ad,res ,kōd }

three-address instruction [COMPUT SCI] In computers, an instruction which includes an operation and specifies the location of three registers. { 'thrē 'ad,res in'strək·shən }

three-dimensional display system [ELECTR] A radar display showing range, azimuth, and elevation simultaneously. { 'thrē di¦men·chən·əl di'splā ,sis·təm }

three-input adder *See* full adder. { 'thrē ¦in,pu̇t 'ad·ər }

three-input subtracter *See* full subtracter. { 'thrē ¦in,pu̇t səb'trak·tər }

three-level subroutine [COMPUT SCI] A subroutine in which a second subroutine is called, and a third subroutine is called by the second subroutine. { 'thrē ¦lev·əl 'səb·rü,tēn }

three-plus-one address [COMPUT SCI] An instruction format containing an operation code, three operand address parts, and a control address. { 'thrē ,pləs ¦wən 'ad,res }

three-pulse canceler [ELECTR] A moving-target indicator technique in which two "two-pulse cancelers" are cascaded together, improving the velocity response by widening the rejection around zero Doppler and, unavoidably, around each associated ambiguity. { 'thrē ¦pəls 'kan·slər }

threshold [ELECTR] In a modulation system, the smallest value of carrier-to-noise ratio at the input to the demodulator for all values above which a small percentage change in the input carrier-to-noise ratio produces a substantially equal or smaller percentage change in the output signal-to-noise ratio. { 'thresh,hōld }

threshold element [COMPUT SCI] A logic circuit which has one output and several weighted inputs, and whose output is energized if and only if the sum of the weights of the energized inputs exceeds a prescribed threshold value. { 'thresh ,hōld ,el·ə·mənt }

thresholding [COMPUT SCI] In machine vision, the comparison of an element's brightness or other characteristic with a set value or threshold. { 'thresh,hōld·iŋ }

threshold signal [ELECTROMAG] A received radio signal (or radar echo) whose power is just above the noise level of the receiver. Also known as minimum detectable signal. { 'thresh,hōld ,sig·nəl }

threshold value [COMPUT SCI] A point beyond which there is a change in the manner a program executes; in particular, an error rate above which the operating system shuts down the computer system on the assumption that a hardware failure has occurred. [CONT SYS] The minimum input that produces a corrective action in an automatic control system. { 'thresh,hōld ,val·yü }

throughput [COMMUN] A measure of the effective rate of transmission of data by a communications system. [COMPUT SCI] The productivity of a data-processing system, as expressed in computing work per minute or hour. { 'thrü ,pu̇t }

through repeater [ELECTR] Microwave repeater that is not equipped to provide for connections to any local facilities other than the service channel. { 'thrü ri,pēd·ər }

thunk [COMPUT SCI] An additional subprogram created by the compiler to represent the evaluation of the argument of an expression in the call-by-name procedure. { thəŋk }

tick [COMMUN] A pulse broadcast at 1-second intervals by standard frequency- and time-broadcasting stations to indicate the exact time. [COMPUT SCI] A time interval equal to 1/60 second, used primarily in discussing computer operations. { tik }

tie line [COMMUN] **1.** A leased communication channel or circuit. **2.** *See* data link. { 'tī ,līn }

TIF *See* telephone influence factor.

TIFF *See* tag image file format. { tif }

tightly coupled computer [COMPUT SCI] A computer linked to another computer in a manner that requires both computers to function as a single unit. { 'tīt·lē ¦kəp·əld kəm'pyüd·ər }

tile painting [COMPUT SCI] **1.** The use of patterns to create shadings that fill shapes and areas on a monochrome display. **2.** The use of very small dots of two or more colors to make blends or shades that fill shapes and areas on a color display. { 'tīl ,pānt·iŋ }

tiling [COMPUT SCI] Dividing an electronic display into two or more nonoverlapping areas that display the outputs of different programs being run concurrently on a computer. { 'tīl·iŋ }

time assignment speech interpolation [COMMUN] Modulation technique based on the fact that speech is never a continuous stream of information, but consists of a large number of short signals; therefore, the period between the speech signals is used for transmitting other data including additional speech signals. { 'tīm ə ¦sīn·mənt 'spēch ,in·tər·pə,lā·shən }

time base [ELECTR] A device which moves the fluorescent spot rhythmically across the screen of the cathode-ray tube. { 'tīm ,bās }

time-base generator *See* sweep generator. { 'tīm ¦bās ,jen·ə,rād·ər }

time-code generator [ELECTR] A crystal-controlled pulse generator that produces a train of pulses with various predetermined widths and spacings, from which the time of day and sometimes also day of year can be determined; used in telemetry and other data-acquisition systems to provide the precise time of each event. { 'tīm ¦kōd ,jen·ə,rā·dər }

time-derived channel [COMMUN] Any of the channels which result from time-division multiplexing of a channel. { 'tīm di¦rīvd ,chan·əl }

time-division data links [COMMUN] Radio communications which use time-division techniques for channel separation. { 'tīm di,vizh·ən 'dad·ə ,liŋks }

time-division multiple access [COMMUN] A technique that allows multiple users who are geographically dispersed to gain access to a communications channel, by permitting each user access to the full pass-band of the channel for a limited time, after which the access right is assigned to another user. Abbreviated TDMA. { ¦tīm də,vizh·ən ,məl·tə·pəl 'ak,ses }

time-division multiplexing [COMMUN] A process for transmitting two or more signals over a common path by using successive time intervals for different signals. Also known as time multiplexing. Abbreviated TDM. [COMPUT SCI] The interleaving of bits or characters in time to compensate for the slowness of input devices as compared to data transmission lines. { 'tīm di ¦vizh·ən ,məl·tə,pleks·iŋ }

time-division multiplier *See* mark-space multiplier. { 'tīm di¦vizh·ən ,məl·tə,plī·ər }

time-division switching system [ELECTR] A type of electronic switching system in which input signals on lines and trunks are sampled periodically, and each active input is associated with the desired output for a specific phase of the period. { 'tīm di¦vizh·ən 'swich·iŋ ,sis·təm }

time-domain reflectometer [ELECTR] An instrument that measures the electrical characteristics of wideband transmission systems, subassemblies, components, and lines by feeding in a voltage step and displaying the superimposed reflected signals on an oscilloscope equipped with a suitable time-base generator. Abbreviated TDR. { 'tīm də¦mān ,rē,flek'täm·əd·ər }

time factor *See* time scale. { 'tīm ,fak·tər }

time gate [ELECTR] A circuit that gives an output only during chosen time intervals. { 'tīm ,gāt }

time hopping [COMMUN] A spread spectrum technique, usually used in combination with other methods, in which the transmitted pulse occurs in a manner determined by a pseudorandom code which places the pulse in one of several possible positions per frame. { 'tīm ,häp·iŋ }

time-invariant system [CONT SYS] A system in which all quantities governing the system's behavior remain constant with time, so that the system's response to a given input does not depend on the time it is applied. { 'tīm in ,ver·ē·ənt ,sis·təm }

time-mark generator [ELECTR] A signal generator that produces highly accurate clock pulses which can be superimposed as pips on a cathode-ray screen for timing the events shown on the display. { 'tīm ¦märk ,jen·ə·rād·ər }

time modulation [COMMUN] Modulation in which the time of occurrence of a definite portion of a waveform is varied in accordance with a modulating signal. { 'tīm ,mäj·ə,lā·shən }

time multiplexing *See* multiprogramming; time-division multiplexing. { 'tīm ,məl·tə,pleks·iŋ }

time-of-day clock [COMPUT SCI] An electronic device that registers the actual time, generally accurate to 0.1 second, through a 24-hour cycle, and transmits its reading to the central processing unit of a computer upon demand. { ¦tīm əv ¦dā ,kläk }

time of delivery [COMMUN] The time at which the addressee or responsible relay agency provides a receipt for a message. { 'tīm əv di 'liv·ə·rē }

time of origin [COMMUN] The time at which a message is released for transmission. { 'tīm əv 'är·ə·jən }

time of receipt [COMMUN] The time at which a receiving station completes reception of a message. { 'tīm əv ri'sēt }

time quantum *See* time slice. { 'tīm ,kwän·təm }

timer [COMPUT SCI] A hardware device that can interrupt a computer program after a time interval specified by the program, generally to remind the program to take some action. [ELECTR] A circuit used in radar and in electronic navigation systems to start pulse transmission and synchronize it with other actions, such as the start of a cathode-ray sweep. { 'tīm·ər }

timer clock [COMPUT SCI] An electronic device in the central processing unit of a computer which times events that occur during the operation of the system in order to carry out such functions as changing computer time, detecting looping and similar error conditions, and keeping a log of operations. { 'tī·mər ,kläk }

time redundancy [COMPUT SCI] Performing a computation more than once and checking the results in order to increase reliability. { 'tīm ri ,dən·dən·sē }

time-reversal reflection *See* optical phase conjugation. { 'tīm ri¦vər·səl ri,flek·shən }

time scale [COMPUT SCI] The ratio of the time duration of an event as simulated by an analog computer to the actual time duration of the event in the physical system under study. Also known as time factor. { 'tīm ,skāl }

time-share [COMPUT SCI] To perform several independent processes almost simultaneously by interleaving the operations of the processes on a single high-speed processor. { 'tīm ,sher }

time-sharing [COMPUT SCI] The simultaneous utilization of a computer system from multiple terminals. { 'tīm ˌsher·iŋ }

time signal [COMMUN] An accurate signal which is broadcast by radio and marks a specified time or time interval, used for setting timepieces and for determining their errors; in particular, a radio signal broadcast at accurately known times each day on a number of different frequencies by WWV and other stations. { 'tīm ˌsig·nəl }

time signal service [COMMUN] Radio communications service for the transmission of time signals of stated high precision, intended for general reception. { 'tīmd 'sig·nəl ˌsər·vəs }

time slice [COMPUT SCI] A time interval during which a time-sharing system is processing one particular computer program. Also known as time quantum. { 'tīm ˌslīs }

time-stamp [COMMUN] A term that indicates the time of a specific action such as the arrival of a byte or the presentation of a presentation unit. { 'tīm ˌstamp }

time-varying system [CONT SYS] A system in which certain quantities governing the system's behavior change with time, so that the system will respond differently to the same input at different times. { 'tīm ¦ver·ē·iŋ ˌsis·təm }

timing-axis oscillator *See* sweep generator. { 'tīm·iŋ ˌak·səs ˌäs·əˌlād·ər }

timing error [COMPUT SCI] An error made in planning or writing a computer program, usually in underestimating the time that will be taken by input/output or other operations, which causes unnecessary delays in the execution of the program. { 'tīm·iŋ ˌer·ər }

timing loop [COMPUT SCI] A set of instructions in a computer program whose execution time is known and whose only function is to cause a delay in processing by causing the loop to be executed an appropriate number of times. { 'tīm·iŋ ˌlüp }

timing signal [COMPUT SCI] A pulse generated by the clock of a digital computer to provide synchronization of its activities. [ELECTR] Any signal recorded simultaneously with data on magnetic tape for use in identifying the exact time of each recorded event. { 'tīm·iŋ ˌsig·nəl }

title bar [COMPUT SCI] An area at the top of a window that contains the name of the file or application in the window. { 'tīd·əl ˌbär }

T junction [ELECTR] A network of waveguides with three waveguide terminals arranged in the form of a letter T; in a rectangular waveguide a symmetrical T junction is arranged by having either all three broadsides in one plane or two broadsides in one plane and the third in a perpendicular plane. { 'tē ˌjəŋk·shən }

T²L *See* transistor-transistor logic.

TM mode *See* transverse magnetic mode. { ¦tē 'em ˌmōd }

TM wave *See* transverse magnetic wave. { ¦tē 'em ˌwāv }

toggle [COMPUT SCI] **1.** To switch back and forth between two stable states or modes of operation. **2.** A hardware or software device that carries out this switching action. [ELECTR] To switch over to an alternate state, as in a flip-flop. { 'täg·əl }

toggle condition [ELECTR] Condition of a flip-flop circuit in which the internal state of the flip-flop changes from 0 to 1 or from 1 to 0. { 'täg·əl kənˌdish·ən }

toggle switch [ELECTR] An electronically operated circuit that holds either of two states until changed. { 'täg·əl ˌswich }

token [COMMUN] A unique grouping of bits that is transmitted as a unit in a communications network and used as a signal to notify stations in the network when they have control and are free to send information or take other specified actions. [COMPUT SCI] **1.** A distinguishable unit in a sequence of characters. **2.** A single byte that is used to represent a keyword in a programming language in order to conserve storage space. **3.** A physical object, such as a badge or identity card, issued to authorized users of a computing system, building, or area. { 'tō·kən }

tokenization [COMPUT SCI] The conversion of keywords of a programming language to tokens in order to conserve storage space. { ˌtō·kən·ə'zā·shən }

token-passing protocol [COMMUN] The assignment of data communications channels to units which communicate according to a fixed priority sequence. { 'tō·kən ¦pas·iŋ 'prōd·əˌkȯl }

token-sharing network [COMMUN] A communications network in which all the stations are linked to a common bus and control is determined by a group of bits (token) that is passed along the bus from station to station. { 'tō·kən ¦sher·iŋ 'netˌwərk }

toll [COMMUN] **1.** Charge made for a connection beyond an exchange boundary. **2.** Any part of telephone plant, circuits, or services for which toll charges are made. { tōl }

toll call [COMMUN] Telephone call to points beyond the area within which telephone calls are covered by a flat monthly rate or are charged for on a message unit basis. { 'tōl ˌkȯl }

toll center [COMMUN] A telephone central office where trunks from end offices are joined to the long-distance system, and operators are present; it is a class-4 office. { 'tōl ˌsen·tər }

toll line [COMMUN] A telephone line or channel that connects different telephone exchanges. { 'tōl ˌlīn }

toll office [COMMUN] A telephone central office which serves mainly to terminate and interconnect toll lines and various types of trunks. { 'tōl ˌȯf·əs }

toll terminal loss [COMMUN] The part of the overall transmission loss on a toll connection that is attributable to the facilities from the toll center through the tributary office, to and including the subscriber's equipment. { 'tōl 'tər·mən·əl ˌlȯs }

tone dialing *See* push-button dialing. { 'tōn ˌdīl·iŋ }

tone generator [ELECTR] A signal generator used to generate an audio-frequency signal suitable for signaling purposes or for testing audio-frequency equipment. { 'tōn ˌjen·əˌrād·ər }

T1 line [COMMUN] High-speed digital connection that transmits data at 1.5 million bits per second through the telephone-switching network. { ˌtēˈwən ˌlīn }

tone-modulated waves [COMMUN] Waves obtained from continuous waves by amplitude-modulating them at audio frequency in a substantially periodic manner. { ˈtōn ¦mäj·əˌlād·əd ˌwāvz }

tone modulation [COMMUN] Type of code-signal transmission obtained by causing the radio-frequency carrier amplitude to vary at a fixed audio frequency. { ˈtōn ˌmäj·əˌlā·shən }

tone-only pager [COMMUN] A receiver in a radio paging system that alerts the user to call a specific telephone number. { ˈtōn ¦on·lē ˈpāj·ər }

tone-operated net-loss adjuster [COMMUN] System for stabilizing the net loss of a telephone circuit by a tone transmitted between conversations. { ˈtōn ¦äp·əˌrād·əd ˈnet ¦lȯs əˌjəs·tər }

tone reversal [COMMUN] Distortion of the recorder copy in facsimile which causes the various shades of black and white not to be in the proper order. { ˈtōn riˌvər·səl }

toolbar [COMPUT SCI] A row or column of on-screen push buttons containing icons that represent frequently accessed commands. { ˈtülˌbär }

top-down analysis [COMPUT SCI] A predictive method of syntactic analysis which, starting from the root symbol, attempts to predict the means by which a string was generated. { ¦täp ¦dau̇n əˈnal·ə·səs }

top-loaded vertical antenna [ELECTROMAG] Vertical antenna constructed so that, because of its greater size at the top, there results modified current distribution, giving a more desirable radiation pattern in the vertical plane. { ¦täp ¦lōd·əd ˈvərd·ə·kəl anˈten·ə }

topology [COMPUT SCI] The physical or logical arrangement of the stations (nodes) in a communications network. { təˈpäl·ə·jē }

torque amplifier [COMPUT SCI] An analog computer device having input and output shafts and supplying work to rotate the output shaft in positional correspondence with the input shaft without imposing any significant torque on the input shaft. { ˈtȯrk ˌam·pləˌfī·ər }

torsional mode delay line [COMPUT SCI] A device in which torsional vibrations are propagated through a solid material to make use of the propagation time of the vibrations to obtain a time delay for the signals. { ˈtȯr·shən·əl ¦mōd diˈlāˌlīn }

TOS *See* tape operating system.

total deadlock [COMPUT SCI] A deadlock that involves all the tasks in a multiprogramming system. { ˈtōd·əl ˈdedˌläk }

total harmonic distortion [ELECTR] Ratio of the power at the fundamental frequency, measured at the output of the transmission system considered, to the power of all harmonics observed at the output of the system because of its nonlinearity, when a single frequency signal of specified power is applied to the input of the system; it is expressed in decibels. { ˈtōd·əl härˈmän·ik diˈstȯr·shən }

touch call *See* push-button dialing. { ˈtəch ˌkȯl }

touchpad [COMPUT SCI] A small, touch-sensitive pad that enables the user to move the pointer on the display screen of a personal computer by moving a finger or other object along the pad, and to click by tapping the pad. { ˈtəch·ˌpad }

touch screen [COMPUT SCI] An electronic display that allows a user to send signals to a computer by touching an area on the display with a finger, pencil, or other object. { ˈtəch ˌskrēn }

tour *See* shift. { tu̇r }

tower [ELECTROMAG] A tall metal structure used as a transmitting antenna, or used with another such structure to support a transmitting antenna wire. { tau̇·ər }

tower case [COMPUT SCI] A system unit that stands in a vertical position. { ˈtau̇·ər ˌkās }

tower radiator [ELECTROMAG] Metal structure used as a transmitting antenna. { ˈtau̇·ər ¦rād·ē ˌād·ər }

Tow-Thomas filter [ELECTR] A multiple-amplifier active filter that has the advantage of ease of design but the disadvantage of lacking a high-pass output in its basic configuration. { ¦tō ˈtäm·əs ˌfil·tər }

TP monitor *See* teleprocessing monitor. { ¦tē ˈpē ˌmän·əd·ər }

trace [COMPUT SCI] To provide a record of every step, or selected steps, executed by a computer program, and by extension, the record produced by this operation. [ELECTR] The visible path of a moving spot on the screen of a cathode-ray tube. Also known as line. { trās }

trace routine [COMPUT SCI] A routine which tracks the execution of a program, step by step, to locate a program malfunction. Also known as tracing routine. { ˈtrās rüˌtēn }

trace statement [COMPUT SCI] A statement, included in certain programming languages, that causes certain error-checking procedures to be carried out on specified segments of a source program. { ˈtrās ˌstāt·mənt }

tracing routine *See* trace routine. { ˈtrās·iŋ rü ˌtēn }

track [ELECTR] **1.** A path for recording one channel of information on a magnetic tape, drum, or other magnetic recording medium; the location of the track is determined by the recording equipment rather than by the medium. **2.** The trace on a plan-position indicator or similar display resulting from the association of successive detections presumed to be from the same moving target; or the same information from an appropriate radar data processor. { trak }

trackball [COMPUT SCI] A ball inset in the console of a video display terminal, the keyboard of a personal computer, or a small box-shaped holder, which can be rotated by the operator, and whose motion is followed by a cursor on the display screen. { ˈtrakˌbȯl }

tracker [COMPUT SCI] An input device used in a virtual environment, which is capable of reporting its location in space and its orientation. { 'trak·ər }

track filtering [ELECTR] In radar data processing, the treatment of each subsequent measurement of a target's position, generally by weighting factors, to reduce the effects of measurement error, resulting in a "smoothing" of the track. { 'trak ,fil·tər·iŋ }

tracking [ELECTR] The condition in which all tuned circuits in a receiver accurately follow the frequency indicated by the tuning dial over the entire tuning range. [ENG] A motion given to the major lobe of a radar or radio antenna such that some preassigned moving target in space is always within the major lobe. { 'trak·iŋ }

Tracking and Data Relay Satellite System [COMMUN] A system providing telecommunication services between low-earth-orbiting user spacecraft and user control centers; it consists of a series of geostationary spacecraft and an earth terminal located at White Sands, New Mexico. Abbreviated TDRSS. { ¦trak·iŋ ən ¦dad·ə ¦rē,lā 'sad·ə·,līt ,sis·təm }

tracking cross [COMPUT SCI] A cross displayed on the screen of a video terminal which automatically follows a light pen. Also known as tracking cursor. { 'trak·iŋ ,krȯs }

tracking cursor *See* tracking cross. { 'trak·iŋ ,kər·sər }

tracking filter [ELECTR] Electronic device for attenuating unwanted signals while passing desired signals, by phase-lock techniques that reduce the effective bandwidth of the circuit and eliminate amplitude variations. { 'trak·iŋ ,fil·tər }

tracking network [ENG] A group of tracking stations whose operations are coordinated in tracking objects through the atmosphere or space. { 'trak·iŋ ,net,wərk }

tracking problem [CONT SYS] The problem of determining a control law which when applied to a dynamical system causes its output to track a given function; the performance index is in many cases taken to be of the integral square error variety. { 'trak·iŋ ,präb·ləm }

tracking radar [ENG] Radar used to monitor the flight and obtain geophysical data from space probes, satellites, and high-altitude rockets. { 'trak·iŋ ,rā,där }

tracking station [ENG] A radio, radar, or other station set up to track an object moving through the atmosphere or space. { 'trak·iŋ ,stā·shən }

tracking system [ENG] Apparatus, such as tracking radar, used in following and recording the position of objects in the sky. { 'trak·iŋ ,sis·təm }

track in range [ELECTR] To adjust the gate of a radar set so that it opens at the correct instant to accept the signal from a target of changing range from the radar. { 'trak in 'rānj }

track pitch [ELECTR] The physical distance between track centers. { 'trak ,pich }

track telling [COMMUN] The process of communicating air surveillance and tactical data information between command and control systems and facilities within the systems. { 'trak ,tel·iŋ }

track-to-track access time [COMPUT SCI] The time required for a read-write head to move between the adjacent cylinders of a disk. { ¦trak tə ¦trak 'ak,ses ,tīm }

track-while-scan [ELECTR] Radar operation used to detect a radar target, compute its velocity, and predict its future position without interfering with continuous radar scanning. { ¦trak ,wīl 'skan }

tractor-feed printer *See* pin-feed printer. { 'trak·tər ¦fēd 'print·ər }

traffic [COMMUN] The messages transmitted and received over a communication channel. { 'traf·ik }

traffic diagram [COMMUN] Chart or illustration used to show the movement and control of traffic over a communications system. { 'traf·ik ,dī·ə ,gram }

traffic distribution [COMMUN] Routing of communications traffic through a terminal to a switchboard or dialing center. { 'traf·ik ,di·strə ,byü·shən }

traffic flow security [COMMUN] Transmission of an uninterrupted flow of random text on a wire or radio link between two stations with no indication to an interceptor of what portions of this steady stream constitute encrypted message text and what portions are merely random filler. { 'traf·ik ¦flō si,kyůr·əd·ē }

traffic forecast [COMMUN] Traffic level prediction on which communications system management decisions and engineering effort are based. { 'traf·ik ,fȯr,kast }

trailer [ELECTR] A bright streak at the right of a dark area or dark line in an analog television picture, or a dark area or streak at the right of a bright part; usually due to insufficient gain at low video frequencies. { 'trā·lər }

trailer label [COMPUT SCI] A record appearing at the end of a magnetic tape that uniquely identifies the tape as one required by the system. { 'trā·lər ,lā·bəl }

trailer record [COMPUT SCI] A record which contains data pertaining to an associated group of records immediately preceding it. { 'trā·lər ,rek·ərd }

trailing antenna [ELECTROMAG] An aircraft radio antenna having one end weighted and trailing free from the aircraft when in flight. { 'trāl·iŋ an ¦ten·ə }

trailing pad [COMPUT SCI] Characters placed to the right of information in a field of data to fulfill length requirements or for cosmetic purposes. { 'trāl·iŋ ,pad }

trainer [ELECTR] A piece of equipment used for training operators of radar, sonar, and other electronic equipment by simulating signals received under operating conditions in the field. { 'trā·nər }

training time [COMPUT SCI] The machine time expended in training employees in the use of the equipment, including such activities as mounting, console operation, converter operation, and printing operation, and time spent in conducting required demonstrations. { 'trān·iŋ ¦tīm }

train printer [COMPUT SCI] A computer printer in which the characters are carried in a track and a hammer strikes the proper character against the paper as it passes the print position. { 'trān ˌprint·ər }

transacter [COMPUT SCI] A system in which data from sources in a number of different locations, as in a factory, are transmitted to a data-processing center and immediately processed by a computer. { tran'sak·tər }

transaction [COMPUT SCI] General description of updating data relevant to any item. { tran'sak·shən }

transaction data [COMPUT SCI] A set of data in a data-processing area in which the incidence of the data is essentially random and unpredictable; hours worked, quantities shipped, and amounts invoiced are examples from, respectively, the areas of payroll, accounts receivable, and accounts payable. { tran'sak·shən ˌdad·ə }

transaction file *See* detail file. { tran'sak·shən ˌfīl }

transaction processing system [COMPUT SCI] A system which processes predefined transactions, one at a time, with direct, on-site entry of the transactions into a terminal, and which produces predefined outputs and maintains the necessary data base. { tran'sak·shən 'präˌses·iŋ ˌsis·təm }

transaction record *See* change record. { tran 'sak·shən ˌrek·ərd }

transaction tape *See* change tape. { tran 'sak·shən ˌtāp }

transceiver [COMPUT SCI] A computer terminal that can transmit and receive information to and from an input/output channel. [ELECTR] A radio transmitter and receiver combined in one unit and having switching arrangements such as to permit both transmitting and receiving. Also known as transmitter-receiver. { tran'sē·vər }

transconductance amplifier [ELECTR] An amplifier whose output current (rather than output voltage) is proportional to its input voltage. { ˌtranz·kənˌduk·təns 'am·pləˌfī·ər }

transconductance-C filter [ELECTR] An integrated-circuit filter that combines the functions of an amplifier and a simulated resistor into a transconductance amplifier. { 'tranz·kən ˌduk·təns 'sē ˌfil·tər }

transconductor *See* transconductance amplifier. { ˌtranz·kən'dək·tər }

transcriber [COMPUT SCI] The equipment used to convert information from one form to another, as for converting computer input data to the medium and language used by the computer. { tranz'krī·bər }

transducer [ENG] Any device or element which converts an input signal into an output signal of a different form; examples include the microphone, loudspeaker, barometer, photoelectric cell, automobile horn, doorbell, and underwater sound transducer. { tranz'dü·sər }

transducer loss [ELECTR] The ratio of the power available to a transducer from a specified source to the power that the transducer delivers to a specified load; usually expressed in decibels. { tranz'dü·sər ˌlȯs }

transfer *See* jump. { 'tranz·fər }

transfer characteristic [ELECTR] **1.** Relation, usually shown by a graph, between the voltage of one electrode and the current to another electrode, with all other electrode voltages being maintained constant. **2.** Function which, multiplied by an input magnitude, will give a resulting output magnitude. **3.** Relation between the illumination on a camera tube and the corresponding output-signal current, under specified conditions of illumination. { 'tranz·fər ˌkar·ik·təˌris·tik }

transfer check [COMPUT SCI] Check (usually automatic) on the accuracy of the transfer of a word in a computer operation. { 'tranz·fər ˌchek }

transfer conditionally [COMPUT SCI] To copy, exchange, read, record, store, transmit, or write data or to change control or jump to another location according to a certain specified rule or in accordance with a certain criterion. { tranz'fər kən'dish·ən·ə·lē }

Transfer Control Protocol *See* Transmission Control Protocol. { ˌtranz·fər kən'trōl ˌprōd·ə ˌkȯl }

transfer function [CONT SYS] The mathematical relationship between the output of a control system and its input: for a linear system, it is the Laplace transform of the output divided by the Laplace transform of the input under conditions of zero initial-energy storage. { 'tranz·fər ˌfəŋk·shən }

transfer-in-channel command [COMPUT SCI] A command used to direct channel control to a specified location in main storage when the next channel command word is not stored in the next location in sequence. { ¦tranz·fər in 'chan·əl kə ˌmand }

transfer instruction [COMPUT SCI] Step in computer operation specifying the next operation to be performed, which is not necessarily the next instruction in sequence. { 'tranz·fər in ˌstrək·shən }

transfer interpreter [COMPUT SCI] A variation of a punched-card interpreter that senses a punched card and prints the punched information on the following card. Also known as posting interpreter. { 'tranz·fər inˌtər·prəd·ər }

transfer matrix [CONT SYS] The generalization of the concept of a transfer function to a multivariable system; it is the matrix whose product with the vector representing the input variables yields the vector representing the output variables. { 'tranz·fər ˌmā·triks }

transfer operation [COMPUT SCI] An operation which moves information from one storage location or one storage medium to another (for example, read, record, copy, transmit, exchange). { 'tranz·fər ˌäp·əˌrā·shən }

transfer rate [COMPUT SCI] The speed at which data are moved from a direct-access device to a central processing unit. { 'tranz·fər ˌrāt }

transferred-electron amplifier [ELECTR] A diode amplifier, which generally uses a transferred-electron diode made from doped *n*-type gallium arsenide, that provides amplification in the gigahertz range to well over 50 gigahertz at power outputs typically below 1 watt continuous-wave. Abbreviated TEA. { 'tranz'fərd i¦lekˌträn 'am·pləˌfī·ər }

transferred-electron device [ELECTR] A semiconductor device, usually a diode, that depends on internal negative resistance caused by transferred electrons in gallium arsenide or indium phosphide at high electric fields; transit time is minimized, permitting oscillation at frequencies up to several hundred megahertz. { 'tranz'fərd i¦lekˌträn di'vīs }

transfer test [COMMUN] Verification of transmitted information by temporary storing, retransmitting, and comparing. { 'tranz·fər ˌtest }

transform [COMPUT SCI] To change the form of digital-computer information without significantly altering its meaning. { tranz'fȯrm }

transformation matrix [ELECTROMAG] A two-by-two matrix which relates the amplitudes of the traveling waves on one side of a waveguide junction to those on the other. { ˌtranz·fər'mā·shən ˌmā·triks }

transformer [ELECTROMAG] An electrical component consisting of two or more multiturn coils of wire placed in close proximity to cause the magnetic field of one to link the other; used to transfer electric energy from one or more alternating-current circuits to one or more other circuits by magnetic induction. { tranz'fȯr·mər }

transformer read-only store [COMPUT SCI] In computers, read-only store in which the presence or absence of mutual inductance between two circuits determines whether a binary 1 or 0 is stored. { tranz'fȯr·mər 'rēd ¦ōn·lē 'stȯr }

transforming section [ELECTROMAG] Length of waveguide or transmission line of modified cross section, or with a metallic or dielectric insert, used for impedance transformation. { tranz'fȯrm·iŋ ˌsek·shən }

transient [PHYS] A pulse, damped oscillation, or other temporary phenomenon occurring in a system prior to reaching a steady-state condition. { 'tranch·ənt }

transient distortion [ELECTR] Distortion due to inability to amplify transients linearly. { 'tranch·ənt diˌstȯr·shən }

transient program [COMPUT SCI] A computer program that is stored in a computer's main memory only while it is being executed. { 'tranch·ənt 'prō·grəm }

transistor [ELECTR] An active component of an electronic circuit consisting of a small block of semiconducting material to which at least three electrical contacts are made, usually two closely spaced rectifying contacts and one ohmic (nonrectifying) contact; it may be used as an amplifier, detector, or switch. { tran'zis·tər }

transistor memory *See* semiconductor memory. { tran'zis·tər ˌmem·rē }

transistor-transistor logic [ELECTR] A logic circuit containing two transistors, for driving large output capacitances at high speed. Abbreviated T^2L; TTL. { tran'zis·tər tran'zis·tər 'läj·ik }

transition [COMMUN] Change from one circuit condition to the other; for example, the change from mark to space or from space to mark. { tran'zish·ən }

transition element [ELECTROMAG] An element used to couple one type of transmission system to another, as for coupling a coaxial line to a waveguide. { tran'zish·ən ˌel·ə·mənt }

transition function [COMPUT SCI] A function which determines the next state of a sequential machine from the present state and the present input. { tran'zish·ən ˌfəŋk·shən }

transition point [ELECTROMAG] A point at which the constants of a circuit change in such a way as to cause reflection of a wave being propagated along the circuit. { tran'zish·ən ˌpȯint }

transit time [ELECTR] The time required for an electron or other charge carrier to travel between two electrodes in an electron tube or transistor. { 'trans·ət ˌtīm }

transit-time microwave diode [ELECTR] A solid-state microwave diode in which the transit time of charge carriers is short enough to permit operation in microwave bands. { 'trans·ət ˌtīm 'mī·krəˌwāv 'dīˌōd }

transit-time mode [ELECTR] A mode of operation of a Gunn diode in which a charge dipole, consisting of an electron accumulation and a depletion layer, travels through the semiconductor at a frequency dependent on the length of the semiconductor layer and the drift velocity. { 'trans·ət ˌtīm ˌmōd }

translate [COMPUT SCI] To convert computer information from one language to another, or to convert characters from one representation set to another, and by extension, the computer instruction which directs the latter conversion to be carried out. { tran'slāt }

translating circuit *See* translator. { tran'slād·iŋ ˌsər·kət }

translation algorithm [COMPUT SCI] A specific, effective, essentially computational method for obtaining a translation from one language to another. { tran'slā·shən 'al·gəˌri<u>th</u>·əm }

translator [COMPUT SCI] A computer network or system having a number of inputs and outputs, so connected that when signals representing information expressed in a certain code are applied to the inputs, the output signals will represent the same information in a different code. Also known as translating circuit. [ELECTR] A combination television receiver and low-power television transmitter, used to pick up television signals on one frequency and retransmit them on another frequency to provide reception in areas not served directly by television stations. { tran'slād·ər }

translator routine [COMPUT SCI] A program which accepts statements in one language and outputs them as statements in another language. { tran'slăd·ər rü,tēn }

transliterate [COMPUT SCI] To represent the characters or words of one language by corresponding characters or words of another language. { tran'slid·ə,rāt }

transmission [ELECTR] **1.** The process of transferring a signal, message, picture, or other form of intelligence from one location to another location by means of wire lines, radio waves, light beams, infrared beams, or other communication systems. **2.** A message, signal, or other form of intelligence that is being transmitted. **3.** *See* transmittance. { tranz'mish·ən }

transmission band [ELECTROMAG] Frequency range above the cutoff frequency in a waveguide, or the comparable useful frequency range for any other transmission line, system, or device. { tranz'mish·ən ,band }

transmission control character [COMMUN] A character included in a message to control its routing to the intended destination. { tranz'mish·ən kən'trōl ,kar·ik·tər }

Transmission Control Protocol [COMMUN] The set of standards that is responsible for breaking down and reassembling the data packets transmitted on the Internet, for ensuring complete delivery of the packets and for controlling data flow. Abbreviated TCP. { tranz,mish·ən kən'trōl ,prōd·ə,kȯl }

Transmission Control Protocol/Internet Protocol [COMPUT SCI] The Internet's principal communication standard, dictating how packets of information are sent and received across multiple networks. TCP breaks down and reassembles packets, and IP ensures that the packets are sent to the correct destination. Abbreviated TCP/IP. { tranz,mish·ən kən'trōl ,prōd·ə,kȯl 'in·tər,net ,prōd·ə,kȯl }

transmission facilities [COMMUN] All equipment and the medium required to transmit a message. { tranz'mish·ən fə,sil·əd·ēz }

transmission gain *See* gain. { tranz'mish·ən ,gān }

transmission gate [ELECTR] A gate circuit that delivers an output waveform that is a replica of a selected input during a specific time interval which is determined by a control signal. { tranz'mish·ən ,gāt }

transmission interface converter [COMPUT SCI] A device that converts data to or from a form suitable for transfer over a channel connecting two computer systems or connecting a computer with its associated data terminals. { tranz'mish·ən 'in·tər,fās kən,vərd·ər }

transmission level [COMMUN] The ratio of the signal power at any point in a transmission system to the signal power at some point in the system chosen as a reference point; usually expressed in decibels. { tranz'mish·ən ,lev·əl }

transmission loss [COMMUN] **1.** The ratio of the power at one point in a transmission system to the power at a point farther along the line; usually expressed in decibels. **2.** The actual power that is lost in transmitting a signal from one point to another through a medium or along a line. Also known as loss. { tranz'mish·ən ,lȯs }

transmission mode *See* mode. { tranz'mish·ən ,mōd }

transmission primaries [COMMUN] The set of three color primaries that correspond to the three independent signals contained in the color signal. { tranz'mish·ən 'prī,mer·ēz }

transmission regulator [ELECTR] In electrical communications, a device that maintains substantially constant transmission levels over a system. { tranz'mish·ən ,reg·yə,lād·ər }

transmission security [COMMUN] Component of communications security which results from all measures designed to protect transmissions from unauthorized interception, traffic analysis, and imitative deception. { tranz'mish·ən si ,kyu̇r·əd·ē }

transmission speed [COMMUN] The number of information elements sent per unit time; usually expressed as bits, characters, bands, word groups, or records per second or per minute. { tranz'mish·ən ,spēd }

transmission time [COMMUN] Absolute time interval from transmission to reception of a signal. { tranz'mish·ən ,tīm }

transmissivity [ELECTROMAG] The ratio of the transmitted radiation to the radiation arriving perpendicular to the boundary between two mediums. { ,tranz·mə'siv·əd·ē }

transmit [COMMUN] To send a message, program, or other information to a person or place by wire, radio, or other means. [COMPUT SCI] To move data from one location to another. { tranz'mit }

transmit-receive module [ELECTR] Microwave circuitry providing signal amplification on transmit, elemental duplexing, receiver functions, and phase control, usually featuring solid-state devices and compact packaging, for use at every element of a phased array radar antenna, the entire assembly constituting an "active" phase array. { tranz'mit ri'sēv ¦mäj·ül }

transmittability [COMMUN] The ability of standard electronic and mechanical elements and automatic communications equipment to handle a code under various signal-to-noise ratios; for example, a code with a variable number of elements such as Morse presents technical problems in automatic interpretation not encountered in a fixed-length code. { tranz ,mid·ə'bil·əd·ē }

transmittance [ELECTROMAG] The radiant power transmitted by a body divided by the total radiant power incident upon the body. Also known as transmission. { tranz'mid·əns }

transmitted-carrier operation [COMMUN] Form of amplitude-modulated carrier transmission in which the carrier wave is transmitted. { tranz'mid·əd ¦kar·ē·ər ,äp·ə,rā·shən }

transmitter [COMMUN] **1.** In telephony, the microphone that converts sound waves into audio-frequency signals. **2.** *See* radio transmitter. { tranz'mid·ər }

transmitter off [COMMUN] A signal sent by a receiving device to a transmitter, directing it to stop sending information if it is doing so, or not to send information if it is preparing to do so. Abbreviated XOFF. { tranz'mid·ər 'ȯf }

transmitter on [COMMUN] A signal sent by a receiving device to a transmitter, directing it to transmit any information it has to send. Abbreviated XON. { tranz'mid·ər 'ȯn }

transmitter-receiver *See* transceiver. { tranz 'mid·ər ri'sē·vər }

transmitting loop loss [COMMUN] That part of the repetition equivalent assignable to the station set, subscriber line, and battery supply circuit which is on the transmitting end. { tranz'mid·iŋ ¦lüp ˌlȯs }

transmitting mode [COMPUT SCI] Condition of an input/output device, such as a magnetic tape when it is actually reading or writing. { tranz'mid·iŋ ˌmōd }

transparent [COMPUT SCI] Pertaining to a device or system that processes data without the user being aware of or needing to understand its operation. { tranz'par·ənt }

transponder [COMMUN] **1.** A transmitter-receiver capable of accepting the challenge of an interrogator and automatically transmitting an appropriate reply. **2.** A receiver-transmitter, such as on satellites, which receives a transmission and retransmits it at another radio frequency. { tranz'pän·dər }

transponder dead time [ELECTR] Time interval between the start of a pulse and the earliest instant at which a new pulse can be received or produced by a transponder. { tranz'pän·dər 'ded ˌtīm }

transponder set [ELECTR] A complete electronic set which is designed to receive an interrogation signal, and which retransmits coded signals that can be interpreted by the interrogating station; it may also utilize the received signal for actuation of additional equipment such as local indicators or servo amplifiers. { tranz'pän·dər ˌset }

transponder suppressed time delay [ELECTR] Overall fixed time delay between reception of an interrogation and transmission of a reply to this interrogation. { tranz'pän·dər sə'prest 'tīm di ˌlā }

transport [COMPUT SCI] **1.** To convey as a whole from one storage device to another in a digital computer. **2.** *See* tape transport. { trans'pȯrt (verb), 'tranzˌpȯrt (noun) }

transportable computer [COMPUT SCI] A microcomputer that can be carried about conveniently but, in contrast to a portable computer, requires an external power source. { tranz'pȯrd·ə·bəl kəm'pyüd·ər }

transport delay unit [COMPUT SCI] A device used in analog computers which produces an output signal as a delayed form of an input signal. Also known as delay unit; transport unit. { 'tranzˌpȯrt di'lā ˌyü·nət }

transport unit *See* transport delay unit. { 'tranz ˌpȯrt ˌyü·nət }

transposition [COMMUN] Interchanging the relative positions of conductors at regular intervals along a transmission line to reduce cross talk. { ˌtranz·pə'zish·ən }

transposition cipher [COMMUN] A cipher in which the order of the characters in the original message is changed. { ˌtranz·pə'zish·ən ˌsī·fər }

transradar [COMMUN] Bandwidth compression system developed for long-range narrow-band transmission of radio signals from a radar receiver to a remote location. { 'tränz¦rāˌdär }

transresistance amplifier [ELECTR] An amplifier whose output voltage is proportional to its input current. { ˌtranz·riˌzis·təns 'am·pləˌfī·ər }

transverse electric mode [ELECTROMAG] A mode in which a particular transverse electric wave is propagated in a waveguide or cavity. Abbreviated TE mode. Also known as H mode (British usage). { trans¦vərs i¦lek·trik ˌmōd }

transverse electric wave [ELECTROMAG] An electromagnetic wave in which the electric field vector is everywhere perpendicular to the direction of propagation. Abbreviated TE wave. Also known as H wave (British usage). { trans ¦vərs i'lek·trik 'wāv }

transverse electromagnetic mode [ELECTROMAG] A mode in which a particular transverse electromagnetic wave is propagated in a waveguide or cavity. Abbreviated TEM mode. { trans ¦vərs i¦lek·trō·mag'ned·ik 'mōd }

transverse electromagnetic wave [ELECTROMAG] An electromagnetic wave in which both the electric and magnetic field vectors are everywhere perpendicular to the direction of propagation. Abbreviated TEM wave. { trans¦vərs i¦lek·trō·mag'ned·ik 'wāv }

transverse magnetic mode [ELECTROMAG] A mode in which a particular transverse magnetic wave is propagated in a waveguide or cavity. Abbreviated TM mode. Also known as E mode (British usage). { trans¦vərs mag'ned·ik 'mōd }

transverse magnetic wave [ELECTROMAG] An electromagnetic wave in which the magnetic field vector is everywhere perpendicular to the direction of propagation. Abbreviated TM wave. Also known as E wave (British usage). { trans ¦vərs mag'ned·ik 'mōd 'wāv }

transverse recording [ELECTR] Technique for recording video signals on magnetic tape using a four-transducer rotating head. { trans¦vərs ri'kȯrd·iŋ }

trap [COMPUT SCI] An automatic transfer of control of a computer to a known location, this transfer occurring when a specified condition is detected by hardware. [ELECTR] **1.** A tuned circuit used in the radio-frequency or intermediate-frequency section of a receiver to reject undesired frequencies; traps in analog television receiver video circuits keep the sound signal out of the picture channel. Also known as rejector. **2.** *See* wave trap. { trap }

trap address [COMPUT SCI] The location at which control is transferred in case of an interrupt as soon as the current instruction is completed. { 'trap 'adˌres }

TRAPATT diode [ELECTR] A *pn* junction diode, similar to the IMPATT diode, but characterized by the formation of a trapped space-charge plasma within the junction region; used in the generation and amplification of microwave power. Derived from trapped plasma avalanche transit time diode. { 'tra,pat ,dī,ōd }

trapezium distortion [ELECTR] A defect in a cathode-ray tube in which the trace is confined within a trapezium rather than a rectangle, usually as a result of interaction between the two pairs of deflection plates. { trə'pē·zē·əm di ,stȯr·shən }

trapezoidal pulse [ELECTR] An electrical pulse in which the voltage rises linearly to some value, remains constant at this value for some time, and then drops linearly to the original value. { ¦trap·ə¦zȯid·əl 'pəls }

trapezoidal wave [ELECTR] A wave consisting of a series of trapezoidal pulses. { ¦trap·ə¦zȯid·əl 'wāv }

trapped plasma avalanche transit time diode *See* TRAPATT diode. { 'trapt 'plaz·mə 'av·ə,lanch 'trans·ət ,tīm 'dī,ōd }

trapping *See* guided propagation. { 'trap·iŋ }

trapping mode [COMPUT SCI] A procedure by means of which the computer, upon encountering a predetermined set of conditions, saves the program in its present status, executes a diagnostic procedure, and then resumes the processing of the program as of the moment of interruption. { 'trap·iŋ ,mōd }

trash heap [COMPUT SCI] An area in a computer's memory that has been assigned to a program but contains data which are no longer useful and are therefore wasteful of storage space. { 'trash ,hēp }

traveling-wave amplifier [ELECTR] An amplifier that uses one or more traveling-wave tubes to provide useful amplification of signals at frequencies of the order of thousands of megahertz. Also known as traveling-wave-tube amplifier (TWTA). { 'trav·əl·iŋ ¦wāv 'am·plə,fī·ər }

traveling-wave antenna [ELECTROMAG] An antenna in which the current distributions are produced by waves of charges propagated in only one direction in the conductors. Also known as progressive-wave antenna. { 'trav·əl·iŋ ¦wāv an'ten·ə }

traveling-wave tube [ELECTR] An electron tube in which a stream of electrons interacts continuously or repeatedly with a guided electromagnetic wave moving substantially in synchronism with it, in such a way that there is a net transfer of energy from the stream to the wave; the tube is used as an amplifier or oscillator at frequencies in the microwave region. { 'trav·əl·iŋ ¦wāv ,tüb }

traveling-wave-tube amplifier *See* traveling-wave amplifier. { ¦trav·əl·iŋ ,wāv ,tüb 'am·plə ,fī·ər }

tree [COMPUT SCI] A data structure in which each element may be logically followed by two or more other elements, there is one element with no predecessor, every other element has a unique predecessor, and there are no circular lists. [ELECTR] A set of connected circuit branches that includes no meshes; responds uniquely to each of the possible combinations of a number of simultaneous inputs. Also known as decoder. { trē }

tree automaton [COMPUT SCI] An automaton that processes inputs in the form of trees, usually trees associated with parsing expressions in context-free languages. { 'trē ,ȯd·ə,mā·shən }

tree diagram [COMPUT SCI] A flow diagram which has no closed paths. { 'trē ,dī·ə,gram }

tree pruning [COMPUT SCI] In computer programming, a strategy for eliminating branches of the complete game tree associated with a given position in a game such as chess or checkers, creating subtrees that explore a limited number of continuations for a limited number of moves. { 'trē ,prün·iŋ }

TRF receiver *See* tuned-radio-frequency receiver. { ,tē,är'ef ri,sē·vər }

triad [COMPUT SCI] A group of three bits, pulses, or characters forming a unit of data. [ELECTR] A triangular group of three small phosphor dots, each emitting one of the three primary colors on the screen of a three-gun color picture tube. { 'trī,ad }

triangular pulse [ELECTR] An electrical pulse in which the voltage rises linearly to some value, and immediately falls linearly to the original value. { trī'aŋ·gyə·lər 'pəls }

triangular wave [ELECTR] A wave consisting of a series of triangular pulses. { trī'aŋ·gyə·lər 'wāv }

tributary station [COMMUN] Communications terminal consisting of equipment compatible for the introduction of messages into or reception from its associated relay station. { 'trib·yə,ter·ē 'stā·shən }

trickling [COMPUT SCI] The temporary transfer of momentarily unneeded data from main storage to secondary storage devices. { 'trik·liŋ }

tricolor picture tube *See* color picture tube. { 'trī,kəl·ər 'pik·chər ,tüb }

trigger [COMPUT SCI] To execute a jump to the first instruction of a program after the program has been loaded into the computer. Also known as initiate. [ELECTR] **1.** To initiate an action, which then continues for a period of time, as by applying a pulse to a trigger circuit. **2.** The pulse used to initiate the action of a trigger circuit. **3.** *See* trigger circuit. { 'trig·ər }

trigger circuit [ELECTR] **1.** A circuit or network in which the output changes abruptly with an infinitesimal change in input at a predetermined operating point. Also known as trigger. **2.** A circuit in which an action is initiated by an input pulse, as in some radar modulators. **3.** *See* bistable multivibrator. { 'trig·ər ,sər·kət }

trimmer capacitor [ELEC] A relatively small variable capacitor used in parallel with a larger variable or fixed capacitor to permit exact adjustment of the capacitance of the parallel combination. { 'tim·ər kə,pas·əd·ər }

triple-conversion receiver [ELECTR] Communications receiver having three different intermediate frequencies to give higher adjacent-channel selectivity and greater image-frequency suppression. { 'trip·əl kən,vər·zhən ri,sē·vər }

triple detection *See* double-superheterodyne reception. { 'trip·əl di,tek·shən }

triple-length working [COMPUT SCI] Processing of data by a computer in which three machine words are used to represent each data item, in order to achieve the desired precision in the results. { 'trip·əl ¦leŋkth 'wərk·iŋ }

triple modular redundancy [COMPUT SCI] A form of redundancy in which the original computer unit is triplicated and each of the three independent units feeds into a majority voter, which outputs the majority signal. { 'trip·əl 'mäj·ə·lər ri'dən·dən·sē }

triple-stub transformer [ELECTROMAG] Microwave transformer in which three stubs are placed a quarter-wavelength apart on a coaxial line and adjusted in length to compensate for impedance mismatch. { 'trip·əl ¦stəb tranz'fȯr·mər }

triplexer [ELECTR] Dual duplexer that permits the use of two receivers simultaneously and independently in a radar system. { 'tri,plek·sər }

triplex system [COMMUN] Telegraph system in which two messages in one direction and one message in the other direction can be sent simultaneously over a single circuit. { 'tri,pleks ,sis·təm }

trip magnet *See* phase magnet. { 'trip ,mag·nət }

tristate logic [ELECTR] A form of transistor-transistor logic in which the output stages or input and output stages can assume three states; two are the normal low-impedance 1 and 0 states, and the third is a high-impedance state that allows many tristate devices to time-share bus lines. { 'trī,stāt 'läj·ik }

Trojan horse [COMPUT SCI] A computer program that has an unannounced (usually undesirable) function in addition to a desirable apparent function. { ,trō·jən 'hȯrs }

trombone [ELECTROMAG] U-shaped, adjustable, coaxial-line matching assembly. { träm'bōn }

troposcatter *See* tropospheric scatter. { 'trōp·ō ,skad·ər }

tropospheric scatter [COMMUN] Scatter propagation of radio waves caused by irregularities in the refractive index of air in the troposphere; used for long-distance communications, with the aid of relay facilities, 180–300 miles (300–500 kilometers) apart. Also known as troposcatter. { ¦trōp·ə¦sfir·ik 'skad·ər }

tropospheric wave [COMMUN] A radio wave that is propagated by reflection from a region of abrupt change in dielectric constant or its gradient in the troposphere. { ¦trōp·ə¦sfir·ik 'wāv }

trouble-location problem [COMPUT SCI] In computers, a test problem used in a diagnostic routine. { 'trəb·əl lō¦kā·shən ,präb·ləm }

troubleshoot [COMPUT SCI] To find and correct errors and faults in a computer, usually in the hardware. { 'trəb·əl,shüt }

true-motion radar presentation [ELECTR] A radar plan-position indicator presentation in which the center of the scope represents the same geographic position, until reset, with all moving objects, including the user's own craft, moving on the scope. { 'trü ¦mō·shən 'rā,där ,pres·ən,tā·shən }

truncated paraboloid [ELECTROMAG] Paraboloid antenna in which a portion of the top and bottom have been cut away to broaden the main lobe in the vertical plane. { 'trəŋ,kād·əd pə'rab·ə,lȯid }

trunk [COMPUT SCI] A path over which information is transferred in a computer. [COMMUN] A telephone line connecting two central offices. Also known as trunk circuit. { trəŋk }

trunk circuit *See* trunk. { 'trəŋk ,sər·kət }

trunk exchange [COMMUN] A telephone exchange whose main function is to interconnect trunks. { 'trəŋk iks'chānj }

trunk group [COMMUN] The collection of trunks of a given type or characteristic that connect two switching points. { 'trəŋk ,grüp }

TSR *See* RAM resident.

T3 line [COMMUN] High-speed digital connection that transmits data at 45 million bits per second through the telephone-switching network. { ,tē'thrē ,līn }

TTL *See* transistor-transistor logic.

TTY *See* teletypewriter.

tube *See* electron tube. { 'tüb }

tunable echo box [ELECTROMAG] Echo box consisting of an adjustable cavity operating in a single mode; if calibrated, the setting of the plunger at resonance will indicate the wavelength. { 'tü·nə·bəl 'ek·ō ,bäks }

Tundra orbit [AERO ENG] An inclined, elliptical, geosynchronous earth satellite orbit designed to provide communications satellite service coverage at high latitudes; it has an inclination of 63.4° degrees and an eccentricity of 0.2684, so that the satellite spends 16 hours each day over the hemisphere (northern or southern) where service coverage is intended and 8 hours over the opposite hemisphere. { 'tən·drə ,ȯr·bət }

tuned cavity *See* cavity resonator. { ¦tünd 'kav·əd·ē }

tuned circuit [ELECTR] A circuit whose components can be adjusted to make the circuit responsive to a particular frequency in a tuning range. Also known as tuning circuit. { ¦tünd 'sər·kət }

tuned-radio-frequency receiver [ELECTR] A radio receiver consisting of a number of amplifier stages that are tuned to resonance at the carrier frequency of the desired signal by a gang capacitor; the amplified signals at the original carrier frequency are fed directly into the detector for demodulation, and the resulting audio-frequency signals are amplified by an audio-frequency amplifier and reproduced by a loudspeaker. Abbreviated TRF receiver. { ¦tünd 'rād·ēo ¦frē·kwən·sē ri,sē·vər }

tuned resonating cavity [ELECTROMAG] Resonating cavity half a wavelength long or some multiple of a half wavelength, used in connection with a waveguide to produce a resultant wave with the amplitude in the cavity greatly exceeding that of the wave in the waveguide. { ¦tünd 'rez·ən,ād·iŋ ,kav·əd·ē }

tuner [ELECTR] The portion of a receiver that contains circuits which can be tuned to accept the carrier frequency of a desired transmitter while rejecting the carrier frequencies of all other stations on the air at that time. { 'tü·nər }

tuning [COMPUT SCI] The use of various techniques involving adjustments to both hardware and software to improve the operating efficiency of a computer system. [ELECTR] The process of adjusting the inductance or the capacitance (or both) in a tuned circuit, for example, in a radio, television, or radar receiver or transmitter, so as to obtain optimum performance at a selected frequency. { 'tün·iŋ }

tuning circuit *See* tuned circuit. { 'tün·iŋ ,sər·kət }

tuning core [ELECTROMAG] A ferrite core that is designed to be moved in and out of a coil or transformer to vary the inductance. { 'tün·iŋ ,kȯr }

tuning indicator [ELECTR] A device that indicates when a radio receiver is tuned accurately to a station; it is connected to a circuit having a direct-current voltage that varies with the strength of the incoming carrier signal. { 'tün·iŋ ,in·də,kād·ər }

tuning range [ELECTR] The frequency range over which a receiver or other piece of equipment can be adjusted by means of a tuning control. { 'tün·iŋ ,rānj }

tuning screw [ELECTROMAG] A screw that is inserted into the top or bottom wall of a waveguide and adjusted as to depth of penetration inside for tuning or impedance-matching purposes. { 'tün·iŋ ,skrü }

tuning stub [ELECTROMAG] Short length of transmission line, usually shorted at its free end, connected to a transmission line for impedance-matching purposes. { 'tün·iŋ ,stəb }

tunnel diode [ELECTR] A heavily doped junction diode that has a negative resistance at very low voltage in the forward bias direction, due to quantum-mechanical tunneling, and a short circuit in the negative bias direction. Also known as Esaki tunnel diode. { 'tən·əl ,dī,ōd }

tunnel junction [ELECTR] A two-terminal electronic device having an extremely thin potential barrier to electron flow, so that the transport characteristic (the current-voltage curve) is primarily governed by the quantum-mechanical tunneling process which permits electrons to penetrate the barrier. { 'tən·əl ,jəŋk·shən }

tuple [COMPUT SCI] A horizontal row of data items in a relational data structure; corresponds to a record or segment in other types of data structures. { 'tü·pəl }

Turing machine [COMPUT SCI] A mathematical idealization of a computing automaton similar in some ways to real computing machines; used by mathematicians to define the concept of computability. { 'tu̇r·iŋ mə,shēn }

turnaround system [COMPUT SCI] In character recognition, a system in which the input data to be read have previously been printed by the computer with which the reader is associated; an application is invoice billing and the subsequent recording of payments. Also known as reentry system. { 'tərn·ə,rau̇nd ,sis·təm }

turnaround time [COMPUT SCI] The delay between submission of a job for a data-processing system and its completion. { 'tərn·ə,rau̇nd ,tīm }

turnkey [COMPUT SCI] A complete computer system delivered to a customer in running condition, with all necessary premises, hardware and software equipment, supplies, and operating personnel. { 'tərn,kē }

turnstile antenna [ELECTROMAG] An antenna consisting of one or more layers of crossed horizontal dipoles on a mast, usually energized so the currents in the two dipoles of a pair are equal and in quadrature; used with television, frequency modulation, and other very-high-frequency or ultra-high-frequency transmitters to obtain an essentially omnidirectional radiation pattern. { 'tərn,stīl an,ten·ə }

turtle [COMPUT SCI] A cursor with the attributes of both position and direction; usually, an arrow that points in the direction it is about to move and generates a line along its path. { 'tərd·əl }

tutorial [COMPUT SCI] A method of computer-assisted instruction that involves a collection of screen formats, generally arranged in sequences that can be selected from a menu, and presented in response to the terminal operator's request. { tü'tȯr·ē·əl }

TV *See* television.

TV camera scanner [COMPUT SCI] In optical character recognition, a device that images an input character onto a sensitive photoconductive target of a camera tube, thereby developing an electric charge pattern on the inner surface of the target; this pattern is then explored by a scanning beam which traces out a rectangular pattern with the result that a waveform is produced which represents the character's most probable identity. { tē'vē ¦kam·rə ,skan·ər }

TVI *See* television interference.

TVRO *See* television receive only antenna.

twin arithmetic units [COMPUT SCI] A feature of some computers where the essential portions of the arithmetic section are virtually duplicated. { 'twin ə'rith·mə,tik ,yü·nəts }

twin axial cable [COMMUN] A transmission line consisting of two coaxial cables enclosed within a single sheath, each used to transmit signals in one direction. { 'twin 'ak·sē·əl 'kā·bəl }

twin check [COMPUT SCI] Continuous check of computer operation, achieved by the duplication of equipment and automatic comparison of results. { 'twin 'chek }

twist [ELECTROMAG] A waveguide section in which there is a progressive rotation of the cross section about the longitudinal axis of the waveguide. { twist }

two-address code [COMPUT SCI] In computers, a code using two-address instructions. { 'tü 'ad ˌres ˌkōd }

two-address instruction [COMPUT SCI] In computers, an instruction which includes an operation and specifies the location of two registers. { 'tü 'adˌres inˌstrək·shən }

two-dimensional electron gas field-effect transistor *See* high-electron-mobility transistor. { ˌtü di'men·shən·əl i¦lekˌträn 'gas ¦fēld i¦fekt tran'zis·tər }

two-dimensional storage [COMPUT SCI] A direct-access storage device in which the storage locations assigned to a particular file do not have to be physically adjacent, but instead may be taken from one or more seek areas. { 'tü ¦di ¦men·shən·əl 'stȯr·ij }

two-gap head [COMPUT SCI] One of two separate magnetic tape heads, one for reading and the other for recording data. { 'tü ¦gap 'hed }

two-hop transmission [COMMUN] Propagation of radio waves in which the waves are reflected from the ionosphere, then reflected from the ground, and then reflected from the ionosphere again before reaching the receiver. { 'tü ¦häp tranz'mish·ən }

two-input subtracter *See* half-subtracter. { 'tü ¦inˌpu̇t səb'trak·tər }

two-level subroutine [COMPUT SCI] A subroutine in which entry is made to a second, lower-level subroutine. { 'tü ¦lev·əl 'səb·rüˌtēn }

two-out-of-five code [COMPUT SCI] An encoding of the decimal digits using five binary bits and having the property that every code element contains two 1's and three 0's. { 'tü au̇d əv ¦fīv 'kōd }

two-part code [COMMUN] Randomized code consisting of an encoding section in which the plain text groups are arranged in alphabetical or other significant order accompanied by their code groups in nonalphabetical or random order, and a decoding section in which the code groups are arranged in alphabetical or numerical order and are accompanied by their meanings given in the encoding section. { 'tü 'pärt 'kōd }

two-pass compiler [COMPUT SCI] A language processor that goes through the program to be translated twice; on the first pass it checks the syntax of statements and constructs a table of symbols, while on the second pass it actually translates program statements into machine language. { 'tü ¦pas kəm'pīl·ər }

two-plus-one address instruction [COMPUT SCI] An instruction in a computer program which has two addresses specifying the locations of operands and one address specifying the location in which the result is to be entered. { 'tü ˌpləs ¦wən 'adˌres inˌstrək·shən }

two-port junction [ELECTROMAG] A waveguide junction with two openings; it can consist either of a discontinuity or obstacle in a waveguide, or of two essentially different waveguides connected together. { 'tü ¦pȯrt 'jəŋk·shən }

two-port system [CONT SYS] A system which has only one input or excitation and only one response or output. { 'tü ¦pȯrt 'sis·təm }

two-pulse canceler [ELECTR] A moving-target indicator canceler which compares the phase variation of two successive pulses received from a target; discriminates against signals with radial velocities which produce a Doppler frequency equal to a multiple of the pulse repetition frequency. { 'tü ¦pəls 'kan·slər }

two-quadrant multiplier [COMPUT SCI] Of an analog computer, a multiplier in which operation is restricted to a single sign of one input variable only. { 'tü ¦kwäd·rənt 'məl·təˌplī·ər }

two-range Decca [NAV] A Decca radio navigation system modified to provide circular lines of position. { 'tü ¦rānj'dek·ə }

two-source frequency keying [COMMUN] Keying in which the modulating wave shifts the output frequency between predetermined values derived from independent sources. { 'tü ¦sȯrs 'frē·kwən·sē ˌkē·iŋ }

two-state Turing machine [COMPUT SCI] A variation of a Turing machine in which only two states are allowed, although the number of symbols may be large. { 'tü ¦stāt 'tu̇r·iŋ mə ˌshēn }

two-symbol Turing machine [COMPUT SCI] A variation of a Turing machine in which only two symbols are permitted, although the number of states may be large. { 'tü ¦sim·bəl 'tu̇r·iŋ mə ˌshēn }

two-tone keying [COMMUN] Keying in which the modulating wave causes the carrier to be modulated with one frequency for the marking condition and modulated with a different frequency for the spacing condition. { 'tü ¦tōn 'kē·iŋ }

two-tone modulation [COMMUN] In teletypewriter operation, a method of modulation in which two different carrier frequencies are employed for the two signaling conditions; the transition from one frequency to the other is abrupt, with resultant phase discontinuities. { 'tü ¦tōn ˌmäj·ə'lā·shən }

TWTA *See* traveling-wave amplifier.

TWX machine *See* teletypewriter. { ˌtē ˌdəb·əl·yü 'eks məˌshēn }

TWX service *See* teletypewriter exchange service. { ˌtē ˌdəb·əl·yü 'eks ˌsər·vəs }

typeahead buffer [COMPUT SCI] A temporary storage device in a keyboard or microcomputer that holds information typed on the keyboard before the central processing unit is ready to accept it. { 'tīp·əˌhed 'bəf·ər }

type A wave *See* continuous wave. { 'tīp ¦ā ˌwāv }

type A1 wave [COMMUN] An unmodulated, keyed, continuous wave. { 'tīp ¦ā¦wən ˌwāv }

type A2 wave [COMMUN] A modulated, keyed, continuous wave. { 'tīp ¦ā¦tü ˌwāv }

type A3 wave [COMMUN] A continuous wave modulated by music, speech, or other sounds. { 'tīp ¦ā¦thrē ˌwāv }

type A4 wave [COMMUN] A superaudio frequency-modulated continuous wave, as used in facsimile systems. { 'tīp ¦ā¦fȯr ˌwāv }

type A9 wave [COMMUN] A composite transmission and continuous wave that is not type A1, A2, A3, A4, or A5 wave. { 'tīp ¦ā¦nīn ˌwāv }

type B wave [COMMUN] A keyed, damped wave. { 'tīp ¦bē ˌwāv }

type drum [COMPUT SCI] A steel cylinder containing 128 to 144 lateral bands, each band containing the alphabet, the digits 0–9, and the standard setof punctuation marks such as commas and periods, and revolving at high speed; printing is achieved by a hammer facing each band and activated at the right time to cause a character to be printed on the paper flowing between hammers and drum. { 'tīp ˌdrəm }

typewriter terminal [COMPUT SCI] An electric typewriter combined with an ASCII or other code generator that provides code output for feeding a computer, calculator, or other digital equipment; the terminal also produces hard copy when driven by incoming code signals. { 'tīpˌrīd·ər ˌtər·mən·əl }

U

UART *See* universal asynchronous receiver transmitter. { 'yü,ärt }

Uda antenna *See* Yagi-Uda antenna. { 'ü·də an ,ten·ə }

UDP *See* user datagram protocol.

U format [COMPUT SCI] A record format which the input/output control system treats as completely unknown and unpredictable. { 'yü ,fȯr ,mat }

UHF *See* ultrahigh frequency.

ULSA *See* ultra-low-side-lobe antenna.

ULSI circuit *See* ultra-large-scale integrated circuit. { ¦yü¦el¦es¦ī 'sər·kət }

ultrahigh frequency [COMMUN] The band of frequencies between 300 and 3000 megahertz in the radio spectrum, corresponding to wavelengths of 10 centimeters to 1 meter. Abbreviated UHF. { ¦əl·trə'hī 'frē·kwən·sē }

ultrahigh-frequency tuner [ELECTR] A tuner in a television receiver for reception of stations transmitting in the ultrahigh-frequency band. { ¦əl·trə'hī 'frē·kwən·sē 'tü·nər }

ultra-large-scale integrated circuit [ELECTR] A complex integrated circuit that contains more than 1,000,000 elements. Abbreviated ULSI circuit. { ¦əl·trə'lärj ¦skāl 'int·ə,grād·əd 'sər·kət }

ultra-low-side-lobe antenna [ELECTROMAG] A radar antenna so carefully designed and constructed that side lobes are all lower than about 45 decibels below the main beam; particularly valuable in modern airborne radar. Abbreviated ULSA. { 'əl·trə ,lō 'sīd ¦lōb an,ten·ə }

ultrashort waves [COMMUN] Radio waves shorter than 10 meters in wavelength; corresponding to frequencies above 30 megahertz. { ¦əl·trə'shȯrt 'wāvz }

ultra-small-aperture terminal [COMMUN] An antenna less than 20 inches (0.5 meter) in diameter that is used for reception of direct broadcasts from geosynchronous satellites. Abbreviated USAT. { ¦əl·trə ¦smȯl ¦ap·ə·chər 'tərm·ə·nəl }

ultrasonic camera [ELECTR] A device which produces a picture display of ultrasonic waves sent through a sample to be inspected or through live tissue; a piezoelectric crystal is used to convert the ultrasonic waves to voltage differences, and the voltage pattern on the crystal modulates the intensity of an electronic beam scanning the crystal; this beam in turn controls the intensity of a beam in a display device. { ¦əl·trə'sän·ik 'kam·rə }

ultrasonic communication [COMMUN] Communication accomplished through water by keying the sound output of echo-ranging sonar on ships or submarines or by using other such devices. { ¦əl·trə'sän·ik kə,myü·nə'kā·shən }

ultraviolet-erasable programmable read-only memory [COMPUT SCI] An integrated-circuit memory chip in which the stored information can be erased only by ultraviolet light and the circuit can be reprogrammed with new information that can be stored indefinitely. Abbreviated UV EPROM; UVPROM. { ¦əl·trə'vī·lət i'rās·ə·bəl 'prō,gram·ə·bəl ¦rēd ,ōn·lē 'mem·rē }

ultraviolet-erasable programmable read-only memory eraser [COMPUT SCI] A device that removes the contents of ultraviolet-erasable programmable read-only memory chips by exposing them to ultraviolet light. { ¦əl·trə'vī·lət i'rās·ə·bəl 'prō,gram·ə·bəl ¦rēd ,ōn·lē 'mem·rē i'rā·sər }

umbrella antenna [ELECTROMAG] Antenna in which the wires are guyed downward in all directions from a central pole or tower to the ground, somewhat like the ribs of an open umbrella. { əm'brel·ə an,ten·ə }

unambiguous name [COMPUT SCI] The name of a file or other data item that completely specifies the item to a computer system. { ¦ən·am'big·yə·wəs 'nām }

unattended operation [COMPUT SCI] An operation in which components in the hardware of a communications terminal or data-processing system operate automatically, allowing handling of signals or data without human intervention. { ¦ən·ə'ten·dəd ,äp·ə'rā·shən }

unattended time [COMPUT SCI] Time during which a computer is turned off but is not undergoing maintenance. Also known as unused time. { ¦ən·ə'ten·dəd ,tīm }

unbreakable cipher [COMMUN] A cipher for which the message or key cannot be obtained through cryptanalysis, even with an unlimited amount of computational power, data storage, and calendar time. { ¦ən'brāk·ə·bəl 'sī·fər }

unbundling [COMPUT SCI] The separate provision of software products and services versus computer hardware (equipment). { ¦ən'bənd·liŋ }

uncatalog [COMPUT SCI] To remove an entry from the system catalog so that the file named in the entry can no longer be accessed by the operating system. { ¦ən'kad·əl,äg }

unconditional [COMPUT SCI] Not subject to conditions external to the specific instruction. { ¦ən·kən'dish·ən·əl }

unconditional jump [COMPUT SCI] A digital-computer instruction that interrupts the normal process of obtaining instructions in an ordered sequence, and specifies the address from which the next instruction must be taken. Also known as unconditional transfer. { ¦ən·kən'dish·ən·əl 'jəmp }

unconditional transfer *See* unconditional jump. { ¦ən·kən'dish·ən·əl 'tranz·fər }

underflow [COMPUT SCI] The generation of a result whose value is smaller than the smallest quantity that can be represented or stored by a computer. { 'ən·dər,flō }

underlap [COMMUN] **1.** In facsimile transmission, the space between the recorded elemental area in one recording line and the adjacent elemental area in the next recording line, when these areas are smaller than normal; or the space between the elemental areas in the direction of the recording line. **2.** The amount by which the effective height of the scanning spot falls short of the nominal width of the scanning line. { 'ən·dər,lap }

undershoot [CONT SYS] The amount by which a system's response to an abrupt change in input falls short of that desired. { 'ən·dər,shüt }

underthrow distortion [COMMUN] Distortion occurring in facsimile when the maximum signal amplitude is too low. { ¦ən·dər¦thrō di'stȯr·shən }

underwater telephone [COMMUN] A method of voice communication using underwater sound as a means of transmission; it functions similarly to a conventional telephone system except that the energy is carried by sound waves in the water rather than by electrical signals through a wire. { ¦ən·dər¦wȯd·ər 'tel·ə,fōn }

undetected error rate [COMMUN] The number of bits (or other units of information) which are received but are not detected or corrected by error-control equipment, divided by the total number of bits (or other units of information) transmitted. Also known as residual error rate. { ¦ən·di'tek·təd 'er·ər ,rāt }

undistorted wave [COMMUN] Periodic wave in which both the attenuation and velocity of propagation are the same for all sinusoidal components, and in which no sinusoidal component is present at one point that is not present at all points. { ¦ən·di'stȯrd·əd 'wāv }

unformatted file [COMPUT SCI] Any data file, such as a text file, that does not have various properties such as a consistent structure with regard to record length and order of data elements. { ¦ən'fȯr,mad·əd 'fīl }

unidirectional antenna [ELECTROMAG] An antenna that has a single well-defined direction of maximum gain. { ¦yü·nə·də'rek·shən·əl an'ten·ə }

unidirectional coupler [ELECTR] Directional coupler that samples only one direction of transmission. { ¦yü·nə·də'rek·shən·əl 'kəp·lər }

unidirectional log-periodic antenna [ELECTROMAG] A broad-band antenna in which the cut-out portions of a log-periodic antenna are mounted at an angle to each other, to give a unidirectional radiation pattern in which the major radiation is in the backward direction, off the apex of the antenna; impedance is essentially constant for all frequencies, as is the radiation pattern. { ¦yü·nə·də'rek·shən·əl 'läg ,pir·ē'äd·ik an'ten·ə }

unidirectional pulse-amplitude modulation [COMMUN] Modulation of pulse-amplitude type in which all pulses rise in the same direction. Also known as single-polarity pulse-amplitude modulation. { ¦yü·nə·də'rek·shən·əl ¦pəls ¦am·plə,tüd ,mäj·ə'lā·shən }

unidirectional pulses [ELECTR] Single polarity pulses which all rise in the same direction. { ¦yü·nə·də'rek·shən·əl 'pəl·səz }

uniformly accessible storage *See* random-access memory. { ,yü·nə'fȯrm·lē ak'ses·ə·bəl 'stȯr·ij }

uniform plane wave [ELECTROMAG] Plane wave in which the electric and magnetic intensities have constant amplitude over the equiphase surfaces; such a wave can only be found in free space at an infinite distance from the source. { 'yü·nə,fȯrm 'plān 'wāv }

uniform resource locator [COMPUT SCI] The unique Internet address assigned to a Web document or resource by which it can be accessed by all Web browsers. The first part of the address specifies the applicable Internet protocol (IP), for example, http or ftp; the second part provides the IP address or domain name of the location. Abbreviated URL. { ,yü·nə,fȯrm ri,sȯrs 'lō·kād·ər }

union [COMPUT SCI] A data structure that can store items of different types, but can store only one item at a time. { 'yün·yən }

union catalog [COMPUT SCI] A merged listing of the contents of two or more catalogs (of libraries, for example). { 'yün·yən 'kad·əl,äg }

unipath *See* nonshared control unit. { 'yü·nə ,path }

unipole [ELECTROMAG] A hypothetical antenna that radiates or receives signals equally well in all directions. Also known as isotropic antenna. { 'yü·nə,pōl }

uniprocessor [COMPUT SCI] A computer that has a single central processing unit and works sequentially on only one program at a time. { 'yü·nə,prä,ses·ər }

uniterm [COMPUT SCI] A word, symbol, or number used as a description for retrieval of information from a collection; especially, such a description used in a coordinate indexing system. { 'yü·nə,tərm }

unit length [COMMUN] Basic element of time used in determining signaling speeds in message transmission. { 'yü·nət 'leŋkth }

unitor [COMPUT SCI] In computers, a device or circuit which performs a function corresponding to the Boolean operation of union. { 'yü·nə·tər }

unit record [COMPUT SCI] Any of a collection of records, all of which have the same form and the same data elements. { 'yü·nət 'rek·ərd }

unit string [COMPUT SCI] A string that has only one element. { 'yü·nət 'striŋ }

unit test [COMPUT SCI] The testing of a module within a computer system. { 'yü·nət ˌtest }

universal asynchronous receiver transmitter [COMPUT SCI] An electronic circuit that converts bytes of data between the parallel format in which bits are stored side by side within a device and the serial format whereby bits are transmitted sequentially over a communications line. Abbreviated UART. { ¦yü·nə¦vər·səl ā'siŋ·krə·nəs ri'sē·vər tranz'mid·ər }

universal language [COMPUT SCI] A programming language that is widely employed to write programs that can be run on a wide variety of computers. { ¦yü·nə¦vər·səl 'laŋ·gwij }

universal product code [COMPUT SCI] **1.** A 10-digit bar code on the outside of a package for electronic scanning at supermarket checkout counters; each digit is represented by the ratio of the widths of adjacent stripes and white areas. **2.** The corresponding combinations of binary digits into which the scanned bars are converted for computer processing that provides continuously updated inventory data and printout of the register tape at the checkout counter. { ¦yü·nə ¦vər·səl 'präd·əkt ˌkōd }

universal serial bus [COMPUT SCI] A serial interface that can transfer data at up to 480 million bits per second and connect up to 127 daisy-chained peripheral devices. Abbreviated USB. { ˌyü·nə ˌvər·səl ˌsir·ē·əl 'bəs }

universal Turing machine [COMPUT SCI] A Turing machine that can simulate any Turing machine. { ¦yü·nə¦vər·səl 'tu̇r·iŋ məˌshēn }

univibrator *See* monostable multivibrator. { ¦yü·nə'vīˌbrād·ər }

Unix [COMPUT SCI] An operating system that was designed for use with microprocessors and with the C programming language, and that has been adopted for use with several 16-bit-microprocessor microcomputers. { 'yü·niks }

unload [COMPUT SCI] To remove or copy data from a computer system. { ən'lōd }

unloading circuit [COMPUT SCI] In an analog computer, a computing element or combination of computing elements capable of reproducing or amplifying a given voltage signal while drawing negligible current from the voltage source. { ¦ən'lōd·iŋ ˌsər·kət }

unloading device [COMPUT SCI] Equipment that holds programs and other data that have been copied or removed from a computer system. { ən'lōd·iŋ diˌvīs }

unmodified instruction *See* basic instruction. { ¦ən'mäd·əˌfīd in'strək·shən }

unpack [COMPUT SCI] **1.** To recover the individual data items contained in packed data. **2.** More specifically, to convert a packed decimal number into individual digits (and sometimes a sign). { ¦ən'pak }

unprotect [COMPUT SCI] To remove restrictions on access to a file so that any computer program can read and alter the data contained in it. { ¦ən·prə'tekt }

unsolicited message [COMPUT SCI] A warning or error message that is automatically issued by a computer program when it detects a problem, and that does not depend on the operator making a query. { ˌən·sə'lis·əd·əd 'mes·ij }

unused time *See* unattended time. { ¦ən'yüzd 'tīm }

unwind [COMPUT SCI] In computers, to rearrange and code a sequence of instructions to eliminate red-tape operations. { ¦ən'wīnd }

up-converter [ELECTR] Type of parametric amplifier which is characterized by the frequency of the output signal being greater than the frequency of the input signal. { 'əp kənˌvərd·ər }

update [COMPUT SCI] **1.** In computers, to modify an instruction so that the address numbers it contains are increased by a stated amount each time the instruction is performed. **2.** To change a record by entering current information; for example, to enter a new address or account number in the record pertaining to an employee or customer. Also known as posting. { ¦əp¦dāt }

update service [COMPUT SCI] A service that guarantees installation of updates to products within a certain period of time after they become available. { 'əpˌdāt ˌsər·vəs }

uplink [COMMUN] The radio transmission path upward from the earth to a communications satellite, or from the earth to aircraft. { 'əpˌliŋk }

upload [COMPUT SCI] To transfer or copy data from a smaller computer, such as a microcomputer, to a larger computer. { 'əpˌlōd }

upper half-power frequency [ELECTR] The frequency on an amplifier response curve which is greater than the frequency for peak response and at which the output voltage is $1/\sqrt{2}$ (that is, 0.707) of its midband or other reference value. { 'əp·ər ¦haf ¦pau̇·ər 'frē·kwən·sē }

upper sideband [COMMUN] The higher of two frequencies or groups of frequencies produced by a modulation process. { 'əp·ər 'sīdˌband }

up time [COMPUT SCI] The time during which equipment is either producing work or is available for productive work. Also known as available time. { 'əp ˌtīm }

upward compatibility [COMPUT SCI] The ability of a newer or larger computer to accept programs from an older or smaller one. { 'əp·wərd kəm ˌpad·ə'bil·əd·ē }

URL *See* uniform resource locator.

usable [COMPUT SCI] Pertaining to a computer system that is easy for all users to work with. { 'yüz·ə·bəl }

USAT *See* ultra-small-aperture terminal. { 'yü ˌsat or ¦yü¦es¦ā'tē }

USB *See* universal serial bus.

UseNet [COMPUT SCI] A global network of newsgroups that is linked by the Internet and other wide-area networks. { 'yüzˌnet }

user [COMMUN] An individual, installation, or activity having access to a switching center through a local private branch exchange, or by dialing an access code. [COMPUT SCI] Anyone who requires the use of services of a computing system or its products. { 'yü·zər }

user datagram protocol [COMMUN] A communications protocol providing a direct way to send and receive datagrams over an IP network but with few error recovery resources, used mainly for broadcasting over a network, for example, with streaming media. Abbreviated UDP. { ˌyüz·ər 'dad·əˌgram ˌprōd·əˌkȯl }

user-defined function [COMPUT SCI] A subroutine written by the programmer to calculate and return the value of a mathematical function. { ¦yü·zər di'fīnd 'fəŋk·shən }

user-defined type [COMPUT SCI] A data type that is not provided by a strongly typed language but is instead created by the programmer for a particular computer program. { ¦yü·zər di'fīnd 'tīp }

user exit [COMPUT SCI] A point in a computer program at which a user can cause control to be transferred outside the program. { 'yü·zər 'eg·zət }

user friendly [COMPUT SCI] Property of a system that is easy for an untrained person to use and sets up an easily understood dialog between the user and the computer. { 'yü·zər 'frend·lē }

user group [COMPUT SCI] An organization of users of the computers of a particular vendor, which shares information and ideas, and may develop system software and influence vendors to change their products. { 'yü·zər ˌgrüp }

userID [COMPUT SCI] The name used to log in to a network, remote server, and so on. { ˌyüz·ər ¦ī'dē }

user interface [COMPUT SCI] **1.** The point at which a user or a user department or organization interacts with a computer system. **2.** The part of an interactive computer program that sends messages to and receives instructions from a terminal user. { 'yü·zər 'in·tərˌfās }

user mode [COMPUT SCI] The mode of operation exercised by the user programs of a computer system in which there is a class of privileged instructions that is not permitted, since these can be executed only by the operating system or executive system. Also known as slave mode. { 'yü·zər ˌmōd }

username *See* mailbox name. { 'yüz·ərˌnām }

user program [COMPUT SCI] A computer program written by the person who uses it or by personnel of the organization that will use it. Also known as roll your own; user-written code. { 'yü·zər 'prōˌgram }

user-programmable memory [COMPUT SCI] That portion of the internal storage of a microcomputer that is available for programs entered or loaded in by the user. { 'yü·zər prō 'gram·ə·bəl 'mem·rē }

user-to-user service [COMMUN] Method of switching that enables direct user-to-user connection which does not include message store-and-forward service. { ¦yü·zər tə ¦yü·zər 'sər·vəs }

user-written code *See* user program. { 'yü·zər ¦writ·ən 'kōd }

utility routine [COMPUT SCI] A program or routine of general usefulness, usually not very complicated, and applicable to many jobs or purposes. { yü'til·əd·ē rüˌtēn }

utilization ratio [COMPUT SCI] The ratio of the effective time on a computer to the total up time. { ˌyüd·əl·ə'zā·shən ˌrā·shō }

uuencode [COMPUT SCI] A protocol for sending binary files in ASCII text format over the Internet, particularly e-mail attachments. { ˌyü'yü·in ˌkōd }

UV EPROM *See* ultraviolet-erasable programmable read-only memory. { ˌyüˌvē 'ēˌpräm }

UVPROM *See* ultraviolet-erasable programmable read-only memory. { ˌyüˌvē'präm }

V

validation [COMPUT SCI] The act of testing for compliance with a standard. { ˌval·əˈdā·shən }

validity check [COMPUT SCI] Computer check of input data, based on known limits for variables in given fields. { vəˈlid·əd·ē ˌchek }

valid program [COMPUT SCI] A computer program whose statements, individually and together, follow the syntactical rules of the programming language in which it is written, so that they are capable of being translated into a machine language program. { ˈval·əd ˈprō ˌgram }

value-added network [COMMUN] A communications network that provides not only communications channels but also other services such as automatic error detection and correction, protocol conversions, and store-and-forward message services. { ˈval·yü ¦ad·əd ˈnetˌwərk }

value parameter [COMPUT SCI] A parameter whose value is copied by a subprogram which can then alter its copy without affecting the original. { ˈval·yü pəˈram·əd·ər }

valve *See* electron tube. { valv }

Van Atta array [ELECTROMAG] Antenna array in which pairs of corner reflectors or other elements equidistant from the center of the array are connected together by a low-loss transmission line in such a way that the received signal is reflected back to its source in a narrow beam to give signal enhancement without amplification. { vaˈnad·ə əˌrā }

vane attenuator *See* flap attenuator. { ˈvān əˈten·yəˌwād·ər }

vanilla [COMPUT SCI] Referring to a generalized system, usually software, that has not been subjected to special modifications, enhancements, or customization. Also known as plain vanilla; pure vanilla. { vəˈnil·ə }

V antenna [ELECTROMAG] An antenna having a V-shaped arrangement of conductors fed by a balanced line at the apex; the included angle, length, and elevation of the conductors are proportioned to give the desired directivity. Also spelled vee antenna. { ˈvē anˌten·ə }

VAR *See* visual-aural range.

varactor [ELECTR] A semiconductor device characterized by a voltage-sensitive capacitance that resides in the space-charge region at the surface of a semiconductor bounded by an insulating layer. Also known as varactor diode; variable-capacitance diode; varicap; voltage-variable capacitor. { vaˈrak·tər }

varactor diode *See* varactor. { vaˈrak·tər ˈdīˌōd }

varactor tuning [ELECTR] A method of tuning in which varactor diodes are used to vary the capacitance of a tuned circuit. { vaˈrak·tər ˈtün·iŋ }

variable [COMPUT SCI] A data item, or specific area in main memory, that can assume any of a set of values. { ˈver·ē·ə·bəl }

variable acceleration *See* dynamic resolution. { ¦ver·ē·ə·bəl ikˌsel·əˈrā·shən }

variable attenuator [ELECTR] An attenuator for reducing the strength of an alternating-current signal either continuously or in steps, without causing appreciable signal distortion, by maintaining a substantially constant impedance match. { ˈver·ē·ə·bəl əˈten·yəˌwād·ər }

variable-bandwidth filter [ELECTR] An electric filter whose upper and lower cutoff frequencies may be independently selected, so that almost any bandwidth may be obtained; it usually consists of several stages of RC filters, each separated by buffer amplifiers; tuning is accomplished by varying the resistance and capacitance values. { ˈver·ē·ə·bəl ¦bandˌwidth ˌfil·tər }

variable bit rate [COMMUN] Operation in a digital system where the bit rate varies with time during the decoding of a compressed bit stream. { ˈvər·ē·ə·bəl ˈbit ˌrāt }

variable-block [COMPUT SCI] Pertaining to an arrangement of data in which the number of words or characters in a block can vary, as determined by the programmer. { ˈver·ē·ə·bəl ˈbläk }

variable-capacitance diode *See* varactor. { ˈver·ē·ə·bəl kə¦pas·əd·əns ˈdīˌōd }

variable carrier modulation *See* controlled carrier modulation. { ˈver·ē·ə·bəl ¦kar·ē·ər ˌmäj·əˈlā·shən }

variable connector [COMPUT SCI] A flow chart symbol representing a sequence connection which is not fixed, but which can be varied by the flow-charted procedure itself; it corresponds to an assigned GO TO in a programming language such as FORTRAN. { ˈver·ē·ə·bəl kəˌnek·tər }

variable-cycle operation [COMPUT SCI] An operation that requires a variable number of regularly timed execution cycles for its completion. { ˈver·ē·ə·bəl ¦sī·kəl ˌäp·əˈrā·shən }

variable field [COMPUT SCI] A field of data whose length is allowed to vary within certain specified limits. { 'ver·ē·ə·bəl 'fēld }

variable-length field [COMPUT SCI] A data field in which the number of characters varies, the length of the field being stored within the field itself. { 'ver·ē·ə·bəl ¦leŋkth 'fēld }

variable-length operation [COMPUT SCI] A computer operation whose operands are allowed to have a variable number of bits or characters. { 'ver·ē·ə·bəl ¦leŋkth ˌäp·ə'rā·shən }

variable-length record [COMPUT SCI] A data or file format that allows each record to be exactly as long as needed. { 'ver·ē·ə·bəl ¦leŋkth 'rek·ərd }

variable-length word [COMPUT SCI] A computer word whose length is determined by the programmer. { 'ver·ē·ə·bəl ¦leŋkth 'wərd }

variable parameter [COMPUT SCI] A parameter whose storage address is passed to a subprogram so that the subprogram can alter its value. { 'ver·ē·ə·bəl pə'ram·əd·ər }

variable point [COMPUT SCI] A system of numeration in which the location of the decimal point is indicated by a special character at that position. { 'ver·ē·ə·bəl 'pȯint }

variable waveguide attenuator [ELECTROMAG] Device designed to introduce a variable amount of attenuation into a waveguide circuit by moving a lossy vane (a component that absorbs electromagnetic energy) either sideways across the waveguide or into the waveguide through a longitudinal slot. { 'ver·ē·ə·bəl 'wāvˌgīd əˌten·yəˌwād·ər }

variable-word-length [COMPUT SCI] A phrase referring to a computer in which the number of characters addressed is not a fixed number but is varied by the data or instruction. { 'ver·ē·ə·bəl 'wərd ˌleŋkth }

variant [COMMUN] **1.** One of two or more cipher or code symbols which have the same plain text equivalent. **2.** One of several plain text meanings that are represented by a single code group. { 'ver·ē·ənt }

variant record [COMPUT SCI] A record variable whose format is made to depend on some circumstance; for example, a record dealing with rates of pay might contain information on hourly rates for some employees and weekly or monthly salaries for others. { 'ver·ē·ənt 'rek·ərd }

varicap *See* varactor. { 'var·əˌkap }

variocoupler [ELECTROMAG] In radio practice, a transformer in which the self-impedance of windings remains essentially constant while the mutual impedance between the windings is adjustable. { ¦ver·ē·ō'kəp·lər }

V-beam radar [ELECTROMAG] A volumetric radar system that uses two fan beams to determine the distance, bearing, and height of a target: one beam is vertical and the other inclined; the beams intersect at ground level and rotate continuously about a vertical axis; the time difference between the arrivals of the echoes of the two beams is a measure of target elevation. { 'vē ¦bēm 'rāˌdär }

VBV *See* video buffering verifier.

V-chip [COMMUN] An electronic device and protocol intended to exert control over program reception based on program content rating. { 'vē ˌchip }

VCO *See* voltage-controlled oscillator.

VCR *See* videocassette recorder.

VCSEL *See* vertical-cavity surface-emitting laser. { ¦vē¦sē¦es¦ē'el *or* ¦vē'sēˌsel }

VDT *See* display terminal.

vector *See* jump vector. { 'vek·tər }

vector canceler [ELECTR] A canceler used in moving-target-indication radar, often using digital methods, operating on both the in-phase and quadrature video signals produced by the phase detectors. { 'vek·tər 'kans·lər }

vectored interrupt [COMPUT SCI] A signal that instructs a computer program to temporarily halt the processing it is doing and transfer control to a routine whose address is given by an entry in a jump vector specified by a value included in the signal. { 'vek·tərd 'in·tə·rəpt }

vector graphics [COMPUT SCI] A computer graphics image-coding technique which codes only the image itself as a series of lines, according to the cartesian coordinates of the lines' origins and terminations. Also known as object-oriented graphics. { 'vek·tər ¦graf·iks }

vector processing [COMPUT SCI] A procedure for speeding the processing of information by a computer, in which pipelined units perform arithmetic operations on uniform, linear arrays of data values, and a single instruction involves the execution of the same operation on every element of the array. { 'vek·tər 'präˌses·iŋ }

vector quantization [COMPUT SCI] A data compression technique in which a finite sequence of values is presented as resembling the template (from among the choices available to a given code book) that minimizes a distortion measure. { ¦vek·tər ˌkwän·tə'zā·shən }

vector table *See* jump vector. { 'vek·tər ˌtā·bəl }

vee antenna *See* V antenna. { 'vē anˌten·ə }

vehicular telephony [COMMUN] The transmission of speech signals to and from mobile radio stations installed in automotive vehicles; typically, each station is equipped with a transmitter and a receiver. { vē'hik·yə·lər tə'lef·ə·nē }

velocity constant [CONT SYS] The ratio of the rate of change of the input command signal to the steady-state error, in a control system where these two quantities are proportional. { və'läs·əd·ē ˌkän·stənt }

velocity control *See* rate control. { və'läs·əd·ē kənˌtrōl }

velocity error [CONT SYS] The difference between the rate of change of the actual position of a control system component and the rate of change of the desired position. { və'läs·əd·ē ˌer·ər }

velocity-modulated oscillator [ELECTR] Oscillator which employs velocity modulation to produce radio-frequency power. Also known as klystron oscillator. { və'läs·əd·ē ˌmäj·əˌlād·əd 'äs·əˌlād·ər }

velocity of light *See* speed of light. { və 'läs·əd·ē əv 'līt }

verb [COMPUT SCI] In COBOL, the action indicating part of an unconditional statement. { vərb }

verification [COMPUT SCI] The process of checking the results of one data transcription against the results of another data transcription; both transcriptions usually involve manual operations. { ˌver·ə·fə'kā·shən }

verify [COMMUN] To ensure that the meaning and phraseology of the transmitted message convey the exact intention of the originator. [COMPUT SCI] To determine whether an operation has been completed correctly. { 'ver·əˌfī }

vertical antenna [ELECTROMAG] A vertical metal tower, rod, or suspended wire used as an antenna. { 'vərd·ə·kəl an'ten·ə }

vertical blanking [ELECTR] Blanking of a video display device during the vertical retrace. { 'vərd·ə·kəl 'blaŋk·iŋ }

vertical-cavity surface-emitting laser [OPTICS] A very small semiconductor laser in which stacks of dielectric mirrors above and below the optically active region form the optical cavity, the active gain medium consists of one or more semiconductor quantum wells placed parallel to the mirrors at an antinode of the cavity resonance, and lasing light emission is from the surface of the semiconductor substrate, normal to the plane of the gain medium. Abbreviated VCSEL. { ¦vərd·i·kəl ˌkav·əd·ē ¦sər·fəs iˌmid·iŋ 'lā·zər }

vertical centering control [ELECTR] The centering control provided in a video display device to shift the position of the entire image vertically in either direction on the screen. { 'vərd·ə·kəl 'sen·tər·iŋ kənˌtrōl }

vertical component effect *See* antenna effect. { 'vərd·ə·kəl kəm'pō·nənt iˌfekt }

vertical definition *See* vertical resolution. { 'vərd·ə·kəl ˌdef·ə'nish·ən }

vertical deflection oscillator [ELECTR] The oscillator that produces, under control of the vertical synchronizing signals, the sawtooth voltage waveform that is amplified to feed the vertical deflection coils on the picture tube of an analog television receiver. Also known as vertical oscillator. { 'vərd·ə·kəl di'flek·shən 'äs·əˌlād·ər }

vertical hold control [ELECTR] The hold control that changes the free-running period of the vertical deflection oscillator in an analog television receiver, so the picture remains steady in the vertical direction. { 'vərd·ə·kəl ¦hōld kənˌtrōl }

vertical instruction [COMPUT SCI] An instruction in machine language to carry out a single operation or a time-ordered series of a fixed number and type of operation on a single set of operands. { 'vərd·ə·kəl in'strək·shən }

vertical interval reference [ELECTR] A reference signal inserted into an analog television program signal every $\frac{1}{60}$ second, in line 19 of the vertical blanking period between NTSC television frames, to provide references for luminance amplitude, black-level amplitude, sync amplitude, chrominance amplitude, and color-burst amplitude and phase. Abbreviated VIR. { 'vərd·ə·kəl ¦in·tər·vəl 'ref·rəns }

vertical linearity control [ELECTR] A linearity control that permits narrowing or expanding the height of the image on the upper half of the screen of a video display, to give linearity in the vertical direction so that circular objects appear as true circles. { 'vərd·ə·kəl ˌlin·ē¦ar·əd·ē kənˌtrōl }

vertically stacked loops *See* stacked loops. { 'vərd·ə·klē ¦stakt 'lüps }

vertical metal oxide semiconductor technology [ELECTR] For semiconductor devices, a technology that involves essentially the formation of four diffused layers in silicon and etching of a V-shaped groove to a precisely controlled depth in the layers, followed by deposition of metal over silicon dioxide in the groove to form the gate electrode. Abbreviated VMOS technology. { 'vərd·ə·kəl ¦medˌəl ¦äkˌsīd ¦sem·i·kənˌdək·tər tek 'näl·ə·jē }

vertical obstacle sonar [ENG] An active sonar used to determine heights of objects in the path of a submersible vehicle; its beam sweeps along a vertical plane, about 30° above and below the direction of the vehicle's motion. Abbreviated VOS. { 'vərd·ə·kəl ¦äb·stə·kəl 'sōˌnär }

vertical oscillator *See* vertical deflection oscillator. { 'vərd·ə·kəl 'äs·əˌlād·ər }

vertical parity check *See* lateral parity check. { 'vərd·ə·kəl 'par·əd·ē ˌchek }

vertical polarization [COMMUN] Transmission of linear polarized radio waves whose electric field vector is perpendicular to the earth's surface. { 'vərd·ə·kəl ˌpō·lə·rə'zā·shən }

vertical recording [ELECTR] Magnetic recording in which bits are magnetized in directions perpendicular to the surface of the recording medium, allowing the bits to be smaller. Also known as perpendicular recording. { 'vərd·ə·kəl ri'kȯrd·iŋ }

vertical redundancy check *See* lateral parity check. { 'vərd·ə·kəl ri'dən·dən·sē ˌchek }

vertical resolution [ELECTR] The number of distinct horizontal lines, alternately black and white, that can be seen in the reproduced image of a video image test pattern; it is primarily fixed by the number of horizontal lines used in scanning. Also known as vertical definition. { 'vərd·ə·kəl ˌrez·ə'lü·shən }

vertical retrace [ELECTR] The return of the electron beam to the top of the screen at the end of each video field. { 'vərd·ə·kəl 'rēˌtrās }

vertical sweep [ELECTR] The downward movement of the scanning beam from top to bottom of the picture being televised. { 'vərd·ə·kəl 'swēp }

vertical synchronizing pulse [ELECTR] One of the six pulses that are transmitted at the end of each field in an analog television system to keep the receiver in field-by-field synchronism with the transmitter. Also known as picture synchronizing pulse. { 'vərd·ə·kəl 'siŋ·krəˌnīz·iŋ ˌpəls }

vertical tab [COMPUT SCI] A control character that causes a computer printer to jump from its current line to another preset line further down the page. { 'vərd·ə·kəl 'tab }

very high frequency [COMMUN] The band of frequencies from 30 to 300 megahertz in the radio spectrum, corresponding to wavelengths of 1 to 10 meters. Abbreviated VHF. { ¦ver·ē ¦hī 'frē·kwən·sē }

very high frequency omnidirectional radio range [NAV] A radio navigation aid operating at very high frequency and supplying bearing information for the entire 360° of azimuth. Abbreviated VOR. { ¦ver·ē ¦hī 'frē·kwən·sē ¦äm·nə·də'rek·shən·əl 'rād·ē·ō,rānj }

very high frequency tuner [ELECTR] A tuner in a television receiver for reception of stations transmitting in the very high frequency band; it generally has 12 discrete positions corresponding to channels 2–13. { ¦ver·ē ¦hī 'frē·kwən·sē 'tün·ər }

very large scale integrated circuit [ELECTR] A complex integrated circuit that contains between 20,000 and 1,000,000 transistors. Abbreviated VLSI circuit. { ¦ver·ē ¦lärj ¦skāl 'int·ə,grād·əd 'sər·kət }

very long range radar [ELECTR] Equipment whose maximum range on a reflecting target of 10.76 square feet (1 square meter) normal to the signal path exceeds 800 miles (1300 kilometers), provided line of sight exists between the target and the radar. { ¦ver·ē ¦lōŋ ¦rānj 'rā,där }

very low frequency [COMMUN] The band of frequencies from 3 to 30 kilohertz in the radio spectrum, corresponding to wavelengths of 10 to 100 kilometers. Abbreviated VLF. { ¦ver·ē ¦lō 'frē·kwən·sē }

very short range radar [ELECTR] Equipment whose range on a reflecting target of 10.76 square feet (1 square meter) normal to the signal path is less than 50 miles (80 kilometers), provided line of sight exists between the target and the radar. { ¦ver·ē ¦shȯrt ¦rānj 'rā,där }

very small aperture terminal [COMMUN] An antenna approximately 6 feet (1.8 meters) or less in diameter, that is used for both broadcast reception and interactive communications via geosynrchronous satellites. Abbreviated VSAT. { ¦ver·ē ¦smȯl ¦ap·ə·chər 'tərm·ə·nəl }

vestigial sideband [COMMUN] The transmitted portion of an amplitude-modulated sideband that has been largely suppressed by a filter having a gradual cutoff in the neighborhood of the carrier frequency; the other sideband is transmitted without much suppression. Abbreviated VSB. { və'stij·ē·əl 'sīd,band }

vestigial-sideband filter [ELECTR] A filter that is inserted between a transmitter and its antenna to suppress part of one of the sidebands. { və'stij·ē·əl ¦sīd,band ,fil·tər }

vestigial-sideband transmission [COMMUN] A type of radio signal transmission for amplitude modulation in which the normal complete sideband on one side of the carrier is transmitted, but only a part of the other sideband is transmitted. Also known as asymmetrical-sideband transmission. { və'stij·ē·əl ¦sīd,band tranz,mish·ən }

VF *See* voice frequency.

V format [COMPUT SCI] A data record format in which the logical records are of variable length and each record begins with a record length indication. { 'vē ,fȯr,mat }

VHF *See* very high frequency.

via point [CONT SYS] A point located midway between the starting and stopping positions of a robot tool tip, through which the tool tip passes without stopping. Also known as way point. { 'vē·ə ,pȯint }

video [ELECTR] **1.** Pertaining to picture signals or to the sections of a television system that carry these signals in either unmodulated or modulated form. **2.** Pertaining to the demodulated radar receiver output that is applied to a radar indicator or otherwise treated in the radar's own data processing. { 'vid·ē·ō }

video adapter [COMPUT SCI] A printed circuit board that is plugged into a computer and generates the text and graphics images on a monitor. Also known as video board; video display board. { 'vid·ē·ō ə,dap·tər }

video board *See* video adapter. { 'vid·ē·ō ,bȯrd }

video buffering verifier [COMMUN] A hypothetical decoder that is conceptually connected to the output of an encoder; provides a constraint on the variability of the data rate that an encoder can produce. Abbreviated VBV. { 'vid·ē·ō 'bəf·ə·riŋ ,ver·ə,fī·ər }

video canceler [ELECTR] A canceler used in moving-target-indication radar operating on the video signals produced by the phase detector of a coherent radar, or on the video signal in a simpler radar, to detect fluctuation of the video signal caused by a target moving over stationary clutter. { 'vid·ē·ō 'kans·lər }

videocassette [ELECTR] A compact plastic case containing a magnetic tape for video recording and playing. { ¦vid·ē·ō,kə'set }

videocassette recorder [ELECTR] A device for video recording and playing of magnetic tapes that are contained in plastic cases. Abbreviated VCR. { ¦vid·ē·ō,kə'set ri,kȯrd·ər }

videoconference [COMMUN] A teleconference that employs some type of video camera or other audio or video equipment to convey pictures and sound from one location to another, either in a one-way or two-way fashion. { ,vid·ē·ō'kän·frəns }

videoconferencing [COMPUT SCI] Two-way interactive, digital communication through video streaming on the Internet, or by communications satellite, video telephone, and so forth. { ,vid·ē·ō'kän·frəns·iŋ }

video correlator [ELECTR] Radar circuit that enhances automatic target detection capability, provides data for digital target plotting, and gives improved immunity to noise, interference, and jamming. { 'vid·ē·ō 'kär·ə,lād·ər }

video dial-tone [COMMUN] A service that enables a subscriber to select a video information provider from among many such providers offering service in a neighborhood, and to access on demand, from this provider, a movie or multimedia content, or similar services such as games and home shopping. { ˌvid·ē·ō 'dīl ˌtōn }

video discrimination [ELECTR] Radar circuit used to reduce the frequency band of the video amplifier stage in which it is used. { 'vid·ē·ō di ˌskrim·ə'nā·shən }

video disk storage See optical disk storage. { 'vid·ē·ō ¦disk ˌstȯr·ij }

video display board See video adapter. { ¦vid·ē·ōdi'splāˌbȯrd }

video display terminal See display terminal. { 'vid·ē·ō di'splāˌtər·mən·əl }

video frequency [COMMUN] One of the frequencies output by a video camera when an image is scanned; for standard-definition systems, it may be any value from almost zero to well over 4 megahertz. { 'vid·ē·ō 'frē·kwən·sē }

video integrator [ELECTR] **1.** Electric counter-countermeasures device that is used to reduce the response to nonsynchronous signals such as noise, and is useful against random pulse signals and noise. **2.** Device which uses the redundancy of repetitive signals to improve the output signal-to-noise ratio, by summing the successive video signals. { 'vid·ē·ō 'int·əˌgrād·ər }

video masking [ELECTR] Method of removing chaff echoes and other extended clutter from radar displays. { 'vid·ē·ō 'mask·iŋ }

video monitor [COMPUT SCI] The cathode-ray-tube screen of a video display terminal. Also known as display screen; monitor. { 'vid·ē·ō 'män·əd·ər }

videophone See video telephone. { 'vid·ē·əˌfōn }

video player [ELECTR] A playback device that converts a video disk, videotape, or other type of recorded video or audio program into signals suitable for driving a home display. { 'vid·ē·ō ˌplā·ər }

video RAM [COMPUT SCI] Dynamic random-access memory optimized for use with video displays. { ˌvid·ē·ō 'ram }

video recorder [ELECTR] A magnetic tape recorder capable of storing video and audio signals for playback at a later time. { 'vid·ē·ō ri ˌkȯrd·ər }

video sensing [COMPUT SCI] In optical character recognition, a scanning technique in which the document is flooded with light from an ordinary light source, and the image of the character is reflected onto the face of a cathode-ray tube, where it is scanned by an electron beam. { 'vid·ē·ō 'sens·iŋ }

video signal [COMMUN] In analog television, the signal containing all of the visual information together with blanking and synchronizing pulses. See target signal. { 'vid·ē·ō ˌsig·nəl }

videotape [ELECTR] A magnetic tape designed primarily for recording of television programs. { 'vid·ē·ōˌtāp }

videotape recorder [ELECTR] A device for video recording and playing of a magnetic tape either in a video cassette or on an open reel. { ¦vid·ē·ō 'tāp riˌkȯrd·ər }

videotape recording [ELECTR] A method of recording video programs on magnetic tape for later rebroadcasting or replay; may also refer to the device that performs this function. Abbreviated VTR. { 'vid·ē·ō¦tāp riˌkȯrd·iŋ }

video telephone [COMMUN] A communication instrument which transmits visual images along with the attendant speech. Also known as videophone. { 'vid·ē·ō 'tel·əˌfōn }

videotex [COMMUN] An electronic home information delivery system, either teletext or videotext. { 'vid·ē·ōˌteks }

videotext [COMMUN] A computer communication service which uses information from a database, and which allows the user, equipped with a limited computer terminal, to interact with the service in selecting information to be displayed, so as to provide electronic mail, teleshopping, financial services, calculation services, and such. { ¦vid·ē·ō'tekst }

video transformer [ELECTR] A transformer designed to transfer, from one circuit to another, the signals containing picture information in a video system. { 'vid·ē·ō tranz'fȯr·mər }

video transmitter See visual transmitter. { 'vid·ē·ō tranz'mid·ər }

vidicon [ELECTR] A camera tube in which a charge-density pattern is formed by photoconduction and stored on a photoconductor surface that is scanned by an electron beam, usually of low-velocity electrons. { 'vid·əˌkän }

Vienna definition language [COMPUT SCI] A language for defining the syntax and semantics of programming languages; consists of a syntactic metalanguage for defining the syntax of programming and data structures, and a semantic metalanguage which specifies programming language semantics operationally in terms of the computations to which programs give rise during execution. { vē'en·ə ˌdef·ə'nish·ən ˌlaŋ·gwij }

viewfinder [ELECTR] An auxiliary optical or electronic device attached to a video camera so the operator can see the scene as the camera sees it. [OPTICS] A device which provides the user of a camera with the view of the subject that is focused by the lens. { 'vyüˌfīn·dər }

viewport See window. { 'vyüˌpȯrt }

viologen display [ELECTR] An electrochromic display based on an electrolyte consisting of an aqueous solution of a dipositively charged organic salt, containing a colorless cation that undergoes a one-electron reduction process to produce a purple radical cation, upon application of a negative potential to the electrode. { vē'äl·ə·jən diˌsplā }

VIR See vertical interval reference.

virgin medium [COMPUT SCI] A material designed to have data recorded on it which is as yet completely lacking any information, such as a paper tape without any punched holes, not even feed holes; in contrast to an empty medium. { 'vər·jən 'mēd·ē·əm }

virtual address [COMPUT SCI] A symbol that can be used as a valid address part but does not necessarily designate an actual location. { 'vər·chə·wəl 'ad,res }

virtual decimal point *See* assumed decimal point. { 'vər·chə·wəl 'des·məl ,pȯint }

virtual direct-access storage [COMPUT SCI] A device used with mass-storage systems, whereby data are retrieved prior to usage by a batch-processing program and automatically transcribed onto disk storage. { 'vər·chə·wəl də¦rekt ¦ak,ses 'stȯr·ij }

virtual environment *See* virtual reality. { ¦vər·chə·wəl in'vī·rən·mənt }

virtual machine [COMPUT SCI] A portion of a computer system or of a computer's time that is controlled by an operating system and functions as though it were a complete system, although in reality the computer is shared with other independent operating systems. { 'vər·chə·wəl mə'shēn }

virtual memory [COMPUT SCI] A combination of primary and secondary memories that can be treated as a single memory by programmers because the computer itself translates a program or virtual address to the actual hardware address. { 'vər·chə·wəl 'mem·rē }

virtual PPI reflectoscope [ENG] A device for superimposing a virtual image of a chart on a plan position indicator (PPI) pattern; the chart is usually prepared with white lines on a black background to the scale of the plan position indicator range scale. { 'vər·chə·wəl ¦pē¦pē'ī ri'flek·tə,skōp }

virtual private network [COMMUN] A wide-area network whose links are provided by a common carrier although they appear to the users to behave like dedicated lines, and whose computers use a common cryptographic key to send messages from one computer in the network to another. Abbreviated VPN. { ¦vər·chə·wəl ,prī·vət 'net,wərk }

virtual reality [COMPUT SCI] A simulation of an environment that is experienced by a human operator provided with a combination of visual (computer-graphic), auditory, and tactile presentations generated by a computer program. Also known as artificial reality; immersive simulation; virtual environment; virtual world. { ¦vər·chə·wəl rē'al·əd·ē }

Virtual Reality Modeling Language [COMPUT SCI] The markup specification for three-dimentional (virtual reality) objects and environments on the Web. Abbreviated VRML. { ,vər·chə·wəl rē,al·əd·ē 'mäd·əl·iŋ ,laŋ·gwij }

virtual world [COMPUT SCI] A navigable visual digital environment. { ¦vər·chə·wəl 'wərld }

virus [COMPUT SCI] A computer program that replicates itself and transfers itself to another computing system. { 'vī·rəs }

visible radiation *See* light. { 'viz·ə·bəl ,rād·ē'ā·shən }

visual-aural range [NAV] A very-high-frequency radio range that provides one course for display to the pilot on a zero-center left-right indicator and another course, at right angles to the first, in the form of aural A-N radio range signals; the A-N aural signals provide a means for differentiating between the two directions of the visual course. Abbreviated VAR. { 'vizh·ə·wəl 'ȯr·əl 'rānj }

visual display unit *See* display tube. { 'vizh·ə·wəl di'splā,yü·nət }

visualization [COMPUT SCI] The process of converting data into a geometric or graphic representation. { ,vizh·ə·lə'zā·shən }

visual radio range [NAV] Any radio range through which aircraft are flown by visual instrumentation not associated with aural reception. { 'vizh·ə·wəl 'rād·ē·ō,rānj }

visual transmitter [COMMUN] The system of devices used to transmit the visual portion of the television signal in an analog television service. { 'vizh·ə·wəl tranz'mid·ər }

Viterbi algorithm [COMMUN] A decoding procedure for convolutional codes that uses the maximum-likelihood method. { vi¦ter·bē 'al·gə ,rith·əm }

VLF *See* very low frequency.

VLSI circuit *See* very large scale integrated circuit. { ¦vē¦el¦es¦ī 'sər·kət }

VMOS technology *See* vertical metal oxide semiconductor technology. { 'vē,mȯs tek,näl·ə·jē }

vocoder [ELECTR] A system of electronic apparatus for synthesizing speech according to dynamic specifications derived from an analysis of that speech. { vō'kōd·ər }

voice call sign [COMMUN] A call sign provided primarily for voice communications. { 'vȯis 'kȯl ,sīn }

voice channel [COMMUN] A communication channel having sufficient bandwidth to carry voice frequencies intelligibly; the minimum bandwidth for an analog voice channel is about 3000 hertz for good intelligibility. { 'vȯis ,chan·əl }

voice coder [ELECTR] Device that converts speech input into digital form prior to encipherment for secure transmission and converts the digital signals back to speech at the receiver. { 'vȯis ,kō·dər }

voice/data system [COMMUN] Integrated communications system for transmitting both voice and digital data. { 'vȯis 'dad·ə ,sis·təm }

voice frequency [COMMUN] An audio frequency in the range essential for transmission of speech of commercial quality, from about 300 to 3400 hertz. Abbreviated VF. Also known as speech frequency. { 'vȯis ,frē·kwən·sē }

voice-frequency carrier telegraphy [COMMUN] Carrier telegraphy in which the carrier currents have frequencies such that the modulated currents may be transmitted over a voice-frequency telephone channel. { 'vȯis ¦frē·kwən·sē ¦kar·ē·ər tə'leg·rə·fē }

voice-frequency dialing [ELECTR] Method of dialing by which the direct-current pulses from the dial are transformed into voice-frequency alternating-current pulses. { 'vȯis ¦frē·kwən·sē ,dī·liŋ }

voice-frequency telegraph system [COMMUN] Telegraph system permitting the use of many channels on a single circuit; a different audio frequency is used for each channel, being keyed in the conventional manner; the various audio frequencies at the receiving end are separated by suitable filter circuits and fed to their respective receiving circuits. { 'vȯis ¦frē·kwən·sē 'tel·ə,graf ,sis·təm }

voice-grade channel [COMMUN] A channel whose bandwidth is large enough to transmit voice-frequency signals. { 'vȯis ¦grād ,chan·əl }

voice mail [COMMUN] A method of storing voice-recorded messages and delivering them electronically to an intended receiver. { 'vȯis ,māl }

voice-operated device [ELECTR] Any of several devices that are brought into operation by a sound signal, or some characteristic of such a signal. { 'vȯis ¦äp·ə,rād·əd di,vīs }

voice recognition unit [COMMUN] A computer peripheral device that recognizes a limited number of spoken words and converts them into equivalent digital signals which can serve as computer input or initiate other desired actions. { 'vȯis ,rek·ig'nish·ən ,yü·nət }

voice response [ENG ACOUS] A computer-controlled recording system in which basic sounds, numerals, words, or phrases are individually stored for playback under computer control as the reply to a keyboarded query. { 'vȯis ri ,späns }

voice store and forward [COMMUN] A computer-supported system that converts spoken messages to digital format, stores them temporarily, and then transmits them to a receiver where they are converted back to sound. { 'vȯis ¦stȯr ən 'fȯr·wərd }

voice synthesizer [ELECTR] A synthesizer that simulates speech in any language by assembling a language's elements or phonemes under digital control, each with the correct inflection, duration, pause, and other speech characteristics. { 'vȯis ,sin·thə,sīz·ər }

void [COMPUT SCI] In optical character recognition, an island of insufficiently inked paper within the area of the intended character stroke. { vȯid }

volatile file [COMPUT SCI] Any file in which data are rapidly added or deleted. { 'väl·əd·əl 'fīl }

volatile memory *See* volatile storage. { 'väl·əd·əl 'mem·rē }

volatile storage [COMPUT SCI] A storage device that must be continuously supplied with energy, or it will lose its retained data. Also known as volatile memory. { 'väl·əd·əl 'stȯr·ij }

voltage-controlled oscillator [ELECTR] An oscillator whose frequency of oscillation can be varied by changing an applied voltage. Abbreviated VCO. { 'vōl·tij kən¦trōld 'äs·ə,lād·ər }

voltage feed [ELECTROMAG] Excitation of a transmitting antenna by applying voltage at a point of maximum potential (at a voltage loop or antinode). { 'vōl·tij ,fēd }

voltage node [ELECTROMAG] Point having zero voltage in a stationary wave system, as in an antenna or transmission line; for example, a voltage node exists at the center of a half-wave antenna. { 'vōl·tij ,nōd }

voltage reflection coefficient [ELECTROMAG] The ratio of the phasor representing the magnitude and phase of the electric field of the backward-traveling wave at a specified cross section of a waveguide to the phasor representing the forward-traveling wave at the same cross section. { 'vōl·tij ri¦flek·shən ,kō·i,fish·ənt }

voltage step electronics [ELECTR] A sudden change in a voltage from one specified value to another at a particular instant in time. { 'vōl·tij ¦step i,lek'trän·iks }

voltage-variable capacitor *See* varactor. { 'vōl·tij ¦ver·ē·əbəl kə'pas·əd·ər }

volume [COMPUT SCI] A single unit of external storage, all of which can be read or written by a single access mechanism or input/output device. { 'väl·yəm }

volume label [COMPUT SCI] A record that contains information about the contents of a particular storage device, usually a disk or magnetic tape, and is written somewhere on that device. { 'väl·yəm ,lā·bəl }

volume table of contents [COMPUT SCI] A list of all the files in a volume, usually with descriptions of their contents and locations. Abbreviated VTOC. { 'väl·yəm ¦tā·bəl əv 'kän,tens }

volume target [ELECTROMAG] A radar target composed of a large number of objects too close together to be resolved. { 'väl·yəm 'tär·gət }

volume test [COMPUT SCI] The processing of a volume of actual data to check for program malfunction. { 'väl·yəm ,test }

volumetric radar [ENG] Radar capable of producing three-dimensional position data on a multiplicity of targets. { ¦väl·yə¦me·trik 'rā,där }

volumetric storage [COMMUN] Any data storage technology in which information is stored throughout a three-dimensional volume rather than merely on a surface. { ¦väl·yə¦me·trik 'stȯr·ij }

von Neumann bottleneck [COMPUT SCI] An inefficiency inherent in the design of any von Neumann machine that arises from the fact that most computer time is spent in moving information between storage and the central processing unit rather than operating on it. { fȯn 'nȯi,män 'bäd·əl,nek }

von Neumann machine [COMPUT SCI] A stored-program computer equipped with a program counter. { fȯn 'nȯi,män mə,shēn }

VOR *See* very high frequency omnidirectional radio range.

Vortac [NAV] A ground radio station consisting of a collocated very-high-frequency omnidirectional radio range (VOR) and Tacan facility; this station permits obtaining polar coordinates by the use of VOR receiver and distance-measuring equipment, or by Tacan equipment alone. { 'vȯr ,tak }

VOS *See* vertical obstacle sonar.

voxel [COMPUT SCI] The smallest box-shaped part of a three-dimensional image or scan. Derived from volume pixel. { 'väks·əl }

VPN *See* virtual private network.

VPR chart [NAV] A type of radar chart for use with VPR (virtual plan position indicator reflectoscope). { ,vē,pē'är ,chärt }

VRC See lateral parity check.
VRML See Virtual Reality Modeling Language.
VSAT See very small aperture terminal. { 'vē ˌsat }
VSB See vestigial sideband.
VTOC See volume table of contents.
VTR See videotape recording.

vulnerability [COMPUT SCI] A weakness in a computing system that can result in harm to the system or its operations, especially when this weakness is exploited by a hostile person or organization or when it is present in conjunction with particular events or circumstances. { ˌvəl·nə·rə'bil·əd·ē }

W

WAAS *See* Wide-Area Augmentation System. { wäs *or* ¦dəb·əl·yü¦ā¦ā'es }

waiting time *See* idle time. { 'wād·iŋ tīm }

wait state [COMPUT SCI] The state of a computer program in which it cannot use the central processing unit normally because the unit is waiting to complete an input/output operation. { 'wāt ˌstāt }

walkthrough [COMPUT SCI] A step-by-step review of a computer program or system during its design to search for errors and problems. { 'wȯk ˌthrü }

wallpaper [COMPUT SCI] The design or image used as a computer monitor background. { 'wȯl ˌpā·pər }

WAN *See* wide-area network.

wand [COMPUT SCI] A hand-held device that contains an optical scanner to sense bar codes and other patterns and transmits the data to a computer. { wänd }

warm boot [COMPUT SCI] To boot a computer system after it has been running. { ¦wȯrm 'büt }

warm start [COMPUT SCI] A resumption of computer operation, following a problem-generated shutdown, in which programs running on the system can resume at the point they were at when the shutdown occurred and data is not lost. { 'wȯrm 'stärt }

warning device [COMPUT SCI] A visible or audible alarm to inform the operator of a machine condition. { 'wȯrn·iŋ diˌvīs }

warning message [COMPUT SCI] A diagnostic message that is issued when a computer program detects an error or potential problem but continues processing. { 'wȯrn·iŋ ˌmes·ij }

watch [COMMUN] The service performed by a qualified operator when on duty in the radio room of a vessel. Also known as radio watch. { wäch }

Waterloo Fortran IV *See* WATFIV. { 'wȯd·ərˌlü 'fȯrˌtran 'fȯr }

WATFIV [COMPUT SCI] A programming language based on FORTRAN that is used in learning environments and is characterized by fast compilation and excellent diagnostic messages and debugging aids. Acronym for Waterloo Fortran IV. { 'wätˌfīv }

WATS *See* Wide Area Telephone Service. { wäts }

wave angle [ELECTROMAG] The angle, either in bearing or elevation, at which a radio wave leaves a transmitting antenna or arrives at a receiving antenna. { 'wāv ˌaŋ·gəl }

wave antenna [ELECTROMAG] Directional antenna composed of a system of parallel, horizontal conductors, varying from a half to several wavelengths long, terminated to ground at the far end in its characteristic impedance. { 'wāv anˌten·ə }

wave clutter *See* sea clutter. { 'wāv ˌkləd·ər }

wave converter [ELECTROMAG] Device for changing a wave of a given pattern into a wave of another pattern, for example, baffle-plate converters, grating converters, and sheath-reshaping converters for waveguides. { 'wāv kənˌvərd·ər }

wave duct [ELECTROMAG] **1.** Waveguide, with tubular boundaries, capable of concentrating the propagation of waves within its boundaries. **2.** Natural duct, formed in air by atmospheric conditions, through which waves of certain frequencies travel with more than average efficiency. { 'wāv ˌdəkt }

waveform-amplitude distortion *See* frequency distortion. { 'wāvˌfȯrm ¦am·pləˌtüd diˌstȯr·shən }

waveform coder *See* waveform compression. { 'wāvˌform ˌkōd·ər }

waveform compression [COMMUN] Any technique for compressing audio or video signals in which a facsimile of the source-signal waveform is replicated at the receiver with a level of distortion that is judged acceptable. Also known as waveform coder. { 'wāvˌfȯrm kəmˌpresh·ən }

wavefront reversal *See* optical phase conjugation. { 'wāvˌfrənt ri'vər·səl }

waveguide [ELECTROMAG] **1.** Broadly, a device which constrains or guides the propagation of electromagnetic waves along a path defined by the physical construction of the waveguide; includes ducts, a pair of parallel wires, and a coaxial cable. Also known as microwave waveguide. **2.** More specifically, a metallic tube which can confine and guide the propagation of electromagnetic waves in the lengthwise direction of the tube. { 'wāvˌgīd }

waveguide bend [ELECTROMAG] A section of waveguide in which the direction of the longitudinal axis is changed; an **E**-plane bend in a rectangular waveguide is bent along the narrow dimension, while an **H**-plane bend is bent along the wide dimension. Also known as waveguide elbow. { 'wāv,gīd ¦bend }

waveguide cavity [ELECTROMAG] A cavity resonator formed by enclosing a section of waveguide between a pair of waveguide windows. { 'wāv,gīd ¦kav·əd·ē }

waveguide connector [ELECTROMAG] A mechanical device for electrically joining and locking together separable mating parts of a waveguide system. Also known as waveguide coupler. { 'wāv,gīd kə¦nek·tər }

waveguide coupler *See* waveguide connector. { 'wāv,gīd ¦kəp·lər }

waveguide critical dimension [ELECTROMAG] Dimension of waveguide cross section which determines the cutoff frequency. { 'wāv,gīd 'krid·ə·kəl də'men·shən }

waveguide cutoff frequency [ELECTROMAG] Frequency limit of propagation along a waveguide for waves of a given field configuration. { 'wāv,gīd 'kəd,ȯf ,frē·kwən·sē }

waveguide elbow *See* waveguide bend. { 'wāv ,gīd ¦el·bō }

waveguide filter [ELECTROMAG] A filter made up of waveguide components, used to change the amplitude-frequency response characteristic of a waveguide system. { 'wāv,gīd ¦fil·tər }

waveguide hybrid [ELECTROMAG] A waveguide circuit that has four arms so arranged that a signal entering through one arm will divide and emerge from the two adjacent arms, but will be unable to reach the opposite arm. { 'wāv,gīd 'hī·brəd }

waveguide junction *See* junction. { 'wāv,gīd ¦jəŋk·shən }

waveguide plunger *See* piston. { 'wāv,gīd ¦plən·jər }

waveguide probe *See* probe. { 'wāv,gīd ¦prōb }

waveguide propagation [COMMUN] Long-range communications in the 10- to 35-kilohertz frequency range that are made possible by the waveguide characteristics of the atmospheric wave duct formed by the ionospheric D layer and the surface of the earth. { 'wāv,gīd ,präp·ə'gā·shən }

waveguide resonator *See* cavity resonator. { 'wāv,gīd ¦rez·ən,ād·ər }

waveguide shim [ELECTROMAG] Thin resilient metal sheet inserted between waveguide components to ensure electrical contact. { 'wāv,gīd ,shim }

waveguide slot [ELECTROMAG] A slot in a waveguide wall, either for coupling with a coaxial cable or another waveguide, or to permit the insertion of a traveling probe for examination of standing waves. { 'wāv,gīd ,slät }

waveguide switch [ELECTROMAG] A switch designed for mechanically positioning a waveguide section so as to couple it to one of several other sections in a waveguide system. { 'wāv ,gīd ,swich }

waveguide window *See* iris. { 'wāv,gīd ¦win·dō }

wave impedance [ELECTROMAG] The ratio, at every point in a specified plane of a waveguide, of the transverse component of the electric field to the transverse component of the magnetic field. { 'wāv im,pēd·əns }

wavelength constant *See* phase constant. { 'wāv,leŋkth ,kän·stənt }

wavelength-division multiplexing [COMMUN] The sharing of the total available pass-band of a transmission medium in the optical portion of the electromagnetic spectrum by assigning individual information streams to signals of different wavelengths. Abbreviated WDM. { ¦wāv,leŋth də¦vizh·ən 'məl·tə,pleks·iŋ }

wave-shaping circuit [ELECTR] An electronic circuit used to create or modify a specified time-varying electrical quantity, usually voltage or current, using combinations of electronic devices, such as vacuum tubes or transistors, and circuit elements, including resistors, capacitors, and inductors. { 'wāv ¦shāp·iŋ ,sər·kət }

wave tail [ELECTR] Part of a signal-wave envelope (in time or distance) between the steady-state value (or crest) and the end of the envelope. { 'wāv ,tāl }

wave tilt [ELECTROMAG] Forward inclination of a radio wave due to its proximity to ground. { 'wāv ,tilt }

wave trap [ELECTR] A resonant circuit connected to the antenna system of a receiver to suppress signals at a particular frequency, such as that of a powerful local station that is interfering with reception of other stations. Also known as trap. { 'wāv ,trap }

way point *See* via point. { 'wā ,pȯint }

WDM *See* wavelength-division multiplexing.

weather radar [ENG] Radar specifically designed to enhance the echo from rain, other precipitates, and weather conditions in general, and with processing techniques by which reflectivity of the scatterers, wind speeds, and direction all can be accurately estimated. Also known as weather observation radar. { 'weth·ər 'rā,där }

Web *See* World Wide Web. { web }

Web browser *See* browser. { 'web ,braůz·ər }

Web page [COMPUT SCI] A document written in HTML and available for viewing on the World Wide Web; it may contain images, sound, video, formatted text, and hyperlinks. { 'web ,pāj }

Web server [COMPUT SCI] A program that processes document requests; it also has a database, which is a repository of data and content. { 'web ,sər·vər }

Web services [COMPUT SCI] A collection of SML-based standards that enable electronic communication and interaction independently of the computer platforms or specific technologies used by the communication parties. { 'web ,serv·ə·səs }

Web site [COMPUT SCI] A collection of thematically related, hyperlinked World Wide Web services, mainly HTML documents, usually located on a specific Web server and reachable through a URL assigned to the site. { 'web ,sīt }

wedge [COMMUN] A convergent pattern of equally spaced black and white lines, used in a television test pattern to indicate resolution. [ELECTROMAG] A waveguide termination consisting of a tapered length of dissipative material introduced into the guide, such as carbon. { wej }

weighted area masks [COMPUT SCI] In character recognition, a set of characters (each character residing in the character reader in the form of weighted points) which theoretically render all input specimens unique, regardless of the size or style. { 'wād·əd 'er·ē·ə ,masks }

weighted code [COMPUT SCI] A method of representing a decimal digit by a combination of bits, in which each bit is assigned a weight, and the value of the decimal digit is found by multiplying each bit by its weight and then summing the results. { 'wād·əd 'kōd }

weighted quasi peak [ELECTR] Referring to a fast attack, slow-decay detector circuit that approximately responds to signal peaks, and that has varying attenuation as a function of frequency so as to produce a measurement that approximates the human hearing system. Abbreviated WQP. { 'wād·əd 'kwä·zē 'pēk }

welcome page *See* home page. { 'wel·kəm ,pāj }

well-regulated system [CONT SYS] A system with a regulator whose action, together with that of the environment, prevents any disturbance from permanently driving the system from a state in which it is stable, that is, a state in which it retains its structure and survives. { 'wel ¦reg·yə,lād·əd ,sis·təm }

wheel printer [COMPUT SCI] A line printer that prints its characters from the rim of a wheel around which is the type for the alphabet, numerals, and other characters. { 'wēl 'print·ər }

whiffletree switch [ELECTR] In computers, a multiposition electronic switch composed of gate tubes and flip-flops, so named because its circuit diagram resembles a whiffletree. { 'wif·əl ,trē ,swich }

WHILE statement [COMPUT SCI] A statement in a computer program that is executed repeatedly, as long as a specified condition holds true. { 'wīl ,stāt·mənt }

whip antenna [ELECTROMAG] A flexible vertical rod antenna, used chiefly on vehicles. Also known as fishpole antenna. { 'wip an,ten·ə }

white compression [COMMUN] In analog television the reduction in picture-signal gain at levels corresponding to light areas, with respect to the gain at the level for midrange light values; the overall effect of white compression is to reduce contrast in the highlights of the picture. { 'wīt kəm,presh·ən }

white level [COMMUN] The carrier signal level corresponding to maximum picture brightness in analog television. { 'wīt ,lev·əl }

white noise [COMMUN] *See* additive white Gaussian noise. [PHYS] Random noise that has a constant energy per unit bandwidth at every frequency in the range of interest. { 'wīt ,nȯiz }

white signal [COMMUN] Signal at any point in a facsimile system produced by the scanning of a minimum density area of the subject copy. { 'wīt ,sig·nəl }

white transmission [COMMUN] **1.** In an amplitude-modulated system, that form of transmission in which the maximum transmitted power corresponds to the minimum density of the subject copy. **2.** In a frequency-modulation system, that form of transmission in which the lowest transmitted frequency corresponds to the minimum density of the subject copy. { 'wīt tranz'mish·ən }

wide-aperature digital VOR [NAV] A very high-frequency omnidirectional radio range in which azimuth angle is determined from a system of crossed wide-baseline interferometers that measure the angles between the line of sight and each of the two baselines. { 'wīd ¦ap·ə·chər 'dij·əd·əl ,vē,ō'är }

Wide-Area Augmentation System [NAV] A satellite-based augmentation system developed by the Federal Aviation Administration in the United States. Abbreviated WAAS. { ¦wīd ¦er·ē·ə ȯg·mən'tā·shən ,sis·təm }

wide-area DGPS [NAV] A version of differential GPS which provides error corrections over a large geographic area, based on the measurements by a widely distributed network of monitor stations that are processed at a centrally located facility. { ¦wīd ¦er·ē·ə ¦dē¦jē¦pē'es }

wide-area network [COMMUN] A computer or telecommunication system consisting of a set of nodes that are interconnected by a set of links, and generally covers a large geographic area, usually on the order of hundreds of miles. Abbreviated WAN. { 'wīd ¦er·ē·ə 'net,wərk }

Wide Area Telephone Service [COMMUN] A special telephone service that allows a customer to call anyone in one or more of six regions into which the continental United States has been divided, on a direct dialing basis, for a flat monthly charge related to the number of regions to be called. Abbreviated WATS. { 'wīd ¦er·ē·ə 'tel·ə,fōn ,sər·vəs }

wideband [ELECTR] Property of a tuner, amplifier, or other device that can pass a broad range of frequencies. { 'wīd¦band }

wideband amplifier [ELECTR] An amplifier that will pass a wide range of frequencies with substantially uniform amplification. { 'wīd¦band 'am·plə,fī·ər }

wideband communications system [COMMUN] Communications system which provides numerous channels of communications on a highly reliable and secure basis which are relatively invulnerable to interruption by natural phenomena or countermeasures; included are multichannel telephone cable, tropospheric scatter, multichannel line-of-sight radio system such as microwave, and satellites. { 'wīd ¦band kə ,myü·nə'kā·shənz ,sis·təm }

wideband ratio [COMMUN] Ratio in a system of the occupied frequency bandwidth to the intelligence bandwidth. { 'wīd ¦band 'rā·shō }

wide-band repeater [ELECTR] Airborne system that receives a radio-frequency signal and relays it to another facility via a separate RF channel. { 'wīd ¦band ri'pēd·ər }

Widrow-Hoff least-mean-squares algorithm [COMMUN] An algorithm that is widely used in adaptive signal processing; for time-discrete analysis with a finite-response filter, it is represented by the first-order difference equation $\mathbf{W}_{k+1} = \mathbf{W}_k + 2\mu e_k \mathbf{X}_k$, where k is a time index that takes on integral values; **W** is a vector whose components are the coefficients of the filter; μ is the convergence coefficient; e_k is the residual signal or error, equal to $y_k - \hat{y}_k$, where y_k and $\hat{y}_k$ are the outputs of the plant (where the unprocessed signal is generated) and the filter, respectively; and **X** is a vector whose components are the present value of the input and L − 1 past values of the input, where L is the number of filter coefficients. { ¦wi·drō ¦häf ˈlēst ˌmēn ˈskwerz ˈal·gəˌrith·əm }

width [COMMUN] **1.** The horizontal dimension of a video picture. **2.** The time duration of a pulse. { width }

width control [ELECTR] Control that adjusts the width of the image on the screen of a video display. { ˈwidth kənˌtrōl }

WiFi™ [COMMUN] An acronym for Wireless Fidelity that denotes equipment conforming to the Institute of Electrical and Electronic Engineers technical specifications for wireless Local Area Networks (IEEE 802.11 standards). Wi-Fi networks utilize unlicensed radio frequencies but are compatible with and may be connected to a wired Ethernet Local Area Network. A typical application is the wireless, high speed connection of a portable computer to the Internet. { ˈwī·ˈfī }

wild card [COMPUT SCI] A symbolic character in a search argument such that any character will satisfy it. { ˈwīld ˌkärd }

Winchester disk [COMPUT SCI] A type of disk storage device characterized by nonremovable or sealed disk packs; extremely narrow tracks; a lubricated surface that allows the head to rest on the surface during start and stop operations; and servomechanisms which utilize a magnetic pattern, recorded on the medium itself, to position the head. { ˈwin·ches·tər ˌdisk }

Winchester technology [COMPUT SCI] Innovations designed to achieve disks with up to 6×10^8 bytes per disk drive; the technology includes nonremovable or sealed disk packs, a read/write head that weighs only 0.25 gram and floats above the surface, magnetic orientation of iron oxide particles on the disk surface, and lubrication of the disk surface. { ˈwin·ches·tər tekˈnäl·ə·jē }

wind [ELECTR] The manner in which magnetic tape is wound onto a reel; in an A wind, the coated surface faces the hub; in a B wind, the coated surface faces away from the hub. { wind }

window [COMPUT SCI] A separate viewing area on a display screen that is established by the computer software. Also known as viewport. [ELECTROMAG] A hole in a partition between two cavities or waveguides, used for coupling. { ˈwin·dō }

window editor [COMPUT SCI] An interactive program that allows the user to view and alter stored information by using the video display. { ˈwin·dōˌed·əd·ər }

windowing [COMPUT SCI] **1.** The procedure of selecting a portion of a large drawing to be displayed on the screen of a computer graphics system, usually by placing a rectangular window over a compressed version of the entire drawing displayed on the screen. **2.** Dividing an electronic display into areas that display the outputs of different programs and can overlap in the same manner as pieces of paper on a desk, partially concealing the contents of pages underneath. { ˈwinˌdō·iŋ }

wired-program computer [COMPUT SCI] A computer in which the sequence of instructions that form the operating program is created by interconnection of wires on a removable control panel. { ˈwīrd ¦prōˌgram kəmˈpyüd·ər }

wire facsimile system [COMMUN] A facsimile system in which messages are sent over wires or cables, rather than by radio. { ˈwīr fakˈsim·ə·lē ˌsis·təm }

wireframe [COMPUT SCI] **1.** In computer-aided design, a line-drawn model. **2.** In computer graphics, an image-rendering technique in which only edges and vertices are shown. { ˌwīrˈfrām }

wireframe model [COMPUT SCI] In computer-aided design, the representation of all surfaces of a three-dimensional object in outline form. { ˈwīrˌfrām ˌmäd·əl }

wiregrating [ELECTROMAG] A series of wires placed in a waveguide that allow one or more types of waves to pass and block all others. { ˈwīrˌgrād·iŋ }

wireless cable [COMMUN] A television broadcasting system in which signals are collected and transmitted to towers for net transmission to homes outfitted with appropriate antennas and receiving equipment. { ˌwīr·ləs ˈkā·bəl }

wireless LAN [COMMUN] A local-area network whose devices or telephones communicate by radio transmissions.s Abbreviated WLAN. { ˈwīr·ləs ˈlan }

wire line [ELECTR] One or more current-conducting wires or cables, used for communication, control, or telemetry. { ˈwīr ˌlīn }

wire-link telemetry [COMMUN] Telemetry in which electric signals are sent over transmission lines, rather than by radio. Also known as hard-wire telemetry. { ˈwīr ¦liŋk təˈlem·ə·trē }

wirephoto [COMMUN] **1.** A photograph transmitted over wires to a facsimile receiver. **2.** *See* facsimile. { ˈwīrˌfōd·ō }

wire printing *See* matrix printing. { ˈwīr ˌprint·iŋ }

wiretap [COMMUN] A concealed connection to a telephone line, office intercommunication line, or other wiring system, for the purpose of monitoring conversations and activities in a room from a remote location without knowledge of the participants, legally or illegally. { ˈwīr ˌtap }

wire telegraphy [COMMUN] Telegraphy in which messages are sent over wires or cables, rather than by radio. { ˈwīr təˈleg·rə·fē }

wiring board *See* control panel. { 'wīr·iŋ ,bȯrd }

WLAN *See* wireless LAN..

word [COMPUT SCI] The fundamental unit of storage capacity for a digital computer, almost always considered to be more than eight bits in length. Also known as computer word. { wȯrd }

word-addressable computer *See* word-oriented computer. { 'wȯrd ə¦dres·ə·bəl kəm'pyüd·ər }

word boundary [COMPUT SCI] A storage address that is a multiple of the word length of a computer. { 'wȯrd ,bau̇n·drē }

word format [COMPUT SCI] Arrangement of characters in a word, with each position or group of positions in the word containing certain specified data. { 'wȯrd 'fȯr,mat }

word length [COMPUT SCI] The number of bits, digits, characters, or bytes in one word. { 'wȯrd ,leŋkth }

word mark [COMPUT SCI] A nondata punctuation bit used to delimit a word in a variable-word-length computer. { 'wȯrd ,märk }

word-oriented computer [COMPUT SCI] A computer in which the locations of words are addressed, and the bits and characters within the words can be addressed only through use of special instructions. Also known as word-addressable computer. { 'wȯrd ¦ȯr·ēn·təd kəm'pyüd·ər }

word processing [COMPUT SCI] The use of computers or computerlike equipment to write, edit, and format text. { 'wȯrd ¦prä,ses·iŋ }

word processor [COMPUT SCI] **1.** A computer that is either dedicated to word processing or is used with a software package that supports word processing, together with a printer. **2.** A person who operates such a device. { 'wərd ,prä,ses·ər }

word rate [COMPUT SCI] In computer operations, the frequency derived from the elapsed period between the beginning of the transmission of one word and the beginning of the transmission of the next word. { 'wȯrd ,rāt }

word time *See* minor cycle. { 'wərd ,tīm }

word wrap [COMPUT SCI] A procedure whereby a word processor automatically ends each line when it is full and starts the next line with the next word, never breaking a word. Also known as wrap mode. { 'wərd ,rap }

work assembly [COMPUT SCI] The clerical activities related to organizing collections of data records and computer programs or series of related programs. { 'wərk ə,sem·blē }

work file [COMPUT SCI] A file created to hold data temporarily during processing. { 'wərk ,fīl }

working program [COMPUT SCI] A valid program which, when translated into machine language, can be executed on a computer. { 'wərk·iŋ 'prō ,gram }

working set [COMPUT SCI] The smallest collection of instruction and data words of a given computer program which should be loaded into the main storage of a computer system so that efficient processing is possible. { 'wərk·iŋ 'set }

working-set window [COMPUT SCI] A fixed time interval during which the working set is referenced. { 'wərk·iŋ ¦set 'win·dō }

working space *See* working storage. { 'wərk·iŋ ,spās }

working storage [COMPUT SCI] **1.** An area of main memory that is reserved by the programmer for storing temporary or intermediate values. Also known as working space. **2.** In COBOL (computer language), a section in the data division used for describing the name, structure, usage, and initial value of program variables that are neither constants nor records of input/output files. { 'wərk·iŋ 'stȯr·ij }

workspace [COMPUT SCI] In a string processing language, the portion of computer memory that contains the string currently being processed. { 'wərk,spās }

work station [COMPUT SCI] A workplace where a person can interact with a computer on a conversational basis, either a microcomputer and printer or a terminal connected to a remote computer. { 'wərk ,stā·shən }

work tape [COMPUT SCI] A magnetic tape that is available for general use during data processing. { 'wərk ,tāp }

World Wide Web [COMPUT SCI] A part of the Internet that contains linked text, image, sound, and video documents. Abbreviated WWW. Also known as Web. { ¦wərld ¦wīd 'web }

worm [COMPUT SCI] A computer program that seeks to replicate itself and to spread, with the goal of consuming and exhausting computer resources, thereby causing computing systems to fail. { wərm }

WORM [COMPUT SCI] Pertaining to a storage device, such as an optical disk, that allows the user to record data only once and to read back the data an unlimited number of times. Abbreviation for write-one, read-many. { wərm }

worst case evaluation [COMPUT SCI] A testing situation in which the most unfavorable possible combination of circumstances is evaluated. { 'wərst ¦kās i,val·yə'wā·shən }

woven-screen storage [COMPUT SCI] Digital storage plane made by weaving wires coated with thin magnetic films; when currents are sent through a selected pair of wires that are at right angles in the screen, storage and readout occur at the intersection of the two wires. { 'wō·vən ¦skrēn 'stȯr·ij }

WQP *See* weighted quasi peak.

wrap mode *See* word wrap. { 'rap ,mōd }

writable control storage [COMPUT SCI] A section of the control storage holding microprograms which can be loaded from a console file or under microprogramming control. { 'rīd·ə·bəl kən'trōl ,stȯr·ij }

write [COMPUT SCI] **1.** To transmit data from any source onto an internal storage medium. **2.** A command directing that an output operation be performed. { rīt }

write enable ring [COMPUT SCI] A file protection ring that must be attached to the hub of a reel of magnetic tape in order to physically allow data to be transcribed onto the reel. Also known as write ring. { 'rīt i¦nā·bəl ,riŋ }

write error [COMPUT SCI] **1.** A condition in which information cannot be written onto or into a storage device, due to dust, dirt, damage to the recording surface, or damaged electronic components. **2.** A condition in which there is an inconsistency between the pattern of bits transmitted to the write head of a magnetic tape drive and the pattern sensed immediately afterward by the read head. { 'rīt ,er·ər }

write head [ELECTR] Device that stores digital information as coded electrical pulses on a storage medium, such as a disk or tape. { 'rīt ,hed }

write inhibit ring [COMPUT SCI] A file protection ring that physically prevents data from being written on a reel of magnetic tape when it is attached to the hub of the reel. { 'rīt in¦hib·ət ,riŋ }

write-once, read-many *See* WORM. { ¦rīt 'wəns ¦rēd 'men·ē }

write protection [COMPUT SCI] **1.** Any procedure used to prevent writing on storage media. **2.** Any software technique that allows a computer program to read from any area in storage but not to write outside its own area. { 'rīt prə,tek·shən }

writer [COMPUT SCI] The part of a job entry system that controls output, in particular, the printer and the spool file. { 'rīd·ər }

write ring *See* write enable ring. { 'rīt ,riŋ }

write time [COMPUT SCI] The time required to transcribe a data item into a computer storage device. { 'rīt ,tīm }

write to operator [COMPUT SCI] A message issued by a computer program and displayed on the system console that provides information or indicates the status of the program and requires no action by the operator. Abbreviated WTO. { 'rīt tü 'äp·ə,rād·ər }

write to operator with reply [COMPUT SCI] A message issued by a computer program and displayed on the system console that requires action by the operator in order for execution of the program to continue. Abbreviated WTOR. { 'rīt tü 'äp·ə,rād·ər wit͟h ri'plī }

WTO *See* write to operator.

WTOR *See* write to operator with reply.

Wullenweber antenna [ELECTROMAG] An antenna array consisting of two concentric circles of masts, connected to be electronically steerable; used for ground-to-air communication at Strategic Air Command bases. { 'wu̇l·ən,web·ər an,ten·ə }

WWV [COMMUN] The call letters of a radio station maintained by the National Institute of Standards and Technology to provide standard radio and audio frequencies and other technical services, such as precision time signals and radio propagation disturbance warnings; the station broadcasts on 2.5, 5, 10, 15, and 20 megahertz 24 hours per day, 7 days per week at various times.

WWVH [COMMUN] Maintained by the National Institute of Standards and Technology, the radio station at Maui, Hawaii, broadcasting services similar to those of WWV on 5, 10, and 15 megahertz.

WWW *See* World Wide Web.

X band [COMMUN] A radio-frequency band extending from 8 to 12 gigahertz. { 'eks ,band }

XML *See* Extensible Markup Language.

XOFF *See* transmitter off. { eks'óf }

XON *See* transmitter on. { eks'ón }

XOR *See* exclusive or. { eks'ór }

X server [COMPUT SCI] Software that draws the screen image and handles standard input in an X Windows System; in contrast to typical usage of the term "server"; an X server is located on the user's computer; the client is the application that is displayed, which may be located on a remote node of the network. { 'eks ,ser·vər }

XS-3 code *See* excess-three code. { 'ek,ses 'thrē ,kōd }

X Windows System [COMPUT SCI] A graphical environment providing window management for computer applications; originally developed to provide a graphical user interface for Unix systems, it has been ported to other platforms. { 'eks 'win,dōz ,sis·təm }

Y

Yagi antenna *See* Yagi-Uda antenna. { 'yäg·ē an,ten·ə }

Yagi-Uda antenna [ELECTROMAG] An end-fire antenna array having maximum radiation in the direction of the array line; it has one dipole connected to the transmission line and a number of equally spaced unconnected dipoles mounted parallel to the first in the same horizontal plane to serve as directors and reflectors. Also known as Uda antenna; Yagi antenna. { 'yäg·ē 'üd·ə an ,ten·ə }

Y circulator [ELECTROMAG] Circulator in which three identical rectangular waveguides are joined to form a symmetrical Y-shaped configuration, with a ferrite post or wedge at its center; power entering any waveguide will emerge from only one adjacent waveguide. { 'wī 'sər·kyə,lād·ər }

yig device [ELECTR] A filter, oscillator, parametric amplifier, or other device that uses an yttrium-iron-garnet crystal in combination with a variable magnetic field to achieve wide-band tuning in microwave circuits. Derived from yttrium-iron-garnet device. { 'yig di,vīs }

yig filter [ELECTR] A filter consisting of an yttrium-iron-garnet crystal positioned in a magnetic field provided by a permanent magnet and a solenoid; tuning is achieved by varying the amount of direct current through the solenoid; the bias magnet serves to tune the filter to the center of the band, thus minimizing the solenoid power required to tune over wide bandwidths. { 'yig ,fil·tər }

yig-tuned parametric amplifier [ELECTR] A parametric amplifier in which tuning is achieved by varying the amount of direct current flowing through the solenoid of a yig filter. { 'yig ¦tünd ¦par·ə¦me·trik 'am·plə,fī·ər }

yig-tuned tunnel-diode oscillator [ELECTR] Microwave oscillator in which precisely controlled wide-band tuning is achieved by varying the current through a tuning solenoid that acts on a yig filter in the tunnel-diode oscillator circuit. { 'yig ¦tünd ¦tən·əl ¦dī,ōd 'äs·ə,lād·ər }

YIQ [COMMUN] The method of representing colors in color video systems, where Y represents the luminosity, which is also the black-and-white signal, I represents red minus the luminosity, and Q represents blue minus the luminosity.

Y junction [ELECTROMAG] A waveguide in which the longitudinal axes of the waveguide form a Y. { 'wī ,jəŋk·shən }

yoke [COMPUT SCI] Two or more read/write heads that are physically joined together and move as a unit over a disk, so that it is possible to read from or write to adjacent tracks without moving the head. [ELECTR] *See* deflection yoke. { yōk }

y parameter [ELECTR] One of a set of four transistor equivalent-circuit parameters, used especially with field-effect transistors, that conveniently specify performance for small voltage and current in an equivalent circuit; the equivalent circuit is a current source with shunt impedance at both input and output. { 'wī pə ,ram·əd·ər }

Y signal *See* luminance signal. { 'wī ,sig·nəl }

yttrium-iron-garnet device *See* yig device. { ¦i·trē·əm ¦ī·ərn ¦gär·nət di,vīs }

Z

Zepp antenna [ELECTROMAG] Horizontal antenna which is a multiple of a half-wavelength long and is fed at one end by one lead of a two-wire transmission line that is some multiple of a quarter-wavelength long. { 'zep an,ten·ə }

zero-access instruction [COMPUT SCI] An instruction consisting of an operation which does not require the designation of an address in the usual sense; for example, the instruction, "shift left 0003," has in its normal address position the amount of the shift desired. { 'zir·ō ¦ak,ses in'strək·shən }

zero-access storage [COMPUT SCI] Computer storage for which waiting time is negligible. { 'zir·ō ¦ak,ses 'stȯr·ij }

zero-address instruction format [COMPUT SCI] An instruction format in which the instruction contains no address, used when an address is not needed to specify the location of the operand, as in repetitive addressing. Also known as addressless instruction format. { 'zir·ō ¦ad ,res in¦strək·shən ,fȯr,mat }

zero-beat reception *See* homodyne reception. { 'zir·ō ¦bēt ri'sep·shən }

zero compression [COMPUT SCI] Any of a number of techniques used to eliminate the storage of nonsignificant leading zeros during data processing in a computer. { 'zir·ō kəm'presh·ən }

zero condition [COMPUT SCI] The state of a magnetic core or other computer memory element in which it represents the value 0. Also known as nought state; zero state. { 'zir·ō kən,dish·ən }

zero error [ELECTR] Delay time occurring within the transmitter and receiver circuits of a radar system; for accurate range data, this delay time must be compensated for in the calibration of the radar display or other time-based determination of range. { 'zir·ō 'er·ər }

zero fill [COMPUT SCI] To place leading zeros in the portion of a field to the left of a numeric value. { 'zir·ō ,fil }

zero flag [COMPUT SCI] A bit in a status register that is set to 1 to indicate that another register in the central processing unit contains all zeros or that two compared values are equal, and is set to 0 to indicate the contrary. { 'zir·ō ,flag }

zero-level address [COMPUT SCI] The operand contained in an instruction so structured as to make immediate use of the operand. { 'zir·ō ¦lev·əl 'ad,res }

zero output [ELECTR] **1.** Voltage response obtained from a magnetic cell in a zero state by a reading or resetting process. **2.** Integrated voltage response obtained from a magnetic cell in a zero state by a reading or resetting process; a ratio of a one output to a zero output is a one-to-zero ratio. { 'zir·ō 'au̇t,pu̇t }

zero state *See* zero condition. { 'zir·ō ,stāt }

zero subcarrier chromaticity [COMMUN] Chromaticity, in color television, which is intended to be displayed when the subcarrier amplitude is zero. { 'zir·ō səb¦kar·ē·ər ,krō·mə'tis·əd·ē }

zero suppression [COMPUT SCI] A process of replacing leading (nonsignificant) zeros in a numeral by blanks; it is an editing operation designed to make computable numerals easily readable to the human eye. { 'zir·ō sə'presh·ən }

zigzag reflections [ELECTROMAG] From a layer of the ionosphere, high-order multiple reflections which may be of abnormal intensity; they occur in waves which travel by multihop ionosphere reflections and finally turn back toward their starting point by repeated reflections from a slightly curved or sloping portion of an ionized layer. { 'zig,zag ri,flek·shənz }

zip [COMPUT SCI] Open standard for file compression and decompression used with personal computers. { zip }

Ziv-Lempel compression [COMPUT SCI] A data compression technique in which data is represented by a sequence of numbers standing for the positions of character strings in a dictionary; this dictionary initially contains every character in the alphabet and is continually enlarged by forming new strings from the string just compressed and the upcoming character in the text. { 'ziv 'lem·pəl kəm'presh·ən }

zone [COMPUT SCI] **1.** One of the top three rows of a punched card, namely, the 11, 12, and zero rows. **2.** *See* storage area. { zōn }

zone bit [COMPUT SCI] One of a set of bits used to indicate some grouping of characters. { 'zōn ,bit }

zone blanking [ELECTR] Method of turning off the cathode-ray tube during part of the sweep of an antenna. { 'zōn 'blaŋk·iŋ }

zoned decimal [COMPUT SCI] A format for use with EBCDIC input and output in which the first four bits of each character are called the zone portion, the second four bits, called the data portion, contain a hexadecimal digit, and the zone portion of the lowest-order character may indicate the sign of an integer; thus, + 1234 (in hexadecimal notation) would be represented as 1111/0001/1111/0010/1111/0011/1100/0100. { 'zōnd 'des·məl }

zoning [ELECTROMAG] The displacement of various portions of the lens or surface of a microwave reflector so the resulting phase front in the near field remains unchanged. Also known as stepping. { 'zōn·iŋ }

Z parameter [ELECTR] One of a set of four transistor equivalent-circuit parameters; they are the inverse of the Y parameters. { 'zē pə,ram·əd·ər }

z-transfer function See pulsed transfer function. { 'zē 'tranz·fər ,fəŋk·shən }

z-transform [MATH] The *z*-transform of a sequence whose general term is f_n is the sum of a series whose general term is $f_n z^{-n}$, where z is a complex variable; n runs over the positive integers for a one-sided transform, over all the integers for a two-sided transform. { 'zē 'tranz ,fȯrm }

Appendix

Appendix

Equivalents of commonly used units for the U.S. Customary System and the metric system

1 inch = 2.5 centimeters (25 millimeters)	1 centimeter = 0.4 inch	1 inch = 0.083 foot
1 foot = 0.3 meter (30 centimeters)	1 meter = 3.3 feet	1 foot = 0.33 yard (12 inches)
1 yard = 0.9 meter	1 meter = 1.1 yards	1 yard = 3 feet (36 inches)
1 mile = 1.6 kilometers	1 kilometer = 0.62 mile	1 mile = 5280 feet (1760 yards)
1 acre = 0.4 hectare	1 hectare = 2.47 acres	
1 acre = 4047 square meters	1 square meter = 0.00025 acre	
1 gallon = 3.8 liters	1 liter = 1.06 quarts = 0.26 gallon	1 quart = 0.25 gallon (32 ounces; 2 pints)
1 fluid ounce = 29.6 milliliters	1 milliliter = 0.034 fluid ounce	1 pint = 0.125 gallon (16 ounces)
32 fluid ounces = 946.4 milliliters		1 gallon = 4 quarts (8 pints)
1 quart = 0.95 liter	1 gram = 0.035 ounce	1 ounce = 0.0625 pound
1 ounce = 28.35 grams	1 kilogram = 2.2 pounds	1 pound = 16 ounces
1 pound = 0.45 kilogram	1 kilogram = 1.1 3 10^{-3} ton	1 ton = 2000 pounds
1 ton = 907.18 kilograms		
°F = (1.8 × °C) + 32	°C = (°F − 32) ÷ 1.8	

Conversion factors for the U.S. Customary System, metric system, and International System

A. Units of length

Units	*cm*	*m*	*in.*	*ft*	*yd*	*mi*
1 cm	= 1	0.01	0.3937008	0.03280840	0.01093613	6.213712×10^{-6}
1 m	= 100.	1	39.37008	3.280840	1.093613	6.213712×10^{-4}
1 in.	= 2.54	0.0254	1	0.08333333 . . .	0.02777777 . . .	1.578283×10^{-5}
1 ft	= 30.48	0.3048	12.	1	0.3333333 . . .	$1.893939... \times 10^{-4}$
1 yd	= 91.44	0.9144	36.	3.	1	$5.681818... \times 10^{-4}$
1 mi	$= 1.609344 \times 10^{5}$	1.609344×10^{3}	6.336×10^{4}	5280.	1760.	1

B. Units of area

Units	*cm*^{2}	*m*^{2}	*in.*^{2}	*ft*^{2}	*yd*^{2}	*mi*^{2}
1 cm^2	= 1	10^{-4}	0.1550003	1.076391×10^{-3}	1.195990×10^{-4}	3.861022×10^{-11}
1 m^2	$= 10^{4}$	1	1550.003	10.76391	1.195990	3.861022×10^{-7}
1 $in.^2$	= 6.4516	6.4516×10^{-4}	1	$6.944444... \times 10^{-3}$	7.716049×10^{-4}	2.490977×10^{-10}
1 ft^2	= 929.0304	0.09290304	144.	1	0.1111111 . . .	3.587007×10^{-8}
1 yd^2	= 8361.273	0.8361273	1296.	9.	1	3.228306×10^{-7}
1 mi^2	$= 2.589988 \times 10^{10}$	2.589988×10^{6}	4.014490×10^{9}	2.78784×10^{7}	3.0976×10^{6}	1

C. Units of volume

Units	m^3	cm^3	liter	$in.^3$	ft^3	qt	gal
1 m^3	= 1	10^6	10^3	6.102374×10^4	35.31467×10^{-3}	1.056688	264.1721
1 cm^3	= 10^{-6}	1	10^{-3}	0.06102374	3.531467×10^{-5}	1.056688×10^{-3}	2.641721×10^{-4}
1 liter	= 10^{-3}	1000.	1	61.02374	0.03531467	1.056688	0.2641721
1 $in.^3$	= 1.638706×10^{-5}	16.38706	0.01638706	1	5.787037×10^{-4}	0.01731602	4.329004×10^{-3}
1 ft^3	= 2.831685×10^{-2}	28316.85	28.31685	1728.	1.	2.992208	7.480520
1 qt	= 9.463529×10^{-4}	946.3529	0.9463529	57.75	0.03342014	1	0.25
1 gal (U.S.)	= 3.785412×10^{-3}	3785.412	3.785412	231.	0.1336806	4.	1

D. Units of mass

Units	g	kg	oz	lb	metric ton	ton
1 g	= 1	10^{-3}	0.03527396	2.204623×10^{-3}	10^{-6}	1.102311×10^{-6}
1 kg	= 1000.	1	35.27396	2.204623	10^{-3}	1.102311×10^{-3}
1 oz (avdp)	= 28.34952	0.02834952	1	0.0625	2.834952×10^{-5}	3.125×10^{-5}
1 lb (avdp)	= 453.5924	0.4535924	16.	1	4.535924×10^{-4}	$5. \times 10^{-4}$
1 metric ton	= 10^8	1000.	35273.96	2204.623	1	1.102311
1 ton	= 907184.7	907.1847	32000.	2000.	0.9071847	1

Conversion factors for the U.S. Customary System, metric system, and International System (*cont.*)

E. Units of density

Units	$g \cdot cm^{-3}$	$g \cdot L^{-1}$, $kg \cdot m^{-3}$	$oz \cdot in.^{-3}$	$lb \cdot in.^{-3}$	$lb \cdot ft^{-3}$	$lb \cdot gal^{-1}$
1 $g \cdot cm^{-3}$	= 1	1000.	0.5780365	0.03612728	62.42795	8.345403
1 $g \cdot L^{-1}$, $kg \cdot m^{-3}$	= 10^{-3}	1	5.780365×10^{-4}	3.612728×10^{-5}	0.06242795	8.345403×10^{-3}
1 $oz \cdot in.^{-3}$	= 1.729994	1729.994	1	0.0625	108.	14.4375
1 $lb \cdot in.^{-3}$	= 27.67991	27679.91	16.	1	1728.	231.
1 $lb \cdot ft^{-3}$	= 0.01601847	16.01847	9.259259×10^{-3}	5.787037×10^{-4}	1	0.1336806
1 $lb \cdot gal^{-1}$	= 0.1198264	119.8264	4.749536×10^{-3}	4.3290043×10^{-3}	7.480519	1

F. Units of pressure

Units	Pa, $N \cdot m^{-2}$	$dyn \cdot cm^{-2}$	*bar*	*atm*	$kgf \cdot cm^{-2}$	*mmHg (torr)*	*in.* Hg	$lbf \cdot in.^{-2}$
1 Pa, 1 $N \cdot m^{-2}$	= 1	10	10^{-5}	9.869233×10^{-6}	1.019716×10^{-5}	7.500617×10^{-3}	2.952999×10^{-4}	1.450377×10^{-4}
1 $dyn \cdot cm^{-2}$	= 0.1	1	10^{-6}	9.869233×10^{-7}	1.019716×10^{-6}	7.500617×10^{-4}	2.952999×10^{-5}	1.450377×10^{-5}
1 bar	= 10^5	10^6	1.	0.9869233	1.019716	750.0617	29.52999	14.50377
1 atm	= 101325	101325.0	1.01325	1	1.033227	760.	29.92126	14.69595
1 $kgf \cdot cm^{-2}$	= 98066.5	980665	0.980665	0.9678411	1	735.5592	28.95903	14.22334
1 mmHg (torr)	= 133.3224	1333.224	1.333224×10^{3}	1.3157895×10^{-3}	1.359510×10^{-3}	1	0.03937008	0.01933678
1 in. Hg	= 3386.388	33863.88	0.03386388	0.03342105	0.03453155	25.4	1	0.4911541
1 $lbf \cdot in.^{-2}$	= 6894.757	68947.57	0.06894757	0.06804596	0.07030696	51.71493	2.036021	1

G. Units of energy

Units	g mass (energy equiv)	J	eV	cal	cal_{IT}	Btu_{IT}	kWh	hp-h	ft-lbf	$ft^3 \cdot lbf \cdot in.^{-2}$	liter-atm
1 g mass (energy equiv)	= 1	8.987552×10^{13}	5.609589×10^{32}	2.148076×10^{3}	2.146640×10^{13}	8.518555×10^{10}	2.496542×10^{7}	3.347918×10^{7}	6.628878×10^{13}	4.603388×10^{11}	8.870024×10^{11}
1 J	$= 1.112650 \times 10^{-14}$	1	6.241510×10^{18}	0.2390057	0.2388459	9.478172×10^{-4}	$2.777777\ldots \times 10^{-7}$	3.725062	0.7375622	5.121960×10^{-3}	9.869233×10^{-3}
1 eV	$= 1.782662 \times 10^{-33}$	1.602176×10^{-19}	1	3.829293×10^{-20}	3.826733×10^{-20}	1.518570×10^{-22}	4.450490×10^{-26}	5.968206×10^{-26}	1.181705×10^{-19}	8.206283×10^{-22}	1.581225×10^{-21}
1 cal	$= 4.655328 \times 10^{-14}$	4.184*	2.611448×10^{19}	1	0.9993312	3.965667×10^{-3}	$1.1622222\ldots \times 10^{-6}$	1.558562×10^{-6}	3.085960	2.143028×10^{-2}	0.04129287
1 cal_{IT}	$= 4.658443 \times 10^{-14}$	4.1868*	2.613195×10^{19}	1.000669	1	3.968321×10^{-3}	1.163×10^{-6}	1.559609×10^{-6}	3.088025	2.144462×10^{-2}	0.04132050
1 Btu_{IT}	$= 1.173908 \times 10^{-11}$	1055.056	6.585141×10^{21}	252.1644	251.9958	1	2.930711 $\times 10^{-}$[illegible]	3.930148×10^{-4}	778.1693	5.403953	10.41259
1 kWh	$= 4.005540 \times 10^{-8}$	3600000.*	2.246944×10^{25}	860420.7	859845.2	3412.142	1	1.341022	2655224.	18349.06	35529.24
1 hp-h	$= 2.986931 \times 10^{-8}$	2384519.	1.675545×10^{25}	641615.6	641186.5	2544.33	0.7456998	1	1980000.	13750.	26494.15
1 ft-lbf	$= 1.508551 \times 10^{-14}$	1.355818	8.462351×10^{18}	0.3240483	0.3238315	1.285067×10^{-3}	3.766161×10^{-7}	$5.050505\ldots \times 10^{-7}$	1	$6.944444\ldots \times 10^{-3}$	0.01338088
1 ft^3 lbf $\cdot$ in.$^{-2}$	$= 2.172313 \times 10^{-12}$	195.2378	1.218579×10^{21}	46.66295.	46.63174	0.1850497	5.423272×10^{-5}	$7.272727\ldots \times 10^{-5}$	144.*	1	1.926847
1 liter-atm	$= 1.127393 \times 10^{-12}$	101.325	6.324210×10^{20}	24.21726	24.20106	0.09603757	2.814583×10^{-5}	3.774419×10^{-5}	74.73349	0.5189825	1

*Numbers followed by an asterisk are definitions of the relation between the two units.

Appendix

Mathematical notation, with definitions

Signs and symbols

+	Plus (sign of addition)
+	Positive
−	Minus (sign of subtraction)
−	Negative
±(∓)	Plus or minus (minus or plus)
×	Times, by (multiplication sign)
·	Multiplied by
+	Sign of division
/	Divided by
:	Ratio sign, divided by, is to
::	Equals, as (proportion)
<	Less than
>	Greater than
≪	Much less than
≫	Much greater than
=	Equals
≡	Identical with
~	Similar to
≈	Approximately equals
≅	Approximately equals, congruent
≤	Equal to or less than
≥	Equal to or greater than
‡ ≠	Not equal to
→ ≐	Approaches
∝	Varies as
∞	Infinity
√	Square root of
∛	Cube root of
∴	Therefore
∥	Parallel to
( ) [ ] { }	Parentheses, brackets and braces; quantities enclosed by them to be taken together in multiplying, dividing, etc.
$\overline{AB}$	Length of line from *A* to *B*
π	pi = 3.14159...
°	Degrees
′	Minutes
″	Seconds
∠	Angle

Mathematical notation, with definitions (*cont.*)

	Signs and symbols (cont.)
dx	Differential of x
Δ	(delta) difference
Δx	Increment of x
$\partial u/\partial x$	Partial derivative of u with respect to x
$\int$	Integral of
$\int_b^a$	Integral of, between limits a and b
$\oint$	Line integral around a closed path
Σ	(sigma) summation of
$f(x), F(x)$	Functions of x
∇	Del or nabla, vector differential operator
∇^2	Laplacian operator
£	Laplace operational symbol
4!	Factorial $4 = 1 \times 2 \times 3 \times 4$
$\|x\|$	Absolute value of x
$\dot{x}$	First derivative of x with respect to time
$\ddot{x}$	Second derivative of x with respect to time
$\mathbf{A} \times \mathbf{B}$	Vector-product; magnitude of **A** times magnitude of **B** times sine of the angle from **A** to **B**; $AB \sin \overline{AB}$
$\mathbf{A} \cdot \mathbf{B}$	Scaler product of **A** and **B**; magnitude of **A** times magnitude of **B** times cosine of the angle from **A** to **B**; $AB \cos \overline{AB}$
	Mathematical logic
$p, q, P(x)$	Sentences, propositional functions, propositions
$-p$, $\sim p$, non p, Np	Negation red "not p" ($\neq$: read "not equal")
$p \vee q$, $p + q$, Apq	Disjunction, read "p or q," "p, q," or both
$p \wedge q$, $p \cdot q$, $p\&q$, Kpq	Conjunction, read "p and q"
$p \rightarrow q$, $p \supset q$, $p \Rightarrow q$, Cpq	Implication, read "p implies q" or "if p then q"
$p \leftrightarrow q$, $p \equiv q$, $p \Leftrightarrow q$, Epq, p iff$q$	Equivalence, read "p is equivalent to q" or "p if and only if q"
n.a.s.c.	Read "necessary and sufficient condition"
( ), [ ], { }, ...,··	Parentheses
$\forall$, Σ	Universal quantifier, read "for all" or "for every"
$\exists$, Π	Existential quantifier, read "there is a" or "there exists"
$\vdash$	Assertion sign ($p \vdash q$: read "q follows from p"; $\vdash p$; read "p is or follows from an axiom," or "p is a tautology"
0, 1	Truth, falsity (values)
=	Identity
$\overset{\mathrm{Df}}{=}$, $\overset{\mathrm{df}}{=}$, $\underset{\mathrm{df}}{=}$, $\equiv$	Definitional identity
■	"End of proof"; "QED"

Appendix

Mathematical notation, with definitions (*cont.*)

	Set theory, relations, functions
X, Y	Sets
$x \in X$	x is a member of the set X
$x \notin X$	x is not a member of X
$A \subset X, A \subseteq X$	Set A is contained in set X
$A \not\subset X, A \not\subseteq X$	A is not contained in X
$X \cup Y, X + Y$	Union of sets X and Y
$X \cap Y, X \cdot Y$	Intersection of sets X and Y
$+, \dotplus, \circ$	Symmetric difference of sets
$\cup X_i, \Sigma X_i$	Union of all the sets X_i
$\cap X_i, \Pi X_i$	Intersection of all the sets X_i
$\varnothing, 0, \Lambda$	Null set, empty set
X', CX, CX	Complement of the set X
$X - Y, X \backslash Y$	Difference of sets X and Y
$x(P(x)); \{x \mid P(x)\}, \{x: P(x)\}$	The set of all x with the property P
$(x,y,z), \langle x,y,z \rangle$	Ordered set of elements x, y, and z; to be distinguished from (x,z,y) for example
$\{x,y,z\}$	Unordered set, the set whose elements are x, y, z and no others
$\{a_1, a_2, \ldots, a_n\}$, $\{a_1\}_{i=1,2,\ldots,n}, \{a_1\}_{i=1}^{n}$	The set whose numbers are a_i, where i is any number whole from 1 to n
$\{a_1, a_2, \ldots\}$, $\{a_1\}_{i=1,2,\ldots}, \{a_1\}_{i=1}^{\infty}$	The set whose members are a_i, where i is any whole positive number
$X \times Y$	Cartesian product, set of all (x,y) such that $x \in X,\ y \in Y$
$\{a_i\}_{i \in I}$	The set whose elements are a_i, where $i \in I$
$xRy, R(x,y)$	Relation
$\equiv, \cong, \sim, \approx$	Equivalence relations, for example, congruence
$\geqq, \geq, \&, \gtrdot, \leqq, \leq, <$	Transitive relations, for example, numerical order
$f \cdot X \to Y, X \xrightarrow{f} Y$, $X \to Y, f \in Y^X$	Function, mapping, transformation
$f^{-1}, \overset{-1}{f}, X \xleftarrow{f^{-1}} Y$	Inverse mapping
$g \circ f$	Composite functions: $(g \circ f)(x) = g(f(x))$
$f(X)$	Image of X by f
$f^{-1}(X)$	Inverse-image set, counter image
1-1, one-one	Read "one-to-one correspondence"
$\begin{array}{ccc} X & \xrightarrow{f} & Y \\ \phi\downarrow & & \downarrow\psi \\ W & \xrightarrow{g} & Z \end{array}$	Diagram: the diagram is commutative in case $\psi \circ f = g \circ \phi$
f/A	Partial mapping, restriction of function f to set A
$\overline{\overline{X}}$, card X, $/X/$	Cardinal of the set A
$\aleph_0$, d	Denumerable infinity

Mathematical notation, with definitions (*cont.*)

	Set theory, relations, functions (cont.)
$c, \mathfrak{c}, 2^{\aleph_0}$	Power of continuum
ω	Order type of the set of positive integers
σ^-	Read "countably"
	Number, numerical functions
1.4; 1,4; 1 · 4	Read "one and four-tenths"
1(1)20(10)100	Read "from 1 to 20 in intervals of 1, and from 20 to 100 in intervals of 10"
const	Constant
$A \geqq 0$	The number A is nonnegative, or, the matrix A is positive definite, or, the matrix A has nonnegative entries
$x \mid y$	Read "x divides y"
$x \equiv y \bmod p$	Read "x congruent to y modulo p"
$a_0 + \frac{1}{a_1 +} \frac{1}{a_2 +} \cdots$ $a_0 + \frac{1\mid}{\mid a_1} + \ldots$	Continued fractions
$[a,b]$	Closed interval
$[a,b)$, $[a,b[$	Half-open interval (open at the right)
(a,b), $]a,b[$	Open interval
$[a,\infty)$, $[a,\rightarrow[$	Interval closed at the left, infinite to the right
$(-\infty,\infty)$, $]\leftarrow,\rightarrow[$	Set of all real numbers
$\max_{x \in X} f(x)$, $\max\{f(x) \mid x \in X\}$	Maximum of $f(x)$ when x is in the set X
min	Minimum
sup, l.u.b.	Supremum, least upper bound
inf, g.l.b.	Infimum, greatest lower bound
$\lim_{x\to a} f(x) = b$, $\lim_{x=a} f(x) = b$, $f(x) \to b$ as $x \to a$	b is the limit of $f(x)$ as x approaches a
$\lim_{x\to a^-} f(x)$, $\lim_{x=a-0} f(x)$, $f(a-)$	Limit of $f(x)$ as x approaches a from the left
lim sup, $\overline{\lim}$	Limit superior
lim inf, $\underline{\lim}$	Limit inferior
l.i.m.	Limit in the mean
$z = x + iy = re^{i\theta}$, $\zeta = \xi + i\eta$, $w = u + iv = \rho e^{i\phi}$	Complex variables
$\bar{z}$, z^*	Complex conjugate
Re, $\Re$	Real part
Im, $\mathfrak{I}$	Imaginary part
arg	Argument

Mathematical notation, with definitions (*cont.*)

	Number, numerical functions (cont.)
$\frac{\partial(u,v)}{\partial(x,y)}$, $\frac{D(u,v)}{D(x,y)}$;	Jacobian, functional determinant
$\int_E f(x)d\mu(x)$	Integral (for example, Lebesgue integral) of function f over set E with respect to measure μ
$f(n) \sim \log n$ as $n \to \infty$	$f(n)/\log n$ approaches 1 as $n \to \infty$
$f(n) = O(\log n)$ as $n \to \infty$	$f(n)/\log n$ is bounded as $n \to \infty$
$f(n) = o(\log n)$	$f(n)/\log n$ approaches zero
$f(n) \nearrow b$, $f(x) \uparrow b$	$f(x)$ increases, approaching the limit b
$f(n) \downarrow b$, $f(x) \searrow b$	$f(x)$ decreases, approaching the limit b
a.e., p.p.	Almost everywhere
ess sup	Essential supremum
C^0, $C^0(X)$, $C(X)$	Space of continuous functions
C^k, $C^k[a, b]$	The class of functions having continuous kth derivative (on $[a,b]$)
C'	Same as C^1
Lip_α, Lip α	Lipschitz class of functions
L^p, L_p, L^p $[a, b]$	Space of functions having integrable absolute pth power (on $[a,b]$)
L'	Same as L^1
$(C, \propto)$, (C, p)	Cesàro summability
	Special functions
$[x]$	The integral part of x
$\binom{n}{k}$, ${}^nC_k$, ${}_nC_k$	Binomial coefficient $n!/k!(n-k)!$
$\left(\frac{n}{p}\right)$	Legendre symbol
e^x, $\exp x$	Exponential function
$\sinh x$, $\cosh x$, $\tanh x$	Hyperbolic functions
sn x, cn x, dn x	Jacobi elliptic functions
$\wp(x)$	Weierstrass elliptic function
$\Gamma(x)$	Gamma function
$J_\nu(x)$	Bessel function
$\mathrm{X}_X(x)$	Characteristic function of the set X; $\chi_X(x) = 1$ in case $x \in X$, otherwise $\chi_X(x) = 0$
sgn x	Signum: sgn = 0, while sgn $x = x/\lvert x\rvert$ for $x \neq 0$
$\delta(x)$	Dirac delta function
	Algebra, tensors, operators
$+, \cdot, \times, \circ, \mathrm{T}, \tau$	Laws of composition in algebraic systems
e, 0	Identity, unit, neutral element (of an additive system)
e, 1, I	Identity, unit, neutral element (of a general algebraic system)

Mathematical notation, with definitions (*cont.*)

	Algebra, tensors, operators (cont.)
e, $\mathfrak{e}$, E, P	Idempotent
a^{-1}	Inverse of a
Hom(M,N)	Group of all homomorphisms of M into N
G/H	Factor group, group of cosets
$[K{:}k]$	Dimension of K over k
$\oplus$, $+$	Direct sum
$\otimes$	Tensor product, Kronecker product
$\wedge$	Exterior product, Grassmann product
$\vec{x}$, $\mathbf{x}$, $\mathfrak{x}$, x	Vector
$\vec{x}\cdot\vec{r}$, $\mathbf{x}\cdot\mathbf{y}$, $(\mathfrak{x},\mathfrak{y})$	Inner product, scalar product, dot product
$\mathbf{x}\times\nabla$, $[\mathfrak{x},\mathfrak{y}]$, $\mathbf{x}\wedge\mathbf{y}$	Outer product, vector product, cross product
$\lvert x\rvert$, $\lvert x\rvert$, $\lVert x\rVert$, $\lVert x\rVert_p$	Norm of the vector x
Ax, xA	The image of x under the transformation A
δ_{ij}	Kronecker delta: $\delta_{ij} = 1$, while $\delta_{ij} = 0$ for $i \neq j$
A', tA, A^t, ${}^tA$	Transpose of the matrix A
A^*, $\bar{A}$	Adjoint, Hermitian conjugate of A
tr A, Sp A	Trace of the matrix A
det A, $\lvert A\rvert$	Determinant of the matrix A
$\Delta^n f(x)$, $\Delta_h{}^n f$, $\underset{h}{\Delta}{}^n f(x)$	Finite differences
$[x_0,x_1]$, $[x_0,x_1x_2]$, $\underset{x_1}{\Delta}u_{x_0}$, $[x_0x_1]f$	Divided differences
∇f, grad f	Read "gradient of f"
$\nabla\cdot\mathbf{v}$, div $\mathbf{v}$	Read "divergence of $\mathbf{v}$"
$\nabla\times\mathbf{v}$, curl $\mathbf{v}$, root $\mathbf{v}$	Read "curl of $\mathbf{v}$"
∇^2, Δ, div grad	Laplacian
$[X,Y]$	Poisson bracket, or commutator, or Lie product
GL(n,R)	Full linear group of degree n over field R
O(n,R)	Full orthogonal group
SO(n,R) O^+ (n,R)	Special orthogonal group
	Topology
E^n	Euclidean n space
S^n	n sphere
$\rho(p,q)$, $d(p,q)$	Metric, distance (between points p and q)
$\overline{X}$, X^-, cl X, X^c	Closure of the set X
FrX, fr$X$, ∂X, bdry X	Frontier, boundary of X
int X, $\mathring{X}$	Interior of X
T_2 space	Hausdoff space

Mathematical notation, with definitions (*cont.*)

	Topology (cont.)
F_σ	Union of countably many closed sets
G_δ	Intersection of countably many open sets
dim X	Dimensionality, dimension of X
$\pi_1(X)$	Fundamental group of the space X
$\pi_n(X)$, $\pi_n(X, A)$	Homotopy groups
$H_n(X)$, $H_n(X,A;G)$, $H_*(X)$	Homology groups
$H^n(X)$, $H^n(X,A;G)$, $H^*(X)$	Cohomology groups
	Probability and statistics
X, Y	Random variables
$P(X \leqq 2)$, $\Pr(X \leqq 2)$	Probability that $X \leqq 2$
$P(X \leqq 2 \mid Y \geqq 1)$	Conditional probability
$E(X)$, $E(X)$	Expectation of X
$E(X \mid Y \geqq 1)$	Conditional expectation
c.d.f.	Cumulative distribution function
p.d.f.	Probability density function
c.f.	Characteristic function
$\overline{x}$	Mean (especially, sample mean)
σ, s.d.	Standard deviation
σ^2 Var, var	Variance
μ_1, μ_2, μ_3, μ_i, μ_{ij}	Moments of a distribution
ρ	Coefficient of correlation
$\rho_{12.34}$	Partial correlation coefficient

Appendix

Schematic electronic symbols*

Ammeter

Amplifier, general

Amplifier, inverting

Amplifier, operational

AND gate

Antenna, balanced

Antenna, general

Antenna, loop

Antenna, loop, multiturn

Battery

Capacitor, feedthrough

Capacitor, fixed

Capacitor, variable

Capacitor, variable, split-rotor

Capacitor, variable, split-stator

Cathode, electron-tube, cold

Cathode, electron-tube, directly heated

Cathode, electron-tube, indirectly heated

Cavity resonator

Cell, electrochemical

Circuit breaker

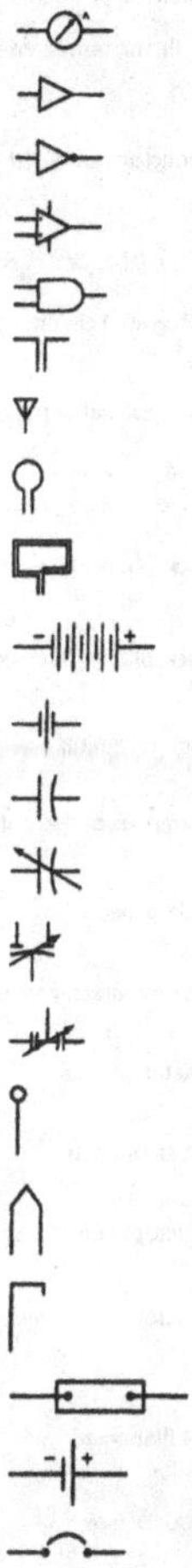

Coaxial cable

Crystal, piezoelectric

Delay line

Diac

Diode, field-effect

Diode, general

Diode, Gunn

Diode, light-emitting

Diode, photosensitive

Diode, PIN

Diode, Schottky

Diode, tunnel

Diode, varactor

Diode, Zener

Directional coupler

Directional wattmeter

Exclusive-OR gate

Female contact, general

Ferrite bead

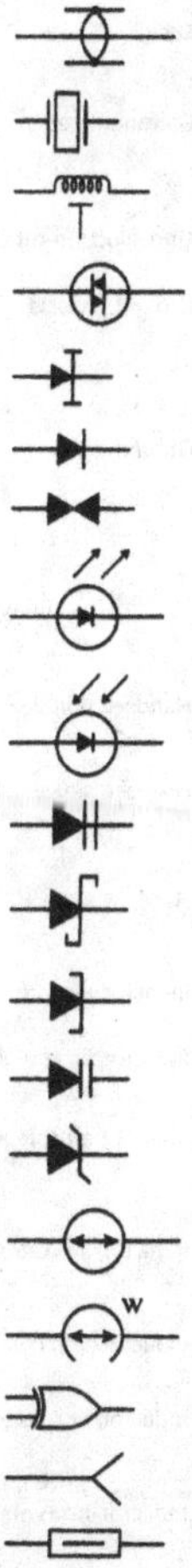

*From S. Gibilisco, *The Illustrated Dictionary of Electronics*, 8th ed., McGraw-Hill, 2001.

Appendix

Filament, electron-tube

Fuse

Galvanometer

Grid, electron-tube

Ground, chassis

Ground, earth

Headset

Handset, double

Headset, single

Headset, stereo

Inductor, air core

Inductor, air core, bifilar

Inductor, air core, tapped

Inductor, air core, variable

Inductor, iron core

Inductor, iron core, bifilar

Inductor, iron core, tapped

Inductor, iron core, variable

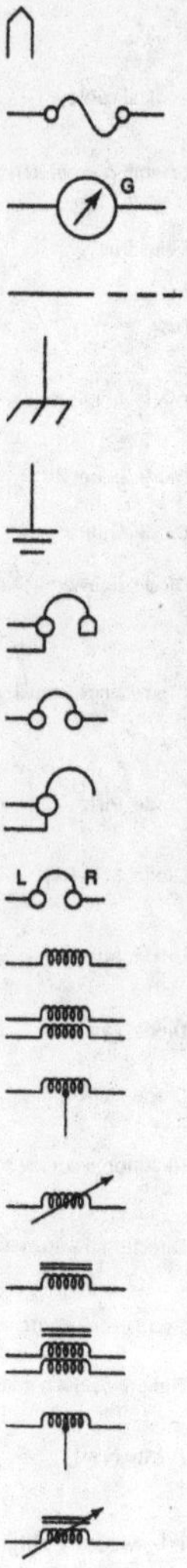

Inductor, powdered-iron core

Inductor, powdered-iron core, bifilar

Inductor, powdered-iron core, tapped

Inductor, powdered-iron core, variable

Integrated circuit, general

Jack, coaxial or photo

Jack, phone, two-conductor

Jack, phone, three-conductor

Key, telegraph

Lamp, incandescent

Lamp, neon

Male contact, general

Meter, general

Microammeter

Microphone

Microphone, directional

Milliammeter

NAND gate

Negative voltage connection

NOR gate

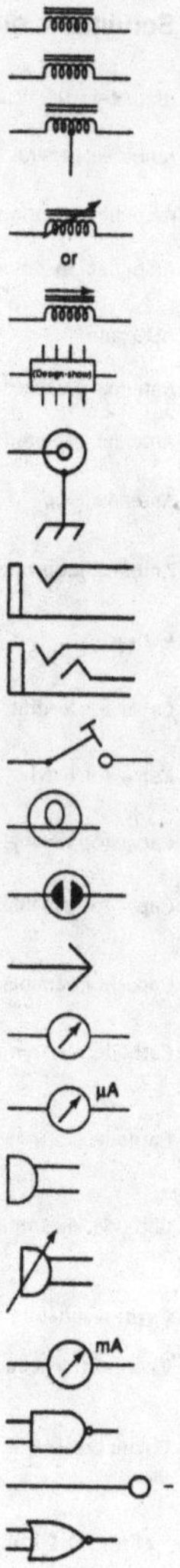

Appendix

NOT gate

Optoisolator

OR gate

Outlet, two-wire, nonpolarized

Outlet, two-wire, polarized

Outlet, three-wire

Outlet, 234-V

Plate, electron-tube

Plug, two-wire, nonpolarized

Plug, two-wire, polarized

Plug, three-wire

Plug, 234-V

Plug, coaxial or phono

Plug, phone, two-conductor

Plug, phone, three-conductor

Positive voltage connection

Potentiometer

Probe, radio-frequency

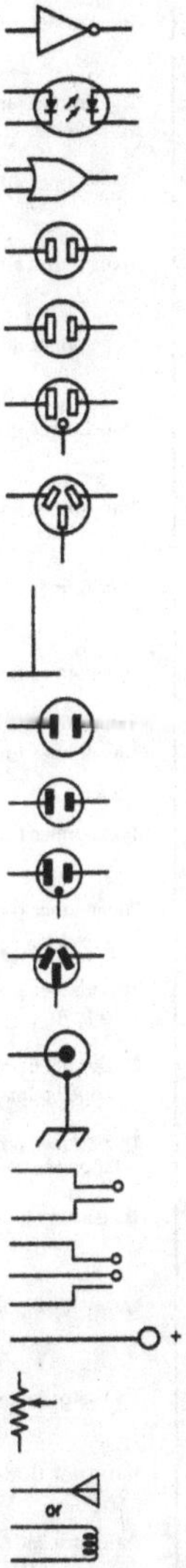

or

Rectifier, gas-filled

Rectifier, high-vacuum

Rectifier, semiconductor

Rectifier, silicon-controlled

Relay, double-pole, double-throw

Relay, double-pole, single-throw

Relay, single-pole, double-throw

Relay, single-pole, single-throw

Resistor, fixed

Resistor, preset

Resistor, tapped

Resonator

Rheostat

Saturable reactor

Signal generator

Solar battery

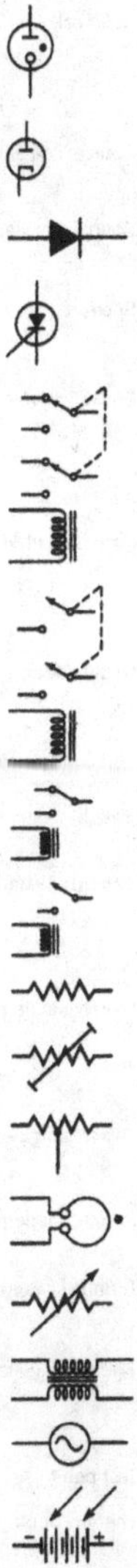

Appendix

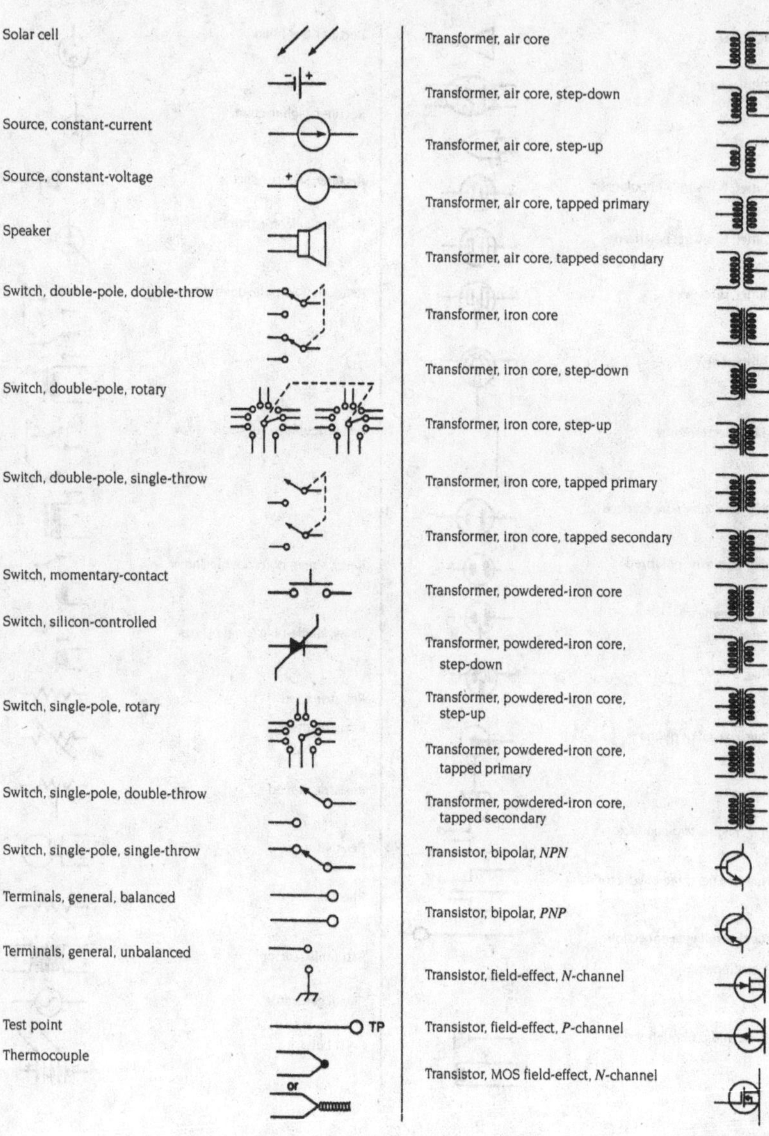
Solar cell
Source, constant-current
Source, constant-voltage
Speaker
Switch, double-pole, double-throw
Switch, double-pole, rotary
Switch, double-pole, single-throw
Switch, momentary-contact
Switch, silicon-controlled
Switch, single-pole, rotary
Switch, single-pole, double-throw
Switch, single-pole, single-throw
Terminals, general, balanced
Terminals, general, unbalanced
Test point
TP
Thermocouple
or
Transformer, air core
Transformer, air core, step-down
Transformer, air core, step-up
Transformer, air core, tapped primary
Transformer, air core, tapped secondary
Transformer, iron core
Transformer, iron core, step-down
Transformer, iron core, step-up
Transformer, iron core, tapped primary
Transformer, iron core, tapped secondary
Transformer, powdered-iron core
Transformer, powdered-iron core, step-down
Transformer, powdered-iron core, step-up
Transformer, powdered-iron core, tapped primary
Transformer, powdered-iron core, tapped secondary
Transistor, bipolar, NPN
Transistor, bipolar, PNP
Transistor, field-effect, N-channel
Transistor, field-effect, P-channel
Transistor, MOS field-effect, N-channel

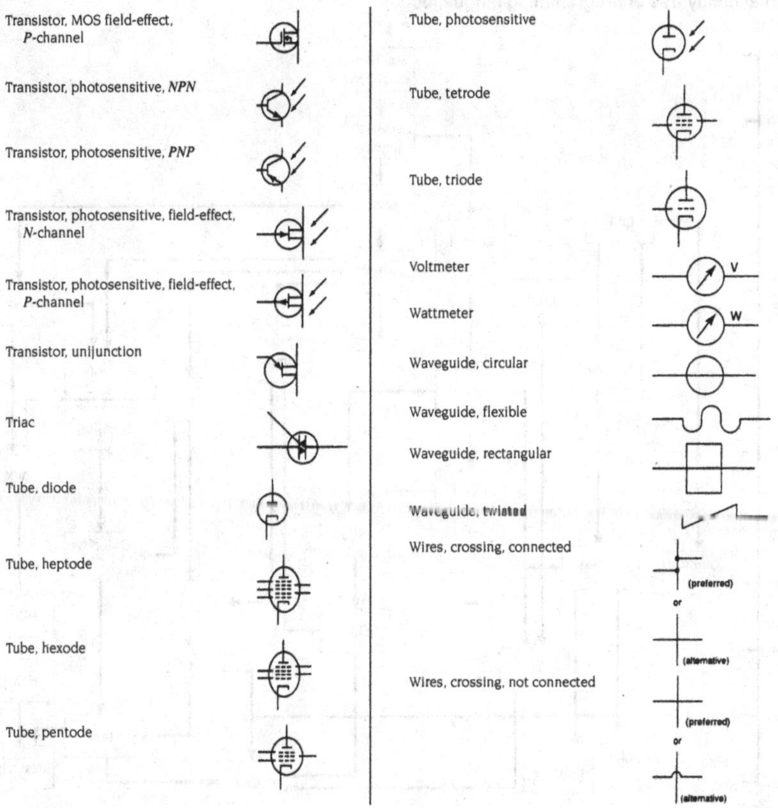
Transistor, MOS field-effect, P-channel
Transistor, photosensitive, NPN
Transistor, photosensitive, PNP
Transistor, photosensitive, field-effect, N-channel
Transistor, photosensitive, field-effect, P-channel
Transistor, unijunction
Triac
Tube, diode
Tube, heptode
Tube, hexode
Tube, pentode
Tube, photosensitive
Tube, tetrode
Tube, triode
Voltmeter
V
Wattmeter
W
Waveguide, circular
Waveguide, flexible
Waveguide, rectangular
Waveguide, twisted
Wires, crossing, connected
(preferred)
or
(alternative)
Wires, crossing, not connected
(preferred)
or
(alternative)

Appendix

Partial family tree of programming languages

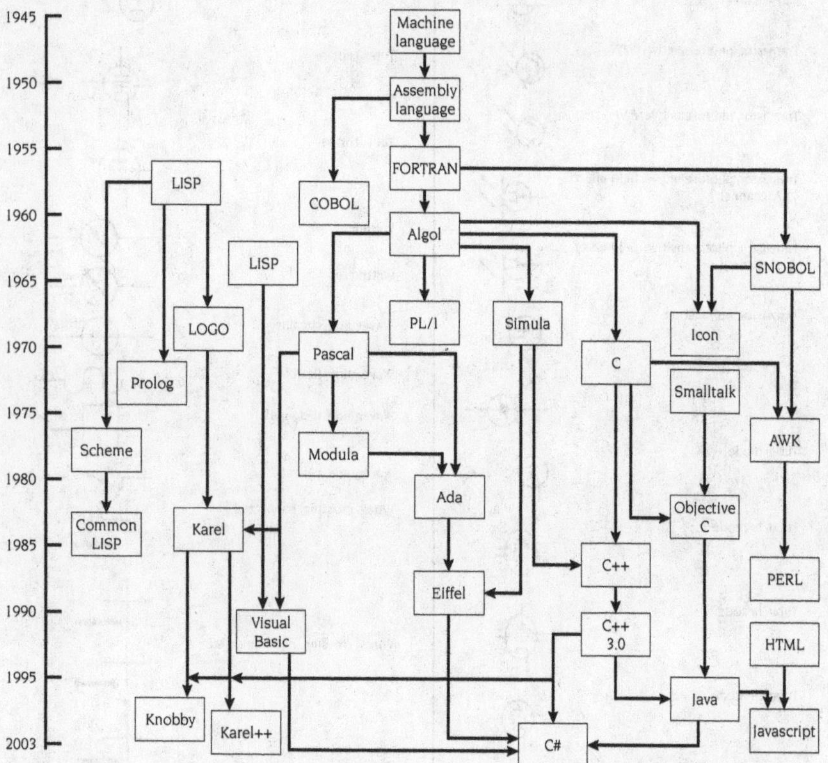

From Glenn D. Blank, Robert F. Barnes, and Edwin J. Kay, *The Universal Computer: Introducing Computer Science with Multimedia*, McGraw-Hill, New York, 2003.

ASCII code*

Order number				Character code	
Decimal	Hexadecimal	Binary	Octal		
0	0	0000000	000	NUL	(Blank)
1	1	0000001	001	SOH	(Start of Header)
2	2	0000010	002	STX	(Start of Text)
3	3	0000011	003	ETX	(End of Text)
4	4	0000100	004	EOT	(End of Transmission)
5	5	0000101	005	ENQ	(Enquiry)
6	6	0000110	006	ACK	(Acknowledge (Positive))
7	7	0000111	007	BEL	(Bell)
8	8	0001000	010	BS	(Backspace)
9	9	0001001	011	HT	(Horizontal Tabulation)
10	A	0001010	012	LF	(Line Feed)
11	B	0001011	013	VT	(Vertical Tabulation)
12	C	0001100	014	FF	(Form Feed)
13	D	0001101	015	CR	(Carriage Return)
14	E	0001110	016	SO	(Shift Out)
15	F	0001111	017	SI	(Shift In)
16	10	0010000	020	DLE	(Data Link Escape)
17	11	0010001	021	DC1	(Device Control 1)
18	12	0010010	022	DC2	(Device Control 2)
19	13	0010011	023	DC3	(Device Control 3)
20	14	0010100	024	DC4	(Device Control 4-Stop)
21	15	0010101	025	NAK	(Negative Acknowledge)
22	16	0010110	026	SYN	(Synchronization)
23	17	0010111	027	ETB	(End of Text Block)
24	18	0011000	030	CAN	(Cancel)
25	19	0011001	031	EM	(End of Medium)
26	1A	0011010	032	SUB	(Substitute)
27	1B	0011011	033	ESC	(Escape)
28	1C	0011100	034	FS	(File Separator)
29	1D	0011101	035	GS	(Group Separator)
30	1E	0011110	036	RS	(Record Separator)
31	1F	0011111	037	US	(Unit Separator)
32	20	0100000	040	SP	(Space)
33	21	0100001	041	!	
34	22	0100010	042	"	
35	23	0100011	043	#	
36	24	0100100	044	$	
37	25	0100101	045	%	
38	26	0100110	046	&	
39	27	0100111	047	'	(Closing Single Quote)
40	28	0101000	050	(	
41	29	0101001	051	)	
42	2A	0101010	052	*	
43	2B	0101011	053	+	
44	2C	0101100	054	,	(Comma)
45	2D	0101101	055	-	(Hyphen)
46	2E	0101110	056	.	(Period)
47	2F	0101111	057	/	
48	30	0110000	060	0	
49	31	0110001	061	1	
50	32	0110010	062	2	
51	33	0110011	063	3	

ASCII code (*cont.*)

Order number					
Decimal	*Hexadecimal*	*Binary*	*Octal*		*Character code*
52	34	0110100	064	4	
53	35	0110101	065	5	
54	36	0110110	066	6	
55	37	0110111	067	7	
56	38	0111000	070	8	
57	39	0111001	071	9	
58	3A	0111010	072	:	
59	3B	0111011	073	;	
60	3C	0111100	074	<	(Less Than)
61	3D	0111101	075	=	
62	3E	0111110	076	>	(Greater Than)
63	3F	0111111	077	?	
64	40	1000000	100	@	
65	41	1000001	101	A	
66	42	1000010	102	B	
67	43	1000011	103	C	
68	44	1000100	104	D	
69	45	1000101	105	E	
70	46	1000110	106	F	
71	47	1000111	107	G	
72	48	1001000	110	H	
73	49	1001001	111	I	
74	4A	1001010	112	J	
75	4B	1001011	113	K	
76	4C	1001100	114	L	
77	4D	1001101	115	M	
78	4E	1001110	116	N	
79	4F	1001111	117	O	
80	50	1010000	120	P	
81	51	1010001	121	Q	
82	52	1010010	122	R	
83	53	1010011	123	S	
84	54	1010100	124	T	
85	55	1010101	125	U	
86	56	1010110	126	V	
87	57	1010111	127	W	
88	58	1011000	130	X	
89	59	1011001	131	Y	
90	5A	1011010	132	Z	
91	5B	1011011	133	[	(Opening Bracket)
92	5C	1011100	134	\	(Reverse Slant)
93	5D	1011101	135	]	(Closing Bracket)
94	5E	1011110	136	^	(Circumflex)
95	5F	1011111	137	_	(Underline)
96	60	1100000	140	`	(Opening Single Quote)
97	61	1100001	141	a	
98	62	1100010	142	b	
99	63	1100011	143	c	
100	64	1100100	144	d	
101	65	1100101	145	e	
102	66	1100110	146	f	
103	67	1100111	147	g	
104	68	1101000	150	h	

ASCII code (*cont.*)

Order number					
Decimal	*Hexadecimal*	*Binary*	*Octal*	*Character code*	
105	69	1101001	151	i	
106	6A	1101010	152	j	
107	6B	1101011	153	k	
108	6C	1101100	154	l	
109	6D	1101101	155	m	
110	6E	1101110	156	n	
111	6F	1101111	157	o	
112	70	1110000	160	p	
113	71	1110001	161	q	
114	72	1110010	162	r	
115	73	1110011	163	s	
116	74	1110100	164	t	
117	75	1110101	165	u	
118	76	1110110	166	v	
119	77	1110111	167	w	
120	78	1111000	170	x	
121	79	1111001	171	y	
122	7A	1111010	172	z	
123	7B	1111011	173	{	(Opening Brace)
124	7C	1111100	174	|	(Vertical Line)
125	7D	1111101	175	}	(Closing Brace)
126	7E	1111110	176	-	(Overline (Tilde))
127	7F	1111111	177	DEL	(Delete/Rubout)

*ASCII (American Standard Code for Information Interchange) is a code for representing English characters as numbers, with each character assigned an order number from 0 through 127. In this table the order numbers are given in decimal, hexadecimal, binary, and octal notations. The first 32 character codes are control codes (non-printable) and the other 96 are representable characters. Since ASCII was developed in the 1960s for use with teletypewriters, the control codes are now rarely used for their original purpose, and their descriptions given here are somewhat obscure.

Appendix

Electromagnetic spectrum

Frequency, Hz	*Wavelength, m*	*Nomenclature*	*Typical source*
10^{23}	3×10^{-15}	Cosmic photons	Astronomical
10^{22}	3×10^{-14}	X-rays	Radioactive nuclei
10^{21}	3×10^{-13}	X-rays	
10^{20}	3×10^{-12}	X-rays	Atomic inner shell, positron-electron annihilation
10^{19}	3×10^{-11}	Soft x-rays	Electron impact on a solid
10^{18}	3×10^{-10}	Ultraviolet, x-rays	Atoms in sparks
10^{17}	3×10^{-9}	Ultraviolet	Atoms in sparks and arcs
10^{16}	3×10^{-8}	Ultraviolet	Atoms in sparks and arcs
10^{15}	3×10^{-7}	Visible spectrum	Atoms, hot bodies, molecules
10^{14}	3×10^{-6}	Infrared	Hot bodies, molecules
10^{13}	3×10^{-5}	Infrared	Hot bodies, molecules
10^{12}	3×10^{-4}	Far-infrared	Hot bodies, molecules
10^{11}	3×10^{-3}	Microwaves	Electronic devices
10^{10}	3×10^{-2}	Microwaves, radar	Electronic devices
10^{9}	3×10^{-1}	Radar	Electronic devices, interstellar hydrogen
10^{8}	3	Television, FM radio	Electronic devices
10^{7}	30	Short-wave radio	Electronic devices
10^{6}	300	AM radio	Electronic devices
10^{5}	3000	Long-wave radio	Electronic devices
10^{4}	3×10^{4}	Induction heating	Electronic devices
10^{3}	3×10^{5}		Electronic devices
100	3×10^{6}	Power	Rotating machinery
10	3×10^{7}	Power	Rotating machinery
1	3×10^{8}		Communicated direct current
0	Infinity	Direct current	Batteries

Radio spectrum

Band number[a]	Frequency range[b]	Approximate wavelength range	Band designation	Some applications[c]	
				Applications	Frequency range[b]
4	3–30 kHz[d]	10–100 km	Very low frequency (VLF)	VLF radio navigation	9–14 kHz[e]
5	30–300 kHz	1–10 km	Low frequency (LF)	Loran C Longwave broadcasting[f]	90–110 kHz 150–290 kHz
6	300 kHz to 3 MHz	100–1000 m	Medium frequency (MF)	AM broadcasting	535–1705 kHz
7	3–30 MHz	10–100 m	High frequency (HF)	Shortwave broadcasting Citizens band (CB)	5.95–26.1 MHz (8 frequency bands) 26.96–27.41 MHz
8	30–300 MHz	1–10 m	Very high frequency (VHF)	Cordless telephones Television channels 2–4 Television channels 5–6 FM broadcasting Instrument landing system (ILS) localizer VOR (VHF omnidirectional range) Television channels 7–13 Terrestrial digital audio broadcasting (DAB)[g]	43.71–44.49 and 46.60–46.98 MHz (base transmitters), 48.75–49.51 and 49.66–50.0 MHz (handset transmitters) 54–72 MHz 76–88 MHz 88–108 MHz 108–112 MHz 108–118 MHz 174–216 MHz 174–240 MHz

Radio spectrum (*cont.*)

Band number[a]	Frequency range[b]	Approximate wavelength range	Band designation	Some applications[e]: Application	Some applications[e]: Frequency range[b]
9	300 MHz to 3 GHz[h]	100–1000 mm	Ultrahigh frequency (UHF)	ILS glide slope	329–335 MHz
				Television channels 14–69	470–806 MHz
				Cellular telephones	824–849 MHz (mobile transmitters), 869–894 MHz (base transmitters)
				"900-MHz" cordless telephones	902–928 Mhz
				Distance measuring equipment (DME), Tacan	960–1215 MHz
				Secondary surveillance radar (SSR)	1030 MHz (beacons), 1090 MHz (transponders)
				Global Positioning System (GPS)	1176.45 MHz (L3), 1227.6 MHz (L2), 1575.42 MHz (L1)
				Radio astronomy[i]	1400–1427 and 1660–1670 MHz
				Terrestrial DAB[g]	1452–1492 MHz
				Inmarsat system	1535–1543.5 MHz (downlink), 1636.5–1645 MHz (uplink)
				Personal communications systems (PCS)	1850–1910 and 1930–1990 MHz (licensed), 1910–1930 (unlicensed)
				Deep space communications	2110–2120 MHz (uplink), 2290–2300 MHz (downlink)
				Satellite DAB	2310–2360 MHz
				Microwave ovens	2400–2500 MHz
10	3–30 GHz[h]	10–100 mm	Superhigh frequency (SHF)	Satellite communications, C band	3.7–4.2 GHz (downlink), 5.9–6.4 GHz (uplink)
				Satellite communications, K_u band	10.7–12.57 GHz (downlink), 12.7–14.8 GHz (uplink)

Radio spectrum (*cont.*)

Band number[a]	Frequency range[b]	Approximate wavelength range	Band designation	Some applications[c]	
				Application	Frequency range[b]
				Direct broadcasting satellite (DBS) systems	12.2–12.7 GHz
				Satellite communications, K_a band	18.55–18.8 and 19.2–20.2 GHz (downlink), 28.35–28.6 and 29.0–30.0 GHz (uplink)
11	30–300 GHz[h,j]	1–10 mm	Extremely high frequency (EHF), or millimeter-wave		
12	300 GHz to 3 THz[j]	0.1–1 mm	Submillimeter		

[a]"Band number N" extends from 0.3 × 10N to 3 × 10N hertz. The upper limit is included in each band; the lower limit is excluded.
[b]kHz = kilohertz (10^3 Hz), MHz = megahertz (10^6 Hz), GHz = gigahertz (10^9 Hz), THz = terahertz (10^{12} Hz).
[c]Radio spectrum allocations can vary in different parts of the world. The allocations given here pertain to the United States unless otherwise noted.
[d]Frequencies below 9 kHz have not been allocated by the International Telecommunications Union (ITU) and the U.S. Federal Communications Commission (FCC).
[e]An important user of this frequency band was the Omega navigation system (4 frequencies, 10.2–13.6 kHz), which was terminated in 1997.
[f]Longwave broadcasting is permitted in Europe.
[g]Terrestrial digital audio broadcasting (DAB) is allocated the174–240-MHz and 1452–1492-MHz bands in Europe, Canada, and other countries. In the United States, these bands are unavailable for this service, and DAB in the AM and FM broadcasting bands is under development.
[h]Microwave frequencies (1–110 GHz) are also classified according to an alternative set of designations given in the table titled "Microwave frequency bands."
[i]Many frequency bands between 13.36 MHz and 275 GHz are allocated to radio astronomy, most of them shared with other services. Two of the most important, listed here, are used to study the hydrogen line at 1420.4 MHz and hydroxyl lines near 1665 and 1667 MHz.
[j]Frequencies above 300 GHz in the United States and above 275 GHz elsewhere have not been allocated by the FCC and the ITU, but allocation of frequencies up to 1000 GHz is anticipated. In the meantime, the ITU and the FCC have urged the protection from harmful interference of observations for radio astronomy, earth exploration satellites, and space research in various frequency bands above 275 GHz around spectral lines that are of interest in these fields.

Appendix

Microwave frequency bands

Microwave band	*Frequency range, GHz*	*Approximate wavelength range, mm*
L	1–2	150–300
S	2–4	75–150
C	4–8	37–75
X	8–12	25–37
K_u	12–18	17–25
K	18–27	11–17
K_a	27–40	7.5–11
V	40–75	4–7.5
W	75–110	2.7–4

www.ingramcontent.com/pod-product-compliance
Lightning Source LLC
Chambersburg PA
CBHW012111070326
40689CB00056B/4951

* 9 7 8 0 0 7 1 4 2 1 7 8 2 *